Encyclopedia of
Plant Physiology

New Series Volume 12 D

Editors

A. Pirson, Göttingen
M.H. Zimmermann, Harvard

Physiological Plant Ecology IV

Ecosystem Processes: Mineral Cycling, Productivity and Man's Influence

Edited by

O.L. Lange P.S. Nobel C.B. Osmond H. Ziegler

Contributors

W.D. Billings W.R. Boynton J.L. Charley J.K. Detling
J. Ehleringer P.G. Falkowski D.M. Gates C.A. Hall
P.J. Harrison P.G. Jarvis C.W. Keefe W.M. Kemp H. Klinge
A. Kohler B.C. Labus K.J. Lendzian J.W. Leverenz
R.S. Loomis E. Medina A. Melzer H.A. Mooney T.R. Parsons
F.W.T. Penning de Vries K. Pfister B.N. Richards
R. Schubert Ch. Steinberg B.R. Strain L.L. Tieszen
M.H. Unsworth W. Urbach J.A. Weber

With 61 Figures

Springer-Verlag Berlin Heidelberg New York 1983

Professor Dr. O.L. LANGE
Lehrstuhl für Botanik II der Universität Würzburg
Mittlerer Dallenbergweg 64, 8700 Würzburg/FRG

Professor P.S. NOBEL
Department of Biology
University of California
Los Angeles, California 90024/USA

Professor C.B. OSMOND
Department of Environmental Biology
Research School of Biological Sciences
Australian National University
Box 475, Canberra City 2601/Australia

Professor Dr. H. ZIEGLER
Institut für Botanik und Mikrobiologie
der Technischen Universität München
Arcisstraße 21, 8000 München 2/FRG

ISBN-13: 978-3-642-68158-5 e-ISBN-13: 978-3-642-68156-1
DOI: 10.1007/978-3-642-68156-1

Library of Congress Cataloging in Publication Data. Main entry under title: Physiological plant ecology. (Encyclopedia of plant physiology; new ser., v. 12A–). Includes bibliographies and indexes. Contents: 1. Responses to the physical environment – – 4. Ecosystem processes – mineral cycling, productivity, and man's influence. 1. Plant physiology – Collected work. 2. Botany – Ecology – Collected work. I. Lange, O.L. (Otto Ludwig) II. Aragno, M. III. Series: Encyclopedia of plant physiology; new ser., v. 12A, etc. QK711.2.E5 vol. 12D 581.1s [581.1] 81-9348

Softcover reprint of the hardcover 1st edition 1983

2131/3130-543210

List of Contributors

W.D. Billings
Department of Botany
Duke University
Durham, North Carolina 27706/USA

W.R. Boynton
University of Maryland
Chesapeake Biological Laboratory
Center for Environmental
and Estuarine Studies
P.O. Box 38
Solomons, Maryland 20688/USA

J.L. Charley
Botany Department
University of New England
Armidale, NSW 2351/Australia

J.K. Detling
Natural Resource Ecology Laboratory
Colorado State University
Ft. Collins, Colorado 80523/USA

J. Ehleringer
Department of Biology
University of Utah
Salt Lake City, Utah 84112/USA

P.G. Falkowski
Brookhaven National Laboratory
Associated Universities, Inc.
Department of Energy and Environment
Upton, New York 11973/USA

D.M. Gates
University of Michigan
Biological Station
4053 Natural Science Building
University of Michigan
Ann Arbor, Michigan 48109/USA

C.A. Hall
Cornell University
Division of Biological Sciences
Section of Ecology and Systematics
Langmuir Laboratory
Ithaca, New York 14850/USA

P.J. Harrison
Departments of Oceanography
and Botany
University of British Columbia
Vancouver BC, V6T 2B1/Canada

P.G. Jarvis
Department of Forestry
and Natural Resources
The University of Edinburgh
King's Buildings
Mayfield Road
Edinburgh, EH9 3JU/United Kingdom

C.W. Keefe
University of Maryland
Chesapeake Biological Laboratory
Center for Environmental
and Estuarine Studies
P.O. Box 38
Solomons, Maryland 20688/USA

W.M. Kemp
University of Maryland
Horn Point Environmental Laboratories
P.O. Box 775
Cambridge, Maryland 21613/USA

H. Klinge
Max-Planck-Institut für Limnologie
D-2320 Plön/FRG

A. Kohler
Institut für Landeskultur und Pflanzen-
ökologie der Universität Hohenheim
Postfach 700562
D-7000 Stuttgart 70/FRG

B.C. Labus
Hoechst AG
Verkauf Landwirtschaft
Beratung K 607
D-6000 Frankfurt-Hoechst/FRG

K.J. LENDZIAN
Institut für Botanik und Mikrobiologie
Technische Universität München
Arcisstraße 21
D-8000 München/FRG

J.W. LEVERENZ
College of Forest Resources – ARIO
University of Washington
Seattle, Washington 98195/USA

R.S. LOOMIS
Department of Agronomy
and Range Science
University of California
Davis, California 95616/USA

E. MEDINA
Centro de Ecología
Instituto Venezolano de Investigaciones
Científicas
Caracas/Venezuela

A. MELZER
Institut für Botanik und Mikrobiologie
der Technischen Universität München
Arcisstraße 21
D-8000 München 2/FRG

H.A. MOONEY
Department of Biological Sciences
Stanford University
Stanford, California 94305/USA

T.R. PARSONS
Department of Oceanography
University of British Columbia
Vancouver, B.C. V6T1W5/Canada

F.W.T. PENNING DE VRIES
Centre for Agrobiological Research
P.O. Box 14
NL-6700 AA Wageningen/
The Netherlands

K. PFISTER
Present address:
CIBA-GEIGY AG
Agrochemical Division
CH-4002 Basle/Switzerland

B.N. RICHARDS
Department of Ecosystem Management
University of New England
Armidale, NSW 2351/Australia

R. SCHUBERT
Martin-Luther-Universität
Sektion Biowissenschaften
Wissenschaftsbereich Geobotanik
u. Botanischer Garten
Neuwerk 21
402 Halle/GDR

CH. STEINBERG
Bayerisches Landesamt
für Wasserwirtschaft
Lazarettstraße 67
D-8000 München 19/FRG

B.R. STRAIN
Department of Botany
Duke University
Durham, North Carolina 27706/USA

L.L. TIESZEN
Department of Biology
Augustana College
Sioux Falls, South Dakota 57197/USA

M.H. UNSWORTH
Department of Physiology
University of Nottingham
School of Agriculture
Loughborough/United Kingdom

W. URBACH
Institut für Botanik
und Pharmazeutische Biologie
Lehrstuhl für Botanik I
der Universität Würzburg
Mittlerer Dallenbergweg 64
D-8700 Würzburg/FRG

J.A. WEBER
Biological Station
The University of Michigan
Ann Arbor, Michigan 48109/USA

Contents Part D

Introduction
O.L. LANGE, P.S. NOBEL, C.B. OSMOND, and H. ZIEGLER 1

1 Nutrient Allocation in Plant Communities: Mineral Cycling in Terrestrial Ecosystems
J.L. CHARLEY and B.N. RICHARDS (With 4 Figures) 5

2 Nutrient Cycling in Freshwater Ecosystems
A. MELZER and CH. STEINBERG (With 6 Figures) 47

3 Nutrient Cycling in Marine Ecosystems
T.R. PARSONS and P.J. HARRISON (With 6 Figures) 85

4 Modeling of Growth and Production
F.W.T. PENNING DE VRIES (With 2 Figures) 117

5 Productivity of Agricultural Systems
R.S. LOOMIS (With 4 Figures) 151

6 Productivity of Grassland and Tundra
L.L. TIESZEN and J.K. DETLING (With 10 Figures) 173

7 Productivity of Desert and Mediterranean-Climate Plants
J. EHLERINGER and H.A. MOONEY (With 6 Figures) 205

8 Productivity of Temperate, Deciduous and Evergreen Forests
P.G. JARVIS and J.W. LEVERENZ (With 10 Figures) 233

9 Productivity of Tropical Forests and Tropical Woodlands
E. MEDINA and H. KLINGE (With 1 Figure) 281

10 Phytoplankton Productivity in Aquatic Ecosystems
W.R. BOYNTON, C.A. HALL, P.G. FALKOWSKI, C.W. KEEFE, and W.M. KEMP (With 2 Figures) 305

11 Effects of Biocides and Growth Regulators: Physiological Basis
K. PFISTER and W. URBACH (With 4 Figures) 329

12 Effects of Biocides and Growth Regulators: Ecological Implications
R. SCHUBERT 393

13 Eutrophication Processes and Pollution of Freshwater Ecosystems Including Waste Heat
A. KOHLER and B.C. LABUS 413

14 Ecophysiological Effects of Atmospheric Pollutants
K.J. LENDZIAN and M.H. UNSWORTH (With 1 Figure) 465

15 Ecophysiological Effects of Changing Atmospheric CO_2 Concentration
D.M. GATES, B.R. STRAIN, and J.A. WEBER (With 4 Figures) 503

16 Man's Influence on Ecosystem Structure, Operation, and Ecophysiological Processes
W.D. BILLINGS (With 1 Figure) 527

Author Index 549

Taxonomic Index 605

Subject Index for Parts A–D 613

Part A: Physiological Plant Ecology I
Responses to the Physical Environment

Introduction
O.L. LANGE, P.S. NOBEL, C.B. OSMOND, and H. ZIEGLER

1 Fundamentals of Radiation and Temperature Relations
G.S. CAMPBELL

2 Photosynthetically Active Radiation
K.J. MCCREE

3 Responses to Different Quantum Flux Densities
O. BJÖRKMAN

4 Non-photosynthetic Responses to Light Quality
D.C. MORGAN and H. SMITH

5 Responses to Photoperiod
F.B. SALISBURY

6 Plant Response to Solar Ultraviolet Radiation
M.M. CALDWELL

7 Responses to Ionizing Radiation
S. ICHIKAWA

8 The Aquatic Environment
W.N. WHEELER and M. NEUSHUL

9 Responses to Light in Aquatic Plants
S.W. JEFFREY

10 Responses of Macrophytes to Temperature
J.A. BERRY and J.K. RAISON

11 Responses of Microorganisms to Temperature
M. ARAGNO

12 Responses to Extreme Temperatures. Cellular and Sub-Cellular Bases
P.L. STEPONKUS

13 Ecological Significance of Resistance to Low Temperature
W. LARCHER and H. BAUER

14 Ecological Significance of Resistance to High Temperature
L. KAPPEN

15 Wind as an Ecological Factor
P.S. NOBEL

16 Fire as an Ecological Factor
P.W. RUNDEL

17 The Soil Environment
P. BENECKE and R.R. VAN DER PLOEG

Part B: Physiological Plant Ecology II
Water Relations and Carbon Assimilation

Introduction
O.L. LANGE, P.S. NOBEL, C.B. OSMOND, and H. ZIEGLER

1 Water in the Soil-Plant-Atmosphere Continuum
J.B. PASSIOURA

2 Water in Tissues and Cells
M.T. TYREE and P.G. JARVIS

3 Water Uptake and Flow in Roots
P.E. WEATHERLEY

4 Water Uptake by Organs Other Than Roots
P.W. RUNDEL

5 Transport and Storage of Water
M.H. ZIMMERMANN and J.A. MILBURN

6 Resistance of Plant Surfaces to Water Loss: Transport Properties of Cutin, Suberin and Associated Lipids
J. SCHÖNHERR

7 Stomatal Responses, Water Loss and CO_2-Assimilation Rates of Plants in Contrasting Environments
E.-D. SCHULZE and A.E. HALL

8 Mathematical Models of Plant Water Loss and Plant Water Relations
A.E. HALL

9 Physiological Responses to Moderate Water Stress
K.T. BRADFORD and T.C. HSIAO

10 Desiccation-Tolerance
J.D. BEWLEY and J.E. KROCHKO

11 Frost Drought and Its Ecological Significance
W. TRANQUILLINI

12 Water Relations in the Germination of Seeds
D. KOLLER and A. HADAS

13 Environmental Aspects of the Germination of Spores
A.S. SUSSMAN and H.A. DOUTHIT, Jr.

14 Physiological Responses to Flooding
R.M.M. CRAWFORD

15 Functional Significance of Different Pathways of CO_2 Fixation in Photosynthesis
C.B. OSMOND, K. WINTER, and H. ZIEGLER

16 Modelling of Photosynthetic Response to Environmental Conditions
G.D. FARQUHAR and S. VON CAEMMERER

17 Regulation of Water Use in Relation to Carbon Gain in Higher Plants
I.R. COWAN

18 Plant Life Forms and Their Carbon, Water and Nutrient Relations
E.-D. SCHULZE

Part C: Physiological Plant Ecology III
Responses to the Chemical and Biological Environment

Introduction
O.L. LANGE, P.S. NOBEL, C.B. OSMOND, and H. ZIEGLER

1 The Ionic Environment and Plant Ionic Relations
M.G. PITMAN and U. LÜTTGE

2 Osmoregulation
R.G. WYN-JONES and J. GORHAM

3 Halotolerant Eukaryotes
R. MUNNS, H. GREENWAY, and G.O. KIRST
4 Halophilic Prokaryotes
A.D. BROWN
5 Physiology and Ecology of Nitrogen Nutrition
M. RUNGE
6 Influence of Limestone, Silicates and Soil pH on Vegetation
H. KINZEL
7 Toxicity and Tolerance in the Response of Plants to Metals
H.W. WOOLHOUSE
8 Ecophysiology of Nitrogen-Fixing Systems
A.H. GIBSON and D.C. JORDAN
9 Ecophysiology of Mycorrhizal Symbioses
M. MOSER and K. HASELWANDTER
10 Ecophysiology of Lichen Symbioses
U. MATTHES and G.B. FEIGE
11 Interactions Between Plants and Animals in Marine Systems
W. HÖLL
12 Ecophysiology of Carnivorous Plants
U. LÜTTGE
13 Host-Parasite Interactions in Higher Plants
P.R. ATSATT
14 Virus Ecology – 'Struggle' of the Genes
A.J. GIBBS
15 Ecophysiology of Zoophilic Pollination
S. VOGEL
16 Physiological Ecology of Fruits and Their Seeds
D.H. JANZEN
17 Physiological and Ecological Implications of Herbivory
S.J. MCNAUGHTON
18 Interactions Between Plants
E.I. NEWMAN

Introduction

O.L. Lange, P.S. Nobel, C.B. Osmond, and H. Ziegler

In the last volume of the series '*Physiological Plant Ecology*' we have asked contributors to address the bases of ecosystem processes in terms of key plant physiological properties. It has often been suggested that it is not profitable to attempt analysis of complex living systems in terms of the properties of component individuals or populations, i.e., the whole is more than the sum of its parts. Nevertheless, assessments of ecological research over the last century show that other approaches are seldom more helpful. Although it is possible to describe complex systems of living organisms in holistic terms, the most useful descriptions are found in terms of the birth, growth and death of individuals. This allows analysis of performance of the parts of the whole considering their synergistic and antagonistic interrelationships and is the basis for a synthesis which elucidates the specific properties of a system. Thus it seems that the description of ecosystem processes is inevitably anchored in physiological understanding. If enquiry into complex living systems is to remain a scientific exercise, it must retain tangible links with physiology. Of course, as was emphasized in Vol. 12A, not all of our physiological understanding is required to explore ecosystem processes. For pragmatic purposes, the whole may be adequantely represented as a good deal less than the sum of its parts. The chapters in this volume help us identify some of the key physiological processes in ecosystem function and help promote those areas of physiological research which are poorly understood in this context.

We have had to single out only a few ecosystem processes for consideration in this volume. This has not been difficult, as physiological understanding of some, such as maintenance of species diversity, can only be perceived at present in dim outlines of plant morphology, metabolism, and resource capture. Recent insights into successional processes and plant competition may provide a path to understanding such questions in physiological terms. It seems, for example, that a relatively small number of combinations of plant life form, reproductive behavior, and longevity have evolved, but present understanding of the physiological bases and significance of these processes in individual plants or populations is meager. Omission of successional processes in this volume is to some extent compensated by excellent recent reviews (F.A. Bazzaz, Ann. Rev. Ecol. Syst. 10, 351–172, 1979; F.A. Bazzaz and S.T.A. Pickett, ibid. 11, 287–310, 1980). The dynamics of plant populations, a key ecosystem process, is now an important research field in itself. Its techniques must be taken into ecophysiology if we are to assess adaptive significance in quantitative terms. Likewise, physiological perspectives must enter population ecology wherever more than numerical satisfaction is sought. These questions are also overlooked in this volume, but have been the subject of recent major works (J.L. Harper, Population biology of plants; Academic Press: London, New York 1977; J.P. GRIME, Plant strategies and vegetation processes; Wiley: London 1979).

All ecosystem processes are consequences of the conversion of solar energy to biologically negotiable chemical energy during photosynthesis, the ecophysiological aspects of which have been dealt with in Vol. 12A and 12B. In ecosystems, the potential of photosynthesis is attenuated by nutritional and biological constraints, some of which are described in Vol. 12C. By beginning with treatments of nutrient cycling as an ecosystem process in this volume, we hope to provide some continuity with the previous volume. Another motive in commencing with nutrient cycling is that in terrestrial (Chap. 1) and aquatic ecosystems (Chaps. 2 and 3), nitrogen and phosphorus nutrition are major limitations of productivity. As shown in these chapters, the gates controlling nutrient cycling involve physiological processes in organisms which have yet to attract the attention of sophisticated laboratory research.

The processes in the ecosystem which attract most attention are productivity and stability, simply because, from an anthropocentric viewpoint, these are of paramount economic and esthetic significance. We have been selective rather than comprehensive in our choice of ecosystems when considering productivity, bearing in mind the wealth of data now available after a decade of research (1964–74) under IBP programs. Much of this has appeared in comprehensive regional reports, such as those of the Japanese IBP Commitee, and in a series of 20 volumes published by Cambridge University Press. Some of these data have been reported in the series Ecological Studies (e.g., Vols. 2, 7, 16, 17, 27, 29, and 32). In spite of this we remain conscious of several oversights. Physiological processes as a basis of ecosystem processes in mangroves, for example, are likely to be more important than in many other ecosystems, but must await further study.

Productivity of ecosystems depends primarily on the rate and duration of photosynthetic activity in individual leaves. It is a product of investments made in the major leaf protein, the enzyme ribulose bisphosphate carboxylase oxygenase, which is by far the most abundant single protein in the biosphere. Although the detailed economics of plant carbon metabolism have now been applied as a basis for modeling plant growth and production (Chap. 4), it has not yet been satisfactorily coupled to nutrient economics, except in agricultural systems where nutrition is a controlled variable (Chap. 5). Yet, as subsequent chapters show, nutrient economics are vital to ecosystem productivity. In herbaceous (Chap. 6) and shrub ecosystems (Chap. 7), leaf nitrogen content is an important index of potential productivity, and of course, an important inducement to herbivory.

In forest ecosystems, characterized by a high proportion of woody, nonproductive biomass, new dimensions must be added to the interactions between nutrition and production (Chap. 8). Here population dynamics clearly intrudes into physiology with complex and as yet unclear relationships emerging between life form, leaf longevity, leaf nitrogen content and photosynthetic activity. It is clearly evident that the ecophysiological understanding of forests at present lags far behind that of herbaceous and shrubby ecosystems, no doubt a consequence of the large size, long life time, and complex structure of trees. The sophistication of forestry practice in temperate environments indicates that economic potential does not have to await ecophysiological understanding. However, our understanding of productivity in tropical forests (Chap. 9), which is perhaps less than in any other ecosystem, is likely to arrive too late to prevent large-scale irreversible change. Productivity of aquatic ecosystems, also limited physiologically by access to nutrients at dilute concentration in solution (Chap. 10), is also poorly understood, even in terms of basic physical processes.

The adverse effects of attempts to promote plant productivity are usually felt at higher tropic levels. Whether one attempts to stimulate plant growth or control competition and herbivory through sophisticated organic chemistry (Chaps. 11 and 12) or simply improves nutrition (Chap. 13), Muir's maxim in ecology is rediscovered: "When we try to pick out anything by itself, we find it hitched to everything else in the Universe". Fortunately, where higher trophic level responses are involved, corrective action tends to be taken relatively rapidly and the effects confined. However, in consuming the fossil fuel residues of several hundred million years plant productivity in the space of a few hundred years, we are perturbing many ecosystem processes in a spectacular way. For example, a minor element, sulfur, has been admitted to the hydraulic cycle in massive proportions (Chap. 14), with many adverse consequences for productivity of aquatic and forest ecosystems. Of much greater significance in the not very distant future is the return of fossil plant biomass to the atmosphere as CO_2 (Chap. 15). Not only will this have important direct effects, increasing the potential productivity of most individual plants, but it will exert major effects on global climate and hydrology, which will have indirect effects, yet to be understood, on productivity and most other ecosystem processes. Without doubt, these effects will be manifest in integrated physiological function of individuals and adaptation will proceed through the filters of performance (productivity) and tolerance (survival).

For good or ill, man is arranging some experiments of grand proportions in ecophysiology (Chap. 16). The dominant mammal and his society will continue to exert pressures, to meet the nutritional needs of many and to preserve the wilderness for a few, which will tax our insight and predictive ingenuity. As suggested in the introduction to Volume 12A, this challenge to our ability to extract meaningful relationships and to contrive achievable experiments underscores the success story of ecological physiology and is one of its special fascinations.

In concluding this endeavor we wish to thank not only the contributors to this volume, but to the previous volumes as well. Aside from the education of the editors, one of our chief rewards has been to have made many new friends, and to have retained so many old friendships while fulfilling our task. We are particularly grateful for the technical support by many colleagues and by many helpers in our laboratories and offices, for support of the publishing house from copy-editing of the manuscripts to production of the volumes, and to Ms. Dipl.-Biol. Maria Braun (Würzburg) for her splendid work in indexing the volumes.

1 Nutrient Allocation in Plant Communities: Mineral Cycling in Terrestrial Ecosystems

J.L. CHARLEY and B.N. RICHARDS

CONTENTS

1.1 Introduction 6
1.1.1 Ecosystem Growth and Mineral Capital 6
1.1.2 Soil Metabolism: Its Central Role in Mineral Cycling 8

1.2 Total Nutrient Capital 9
1.2.1 Sources of Nutrients 9
1.2.1.1 Atmospheric 9
1.2.1.2 Parent Material 12
1.2.1.3 Relative Importance of Atmospheric and Parent Material Sources 13
1.2.2 Nutrient Stocks 14
1.2.2.1 Soils 14
1.2.2.2 Vegetation and Litter 16

1.3 Mineral Cycling Processes 18
1.3.1 The Biochemical Cycle 18
1.3.2 Biogeochemical Cycling 20
1.3.2.1 Nutrient Uptake 20
1.3.2.2 Nutrient Return 23
1.3.3 Nutrient Release from Residues: Mineralization of Litter and Soil Organic Matter 27
1.3.3.1 Biological Determinants of Organic Matter Decomposition . . 28
1.3.3.2 Distribution of Microbiological Activity in Space and Time . . 30
1.3.3.3 Measurement of Decomposition Rate 32
1.3.3.4 Extrinsic and Intrinsic Factors Affecting Decomposition of Detritus 33
1.3.3.5 Regulation of Decomposition Processes 35
1.3.4 The Geochemical Cycle: Nutrient Losses 36
1.3.4.1 Gaseous Losses 36
1.3.4.2 Outgoes in Drainage Waters 37

1.4 Conclusions 39

References 39

1.1 Introduction

In recent years the literature on mineral cycling in terrestrial ecosystems has expanded greatly, and few if any of the major biomes have escaped attention. We do not propose to review this large body of information in detail. Rather, it is our intention to outline the broad features of mineral acquisition, distribution and cycling that apply generally in terrestrial ecosystems, and to highlight several facets of the internal cycle that are vital for community nutrition but which seem to have been rather neglected. Inevitably, we cannot treat all essential elements equally, nor can we deal with the functional aspects of elemental storage and flux in all their complexity. The aboveground components of ecosystems have been studied in great detail, whereas spatial and temporal patterns in soil are but poorly understood. The account that follows is in part, therefore, an attempt to redress this imbalance and consequently is uneven in its treatment of both the static and dynamic attributes of ecosystems.

1.1.1 Ecosystem Growth and Mineral Capital

Ecosystems are open systems subject to inputs from the substratum and the aerial environment, as well as outputs of dissolved, particulate and gaseous materials to drainage sinks and the atmosphere. Elements accumulate within the system as a result of successional processes that result in increased biomass. The course of this developmental sequence is determined by the interaction of the independent state factors, climate, organisms, parent material and relief, that operate during the developmental interval (JENNY 1961). So long as biomass increases, i.e. provided net primary production occurs, the system's content of minerals will increase. Ultimately, in the absence of disturbance, and provided the state factors remain effectively constant, a condition of quasi-equilibrium will be reached. Invariably, however, there will be fluctuations about this steady state due to short-term environmental change.

According to ODUM (1969), ecosystems conserve essential elements by evolving mechanisms that promote internal recycling, so that succession may be viewed as the process by which plant communities accumulate sufficient nutrient capital to allow the rise of succeeding ones, until such time as the steady state is reached. Having attained this mature condition, the system is held to exercise the maximum degree of control over its accumulated store of nutrients and the maximum ability to entrap fresh inputs. Recently, however, evidence has accumulated to support an alternative hypothesis that challenges crucial aspects of the traditional ecosystem growth model and the successional pattern of element accumulation and retention (VITOUSEK and REINERS 1975; VITOUSEK 1977; BORMANN and LIKENS 1979). This evidence suggests that maximum capacity to entrap inputs of elements and regulate their export coincides with that phase of succession, primary or secondary, where net primary production (and thus rate of accumulation of nutrients in the biomass) is greatest. Thus in terms of the balance between inputs and outputs, the "leakiness" of an ecosystem is high as succession is initiated, rapidly declines as growth rate increases, then rises again in the later stages of development as net community production falls to zero at system maturity. These critical reassessments of the "strategy" of ecosystem development have important implications for the acquisition and retention of nutrient capital.

VITOUSEK'S (1977) model is based on mineral input–output balance in relation to net community productivity during the course of primary and secondary succession. Inputs are assumed to be constant over time, and biomass to increase asymptotically to the mature condition. Storage of nutrients regulates output, storage being determined by net productivity and the supply of individual chemical species. Output equals input at the initiation of primary succession (time zero) and again when the steady state is reached in both primary and secondary successions. Following major disturbance of the ecosystem, net productivity is negative and, for a brief period, rundown of previously accumulated nutrient capital takes place until such time as the site is revegetated and the growing plant community provides an expanding, internal nutrient sink. BORMANN and LIKENS (1979), in their recent account of the Hubbard Brook study, convincingly argue a similar proposition of mineral loss regulation by growth processes, but contend that a simple comparison of the balance between mineral inputs and outputs at different successional stages is insufficient to account fully for the complex changes in mineral cycling that take place during ecosystem development. Their models of biomass accumulation and elemental storage relate specifically to secondary succession after a major disturbance such as clearcutting. However, they have much wider relevance, for they raise issues concerning the validity of the commonly accepted community growth curve, viz. that which shows asymptotic growth towards a steady state.

BORMANN and LIKENS (1979) propose four principal phases of ecosystem development after clearcutting and three, by implication, in primary succession. There is a relatively short reorganization interval (R) following disturbance, an aggradation phase (A) of major growth in biomass and nutrient capital, then a transition phase (T) of varying duration when biomass and nutrient content decline to a condition of steady state (SS).

From their analysis these authors conclude that the tightness of nutrient cycles does not develop in parallel with biomass accretion; rather it is greatest in the A phase and least during the R and T intervals. They visualise the aggradation phase, which has the greatest degree of biogeochemical stability, as culminating in a biologically unstable condition, and attribute this instability to its even-agedness which is subsequently destroyed by random mortality of individuals or subpopulations. In consequence, there is a decline in total biomass and a concomitant loss of nutrients as community respiration exceeds production. Over a protracted period of such transition, the ecosystem stabilises in terms of biomass and nutrient storage. At such time it will consist of a structural and floristic mosaic of varying complexity comprised of individual areas representative of all four developmental stages. Consequently, they designate the ultimate condition of the whole ecosystem as the shifting mosaic steady state and ascribe to it a degree of regulatory control over biogeochemical export which reflects the relative proportions of its constituent phases.

There is a growing body of evidence underlining the importance of natural disturbance, such as fire and wind, as causes of structural and functional patterns in plant communities (BORMANN and LIKENS 1979; WHITE 1979; see also Chaps. 15 and 16, Vol. 12A). Accordingly, the probability of autogenic succession proceeding uninterrupted to a steady state may be lower than commonly supposed, and the shifting mosaic steady state may therefore be a more generally applicable model of spatial and temporal variation in terrestrial ecosystems. If so, a re-examination and re-interpretation of much mineral cycling work may be required.

1.1.2 Soil Metabolism: Its Central Role in Mineral Cycling

Except in special circumstances, the solar energy captured by photosynthesis (gross primary production) is the total amount of energy available for all the energy-consuming processes of ecosystems. These include biosynthesis, growth, nutrient uptake and transport, the energy needed for these purposes being provided by the oxidation of photosynthate during respiration. The soil, or soil–litter subsystem, is deficient in photosynthetic organisms and so does not have the capacity to capture significant quantities of solar energy itself. Instead, it depends on energy-rich substances produced elsewhere, viz. plant and animal residues or "litter". Litter decomposition is complex and involves both microbes and the soil fauna, the two major microbial groups participating being fungi and bacteria. The relative contributions of the major categories of decomposer organisms has yet to be determined for most terrestrial ecosystems.

The soil–litter subsystem acts as a control valve (the "work gate" of ODUM 1970) through which must pass virtually all the nutrients taken up by the primary producers and most of the energy fixed by them as well (Fig. 1.1). Not only this, but the rate at which energy and nutrients flow through the gate frequently governs the productivity of the whole ecosystem. This regulatory role (see Sect. 1.3.3.5) is particularly important in systems which operate on a very low nutrient capital, and in those where decomposition processes are limited by environmental extremes. In such situations the level of community productivity is likely to be determined by organic matter turnover in the soil–litter subsystem.

In this context, the operation of the soil–litter subsystem is the key to understanding total ecosystem function. It is the site of mineralization processes that release nutrients previously bound in the plant biomass, so that they are once again available for re-cycling through aboveground portions of the ecosystem. Mineralization is brought about by soil animals and microorganisms which depend upon a regular accession of plant and animal residues to provide them with food and energy. Thus it is in the soil–litter subsystem that the two major functional processes of energy flow and nutrient cycling interact most strongly, and neither process can be fully understood without adequate consideration of the other. A detailed examination of soil metabolism is therefore an essential feature of any comprehensive programme of ecosystem analysis.

While much has been written about nutrient cycles in general, particularly at the total system level, there has been little emphasis on the mechanisms of nutrient interchange that operate at the interface between the soil and vegetation subsystems. There

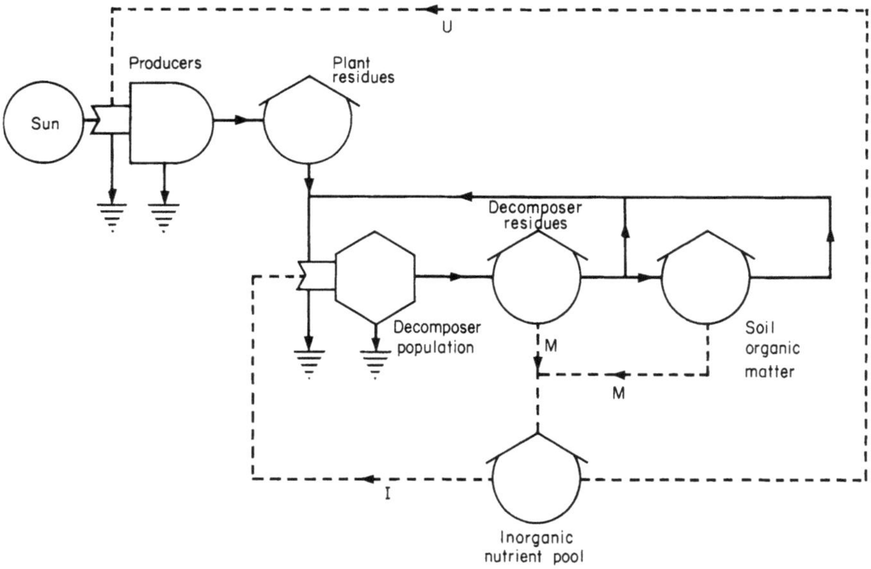

Fig. 1.1. Simplified energy circuit diagram (sensu ODUM 1970) showing how the processes of organic matter decomposition in the soil-litter subsystem affect community productivity overall. The plant-herbivore subsystem has been omitted for the sake of clarity. Major energy flows are shown as *solid lines*, pathways of inorganic nutrients as *dotted lines*. *M* mineralization; *I* immobilization; *U* plant uptake. (Redrawn from RICHARDS 1974)

is a large gap in our understanding of mineral cycling at this process level. This is especially true when one considers day-to-day variations in nutrient flux, yet such short-term changes might be of considerable significance in determining system function overall.

1.2 Total Nutrient Capital

The potential susceptibility of an ecosystem to major nutrient depletion, and its capacity to recover quickly from disturbance can depend on the amount of total system capital that is held in aboveground biomass relative to the soil. In the same way, the gross quantities, kinds and distribution of nutrient elements in the solum serve as a basis for judgments about the potential for nutrient depletion should surface truncation of the profile occur. For these reasons, it is useful to consider briefly the principal sources of nutrients for ecosystem development and the static picture of elemental distributions among the several compartments.

1.2.1 Sources of Nutrients

1.2.1.1 Atmospheric

Nutrient inputs from the atmosphere include dry sedimentation of aerosols, their scrubbing from the air by rain, collection of the same materials on surfaces by impaction from moving air masses, adsorption onto charged surfaces, and the fixation of gases by biological and non-biological mechanisms.

Accessions in rainfall have been extensively studied and fair estimates of their annual contribution can be had for most parts of the world (STEINHARDT 1973). With few exceptions values obtained refer to bulk precipitation, which includes sedimentation into gauges between rain events as well as what is collected during rain or snowfall.

Much higher inputs of chloride, sulphur and the metallic cations are often recorded in coastal areas, and sulphur accessions in areas subject to industrial pollution are commonly greater again.

The atmospheric loadings of nitrogen and phosphorus measured by NICHOLLS and COX (1978) at a small lake in Ontario are probably fair estimates of what might be expected generally in forested areas away from significant dust and industrial influence. For phosphorus they found an input of 0.74 kg ha^{-1} yr^{-1}, 20% of which was from locally derived pollen, while for total nitrogen the income amounted to 16 kg ha^{-1} yr^{-1}, 63% of which was in dissolved inorganic form.

Bulk precipitation is the geochemically significant input, not merely the chemical load in rainwater. Even so, there remains a further accession that is not collected by bulk precipitation gauges located in the open. This is the elemental content of aerosols impacted on vegetation surfaces. Approximation of impaction input to forested ecosystems can be had by determining the difference between the quantities of elements collected by gauges under canopies and those in the open. In these circumstances the catch of most elements under canopies is the greater since it includes a component scavenged by foliage from moving air masses, plus whatever may be leached or exuded from leaves. An example of the potential significance of this nutrient source is provided by NIHLGARD (1970), whose results are shown in Table 1.1.

Table 1.1. Nutrient catch by aerosol impaction on plastic screens in a forest clearing in Sweden[a]

Element	Elemental content (kg ha^{-1})		
	In rainfall	Below plastic nets	Aerosol inputs
Na	0.9	8.9	8.0
K	0.3	1.0	0.7
Ca	1.0	5.5	4.5
Mg	0.2	1.6	1.4
Cl	2.1	11.6	9.5
SO_4-S	2.2	8.4	6.2
PO_4-P	0.02	0.07	0.05
Total N	2.1	5.2	3.1
Rain (mm)	276	291	

[a] Data from NIHLGARD (1970) for a 7-month collection period

In addition to contributions of particulate matter, the atmosphere is a source of gaseous forms of S and N. The first occurs principally as SO_2, while the latter is represented by N_2 and trace amounts of NH_3 and nitrogen oxides (see Chap. 14, this Vol.).

Sulphur dioxide absorption by soils (ALWAY et al. 1937) and leaves (THOMAS and HILL 1935) has been recognised for some time and has attracted considerable attention in recent years. The capacity of soil to absorb SO_2 is potentially very high, as demonstrated by the experiments of FUHR et al. (1948) in which gas concentrations as high as 1×10^7 µg m^{-3} SO_2 were reduced to trace levels after 40 min over soil. Similarly, foliar absorption as demonstrated by OLSEN (1957) for healthy cotton plants was estimated to provide about 30% of the plants' needs. It appears, therefore, that atmospheric SO_2 is a normal, and perhaps necessary, source of S in many situations.

Early reports of ammonia absorption (INGHAM 1950a, b) have been substantiated by subsequent work with soils and plants. MALO and PURVIS (1964), for example, have suggested gains from this source ranging from 57 to 225 g ha^{-1} day^{-1} for agricultural soils in New Jersey. Plant absorption (HUTCHINSON et al. 1972) is also sufficiently effective for ammonia to be considered a potentially important source of N, particularly as concentrations in air average between 30 and 40 µg m^{-3} (MALO and PURVIS 1964). However, as with SO_2, extrapolation from experimental laboratory conditions to actual field situations is most uncertain and the magnitude of real input from this source must remain a matter for speculation.

Nitrogen Fixation. Biological nitrogen fixation involves either symbiotic or free-living soil microbes. The former include root nodule symbioses and some lichens. The ecophysiology of nitrogen fixers is discussed elsewhere in this series (Chaps. 8 and 10, Vol. 12 C), but brief mention must be made here of those aspects pertinent to mineral cycling.

While none would deny the agronomic importance of the legume–rhizobium symbiosis, its significance from the geochemical viewpoint is another matter. Following a critical review of the data then available, HUTCHINSON (1954) concluded that legumes are responsible for only 3%–4% of the nitrogen fixed at the surface of the earth each year. The cultivated herbaceous legumes of the family Fabaceae are capable of fixing 100 kg ha^{-1} or more nitrogen annually, but it would be unwise to assume that fixation rates attained by these herbaceous species in cultivation apply also to wild legumes. Vast areas of the earth's surface however, especially in the tropics and subtropics, are populated by woody perennial legumes, so that the contribution of the Leguminosae as a whole to the geochemical cycle of nitrogen may be greater than Hutchinson was prepared to concede. None the less, it is worth noting that many tropical legumes apparently do not nodulate (NORRIS 1956) and therefore presumably do not fix nitrogen at all.

Perhaps of greater geochemical significance than legumes are the actinorrhiza symbioses of non-leguminous angiosperms. Under appropriate ecological conditions, nodulated non-legumes may achieve quite high fixation rates, for example 62–164 kg N ha^{-1} yr^{-1} for *Alnus crispa* in Alaska (CROCKER and MAJOR 1955), and 58 kg N ha^{-1} yr^{-1} for *Casuarina equisetifolia* in Africa (DOMMERGUES 1963). The contribution of actinorrhizas to the nitrogen economy of many forest ecosystems is no doubt considerable, and bearing in mind their broad distribution, it is likely they will prove to be among the major contributors to nitrogen cycling in the biosphere.

Nitrogen fixation also occurs in the coralloid roots of some gymnosperms (BERGERSEN et al. 1965), the endophyte being a blue-green alga. These plants may be important in the nitrogen economy of particular ecosystems, but their geochemical significance is slight. The same is true of lichen symbioses involving blue-green algae, although lichens have an additional role in initiating mineral cycling by colonizing bare rock surfaces.

The organisms responsible for non-symbiotic nitrogen fixation are either bacteria or blue-green algae, i.e. prokaryotes. Because of the limitations imposed by their energy requirements, and their preferential use of mineral nitrogen, the agronomic value of heterotrophic nitrogen-fixing bacteria in cultivated soils is thought to be slight, and the same may be true of their geochemical significance. It has been suggested (JENSEN 1950) that their efficiency may be greater in uncultivated soils, where crops are not removed but where plants are allowed to die and decompose in situ, but available evidence points to a fixation rate of less than 5 kg N ha^{-1} yr^{-1}. Fixation by free-living nitrogen fixers is, however, greatly stimulated in the rhizosphere, especially in grassland ecosystems where it may be an order of magnitude higher (JORDAN 1981). Nitrogen fixation occurring in association with mycorrhizal root systems should also be ascribed to the activities of free-living bacteria in the rhizosphere (BEVEGE et al. 1978).

The blue-green algae (cyanobacteria) are autotrophic microorganisms whose geochemical significance as nitrogen fixers is greater than that of any of the free-living bacteria, since they are independent of exogenous organic carbon sources and occur over a wide range of climates, from tropical to polar (JORDAN 1981). They have a recognized role in wetland rice cultivation where they may fix perhaps 50 kg N ha^{-1} yr^{-1}, although the magnitude of their contribution relative to that of free-living and associative heterotrophic bacteria has not yet been satisfactorily quantified (WATANABE 1981).

1.2.1.2 Parent Material

All essential micro- and macro-nutrient elements other than N may be derived principally from soil parent material. During soil formation, ions released by weathering may be removed by leaching, accumulated in biomass or incorporated in secondary minerals and retained in the solum. However, accurate determination of the fraction of minerals released by weathering that is retained in the functional volume of the ecosystem is complicated because of the simultaneous accumulation of various atmospheric inputs. On the other hand, it is not a particularly difficult task to determine with fair precision how much is lost from the elemental content of a given volume of rock when it has weathered to soil (SIMONSON 1970). Where precipitation is high and weathering intense, soluble products of parent material decay are prone to removal in drainage, but towards more arid conditions such loss decreases and is effectively zero where precipitation is insufficient to promote periodic leaching of the profile. Here chemical constituents of rock decay remain to contribute to total stocks.

Because of its central importance in biological function, and because parent material is the only significant source, the fate of P during pedogenesis and ecosystem development is of special significance. Several of the trace elements may be included in this category as well. Relative to the common cations and S, for example, the persistence of P in soils is marked, provided Fe and Al are present in significant quantities. Even so, studies such as that of WILD on granitic soils in temperate and subtropical Australia (WILD 1958) show that losses during soil formation may be substantial. Notwithstanding this, elements which form highly insoluble secondary compounds under oxidizing and reducing conditions are the ones best able to resist the influence of leaching and by their soil content most closely reflect their original status in parent material. The essential elements amongst these, provided they are not significant in atmospheric inputs, are those which may limit ecosystem development most by virtue of deficiencies in parent material, a thesis well illustrated by the work of BEADLE (1966) on nutrient factors moulding the development of the Australian vegetation.

Weathering. As GORHAM et al. (1979) point out, soluble products of weathering are best defined as potential inputs to the ecosystem, provided they are generated within the root zone and thus capable of being incorporated into the organic regime. However, there is presently no reliable method for determining either natural rates of release of nutrients from resistates or the efficiency of their capture by the biota. Accordingly, most estimates of the contribution of weathering are arrived at by difference and are thus subject to the uncertainties associated with determination of compartment nutrient stocks.

That biological agencies promote solubilization of primary minerals has been known ever since Julius von Sachs demonstrated the ability of plant roots to etch the surface of marble. There is considerable uncertainty, however, as to the quantitative significance in mineral cycling of such acid production by roots and microorganisms, since carbonic acid is weak and many of the organic acids of microbial and plant origin occur only at low concentrations. On the other hand, many organic acids chelate readily with metallic ions found in silicate minerals and chelation is known to increase nutrient availability. Microorganisms capable of producing 2-ketogluconic acid, a powerful chelating agent, are preferentially stimulated in the rhizosphere (WEBLEY and DUFF 1965), utilizing glucose exuded by the roots as a substrate. If solubilization of primary minerals is thereby facilitated, this phenomenon might be significant in the initiation of nutrient cycles, or in the aggradation phase of ecosystem development generally.

1.2.1.3 Relative Importance of Atmospheric and Parent Material Sources

In primary succession, where soil and vegetation are developing together, judgements about the relative importance of the two principal nutrient sources may well be influenced by considerations different from those applied when secondary succession is the basis of discussion. In primary succession, the initial capture of elements is taking place and nutrient stocks are being accumulated for posterity, as it were. With the passage of time, however, the matters of central importance to continuation of the ecosystem will become the retention of what has been accumulated and an increasing dependence on atmospheric inputs. The older and more deeply weathered a soil, the less its residue of primary minerals and the less the ability of root systems to recover nutrient ions from depth. In the extreme cases of lowland tropical rainforests on deep soils derived from nutritionally poor parent materials, the ecosystem may well be virtually dependent on atmospheric inputs if it is to make good those small losses to drainage that inevitably occur, and so maintain its steady state.

1.2.2 Nutrient Stocks

1.2.2.1 Soils

Two considerations limit the usefulness of studies of mineral capital in ecosystems. The first is that a significant number deal with biomass mineral capital alone; the second is that where soil analyses are given they commonly refer only to part of the profile and not to the entire depth that is accessible to roots. If an adequate appraisal of the nutritional physiology of an ecosystem is to be made, omissions of the first kind are serious. As to the second matter, limited sampling depth may not be a major deficiency because in the majority of situations the greater part of the significant, short-term nutrient flux takes place in a relatively shallow zone near the profile surface where roots, organic matter and microbiological activity are concentrated. It should also be kept in mind that only a small fraction of the total soil volume above the lower limit of root penetration may be utilized effectively for nutrient uptake (VOIGT et al. 1964).

For present purposes our concern is primarily to indicate the range of nutrient element stocks that may be encountered across a spectrum of soils representing a broad range of climatic conditions and parent materials ranging from nutrient-rich to nutrient-poor. Accordingly, we have calculated volumetric contents of several major elements (Table 1.2) for a suite of Australian soils that has been sampled to at least 1 m, and where standard methods of analysis have been employed. Higher upper limits for stocks of each element do occur. Nevertheless, the values given are probably quite sufficient to serve as a general indication of soil nutrient capital for comparison with storage in biomass of widely differing vegetation types throughout the world.

Cursory examination of the data in Table 1.2 shows the influence of parent material and climate on nutrient reserves generally, and the interaction of these variables on reserves of N and the exchangeable ions. Phosphorus levels, however, do not correlate with climate but are strongly influenced by the nature of the parent material. For example, it is instructive to compare samples 6 and 7, which represent very similar climates but significantly different vegetation; also soils 8, 9 and 10, which developed on moderate, poor and very

Table 1.2. Nutrient element contents in the 0–100 cm zone of a range of Australian soils[a]

Soil	Annual rainfall (mm)	Parent material	Vegetation	Elemental content ($g\ m^{-3}$)				
				Total		Exchangeable		
				N	P	Ca	Mg	K
1	330	Quaternary alluvium	Eucalypt woodland	1,490	740	4,530	1,140	450
2	380	Quaternary alluvium	Shrub steppe	770	290	1,710	1,900	410
3	380	Quaternary alluvium	Tall eucalypt woodland	680	870	3,420	1,450	840
4	635	Siltstone	Dwarf woodland	880	170	1,380	2,020	460
5	660	Basalt colluvium	Grassland	1,490	440	8,380	2,760	300
6	1,090	Sandstone	Eucalypt forest	600	320	300	220	60
7	1,070	Basalt	Rainforest	1,940	3,310	1,390	270	1,150
8	1,575	Shale	Tall eucalypt forest	1,370	230	40	80	40
9	1,625	Sandstone	Eucalypt forest	350	110	120	120	70
10	1,600	Siliceous sand	Rainforest	120	20	50	30	90
11	2,135	Basalt colluvium	Alpine woodland	1,620	4,920	100	30	60

[a] Soil 10, data of authors; remainder calculated from analytical data of STACE et al. (1968)

poor parent materials all under fairly high rainfall, support forests the largest of which occurs on the poorest soil. Such apparent anomalies are not uncommon and merely highlight the inadequacy of dealing with biomass nutrient content alone when attempting to understand the overall functional processes of terrestrial ecosystems. Clearly, the contrast between samples 6 and 7 or 8 and 10 would be of major significance in assessing the relative consequences of disruption to the nutrient cycle should the forests be burnt, selectively logged or clearcut. They may also serve as an indication of the relative importance of atmospheric and parent material sources for the maintenance of steady state conditions in the different ecosystems.

Anisotropic Nutrient Distribution. An aspect of soil nutrient capital seldom commented on is its spatial heterogeneity, yet this is one of its most consistent and significant features. In forest soils and many others, elements such as N and S, which occur mainly in organic combination, commonly show decreased concentration from the surface to subsoil. Much the same pattern occurs with organically bound P and the exchangeable reserves of Ca and K. Examples from a collection of acid forest soils are shown in Fig. 1.2.

A horizontal pattern of nutrient distribution may also be evident in surface soil. ZINKE (1962) was the first to show clearly the concordance between soil chemical patterning and litter "haloes" of individual trees. Thereafter, various workers have shown that diversion of precipitation to stemflow affects soil chemical and physical properties around the boles of trees (GERSPER and HOLOWAYCHUK 1970a, b, 1971), and that litter concentration under shrubs enhances soil nutrient concentrations in this microhabitat (TIEDEMANN and KLEMMEDSON 1973; CHARLEY and WEST 1975).

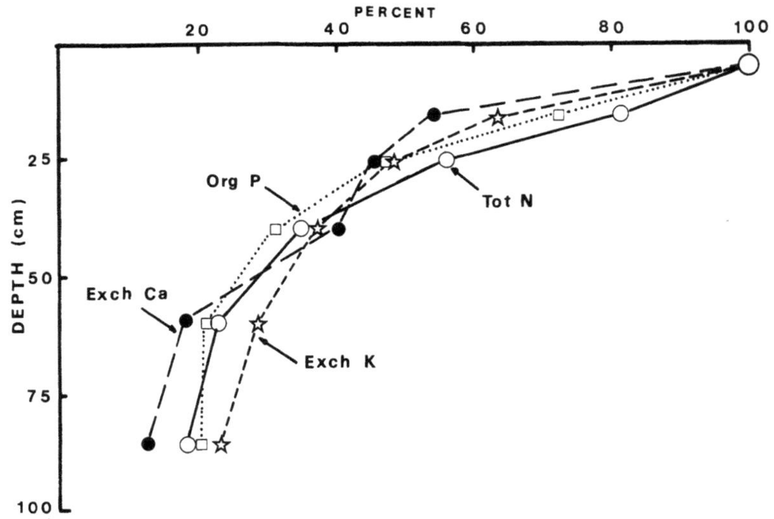

Fig. 1.2. Relative changes in soil chemical characteristics with depth in some Australian forest soils. Within profiles, values are plotted as percentages of the highest concentration recorded for the profile

Demonstration of small-scale horizontal patterns serves as a reminder that average values of nutrient status in surface soils obtained from randomly chosen samples may be misleading. If enough samples are collected in this manner they may be quite adequate for calculation of nutrient capital, but they will not reveal whether horizontal concentration gradients are present. Significant functional characteristics of the soil body could therefore be overlooked.

1.2.2.2 Vegetation and Litter

Estimates of biomass and litter layer mineral contents (Table 1.3) are available from most major vegetation types. These span climatic regimes from the arctic to the tropics and from the wettest environment to the very dry. Since such information was assembled and reviewed by RODIN and BASILEVICH (1967) there have been many additions to the literature, but the summary picture drawn by those authors remains a useful one for illustrating world-wide trends. The profound effect of environmental conditions on decomposition of organic debris and the circulation of elements is well illustrated by the striking contrast between annual litter fall and floor litter mineral stocks of tropical and high-latitude vegetation. Tropical rainforest, the largest and most mineral-rich vegetation formation, cycles by far the greatest quantity of elements annually and does this through a very small surface litter pool. On the other hand, in northern areas, and shrub tundra in particular, the relatively small yearly mineral input of litter is processed slowly by decomposition agents and the result is a very substantial accumulation of mineral matter in the floor compartment. In this contrast the predominant control is undoubtedly temperature; towards arid

Table 1.3. Standing stocks and some components of turnover of organic matter and minerals in selected world vegetation types (RODIN and BASILEVICH 1967)

Component	Shrub tundra	Southern taiga: Pine	Southern taiga: Spruce	Beech	Sub-trop. decid-uous	Trop-ical rain forest	Savanna	Meadow steppe	Semi-shrub desert
Dry weight (Mg ha^{-1})									
Biomass	28	280	330	370	410	>500	67	25	4.3
Litterfall (yr^{-1})	2.4	4.7	5.5	9.0	21	25	11.5	13.7	1.2
Floor litter	84	45	35	15	10	2	1.3	12	tr
Mineral content (kg ha^{-1})									
Biomass	901	1,878	2,700	4,196	5,283	11,081	727	1,183	185
Green parts	110	358	550	306	770	2,924	152	431	15
Litterfall (yr^{-1})	107	58	120	352	795	1,540	–	682	59
Floor litter	4,200	1,730	1,300	1,000	600	178	–	800	tr

Table 1.4. Distributions of macroelements amongst vegetation (V), litter (L) and soil (S) compartments in a range of forest ecosystems

Forest type[a]	Soil depth sampled (cm)	Nutrient element and storage compartment (%) N			P			K			Ca			Mg		
		V	L	S	V	L	S	V	L	S	V	L	S	V	L	S
Yellow birch (1)	$-B_2$				21	60	19	30	40	30	43	49	8	28	60	12
Spruce-Fir (1)	$-B_2$				29	64	7	37	50	13	51	40	9	27	63	10
Fir (2)	70	8	18	74	21	77	2	29	25	46	17	22	61	22	21	57
Oak-hickory (3)	60	8	1	91	20	4	76	32	1	67	31	2	67	9	1	90
Mixed oak (4)	80	10	1	89	5	1	94	1	1	98	1	1	98	2	1	97
Mixed oak (5)	45	10	1	89	3	1	96	27	3	70	32	2	66	5	1	94
Eucalypt (6)	200	1	1	98	17	7	76	41	1	58	40	2	58	13	1	86
Rainforest (7)	70	8	1	91	71	7	22	31	1	68	44	7	49	59	3	38
Rainforest (8)	30				79	8	13	89	1	10	14	1	85	15	1	84
Rainforest (9)	25	30	1	69				58	1	41	50	1	49	50	1	49

[a] (1) WEAVER (1975); (2, 7) TSUTSUMI (1971); (3) ROLFE et al. (1978); (4) DUVIGNEAUD and DENAEYER – DE SMET (1970); (5) JOHNSON and RISSER (1974); (6) WESTMAN and ROGERS (1977); (8, 9) GOLLEY et al. (1975)

conditions, however, moisture limitation is the chief regulator of biomass, litter fall and litter layer mineral stocks.

The distributions of total ecosystem nutrient capital between vegetation, litter and soil that are shown in Table 1.4 indicate in a general way what might be expected in forest ecosystems. In the order in which the forest types are listed, the environment ranges from cold through temperate to subtropical and tropical. Weathering and leaching regimes therefore vary greatly; so too does soil parent material, which is unfortunate but unavoidable. Nevertheless, certain broad trends are evident, the principal ones being that the proportions of nutrients retained in litter decline from cold climates, and that vegetation storage of most elements is relatively important at high latitudes, falls somewhat towards mid-latitudes and then becomes more significant again towards the tropics. The importance of soil as a reservoir is most striking in the case of N, but with only a few exceptions, notably in the case of P, it predominates generally. Even in the tropical rainforests (Table 1.4: 8, 9) the soil is a significant store, despite underestimation due to the clearly inadequate sampling depth. Similar distributions for the mean of 16 of the 32 forest communities listed by COLE and RAPP (1981) show the soil pool to be 81%, 96%, 64%, 62%, and 67% of total stocks for N, P, K, Ca, and Mg, respectively. The ranges about these mean values are N 53%–97%, P 94%–98%, K 8%–99%, Ca 20%–91%, and Mg 26%–99%.

These data are too limited to allow more than a tentative appraisal of how nutrient capital of ecosystems may be apportioned among compartments. Nevertheless, they do serve to indicate the variability that can occur and in so doing they caution against generalization on matters such as the effect of harvest removals on nutrient capital.

1.3 Mineral Cycling Processes

Three principal mineral flow pathways affect the nutrition of terrestrial communities (SWITZER and NELSON 1972). In order of increasing degree of internalization they are geochemical, biogeochemical and biochemical cycling. The first links the external environment to the ecosystem, the second is the circulation of nutrient capital between soil, standing crop and litter subsystems, while the third includes nutrient redistributions in the living biomass that act to conserve elements within the standing crop. All are coupled in overall community nutrition, but the relative significance of one or other of these major pathways differs from element to element. So far the inputs of the geochemical pathway have been mentioned (Sect. 1.2.1) and losses along this route will be discussed later (Sect. 1.3.4). We turn now to the biochemical cycle before considering the main features of biogeochemical circulation within the ecosystem and the processes that link its separate parts. The conceptual model here is the much simplified one shown in Fig. 1.3.

1.3.1 The Biochemical Cycle

During leaf development and ageing total nutrient content increases to a maximum and then declines, often to quite low levels of certain elements. As will be apparent from our treatment of throughfall (Sect. 1.3.2.2), part of this decline may be accounted for by leaching. But elements such as N and P, which are generally minor constituents of canopy washings, are two whose content falls most during ageing and final senescence. While the leaf remains metabolically

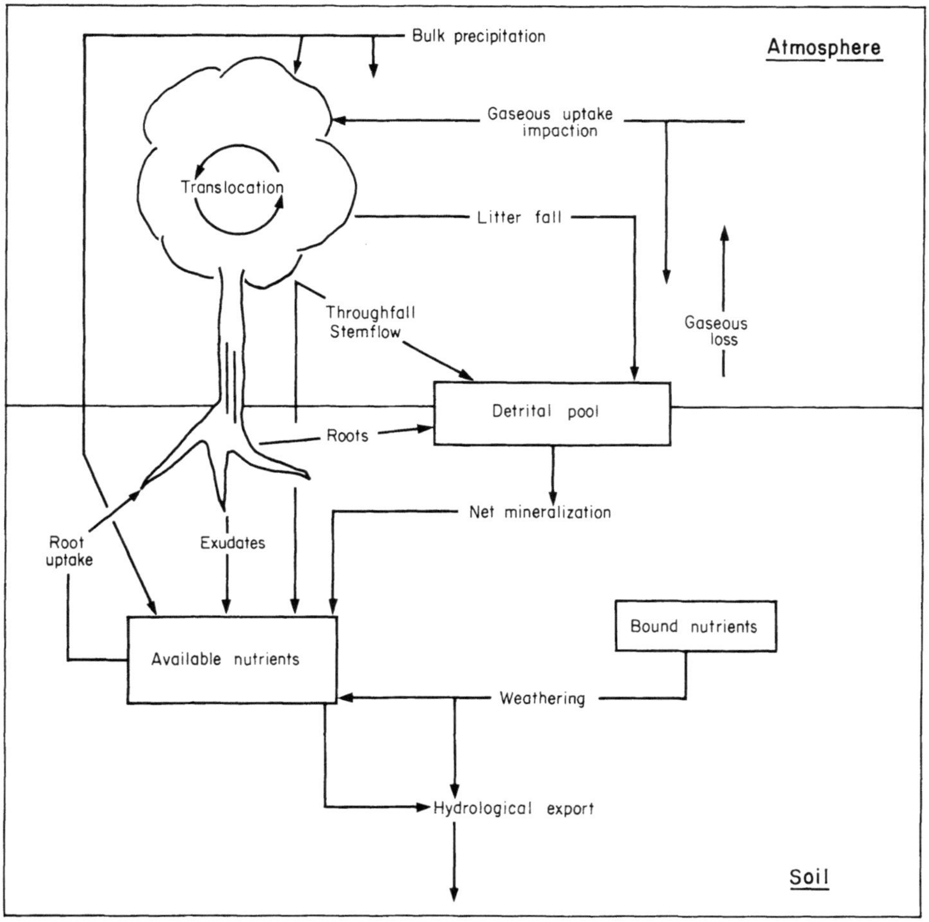

Fig. 1.3. Generalized picture of mineral cycling pathways in terrestrial ecosystems

active it will continue to be resupplied even though its ability to retain elements may deteriorate from quite an early age. Once biosynthetic activity begins to fall, mobilization of many nutrient elements increases significantly and moderate to high proportions of the total contents of the lamina can be translocated elsewhere within the plant.

Such withdrawal prior to leaf shedding has been noted in a wide range of species. Most commonly there is extensive depletion of N and P, little if any removal of Ca and only minor to moderate changes in Mg content. Sulphur may be withdrawn in moderate amounts, so too Cu and Zn, whereas K has high mobility, but its degree of reutilization is difficult to assess because of the extent to which it can be leached.

The significance of nutrient withdrawal in the nutrition of plant communities was given scant attention prior to publication of SWITZER and NELSON'S (1972) paper on elemental stocks and cycling in an age sequence of *Pinus taeda* stands. Their partitioning of nutrient flux led these authors to propose that an internal, biochemical subcycle should be recognized in addition to the well-known import–export geochemical and soil–plant biogeochemical cycles. Of particular interest is their finding that after 20 years of biomass and nutrient accumulation the pine ecosystem drew very little of its annual nutrient requirements from soil reserves. Instead it obtained most of its needs (especially for Ca) from the established, external litter decay (biogeochemical) cycle and what was retained internally by withdrawal from ageing tissues. Internal transfer via the biochemical cycle varied greatly depending on the element, being 39%, 60%, 22%, trace, 24% and 22% of annual requirement for N, P, K, Ca, Mg and S, respectively. BORMANN and LIKENS (1979, p. 75) provide additional evidence of the ecological significance of withdrawal with a value of 33% internal reutilization of the N used in growth processes each year by a 55-year-old deciduous, broadleaf forest ecosystem at Hubbard Brook. Similarly, a recent detailed study by ATTIWILL (1980) confirms the significance of internal conservation of P. In this case, gross annual demand of a mature *Eucalyptus obliqua* stand was 0.3 g m^{-2} of which 46% was supplied by biochemical cycling (not all from ageing leaves), 36% by biogeochemical cycling and only 18% from soil reserves. In contrast, only 2% of the annual Ca demand of 3.4 g m^{-2} was met by biochemical cycling, whereas 82% came from biogeochemical cycling and the remainder from soil reserves and atmospheric sources.

A phenomenon similar to leaf nutrient withdrawal has been observed in the ageing of wood, specifically in the transition from sapwood to heartwood. ORMAN and WILL (1960) analyzed radial nutrient gradients in bark and wood of *Pinus radiata* at different positions along the stem and found clear evidence of N, P and K withdrawal across the sapwood–heartwood transition. In contrast, Ca was shown to be sequestered in heartwood. BAMBER (1976) assembled further evidence of P and K depletion in the heartwood of six *Eucalyptus* species and suggested that storage in heartwood may represent a nutrient conservation mechanism that reduces the tree's dependence on external sources. In ATTIWILL's (1980) analysis of an *E. obliqua* forest, 31% of the total biochemical cycle of P was supplied by withdrawal from sapwood during its transition to heartwood.

1.3.2 Biogeochemical Cycling

1.3.2.1 Nutrient Uptake

Logically, cycling of elements begins at the root–soil interface where ions enter the root by active and passive processes. The adequacy of uptake in relation to the requirements of growth is dependent on ion concentrations at the absorbing surface and the rate of replenishment of soil solution from organic and inorganic solid phase reserves. Accordingly, desorption and dissolution reactions at inorganic surfaces, net release of ions from organic substrates as a consequence of microbial activity, and subsequent transport of ions from the various sources to the root sink are central considerations in community nutrition. Transport mechanisms were reviewed by BARLEY (1970) and are treated in detail by NYE and TINKER (1977).

Table 1.5. Ionic composition of some solutions from surface soils of forest ecosystems

Forest	Climate	Soil parent material	Ion concentrations ($\mu g\ ml^{-1}$)									
			Na	K	Ca	Mg	Min. N	P	S	Cl	HCO_3	pH
			Field lysimeter extracts [a]									
(1)	Temperate	Quartz diorite	0.8	0.4	1.8	0.7	0.4	0.03	5.0	1.6	5.0	6.2
(2)	Temperate	Basic igneous	8.6	13.8	41.9	8.0	3.4	0.03	8.6	17.0	68.6	7.0
(3)	Temperate	Sediments	0.7	0.9	2.5	2.3	0.3				4.9	5.9
			Laboratory extractions [b]									
(4)	Tropical	Basalt	5.8	10.5	2.2	1.8	85.6	<0.005	40.1	75.3	46.4	5.9
(5)	Tropical	Dacite	19.8	30.8	15.8	17.4	80.5	<0.005	49.0	163.3	98.8	5.0
(6)	Tropical	Sediments	6.7	17.6	8.8	5.0	2.9	<0.005	7.2	28.4	28.1	7.4

[a] (1) FELLER (1977); (2) McCOLL (1978); (3) McCOLL (1970)
[b] (4), (5), (6) GILLMAN and BELL (1978)

In all but a few soils, notably those of arid regions, soil water is a very dilute solution of electrolytes. The data of Table 1.5 are probably representative of the general range to be expected, although it should be pointed out that the results of GILLMAN and BELL (1978) for weathered tropical soils are unusually high in mineral N and S because samples were incubated prior to solution extraction. More commonly, solution concentrations at root surfaces in forested ecosystems would be closer to the lysimeter leachate values tabulated. What these solution data emphasize is the efficiency of uptake mechanisms at low concentrations (for example CHAPIN 1974, for P) and the ability of transport processes to resupply the absorbing surface.

Disposition of the absorption sink, i.e. the root system, is uneven in most soils, the common pattern of distribution being concentration of a large proportion of total root biomass in the first 20–30 cm of the profile (SANTANTONIO et al. 1977). Accession of organic residues on and in surface soil, and the attendant stimulus to mineralization processes here no doubt explain much of this profuse root ramification near the soil surface. Limited root development at greater depth may result from low chemical fertility (SUTTON 1969), poor aeration, high bulk density or, in low rainfall areas, limited depth of water penetration.

Not only are roots unevenly distributed in the soil both vertically and horizontally, but their activity varies in time (LYR and HOFFMANN 1967; HOFFMANN 1972), one or more cycles of endogenously controlled activity and quiescence occurring over a growing season. Furthermore, fluctuations in soil moisture and temperature exercise control on endogenously governed activity (DEANS 1979).

Changes in roots with age, such as an increasing degree of suberization, also affect their absorption capacity. KRAMER and BULLOCK (1966) have shown for loblolly pine that this can proceed to the point where less than 7% of total root surface remains unaffected. Suberization restricts mycorrhizal infection, and the significance of non-suberized, ectomycorrhizal roots in promoting uptake of $H_2PO_4^-$ (BOWEN and THEODOROU 1967) is well established.

As HERMANN (1977) makes clear, data on root systems in natural communities are limited and information on their physiological characteristics particularly so, yet the need for such information is becoming increasingly apparent as studies of community nutrition shift their focus from the gross features of the nutrient turnover cycle to a more detailed analysis of processes that occur in the soil body, particularly near the surface.

Effect of Mycorrhizas on Nutrient Uptake. Because mycorrhizal associations have evolved concurrently with the evolution of root systems (NICHOLSON 1975), studies of ion uptake by roots must take account of mycorrhizas. Several different kinds of mycorrhiza are recognized but only two – ectomycorrhiza and vesicular-arbuscular (endotrophic) mycorrhiza – will be considered. The former is characteristic of forest trees of the families Pinaceae, Betulaceae, Fagaceae, Dipterocarpaceae and Myrtaceae, while vesicular-arbuscular (V-A) mycorrhizas are widespread throughout the plant kingdom. Most ectomycorrhizal fungi are unable to utilize the complex polysaccharides found in plant residues, and so are unlikely to promote rapid breakdown of litter and soil organic matter. Probably this is true of V-A endophytes also. Hence there is no reason to suppose that trees harbouring such symbionts might be independent of normal mineralization processes. On the other hand, there is much evidence to indicate that both kinds of mycorrhiza function very efficiently as nutrient absorbing organs (HARLEY 1969; Chap. 9, Vol. 12 C).

Role of Mycorrhizas in Biogeochemical Cycling. An energy circuit diagram (sensu ODUM 1970) showing the potential role of mycorrhizas in nutrient cycling is given in Fig. 1.4. Plants and soil microorganisms are shown as drawing on

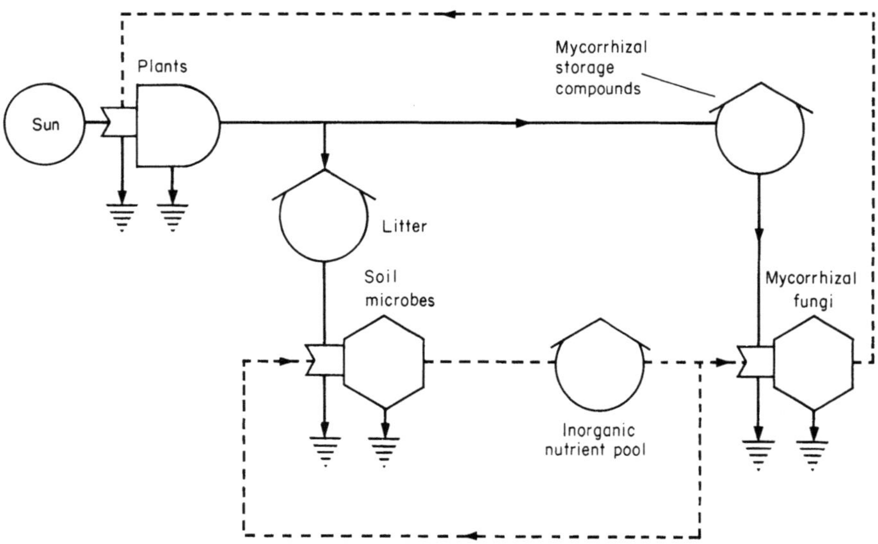

Fig. 1.4. Energy circuit diagram showing the postulated role of mycorrhizas in biogeochemical cycling. (Redrawn from RICHARDS 1974)

the same inorganic nutrient pool, and the two nutrient cycles are assumed to be competitive. The mycorrhizal fungus is interposed between the nutrient pool and the plant, and primary productivity is determined by the operation of a "work gate" through which energy flow from the sun is controlled by mineral nutrients drawn from the inorganic pool in the soil. The competitive flow of nutrients towards the heterotrophic soil microbes (i.e. the immobilization rate) is governed at another work gate by the energy input from plant residues. This is a function of the rate at which the litter reservoir is replenished, which in turn depends on the rate of energy capture by the producers.

The fungus acts as a physiological sink for plant assimilates (HARLEY and LEWIS 1969; BEVEGE et al. 1975), converting soluble sugars into compounds which the host is unable to utilize. This maintains a concentration gradient ensuring the continual movement of sugars into the fungal symbiont, which in turn activates the work gate that regulates nutrient flow into the mycorrhiza and thence to the host. Being independent of plant residues as an energy source, the mycorrhizal fungus competes successfully with free-living soil microorganisms for mineral nutrients. In low-fertility soils the role of mycorrhizas is seen as speeding the passage of essential elements through the biogeochemical cycle, and minimizing losses from the system through leaching. Productivity is increased by mycorrhizas through their ability to increase the supply of limiting nutrients at the work gate which controls the rate of capture of solar energy by the plants. The mechanism which initiates this increase in ecosystem productivity is the diversion of photosynthate into fungal storage carbohydrates or other carbon compounds. Just how much is diverted is a matter of some dispute (HARLEY 1975), but it is likely to be only a small fraction of the total produced.

1.3.2.2 Nutrient Return

Following uptake and internal utilization, nutrients are returned eventually to the detrital pool either as ions or in organic combination. The principal agencies involved are leaf consumption by grazing animals, canopy leaching, senescence and litter fall, shedding of roots, and exudation from roots. The first of these, which concerns the role of phytophagous insects as regulators of mineral cycling, is not discussed here, although it has potential significance in some ecosystems (SPRINGETT 1978).

Throughfall and Stemflow. Rain striking plant surfaces either drips to the soil as throughfall or is channelled to the ground as stemflow. In most situations 85% or more of input is throughfall and something less than 10% is stemflow (HELVEY and PATRIC 1965). The contribution of stemflow may be higher in tropical forests; between 17.5% and 22% according to JORDAN (1978).

Compared to rainfall, both throughfall and stemflow are enriched in nutrient elements and their combined contribution to the annual mineral cycle may be substantial. Stemflow is frequently more enriched than throughfall, but its smaller volume means that throughfall is the more important nutrient-transporting process. Nevertheless, because stemflow input to the ground is very localized, it can have a relatively powerful influence on soil chemical characteristics immediately about the boles of trees (PATTERSON 1975; GERSPER and HOLOWAYCHUK 1970b, 1971).

Nutrient enrichment of throughfall and stemflow includes contributions from particulate matter deposited on leaves by sedimentation and impaction and from soluble materials exuded from or leached out of leaves. Unfortunately, however, there is no precise method for separating these sources to determine how much of the nutrient input to the floor of the community is true addition from external sources (the geochemical cycle) and how much represents nutrients cycling within the system (the biogeochemical cycle). In consequence, leaf leaching must be assessed using artificial leaching procedures as in the study of GOSZ et al. (1975). These authors found that early summer sun leaves lost more nutrients than shade leaves, but both lost very little. Later collections lost substantially more nutrient, but differences between sun and shade leaves were minor. Very large losses occurred from old leaves, particularly those collected after abscission, with K removal amounting to 80%–90% of total leaf content of the element. The common pattern of leachability for the three species studied was K > Mg > Ca > N > Na and this was in accordance with that shown by decomposing litter.

These results, together with the information assembled in TUKEY'S (1970) review, indicate that the principal factors affecting leaf leaching are species, integrity of the cuticle, leaf age, wettability, metabolic activity of the leaf and duration of rain.

If we consider the relative importance of the major nutrients in throughfall, the ordering of elements is generally similar to that found in artificial leaching experiments: thus K > Ca = Mg > P = N in four forest types in Tennessee (HENDERSON et al. 1977), K > Ca > Mg > Na for two Australian rainforests (BRASELL and SINCLAIR pers comm), K > Ca > Mg > Na in a New Hampshire hardwood forest (EATON et al. 1973), and K > N > S > Ca > Mg > P in a loblolly pine plantation, North Carolina (WELLS et al. 1975). It would seem therefore, that where maritime influences and dust deposition are minor, the bulk of the nutrients in throughfall (and stemflow) probably derives from leaching and thus represents a component of the internal or biogeochemical nutrient cycle.

Annual nutrient load in throughfall varies greatly by forest type and geographic location. There do not appear to be consistent differences between conifers and broadleaf species, although the former may contribute lesser quantities of elements, but there is little doubt that throughfall nutrients in tropical forests are the largest overall. The selection of data in Table 1.6 gives an approximate picture of the range of nutrient inputs that might be expected in various communities.

Plant Residues: Litter Fall. According to RODIN and BASILEVICH (1967), mineral return in annual litter fall may exceed 2,000 kg ha^{-1} in tropical rainforests; is between 250 and 500 kg ha^{-1} in temperate deciduous forests, steppe, savanna and some desert shrub communities with annuals; from 100 to 200 kg ha^{-1} in coniferous forests of the temperate zone; then declines through shrub tundra and arid steppe to values between 40 and 60 kg ha^{-1} in arctic tundra, northern taiga spruce forests and some ephemeral desert communities. Absolute values

Table 1.6. Elemental inputs to the forest floor in throughfall for a range of forest types.[a] Values do not include accessions in rainfall

Forest type	Location	Annual elemental input (kg ha^{-1})							
		Na	K	Ca	Mg	Tot. N	Cl	S	P
Rainforest (1)	Ivory Coast		177	28	50	64			9.1
Rainforest (2)	Panama		50	40	10				0.6
Rainforest (3)	Ghana		202	16	7				3.3
Beech (4)	Sweden	8	8	6	2	0.3	20	7	0.05
Beech (5)	New Zealand	11	25	6	2	0.6	44	2	0.3
Spruce (6)	Sweden	17	21	11	4	13	35	34	0.04
Oak (7)	England	34	25	12	7	0.6			0.6
Oak (8)	U.S.A.			8	3	5.5			0.7
Pine (9)	New Zealand	8	17	1					2.1
Pine (10)	U.S.A.		9	2	1	2.8		3	0.25
Eucalypt (11)	Australia	44	9	14	7	35	85	17	
Eucalypt (12)	Australia	17	36	6	5		15		

[a] (1) Bernhard-Reversat (1975); (2) Golley et al. (1975); (3) Nye (1961); (4), (6) Nihlgard (1970); (5) Miller (1963); (7) Carlisle et al. (1967); (8) Reiners (1972); (9) Will (1959); (10) Wells et al. (1975); (11) Westman (1978); (12) Turvey (1979)

of particular elements vary a great deal between and within ecosystem types, yet certain patterns are evident. Thus N in annual litter fall of temperate deciduous and coniferous forests is fairly constant at 50–70 kg ha^{-1}, whereas Ca return is 50–150 kg ha^{-1} in the former but only 20–50 kg ha^{-1} in the latter. In subtropical and tropical forests N return is 250–325 kg ha^{-1} while Ca in litter fall is 200–300 kg ha^{-1}. Silicon is outstanding in that it is cycled in large amounts (>700 kg ha^{-1}) in tropical rainforests and in substantial quantities in grass steppe, savanna and some desert shrub communities, but it is a very small component of mineral turnover in most forests of the temperate zone. As for K, Mg, P and S, the summary data of Rodin and Basilevich (1967) show no unusual trends, only the kind of variation that would be expected on the basis of differences in plant biomass and the amount of annual litter fall.

Bray and Gorham's (1964) review and numerous studies since show that moist tropical forests shed litter at a fairly steady rate throughout the year, whereas deposition in arid-zone ecosystems is unpredictable because of the large random element in the timing and magnitude of precipitation events. Between these extremes the annual pattern of litterfall is usually composed of a single major wave from maximum to minimum deposition, though the positions of the stationary points and the slopes of the curves about them differ appreciably, particularly between evergreen and deciduous vegetation.

The proportions of the dominant elements in litter fall have been used by Rodin and Basilevich (1967) to categorize some of the major vegetation types. Thus nitrogen clearly dominates the mineral content of litter fall in tundra and is also prominent in subtropical forests and coniferous and deciduous forests of the temperate zone. On the other hand, Ca is predominant in broadleaf forests of the temperate zone and in semi-shrub desert communities. Silicon predominates in savanna, steppe, semi-shrub desert and tropical rainforest communities, while Cl dominates the annual mineral cycle of communities on desert solonchaks.

A great deal of what is known about annual nutrient return in litter comes from measurements made over 1 or 2 years. Few studies have extended beyond this, yet it has long been known that year-to-year variability can be large (BRAY and GORHAM 1964). Our own data for a mixed *Eucalyptus* forest under warm temperate conditions demonstrate this variation and show also that there can be an appreciable shift from year to year in the timing of the major period of litter fall.

Windthrown trees bring large quantities of fresh foliage of high nutrient content to the forest floor. This is supplemented by the wind-pruning of foliage-bearing branches and branchlets, which itself is a major factor in returning organic matter to the soil subsystem in storm-prone forest ecosystems. GRIER (1977) found that the dry weight of green leaves removed from trees by wind during a 3-year period was equivalent to more than one-third of annual foliage production in this hemlock-spruce stand. Similarly, green branches, presumably broken by wind and snow, make a large contribution to litter fall in an *Abies amabilis* forest in Washington (TURNER and SINGER 1976). It is not known how much these sudden inputs of nutrient-rich leaves change decomposition patterns on the forest floor, but a priming effect might be expected.

In measuring litter fall large numbers of traps are usually necessary if variance is to be reduced to acceptable levels; this is well known and requires no further comment. However, the extent to which such spatial heterogeneity reflects fixed patterns of litter deposition that differ in quantity and quality is not clear. Several striking examples of patterning in semi-arid ecosystems have been noted, but there is little information for forested areas. There is ample evidence to show that interspecific differences in leaf nutrient content occur within plant communities and one would expect these to be reflected in the quality of litter fall. In this respect TAPPEINER and ALM (1975) found that abrupt changes in turnover rates of the forest floor coincided with changes in the type of undergrowth in a *Pinus resinosa* stand.

The contributions of different plant parts to litter fall are variable but leaves generally comprise between 50% and 70% of the annual total. Because their nutrient concentrations are relatively high, they also add most of the inputs of N, P, K, Ca, Mg and S that reach the floor in organic debris.

Plant Residues: Root Detritus. Belowground additions to the detrital pool remain one of the least-studied aspects of mineral cycling. It is evident however, that the quantities of roots entering the decomposer pathway can be very large, as evidenced by the 9 Mg ha^{-1} yr^{-1} turnover measured by HARRIS et al. (1977) in deciduous forest and pine plantations, and the 21 to 10 Mg ha^{-1} total root biomass change that occurred over a 4 months' period in an oak wood in Minnesota (OVINGTON et al. 1963).

What such large and rapid change in root biomass means in terms of nutrient flux to the detrital pool must be largely inferred at present, since few mineral analyses are available. Even so, the recent data of COX et al. (1978) for N and K turnover suggest that perhaps too much emphasis has been placed on aboveground litter fall. In the *Liriodendron* forest studied by these authors root death added 6.7 Mg ha^{-1} of organic matter to the soil during the year and this detritus contained 76 kg of N and 128 kg of K. Aboveground litter fall, on the other hand, was only 3.3 Mg ha^{-1} and included 42 kg of N and 10 of K. About 9 kg of N and 14 of K went to the decomposition pool by way of root consumers and a further 128 kg of K was estimated to have come from exudation and leaching of live roots; N additions from this source were assumed to be minor relative to K release. Of the total annual turnover in the ecosystem, roots therefore contributed about 70%, 60% and 85% of dry matter, N and K respectively. Pronounced seasonal peaks of N and K release from roots <0.5 cm in diameter occurred in spring and winter.

Root Exudates. Exudates from healthy roots of agricultural plants have been extensively studied and their composition is well known (ROVIRA 1969). Research on tree root exudates is limited by comparison (BOWEN and THEODOROU 1973), but at least 21 amino acids/amides, 10 carbohydrates and 14 organic acids have been identified in studies of 10 species of forest tree seedlings. Inorganic constituents have been largely neglected since they do not usually serve as energy sources for microorganisms. Latterly, however, SMITH (1976) has collected exudates from attached roots of three forest tree species and identified Na^+, K^+, NH_4^+, Ca^{2+}, Mg^{2+}, Cl^-, NO_3^-, SO_4^{2-} and PO_4^{3-} in the net release to vials of solution in which tips of new woody roots were inserted for 14 days. As Smith acknowledged, his collection technique (SMITH 1970) is somewhat artificial. Nevertheless, the interspecific differences demonstrated are probably real even if the quantitative information is not readily convertible to real outgoes from roots in natural contact with their nutrient medium. If the net release rates measured are an approximation of what happens in the field, then clearly the inorganic fluxes are substantial; for example, *Betula allegheniensis* released about 13 μg of K per mg of oven dry root in 14 days and about 5, 11 and 1.6 μg of Ca^{2+}, SO_4^{2-} and PO_4^{3-} respectively in the same period. *Fagus grandifolia* and *Acer saccharum* released roughly 25%–75% of these amounts. Organic acids dominated the organic fraction, while Na^+, K^+, and SO_4^{2-} were the major inorganic constituents.

1.3.3 Nutrient Release from Residues: Mineralization of Litter and Soil Organic Matter

The key role played by microbiological processes in mineral cycling has already been emphasized (Sect. 1.1.2). Regulation of plant growth by the heterotrophic soil microflora may be achieved through a number of elements, but nutrients occurring principally in organic combination are the ones most likely to be implicated. In this regard, nitrogen mineralization reactions are paramount in most terrestrial communities; sulphur transformations are also significant, especially in humid environments. Chapter 5 of Volume 12 C gives a detailed account of nitrogen mineralization. Here, however, we are concerned more with how such processes affect the timing and level of supply of nutritionally important elements, in relation to total ecosystem reserves.

1.3.3.1 Biological Determinants of Organic Matter Decomposition

While the softer tissues of plants and small animals are thought to be decomposed principally by microbes, the breakdown of more refractory substances involves the complementary activities of both soil flora and fauna. Some tissues decompose at or on the soil surface, releasing soluble products which move into the soil in aqueous solution. In other instances, fresh organic matter may be directly incorporated in the soil (by earthworms for example) before any decomposition occurs. The effects of invertebrates on the decomposition of woody substrates may be one of the more significant examples of fauna-microflora interactions, according to WITKAMP and AUSMUS (1976). They suggest that the microbial succession which follows the tunnelling activities of termites and wood borers in fallen tree trunks enhances the incorporation of wood into soil organic matter by translocating nitrogen to the sites of decay through fungal mycelia and perhaps by stimulating bacterial nitrogen fixation in situ.

The main contribution of soil animals has been thought to be the increase in surface area available for microbial colonization through comminution of the litter, since the chemical composition of arthropod faeces is generally very similar to that of food before ingestion (EDWARDS et al. 1970), and because physically excluding certain groups of the macro- and mesofauna greatly retards decomposition of leaves (EDWARDS and HEATH 1963). Faecal material is thought to be a very suitable substrate for bacterial (though perhaps not fungal) exploitation. However, this conclusion was questioned by SATCHELL (1974), who drew attention to evidence for the resistance of faecal pellets to decomposition and their consequent persistence in soil and litter. Reingestion of microbially enriched faeces is seen as a means of improving the digestibility of poor food and conserving nutrients (MCBEE 1971), but whether coprophagy is of any real significance in mineral cycling, as suggested by WEIGERT et al. (1970), remains an open question.

SATCHELL (1974) emphasized one aspect of litter comminution by arthropods that is frequently overlooked, viz. its effect on the topography of the litter layer. Freshly fallen litter (the L-layer or O_1 horizon) is moist for only short periods after rain because of its looseness and consequent rapid drying, whereas on fragmentation the litter becomes more closely appressed to the soil surface, creating a zone where the humidity is higher and more stable. The increase in water-holding capacity and decrease in evaporation brought about by flattening the litter layer may prove to be a more important outcome of litter fragmentation by invertebrates than the greater surface area it provides for microbial colonization.

Despite the strength of argument in support of physical interpretations of the effect of soil and litter arthropods, there has recently been a reassertion of earlier views that the fauna influences decomposition and nutrient flux directly (CROSSLEY 1977). Although the elemental concentrations of ingested and egested material differ only slightly, the molecular forms of the elements may be quite distinct (WITKAMP and AUSMUS 1976) and this may consequently modify the rate of mineralization. Thus the presence of invertebrates doubled the loss of ^{134}Cs during litter breakdown in the field (CROSSLEY and WITKAMP 1966), even though the litter fauna accounted for only 1% of the energy expended by the decomposer group as a whole. A small initial energy expenditure by invertebrates may therefore greatly increase mineralization of litter.

The Rhizosphere Effect. Microbial activity is greatly modified by proximity to plant roots, and the soil in the immediate vicinity of plant roots, i.e. in the rhizosphere, is a highly favourable habitat for microorganisms. The high bacterial density in the rhizosphere produces intense competition and the selection pressures arising from this tend to favour rapidly growing and biochemically versatile microorganisms over slower-growing and less versatile strains. This implies that the rhizosphere microflora has a greater ability than the general soil population to effect rapid biochemical changes and there is some evidence to support this contention. ESTERMANN and MCLAREN (1961) have shown that the enzymatic activity of barley roots is increased by the presence of a root surface microflora, and as a result of this the plant's ability to utilize urea and some organic phosphates is increased. Ammonifying bacteria are greatly stimulated by proximity to living roots, and this should lead to greater mineralization of organic N in the rhizosphere. Field evidence, however, indicates that less mineral N is present in soil under crops than in fallow, even when allowance is made for the amount of N taken up by the plants. This apparent anomaly is due to the fact that most of the N mineralized under a plant cover is rapidly assimilated again by the microflora, as was shown by BARTHOLOMEW and CLARK (1950) using ^{15}N as a tracer. Since the availability of plant nutrients is the resultant of the opposing processes of mineralization and immobilization, the overall rhizosphere effect could be detrimental as often as it is beneficial. On balance however, the effect is more likely to be beneficial, for the nutrients mainly involved in these microbial transformations are the anions nitrate, sulphate and phosphate, which are normally absorbed from the soil solution. It is argued that the cyclic turnover through organic and inorganic forms, which is intensified by the rhizosphere microflora, minimizes loss through leaching and provides a continual supply of anions in the root absorption zone.

Soil Enzymes. An important reaction in soils which liberates mineral nitrogen is urea hydrolysis. Large amounts of urea are continually added to the soil–litter subsystem in animal excreta; smaller quantities derive from the oxidation of nucleic acids. Urea production, while greatest in the specialized urea bacteria, is not restricted to these microbes but is a property of many bacteria and fungi. Urea hydrolysis is not confined to living microorganisms either; extracellular urease activity remains a relatively permanent and almost ubiquitous feature of soils even in the absence of microbial activity.

Soil enzyme activity contributes to the biogeochemical cycling of P but to what extent is unknown. About 1% of the organic P in soils is phospholipid, 5%–10% is nucleic acid, and usually less than 50% (though up to 60%) is inositol phosphate (HALSTEAD and MCKERCHER 1975). Hydrolysis of inositol phosphates is achieved by phosphatases which are produced by both bacteria and fungi, including mycorrhizal fungi (see MOSER and HASELWANDTER, Chap. 9 of Vol. 12 C). Indirect evidence for the existence of extracellular phosphatases was reviewed by MCLAREN and SKUJINS (1971). GREAVES and WEBLEY (1969) showed that microorganisms capable of hydrolysing organic P were more abundant in the rhizosphere than in non-rhizosphere soil and that many of them produced extracellular phosphatases. This may not necessarily lead to increased P uptake by plants, since BARBER and LOUGHMAN (1967) found that at low phosphate levels rhizosphere microorganisms accumulated P from organic sources in competition with plant roots.

Soluble organic P is a recognized source of phosphate for plants, but according to HALSTEAD and MCKERCHER (1975) it is uncertain whether it must be mineralized prior to plant uptake. Laboratory mineralization trials suggest that some soils might release sufficient P from organic forms to supply crop requirements, although the actual source of the mineralized P has not been established. Studies in native grasslands have shown parallel fluctuations in $NaHCO_3$-extractable organic P and phosphatase activity and doubtless phosphorus mineralization processes are relatively more important in natural communities than in agricultural ecosystems.

1.3.3.2 Distribution of Microbiological Activity in Space and Time

The soil–litter subsystem is not a uniform milieu for microorganisms but varies considerably from place to place and from time to time. No discussion of spatial and temporal variability in microbiological activity would be complete without reference to the microscopic dimensions of the habitat. Thus "a single cellulose fibre provides a specialized environment with its own characteristic microflora" (STANIER 1953). While in a mature ecosystem the gross pattern of distribution of substrates changes only slowly with the passage of time, on the minute scale of microbial habitats changes are vast and rapid. To understand the functional aspects of mineral cycling it is therefore necessary to appreciate that substrates for chemoheterotrophs in soil are ephemeral. This is especially true of the zymogenous organisms: these must either disperse to fresh substrates, or enter a quiescent stage until new substrates are brought to them by mechanisms such as leaf fall and root decay of plants, and by death or defaecation of soil animals. The growth pattern of many chemoheterotrophic soil microbes is thus one of successive cycles of substrate colonization, exploitation and exhaustion, separated by periods of migration or quiescence (GARRETT 1951). In this context the soil is seen as a matrix of discrete and transient micro-habitats.

Notwithstanding the conceptual significance of this view of the soil environment, there are readily discernible and more or less permanent gross patterns of decomposer activity which have much relevance for mineral cycling. The anisotropic distribution of elements previously discussed (Sect. 1.2.2.1) reflects changes in the intensity of microbial activity and the tempo of mineralization reactions is governed by the interaction of soil moisture and temperature on microbiological processes (Chap. 5, Vol. 12 C).

Spatial Variability. There is little doubt that the primary factor limiting the activity of chemoheterotrophic microbes is substrate availability. Substrates for soil microorganisms have been defined by GARRETT (1951) as "living or dead, virgin or partially decomposed plant or animal tissues lying in or upon the soil, or soluble products diffusing therefrom".

An important class of substrates in soil emanates from the roots of vascular plants, in the form of sloughed-off roots or root exudates. These substrates give rise to the rhizosphere effect, the potential significance of which was discussed above in relation to mineralization processes. Laboratory evidence points to the existence of a specialized rhizosphere microflora as being potentially significant in the initiation of nutrient cycling in skeletal soils (Sect. 1.2.1.2) and it is conceivable that rhizosphere microbes affect the rate and pattern of nutrient cycling generally.

Within the soil, chemoheterotrophic microbes are characteristically concentrated in the upper part of the profile. Since this pattern of distribution exists even in soils where organic matter does not tend to concentrate in the surface horizon, factors other than energy supply, e.g. oxygen and carbon dioxide levels, may become limiting. Nevertheless, organic matter distribution remains the primary determinant of spatial pattern in soil microbial acitvity. Within the surface litter, studies of fungal succession (KENDRICK and BURGES 1962) clearly show vertical patterning of the microflora, the genera which dominate the L-layer on the forest floor in a scots pine stand being quite distinct from those of the F_1-layer, and these in turn differing from the dominant fungi of the F_2-layer.

Spatial heterogeneity in the distribution of soil and litter fauna is also well documented. For the larger invertebrates, non-random horizontal variation may be related to the presence of suitable cover (WALLWORK 1970). Vertical zonation is a general characteristic of the soil fauna, often to the extent that specialized life forms typify successive layers in litter and soil. This zonation is quite evident despite the fact that there is considerable vertical migration in most soils, and leads to speculation that it reflects quite distinct functional processes in the several strata.

Since spatial heterogeneity in the distribution of the soil biota is clearly discernible, organic matter decomposition processes should show similar variation. Depth functions of nitrogen mineralization are well known in cultivated soils and have been recognized more recently in natural ecosystems. Horizontal gradients of net mineralization are also evident in rangeland communities characterized by perennial shrubs (CHARLEY and WEST 1977) and in forests dominated by long-lived, large trees (WARD personal communication). Horizontal variation of a different kind is found in some tropical and subtropical communities, where much of the organic matter which would be subject to mineralization in soils of temperate regions is sequestered in termite mounds (LEE and WOOD 1971).

Temporal Variability. Litter deposition, or any other process whereby organic residues are added to the soil, does not take place at a constant rate. The seasonal pattern of organic matter addition to the soil–litter subsystem is most apparent in deciduous forests, but is recognizable in other communities as well (Sect. 1.3.2.2). The introduction of carbonaceous materials to soil stimulates the zymogenous microflora, which is normally present in a resting state. Temporal variation in substrate availability for these microbes is likely to have a marked effect on the cycling of nutrients such as N which are found principally in organic form. It seems that only a small proportion of the total soil organic matter participates in the mineralization-immobilization cycle at any one time, and this fraction contains a ready supply of available substrates. Nitrogen mineralized during the course of any one growing season may amount to no more than 3% of the soil's organic N (SATCHELL 1974). The greater proportion of organic matter that enters the soil becomes more or less stabilized against further microbial attack, once it has been partially decomposed. It therefore constitutes a passive organic phase existing contemporaneously with, but outside, the mineralization-immobilization cycle (JANSSON 1958).

In communities whose productivity is closely linked to low and unreliable rainfall, such as the rangelands of semi-arid regions, substrates accumulate during dry periods and a pulse of mineralization occurs following rain. Falls of rain too small to elicit a plant response may result in the build-up of a pool of inorganic nitrogen which is available to be drawn on by plants as soon as more substantial rains occur. Similar temporal patterns of mineralization are found in monsoon climates which experience a marked seasonal pattern of precipitation, as shown by GREENLAND (1958) for soil from tropical high forest in Ghana. Less pronounced temporal variation in the mineralization of soil N occurs in more equable climates. In three contrasting high forest communities in eastern Australia, significant seasonal differences were found in the standing crop of mineral N in the topsoil (RICHARDS et al. unpublished), the lowest values for both NH_4-N and NO_3-N being found in late winter (the driest season) and the highest in late summer (the period of maximum rainfall). Seasonal variation in mineralization potential, as measured by laboratory incubation of intact soil cores, was also evident, tending to be greatest in late winter when the standing crop of mineral N was lowest: this may again reflect the accumulation of available substrates during periods when field conditions for mineralization are least favourable.

For reasons that are not entirely clear (see Chap. 5, Vol. 12 C), there is rapid oxidation of organic matter in soil that has been remoistened after drying, the flush of decomposition lasting several days. This phenomenon was first recorded by LEBEDJANTZEV (1924). Generally, the longer the drying phase and the higher the temperature during drying, the greater the mineralization rate after wetting (BIRCH 1958, 1960).

In addition to those temporal changes in microbial activity and concomitant mineralization that are associated with variations in soil moisture status, and therefore rainfall distribution, there are daily and seasonal trends determined by prevailing temperatures during periods when soil moisture is readily available. Diurnal rhythms of microbial activity have been shown by measurements of soil respiration in semi-arid rangelands, and in more humid environments by WITKAMP (1969) and others. Seasonal trends are also clearly evident (RICHARDS and CHARLEY 1977). Short term patterns of mineral transformation are not so readily detected but must occur none the less.

1.3.3.3 Measurement of Decomposition Rate

Spatial and temporal patterns in substrate availability and microbial activity greatly complicate the measurement of mineralization rates and, consequently, nutrient fluxes in the soil–litter subsystem. If there was no pattern to the variability, then unrestricted random sampling could (subject to certain provisos related to the effect of sampling per se on microbiological processes) provide a reliable estimate of the rate of organic matter decomposition. However, given the existence of spatial and temporal patterns such as those described above, some form of stratified random sampling is necessary, and this pre-supposes a knowledge of substrate and activity patterns. To the best of our knowledge, no one has addressed this key problem in the study of soil metabolism.

Nutrient fluxes associated with organic matter decomposition may be estimated by a variety of methods. Five commonly used approaches are: difference; litter fall; litter decomposition (litter bags); soil respiration; and soil mineralization. All provide relatively crude measures of carbon and nutrient throughput in the soil–litter subsystem. Soil respiration techniques tend to over-estimate soil metabolism, while the other methods in general underestimate the activities of the soil and litter microbiota. This is because none takes into account the contribution of root detritus to the soil subsystem and this may be a significant source of substrates for decomposer organisms (AUSMUS and WITKAMP 1974; WAID 1974; COLEMAN 1976).

Space does not permit detailed consideration of the several methodologies, however in Chap. 5 of Vol. 12 C N mineralization techniques are discussed. In passing, it should be noted that the latter are applicable to other elements for which organic compounds constitute significant ecosystem reserves, e.g. S in forests developed on high dune systems (HURDITCH et al. 1980).

1.3.3.4 Extrinsic and Intrinsic Factors Affecting Decomposition of Detritus

Decomposition rates reflect the combined influence of environmental factors, especially climate, and biological factors, chief of which is the inherent susceptibility of the substrate to microbial attack.

Substrate Quality. It is a central concept of soil microbiology that mineralization and immobilization of nutrients proceed concurrently with nutrients cycling between an organic and inorganic phase continuously: the supply of readily decomposable substrates generally determines which phase predominates at any particular time. Implicit in this concept of a mineralization-immobilization cycle is the assumption that plants cannot compete successfully with microorganisms for inorganic nutrients so that only when there is net mineralization can they acquire nutrients. There is a large body of evidence to support this contention for N which is the element in greatest demand by both plants and microorganisms, and for which there is no primary mineral source in the soil. Whether microbial immobilization ever restricts primary production by causing other nutrients to become limiting for plant growth is much more conjectural. Even with N, a fresh input of substrates will result in immobilization only if these are high in energy in relation to their N status: thus largely carbonaceous litter will cause greater immobilization than highly proteinaceous materials. The period of net immobilization depends on the relative proportions of the two factors, viz. energy and nitrogen, and this is roughly approximated by the C/N ratio of the residues undergoing decomposition. Agronomic experience in the northern temperate region indicates that only those residues with C/N ratios of less than 20:1 are likely to provide available nitrogen for crop growth when they are incorporated in soils at the beginning of a growing season. This concept has limited value for other ecosystems however; for example, the C/N ratio of litter in forest communities is often greater than 100:1 when it reaches the ground and may still exceed 40:1 before the comminuted fragments are physically incorporated into the mineral soil, yet 5 to 10 Mg ha^{-1} of such litter are decomposed in productive forests each year. Clearly factors other than C/N ratio affect substrate decomposability.

Physiological adaptation of decomposer organisms is one such factor, for as GRIFFIN (1972) has emphasized, the growth of fungi in heartwoods, which have C/N ratios of the order of 1,000:1, requires great efficiency in nitrogen metabolism. This is achieved in a number of ways: re-use of mycelial nitrogen by internal translocation of cytoplasmic contents from old to young cells, or their autolytic products; production of extracellular lytic enzymes which render cell walls available for assimilation by younger cells; and preferential allocation of nitrogen to metabolically active systems. Griffin noted, however, that wood-decaying fungi can only re-use nitrogen in the presence of exogenous carbohydrate. In fact, most cellulolytic fungi show greatly reduced production of cellulases as substrate C/N ratio increases. The white rot fungi are exceptional in this regard, being able to produce cellulases at a C/N ratio of 2,000:1.

So far as inherent decomposability goes, the chemical nature of organic residues seems to be more important than their C/N ratio. In general, hemicelluloses decompose more rapidly and are utilized as substrates by a wider range of microorganisms than cellulose, and lignin is even more refractory. As the lignin content of plant tissues increases, so too does the ability of the litter microflora to decompose them decline, and the capacity to use lignin as a substrate is not widespread among microorganisms. Several authors have shown lignin content to be a good index of substrate quality for the prediction of litter decomposition rates. For various litter components in old-growth Douglas fir forests in Oregon, FOGEL and CROMACK (1977) found that weight loss was better correlated with lignin concentration than with C/N ratio. Changes in log biomass were also found to be much better correlated with lignin content than with cellulose, acid-detergent cell wall constituents, or C/N ratio (see RICHARDS et al. 1981). These findings support the contention that structural complexity is a more important determinant of substrate decomposability than C/N ratio or nitrogen status per se.

To complete this discussion of substrate quality, brief reference is made to the role of polyphenols and other water-soluble extractives of litter. Polyphenols in conifer needles are thought to react with proteins of the mesophyll cells just prior to leaf abscission, forming complexes which are quite resistant to decay. This results in mesophyll cells becoming a major recognizable constituent of mor humus in coniferous forests (HANDLEY 1954). Polyphenols also combine with proteins of plant and microbial origin in the litter layer and there is good evidence that they are important in determining rates of litter decomposition (WILLIAMS and GRAY 1974).

Climate. Decomposition proceeds more rapidly in tropical than temperate climates (JENNY et al. 1949; OLSON 1963): decay constants (k values) range from about 0.1 in arctic and subalpine communities to 4.0 in tropical lowland rainforests. Decomposition rates based on litter bag studies were compared along environmental gradients of temperature, moisture and altitude in the western coniferous forest biome project of the IBP. Decay constants in old-growth Douglas fir forests aged about 450 years decreased from 0.29 at cool, moist sites to 0.22 in warmer, drier habitats (FOGEL and CROMACK 1977). Although decay rates generally decrease with increasing altitude, decomposition proceeds quite rapidly under heavy snow in a subalpine *Abies amabilis* community (see RICHARDS et al. 1981).

Interaction of Climate and Substrate Quality. When the decay constant (k) is plotted against time, the shape of the resulting curve suggests that decomposition proceeds as a series of exponential steps, despite the fact that one exponential function can sometimes be fitted to the data quite successfully. In other words, it seems certain that k values are not constant throughout time. This is to be expected, since litter layers in most plant communities normally experience alternate periods of wetting and drying and cycles of fluctuating temperature. Even if environmental conditions did not vary, decomposition would not proceed at a constant rate because of differences in the decomposability of different substrates. Thus the course of litter decomposition might best be represented as a combination of the decay curves of the several constituents, sugars, hemicelluloses, cellulose, lignin, waxes and polyphenols etc., in proportion to the amounts in which they occur (MINDERMAN 1968).

MEENTEMEYER (1978) set out specifically to examine the dual control of regional climate (subpolar to warm temperate) and substrate quality on litter decomposition, using lignin concentration as an index of substrate quality and actual evapotranspiration (AET) as an integrator of climatic variables. Regression analysis revealed that 51% of the variance in the annual percentage weight loss from litter bags was accounted for by AET and a further 19% by the AET/lignin ratio. On a global scale, climatic control should be even more dominant. Within any region of uniform macroclimate, at sites of comparable topography and aspect, one would expect substrate quality as measured by lignin status to govern the rate of litter decomposition. The relative control by lignin over decomposition rate is not uniform over different climatic regions, however: the more favourable the temperature and moisture regime, the more rapid the decay for a given lignin content, but the higher the lignin content, the more favourable the climatic conditions needed for litter breakdown.

1.3.3.5 Regulation of Decomposition Processes

There is evidence that leaves of woody perennials may be largely depleted of simple carbohydrates before abscission and that nutrient withdrawal from leaves occurs during senescence. The extent to which the processes of autolysis and translocation within the leaf influence primary colonization is an unknown factor, but both leaf physiology and anatomy are certain to have some effect on the pattern of microbial succession in decomposing leaf litter. However, the degree of physical breakdown by the mesofauna depends not only on leaf anatomy but also on the kinds of animals involved. The various litter-feeding and phytophagous invertebrates have their own distinctive patterns of leaf consumption, some mining the internal parenchyma tissues, others consuming the cuticle and epidermis, and still others reducing the leaf to a skeleton of vascular tissue.

Root uptake, exchange reactions in the soil, and recycling by decomposers are processes which utilize the products of net mineralization and "minimize losses from the ecosystem of elements whose absence would severely limit primary production" (WITKAMP and AUSMUS 1976). This is an acceptable viewpoint, but it is a large step from here to propose that mature ecosystems are so closely integrated that decomposition processes are "programmed" by plant physiology and anatomy so as to conserve nutrients.

1.3.4 The Geochemical Cycle: Nutrient Losses

When elements escape from the biogeochemical cycle they may enter the geochemical cycle. Some aspects of how this cycle impinges on nutrient dynamics were touched on previously when considering inputs from atmospheric and parent material sources. Here we deal very briefly with several of the losses of nutrient elements to sinks external to the ecosystem.

1.3.4.1 Gaseous Losses

The Nitrogen Cycle. When the decomposition of nitrogenous organic residues is accompanied by a marked increase in soil reaction, as in the hydrolysis of urea (Sect. 1.3.3.1), substantial amounts of ammonia may escape to the atmosphere if the pH of the micro-habitat rises above 8. Considerable quantities of N are also volatilized by fire, which is a normal environmental variable in many terrestrial ecosystems (see also Chap. 16, Vol 12A). In most circumstances, however, the N incorporated by organisms is eventually returned to the inorganic state by mineralization processes involving its stepwise oxidation from NH_4^+ to NO_2^- and NO_3^-. In terrestrial ecosystems the ammonium ion is normally adsorbed on colloidal surfaces in the upper part of the soil profile where it is available for uptake by plants and microbes or for further oxidation by the nitrifying bacteria. The nitrate ion is mobile and hence may be lost to groundwater by leaching and find its way into streams and, eventually, into the ocean. Nitrate which is not leached is either assimilated by organisms or else denitrified, i.e. reduced to gaseous form, and this is thought also to be the ultimate fate of most of the nitrate that reaches the sea.

Denitrification is thus the major process by which the nitrogen that has been fixed biologically (Sect. 1.2.1.1) is returned to the atmosphere. It is an anaerobic process, the end products of which are nitrous oxide or molecular nitrogen that diffuse out of the soil into the atmosphere, and so are lost from the ecosystem. While denitrification dominates the N cycle in waterlogged soils, it is not confined to such environments. Rather, it is a normal and regular reaction in all terrestrial ecosystems, since anaerobic microhabitats exist temporarily even in the best-aerated soils. In mature communities approaching the steady-state condition, the rate of denitrification presumably balances that of biological N fixation.

The Sulphur Cycle. Plant litter and soil organic matter contain a variety of compounds in which sulphur is present in the sulfhydryl (–SH) form. Since plants take up S mainly as sulphate the mineralization of organic residues is an essential component of the biogeochemical cycle (Sect. 1.3.3.3). The geochemical cycle also plays an important role in the S nutrition of plant communities, for much of the S used in terrestrial ecosystems is volatilized from marine sources. Volatilization of S also occurs, along with N, during fires. An anaerobic process analogous to denitrification which occurs in soils, but is much more characteristic of estuarine muds, is sulphate reduction. The H_2S produced diffuses into the atmosphere where it is oxidized spontaneously to SO_2, dissolved in water to form finally sulphate and deposited on land surfaces in rain. Direct absorption of atmospheric SO_2 by plants and soil can also occur (Sect. 1.2.1.1, see also Chap. 14, this Vol.).

1.3.4.2 Outgoes in Drainage Waters

Wherever rainfall is sufficient to through-leach the soil profile periodically, nutrient ions are carried away to stream drainage. The evidence suggests, however, that in steady-state ecosystems these losses are essentially in balance with inputs and that net loss from the functional volume of the system does not occur. Indeed, one of the more remarkable characteristics of such systems is the efficiency with which they retain their nutrient capital. In aggrading ecosystems this ability is even more pronounced, as they possess also a powerful capacity for recruitment of additional nutrient capital from atmospheric and weathering inputs.

The ability to retain what has been accumulated, and to add to it under conditions of net growth, results from a number of physicochemical and biological mechanisms that buffer the ecosystem against leaching. Broadly speaking, these comprise a chain of sinks or filters between input at canopy level and the loss point located approximately at the lowest depth of effective root development. Of particular note are active uptake by roots and microorganisms, and cation exchange and sorption reactions at solution-solid-phase surfaces generally. Collectively, these several mechanisms act to sustain essentially closed internal nutrient cycles in undisturbed systems.

Ion concentration increases observed in throughfall and litter leachates frequently undergo attenuation as the solution passes through the solum, so that quality of drainage collected at depth, or streamwater, often bears little relationship to that of water entering the profile (Table 1.7). When water volume is taken into account, and the observed concentration changes translated into annual elemental fluxes, as in the studies of Heinrichs and Mayer (1977) and Best and Monk (1975), the location and strength of cation retention mechanisms is evident (Table 1.8). Similarly, $H_2PO_4^-$ ions are retained with great tenacity and NH_4^+ also (Henderson et al. 1978; Stark and Jordan 1978; Jordan et al. 1979; Lewis and Grant 1979). For NO_3^- and SO_4^{2-}, however, the response in undisturbed ecosystems seems more uncertain (Johnson et al. 1979; Lewis and Grant 1979; Martin 1979), net retention or loss being more dependent upon the particular balance between decomposition processes and active absorption at the time than is the case with the common cations and phosphate. Even so, net loss of these elemental species seems not to occur in undisturbed systems, except perhaps as a transient phenomenon.

Table 1.7. Changes in ionic composition of water during passage through two forest ecosystems

Solution	Ion concentrations (μg ml^{-1})								
	Na	K	Ca	Mg	Min. N	P	S	HCO_3^-	pH
Tropical forest[a]									
Rainfall	0.5	0.2	0.2	0.1	0.1	0.00		1.2	5.8
Throughfall	0.4	0.7	0.3	0.1	0.2	0.03		3.7	5.9
Litter leachate	0.6	1.1	1.3	1.1	0.8	0.00		14.6	6.0
Soil leachate (8 cm)	0.7	0.9	2.5	2.3	0.3	0.00		4.9	5.9
Stream water	1.6	0.4	0.8	0.3	0.1	0.00		4.9	6.5
Temperate forest[b]									
Rainfall	0.3	0.1	0.2	0.1	0.8	0.01	1.3	0.3	4.5
Throughfall	0.6	1.1	1.0	0.3	1.1	0.13	4.9	0.2	4.4
Litter leachate	1.0	1.7	2.9	1.8	0.2	0.10	13.9	6.0	5.4
Soil leachate (A_1 hor)	0.8	0.4	1.8	0.7	0.4	0.03	5.0	5.0	6.2
Stream water	1.0	0.1	1.5	0.3	0.3	0.01	2.1	7.8	6.8

[a] MCCOLL (1970)
[b] FELLER (1977)

Table 1.8. Yearly fluxes of cations in water flow through mixed hardwood and pine watersheds in North Carolina. (BEST and MONK 1975)

Flux component	Hardwood					White pine				
	Water	Na	K	Ca	Mg	Water	Na	K	Ca	Mg
	(mm)	(kg ha^{-1} yr^{-1})				(mm)	(kg ha^{-1} yr^{-1})			
Rainfall	1,704	6.3	4.7	4.2	0.8	1,760	6.8	4.9	4.4	0.8
Throughfall	1,380	8.6	30.5	8.1	3.1	1,328	7.2	30.5	6.3	2.0
Litter leachate	906	8.5	31.7	23.4	5.9	1,045	10.2	34.0	20.9	5.4
Soil percolate	172	1.9	2.1	1.9	0.6	215	2.6	4.1	3.6	0.9
Stream discharge	893	7.8	4.2	5.7	2.5	657	4.7	2.8	3.2	1.4

In the case of those nutrient elements, such as N and S, which occur predominantly in organic combination, measurement of the balance between atmospheric inputs and drainage outgoes may indicate the degree of control or relative leakiness of the ecosystem, provided due allowance is made for short-term fluctuations that could be meaningless. But for those such as Ca and K, streamwater outgoes cannot be expected to reveal the true situation regarding internal nutrient dynamics of an ecosystem if weathering below the root zone is contributing much of the dissolved load (HENDERSON et al. 1978). In instances such as these, where streamwater quality is partly or largely under geological control, other investigative procedures are called for if total ion load of drainage is to be partitioned into real and apparent outgoes from the ecosystem. Here tension lysimeters installed at several depths in and below the root zone are appropriate. However, there can be problems associated with the use of these devices, such as their water collection efficiency, their possible effects on water quality, and the degree of replication that may be called for in laterally variable soils. Nevertheless, as the data in Tables 1.7, 1.8 and other studies (HEINRICHS and MAYER 1977; JOHNSON et al. 1977; MCCOLL 1978) indicate, they can provide useful insights into where particular ionic species enter percolating solution in greatest amounts, whether or not they are subsequently withdrawn at greater depth, and the effectiveness of their recapture.

1.4 Conclusions

Our brief treatment of mineral cycling purposely omits any consideration of how nutrient capital and nutrient flux in terrestrial ecosystems are affected by disruptive factors such as acid rain, fire, and logging in its various forms. Space does not permit adequate coverage of these important matters, therefore we have merely attempted to draw attention to several common features of nutrient distribution and conservation which determine susceptibility of ecosystems to elemental loss. In the main these relate to the capacity of soil to retain nutrient capital against leaching, and to refurnish the elements sequestered in biomass during secondary succession.

It is fair to say that most nutrient cycling studies of the last decade have been concerned primarily with aboveground nutrient capital and flux, together with measurements of gross atmospheric inputs and streamwater outputs. Obviously such information is vital, but much more than this will be required in a future that will demand perhaps more intensive, and certainly more careful, utilisation of vegetation resources than hitherto. In particular, there will be a greater need for information on soil-based processes if response to specific long- and short-term disturbance is to be reliably predicted. Fortunately, there has been increasing interest in these processes in recent years, and it is to be hoped that this trend will not only continue but intensify.

References

Alway FJ, Marsh AW, Methley WJ (1937) Sufficiency of atmospheric sulphur for maximum crop yields. Soil Sci Soc Am Proc 2:229–238

Attiwill PM (1980) Nutrient cycling in a *Eucalyptus obliqua* (L'Hérit.) forest. IV Nutrient uptake and nutrient return. Aust J Bot 28:199–222

Ausmus BS, Witkamp M (1974) Litter and soil microbial dynamics in a deciduous forest stand. East Deciduous For Biome – IBP – 73–10

Bamber RK (1976) Heartwood, its function and formation. Wood Sci Technol 10:1–8

Barber DA, Loughman BC (1967) The effect of micro-organisms on the absorption of inorganic nutrients by intact plants. J Exp Bot 18:170–176

Barley KP (1970) The configuration of the root system in relation to nutrient uptake. Adv Agron 22:159–201

Bartholomew WV, Clark FE (1950) Nitrogen transformations in soil in relation to the rhizosphere microflora. Int Congr Soil Sci Trans 4(2):112–113
Beadle NCW (1966) Soil phosphate and its role in molding segments of the Australian flora and vegetation, with special reference to xeromorphy and sclerophylly. Ecology 47:992–1007
Bergersen FJ, Kennedy GS, Wittman W (1965) Nitrogen fixation in the coralloid roots of *Macrozamia communis* L. Johnson. Aust J Biol Sci 18:1135–1142
Bernhard-Reversat F (1975) Nutrients in throughfall and their quantitative importance in rain forest mineral cycles. In: Golley FB, Medina E (eds) Tropical ecological systems. Springer, Berlin Heidelberg New York
Best GR, Monk CD (1975) Cation flux in hardwood and white pine watersheds. In: Howell FG, Gentry JB, Smith MH (eds) Mineral cycling in southeastern ecosystems. US Energ Res Dev Admin, Tech Inf Center, Washington DC
Bevege DI, Bowen GD, Skinner MF (1975) Comparative carbohydrate physiology of ecto- and endo-mycorrhizas. In: Sanders FE, Mosse B, Tinker PB (eds) Endomycorrhizas. Proceedings of a symposium held at the University of Leeds, 22–25 July 1974. Academic Press, London New York
Bevege DI, Richards BN, Moore AW (1978) Fixation of atmospheric nitrogen in association with conifers. CSIRO Aust Div Soils Div Rep No 26
Birch HF (1958) The effect of soil drying on humus decomposition and nitrogen availability. Plant Soil 10:9–31
Birch HF (1960) Nitrification in soil after different periods of dryness. Plant Soil 12:81–96
Bormann FH, Likens GE (1979) Pattern and process in a forested ecosystem. Springer, Berlin Heidelberg New York
Bowen GD, Theodorou C (1967) Studies on phosphate uptake by mycorrhizas. Proc 14th IUFRO Congr, Munich 24:116–138
Bowen GD, Theodorou C (1973) Growth of ectomycorrhizal fungi around seeds and roots. In: Marks GC, Kozlowski TT (eds) Ectomycorrhizas: their physiology and ecology. Academic Press, London New York
Bray JR, Gorham E (1964) Litter production in forests of the world. Adv Ecol Res 2:101–157
Carlisle A, Brown AHF, White EJ (1967) The nutrient content of tree stem flow and ground flora litter and leachates in a sessile oak (*Quercus petraea*) woodland. J Ecol 55:615–627
Chapin FS (1974) Morphological and physiological mechanisms of temperature compensation in phosphate absorption along a latitudinal gradient. Ecology 55:1180–1198
Charley JL, West NE (1975) Plant-induced soil chemical patterns in some desert shrub-dominated ecosystems of Utah. J Ecol 63:945–964
Charley JL, West NE (1977) Micro-patterns of nitrogen mineralization activity in soils of some shrub-dominated semi-desert ecosystems of Utah. Soil Biol Biochem 9:357–365
Cole DW, Rapp M (1981) Elemental cycling in forest ecosystems. In: Reichle DE (ed) Dynamic properties of forest ecosystems. Cambridge Univ Press, Cambridge
Coleman DC (1976) A review of root production processes and their influence on soil biota in terrestrial ecosystems. In: Anderson JM, Macfadyen A (eds) The role of terrestrial and aquatic organisms in decomposition processes. Blackwell, Oxford
Cox TL, Harris WF, Ausmus BS, Edwards NT (1978) The role of roots in biogeochemical cycles in an eastern deciduous forest. Pedobiologia 18:264–271
Crocker RL, Major J (1955) Soil development in relation to vegetation and surface age at Glacier Bay, Alaska. J Ecol 43:427–448
Crossley DA Jr (1977) The roles of terrestrial saprophagous arthropods in forest soils. In: Mattson WJ (ed) The role of arthropods in forest ecosystems. Springer, Berlin Heidelberg New York
Crossley DA Jr, Witkamp M (1966) The role of arthropods and microflora in breakdown of white oak litter. Pedobiologia 6:293–303
Deans JD (1979) Fluctuations of the soil environment and fine root growth in a young sitka spruce plantation. Plant Soil 52:195–208

Dommergues Y (1963) Evaluation du taux de fixation de l'azote dans un sol dunaire reboisé en filao (*Casuarina equisetifolia*). Agrochimica 7:335–340
Duvigneaud P, Denaeyer-De Smet S (1970) Biological cycling of minerals in temperate deciduous forests. In: Reichle DE (ed) Ecol Stud, vol I. Analysis of temperate forest ecosystems. Springer, Berlin Heidelberg New York
Eaton JS, Likens GE, Bormann FH (1973) Throughfall and stemflow chemistry in a northern hardwood forest. J Ecol 61:495–508
Edwards CA, Heath GW (1963) The role of soil animals in breakdown of leaf material. In: Doeksen J, Drift Van der J (eds) Soil organisms. Elsevier/North Holland Biomedical Press, Amsterdam New York
Edwards CA, Reichle DE, Crossley DA Jr (1970) The role of soil invertebrates in turnover of organic matter and nutrients. In: Reichle DE (ed) Analysis of temperate forest ecosystems. Springer, Berlin Heidelberg New York
Estermann EF, McLaren AD (1961) Contributions of rhizosphere organisms to the total capacity of plants to utilize organic nutrients. Plant Soil 15:243–260
Feller MC (1977) Nutrient movement through western hemlock – western red cedar ecosystems in southwestern British Columbia. Ecology 58:1269–1283
Fogel R, Cromack K Jr (1977) Effect of habitat and substrate quality on douglas fir litter decomposition in western Oregon. Can J Bot 55:1632–1640
Fuhr I, Bransford AV, Silver SD (1948) Sorption of fumigant vapors by soil. Science 107:274–275
Garrett SD (1951) Ecological groups of soil fungi: a survey of substrate relationships. New Phytol 50:149–166
Gersper PL, Holowaychuk N (1970a) Effect of stemflow water on a Miami soil under a beech tree: 1. Morphological and physical properties. Soil Sci Soc Am Proc 34:779–786
Gersper PL, Holowaychuk N (1970b) Effect of stemflow water on a Miami soil under a beech tree: 2. Chemical properties. Soil Sci Soc Am Proc 34:786–794
Gersper PL, Holowaychuk N (1971) Some effects of stem flow from forest canopy trees on chemical properties of soils. Ecology 52:691–702
Gillman GP, Bell LC (1978) Soil solution studies on weathered soils from tropical North Queensland. Aust J Soil Res 16:67–77
Golley FB, McGinnis JT, Clements RG, Child GI, Duever MJ (1975) Mineral cycling in a tropical moist forest ecosystem. Univ Georgia Press, Athens, Georgia
Gorham E, Vitousek PM, Reiners WA (1979) The regulation of chemical budgets over the course of terrestrial ecosystem succession. Annu Rev Ecol Syst 10:53–84
Gosz JR, Likens GE, Eaton JS, Bormann FH (1975) Leaching of nutrients from leaves of selected tree species in New Hampshire. In: Howell FG, Gentry JB, Smith MH (eds) Mineral cycling in southeastern ecosystems. US Energ Res Dev Admin, Tech Inf Center, Washington DC
Greaves MP, Webley DM (1969) The hydrolysis of myoinositol hexaphosphate by soil microorganisms. Soil Biol Biochem 1:37–43
Greenland DJ (1958) Nitrate fluctuations in tropical soils. J Agric Sci 50:82–92
Grier CC (1977) Biomass, productivity and nitrogen-phosphorus cycles in hemlock-spruce stands of the central Oregon coast. In: Zososki RJ, Atkinson WA (eds) Proceedings conference on intensive management of western hemlock. Inst For Prod Bull 21. Coll For Res, Univ Washington, Seattle
Griffin DM (1972) Ecology of soil fungi. Chapman & Hall, London
Halstead RL, McKercher R (1975) Biochemistry and cycling of phosphorus. In: Paul EA, McLaren AD (eds) Soil biochemistry, vol IV. Dekker, New York
Handley WRC (1954) Mull and mor formation in relation to forest soils. Bull For Comm London 23:iv+115
Harley JL (1969) The biology of mycorrhiza, 2nd edn. Hill, London
Harley JL (1975) Problems of mycotrophy. In: Sanders FE, Mosse B, Tinker PB (eds) Endomycorrhizas. Proceedings of a symposium held at the University of Leeds, 22–25 July 1974. Academic Press, London New York
Harley JL, Lewis DH (1969) The physiology of ectotrophic mycorrhizas. Adv Microb Physiol 3:53–81

Harris WF, Kinerson RS, Edwards NT (1977) Comparison of belowground biomass of natural deciduous forests and loblolly pine plantations. In: Marshall JK (ed) The belowground ecosystem: a synthesis of plant-associated processes. Range Sci Dep Sci Ser 26. Colorado State Univ, Fort Collins, Colorado
Heinrichs H, Mayer R (1977) Distribution and cycling of major and trace elements in two Central European forest ecosystems. J Environ Qual 6:402–407
Helvey JD, Patric JH (1965) Canopy and litter interception of rainfall by hardwoods of eastern United States. Water Resour Res 1:193–206
Henderson GS, Harris WF, Todd DE, Grizzard T (1977) Quantity and chemistry of throughfall as influenced by forest type and season. J Ecol 65:365–374
Henderson GS, Swank WT, Waide JB, Grier CC (1978) Nutrient budgets of Appalachian and Cascade watersheds: a comparison. For Sci 24:385–397
Hermann RK (1977) Growth and production of tree roots: a review. In: Marshall JK (ed) The belowground ecosystem: a synthesis of plant-associated processes. Range Sci Dep Sci Ser 26. Colorado State Univ, Fort Collins, Colorado
Hoffmann G (1972) Wachstumsrhythmik der Wurzeln und Sproßachsen von Forstgehölzen. Flora 161:303–319
Hurditch WJ, Charley JL, Richards BN (1980) Sulfur cycling in Fraser Island and New South Wales coastal forests. In: Freney JR, Nicolson AJ (eds) Sulfur in Australia. Aust Acad Sci, Canberra
Hutchinson GE (1954) The biogeochemistry of the terrestrial atmosphere. In: Kuiper GP (ed) The earth as a planet. Univ Chicago Press, Chicago
Hutchinson GL, Millington RJ, Peters DB (1972) Atmospheric ammonia: absorption by plant leaves. Science 175:771–772
Ingham G (1950a) The mineral content of air and rain and its importance to agriculture. J Agric Sci 40:55–61
Ingham G (1950b) Effect of materials absorbed from the atmosphere in maintaining soil fertility. Soil Sci 70:205–212
Jansson SL (1958) Tracer studies on nitrogen transformations in soil with special attention to mineralization-immobilization relationships. K Landbouwhogesch Annl 24:101–361
Jenny H (1961) Derivation of state factor equations of soils and ecosystems. Soil Sci Soc Am Proc 25:385–388
Jenny H, Gessel SP, Bingham FT (1949) Comparative study of decomposition rates of organic matter in temperate and tropical regions. Soil Sci 68:419–432
Jensen HL (1950) A survey of biological nitrogen fixation in relation to the world supply of nitrogen. Int Congr Soil Sci Trans 4(1):165–172
Johnson DW, Cole DW, Gessel SP (1979) Acid precipitation and soil sulphate absorption properties in a tropical and in a temperate forest soil. Biotropica 11:38–42
Johnson DW, Cole DW, Gessel SP, Singer MJ, Mind RV (1977) Carbonic acid leaching in a tropical, temperate, subalpine and northern forest soil. Arctic Alp Res 9:329–343
Johnson DW, Cole DW, Bledsoe CS, Cromack K, Edmonds RL, Gessel SP, Grier CC, Richards BN, Vogt KA (1982) Nutrient cycling in forests of the Pacific northwest. In: Edmonds RL (ed) Analysis of coniferous forest ecosystems in the western United States. Hutchinson Ross, Stroudsburg Pa
Johnson FL, Risser PG (1974) Biomass, annual net primary production, and dynamics of six mineral elements in a post oak-blackjack oak forest. Ecology 55:1246–1258
Jordan CF (1978) Stem flow and nutrient transfer in a tropical rain forest. Oikos 31:257–263
Jordan CF, Todd RL, Escalante G (1979) Nitrogen conservation in a tropical rain forest. Oecologia 39:123–128
Jordan DC (1981) Nitrogen fixation by selected free-living and associative microorganisms. In: Gibson AH, Newton WE (eds) Current perspectives in nitrogen fixation. Aust Acad Sci, Canberra
Kendrick WB, Burges A (1962) Biological aspects of the decay of *Pinus sylvestris* leaf litter. Nova Hedwigia Z Kryptogamenkd 4:313–342
Kramer PJ, Bullock HC (1966) Seasonal variations in the proportions of suberized and unsuberized roots of trees in relation to the absorption of water. Am J Bot 53:200–204

Lebedjantzev AN (1924) Drying of soil as one of the natural factors in maintaining soil fertility. Soil Sci 18:419–447
Lee KE, Wood TG (1971) Physical and chemical effects on soils of some Australian termites, and their pedological significance. Pedobiologia 11:376–409
Lewis WM, Grant MC (1979) Relationships between stream discharge and yield of dissolved substances from a Colorado mountain watershed. Soil Sci 128:353–363
Lyr H, Hoffmann G (1967) Growth rates and growth periodicity of tree roots. In: Romberger JA, Mikola P (eds) International review of forestry research, vol II. Academic Press, London New York
Malo BA, Purvis ER (1964) Soil absorption of atmospheric ammonia. Soil Sci 97:242–247
Martin CW (1979) Precipitation and steamwater chemistry in an undisturbed forested watershed in New Hampshire. Ecology 60:36–42
McBee RH (1971) Significance of intestinal microflora in herbivory. Annu Rev Ecol Syst 2:165–176
McColl JG (1970) Properties of some natural waters in a tropical wet forest of Costa Rica. Bioscience 20:1095–1100
McColl JG (1978) Ionic composition of forest soil solution and effects of clearcutting. Soil Sci Soc Am J 42:358–363
McLaren AD, Skujins J (1971) Trends in the biochemistry of terrestrial soils. In: McLaren AD, Skujins J (eds) Soil biochemistry, vol II. Dekker, New York
Meentemeyer V (1978) Macroclimate and lignin control of litter decomposition rates. Ecology 59:465–472
Miller RB (1963) Plant nutrients in hard beech III. The cycle of nutrients. NZ J Sci 6:388–413
Minderman G (1968) Addition, decomposition and accumulation of organic matter in forests. J Ecol 56:355–362
Nicholls KH, Cox MC (1978) Atmospheric nitrogen and phosphorus loading to Harp Lake, Ontario, Canada. Water Resour Res 14:589–592
Nicholson TH (1975) Evolution of vesicular-arbuscular mycorrhizas. In: Sanders FE, Mosse B, Tinker PB (eds) Endomycorrhizas. Proceedings of a symposium held at the University of Leeds, 22–25 July 1974. Academic Press, London New York
Nihlgard B (1970) Precipitation, its chemical composition and effect on soil water in a beech and a spruce forest in south Sweden. Oikos 21:208–217
Norris DO (1956) Legumes and the rhizobium symbiosis. Emp J Exp Agric 24:247–270
Nye PH (1961) Organic matter and nutrient cycles under moist tropical forest. Plant Soil 13:333–346
Nye PH, Tinker PB (1977) Solute movement in the soil-plant system. Univ California Press, Berkeley
Odum EP (1969) The strategy of ecosystem development. Science 164:262–270
Odum HT (1970) The rain forest and man: an introduction. In: Odum HT, Pigeon RF (eds) A tropical rain forest. A study of irradiation and ecology at El Verde, Puerto Rico. Div Tech Inf US At Energ Comm. USAEC, Washington
Olsen RA (1957) Absorption of SO_2 from the atmosphere by cotton plants. Soil Sci 84:107–111
Olson JS (1963) Energy storage and the balance of producers and decomposers in ecological systems. Ecology 44:322–331
Orman HR, Will GM (1960) The nutrient content of *Pinus radiata* trees. NZ J Sci 3:510–522
Ovington JD, Heitkamp D, Lawrence DB (1963) Plant biomass and productivity of prairie, savanna, oakwood and maize field ecosystems in central Minnesota. Ecology 44:52–63
Patterson DT (1975) Nutrient return in the stemflow and throughfall of individual trees in the Piedmont deciduous forest. In: Howell FG, Gentry JB, Smith MH (eds) Mineral cycling in southeastern ecosystems. US Energ Res Dev Admin Tech Inf Center, Washington DC
Reiners WA (1972) Nutrient content of canopy throughfall in three Minnesota forests. Oikos 23:14–22
Richards BN (1974) Introduction to the soil ecosystem. Longman, London New York

Richards BN, Charley JL (1977) Carbon and nitrogen flux through native forest floors. In: Hingston FJ (ed) Symposium on nutrient cycling in indigenous forest ecosystems. Inst For Res Protect, Como, W Austr, CSIRO Div Land Res Managem, Perth
Rodin LE, Basilevich NI (1967) Production and mineral cycling in terrestrial vegetation. In: Fogg GE (ed) Oliver and Boyd, Edinburgh
Rolfe GL, Akhtar MA, Arnold LE (1978) Nutrient distribution and flux in a mature oak-hickory forest. For Sci 24:122–130
Rovira AD (1969) Plant root exudates. Bot Rev 35:35–57
Santantonio D, Hermann RK, Overton WS (1977) Root biomass studies in forest ecosystems. Pedobiologia 17:1–31
Satchell JE (1974) Litter-interface of animate/inanimate matter. In: Dickinson CH, Pugh GJF (eds) Biology of plant litter decomposition, vol I. Academic Press, London New York
Simonson RW (1970) Loss of nutrient elements during soil formation. In: Engelstad PO (ed) Nutrient mobility in soils: accumulation and losses. Soil Sci Soc Am Spec Publ 4, Madison, Wisconsin
Smith WH (1970) Technique for collection of root exudates from mature trees. Plant Soil 32:238–241
Smith WH (1976) Character and significance of forest tree root exudates. Ecology 57:324–331
Springett BP (1978) On the ecological role of insects in Australian eucalypt forests. Aust J Ecol 3:129–139
Stace HCT, Hubble GD, Brewer R, Northcote KH, Sleeman JR, Mulcahy MJ, Hallsworth EG (1968) A handbook of Australian soils. Rellim Technical Publ, Glenside, South Australia
Stanier RY (1953) Adaptation, evolutionary and physiological: or Darwinism among the microorganisms. In: Davies R, Gale EF (eds) Adaptation in micro-organisms. 3rd Symp Soc Gen Microbiol. Cambridge Univ Press, London
Stark N, Jordan CF (1978) Nutrient retention by the root mat of an Amazonian rain forest. Ecology 59:434–437
Steinhardt U (1973) Input of chemical elements from the atmosphere. A tabular review of literature. Göttinger Bodenkd Ber 29:93–132
Sutton RF (1969) Form and development of conifer root systems. Tech Commun 7 For Bur, Oxford
Switzer GL, Nelson LE (1972) Nutrient accumulation and cycling in loblolly pine (*Pinus taeda* L.) plantation ecosystems: the first twenty years. Soil Sci Soc Am Proc 36:143–147
Tappeiner JC, Alm AA (1975) Undergrowth vegetation effects on the nutrient content of litterfall and soils in red pine and birch stands in northern Minnesota. Ecology 56:1193–1200
Thomas WD, Hill GW (1935) Absorption of sulphur dioxide by alfalfa and its relation to leaf injury. Plant Physiol 10:291–307
Tiedemann AR, Klemmedson JO (1973) Effect of mesquite on physical and chemical properties of the soil. J Range Manage 26:27–29
Tsutsumi T (1971) Accumulation and circulation of elements in forest ecosystems. In: Duvigneaud P (ed) Productivity of forest ecosystems. Ecology and conservation 4, UNESCO, Paris
Tukey HB (1970) The leaching of substances from plants. Annu Rev Plant Physiol 21:305–324
Turner J, Singer MJ (1976) Nutrient distribution and cycling in a sub-alpine coniferous forest ecosystem. J Appl Ecol 13:295–301
Turvey ND (1979) Pathways of mineral transfer through selected forest floors in northeastern New South Wales. Ph D thesis, Univ New England, Armidale NSW, Australia
Vitousek PM (1977) The regulation of element concentrations in mountain streams in the northeastern United States. Ecol Monogr 47:65–87
Vitousek PM, Reiners WA (1975) Ecosystem succession and nutrient retention: a hypothesis. BioScience 25:376–381

Voigt GK, Richards BN, Mannion EC (1964) Nutrient utilization by young pitch pine. Soil Sci Soc Am Proc 28:707–709
Waid JS (1974) Decomposition of roots. In: Dickinson CH, Pugh GJF (eds) Biology of plant litter decomposition, vol I. Academic Press, London New York
Wallwork JA (1970) Ecology of soil animals. McGraw-Hill, London
Watanabe I (1981) Biological nitrogen fixation associated with wetland rice. In: Gibson AH, Newton WE (eds) Current perspectives in nitrogen fixation. Aust Acad Sci, Canberra
Weaver GT (1975) The quantity and distribution of four nutrient elements in high-elevation forest ecosystems, Balsam Mountains, North Carolina. In: Howell FG, Gentry JB, Smith MH (eds) Mineral cycling in southeastern ecosystems. US Energ Res Dev Admin Tech Inf Center, Washington DC
Webley DM, Duff RB (1965) The incidence, in soils and other habitats, of microorganisms producing 2-ketogluconic acid. Plant Soil 22:307–313
Weigert RG, Coleman DC, Odum EP (1970) Energetics of the litter-soil subsystem. In: Phillipson J (ed) Methods of study in soil ecology. UNESCO, Paris
Wells EG, Nicholas AK, Bual SW (1975) Some effects of fertilization on mineral cycling in loblolly pine. In: Howell FG, Gentry JB, Smith MH (eds) Mineral cycling in southeastern ecosystems. US Energ Res Dev Admin, Tech Inf Center, Washington DC
Westman WE (1978) Inputs and cycling of mineral nutrients in a coastal subtropical eucalypt forest. J Ecol 66:513–531
Westman WE, Rogers RW (1977) Nutrient stocks in a subtropical eucalypt forest, North Stradbroke Island. Aust J Ecol 2:447–460
White PS (1979) Pattern, process, and natural disturbance in vegetation. Bot Rev 45:229–299
Wild A (1958) The phosphate content of Australian soils. Aust J Agric Res 9:193–204
Will GM (1959) Nutrient return in litter and rainfall under some exotic conifer stands in New Zealand. NZ J Agric Sci 2:719–724
Williams ST, Gray TRG (1974) Decomposition of litter on the soil surface. In: Dickinson CH, Pugh GJF (eds) Biology of plant litter decomposition, vol II. Academic Press, London New York
Witkamp M (1969) Cycles of temperature and carbon dioxide evolution from litter and soil. Ecology 50:922–924
Witkamp M, Ausmus BS (1976) Processes in decomposition and nutrient transfer in forest systems. In: Anderson JM, Macfadyen A (eds) The role of terrestrial and aquatic organisms in decomposition processes. Blackwell, Oxford
Zinke PJ (1962) The pattern of influence of individual forest trees on soil properties. Ecology 43:130–133

2 Nutrient Cycling in Freshwater Ecosystems

A. MELZER and CH. STEINBERG

CONTENTS

2.1 Introduction 48

2.2 Carbon 48
2.2.1 Inorganic Carbon and Carbon Uptake 48
2.2.2 Organic Carbon 49
2.2.2.1 Release of Dissolved Organic Carbon (DOC) 50
2.2.2.2 Some Problems Arising with Measuring Heterotrophic Activity and Bacterial Production 51
2.2.2.3 Role of Detritus 52
2.2.2.4 Methane 54

2.3 Nitrogen 55
2.3.1 Nitrogen Input 55
2.3.1.1 N_2 Fixation 56
2.3.2 Nitrogen Transformations by Aquatic Microorganisms 57
2.3.2.1 Nitrification 57
2.3.2.2 Denitrification 58
2.3.2.3 Ammonia Mineralization 59
2.3.3 Nitrogen Nutrition of Primary Producers 59
2.3.3.1 Aquatic Vascular Plants 59
2.3.3.2 Algae 60
2.3.4 Organic Nitrogen 61

2.4 Phosphorus 62
2.4.1 Forms of Phosphorus and Their Dynamics 62
2.4.2 Role of Pelagic Consumers 63
2.4.3 Phytoplankton and Bacteria 63
2.4.3.1 Algae 63
2.4.3.2 Algae Versus Bacteria 64
2.4.4 Phosphorus in the Littoral Zone 64
2.4.5 Phosphorus in Seston and Sediments 64
2.4.5.1 Precipitation and Sorption of Phosphorus 65
2.4.5.2 Release from the Sediments 65
2.4.6 Phosphorus Retention in Flowing Waters 65

2.5 Sulphur 66
2.5.1 General Ecological Aspects 66
2.5.2 Microbial Transformations of Inorganic Sulphur 66
2.5.2.1 Sulphur-Reducing Processes 68
2.5.2.2 Sulphur-Oxidizing Processes 68
2.5.3 Volatile Organic Sulphur Compounds 69

2.6 Silica 69
2.6.1 Ecosystem Approach 69
2.6.2 Silica Within Diatoms 70

2.7 Heavy Metals, with Special References to Iron and Manganese 70
2.7.1 Forms of Metals in Natural Waters 71
2.7.2 Cycling of Iron and Manganese 72
2.7.3 Iron and Manganese in Sediments 74

2.8 Conclusions 75

References 75

2.1 Introduction

Limnology has been developed mainly from biological sciences. Limnochemistry, a part of limnology, has long been neglected by researchers and was thus referred to as "illustrative tapestry" by SCHINDLER et al. (1975). The stimulating works of OHLE (e.g. 1954, 1958, 1962) and the investigations on the problem of eutrophication, by shedding light on the importance of the different nutritive substances in freshwaters, changed the role of limnochemistry substantially.

The most important nutrients for primary producers in freshwater ecosystems are phosphorus, nitrogen, sulphur, silica, trace metals, as well as some organic compounds. Each of these nutrients can limit the primary production in running and standing waters. It seems by and large that phosphorus is the most frequently limiting factor (cf. LIKENS 1972). In a number of different freshwater types (alkalinic–acid; Ca-rich–Ca-poor; running–standing; temperate–tropic), however, nitrogen, silica and even sulphur and inorganic carbon can be the limiting factors under certain circumstances.

In this chapter an attempt is made to summarize the current literature on nutrient cycling in freshwater. We emphasize the biological and biogeochemical connections which are part of the nutrient cycling. During the course of searching through the literature, it became evident that some of the views on nutrient cycling in the textbooks are in conflict with those found in the literature. These textbooks (e.g. HUTCHINSON 1957, 1967, 1975; WETZEL 1975), however, are in general excellent and highly recommended to those who are interested in limnological aspects. It is believed that further research aimed at the nutrient cycling is needed and it is hoped that this review will, in part, pave the way for such investigations.

2.2 Carbon

2.2.1 Inorganic Carbon and Carbon Uptake

The calcium carbonate-carbonic acid equilibrium can be specified sufficiently in artificial solutions. However, it is not entirely possible to transfer the known equation to natural water. It appears that at pH 8 (well within the pH range of hard waters), calculated values of total CO_2 greatly exceed analytical values; the same is true at low pH, e.g. in soft waters or bogs (WETZEL and RICH 1973). These differences are in part a result of interferences by organic acids and other organic water constituents: the organic water constituents slow down the formation of calcite considerably (CHAVE 1970; REDDY 1975) and presumably also the dissociation of HCO_3^- to CO_2 (WETZEL 1972). Furthermore the coprecipitation of organic and inorganic trace nutrients at autochthonous calcite may have negative effects for the maintainance of the primary production.

WETZEL and OTSUKI (1974) observed a coprecipitation of fulvic acids and presume that a similar elimination mechanism is also effective for vitamins and similar substances. A limitation of the aquatic production by carbon very probably only occurs (if at all) via the described coprecipitation mechanism and the retarded dissociation, as is indicated by the symposium *Nutrient and eutrophication: The limiting-nutrient controversy* (LIKENS 1972). Even in soft-water lakes limitation does not come about primarily through too low a supply of CO_2 (SCHINDLER et al. 1972). SCHINDLER et al. (1975) stress that in the few natural waters where carbon is "limiting" to plankton, it must be regarded as a symptom, rather than a cause, of eutrophication (by P- and N compounds).

The question which CO_2 uptake and fixation mechanism is used by algae and aquatic macrophytes has only recently been followed up with particular interest, although three classic types have been known for a long time:

Fontinalis type (uptake of free CO_2), *Elodea* type (uptake of CO_2 and HCO_3^- in varying proportions) and *Scenedesmus* type (uptake of HCO_3^- and little CO_2) (from ZIEGLER 1979 with references to the original works). Further details are treated in Chap. 15, Vol. 12B.

The contribution of organs other than leaves to CO_2/HCO_3^- exchange by aquatic plants needs to be clarified. In some soft water plants CO_2/HCO_3^- is exchanged through the roots rather than through the leaves (WIUM-ANDERSEN 1971; SØNDERGAARD and SAND-JENSEN 1979) although of course, carbon reduction takes place in leaf photosynthetic tissue. On the other hand, when the HCO_3^- concentration of water is high (e.g. in sea water) most of the CO_2/HCO_3^- exchange occurs across leaf tissue itself (WETZEL and PENHALE 1979).

2.2.2 Organic Carbon

The circulation of organic carbon in a freshwater ecosystem is represented in a simplified form in Fig. 2.1 (see also Chap. 10, this Vol.). In principle, it is true of both still and flowing waters, however there are great differences in the proportion of autochthonous and allochthonous parts in the individual water types. In rivers and stagnant bog water, the organic carbon of the external humic substances and of the dead foliage plays a main role, and the primary production in the water is quantitatively less important. In still clear waters the autochthonous predominates over the allochthonous input of carbon.

In Fig. 2.1 litoral flora and phytoplankton count as autochthonous primary producers, along with the photoautotrophic and chemoautotrophic bacterial flora. It must be stressed that the differentiation between dissolved organic carbon (DOC) and particulate

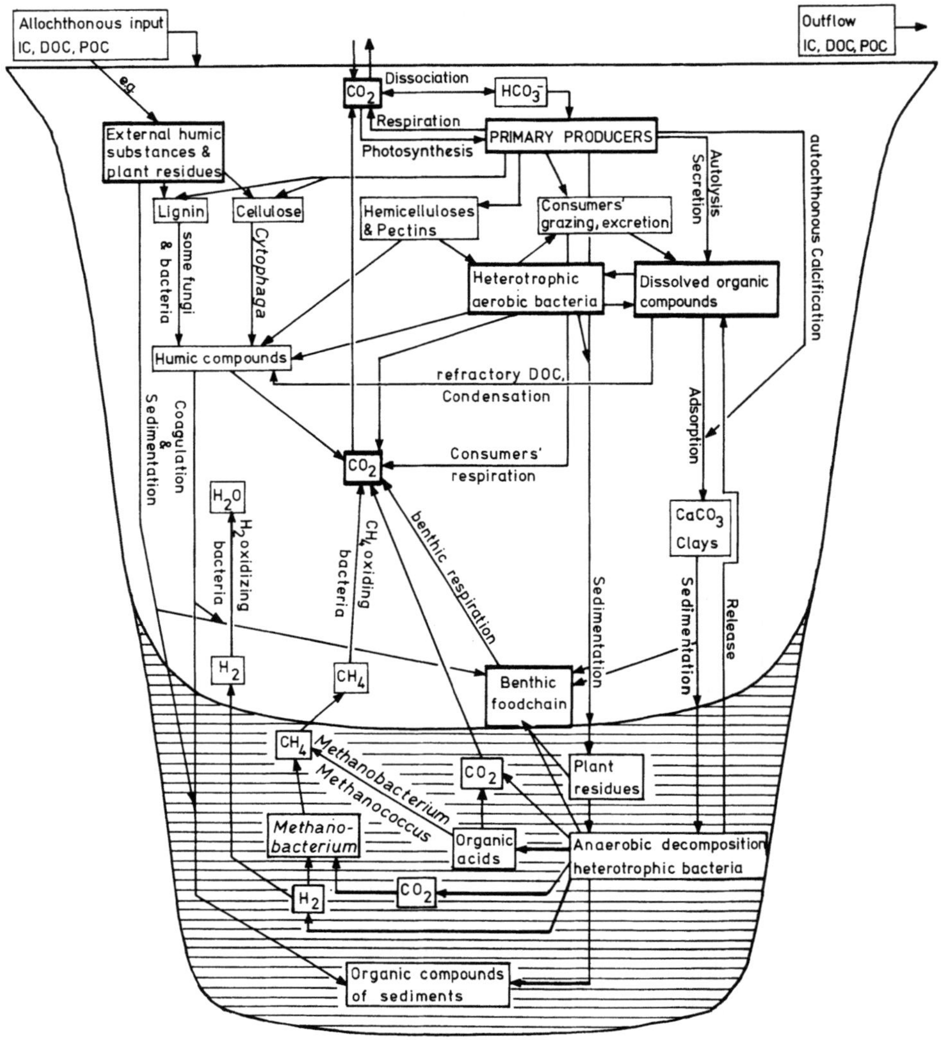

Fig. 2.1. Carbon cycling. *IC* inorganic carbon; *DOC* dissolved organic carbon; *POC* particulate organic carbon. (Modified after KUSNEZOW 1959 and WETZEL 1975, supplemented after several authors)

organic carbon (POC) in Fig. 2.1 follows the analytical methods and does not reflect the actual conditions. Both forms are dead organic matter, therefore detritus. From a functional point of view, the energetic transformations of dissolved and particulate organic carbon are similar; only the rates of transformation differ (WETZEL 1975). Also a distinction between humic and dissolved organic compounds is only justified, when a different origin and possibly a varying degree of bacterial transformation is meant.

2.2.2.1 Release of Dissolved Organic Carbon (DOC)

In spite of some errors in the methods and of some false interpretations (cf. SHARP 1977) it appears to be unquestionable that healthy phytoplankton and submersed macrophytes excrete a part of the photosynthetic products to the medium (FOGG 1977). In the case of phytoplankton the proportion normally lies below 5% (for example BERMAN 1976; NALEWAJKO and SCHINDLER 1976; LEE and NALEWAJKO 1978; HARRIS 1978) and in the case of macrophytes between 4% and 10% (HOUGH and WETZEL 1975; WETZEL 1979), whereas epiphytized plants excrete less than those with no epiphytes (PENHALE and SMITH 1977). The release through autolysis or rotting biomass of macrophytes considerably exceeds this value. The substances released from healthy cells can be divided into two classes: metabolic intermediates of low molecular weight released by diffusion and end products of metabolism, the liberation of which from healthy cells is more or less an active process (FOGG 1971). These two types of substances also appear in natural plankton populations (STEINBERG 1978).

Two reviews discuss the nature of the released products (FOGG 1971; HELLEBUST 1974). Glycollic acid is always detected in the excretions from healthy cells and is taken as evidence of excretion (FOGG 1977). During an *Oscillatoria redekei* bloom, JÜTTNER and MATUSCHEK (1978) found glycollic acid to account for up to 34% of the low molecular weight compounds released. The extracellular products include carbohydrates (cf. STABEL 1977), peptides and proteins (cf. STEINBERG 1977), volatile substances (cf. JÜTTNER 1981), vitamins and other growth-promoting substances, substances having specific effects in reproduction, auto-inhibitors, antibiotics and toxins (cf. MURPHY et al. 1976; WOLFE and RICE 1979; MCCRACKEN et al. 1980).

The release of biologically active substances seems to be one mechanism among many, by which certain species of algae are able to suppress others. Cyanophytes, for example, can excrete iron-selective chelators (hydroxamic acids) so that iron is no longer available for other algae (MURPHY et al. 1976). LAMPERT (1978) describes a particular detail of the DOC release, namely the release of DOC by grazing zooplankton. Up to 17% of the algal carbon ingested was initially lost as DOC from algae damaged during feeding of *Daphnia pulex* (Cladocera). Only about 4% of the carbon from algae swallowed whole is released during ingestion. Additional DOC is produced by secretion from *Daphnia* and by leaching from their faeces. We can estimate that during those periods of the year when phytoplankton primary production is totally consumed by the zooplankton, more than 10% of the carbon fixed in photosynthesis is added to the DOC pool through grazing zooplankton. This is in the same range as the quantity excreted directly by phytoplankton.

2.2.2.2 Some Problems Arising with Measuring Heterotrophic Activity and Bacterial Production

The role of bacterioplankton in the cycling of organic matter has long been recognized as a pathway in the organic carbon cycle of natural waters, but has not yet been fully characterized. Whereas the significance of bacterial catabolic activity has been amply demonstrated, methodological problems have made it difficult to quantify the rates of biosynthesis and growth in the same manner as primary production. For this reason, the method rather than the subject matter is important in the following discussions.

ROMANENKO (1964) estimated the bacterial production of intact populations by dark uptake of $^{14}CO_2$, assuming this uptake to be constantly 6% of the total bacterial production. But in detailed ecological studies the rate was shown to vary from 1% to 12% and more (OVERBECK and DALEY 1973). Apart from that, the ecological conditions for heterotrophic CO_2 fixation are not sufficiently known (cf. OVERBECK 1979 vs. MARXSEN 1980).

Another approach in measuring overall heterotrophic activity and bacteria production is the use of radioactively labelled monomeric substrates (glucose, acetate) as pioneered by PARSONS and STRICKLAND (1962) and by WRIGHT and HOBBIE (1966). It yields a *relative* measure of microbial heterotrophic activity only, which may be useful in comparing one water body with another. Apart from the fact that the uptake of labelled substrates by a heterogenic aquatic micro-flora does not necessarily follow the single-enzyme kinetics relied on for calculation (KRAMBECK 1979), such results cannot be converted into a measure of total microbial heterotrophic activity, chiefly because of the diversity within the DOC pool. A further great difficulty is that algae also are able to compete effectively with the bacteria for dissolved organic substrates (WHEELER et al. 1977). The use of labelled polymeric amino acids or of lysis products from labelled algal cultures is in principle superior to the use of monomers (cf. SAUNDERS 1972). In a more recent paper, STABEL (1981) compared monomeric and polymeric carbon mineralization calculations. He found that the heterotrophic uptake of glucose accounts for about 20%, the use of lysis products, however, for about 50% of the mineralization measured by seston trapping. Some further attempts to evaluate production (of bacteria), e.g. by the RNA- or DNA synthesis (FUHRMANN and AZAM 1980) or by the uptake of SO_4^{2-} (MONHEIMER 1975; CAMPBELL and BAKER 1978), have the disadvantage that they include many generalized assumptions, among other things again that of a constant conversion factor of uptake or (rather) synthesis to the total heterotrophic production. There is a very laborious alternative in the possibility of combining several techniques, as has been attempted in recent works (CAMPBELL and BAKER 1978; FUHRMANN and AZAM 1980; JORDAN and LIKENS 1980).

2.2.2.3 Role of Detritus

The most useful definition of "detritus" is given by RICH and WETZEL (1978), who define detritus as any non-predatory loss of organic carbon from any trophic level or inputs from sources external to the ecosystem that enter and cycle in the system (allochthonous detritus = empneuston). This definition does not distinguish between particulate and dissolved organic matter. In terms of the ecosystems there is no functional difference, energetically, between dissolved organic carbon lost from a phytoplankter and faeces or exuviae from an animal (RICH and WEZEL 1978). These authors also stress the significance of a detrital foodchain by which the chemical energy of detritial organic matter becomes available to the biota. Many significant examples of detritus foodchains are given in MELCHIORRI-SANTOLINI and HOPTON (1972).

Particulate Detritus. As stated above, in running waters detritus plays a greater role in supplying energy to the systems than primary production does in general. A study on Bear Brook, New Hampshire, one of the few studies that has itemized the nature of input in a stream, has shown that allochthonous organic matter accounted for 99.8% of the energy available at primary producer level, and that 52.9% of this was in the form of particulate organic litter (FISHER and LIKENS 1972, quoted in KAUSHIK 1975). The particulate detritus becomes available to stream invertebrates after colonization by microorganisms, especially by hyphomycete fungi (BÄRLOCHER et al. 1978) and/or cellulolytic bacteria (ROBB et al. 1979) as primary colonizers. Thus the quality of food available to benthos largely depends upon the capacity of plant tissue to support microbial populations. For example, elm and maple leaves provide better-quality food than oak or beech leaves, the latter even better than *Eucalyptus* or *Nothofagus* leaves or coniferous needles (KAUSHIK 1975; BLACKBURN and PETR 1979). Multiple-choice feeding experiments showed that caddis fly larvae selected detritus having the most microbial detritus (BÄRLOCHER et al. 1978). Some more chemical and physical parameters influencing the decay of leaf litter are summarized in KAUSHIK (1975); MEYER (1980) stresses the influence of velocity and turbulence of running waters. The great influence of temperature was shown, for example, by SUBERKROPP et al. (1975).

This mechanism of detritus processing is principally also valid for lakes (BARNES et al. 1978). GODSHALK and WETZEL (1978) tested the conditions of conversion of autochthonous particulate detritus of aquatic angiosperms into dissolved organic matter (DOM) and stated that temperature was the most important factor influencing the rates of decomposition and conversion into DOM, while O_2 concentrations controlled the efficiency of decomposition and the conversion of DOM to CO_2.

Dissolved Detritus. In general, dissolved detritus exceeds concentrations of particulate organic carbon, dead or alive, by an order of magnitude, the main part of which is due to fulvic acids (=water-soluble fraction of humic acids). Some properties of the dissolved detritus are essential in the functioning of the aquatic ecosystem, especially the property of metal–organic interactions.

Although the general character of fulvic acids can be clearly defined as phenolic, they also include certain functional subsidiary groups (hydroxyl-, carbonyl- and carboxyl-groups among others; GAMBLE and SCHNITZER 1974). It is, however, difficult in a particular case to define the structural formula even for single humus isolates. Thus, great variations exist in the data about the ratio of the phenolic groups in a fulvic acid association (cf. GAMBLE and SCHNITZER 1974; BERGMANN 1978). Furthermore, there is no standardized pattern for the position of the functional subsidiary groups, which also include, apart from those already mentioned, carbohydrates, peptides, amino acids and ethylene-oxide (GJESSING 1976; STEINBERG 1977; BERGMANN 1978): *the* "fulvic acid" does not exist. Fulvic acids are always a mixture of diverse polymers. BERGMANN (1978), who isolated and examined humic acids from sewage attributed the complexing and decomplexing of metals to the existence of flexible pseudo-cavities, rather than rigid cavities, in the molecules. The latter, however, were assumed by GJESSING (1976). These metal–organic interactions are in principle effective for both the transition metals and the metals of the main groups. Fractions of the dissolved organic substances with various molecular weight have very different uptake capacities and also "preferences" for some metals (STEINBERG 1980). Lead, for example, was only found in the macromolecular fraction (M.W.$\geqq$1,500). Some metals, e.g. copper, show their most toxic effects towards aquatic microorganisms when in an ionic state. The toxic effect decreases when the strength of the organic complex of these metals increases (SUNDA and GUILLARD 1976). Some algae are capable of producing effective, specific, detoxifying exosubstances (SWALLOW et al. 1978), compared with a greater number capable of producing weak copper-complexing agents, which cannot sufficiently eliminate the toxicity (MCKNIGHT and MOREL 1979). The iron-specific chelators of cyanophytes also belong to these weak copper-complexing agents (MCKNIGHT and MOREL 1980). BACCINI and SUTER (1979) isolated an organic ligand from lake water, the conditional formation constant of which was at least two orders of magnitude greater than the ones of well-described bidentate organic ligands.

The fulvic acid–metal compounds promote microbial growth by making trace metals available (cf. Barber 1974). Some results of Steinberg and Herrmann (1980) indicate that the oligomeric fractions of molecular weights from 300 to 700 are most effective in this connection. A further indirect function of the fulvic acids in promoting microorganisms lies in their property as redox catalyst (Schindler et al. 1976). They are capable of catalyzing the transport of electrons from reduced materials in the sediments to electron acceptors in the water.

Much less important than the indirect function of fulvic acids for microorganisms is the direct effect as nutrients. Haan (1977) reports on the possibility of cometabolism of an *Arthrobacter* strain. Steinberg (1977) and Stabel et al. (1979) discovered increased catabolism of dissolved detritus after the addition of organic nutrients to microorganism cultures. Finally Giesy (1976) found that a *Scenedesmus* strain is able to live (photoheterotrophically) on fulvic acids, and is therefore also able to make use of these substances as a C source during photosynthesis.

2.2.2.4 Methane

Methane gas is produced by bacteria in anaerobic environments where organic matter is being decomposed. The rate of methane production may be an indicator of the overall productivity of the aquatic ecosystem (Ohle 1958), as methane production is a direct function of the rate of organic input (Kelly Robertson 1979). Methane production is an important factor in regenerating carbon from sediments and maintaining pelagic production; as much as 55% of the total carbon input was regenerated as methane in a soft water lake during 1 year, and 36% of this total carbon input was recycled by methane oxidation (Rudd and Hamilton 1978). Sediments of dammed rivers, suffering from heavy pollution with organic wastes, may show methanogenesis exceeding by far that of eutrophic stratified lakes (Zaiss and Kaltwasser 1979).

The distribution of methanotrophs and the localization of methane oxidation is determined by the gradient of methane concentrations in aerobic water. When the concentration of methane decreases, e.g. to <5 μM, the rate of population growth is limited (Harrits and Hanson 1980). Yeast-like methanotrophs, which are capable of oxidizing nitrite, may also take part in the oxidation process (Rudd et al. 1976). During some winters, production of particulate carbon by CH_4 oxidizers may be an important source for grazers, as stated by Rudd and Hamilton (1978) – another example of the detritus food chain. Methane oxidation was the most significant contributor to the development of total lake anoxia under ice cover when large amounts of methane are trapped by an early freeze-up (Rudd and Hamilton 1978). In some case studies authors calculated that methane loss to the atmosphere can be an important loss in the carbon cycle of a lake (Barber and Ensign 1979), and in this water lilies (*Nuphar luteum*) can play a substantial role (Dacey and Klug 1979).

2.3 Nitrogen

2.3.1 Nitrogen Input

Nitrogen inputs to standing and running waters originate through above- or underground sources, or they arrive through aerial means, by precipitation, aerosols, fallen leaves and the like. A characteristic of nitrogen input is the process of N_2 fixation (cf. Chap. 8, Vol. 12C). In this chapter the aspects of N_2 fixation as a possibility of nitrogen input will be exclusively examined. A generalized scheme of the nitrogen cycle in a simplified lake system is given in Fig. 2.2.

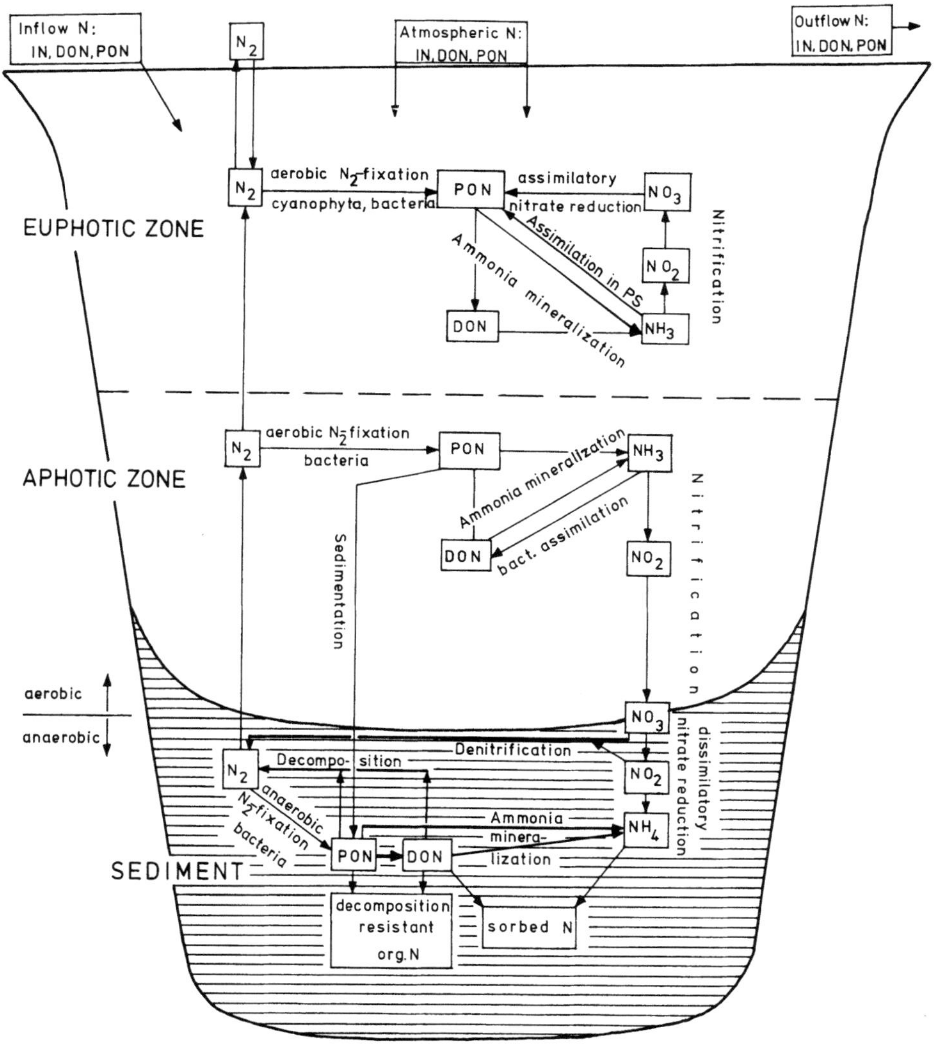

Fig. 2.2. Nitrogen cycling. *IN* inorganic nitrogen; *DON* dissolved organic nitrogen; *PON* particulate organic nitrogen; *PS* photosynthesis. (Modified after KUSNEZOW 1959 and WETZEL 1975, supplemented after several authors)

2.3.1.1 N_2 Fixation

Blue-Green Algae. The physiology and biochemistry of N_2 fixation is fully detailed in other contributions to this series (STEWART 1979; see also the forthcoming Vol. 15). What will be considered here are the most important ecological aspects of the nitrogen cycle. The N_2 fixation in lakes caused by blue-green algae quantitatively outweighs by far the N_2 fixation by bacterial groups. One recognizes among the *Hormogonales* a whole series of heterocystous genera, such as *Anabaena, Aphanizomenon, Gloeotrichia,* which can fix dinitrogen under aerobic conditions. Close relationships exist between the number of heterocysts and the rates of N_2 fixation (cf. HORNE and FOGG 1970; HORNE and GOLDMAN 1972; HORNE et al. 1979). The inhibition of the enzyme nitrogenase responsible for the N_2-fixing process and the repression of the heterocyst formation through inorganic nitrogen have been experimentally proven (OGAWA and CARR 1969). In natural water, however, these interactions are not always realized (GRANHALL and LUNDGREN 1971; LÄNNERGREN and LUNDGREN 1974). N_2 fixation like photosynthesis is a light-dependent process. The competition for photoreductants between these two processes can be bypassed through a temporal transfer of these two synthetic processes, as PAERL (1979) proved with *Anabaena* in Lake Rotongaio (New Zealand). The activity of photosynthesis prevails during the first part of the light phase. Later, however, it is curtailed because of the increased O_2 concentration, due to photosynthesis, in the water. Thus photoreductants are freed for the N_2 fixation process, which has its maximum activity during early afternoon. Although photosynthesis ceases on darkness, stored reduced photosynthetic products continue to be used for N_2 fixation. The amount of N_2 fixation during the dark phase is therefore often one-third to one-fifth the amount during the light phase (HORNE 1979). Correlations between the N_2 fixation capacity in lakes and the degree of eutrophication are discussed in Chap. 13 of this Volume and by HORNE (1977).

Without doubt, the N_2 fixation capacity of the planktonic blue-green algae is in general greater than that of the species which colonize other locations in the water. It has been reported, however, that rates of fixation of blue-green algae, living epiphytically on water plants (FINKE and SEELEY 1978), benthically on lake bottoms (LAM et al. 1979), as well as symbiotically in water ferns of the genus *Azolla* (KELLAR and GOLDMAN 1979), can sometimes exceed the rates of N_2 fixation of the planktonic types. The amount of N fixed from the *Anabaena azollae* in Lake Ngahewa (New Zealand) was 16.4 g m^{-2} a^{-1} and thus approximately five times as much as was produced by the planktonically living *Anabaena* (KELLAR and GOLDMAN 1979). In smaller bodies of running water, the epiphytically growing N_2-fixing species (e.g. *Nostoc*) prevail (HORNE and CARMIGGELT 1975). These algae endure even high water velocities.

Bacteria. The N_2 fixation produced by carbon-heterotrophic bateria is, first of all, coupled with the availability of organic substances. A maximum amount of fixation in the sediment always takes place in the top centimeters (BROOKS et al. 1971; KEIRN and BREZONIK 1971; MCGREGOR et al. 1973). At any given time the rate of fixation per g sediment amounts thereby to only a few ng N h^{-1}. Species of *Clostridium* and *Azotobacter* are mainly involved in the N_2-fixing process in sediment (KUSNETSOV 1959). Photoautotrophic, planktonic sulfur bacteria fix over 3 μg N l^{-1} h^{-1} in Lake Mize (Florida) (KEIRN and BREZONIK 1971). The N_2 fixation in the rhizosphere of submersed and emergent waterplants is especially high. In a *Glyceria borealis* stand of 1 ha approximately 60 kg N per year were fixed (BRISTOW 1974).

2.3.2 Nitrogen Transformations by Aquatic Microorganisms

2.3.2.1 Nitrification

In a broad sense, by nitrification one understands the biological oxidation of reduced inorganic and organic nitrogen compounds to nitrate (see also Chap. 5, Vol. 12C). The carbon-autotrophic nitrifiers derive their energy exclusively from the oxidation of ammonia and nitrite, while for the carbon-heterotrophic nitrifiers the oxidation of ammonia (and organic-N) is not the sole energy source. With carbon-autotrophic nitrification ammonia is oxidized by bacteria such as *Nitrosomonas* to nitrite, while others such as *Nitrobacter* effect the further oxidation to nitrate. However, other bacteria are also known to produce these reaction steps (see review by PAINTER 1977). Recently W. JUNG (personal communication) found *Acinetobacter* species as well as the Gram-negative nitrifiers of the genus *Pseudomonas* in the discharge of purification plants. Because oxygen must be present, nitrification is successful only in the top oxidized sediment layers. When the sediment in incubation experiments was artificially disturbed (CHEN et al. 1972) or the water layer was artificially aerated (GRAETZ et al. 1973), high rates of nitrification of approximately 1–4 mg NO_3 l^{-1} d^{-1} could be measured. As mentioned above, the nitrification and denitrification processes in lake and river sediment stand in close connection with one another. Through wind-induced currents, O_2 arrives in the top sediment layers, where the accumulated ammonium is nitrified. After oxygen is depleted, denitrification commences.

The nitrification process plays a particularly important role in waste water technology. The microbial background, as well as growth requirements and inhibition processes, are described by PAINTER (1977). If ammonia, which is formed during the biological cleaning of waste water, is not brought immediately into running waters but rather is oxidized to nitrate through nitrifying bacteria, this has the following consequences for the waters concerned: (1) nitrate is, in natural concentrations, untoxic for plants and animals, while ammonia can be toxic to fish with concentrations far under 1 mg N l^{-1} depending on the pH value of the water; (2) because of the nitrification of ammonia in flowing water, a higher biological oxygen demand would appear, since for the bacterial oxidation from 1 mg NH_3–N approximately 4.33 mg of oxygen are needed (WERZENAK and GANNON 1968); (3) through the oxidation of ammonium larger quantities of nitrite would appear in the water, which is likewise toxic in small quantities. For the rainbow trout (*Salmo gairdneri*), the LC 50, depending on fish weight amounts to approximately 0.2 mg NO_2–N l^{-1} (RUSSO et al. 1974).

The estimates which GARLAND (1973) employed for the nitrification in the River Trent give an insight into the quantitative proportions of the process. In 1970, an average of 6.9 t NH_3–N d^{-1} was nitrified. For this process approximately 30 t O_2 were necessary, which accounts for 75% of the total oxygen demand. In this river, 80% of the nitrifying bacteria were in the sediment, and indeed were to be found in the top 1 cm only (CURTIS et al. 1975). In other running waters, especially in rocky, pebbly-bottomed and shallow brooks, the entire nitrification production can be produced by non-planktonic cells (FINSTEIN et al. 1978).

Next to fungi, the bacteria are important carbon-heterotrophic nitrifiers. In an examination of lakes in northern Germany WITZEL (1973) found that heterotrophic nitrifiers dominate in lakes with high nutrient concentrations but with an unfavourable oxygen relationship, while the autotrophic nitrifiers dominate in lakes with well-aerated conditions. With respect to their nitrifying capacity in natural waters, the heterotrophic nitrifiers are at a great disadvantage to the autotrophic types (FOCHT and VERSTRAETE 1977). The release of nitrite into the surrounding milieu is in a distinct metabolic feature, as WITZEL and OVERBECK (1979) proved with *Arthrobacter* sp., and is not brought about through a lysis of cells, as for example ALEXANDER et al. (1960) hypothesized.

2.3.2.2 Denitrification

Denitrification means the reduction of nitrate to nitrous oxide and, more important, dinitrogen; this reduction is due to bacteria. Nitrate thereby serves these bacteria as a final electron acceptor. This phenomenon is thus called nitrate dissimilation. Among various groups of bacteria, the end products of the dissimilative nitrate reduction can also be nitrite or ammonia; however, one cannot speak then of denitrification which always leads to gaseous end products (cf. PAINTER 1977). According to PAYNE (1973), bacteria and fungi from over 60 genera can reduce nitrate dissimilatively, but only species of 17 genera (e.g. *Achromobacter, Bacillus, Micrococcus, Pseudomonas*) are capable of the true denitrification process.

The denitrification process takes place primarily under anaerobic conditions; at least in the immediate surroundings of the cells microanaerobic conditions must prevail (PAINTER 1970; KEENEY 1973). With regard to sensitivity to O_2, however, there appear to be additional differences among various bacteria (PAINTER 1977). Of special note is denitrification in sediments, where enough organic substance is at the disposal of the bacteria. ANDERSEN (1977) found a positive correlation between the content of organic substance in the sediment of six Danish lakes and the rate of denitrification. In experiments with lake sediment, up to 95% of the added nitrate was eliminated in the N_2 form (VAN KESSEL 1977). Under natural conditions the loss was only 60%–80% in the hypolimnion (BREZONIK and LEE 1968) as well as in the sediment of lakes (CHEN et al. 1972). The denitrification process in sediments is limited mostly by lack of nitrate, the concentrations of which also determine the rate of denitrification. It is not surprising, then, that one finds the highest rates in the sediment of running waters with up to 1,500 mg N m^{-2} d^{-1} (EDWARDS and ROLLEY 1965), since through the movement of the water new nitrate is continually supplied to the top sediment layer. For the same reason, the denitrification process in flat estuaries (NEDWELL 1975) and lakes (ANDERSEN 1974) considerably exceeds that of deep waters (VOLLENWEIDER 1968; KEENEY et al. 1971). As has been shown experimentally, denitrification may be enhanced by benthic living organisms, especially tubificid worms (CHATARPAUL et al. 1979).

2.3.2.3 Ammonia Mineralization

When the N-containing compounds, either as particulate or dissolved organic substance, are finally reduced to ammonia through the action of heterotrophic organisms, then one can speak of an ammonia mineralization process, or ammonification. In freshwaters, above all, the action of bacteria (KUSNETSOV 1968; KAUSHIK and HYNES 1971) is of the highest importance. The highest cell count is usually found in the top centimeters of sediments. NIEWOLAK et al. (1978) found up to 8×10^{12} cells g^{-1} dry wt. of bottom sediment in fertilized Masurian lakes and the highest cell count was measured in summer.

Ammonia mineralization is high when the sediment contains a high percentage of organic substance. Thus a correlation exists between this process and the productivity of waters. KAMP-NIELSEN (1974) proved this by studying the rates of ammonia mineralization of Danish lakes of different trophic status. The rates oscillated between ca. 3 and 14 mg NH_3–N m^{-2} d^{-1}. In experiments with the decomposition of algae, larger rates of ammonia mineralization were found under anaerobic than under aerobic conditions. In the sediment of freshwaters, the correlations differ somewhat, as AUSTIN and LEE (1973) showed with Lake Mendota.

TOETZ and COLE (1980) give detailed information about the rate of ammonia mineralization in the euphotic zone of a lake. Such examinations are of great ecological importance, since the heterotrophic ammonia mineralization and the assimilation by autotrophs in this zone are narrowly coupled in space and time. An ammonia mineralization between 93 and 412 μg NH_4^+–N l^{-1} d^{-1} was determined, and the turnover times of the ammonia pool and the particulate nitrogen were between 0.2 and 0.9 day, respectively. In contrast, the rates determined for oceans (HARRISON 1978) were, for the most part, several days. As the examinations of TOETZ and COLE (1980) showed, the NH_4^+ uptake of primary producers outweighed the uptake of available NO_3^-, which re-emphasizes the importance of ammonia mineralization in this zone of the lake.

In Clear Lake (California), HORNE and ROTH (1980) observed an especially interesting and previously unattended connection between the ammonia excretion of zooplankton and the physiology of N_2-fixing algae. The suppression of N_2 fixation in midsummer was traced back to the ammonia excretion of the *Daphnia*. The observed suppression of N_2 fixation was induced when lakewater contained more than 100 μg NH_3–N l^{-1}. With experimental insertion of an insecticide against *Daphnia*, the N_2-fixing capacity ceased.

2.3.3 Nitrogen Nutrition of Primary Producers

2.3.3.1 Aquatic Vascular Plants

Nitrate, ammonia and nitrite are available as inorganic N sources for use by the aquatic vascular plants. Whether the nutritive absorption of water plants occurs through the roots and/or leaves is a long discussed question (cf. Sect. 2.4: phosphorus) which has been solved – at least partially – by the use of isotopes.

TOETZ (1974) followed the absorption and the translocation of $^{15}NH_4^+$ in aquatic vascular plants. Such experiments with ammonia are especially informative, since in sediments of freshwater, NH_4^+ (in contrast to nitrate) continually occurs in large quantities. As is shown in the experiments of TOETZ (1974), the absorption of labelled NH_4^+ through the roots and the basal plant parts, respectively, and the subsequent transport in the apical tissue is not the same for all water plants. The absorption and transport was observed in *Elodea densa* and *Ceratophyllum demersum* but not for a *Scirpus* species. In *Potamogeton pectinatus,* labelled N was translocated only after 3 days of incubation. Additional examinations are necessary to reveal whether the other water plants fit in these three categories, and which ecophysiological explanation for this phenomenon can be found.

The absorption of nutrients (nitrogen, among others) through the leaves of submerged plants is not questioned. However, the water surrounding the leaves of submerged macrophytes is clearly different from the interstitial water in the root area in its concentrations of inorganic nitrogen. The ammonia concentrations are much smaller, while nitrate dominates. Ammonia is absorbed preferentially. This has been proven for a series of water plants (cf., for example, FERGUSON and BOLLARD 1969; SCHWOERBEL and TILLMANNS 1972, 1977). With *Myriophyllum spicatum,* NICHOLS and KEENEY (1976) could show, for example, that nitrate was not absorbed until NH_4^+–N-levels had decreased to less than 30 $\mu g\ l^{-1}$. Certainly, the nitrate assimilation in this plant can be clearly stimulated through increasing P concentrations in the water (BEST and MANTAI 1978). The favoured absorption of ammonia is a great advantage, since the reduction of nitrate to ammonia needs a great amount of energy. The preferential use of NH_4^+ in the cell metabolism is encouraged by the fact that NH_4^+ inhibits (among others) the activity of nitrate reductase in water plants (HERRERA et al. 1972; OREBAMJO and STEWART 1975).

2.3.3.2 Algae

The favoured assimilation of ammonia in phytoplankton has long been known (see review of FORSBERG 1977). The kinetics of nitrogen uptake by planktonic algae can be described in more detail than that of aquatic vascular plants. A theoretical discussion is given in the next chapter of this volume.

For *Oscillatoria agardhii,* one of the important planktonic blue-green algae of eutrophic waters, ZEVENBOOM and MUR (1978) showed the dependency of nitrate absorption on nutritive status and light. The highest rates of uptake were exhibited by N- and P-deficient cultures ($V_{max\,NO_3}$: 0.82, respectively 0.76 $\mu mol\ mg^{-1}$ dry wt. h^{-1}) and the lowest by light-limited cultures (0.1 $\mu mol\ mg^{-1}$ dry wt. h^{-1}). Simultaneous examinations of a Dutch lake indicated the seasonal variation of $V_{max\,NO_3}$, as well as the half-saturation constant of nutrient uptake. With the help of kinetic data, which was ascertained through chemostat cultures, ZEVENBOOM and MUR (1978) demonstrated the actual growth-limiting factors.

The ^{13}N-radioisotope method must be applied to estimate N absorption of natural N-limited phytoplankton, since with the ^{15}N method the concentrations needed are too high and unnatural. Thus, in Castle Lake (California), GERSBERG et al. (1978) measured a nitrogen uptake of only 4.5 ng N l^{-1} h^{-1}, with a nitrate and ammonia content of lake water from 5.4 and 1.7 μg N l^{-1}, respectively.

2.3.4 Organic Nitrogen

In this section, the dissolved organic nitrogen (DON) will be examined, since the living and dead particulate organic nitrogen (PON), when it is not withdrawn from nitrogen circulation (for example, by sedimentation), is finally transformed into DON or into NH_4^+. Because of various similarities between DON and DOC, Sect. 2.2.2 on organic carbon should also be referred to.

The capability of DON utilization has been experimentally proven for several green algae and flagellates, e.g. *Chlorella, Scenedesmus, Chlamydomonas* and *Euglena* (for references see SYRETT and LEFTLEY 1976; FORSBERG 1977). Usually urea, uric acid, amino acids and amides are best utilized. In natural waters, the N nourishment of algae brought about in this way is hypothesized to be of subordinate importance; so that according to VOLLENWEIDER (1968), such inquiries are of greater physiological than ecological interest. Only when the contents of inorganic N compounds are very low or consumed can DON serve as the N source for algae (GOERING 1972). Some DON compounds, e.g. polypeptides, amino acids and other N-containing organic substances have the ability to form chelates and may become important in this indirect way for algal nourishment. The availability of macro- and mironutrients is either improved or impaired through chelation (e.g. WETZEL 1968). Through their capability of excreting specific chelators, algae can themselves regulate nutrient-availability (MURPHY et al. 1976; MCKNIGHT and MOREL 1979). Along with high molecular weight DON, algae also release peptides and amino acids (cf. GOCKE 1970 and STEINBERG 1977) and amines such as methylamine, trimethylamine, 1-aminopropane and others. HERRMANN and JÜTTNER (1977) established this in both *Chlamydomonas* cultures and in a *Chlamydomonas* bloom in the River Ruhr. The same authors reported the release of amines by other algae and a characteristic composition of the amines released by algae of different systematic classes was found. DON is excreted not only by algae and aquatic macrophytes (WETZEL and MANNY 1972a), but also by bacteria and fungi, which are able to dissolve organic substances from particulate organic matter through the release of exoenzymes (WETZEL and MANNY 1972b). As was shown with experimentally obtained excretion products of algae, as well as with the artificially enriched organic substances of a lake, DON is by no means fully utilized by microorganisms; rather a scarcely degradable N-containing organic fraction always remains, due to complexing by fulvic acids (STEINBERG 1977).

DON is also released from the other members of the aquatic food chain. Zooplankton excretes actively into the medium (BERNICE 1972) and DON is released passively by the grazing activity and the digestive systems. NOWAK (1975) found that the zooplankton of an euproductive lake eliminated up to 20% of the consumed undigested algae biomass and through this means considerably elevated the DON content of the water.

2.4 Phosphorus

2.4.1 Forms of Phosphorus and Their Dynamics

The fraction of molybdenium blue-reactive phosphorus in natural waters, which is often incorrectly called "orthophosphate", contains, apart from real orthophosphate, much phosphorus which is linked to a colloidal (M.W. $\geqq$ 5,000,000) and a low molecular weight organic fraction (M.W. $\sim$ 250) (cf. LEAN 1973 and many others). Real orthophosphate is usually quantitatively insignificant and its ratio, in relation to the remaining dissolved reactive phosphorus, is specific to the particular water (WHITE and PAYNE 1980). There is an intensive dynamic interaction between the individual fractions of phosphorus, including the phosphorus bound in plankton organisms and in seston which is exemplified in Fig. 2.3. Compared with the interaction among the individual P fractions, which is almost entirely cyclic, the allochthonous P supply, which is very significant in the eutrophication process, and the loss through sedimentation are both very small. However, since the supply and the sedimentation take place at a steady rate, these processes become more important in the overall metabolism and the ontogeny of the water. The model presented here is also valid in principle for kinetics in flowing water, however, the exchange rates between the individual P compartments are clearly lower than those of lakes (PETERS 1978).

A part of the particulate phosphorus (e.g. minerals like apatite) does not participate in the exchange. Since the macromolecular phosphorus fraction is

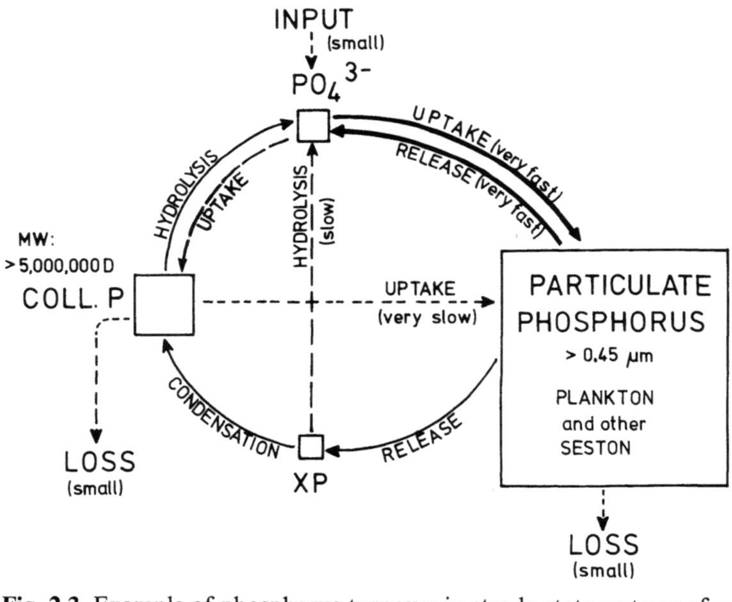

Fig. 2.3. Example of phosphorus turnover in steady-state systems of epilimnetic freshwaters. *XP* low molecular weight (ca. 250) organic phosphorus compound. (Modified after LEAN 1973, supplemented after LEVINE 1975 and PETERS 1979)

present in river water but not in rain, it must be formed in the soil or in the river itself (PETERS 1977). According to PETERS (1979) the dynamics of P between the various trophic gradients is probably higher in oligotrophic than in eutrophic waters.

The condensation process, which leads to the colloidal P fraction, can be assisted by microorganisms (JACKSON and SCHINDLER 1975) and probably also abiotically (STEINBERG and SCHRIMPF 1980).

The process of hydrolytic release of orthophosphate from extracellular reserves is controlled by alkaline and, to a lesser degree, by acid phosphatases (REICHARDT et al. 1967; BERMAN 1970). The enzymes can exist in free solution in water (REICHARDT et al. 1967) or on the cell walls of organisms (GOLTERMAN 1975; JANSSON 1976; KUENZLER and PERRAS 1965). The production of phosphatases can be considered as a means of extending the ecological potency of the particular species, making otherwise unavailable P reserves accessible to the population (HEATH and COOKE 1975).

2.4.2 Role of Pelagic Consumers

Work by HARGRAVE and GREEN (1968), JOHANNES (1964) and RIGLER (1973) point out that zooplankton feeds on particulate P components and afterwards releases dissolved P compounds. The majority (85%–90%) of the released phosphorus consists of orthophosphate (PETERS and LEAN 1973; FERRANTE 1976). Generally the effect of zooplankton on phytoplankton development appears to be positive when the algal density is relatively small: in this case the P-remobilization stimulates the growth of algae and provides up to 44% of the phosphorus needed by the phytoplankton (LAROW and MCNAUGHT 1978). During algal blooms the effect is reversed, since the remobilization rates are very small and cannot cover the algal requirement. Fish also appear to play a similar role to that of zooplankton (LAMARRA 1975; STERNIK 1978).

2.4.3 Phytoplankton and Bacteria

2.4.3.1 Algae

Algae can make use of the prefered orthophosphate and frequently also other P sources, such as polyphosphates. At times of optimal P supply very many algae are able to take up phosphorus in excess and store it as polyphosphate. Intensive examination of the cyanophyte *Oscillatoria redekei* by ZAISS (1976) showed that RNA (but not DNA) and, for a short period, also ATP can serve as reservoirs. The internal reservoirs make population growth possible even when external nutritional sources are exhausted. This signifies a considerable competitive advantage for cyanophytes, which are often able to migrate into nutrient rich parts of a lake (hypolimnion) (STEWART et al. 1978).

Various culture studies of P-uptake kinetics show that the maximum uptake rate is a function of the external substrate concentration as well as the concentration of the internal P components (RHEE 1973), with an inverse relation between intracellular P and the uptake rate (BROWN and HARRIS 1978). Since more than just one P-uptake system can be active, the uptake kinetics are not necessarily simple (BROWN et al. 1978). On the other hand, the uptake process can be counteracted by simultaneous excretion (NALEWAJKO and LEAN 1978), including compounds such as cyclic adenosine-monophosphate (FRANCKO and WETZEL 1980).

2.4.3.2 Algae Versus Bacteria

FUHS et al. (1972) examined the P uptake in two species of diatoms and three bacterial species. Surprisingly, the maximum uptake rates per unit area of the cell surface were very similar in all the organisms tested. Bacteria showed a lower affinity to orthophosphate than algae, but had a higher ratio between surface area and volume, and therefore were able to grow faster than algae.

BURNISON (1975) found that ^{32}P is immediately taken up by algae and bacteria. PAERL and LEAN (1976) established that bacteria were more densely labelled with radioactive phosphorus than algae. However, since the algal biomass greatly exceeds that of the bacteria, it is very probable that the algal populations develop the greater activity in P turnover in free waters.

2.4.4 Phosphorus in the Littoral Zone

For a long time it was not clear which organs of macrophytic water plants perform P uptake. LITTLEFIELD and FORSBERG (1965) found an equal uptake through all parts in *Chara*. The angiosperms are supplied predominantly via the root system (MCROY and BARSDATE 1970; BRISTOW and WITHCOMBE 1971; DENNY 1972; TWILLEY et al. 1977). In almost all the plants examined there existed phosphorus fluxes in two directions: from the roots to the leaves and vice versa. Part of the P taken up is excreted into the medium (free water, sediment) and supports growth of epiphytic algae and bacteria.

Some authors contradict the opinion that macrophytes secrete P actively into the medium. They also observe a release of P from macrophytes, but attribute this to the decomposition of decaying plant biomass only (PRENTKI and ADAMS 1979; KIRKMAN et al. 1979; BARKO and SMART 1980).

2.4.5 Phosphorus in Seston and Sediments

To the inorganic forms of phosphorus in seston and sediments belong compounds with iron, aluminium, calcium, sometimes combined with fluorides, and in addition adsorption complexes with anions in clay minerals or organic complexes. Regarding organic phosphorus in the various sediment layers in lakes, different kinds of production can be distinguished. WILLIAMS et al. (1976) found in Lake Erie a significant increase of the organic P fraction with increasing eutrophication of the lake.

2.4.5.1 Precipitation and Adsorption of Phosphorus

A formation of defined iron, aluminium and calcium phosphates is possibly conceivable; however, on account of the solubility products a high phosphorus concentration must be present, which seldom appears in free waters. According to MORGAN and STUMM (1965) the essential process in precipitation of P is the formation of iron-III-hydroxophosphates, which also includes a chemosorption of phosphate. Former opinions of OHLE (1937) and EINSELE (1938) should be revised. They presumed a formation of defined iron-phosphate from iron-III-hydroxidegels in the sediment–water contact zone in the presence of oxygen.

Fe- and Al-hydroxides and hydrated oxides appear as the most important binding groups of phosphorus in the sediment (LI et al. 1972). It has been established that between 19% and 43% of the inorganic phosphorus found in sediment is exchangeable, and the exchange is greater in calcium-rich sediment than in calcium-deficient sediment (WILLIAMS et al. 1970).

The adsorption of P to sediment particles can be characterized by FREUNDLICH'S as well as by LANGMUIR'S adsorption isotherms (FISCHER and BAUMANN 1978; KU et al. 1978, among others). It is generally true that after a rapid adsorption reaction, a considerably slower one follows, in which the phosphate ions presumably diffuse inside the adsorber and cause chemical reactions.

2.4.5.2 Release from the Sediments

In classical studies the mechanism via reduction of FeIII and MnIV was considered to be the only mechanism of P-release. Works by OLSEN (1958), TESSENOW (1972), KAMP-NIELSEN (1974) and ANDERSEN (1975) brought the first clear result, which recorded P release from aerobic sediments as well. Generally, the recycling of phosphorus from the solid phase of the sediment into the sediment-water interface takes place via the pore waters by two interconnected processes (TESSENOW 1979):

1. dissolution in the sediment by physicochemical (e.g. redox) reactions and/or biochemical reactions, for example by the activity of chironomid larvae (GALLEPP et al. 1978) and

2. migration into the overlying water through diffusion and exchange according to a concentration gradient. Also P release from freshly sedimented seston does not depend to any great extent on the redox potential (TESSENOW 1972), since the ions released from the seston preferably diffuse into the free water and not into the sediment.

2.4.6 Phosphorus Retention in Flowing Waters

Little attention has been paid to phosphorus retention in flowing waters. Dependent on the considerable amount of experimentation necessary, most works only describe this phenomenon more or less exactly; detailed examinations of the mechanism come from KAPLAN et al. (1975), GREGORY (1978) and MEYER (1979). Retention can be caused biologically by aquatic macrophytes (SMITH et al. 1978) or by micro-organisms and/or purely physicochemically. MEYER

(1979) describes chemosorption to organic and mineral sediment particles as the most important mechanism for a clean woodland stream. The adsorption reaction occurs quickly and depends on the P concentration in the water. Adsorption through predominantly physicochemical processes is reversible: if the surrounding P concentration falls under the specific equilibrium concentration the sediment releases phosphorus again.

In contrast, work by Gregory (1978) shows a P adsorption dominated by biological processes in the substrate. The biomasses of the algae, bacteria and fungi appear as regulating factors, whereby the epilithic algae most effectively eliminate P. The quality of the allochthonous detritus plays a considerable role in the biotic P uptake (cf. C cycle). Therefore slowly decomposable detritus does not cause such rich microbial growth and at the same time not such high P elimination from the water, as does more easily usable detritus. It must be noted that the evaluation of the retention factors is specific to each individual running water and therefore different studies cannot produce identical results.

2.5 Sulphur

2.5.1 General Ecological Aspects

The following important characteristics, among others, can be ascertained for the sulphur cycle: (1) in freshwater ecosystems, sulphur scarcely becomes a growth-limiting factor in the sense of Liebig for photoautotrophic organisms. Sulphur-deficient situations occur mainly in tropical waters (Amazon region: Braun 1952; Lake George, Lake Victoria – Africa: Beauchamp 1953); bacteria are the principal participants in the transformation of inorganic sulphur from more oxidized to more reduced compounds and vice versa. There exist, therefore, subtle similarities and connections with the bacterial transformation processes in the N cycle; (3) through the anaerobic reduction of sulfate and the decomposition of S-containing organic substances, the solubility and availability of other nutrients and especially the living conditions for all aquatic organisms are fundamentally altered; (4) the sulphur concentrations in surface waters have increased considerably not only in highly developed countries, but also in abandoned or thinly populated regions through atmospheric fallout (cf. Chap. 14, this Vol.). For an illustration of the major pathways of sulphur in a lake system see Fig. 2.4.

2.5.2 Microbial Transformations of Inorganic Sulphur

In this section the transformations by microorganisms and their important consequences, upon which the ecology of waters is based, will be discussed. The stoichiometry of the corresponding reactions, as well as information to the systematics of the bacteria participating in the sulphur cycle, are described in Wetzel (1975) and the references quoted there.

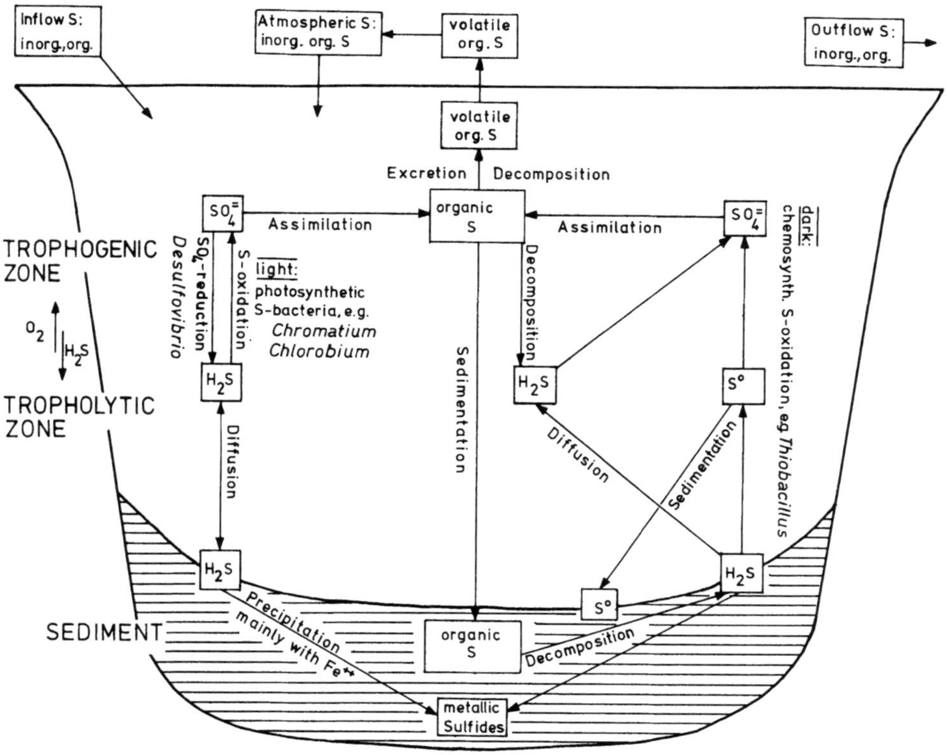

Fig. 2.4. Sulphur cycling. (Modified after KUSNEZOW 1959 and WETZEL 1975)

2.5.2.1 Sulphur-Reducing Processes

Of special importance in waters is the microbial reduction of sulphate and other inorganic S compounds such as sulphite, thiosulphate, hyposulphite and elemental sulphur to hydrogen sulphide. As in denitrification, the highest sulphur-reducing activity is in the top sediment layers and can, in standing waters, proceed from the sediment into free waters. Under these conditions of low redox potentials, the reactivation of phosphorus absorbed and immobilized upon and in sediment into dissolved forms is substantially increased (cf. Sect. 2.4: phosphorus). Since the intensity of the bacterial H_2S formation directly depends on the sulphate quantities available for use, OHLE (1954) coined the term: sulphate as "catalyst" for the limnetic nutrient cycling. While sulphur-reduction is chiefly accomplished by bacteria of the genus *Desulfovibrio* and *Desulfotomaculum,* various species can participate in the H_2S formation of S-containing organic substance according to the type of lake and trophic status.

A release of H_2S based on a sulphur reduction has been observed under anaerobic conditions as well as with the isolated thermophilic blue-green alga *Synechococcus lividus* of the Lower Geyser Basin (Yellowstone National Park, SHERIDAN 1973).

CAPPENBERG (1975a, b, 1978) determined an ecologically interesting connection between sulphate-reducing and methane-producing bacteria in the mud of Lake Vechten (Netherlands), as well as in continuous culture studies. The two bacterial groups occur with maximum numbers at different depths in the mud of the lake, the sulphate-reducing *Desulfovibrio* at a depth of 0–2 cm and the methane-producing *Methanobacterium* sp. at a depth of 4–6 cm. Experiments with selective inhibition of the sulphate-reducing and the methane-producing processes indicated that lactate is the main source of energy for the first and acetate for the second process. Thereupon a commensalism between the two groups could be described using ^{14}C-labelled substrates. The *Methanobacterium* which ferments acetate benefits from the acetate released by *Desulfovibrio*. This bacterium, in turn, is not affected in the presence of the former. Its abundance depends rather on the concentration of sulphate ions in the mud.

As was recently established (NRIAGU et al. 1979), the enzyme rhodanese (thiosulphate: cyanide sulphur transferase), which participates in the oxidation and reduction of unstable S compounds like thiosulphate and polythionates, occurs in lake sediments in free form. In the greatly reduced sediment of Lake Erie, the rhodanese activity as well as the thiosulphate concentrations were lower than in the heavily oxidized sediment of Lake Ontario.

2.5.2.2 Sulphur-Oxidizing Processes

The bacteria which participate in the oxidation of H_2S, $S_2O_3^{2-}$ and S^0 can be divided, according to their physiology, into two main groups (cf. WETZEL 1975); (1) the mostly aerobic, chemosynthetic (colourless) sulphur-oxidizing bacteria (of the genera *Thiobacillus, Thiothrix, Beggiatoa*), and (2) the anaerobic, photosynthetic (coloured) sulphur bacteria, that can be divided into the green sulphur bacteria (Chlorobacteriaceae) and the purple sulphur bacteria (Thiorhodaceae). Simply stated, the energy achieved by the oxidation of sulphur serves the first group in the CO_2-fixing process, while sulphur serves the latter group as the electron donor in the photosynthetic processes.

Corresponding to their different demands towards the inorganic sulphur components, to oxygen and light, the redox- and pH conditions are also apportioned in characteristic ways to the various sulphur-oxidizing bacteria in waters. The following simplified scheme is valid for the occurrence of both bacterial groups in lakes with a hypolimnetic, H_2S-containing water body: if the top H_2S layer boundary is located in the euphotic zone, photoautotrophic sulphur bacteria appear, and the Chromatiacea *Chromatium* especially can build sharply delineated horizons with very high cell counts (KUSNEZOW 1959). In contrast, when the boundary between O_2 and H_2S is found in the aphotic zone, then the aerobic, chemosynthetic sulphur-oxidizing bacteria (e.g. *Thiobacillus*) can dominate. In meromictic lakes the H_2S-occurrence lies sometimes in, at other times beneath the euphotic zone (SOROKIN 1970). Lake Belovod belongs to the first type, where *Thiobacilli* additionally appear among a dense population of purple sulphur bacteria. The biomass of bacteria produced in this zone represents an important nutritive source for the filtering zooplankton.

2.5.3 Volatile Organic Sulphur Compounds

It was hypothesized by JUNGE (1960) for marine ecosystems that organic sulphur, not H_2S, is the principal atmospheric carrier for sulphur. Later, LOVELOCK et al. (1972) found dimethylsulphide (DMS) to be the principal sulphur gas in oceans. Many planktonic and benthic algae can participate in the formation of DMS. Recently BECHARD and RAYBURN (1979) showed with a eutrophic pond, that in freshwaters as well, DMS can be the principal biogenic sulphur compound. Previously (JENKINS et al. 1967) the presence of volatile sulphides (dimethylsulphide, dimethyldisulphide, methylmercaptane), which had been detected in algal cultures, had been attributed to the action of bacteria. BECHARD and RAYBURN (1979), through investigations with axenic and non-axenic algal cultures, could demonstrate that blue-green algae produced DMS, while representatives of the Chlorophyceae, Xanthophyceae and Bacillariophyceae did not.

2.6 Silica

2.6.1 Ecosystem Approach

TESSENOW (1966) studied silica cycling in freshwaters intensively. He states that almost all freshwaters are strongly undersaturated in silica. There exist two forms of silicon in freshwaters: dissolved silicic acid and amorphous silica in diatom frustules and adsorbed to inorganic complexes with iron and aluminium hydroxides. Silica cycling is due to biogenic incorporation and subsequent sedimentation. Abiogenic precipitation in the pelagic regions is unimportant. Utilization of silica by diatoms occurs during photosynthesis and even increases in darkness in living cells. One main source for silica recycling is the sediment. The interstitial concentrations in the sediment increase as the pH is reduced below 7, decrease between 7 to 9 and greatly increase above pH 9. The dissolution is a temperature-dependent, but redox-independent, diffusion process from the sediment involving eddy diffusion. Movement in the sediment–water interface produced by benthic organisms (e.g. tubificid worms or larvae of chironomids) or methane ebullition under anaerobic conditions supports the recycling of silica into the euphotic zone. KAMP-NIELSEN (1974) quantified the release of silica in a eutrophic as well as an oligotrophic lake. The release in eutrophic Furesø was about 95 mg Si m^{-2} d^{-1} and only 5 mg Si m^{-2} d^{-1} in oligotrophic Grane Langsø. These figures are valid under both aerobic and anaerobic conditions.

Another mechanism of silica recycling may be the dissolution of diatom frustules during sedimentation. PARKER et al. (1977) found that the major fraction ($\sim$98%) of amorphous silicon produced annually as diatom frustules was decomposed before incorporation in the permanent sediment (of Lake Michigan and perhaps in some others, but not all lakes, TESSENOW 1966).

The concentration of external silica strongly influences growth of diatoms. For example, *Asterionella formosa* populations decrease below 0.5 mg Si l^{-1}

and can be displaced by *Cyclotella meneghiniana* in a resource-based competition experiment, whereas *Asterionella* itself displaced *Cyclotella* when both are limited by phosphorus (TITMAN 1976). In general, decreasing silica concentrations during diatom blooms is one regulating factor in the diatom succession. No diatom growth at all can be observed at concentrations below 0.1 mg SiO_2 l^{-1} (MÜLLER 1977). In cultures and freshwater ecosystems, many diatom species react faster to SiO_2 deficiency than on removal of any other nutrient (except for water and carbon sources) and diatom bloom breakdown, e.g. in late spring, is mainly caused by lack of silica rather than phosphorus or nitrogen (MÜLLER 1977).

2.6.2 Silica Within Diatoms

Besides diatoms, some chrysophytes (e.g. *Mallomonas, Synura, Ochromonas, Spumella*) and some macrophytes (*Phragmites australis, Equisetum sp., Lemna minor* and others) require silicon as essential nutrient.

Intracellular silica metabolism is a silicate metabolism for no bonds other than Si-O have been demonstrated in biological systems – so far. Silica metabolism is well documented in diatoms. Obviously the formation of frustules can be seen. But besides frustule formation there are at least four different interactions of $Si(OH)_4$ biochemistry with the general metabolism of cells, as described by WERNER (1977):

1. in the citrate cycle between acetyl-CoA and 2-oxoglutarate;
2. in the synthesis of proteins, localized in cell organelles (chromosomes, chloroplasts);
3. in the regulation of respiration and chrysolaminarin utilization;
4. in the regulation of chlorophyll synthesis.

There exist uncertainties about the character of the soluble silicon pool in diatoms. SULLIVAN (1979) postulates the existence of colloidal silica polymeric chains, organo-silicon complexes and monosilicic acid, whereas BINDER and CHISHOLM (1980) found monosilicic acid and short polymers only.

Compared to other cell wall materials, such as cellulose, pectin, chitin, mannans, xylans or alginic acids, polycondensation of $Si(OH)_4$ might be energetically more economical by a factor of 5 to 10. Considering that silicates can contribute up to 30% of the dry weight of cells, this could have been of decisive ecological and evolutionary advantage (WERNER 1977).

2.7 Heavy Metals, with Special References to Iron and Manganese

Some metals of the first main group of the periodic table belong to the essential macro-nutrients (K, Mg). Except for chromium and nickel all metals of the first transition series play a decisive role as trace metals in biological processes. In recent papers, nickel is shown to be an essential metal for some bacteria (DIECKERT et al. 1979; SCHÖNHEIT et al. 1979). Most recentely, a nutrient requirement has been found for the half-metal selenium in a pyrrophyte (*Peridinium cinctum*) by LINDSTRÖM (1980). Beside their role as nutrients, various metals

Table 2.1. Types of metal species in water. (Wilkins 1979, based on Stumm and Bilinski 1972)

	Metal species	Example	
Increasing size of metal species ↓	1. Free metal ion	$Cu(H_2O)_6^{2+}$	True solution
	2. Simple radicals	UO_2^{2+}, VO_3^-	
	3. Inorganic complexes	$CdCl^+$, $CuCO_3^0$, $Pb(CO_3)_2^{2-}$	
	4. Organic complexes chelates and compounds	$Cu\text{-}OOC\text{-}CH_3^+$, $Hg(CH_3)_2$	
	5. Metals bound to organic materials of high molecular mass	Pb-fulvic acid polymers	Transitional
	6. Colloids	$Fe(OH)_n$	
	7. Metals sorbed on colloids	Pb on clays	Transitional / Particulate
	8. Precipitates, mineral particles	$PbCO_3$, PbS	Particulate
	9. Metals in living and dead biota (bacteria, algae, etc.)		

of the main groups and other transition series are important because of their potential toxic effects, when they are present in elevated concentrations and in free ionic state (see also carbon-cycling the discussion in Chap. 7 of Vol. 12 C.)

2.7.1 Forms of Metals in Natural Waters

The types of metal species which can occur in natural waters are listed in Table 2.1. In addition to these species, many metals can exist in different oxidation states that commonly show markedly different features. Thus, the oxidation state may govern whether a metal is present as an anionic or cationic species, for example Mo, V, and Cr. The solubility and tendency to complexing of a metal (e.g. Mn and Fe) may also depend heavily on its oxidation state (Wilson 1979).

Groth (1971) investigated the significance of the trace metals for primary production in some north German lakes, and discovered that the proportion of the metal which is held in solution by organic ligands is almost always sufficient for phytoplankton production. However, precipitation or co-precipitation of metals with clay or calcite can sometimes lead to a decrease in production (Wetzel 1975). The enrichment of the metals in plankton organisms follows the series:

$$Fe > Zn > Cu > Co > Mn > Mo;$$

the release in plankton decomposition the series:

$$Fe > Mn > Co > Zn > Cu > Mo \quad (\text{Groth } 1971).$$

During metal absorption aquatic microorganisms are not able to differentiate between essential and non-essential metals (STEINBERG and HERRMANN 1980). In the following, the cycles of iron and manganese will be discussed in more detail. These are two of the best-known and most important metal cycles, although a few important facts are still not clear. For the other metallic cycles see NRIAGU (1980a, b).

2.7.2 Cycling of Iron and Manganese

The stability of iron and manganese minerals in relation to Eh-pH conditions is shown in Fig. 2.5 (Eh = redox potential measured against standard hydrogen electrode). Shaded areas indicate natural Eh-pH conditions in the free water region of a hardwater lake. The conditions in acidic lakes or in sediments are altered to low pH and Eh values. Under natural conditions and in the absence of organic ligands, iron exists as hydrated ferric hydroxide. In the presence of organic ligands, the organically complexed iron may be predominant. Phosphate does not influence the solubility of Fe^{3+} when inorganic P concentrations are less than 10^{-4} mol l^{-1}, as is usually the case.

With manganese the redox equilibrium Eh values are higher and the rates of oxidation slower than for iron. At Eh 0.4 V and above a pH of 8.0, an intermediate oxide complex is formed (in the absence of organic ligands) in which Mn^{2+} is adsorbed onto Mn(IV)-oxides. Under natural conditions, more chemical species can exist with manganese than with iron. WETZEL (1975) concluded that the direct utilization of iron and manganese bacteria in energetic transformation occurs, but is minor in comparison to heterotrophic metabolism of organic substrates in most natural waters. Chemosynthetic utilization of energy from inorganic oxidation for CO_2 fixation is relatively inefficient, especially in the case of the oxidation of iron and manganese. Sheathed and stalked bacteria, algae, protozoans, flagellates, and specific true bacteria precipitate ferric and manganese oxides in their cells.

Ionic forms of iron readily serve *Thiobacillus* type of iron oxidizing organisms as an energy source (LUNDGREN and DEAN 1979). Iron carbonates and sulphides are substrates for *Leptothrix* and *Gallionella* organisms. Relatively low levels of these iron-containing compounds can serve as nutrients under conditions of neutral pH, low oxygen pressure and low Eh; a continuous iron supply

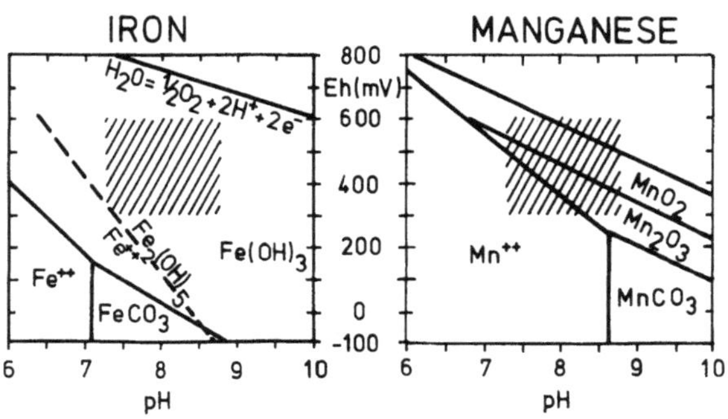

Fig. 2.5. Stability of iron and manganese minerals in relation to Eh-pH conditions. *Shaded areas* indicate natural Eh-pH conditions in the free water of a hardwater system. (After STUMM and MORGAN 1970; BURNS and NRIAGU 1976)

is necessary to support microbial growth. Some species of *Leptothrix* are facultative iron bacteria that can oxidize both ferrous and manganese salts, whereas *Gallionella* is restricted obligately to iron. Other organisms that function within a short distance of the Fe^{2+}/Fe^{3+} borderline are species of *Siderocapsa* which oxidize complexes of iron. It is thought that *Siderocapsa* utilizes the organic part of the complex for nutrition, liberating the ferrous iron. A rapid oxidation of the liberated iron occurs, resulting in the formation of ferric hydroxide precipitates which accumulate upon the cell surface (LUNDGREN and Dean 1979). In a recent study, coprecipitation of Mn, Al, SiO_2, Ca, K, Ti, S and P in Lake Washington by iron bacteria was detected. Coprecipitation of P accounted for up to 8% of the epilimnic P content (GREGORY et al. 1980).

In various Mn oxidizers (e.g. *Pseudomonas manganoxidans*) the Mn oxidation takes place intracellularly with the help of a Mn oxidizing protein. Proteins also play a part in extracellular Mn oxidation, as was determined in culture filtrates of *"Metallogenium" symbioticum* (SCHWEISFURTH et al. 1978). SCHMIDT (1979) found that *Siderocapsa geminata* in Lake Plußsee (Holstein), apart from being able to oxidize reduced Fe- and Mn components during the station-

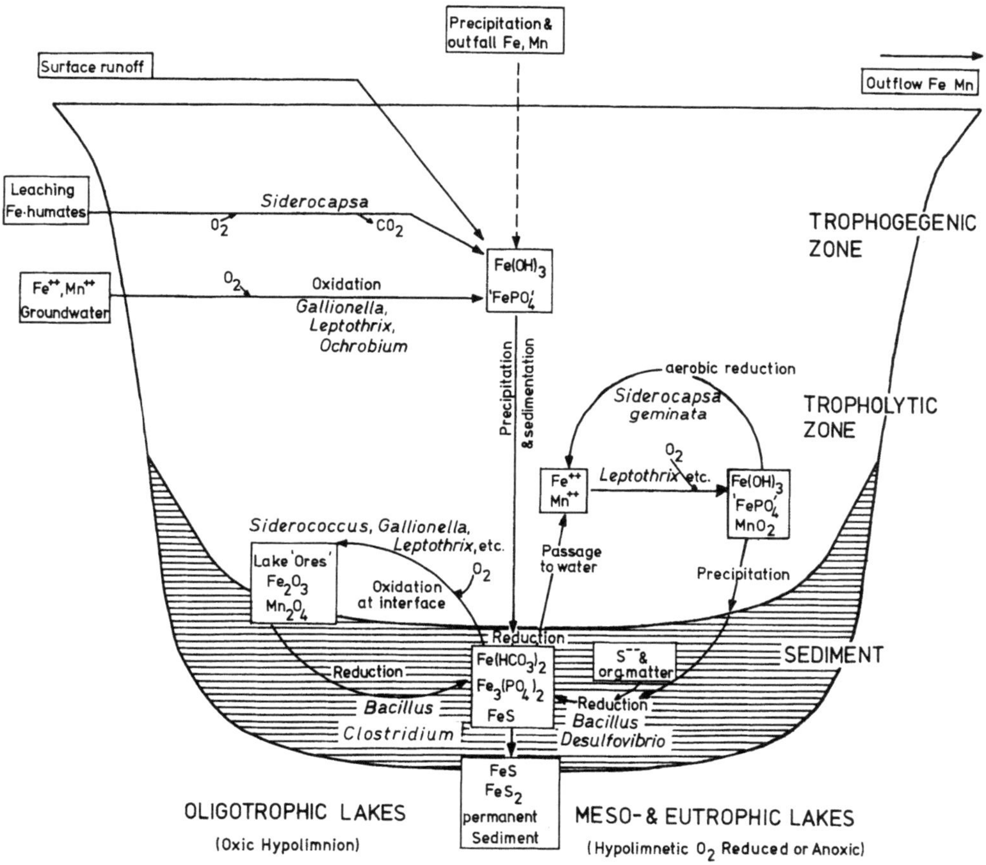

Fig. 2.6. Cycling of iron and manganese. (Modified after KUSNEZOW 1959 and WETZEL 1975, supplemented after SCHMIDT 1979)

ary growth phase, is also capable of reducing oxidized components. According to SCHMIDT (1979), the organic ligand seems not to be utilized in Mn oxidation, contrary to the descriptions of the iron cycle. Generally, the ecological significance of this bacterium is that, in lakes, it catalyses an oxidation of dissolved and a reduction of particulate metal compounds, namely under conditions in which a corresponding chemical transformation is not possible (SCHWEISFURTH 1973; SCHMIDT 1979). The cycling of iron and manganese is summarized in Fig. 2.6.

At this stage it may be pointed out that the biology and above all the systematics of iron bacteria belong to the most obscure fields of aquatic microbiology. Thus, SCHMIDT (1979) questions the taxonomy of *Siderocapsa* and regards some descriptions of species as those of various environmental forms. Further, the opinion of RUTTNER (1938!) concerning his so-called *Metallogenium* is gradually being accepted, that this most frequently occurring "iron bacterium" is not a living structure (KLAVENESS 1977; SCHMIDT 1979; GREGORY et al. 1980), especially as MEREK (1973) was able to provide evidence of various filiform structures that closely resembled "*Metallogenium*" by chemically agitating a silicate solution containing ferrous sulphate and ferric chloride.

The concentration of Mn is one controller of algal succession: low concentrations of Mn favour the development of blue-greens, whereas concentrations of $\geqq 40\ \mu g\ l^{-1}$ more favour the development of diatom communities (PATRICK 1978). In higher concentrations other metals, e.g. V, Se, however, support cyanophytes rather than green and other algae.

2.7.3 Iron and Manganese in Sediments

The release of iron and manganese occurs chiefly under anaerobic conditions in the sediment–water contact zone, but also takes place under aerobic conditions, although greatly reduced. TESSENOW (1972) observed that under aerobic conditions a constant, low iron release took place, which increased erratically after the oxidized upper layer had been completely reduced after plankton sedimentation. The reduction of the upper layer of the sediment can also take place when the oxygen content in the overlying water is high, namely when a layer of fungus develops on a plankton sediment, which separates the sediment from O_2-rich water. In a sediment with an oxidized surface, an intensive migration of reduced iron and manganese components, caused by diffusion, takes place into the upper layer, where these components are oxidized and therefore are largely fixed.

The redox-dependent release of metals from sediments generally gains an increasing environmental-toxicological importance in flowing waters regulated by dams (cf. NEULAND et al. 1978). Under aerobic conditions the trace metals are normally bound in the sediments by precipitation and coprecipitation with Fe- and Mn-hydroxides, a few hundred metres after being introduced into the flowing water. In waters with a high calcium carbonate content additional mechanisms hold true. However, if anaerobic conditions occur in river impoundments, the linked heavy metals are released after the dissolution of Mn- and Fe hydroxide. The erratic release of metals can then lead to dangerous toxification in the water.

2.8 Conclusions

In this chapter, several open questions remain, together with contradictory results which are briefly summarized here. For example, the question of which inorganic and organic carbon compounds are taken up by aquatic primary producers and the related ecological significance needs to be further clarified. Similarly, the role of particulate and soluble dead organic material in the aquatic ecosystem must be investigated: are these compounds mainly inert or do they exert a positive effect on the availability of micronutrients?

Some of these open questions concerning the ecology of phytoplankton are treated and partly answered in the recently published book, *The physiological ecology of phytoplankton* (MORRIS 1980). Until now, a convenient method for the determination of bacterial productivity in standing and running waters has been lacking. Furthermore, knowledge concerning the "metabolic coupling" between individual trophic levels needs to be increased, as well as of the role of heterotrophic nitrification in standing and running waters. However, the effect of aquatic macrophytes on the open water of lakes is difficult to determine: are nutrients released by living or decaying littoral plants into the aquatic environment?

In addition to these detailed problems, certain general aspects must be considered. Studies in running water are rarer than comparable investigations in standing water, for example, due to practical difficulties less is known of nutrient cycling of the rates of primary production. In the future this deficiency needs to be rectified.

References

Alexander M, Marshall KC, Hirsch P (1960) Autotrophy and heterotrophy in nitrification. Proc Comm III, 7th Int Congr Soil Sci, Madison, Wis

Andersen JM (1974) Nitrogen and phosphorus budgets and the role of sediments in six shallow Danish lakes. Arch Hydrobiol 74:528–550

Andersen JM (1975) Influences of pH on release of phosphorus from lake sediments. Arch Hydrobiol 76:411–419

Andersen JM (1977) Rates of denitrification of undisturbed sediment from six lakes as a function of nitrate concentration, oxygen and temperature. Arch Hydobiol 80:147–159

Austin ER, Lee FG (1973) Nitrogen release from lake sediments. J Water Pollut Control Fed 45:870–879

Baccini P, Suter U (1979) Chemical speciation and biological availability of copper in lake water. Schweiz Z Hydrol 41:291–314

Barber LE, Ensign JC (1979) Methane formation and release in a small Wisconsin lake. Geomicrobiol J 1:341–353

Barber RT (1974) Organic ligands and phytoplankton growth in nutrient rich seawater. In: Singer P (ed) Trace metals and metal-organic interactions in natural waters. Ann Arbor Science, Ann Arbor

Barko JW, Smart RM (1980) Mobilization of sediment phosphorus by submersed freshwater macrophytes. Freshwater Biol 10:229–238

Bärlocher F, Mackay RJ, Wiggins GB (1978) Detritus processing in a temporary vernal pool in southern Ontario. Arch Hydrobiol 81:269–295

Barnes JR, Ovink R, Cummins KW (1978) Leaf litter processing in Gull Lake, Michigan, USA. Verh Int Ver Limnol 20:475–479
Beauchamp RSA (1953) Sulphates in African inland waters. Nature (London) 171:769–771
Bechard MJ, Rayburn WR (1979) Volatile organic sulfides from freshwater algae. J Phycol 15:379–383
Bergmann W (1978) Zur Strukturaufklärung von Huminsäuren aus Abwasser. Dissertation, Univ Tübingen
Berman T (1970) Alkaline phosphatases and phosphorous availability in Lake Kinneret. Limnol Oceanogr 15:663–674
Berman T (1976) Release of dissolved organic matter by photosynthesizing algae in Lake Kinneret, Israel. Freshwater Biol 6:13–18
Bernice R (1972) Nitrogen excretion in *Streptocephalus dichotomus* BAIRD (Crustaceae: Anostraca). Hydrobiologia 39:449–456
Best D, Mantai KE (1978) Growth of *Myriophyllum:* Sediment or lake water as the source of nitrogen and phosphorous. Ecology 59:1075–1080
Binder BJ, Chisholm SW (1980) Changes in the soluble silicon pool size in the marine diatom *Thalassiosira weisflogii.* Mar Biol Lett 1:205–212
Blackburn WM, Petr T (1979) Forest litter decomposition and benthos in a mountain stream in Victoria, Australia. Arch Hydrobiol 86:453–498
Braun R (1952) Limnologische Untersuchungen an einigen Seen im Amazonasgebiet. Schweiz Z Hydrol 14:1–128
Brezonik PL, Lee GF (1968) Denitrification as a nitrogen sink in Lake Mendota, Wis. Environ Sci Technol 2:120–125
Bristow JM (1974) Nitrogen fixation in the rhizosphere of freshwater angiosperms. Can J Bot 52:217–221
Bristow JM, Whitcombe M (1971) The role of roots in the nutrition of aquatic vascular plants. Am J Bot 58:8–13
Brooks RH, Brezonik PL, Putnam HD, Keirn MA (1971) Nitrogen fixation in an estuarine environment: the Waccasassa on the Florida Gulf Coast. Limnol Oceanogr 16:701–711
Brown EJ, Harris RF (1978) Kinetics of algal transient phosphate uptake and the cell quota concept. Limnol Oceanogr 23:35–40
Brown EJ, Harris RF, Koonce JK (1978) Kinetics of phosphate uptake by aquatic microorganisms: deviation from a simple Michaelis-Menten equation. Limnol Oceanogr 23:26–34
Burnison BK (1975) Microbial ATP studies. Verh Int Ver Limnol 19:286–290
Burns NM, Nriagu JO (1976) Forms of iron and manganese in Lake Erie waters. J Fish Res Board Can 33:463–470
Campbell PGC, Baker JH (1978) Estimation of bacterial production in freshwaters by the simultaneous measurement of [^{35}S]sulphate and D-[^{3}H]glucose uptake in the dark. Can J Microbiol 24:939–946
Cappenberg ThE (1975a) A study of mixed continuous cultures of sulfate-reducing and methane-producing bacteria. Microb Ecol 2:60–72
Cappenberg ThE (1975b) Relationships between sulfate-reducing and methane-producing bacteria. Plant Soil 43:125–139
Cappenberg ThE (1978) Microenvironments for sulfate reduction and methane production in freshwater sediments. In: Krumbein WE (ed) Environ Biogeochem Geomicrobiol, vol. I. Ann Arbor Science, Ann Arbor
Chatarpaul L, Robinson JB, Kaushik NK (1979) Role of tubificid worms on nitrogen transformations in stream sediment. J Fish Res Board Can 36:673–678
Chave KE (1970) Carbonate-organic interactions in sea water. In: Hood DW (ed) Organic matter in natural waters. Inst Mar Sci Occas Publ 1, pp 373–385
Chen RL, Keeney DR, Graetz DA (1972) Nitrification in sediments of selected Wisconsin lakes. J Environ Qual 1:151–154
Curtis EJC, Durrant K, Harman MMI (1975) Nitrification in rivers in the Trent basin. Water Res 9:255–268

Dacey JWH, Klug MJ (1979) Methane efflux from lake sediments through water lilies. Science 203:1253–1255
Denny P (1972) Sites of nutrient absorption in aquatic macrophytes. J Ecol 60:819–829
Dieckert GB, Graf EG, Thauer RK (1979) Nickel requirement for carbon monoxide dehydrogenase formation in *Clostridium pasteurianum*. Arch Microbiol 122:117–120
Edwards RW, Rolley HLJ (1965) Oxygen consumption in river muds. J Ecol 59:1–19
Einsele W (1938) Über chemische und kolloidchemische Vorgänge in Eisen-Phosphat-Systemen unter limnochemischen und limnogeologischen Gesichtspunkten. Arch Hydrobiol 33:361–387
Ferguson AR, Bollard EG (1969) Nitrogen metabolism of *Spirodela oligorrhiza*. I. Utilization of ammonium, nitrate and nitrite. Planta 88:344–352
Ferrante JG (1976) The characterization of phosphorus excretion products of a natural population of limnetic zooplankton. Hydrobiologia 50:11–15
Finke R, Seeley HW Jr (1978) Nitrogen fixation (acetylene reduction) by epiphytes of freshwater macrophytes. Appl Environ Microbiol 36:129–138
Finstein MS, Strom PF, Matulewich VA (1978) Discussion of: R.J. Courchaine (1968) Significance of nitrification in stream analysis-effects on the oxygen balance. J Water Pollut Control Fed 40:835–847. J Water Pollut Control Fed 50:2055–2057
Fischer WR, Baumann G (1978) Die Sorption von Phosphat an carbonatreichen Unterwasserböden. Z Pflanzenernaehr Bodenkd 141:607–620
Fisher SG, Likens GE (1972) Stream ecosystems: organic energy budget. BioScience 22:33–35
Focht DD, Verstraete W (1977) Biochemical ecology of nitrification and denitrification. In: Alexander M (ed) Advances in microbial ecology, vol I. Plenum, New York London
Fogg GE (1971) Extracellular products of algae in freshwater. Arch Hydrobiol Beih Ergebn Limnol 5:1–25
Fogg GE (1977) Excretion of organic matter by phytoplankton. Limnol Oceanogr 22:576–577
Forsberg C (1977) Nitrogen as a growth factor in fresh water. Prog Water Tech 8:275–290
Francko DA, Wetzel RG (1980) Cyclic adenosine-3′:5′-monophosphate: production and extracellular release from green and blue-green algae. Physiol Plant 49:65–67
Fuhrmann JA, Azam F (1980) Bacterioplankton secondary production estimates for coastal waters of British Columbia, Antarctica, and California. Appl Environ Microbiol 39:1085–1095
Fuhs GW, Demmerle SD, Canelli E, Chen M (1972) Characterization of phosphorus-limited plankton algae (with reflections on the limiting nutrient concept). In: Likens GE (ed) Nutrients and eutrophication: the limiting nutrient controversy. Spec Symp Am Soc Limnol Oceanogr 1:113–133
Gallepp GW, Kitchell JF, Bartell SM (1978) Phosphorus release from lake sediments as effected by chironomids. Verh Int Ver Limnol 20:458–465
Gamble DS, Schnitzer M (1974) The chemistry of fulvic acid and its reactions with metal ions. In: Singer P (ed) Trace Metals and Metal-Organic Interactions in Natural Waters. Ann Arbor Science, Ann Arbor
Garland JHN (1973) Nitrification in rivers: studies in the Trent basin. In: James A (ed) Symposium on the use of mathematical models in water pollution control, vol II. Wiley-Interscience, Newcastle
Gersberg RM, Axler RP, Krohn K, Peek N, Goldman CR (1978) Nitrate uptake by phytoplankton: measurements utilizing the radiotracer ^{13}N. Verh Int Ver Limnol 20:388–392
Giesy JP Jr (1976) Stimulation of growth in *Scenedesmus obliquus* (Chlorophyceae) by humic acids under iron-limited conditions. J Phycol 12:172–179
Gjessing ET (1976) Physical and chemical characteristics of aquatic humus. Ann Arbor Science, Ann Arbor
Gocke K (1970) Untersuchungen über Abgabe und Aufnahme von Aminosäuren und Polypeptiden durch Planktonorganismen. Arch Hydrobiol 67:285–367
Godshalk GL, Wetzel RG (1978) Decomposition in the littoral zone of lakes. In: Good

RE, Whigham DF, Simpson RL (eds) Freshwater wetlands: ecological processes and management potential. Academic Press, London New York
Goering JJ (1972) The role of nitrogen in eutrophic processes. In: Mitchel R (ed) Water pollution microbiology. Wiley-Interscience, New York
Golterman HL (1975) Physiological limnology. Elsevier, Amsterdam
Graetz DA, Keeney DR, Aspiras RB (1973) Eh status of lake sediment–water systems in relation to nitrogen transformations. Limnol Oceanogr 18:908–917
Granhall U, Lundgren A (1971) Nitrogen fixation in Lake Erken. Limnol Oceanogr 16:711–719
Gregory E, Perry RS, Staley JT (1980) Characterization, distribution and significance of *Metallogenium* in Lake Washington. Microb Ecol 6:125–140
Gregory SV (1978) Phosphorus dynamics on organic and inorganic substrates in streams. Verh Int Ver Limnol 20:1340–1346
Groth P (1971) Untersuchungen über einige Spurenmetalle in Seen. Arch Hydrobiol 68:305–375
Haan de H (1977) Effect of benzoate on microbial decomposition of fulvic acids in Tjeukemeer (The Netherlands). Limnol Oceanogr 22:38–44
Hargrave BT, Green GH (1968) Phosphorus excretion by zooplankton. Limnol Oceanogr 13:332–342
Harris GP (1978) Photosynthesis, productivity and growth: the physiological ecology of phytoplankton. Arch Hydrobiol Beih Ergebn Limnol 10:1–171
Harrison W (1978) Experimental measurement of nitrogen mineralization in coastal waters. Limnol Oceanogr 23:659–683
Harrits SM, Hanson RS (1980) Stratification of aerobic methane-oxidizing organisms in Lake Mendota, Madison, Wis. Limnol Oceanogr 25:412–421
Heath RT, Cooke GD (1975) The significance of alkaline phosphatase in a eutrophic lake. Verh Int Ver Limnol 19:959–965
Hellebust JA (1974) Extracellular products. In: Stewart WDP (ed) Algal physiology and biochemistry. Bot Monogr 10:838–863
Herrera J, Paneque A, Maldonado JM, Barea JL, Losada M (1972) Regulation by ammonia of nitrate reductase synthesis and activity in *Chlamydomonas reinhardi*. Biochem Biophys Res Commun 48:996–1003
Herrmann V, Jüttner F (1977) Excretion products of algae. Anal Biochem 78:365–373
Horne AJ (1977) Nitrogen fixation – a review of this phenomenon as a polluting process. Prog Water Tech 8:359–372
Horne AJ, Carmiggelt CJW (1975) Algal nitrogen fixation in California streams. I. Seasonal cycles. Freshwater Biol 5:461–470
Horne AJ, Fogg GE (1970) Nitrogen fixation in some English lakes. Proc R Soc London Ser B 175:351–366
Horne AJ, Goldman CR (1972) Nitrogen fixation in Clear Lake, California. I. Seasonal variation and the role of heterocysts. Limnol Oceanogr 17:693–703
Horne AJ, Sandusky JC, Carmiggelt CJW (1979) Nitrogen fixation in Clear Lake, California. 3. Repetitive synoptic sampling of the spring *Aphanizomenon* blooms. Limnol Oceanogr 24:316–328
Hough RA, Wetzel RG (1975) The release of dissolved organic carbon from submersed aquatic macrophytes: diel, seasonal, and community relationships. Verh Int Ver Limnol 19:939–948
Hutchinson GE (1957–75) A treatise on limnology, vol I–III. Wiley and Sons, New York
Jackson TA, Schindler DW (1975) The biogeochemistry of phosphorus in an experimental lake environment: evidence for the formation of humic-metal-phosphate complexes. Verh Int Ver Limnol 19:211–221
Jansson M (1976) Phosphatases in lake water: characterization of enzymes from phytoplankton and zooplankton by gel filtration. Science 194:320–321
Jenkins D, Medsker LL, Thomas JF (1967) Odorous compounds in natural waters. Some sulfur compounds associated with blue-green algae. Environ Sci Technol 1:731–735

Johannes RE (1964) Phosphorus excretion and body size in marine animals: microzooplankton and nutrient regeneration. Science 146:923–924
Jordan MJ, Likens GE (1980) Measurement of planktonic bacterial production in an oligotrophic lake. Limnol Oceanogr 25:719–732
Jüttner F (1981) Biologically active compounds released during algal blooms. Paper presented at: Verh Int Ver Limnol, Kyoto 21:227–230
Jüttner F, Matuschek T (1978) The release of low molecular weight compounds by the phytoplankton in a eutrophic lake. Water Res 12:251–255
Junge CE (1960) Sulfur in the atmosphere. J Geophys Res 69:227–237
Kamp-Nielsen L (1974) Mud-water exchange of phosphate and other ions in undisturbed sediment cores and factors affecting the exchange rates. Arch Hydrobiol 73:218–237
Kaplan LA, Goldman CR, Knight AW (1975) Phosphorus sorption in a phosphorus poor stream. Bull Ecol Soc Am 56:24
Kaushik NK (1975) Decomposition of allochthonous organic matter and secondary production in stream ecosystems. In: Productivity of world ecosystems. Nat Acad Sci, Washington DC
Kaushik NK, Hynes HBN (1971) The fate of dead leaves that fall into streams. Arch Hydrobiol 68:465–515
Keeney DR (1973) The nitrogen cycle in sediment water systems. J Environ Qual 2:15–29
Keeney DR, Chen RI, Graetz DA (1971) Importance of denitrification and nitrate reduction in sediments to nitrogen budgets of lakes. Nature (London) 233:66–67
Keirn MA, Brezonik PL (1971) Nitrogen fixation by bacteria in Lake Mize, Florida, and in some lacustrine sediments. Limnol Oceanogr 16:720–731
Kellar PE, Goldman CR (1979) A comparative study of nitrogen fixation by the *Anabaena–Azollae* symbiosis and free-living populations of *Anabaena* spp. in Lake Ngahewa, New Zeeland. Oecologia 43:269–281
Kelly Robertson C (1979) Quantitative comparison of the significance of methane in the carbon cycles of two small lakes. Arch Hydrobiol Beih Ergebn Limnol 12:123–135
Kessel van JF (1977) Factors affecting the denitrification rate in two water-sediment systems. Water Res 11:259–267
Kirkman H, Griffith FB, Parker RR (1979) The release of reactive phosphate by a *Posidonia australis* seagrass community. Aquat Bot 6:329–337
Klaveness D (1977) Morphology, distribution and significance of the manganese-accumulating microorganism *Metallogenium* in lakes. Hydrobiologia 56:25–33
Krambeck C (1979) Application and limitations of the Michaelis-Menten equation in microbial ecology. Arch Hydrobiol Beih Ergebn Limnol 12:64–76
Ku WC, Digiano FA, Feng TH (1978) Factors affecting phosphate adsorption equilibria in lake sediments. Water Res 12:1069–1974
Kuenzler EJ, Perras JP (1965) Phosphatases of marine algae. Biol Bull Woods Hole Mass 128:271–284
Kusnetsov SI (1968) Recent studies on the role of microorganisms in the cycling of substances in lakes. Limnol Oceanogr 13:211–224
Kusnezov SI (1959) Die Rolle der Mikroorganismen im Stoffkreislauf der Seen. Deutscher Verlag der Wissenschaften, Leipzig
Lännergren C, Lundgren A (1974) Acetylene reduction and primary production in Lake Erken. Oikos 25:365–369
Lam CWY, Vincent WF, Silvester WB (1979) Nitrogenase activity of nitrogen fixation by freshwater benthic blue-green algae. NZ J Mar Freshwater Res 13:187–192
Lamarra VA Jr (1975) Digestive activities of carp as a major contributor to the nutrient loading of lakes. Verh Int Ver Limnol 19:2461–2468
Lampert W (1978) Release of dissolved organic carbon by grazing zooplankton. Limnol Oceanogr 23:831–834
Larow EJ, McNaught DC (1978) Systems and organismal aspects of phosphorus remineralization. Hydrobiologia 59:151–154
Lean DRS (1973) Phosphorus dynamics in lake water. Science 179:678–680
Lee K, Nalewajko C (1978) Photosynthesis, extracellular release and glycollic acid uptake by plankton: fractionation studies. Verh Int Ver Limnol 20:257–262

Levine S (1975) Orthophosphate concentration and flux within the epilimnion of two Canadian Shield lakes. Verh Int Ver Limnol 19:624–629
Li WC, Armstrong DE, Williams JDH, Harris RF, Syers JK (1972) Rate and extent of inorganic phosphate exchange in lake sediments. Soil Sci Am Proc 36:279–285
Likens GE (ed) (1972) Nutrients and eutrophication: the limiting nutrient controversy, Spec Symp. Am Soc Limnol Oceanogr 1:1–328
Lindström K (1980) *Peridinium cinctum* bioassay of Se in Lake Erken. Arch Hydrobiol 89:110–117
Littlefield L, Forsberg C (1965) Absorption and translocation of phosphorus-32 by *Chara globularis* (Thuill.). Physiol Plant 18:291–296
Lovelock JE, Maggs RJ, Rasmussen RA (1972) Atmospheric dimethyl sulfide and the natural sulfur cycle. Nature (London) 237:252–253
Lundgren DG, Dean W (1979) Biogeochemistry of iron. In: Trudinger PA, Swaine DJ (eds) Biogeochemical cycling of mineral-forming elements. Elsevier, Amsterdam
Marxsen J (1980) Untersuchungen zur Ökologie der Bakterien in der fließenden Welle von Bächen. I. Chemismus, Primärproduktion, CO_2-Dunkelfixierung und Eintrag von partikulärem organischem Material. Arch Hydrobiol Suppl 57:461–533
McCracken MD, Middaugh RE, Middaugh RS (1980) A chemical characterization of an algal inhibitor obtained from *Chlamydomonas*. Hydrobiologia 70:271–276
McGregor AN, Keeney DR, Chen KL (1973) Nitrogen fixation in lake sediments: contribution to nitrogen budget of Lake Mendonta. Environ Lett 4:21–26
McKnight DM, Morel FMM (1979) Release of weak and strong copper complexing agents by algae. Limnol Oceanogr 24:823–838
McKnight DM, Morel FMM (1980) Copper complexation by siderophores from filamentous blue-green algae. Limnol Oceanogr 25:62–71
McRoy CP, Barsdate RJ (1970) Phosphate absorption in eelgrass. Limnol Oceanogr 15:6–13
Melchiorri-Santolini U, Hopton JW (eds) (1972) Detritus and its role in aquatic ecosystems. Mem Ist Ital Idrobiol Suppl 29:
Merek EL (1973) Imaging and life detection. Bio Sci 23:364–371
Meyer JL (1979) The role of sediments and bryophytes in phosphorus dynamics in a headwater stream ecosystem. Limnol Oceanogr 24:365–375
Meyer JL (1980) Dynamics of phosphorus and organic matter during leaf decomposition in a forest stream. Oikos 34:44–53
Monheimer RH (1975) Planktonic microbial heterotrophy: its significance to community biomass production. Verh Int Ver Limnol 19:2658–2663
Morgan JJ, Stumm W (1965) The role of multivalent metal oxides in limnological transformations as examplified by iron and manganese. Proc 2nd Int Water Pollut Res Conf Tokyo, Japan
Morris I (ed) (1980) The physiological ecology of phytoplankton. Studies in ecology, vol VII. Blackwell, Oxford
Müller U (1977) Stoffhaushalt, Phytoplankton und Primärproduktion in drei ostholsteinischen Seen unterschiedlichen Trophiegrades. Dissertation, Univ Kiel
Murphy TP, Lean DRS, Nalewajko C (1976) Blue-green algae: Their excretion of iron-selective chelators enables them to dominate other algae. Science 192:900–902
Nalewajko C, Lean DRS (1978) Phosphorus kinetics – algal growth relationships in batch cultures. Mitt Int Ver Limnol 21:184–192
Nalejwajko C, Schindler DW (1976) Primary production, extracellular release, and heterotrophy in two lakes in the ELA, Northwestern Ontario. J Fish Res Board Can 33:219–226
Nedwell DB (1975) Inorganic nitrogen metabolism in a eutrophicated tropical mangrove estuary. Water Res 9:221–232
Neuland H, Schrimpff E, Herrmann R (1978) Zur Änderung der Spurenmetallgehalte im fließenden Wasserkörper und in den Sedimenten entlang eines Flußabschnittes des Roten Mains in Abhängigkeit von Redoxpotential, pH und anderen Einflußgrößen. Catena 5:19–31
Nichols DS, Keeney DR (1976) Nitrogen nutrition of *Myriophyllum spicatum*: uptake and translocation of ^{15}N by shoots and roots. Freshwater Biol 6:145–154

Niewolak S, Korycka A, Potocka E (1978) Ammonification processes in fertilized lakes. Ekol Pol 26:555–572
Nowak KE (1975) Die Bedeutung des Zooplanktons für den Stoffhaushalt des Schierensees. Arch Hydrobiol 75:149–224
Nriagu JO (ed) (1980a) Cadmium in the environment. Wiley and Sons, New York
Nriagu JO (ed) (1980b) Zinc in the environment. Wiley and Sons, New York
Nriagu JO, Coker RD, Kemp ALW (1979) Thiosulfate, polythionates and rhodanese activity in Lakes Erie and Ontario sediments. Limnol Oceanogr 24:383–389
Ogawa RE, Carr JF (1969) The influence of nitrogen on heterocyst production in blue-green algae. Limnol Oceanogr 14:342–351
Ohle W (1937) Kolloidgele als Nährstoffregulanten der Gewässer. Naturwissenschaften 25:471–474
Ohle W (1954) Sulfat als „Katalysator" des limnischen Stoffkreislaufes. Vom Wasser 21:13–32
Ohle W (1958) Die Stoffwechseldynamik der Seen in Abhängigkeit von der Gasausscheidung ihres Schlammes. Wasser 25:127–149
Ohle W (1962) Der Stoffhaushalt der Seen als Grundlage einer allgemeinen Stoffwechseldynamik der Gewässer. Kiel Meeresforsch 18:107–120
Olsen S (1958) Phosphate adsorption and isotopic exchange in lake muds. Experiments with P-32. Preliminary report. Verh Int Ver Limnol 13:915–922
Orebamjo TO, Stewart GR (1975) Ammonium inactivation of nitrate reductase in *Lemna minor* L. Planta 122:37–44
Overbeck J (1979) Dark CO_2 uptake – biochemical background and its relevance to in situ bacterial production. Arch Hydrobiol Beih Ergebn Limnol 12:38–47
Overbeck J, Daley RJ (1973) Some precautionary comments on the Romanenko technique for estimating heterotrophic bacterial production. Bull Ecol Res Commun (Stockholm) 17:342–344
Paerl HW (1979) Optimization of carbon dioxide and nitrogen fixation by the blue-green alga *Anabaena* in freshwater blooms. Oecologia 38:275–290
Paerl HW, Lean DRS (1976) Visual observations of the uptake of phosphorus by lakewater plankton. J Fish Res Board Can 33:2805–2813
Painter HA (1970) A review of literature of inorganic nitrogen metabolism in microorganisms. Water Res 4:393–450
Painter HA (1977) Microbial transformations of inorganic nitrogen. Prog Water Technol 8:3–29
Parker JI, Conway HL, Yaguchi EM (1977) Dissolution of diatom frustules and recycling of amorphous silicon in Lake Michigan. J Fish Res Board Can 34:545–551
Parsons TR, Strickland JDH (1962) On the production of particulate organic carbon by heterotrophic processes in sea water. Deep Sea Res 8:211–222
Patrick R (1978) Effects of trace metals in the aquatic ecosystem. Am Sci 66:185–191
Payne WJ (1973) Reduction of nitrogenous oxides by microorganisms. Bacteriol Rev 37:410–452
Penhale PA, Smith WO Jr (1977) Excretion of dissolved organic carbon by eelgrass (*Zostera marina*) and its epiphytes. Limnol Oceanogr 22:400–403
Peters RH (1977) Availability of atmospheric orthophosphate. J Fish Res Board Can 34:918–924
Peters RH (1978) Concentrations and kinetics of phosphorus fractions in water from streams entering Lake Memphremagog. J Fish Res Board Can 35:315–328
Peters RH (1979) Concentrations and kinetics of phosphorus fractions along the trophic gradient of Lake Memphremagog. J Fish Res Board Can 36:970–979
Peters RH, Lean DRS (1973) The characterization of soluble phosphorus released by limnetic zooplankton. Limnol Oceanogr 18:270–279
Prentki RT, Adams MS (1979) The phosphorus budget of the littoral zone of Lake Wingra, Wisconsin. 42nd Meet Am Soc Limnol Oceanogr, Stony Brook, New York
Reddy MM (1975) Kinetics of calcium carbonate formations. Verh Int Ver Limnol 19:429–438
Reichardt W, Overbeck J, Steubing L (1967) Free dissolved enzymes in lake waters. Nature (London) 216:1345–1347

Rhee GY (1973) A continuous culture study of phosphate uptake, growth rate and polyphosphate in *Scenedesmus* sp. J Phycol 9:495–506
Rich PH, Wetzel RG (1978) Detritus in the lake ecosystem. Am Nat 112:57–71
Rigler FH (1973) A dynamic view of the phosphorus cycle in lakes. In: Griffith et al. (eds) Environmental phosphorus handbook. Wiley and Sons, New York
Robb F, Davies BR, Cross R, Kenyon C, Howard-Williams C (1979) Cellulolytic bacteria as primary colonizers of *Potamogeton pectinatus* L. (sago pond weed) from a brackish south-temperate coastal lake. Microb Ecol 5:167–177
Romanenko VI (1964) Heterotroph CO_2 assimilation by bacterial flora of water. Mikrobiol 33:679–683 (in Russian), quoted in Overbeck and Daley (1973)
Roth JC, Horne AJ (1981) Algal nitrogen fixation and microcrustacean abundance: an unregarded interrelationship between zoo- and phytoplankton. Verh Int Ver Limnol, Kyoto 21:333–338
Rudd JWM, Hamilton RD (1978) Methane cycling in a eutrophic Shield lake and its effect on whole lake metabolism. Limnol Oceanogr 23:337–348
Rudd JWM, Furutani A, Flett RJ, Hamilton RD (1976) Factors controlling methane oxidation in Shield lakes. The role of nitrogen fixation and oxygen concentration. Limnol Oceanogr 21:357–364
Russo RC, Smith CF, Thurston RU (1974) Acute toxicity of nitrite to rainbow trout (*Salmo gairdneri*). J Fish Res Board Can 31:1653–1655
Ruttner F (1938) Limnologische Studien an einigen Seen der Ostalpen. Arch Hydrobiol 32:167–319
Saunders GW (1972) The transformation of artificial detritus in lake water. Mem Ist Ital Idrobiol Suppl 29:261–288
Schindler DW, Brunskill GJ, Emerson S, Broecker WS, Peng TH (1972) Atmospheric carbon dioxide: its role in maintaining phytoplankton standing crops. Science 177:1192–1194
Schindler DW, Lean DRS, Fee EJ (1975) Nutrient cycling in freshwater ecosystems. In: Productivity of world ecosystems. Natl Acad Sci, Washington DC
Schindler JE, Williams DJ, Zimmerman AP (1976) Investigation of extracellular electron transport by humic acids. In: Nriagu JO (ed) Environmental biogeochemistry, vol I. Carbon, nitrogen, phosphorus, sulfur and selenium cycles. Ann Arbor Science, Ann Arbor
Schmidt WD (1979) Morphologie und Physiologie manganoxidierender Mikroorganismen. Kultur und in situ Untersuchungen zur ökologisch-mikrobiologischen Charakterisierung von *Metallogenium* sp. und *Siderocapsa geminata* im Plussee. Dissertation, Univ Kiel
Schönheit P, Moll J, Thauer RK (1979) Nickel, cobalt and molybdenum requirement for growth of *Methanobacterium thermoautotrophicum*. Arch Microbiol 123:105–107
Schweisfurth R (1973) Manganoxidierende Bakterien. Z Bakteriol I 233:257–270
Schweisfurth R, Eleftheriadis D, Gundlach H, Jacobs M, Jung W (1978) Microbiology of the precipitation of manganese. In: Krumbein WE (ed) Environmental biogeochemistry and geomicrobiology, vol III. Methods, metals and assessment. Ann Arbor Science, Ann Arbor
Schwoerbel J, Tillmanns GC (1972) Ammonium-Adaptation bei submersen Phanerogamen in situ. Arch Hydrobiol Suppl 42:139–141
Schwoerbel J, Tillmanns GC (1977) Nitrataufnahme aus dem Wasser und Nitratreduktase-Aktivität bei *Fontinalis antipyretica* L. im Hell-Dunkel-Wechsel. Arch Hydrobiol Suppl 48:412–423
Sharp JH (1977) Excretion of organic matter by marine phytoplankton: do healthy cells do it? Limnol Oceanogr 22:381–399
Sheridan RP (1973) Hydrogen sulfide production by *Synechococcus lividus* Y 52-s[1]. J Phycol 9:445–457
Smith CS, Adams MS, Schmitt MR, Adams SS (1978) Phosphorus in the water, sediment, and vegetation of the Crnojvića River, Montenegro, Yugoslavia. Verh Int Ver Limnol 20:1536–1542
Søndergaard M, Sand-Jensen K (1979) Carbon uptake by leaves and roots of *Littorella uniflora* (L.) Aschers. Aquat Bot 6:1–12

Sorokin YuI (1970) Interrelations between sulphur and carbon turnover in meromictic lakes. Arch Hydrobiol 66:391–446

Stabel HH (1977) Gebundene Kohlenhydrate als stabile Komponenten im Schöhsee und in *Scenedesmus*-Kulturen. Arch Hydrobiol Suppl 53:159–254

Stabel HH (1981) In situ studies on the estimation of the heterotrophic bacterial activity using two types of substrates. Verh Int Ver Limnol, Kyoto 21:1359–1364

Stabel HH, Moaledj K, Overbeck J (1979) On the degradation of dissolved organic molecules from Plußsee by oligocarbophilic bacteria. Arch Hydrobiol Beih Ergebn Limnol 12:95–104

Steinberg C (1977) Schwer abbaubare, stickstoffhaltige gelöste organische Substanzen im Schöhsee und in Algenkulturen. Arch Hydrobiol Suppl 53:48–158

Steinberg C (1978) Freisetzung gelösten organischen Kohlenstoffs (DOC) verschiedener Molekülgrößen in Planktongesellschaften. Arch Hydrobiol 82:155–165

Steinberg C (1980) Species of dissolved metals derived from oligotrophic hard water. Water Res 14:1239–1250

Steinberg C, Herrmann A (1981) Utilization of dissolved metal organic compounds by freshwater microorganisms. Verh Int Ver Limnol, Kyoto, 21:231–235

Steinberg C, Schrimpf A (1980) Phosphoranalytik – ein gelöstes Problem? Vom Wasser 55:295–302

Sternik KH (1978) Eine Methode zur kontinuierlichen Bestimmung der Phosphatexkretion von Fischen unter Verwendung eines Anionenaustauscherharzes. Arch Hydrobiol Suppl 55:24–61

Stewart WDP (1979) N_2-Fixation and photosynthesis in microorganisms. In: Gibbs M, Latzko E (eds) Photosynthesis II. Encyclopedia of plant physiology new ser vol VI. Springer, Berlin Heidelberg New York

Stewart WDP, Pemble M, Al-Ugaily L (1978) Nitrogen and phosphorus storage and utilization in blue-green algae. Mitt Int Ver Limnol 21:224–247

Stumm W, Bilinski H (1972) Trace metals in natural waters. Difficulties of interpretation arising from our ignorance on their speciation. In: Jenkins SH (ed) Advances in water pollution research. Pergamon, Oxford New York

Stumm W, Morgan JJ (1970) Aquatic chemistry. Wiley and Sons, New York

Suberkropp K, Klug MJ, Cummins KW (1975) Community processing of leaf litter in woodland streams. Verh Int Ver Limnol 19:1653–1685

Sullivan CW (1979) Diatom mineralization of silicic acid. IV. Kinetics of soluble Si pool formation in exponentially growing and synchronized *Navicula pelliculosa*. J Phycol 15:210–216

Sunda W, Guillard RRL (1976) The relationship between cupric ion activity and the toxicity of copper to phytoplankton. J Mar Res 34:511–529

Swallow KC, Westall JC, McKnight DM, Morel NML, Morel FMM (1978) Potentiometric determination of copper complexation by phytoplankton exudates. Limnol Oceanogr 23:538–542

Syrett PJ, Leftley JW (1976) Nitrate and urea assimilation by algae. In: Sunderland N (ed) Perspectives in experimental biology, vol II. Pergamon, Oxford New York

Tessenow U (1966) Untersuchungen über den Kieselsäurehaushalt der Binnengewässer. Arch Hydrobiol Suppl 32:1–136

Tessenow U (1972) Lösungs-, Diffusions- und Sorptionsprozesse in der Oberschicht von Seesedimenten. I. Ein Langzeitexperiment unter aeroben and anaeroben Bedingungen im Fließgleichgewicht. Arch Hydrobiol Suppl 38:353–398

Tessenow U (1979) Die Wechselwirkungen zwischen Sediment und Wasser in ihrer Bedeutung für den Nährstoffhaushalt von Seen. Z Wasser Abwasserforsch 12:29–36

Titman D (1976) Ecological competition between algae. Experimental confirmation of resource-based competition theory. Science 192:463–465

Toetz DW (1974) Uptake and translocation of ammonia by freshwater hydrophytes. Ecology 55:199–201

Toetz DW, Cole B (1980) Ammonia mineralization and cycling in Shagawa Lake, Minnesota. Arch Hydrobiol 88:9–23

Twilley RR, Brinson MM, Davis GJ (1977) Phosphorus absorption, translocation, and secretion in *Nuphar luteum*. Limnol Oceanogr 22:1022–1032

Vollenweider RA (1968) Scientific fundamentals of the eutrophication of lakes and flowing waters with particular reference to nitrogen and phosphorous as factors in eutrophication. Tech Rep DAS/CSI/68.27. Organ Econ Coop Dev (OECD) Paris
Werner D (1977) Silicate metabolism. In: Werner D (ed) The biology of diatoms. Botanical monographs, vol XIII. Blackwell, Oxford
Werzenak CT, Gannon JJ (1968) Evaluation of nitrification in streams. J Sanit Eng Div Am Soc Div Eng 94:883–995
Wetzel RG (1968) Dissolved organic matter and phytoplankton productivity in marl lakes. Mitt Int Ver Limnol 14:261–270
Wetzel RG (1972) The role of carbon in hard-water marl lakes. In: Likens GE (ed) Nutrients and eutrophication: the limiting nutrient controversy. Spec Symp. Am Soc Limnol Oceanogr 1:84–97
Wetzel RG (1975) Limnology. Saunders, Philadelphia London Toronto
Wetzel RG (1979) The role of the littoral zone and detritus in lake metabolism. Arch Hydrobiol Beih Ergebn Limnol 13:145–161
Wetzel RG, Manny BA (1972a) Secretion of dissolved organic carbon and nitrogen by aquatic macrophytes. Verh Int Ver Limnol 18:162–170
Wetzel RG, Manny BA (1972b) Decomposition of dissolved organic carbon and nitrogen compounds from leaves in an experimental hard-water stream. Limnol Oceanogr 17:927–931
Wetzel RG, Otsuki A (1974) Allochthonous organic carbon of a marl lake. Arch Hydrobiol 73:31–56
Wetzel RG, Penhale PA (1979) Transport of carbon and excretion of dissolved organic carbon by leaves and roots/rhizomes in seagrasses and their epiphytes. Aquat Bot 6:149–158
Wetzel RG, Rich PH (1973) Carbon in freshwater systems. In: Woodwell GM, Pecan EV (eds) Carbon in the biosphere. Brookhaven Symp Biol 24. Springsfield, Virginia
Wheeler P, North B, Littler M, Stephens G (1977) Uptake of glycine by natural phytoplankton populations. Limnol Oceanogr 22:900–910
White E, Payne G (1980) Distribution and biological availability of reactive high molecular weight phosphorus in natural waters in New Zealand. Can J Fish Aquat Sci 37:664–669
Williams JD, Syers JK, Harris RF, Armstrong DE (1970) Adsorption and desorption of inorganic phosphorus by lake sediments in a 0.1 M NaCl system. Environ Sci Technol 4:517–519
Williams JDH, Murphy TP, Mayer T (1976) Rates of phosphorus forms in Lake Erie sediments. J Fish Res Board Can 33:430–439
Wilson AL (1979) Trace metals in waters. Philos Trans R Soc London Ser B 228:25–39
Witzel KP (1973) Untersuchungen zur Physiologie heterotropher Nitrifikanten und ihr Vorkommen in ostholsteinischen Seen. Dissertation, Univ Kiel
Witzel KP, Overbeck J (1979) Heterotrophic nitrification by *Arthrobacter* sp. (strain 9006) as influenced by different cultural conditions, growth state and acetate metabolism. Arch Microbiol 122:137–143
Wium-Andersen S (1971) Photosynthetic uptake of free CO_2 by roots of *Lobelia dortmanna*. Physiol Plant 25:245–248
Wolfe JM, Rice EL (1979) Allelopathic interactions among algae. J Chem Ecol 5:533–542
Wright RT, Hobbie JE (1966) Use of glucose and acetate by bacteria in aquatic ecosystems. Ecology 47:447–464
Zaiss U (1976) Physiologische und ökologische Untersuchungen zur Regulation der Phosphatspeicherung bei *Oscillatoria redekei*. Dissertation, Univ Kiel
Zaiss U, Kaltwasser H (1979) Über den Einfluß wasserbaulicher Maßnahmen auf die mikrobiologische Gasproduktion in Fließgewässersedimenten. Arch Hydrobiol 87:314–326
Zevenboom W, Mur LR (1978) On nitrate uptake by *Oscillatoria agardhii*. Verh Int Ver Limnol 20:2302–2307
Ziegler H (1979) Diskriminierung von Kohlenstoff- und Wasserstoffisotopen: Zusammenhänge mit dem Photosynthesemechanismus und den Standortbedingungen. Ber Dtsch Bot Ges 92:162–184

3 Nutrient Cycling in Marine Ecosystems

T.R. PARSONS and P.J. HARRISON

CONTENTS

3.1 Introduction 85

3.2 The Nitrogen Cycle 86
3.2.1 Physical Advection of Nutrients 88
3.2.2 Nutrient Input from Terrestrial and Atmospheric Sources 89
3.2.3 Nutrient Uptake 91
3.2.4 Nutrient Regeneration in the Water Column 95
3.2.5 Nutrient Regeneration from Sediments 97

3.3 The Phosphorus Cycle 98

3.4 The Silicon Cycle 100

3.5 The Transfer of Trace Elements and Compounds in the Nutrient Cycle of the Sea 102

3.6 The Role of Macrophytes in Nutrient Cycles 104

3.7 Conclusions: Biological Consequences of Nutrient Ratios and Flux – Phytoplankton Species Succession and Coexistence 105

References 108

3.1 Introduction

Since seawater contains abundant supplies of certain plant nutrients it may be considered that some nutrients are never limiting to plant growth in the sea. These include elements and compounds such as Na^+, K^+, Ca^{2+}, Mg^{2+}, Cl^-, SO_4^{2-}, as well as water itself, and carbon dioxide (mostly as bicarbonate). In general, it is the concentration of such nutrients as nitrogen, phosphorus, silicon, iron, and a number of organic substances (e.g. vitamins) that restrict plant growth either because of the low concentration of these elements and compounds or due to their unavailability for algal metabolism. Nitrogen, for example, is present in seawater in large amounts as dissolved N_2, but only certain blue-green algae or bacteria can reduce this gas to a metabolic form (e.g. ammonium). Similarly, iron may be present, but in the absence of an organic chelator it may not be readily available to some phytoplankton. The nutrient which is most often cited as causing a limitation of phytoplankton growth in the sea is nitrogen (e.g. DUGDALE and GOERING 1967; EPPLEY et al. 1969b). Because of the importance of nitrogen in the marine environment, particular attention has been given to processes leading to its availability and recycling. In starting with this element, much of the discussion regarding upwelling, terrestrial input, sediment regeneration, etc. may also apply to other elements, particularly phosphorus and silicon. Consequently, the latter elements are considered in less detail at the end of this chapter, in order to avoid repetition.

For the purposes of this report, "nutrients" have been taken to include all autotrophic and auxotrophic requirements of marine plants which may be considered at times to be growth-rate limiting. The discussion will not, however, include the photosynthetic uptake of carbon itself (e.g. CO_2 and HCO_3^-) due to space considerations (see Chap. 10, this Vol. and Chap. 15, Vol. 12B). Further, emphasis has been placed on the biological aspects of the nutrient cycles and less attention has been given to their geochemistry. In discussing these nutrient requirements, it is apparent that less is known about some of the auxotrophic requirements of phytoplankton and seaweeds than about the major elements for autotrophic production. One further limitation regarding the discussion of marine nutrient cycles in the sea has been imposed by consideration of the relative spatial importance of phytoplankton and seaweeds. Since the former are present in seven-tenths of the world's surface waters, while the latter only occupy a small fringe of coastal waters, greater space has been given to the discussion of phytoplankton nutrient cycles than to the role of seaweeds in the same processes. Various aspects of nutrient cycles have been reviewed previously and additional information is available in these reviews (BARNES 1957; REDFIELD et al. 1963; RILEY and SKIRROW 1965; RILEY and CHESTER 1971; ALLEN and KRAMER 1972; SPENCER 1975; DUGDALE 1976; SÖDERLUND and SVENSSON 1976). Nutrient cycling in freshwater ecosystems is treated in Chap. 2, this Vol.

3.2 The Nitrogen Cycle

Nitrogen (Fig. 3.1 A) has the most complex cycle (compared to P or Si), partially because of the numerous forms in which it can exist. The most abundant form of nitrogen in the sea is dissolved nitrogen gas. Saturation concentrations are in the range of 370 to 800 μg-at l^{-1}, depending upon the salinity of the water (RILEY and CHESTER 1971). Seawater also contains much lower concentrations (see Table 3.1) of a wide range of inorganic and organic nitrogen compounds, in which nitrogen occurs in as many as nine different oxidation states ranging from -3 to $+5$. Of this group of compounds, nitrate and ammonium are extremely important because of their utilization by the dominant primary producers in the sea, the phytoplankton, but their concentration is an order of magnitude less than that for the dissolved gas.

The processes that are important to this cycle are either those that account for input of nitrogen into the food chain, or those that contribute to the loss of nitrogen into various sinks from which it may be eventually recycled. Processes which bring nitrogen into the euphotic zone where it can be used directly in autotrophic production by plants include physical advection of primarily

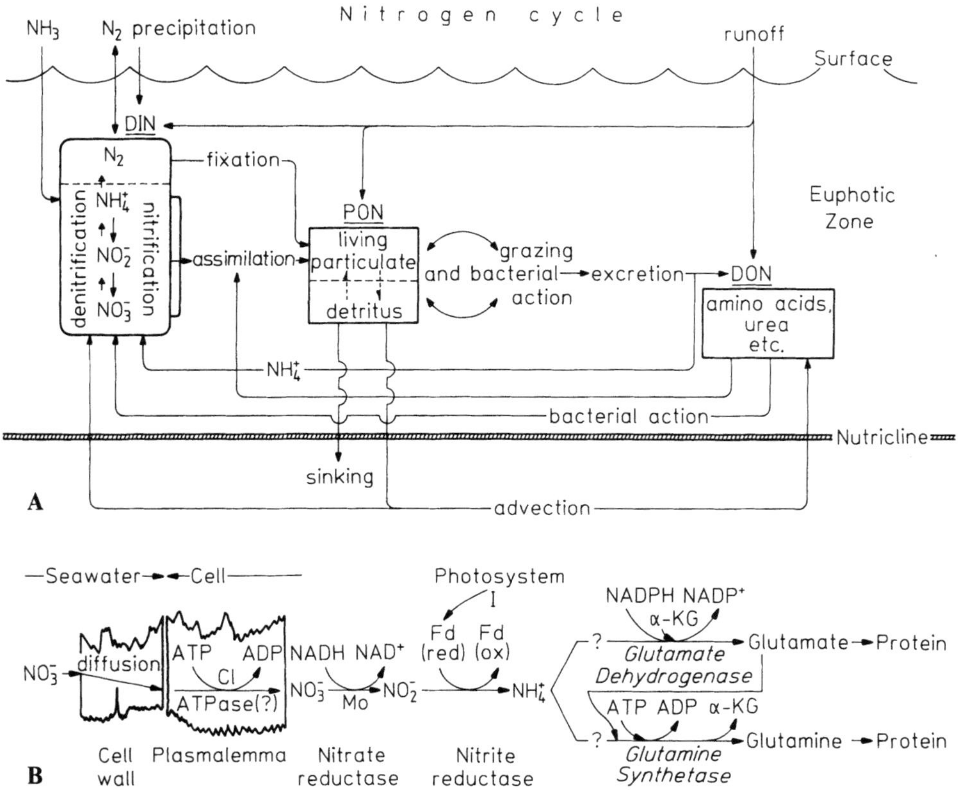

Fig. 3.1. A The nitrogen cycle in the euphotic zone of the sea as given by TURPIN (1980). *DIN* dissolved inorganic nitrogen; *PON* particulate organic nitrogen; *DON* dissolved organic nitrogen. **B** Pathway for nitrate uptake and assimilation in phytoplankton. *ATP* adenosine triphosphate; *ADP* adenosine diphosphate; NAD^+ and *NADH* nicotinamide adenine dinucleotide (oxidized and reduced forms); *Fd* ferredoxin; $NADP^+$ and *NADPH* nicotinamide adenine dinucleotide phosphate (oxidized and reduced forms); *α-KG* alpha ketoglutaric acid (PACKARD 1979)

Table 3.1. Typical nitrate, nitrite, ammonium, phosphate and silicate concentrations at various depths in the three major oceans

Ocean and depth (m)		Nutrient concentration (μg-at/l)				
		NO_3^-	NO_2^-	NH_4^+	PO_4^{3-}	SiO_4^{4-}
Atlantic	5	0–20	0–0.5	0–3	0.1–1	2–30
	4,000	~20	<0.5	~4	~2	~50
Indian	5	0–20	0–0.5	0–3	0.1–1	2–30
	4,000	~35	<0.5	~4	~3	~100
Pacific	5	0–20	0–0.5	0–3	0.1–1	2–30
	4,000	~40	<0.5	~4	~3	~150

Table 3.2. Nutrient budget of world's oceans. (EMERY et al. 1955)

	Millions of metric tons		
	Nitrogen	Phosphorus	Silicon
Reserve in ocean	920,000	120,000	4,000,000
Annual use by phytoplankton	9,600	1,300	Insignificant
Annual contribution by rivers	19	14	4,300
Dissolved	19	2	150
Suspended	0	12	4,150
Annual contribution by rain	59	0	0
Annual loss to sediments	9	13	3,800

nitrate from below the nutricline, atmospheric input of ammonia either in rain or through N_2 fixation by bacteria and blue-green algae or, in coastal areas, the terrigenous input of nitrogen either from land drainage or due to eutrophication from sewage and agricultural fertilizers. All these sources of nitrogen, including both inorganic and organic forms, can be utilized by many species of phytoplankton and also by bacteria if an organic substrate is present for heterotrophic production.

The loss of nitrogen from the food chain takes place when plants and animals die, or as a result of metabolic excretion by animals in the benthic, zooplankton or nekton communities. The decomposition of nitrogen products in the food chain may result in a net loss if nitrogen is lost from the euphotic zone through sedimentation, denitrification, or accumulation as nitrate in the deep ocean waters. Alternatively, many of the decomposed nitrogen products are recycled within the water column.

In the following discussion each of the above processes will be discussed in relation to the distribution of the nitrogen in the sea as summarized in the nitrogen budget (Table 3.2).

3.2.1 Physical Advection of Nutrients

Large ocean areas north and south of the equator are dominated by anticyclonic gyres in which the water circulation tends to suppress net vertical transport of deeper waters towards the surface. In such areas (e.g. the Sargasso Sea) the amount of nitrate-nitrogen found in the surface layers is generally below the limit of analytical detection (i.e. <0.1 µg-at l^{-1}) and these areas are generally considered to be nitrogen-limited. Supplies of nitrogen primarily come from recycling of ammonium and N_2 fixation and these are the determinant factors in phytoplankton growth. Such areas are marked by very slow primary productivities (<50 g C m^{-2} yr^{-1}) and constitute approximately half of the total ocean surface area (KOBLENTZ-MISHKE et al. 1970).

In regions of the world's oceans where there is a circulation pattern which results in a net vertical transport of deep water into the euphotic zone, primary productivity values are considerably higher than in anticyclonic gyres. Such areas are generally classified as regions of upwelling if they are fairly large (e.g. the Equatorial and Antarctic divergences, as well as the Peru, Benguela and Canary currents). The upwelling in these areas is primarily wind-driven and WALSH et al. (1977) have shown that phytoplankton growth responses are closely coupled to the relaxation of the winds. The nitrate which is brought into the euphotic zone by the winds is quickly used by the phytoplankton as the water column stratifies after wind relaxation. Smaller regions of turbulent upwelling are referred to as fronts and BOWMAN and ESAIAS (1977) describe several forms of the latter. These are boundary regions of current, shelf break fronts at the boundary of shelf and slope waters, localized upwelling (coastal), plume fronts where rivers discharge into the oceans and shallow sea fronts around islands, and banks or shoals where tidally mixed near shore water exchanges with deeper stratified offshore water. All these areas tend to transport nutrients into the euphotic zone (WALSH et al. 1978) and hence the productivity of such regions is high (i.e. >200 g C m^{-2} yr^{-1}) although in area they are less than 25% of the total ocean and therefore are less significant on a global basis. Between the two extremes of anticyclonic gyres and upwelling regions, the remainder of the world's oceans generally show seasonally high productivity due to the presence of nutrients in the spring, followed by depressed summer rates due to stratification of the water column when vertical nutrient transport is restricted until winter storms again mix the water mass (FINENKO 1978).

In summarizing many of these observations, DUGDALE and GOERING (1967) have suggested that the level of primary productivity in any area could be evaluated on the basis of the ratio, ρ_{NO_3}/ρ_{NH_4}, where ρ represents phytoplankton assimilation rates of nitrate and ammonium, respectively. This ratio gives an ecological measure of how much *new* primary production is generated by the advection of deep water containing high nitrate versus how much is *old* production from recycling processes. EPPLEY et al. (1979a) showed that in very stable, North Pacific Central Gyre waters, this ratio was ca. 0.1, indicating the importance of ammonium recycling, but that in coastal upwelling water off California it was closer to 0.8, indicating that more of the primary production was *new*. However, the ratio may be changed by other processes, such as sewage in coastal areas, or regenerative processes at the sediment–water interface. In the former case, EPPLEY et al. (1979b) found that 80% of the nitrogen for phytoplankton production may come from ammonium near a sewage outfall. In the latter case, DUGDALE and GOERING (1970) showed that 50% of the euphotic nitrogen supply off the coast of Peru came from ammonium-based regeneration at the sediment–water interface. Dissolved organic nitrogen compounds such as amino acids may also be advected into the euphotic zone and used for phytoplankton growth.

3.2.2 Nutrient Input from Terrestrial and Atmospheric Sources

Rivers associated with land drainage in which there is little or no pollution may contain much lower quantities of nutrients than are generally found in deep oceanic seawater. For example, NAIMAN and SIBERT (1978) report that the Nanaimo River in British Columbia contains ca. 3 µg-at NO_3–N l^{-1} while rivers draining from glacial runoff contain practically no nutrients (e.g. the Skeena River, B.C. – Water Quality data, B.C. 1961–1971). On the other hand there are certain rivers, such as the Nile (prior to the construction of the Aswan dam) which appear to be naturally rich in nutrients relative to the Mediterranean Sea, even prior to recent trends in the eutrophication of freshwater systems (HALIM and MORCOS 1965). The Amazon River, which supplies ca. 20% of the entire world-wide river runoff, carries less than half the load of major nutrients found in surface waters of the Atlantic (WILLIAMS 1968).

In regions of high population and intensive land use for agriculture, the concentration of nitrate, ammonium, phosphate and dissolved organic nutrients in river water may greatly exceed that in the surrounding sea (see Chap. 13, this Vol.). In 1965, for example, levels of over 100 µg-at NO_3–N l^{-1} were recorded on the Potomac River, Washington, D.C. (CARPENTER et al. 1969). KETCHUM (1969) showed a general relationship between the amount of chlorophyll a in seawater and the phosphate concentration; in this relationship, waters enriched by land drainage had chlorophyll a values of >10 µg l^{-1} and phosphate concentrations >2.8 µg-at l^{-1}. In the Hudson River estuary and the adjacent New York Bight, the quantity of nitrogen supplied by sewage greatly exceeds the ability of photosynthetic organisms to assimilate the dissolved nitrogen and concentrations of ammonium and nitrate in excess of 20 µg-at l^{-1} are found at many times of the year (GARSIDE et al. 1976). In contrast, off the coast of southern California where nitrogen is generally in short supply (THOMAS 1969), extensive eutrophication due largely to ammonium occurs in the vicinity of sewage outfalls (THOMAS et al. 1974) without any sustained increase in the concentration of ammonium.

Atmospheric sources of nitrogen include ammonia and dissolved nitrogen gas. In the latter example the gas dissolves in seawater where it is utilized as a nitrogen source in a process termed nitrogen fixation, which reduces the gas to a metabolic form, ammonium. Nitrogen fixation in the marine environment has been reported to occur as a result of bacterial activity (e.g. BROOKS et al. 1971), in attached blue-green algae (e.g. STEWART 1964) and by blue-green algae in the pelagic environment (DUGDALE et al. 1961). In particular the genus *Oscillatoria* spp. (*Trichodesmium*) has been shown to be widely distributed in tropical and subtropical pelagic environments (see SOURNIA 1970, and references cited therein). It has also been reported (e.g. MAGUE et al. 1974) that intracellular blue-green algae may be associated with senescent diatoms, particularly the genus *Rhizosolenia*.

The quantity of ammonia measured in the atmosphere away from continents appears to be about 0.06 µg m^{-3}, increasing by at least two orders of magnitude in coastal environments, particularly under conditions of atmospheric pollution (AYERS and GRAS 1980). Small amounts of nitrate may also be present in the atmosphere. MENZEL and SPAETH (1962b) determined the quantity of ammonium in rainwater that could be transferred to the ocean. From an average ammonium content of 4.97 µg-at l^{-1} in rainwater, it was shown that sea surface ammonium concentrations were closely correlated with the amount of rainfall in Sargasso Sea waters. A much less significant effect of atmospheric ammonia on the supply of nitrogen to surface waters would be expected in less oligotrophic waters.

3.2.3 Nutrient Uptake

From a historical point of view, the earliest nutrient kinetic experiments with marine phytoplankton showed firstly that at low nutrient concentrations uptake rate was proportional to concentration, and secondly, that the total yield of phytoplankton standing stock was proportional to the total amount of rate-limiting nutrient (KETCHUM 1939; SPENCER 1954). A formulation of nutrient uptake kinetics for phytoplankton was later (CAPERON 1967; DUGDALE 1967) given in terms similar to Michaelis-Menten kinetics for enzymes, viz:

$$V = \frac{V_{max} S}{K_s + S} \tag{3.1}$$

where V and V_{max} are the nutrient uptake and the maximum rate of uptake, respectively, measured in units of $time^{-1}$, and K_s, the saturation constant, is the substrate concentration (S) at which $V = V_{max}/2$. Under steady-state conditions of nutrient limitation, it may often be possible to show that the K_s for growth is the same as the K_s for nutrient uptake (EPPLEY and THOMAS 1969). Growth is also a function of limiting nutrient concentration and the relationship is described by a rectangular hyperbola as follows:

$$\mu = \mu_{max} \frac{S}{K_s + S} \tag{3.2}$$

where μ and μ_{max} are the growth rate and maximum growth rate, respectively, measured in units of $time^{-1}$. Although K_s has been traditionally regarded as an important indicator of advantage in nutrient competition, a change solely in V_{max} or μ_{max} produces an "apparent" change in K_s. To overcome this problem, it has been recently suggested that the ratio V_{max}/K_s, which is the slope of the uptake equation at low nutrient concentrations, offers a simple way of emphasizing both parameters (HEALEY 1980).

Departures from Eq. (3.2) are known to occur, particularly if it can be shown that nutrients are taken up without cell division occurring (e.g. KUENZLER and KETCHUM 1962; FITZGERALD 1968). This reflects a change in the cell quota (Q), or the nutrient concentration per cell, such that:

$$\frac{\text{Uptake rate}}{\text{cell division rate}} = Q. \tag{3.3}$$

EPPLEY and STRICKLAND (1968) suggested that phytoplankton requires a certain minimum content of nitrogen (or any other nutrient) before cell division can proceed; for *Dunaliella tertiolecta* the minimal nitrogen per cell (Q_0) was found to be around 1×10^{-15} mol $cell^{-1}$. The uptake of nitrate by the cell appears to follow Michaelis-Menten kinetics (CAPERON 1968) and the maximum growth rate can be related to the observed growth rate through the cell quota (DROOP 1970), as follows:

$$\mu = \mu_{max} \left(1 - \frac{Q_0}{Q}\right). \tag{3.4}$$

Physiological and morphological consequences of suboptimal concentrations of nutrients have been well reviewed (HEALEY 1973).

The processes involved in the use of nitrate by phytoplankton thus appear to involve two steps; the first step is the uptake of nitrate from outside the cell, and the second involves the utilization of nitrate within the cell (Fig. 3.1 B). These two processes involve different enzyme systems. The first involves the active uptake of NO_3^- from the water and the translocation across the cell membrane. FALKOWSKI (1975) showed that in the marine diatom, *Skeletonema costatum,* an enzyme (a NO_3^-, Cl^--activated ATPase) required ATP and had a K_m of 0.9×10^{-6} M which was close to the K_s generally found for nitrate uptake by the whole cell; however, no evidence was presented which demonstrated that this enzyme was specifically associated with the cell membrane or plasmalemma. Diffusion of nitrate across the cell membrane is generally discounted because it cannot maintain the intracellular nitrate levels (e.g. ~ 100 μM) required to insure the effective action of nitrate reductase (PACKARD 1979).

Within the cell, the enzyme, nitrate reductase, reduces nitrate to nitrite, a process which also requires energy in the form of a photosynthetic reductant. The synthesis of nitrate reductase is inhibited by the presence of ammonium at concentrations above 0.5 to 1 μM (EPPLEY et al. 1969a). The observed preferential uptake of nitrogen is inversely proportional to the degree to which the nitrogen is oxidized, with ammonium being taken up approximately twice as fast as nitrite or nitrate, depending on the nutrient past history of the cell (CONWAY 1977). In the presence of nitrite, only nitrite reductase is synthesized, while in the presence of nitrate the synthesis of both nitrate and nitrite reductase is necessary (LUI and ROELS 1972).

The high requirement for ATP in the nitrate assimilation process accounts for a number of other observations regarding nitrate uptake by phytoplankton in the marine environment. These include: the observed uptake of nitrate in the dark (since ATP can be made available by oxidative phosphorylation), temporary inhibition of carbon dioxide uptake, through competition for ATP, and increased photosynthetic pigment synthesis, together with enhanced longer-term carbon dioxide uptake through the supply of more energy and eventual protein synthesis from nitrate incorporation (for discussion see FALKOWSKI and STONE 1975).

Ammonium, like nitrate and nitrite, appears to be transported across the cell membrane by an active process, although no specific enzyme(s) has been associated with this process. After ammonium is taken up by the cell or is produced from nitrate reduction as described above, there are two possible pathways by which ammonium can be assimilated into amino acids (Fig. 3.1 B). The major route of ammonium assimilation has been considered for many years to occur via the reductive amination of α-ketoglutarate, catalyzed by the enzyme, glutamate dehydrogenase (GDH) (AHMED et al. 1977). However, recent work has shown the incorporation of ammonium into the amino acid pool occurs under most conditions via glutamine synthetase where ammonium combines with glutamate with the use of ATP (FALKOWSKI and RIVKIN 1976; TURPIN and HARRISON 1978).

Phytoplankton can also utilize organic nitrogen compounds such as urea as a nitrogen source and the K_s value for uptake appears to be similar to that found for ammonium (i.e. ca. 0.5–1.5 µg-at l^{-1}; McCarthy 1972). Several common amino acids have also been shown to be taken up by marine phytoplankton when supplied at near natural concentrations; lysine was taken up most readily (Wheeler et al. 1974). Uptake rates were higher in cells that had been deprived of nitrogen during growth, suggesting that these organic nitrogen compounds may become increasingly important during summer months when inorganic nitrogen concentration is low. Data from an 11-year study in the English Channel show that even though nitrate becomes exhausted in the summer months, there were always significant amounts (e.g. 6–10 µg-at l^{-1}) of dissolved organic nitrogen, suggesting that this form of nitrogen may be important for some species of phytoplankton (Butler et al. 1979).

Differences in K_s values for different substrates as well as for different species have been recorded by several workers. Eppley et al. (1969b) have summarized some of these observations into a number of generalizations. These are: (1) that larger species of phytoplankton tend to have higher K_s values; (2) that high K_s values for nitrate are often associated with high K_s values for ammonium; (3) that K_s values for oceanic phytoplankton are lower than K_s values for coastal phytoplankton; (4) that K_s values are temperature dependent. From Carpenter and Guillard (1971) another point which can be added is that (5) K_s values for different clones of the same species may vary appreciably. A summary of some values for K_s and V_{max} of phytoplankton [Eq. (3.1)] are given in Table 3.3. It should be noted that while these parameters are called "constants", in fact they do vary considerably, particularly with the nutrient past history of the organism (Perry 1976; Dugdale 1977).

The uptake of carbon, nitrogen and phosphorus by marine phytoplankton is generally found to be in the ratio 106:16:1 (Redfield 1934). In a recent discussion of this ratio, Goldman et al. (1979) have pointed out that this ratio is often maintained even in areas of apparent low phytoplankton productivity due to nutrient limitation. Their explanation for this is that under steady-state conditions, such as may persist over extended periods of time in both low and high nutrient regimes, the phytoplankton organisms are actually growing near their maximum growth rate and are being controlled by grazers. In this case under oligotrophic conditions, the herbivorous zooplankton must supply sufficient regenerated nitrogen and phosphorus to sustain the near maximal growth rate of the phytoplankton. However, if these oligotrophic phytoplankton are slow-growing species, as recently suggested by Jackson (1980), then regenerated nutrients may be adequate to balance nutrient uptake demands. Under conditions in which there is an imbalance of N/P in the water (such as due to sewage outfalls, faster regeneration rate of phosphorus than nitrogen, or under some conditions of laboratory cultures), the CNP cellular composition ratio may change by several hundred percent (e.g. Banse 1974a, b, and references cited therein).

Various strategies for increasing the uptake of nutrients under rate-limiting conditions are apparent among different members of the phytoplankton community. Thus the increase in the ratio of surface area to volume will encourage the presence of small phyto-

Table 3.3. Some representative values for K_s and V_{max} among marine phytoplankton

Species or region	Substrate	K_s (μg-at l^{-1})	V_{max} (h^{-1})	Reference
Oligotrophic	Nitrate	0.01–0.14	0.001–0.007	MACISAAC and DUGDALE (1969)
Tropical Pacific	Ammonium	0.01–0.62	0.006–0.009	MACISAAC and DUGDALE (1969)
Eutrophic, tropical Pacific	Nitrate	0.98	0.036	MACISAAC and DUGDALE (1969)
Eutrophic, Subarctic Pacific	Nitrate	4.21	0.016	MACISAAC and DUGDALE (1969)
Skeletonema costatum	Urea	1.41	0.0151	MCCARTHY (1972)
Ditylum brightwellii	Urea	0.41	0.0109	MCCARTHY (1972)
Thalassiosira fluviatilis (Actin)	Urea	1.66	0.0239	MCCARTHY (1972)
Thalassiosira fluviatilis (T. Flur.)	Urea	0.45	0.0303	MCCARTHY (1972)
Ditylum brightwellii	Nitrate	4	0.14	EPPLEY and COATSWORTH (1968)
Ditylum brightwellii	Nitrate	2	–	EPPLEY and COATSWORTH (1968)
Chaetoceros gracilis	Phosphate	0.1	0.13	THOMAS and DODSON (1968)
Thalassiosira pseudonana	Phosphate	0.6–0.7	–	PERRY (1976)
Thalassiosira pseudonana	Silicate	0.19–0.98	0.14	GUILLARD et al. (1973)
Skeletonema costatum	Silicate	0.5	–	GUILLARD et al. (1973)
Chaetoceros gracilis	Silicate	0.47	0.12	THOMAS and DODSON (1975)
Leptocylindrus	Silicate	0.33–0.74	0.1–0.15	THOMAS and DODSON (1975)
Ditylum brightwellii	Silicate	2.96	–	PAASCHE (1973)
Peru upwelling	Silicate	2.93	0.075	GOERING (1974)

plankton in oligotrophic waters since they are less prone to diffusion limitation (PASCIAK and GAVIS 1974). However, surface area can also be increased by the presence of cytoplasm-filled spines or any distortion of shape away from the spherical. Sinking of phytoplankton out of a region of low nutrients to deeper water where nutrients are more abundant is another mechanism facilitating increased nutrient uptake. Sharp changes in the water density near the pycnocline plus an increase in buoyancy resulting from nutrient uptake (e.g. STEELE and YENTSCH 1960) is believed to retard the cells from sinking out of the euphotic zone when they have sunk to the depth of the nutricline. Alternatively, for flagellates, there is an obvious ability to increase the flow of the water over the cell surface or to migrate vertically to the nutricline in order to increase nutrient uptake (HARRISON 1973). Physiological adaptations through unknown mechanisms to low nutrient concentrations leads to an enhanced uptake ability (an increase in V_{max}) when high nutrient concentrations are encountered (CONWAY and HARRISON 1977; MCCARTHY and GOLDMAN 1979). There is some suggestion that diatoms may have some advantage in this regard.

3.2.4 Nutrient Regeneration in the Water Column

Regeneration of nitrogen in the water column occurs as a result of two largely separate processes, the one involves bacteria and the other results from the excrement of marine fauna, particularly the excretion of ammonium by the very large zooplankton community. The regeneration process is reversible and may be represented as follows:

$$\text{Organic N} \leftrightarrow NH_4^+ \leftrightarrow NO_2^- \leftrightarrow NO_3^-.$$

The oxidation of ammonium to nitrate, generally referred to as nitrification, was demonstrated to occur in classical experiments by RAKESTRAW and VON BRAND (1947, and references cited therein). These reactions are exothermic and only require activation to make them proceed. The process appears to be performed in part by bacteria (CARLUCCI and STRICKLAND 1968) which may be free-floating or attached to organic and inorganic suspended matter. The bacteria satisfy their energy requirements through oxidation of organic matter to carbon dioxide according to the following generalized stoichiometric relationship (REDFIELD et al. 1963):

$$(CH_2O)_{106}(NH_3)_{16}\,H_3PO_4 + 138\,O_2 \rightarrow$$
$$106\,CO_2 + 122\,H_2O + 16\,HNO_3 + H_3PO_4.$$

The presence of a small nitrite maximum in stratified water below the euphotic zone (50–125 m) is generally taken as in situ evidence of the first step in the oxidation of ammonium as given above (e.g. VACCARO and RYTHER 1960). Recent laboratory experiments indicate that the nitrite maximum may also be formed by phytoplankton excretion (LAWS and WONG 1978; SERRA et al. 1978). When nitrogen-limited phytoplankton were given nitrate in the dark, it was taken up and nitrite was excreted (CARLUCCI et al. 1970). In some low-oxygen waters, such as off Peru, and in the Benguela current (CARLUCCI and STRICKLAND 1968; CALVERT and PRICE 1971), nitrate reduction to nitrite by bacteria may give rise to a secondary, deep nitrite maximum (HATTORI and WADA 1971).

The breakdown of organic nitrogen and the release of ammonium (generally referred to as remineralization) has been suggested by a number of authors as a mechanism sustaining phytoplankton production in the euphotic zone, in the absence of any appreciable new nitrate-nitrogen from deeper water (WILLIAMS et al. 1976; HOLLIBAUGH 1978). HARRISON (1978) found a strong correlation between remineralization and nitrogen assimilation by phytoplankton. In his experiments, remineralization was largely accounted for by the microzooplankton (<183 µm) and the turnover time was from 3 to 16 days. Since the amount of nitrogen turned over per day was usually of the order 0.1 µg-at l^{-1},

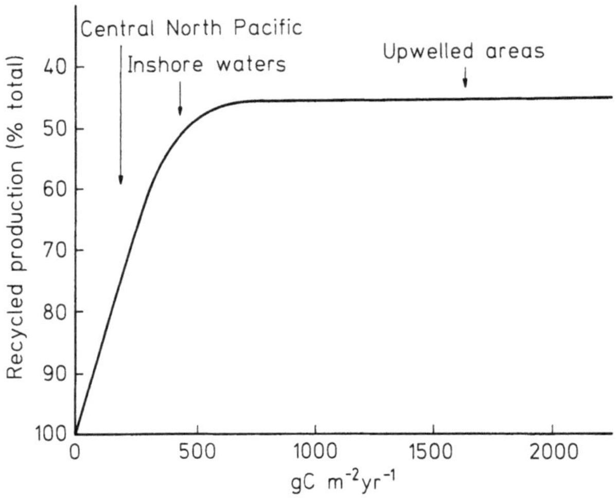

Fig. 3.2. Percent of production coming from recycled nutrients in different waters. (Adapted from EPPLEY and PETERSON 1979)

the total seasonal turnover for a period of 200 days would be equivalent to a nitrogen supply of 20 µg-at l^{-1}. By using an entirely different approach based on phytoplankton assimilation, a similar value was obtained by PARSONS et al. (1977). In the latter experiments the nitrogen turnover of the water column through remineralization amounted to ca. 1.52 µg-at m^{-2} day^{-1} which was equivalent in toto to an approximate doubling of the nitrogen supply in the water column compared with winter levels. EPPLEY and PETERSON (1979) have related the proportion of recycled nitrogen to new nitrogen for environments of different productivities (Fig. 3.2). From these results it may be shown that in highly oligotrophic environments (e.g. primary productivity ca. 25 g C m^{-2} yr^{-1}) nitrogen may be recycled ca. 20 times before sinking out of the euphotic zone. However, for coastal and upwelled areas, nitrogen may be recycled approximately once per year.

The rate at which zooplankton remineralize phytoplankton organic nitrogen back to ammonium appears to be related to the quantity of phytoplankton consumed. BUTLER et al. (1970) found 10% of zooplankton body nitrogen to be excreted per day during the spring bloom but only 2% during the winter. TAKAHASHI and IKEDA (1975) found an almost ten fold increase in the excretion rate of ammonium from *Euphausia pacifica* when the phytoplankton cell density was increased from zero to 70 µg chl a l^{-1}. Zooplankton have also been shown to release dissolved free amino acids during grazing (WEBB and JOHANNES 1967). CORNER and DAVIES (1971) have summarized nitrogen excretion rates of different zooplankton; the rate for many animals and diets fell in the approximate range 1–10 µg N mg^{-1} body wt d^{-1}.

Loss of nitrogen by denitrification of nitrate to gaseous nitrogen can occur under anoxic conditions in the marine environment (GOERING and DUGDALE 1966). From detailed chromatographic studies using the marine bacterium *Pseudomonas perfectomarinus*, BARBAREE and PAYNE (1967) showed that intermediary products were nitric and nitrous oxides, giving rise to the following sequence:

$$NO_3 \rightarrow NO_2 \rightarrow NO \rightarrow N_2O \rightarrow N_2.$$

Since N_2 is unavailable to phytoplankton other than nitrogen fixers, the occurrence of the reaction given above must represent an appreciable nitrogen loss occurring in anoxic sediments (GOERING and PAMATMAT 1971).

3.2.5 Nutrient Regeneration from Sediments

Nitrogen-containing particles, produced primarily in the euphotic zone, reach the sediments by sinking. Sinking rates vary with particle size and whether the particle is living or dead. Living phytoplankton cells sink as slowly as 0.2 to $2\ m\ d^{-1}$ (SMAYDA 1970) while larger particles such as faecal pellets from zooplankton or euphausiids sink at rates up to 50 to $200\ m\ d^{-1}$ (SMAYDA 1969).

Inorganic nitrogen in interstitial waters is generally in the form of ammonium due to the reducing environment of anoxic sediments. In these sediments anaerobic bacteria must obtain oxygen from nitrate and sulphate compounds to oxidize organic matter, yielding large amounts of sulphides, nitrogen gas and ammonia according to the following generalized equation for nitrate reduction (REDFIELD et al. 1963):

$$(CH_2O)_{106}(NH_3)_{16}\,H_3PO_4 + 84.8\,HNO_3 \rightarrow$$
$$106\,CO_2 + 42.4\,N_2 + 148.4\,H_2O + 16\,NH_3 + H_3PO_4 .$$

Further, the total concentration of all dissolved nutrients in sediment waters is much higher than in the overlying seawater. MARTENS et al. (1978) showed that in Long Island Sound sediments, ammonium concentrations in 100 cm cores may reach 10 mM and reactive phosphate, >1.0 mM. If the circulation of seawater is such as to entrain some fraction of the interstitial sediment water into the surface layers, then nutrient enrichment of the water column results from these regenerated nutrients. SMETACEK et al. (1976) have demonstrated this point in association with high salinity seawater intrusions in the Kiel Bight. The authors showed that an intrusion of high salinity water could displace low salinity interstitial water and that the resulting input of ammonium, phosphate and silicate to the water column was approximately ten times the concentration prior to the intrusion.

In Narragansett Bay it is estimated that up to 80% of the nutrients entering the bay at certain times of the year come from the sediments. The transport of nutrients into the water column was reported to occur as a result of two mechanisms which were approximately equal in magnitude. The one resulted from the activity of burrowing benthic organisms and the other was by diffusion (NIXON et al. 1976; MCCAFFNEY et al. 1980). In another study, HARTWIG (1976) determined the nitrogen and phosphorus released from a subtidal (18.3 m depth) siliceous sediment off La Jolla, California, as $829\ \mu mol\ m^{-2}\ d^{-1}$ mainly as ammonium and $77\ \mu mol\ m^{-2}\ d^{-1}$ as phosphate, respectively. From current velocity studies this release was believed to be transported out of the area before it could be utilized by the endemic phytoplankton population. In upwelled areas it appears that nutrients entrained after being regenerated in the sediments are much more important in the total photosynthesis of surface phytoplankton. Off the Spanish Sahara, ROWE et al. (1977) reported that the elements Si/N/P were regenerated in the ratio 11:8:1 and that the nitrogen flux out of the sediments was 410 μmol m^{-2} h^{-1} which accounted for 30% to 40% of the nitrogen presumed to be required for photosynthesis in the same area. Similar results were found by DUGDALE and GOERING (1970) for nitrogen off the coast of Peru.

3.3 The Phosphorus Cycle

Since phosphate is widely used in many essential life processes, including energy transfer, the element occupies an important position in the productivity of the sea. However, unlike nitrogen, phosphorus cycles rapidly through the food chain and consequently it is seldom limiting in the marine environment. The phosphorus cycle is simpler than nitrogen, primarily due to the fact that dissolved inorganic phosphorus exists in one form, orthophosphate, and that phosphorus is used in the cell in the form of unreduced phosphate; hence, there are no processes comparable to nitrification, and denitrification which are found in the nitrogen cycle.

The seasonal cycle of phosphorus compounds in temperate latitudes is shown in Fig. 3.3 and representative concentration values are reported in Table 3.1. The cycle is similar to that for other elements (e.g. nitrogen) in that much of the element is tied up as soluble organic and particulate form during the summer months, while inorganic orthophosphate is high during the winter. It should be noted that it is difficult to measure the orthophosphate concentration accurately because of the interference from labile esters of the dissolved organic phosphate fraction which may be hydrolyzed by the acid conditions used in the analysis.

Phosphorus is advected into the euphotic zone largely as orthophosphate, primarily through wind-induced mixing or upwelling as described for nitrogen. Contributions by rivers are important to the phosphorus budget (PIERROU 1976) with the largest portion coming into the ocean as particulate rather than dissolved phosphorus (Table 3.2). Unlike nitrogen, little phosphorus appears to be added to the oceans from rainfall, although recent estimates indicate that this may be more (e.g. 20 μg l^{-1}) than previously thought (PETERS 1977).

Phosphate, like nitrate, is taken up in the sea by both autotrophic and heterotrophic marine microorganisms. In a comparison of the relative proportions of bacterial uptake versus uptake by phytoplankton, HARRISON et al. (1977) found that in coastal waters (where organic substrates for heterotrophic production may be high), at least 50% of the inorganic phosphorus uptake could be attributed to bacteria.

Short-term phosphate uptake experiments generally show Michaelis-Menten kinetics, similar to those given for nitrate (e.g. PERRY 1976; RESHKIN and KNAUER 1979). Representative values of the uptake parameters V_{max} and K_s are given in Table 3.3. As in the case of nitrate, uptake may also be governed

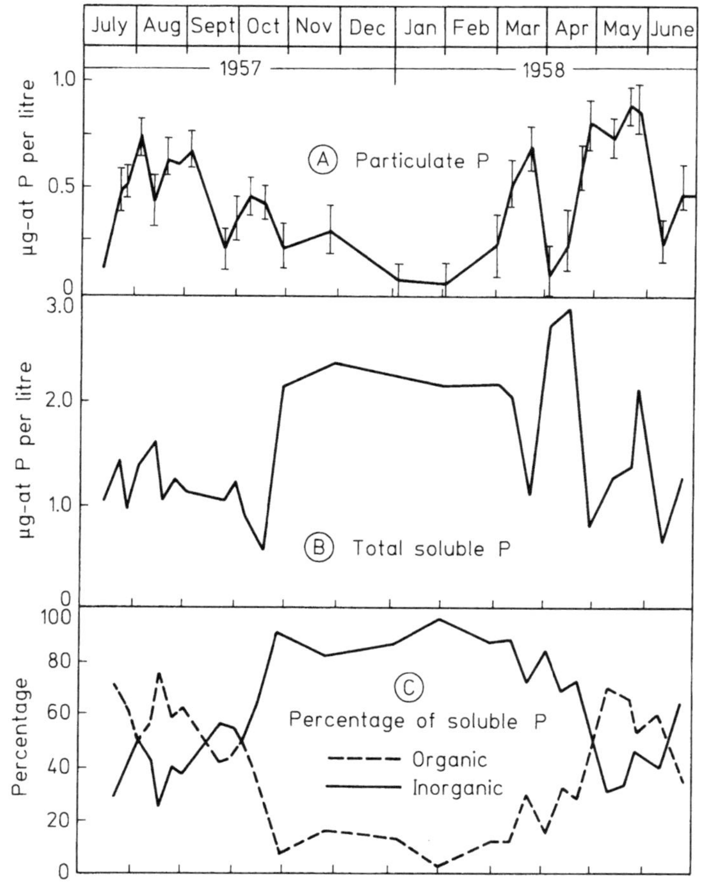

Fig. 3.3. Annual cycle of phosphorus at 5 m in a temperate coastal environment. (From STRICKLAND and AUSTIN 1960)

by the cell quota (e.g. SENFT 1978). GOLDMAN (1979) has indicated, however, that the cell quota concept may be difficult to apply as described previously for nitrate [Eq. (3.4)], since the minimum cell quota appears to be temperature-dependent and at low temperatures (suboptimum for the species) the minimum cell quota was found to double.

Phytoplankton are generally unable to directly take up even simple, dissolved phosphate esters. Instead they obtain the phosphorus from these compounds by producing a membrane-bound or extracellular enzyme, alkaline phosphatase. This enzyme cleaves off the phosphate group on the compound and then the free phosphate ion can be taken in by the regular phosphate transport system. Alkaline phosphatase may be induced in some phytoplankton at orthophosphate concentrations <0.1 μg-at l^{-1}, particularly when the organic phosphate concentration is appreciably higher than the inorganic. This enzyme has been detected in oligotrophic waters such as the Sargasso Sea and the central North Pacific gyre (KUENZLER and PERRAS 1965; PERRY 1976).

Phosphorus is recycled in the sea between inorganic phosphate dissolved in seawater and organic phosphate produced as a result of biological activity of marine organisms. Even during the spring phytoplankton bloom in the North Sea, the phosphate concentration did not decrease below 0.6 μg-at l^{-1} (CUSHING 1964). This was in part due to the recycling of phosphorus by zooplankton since in the absence of zooplankton, ANTIA et al. (1963) showed that phosphate could decrease to 0.1 μg-at l^{-1} during a phytoplankton bloom in similar latitudes. The excretion of phosphorus by zooplankton has been studied by a number of authors (e.g. POMEROY et al. 1963; BUTLER et al. 1970). In the last reference, the authors report that the grazing activity of *Calanus* effectively returned more than 80% of the phytoplankton phosphorus to the environment, either as particulate organic phosphorus in faecal material (ca. 23%) or as soluble organic phosphorus from the animal's metabolism (ca. 60%). Some of the organic phosphorus released into seawater from herbivorous grazers is readily hydrolyzed back to inorganic phosphate at the alkaline pH of seawater. Other organic phosphorus compounds are attacked by free enzymes in the water, phosphatases, which cleave off the phosphate ion. These phosphatases are produced by the phytoplankton or by zooplankton, as recently shown in lake studies (JANSSON 1976). Small amounts of dissolved organic phosphorus can be excreted by phytoplankton but it is generally minor compared to zooplankton production. In fact, some of the high estimates of excretion and reutilization (KUENZLER 1970) may be due to fibril production by some freshwater phytoplankton species growing under phosphorus limitation; the fibrils appear to be broken off during filtration, giving rise to "apparent" excretion and some are caught by the filter during refiltration to measure uptake by the cells (LEAN 1976).

POSTMA (1971) has calculated that advection transports about half the amount of phosphorus that is required to maintain the estimated global primary productivity. This suggests that, on average, an atom of phosphorus is cycled twice through a cycle of assimilation and regeneration before it is lost to deep water and sediments. However, in special situations and in the tropics the number of cycles could be much greater.

Particulate organic phosphorus which sinks to the sediment can be recycled (e.g. HARTWIG 1976) but may also be permanently retained in the sediments as a variety of inorganic minerals. The latter include phosphorites (e.g. VEEH et al. 1973), vivianite (e.g. BRAY et al. 1973), struvite (e.g. MALONE and TOWE 1970) and apatite (e.g. RILEY and CHESTER 1971).

3.4 The Silicon Cycle

Silicon uptake by phytoplankton is an important factor in the growth of diatoms, although a few other classes of phytoplankton (e.g. silicoflagellates) and protozoa (e.g. radiolarians) also require silicon. Silicon is present in seawater generally either as soluble inorganic silicate [silicic acid, $Si(OH)_4$] or particulate silica; the latter occurs as a hydrated form of silica (opal) in diatom frustules or in coastal areas, as various forms of mineral silicates. Silicon can exist as polymers of silicic acid, but it is the monomeric form that is thought to be biologically available. Very little soluble polymerized silicate exists in seawater, although PAASCHE (1973) has reported a non-biologically available form of inorganic silicon which generally amounted to <1 µg-at Si l^{-1}.

From studies on the silicon balance in the sea (CALVERT 1968; BURTON and LISS 1968; HURD 1977) it appears that approximately 4×10^{14} g SiO_2 yr^{-1} is contributed by rivers entering the ocean while $<0.1 \times 10^{14}$ g SiO_2 yr^{-1} may be contributed by submarine weathering and vulcanism. Approximately 1.9–3.6×10^{14} g SiO_2 yr^{-1} can be accounted for in terms of loss from the oceans as diatomaceous ooze. If these figures are correct, they leave a silica imbalance of around 0.5–2×10^{14} g SiO_2 yr^{-1} which BURTON and LISS (1968) suggest is removed by non-biological processes including adsorption of particulate matter and formation of aluminosilicates.

Sources of silicon are from advection of deep water into the surface waters by wind events, and dissolved and particulate silicon in river outflow; the latter is particularly important (Table 3.2). There is little atmospheric contribution to the ocean from rain or dust. This information has been summarized in a mathematical model (GRILL 1970).

The silicon cycle is the simplest of the three nutrient cycles considered here. It has only one inorganic form, silicic acid or silicate and no known organic forms of biological importance. Hence the main cycling pathway is from dissolved inorganic to particulate and back to the inorganic form.

The uptake of silicate by phytoplankton under steady-state conditions generally follows Michaelis-Menten kinetics (e.g. PAASCHE 1973). However, some anomalies in the uptake and growth responses have been observed (HARRISON et al. 1976; CONWAY et al. 1976). According to DAVIS et al. (1978), steps in the uptake of silicate are governed by the external substrate concentrations, the internal pool of dissolved silicate and the amount of silica laid down in the frustule. Since growth rate was related to frustule formation, which was not always directly related to the amount of dissolved silicate in the medium, it is apparent that a single equation describing silicate-dependent growth might not be an appropriate description of this process during transient conditions. Representative values of the uptake parameters are given in Table 3.3.

Considering the importance of the diatom ecology in the sea in maintaining high energy food chains (e.g. PARSONS 1979), it may be a critical factor in sustaining the high yield of fish in areas such as the Northwest African upwelling region. However, NELSON and GOERING (1978) found that silicate assimilation by phytoplankton extended to about twice the depth to which carbon and nitrogen assimilation was observed. Further, these authors found off the coast of California that upwelling and remineralization of silicate were sufficient to meet the demand by phytoplankton and that consequently production was not limited by the concentration of silicate. Low silicate concentrations (<2 µg-at l^{-1}) have been reported in large areas of the eastern tropical Pacific Ocean, covering areas to the north and south of the equatorial upwelling zone and away from the Peru Current and Costa Rica Dome. THOMAS and DODSON (1975) reported that while silicate may limit some diatom species in this area, it is probably secondary to nitrogen in controlling productivity.

The rate of silicate remineralization is slow compared with nitrogen and phosphorus (DUGDALE 1972). For this reason, it is possible that while nitrogen is the limiting nutrient during the spring bloom, silicate may become limiting late in the summer in estuarine areas, due to the rapid sedimentation of particulate silica via diatom frustules packaged together in relatively large faecal pellets.

In experiments involving the decay of a diatom crop, GRILL and RICHARDS (1964) found inorganic ammonium increased 30 times, phosphate 8 times and silicate 5 times over a 40-day period in the dark. The frustules of dead cells are known to dissolve (LEWIN 1961) and even the frustules of living cells can show measurable dissolution (NELSON et al. 1976). However, unlike nitrogen and phosphorus remineralization, silicate regeneration is not greatly enhanced by herbivorous grazers. WHITLEDGE and PACKARD (1971) found that silicate was excreted by herbivorous anchovies feeding on diatoms at ca. 15% and 20% the rate at which inorganic nitrogen and phosphorus were excreted, respectively.

3.5 The Transfer of Trace Elements and Compounds in the Nutrient Cycle of the Sea

While nitrogen, phosphorus and silicon occupy a central position in both regulating the rate and total standing stock of primary producers, other elements and compounds are known to influence phytoplankton growth, particularly with respect to species diversity and succession. In general, minor metals required by phytoplankton include Fe, Mn, Cu, Zn, Bo, Mo and Co. Organics may include such vitamins as biotin, thiamine and B_{12}, as well as chelators, which either make some metals more available (e.g. iron) or may protect the alga from high concentrations of heavy metals (e.g. copper) by forming a complex with the metal and making it non-toxic.

The speciation of *iron* in seawater is complex (BYRNE and KESTER 1976). At the pH of seawater the ferric ion combines with hydroxyl ions forming ferric hydroxide which is relatively insoluble ($K_{sp} = 10^{-38}$ M). Therefore much of the iron is maintained in solution only by the formation of a complex with natural chelators or ligands such as humic acid. Recent measurements in the western North Pacific indicate that dissolved iron concentrations are 2.2 to 3.8 $\mu g\ l^{-1}$, of which 80% to 90% is organic iron (SUGIMURA et al. 1978). The chelators not only serve to increase the solubility of iron, but also make the iron biologically available. It is not clear whether the chelated iron is taken across the cell membrane or is exchanged at the cell surface. There is recent evidence that these natural chelators are produced by marine microorganisms (GONYE and CARPENTER 1974; ESTEP et al. 1975). Specific iron-binding compounds called siderophores or siderochromes have been chemically identified from freshwater blue-green algae (SIMPSON and NEILANDS 1976) and their ecological role elucidated (MURPHY et al. 1976), but so far no specific compounds have been examined in the marine environment.

Laboratory studies with iron and its possible limiting concentrations are difficult to conduct because of the adsorption of iron onto the glass walls of the container and the rapid conversion (30–40 h) of soluble iron to particulate (LEWIN and CHEN 1973).

PAASCHE (1977) found that the adsorption of iron onto glass could be markedly decreased by treating the glass with a silicone coating. When iron in the culture medium adsorbs onto glass, the growth rate of phytoplankton culture is decreased; culture growth can be stimulated by adding more iron or especially an artificial chelator, EDTA (BARBER et al. 1971; BARBER 1973). However, caution must be exercised in interpreting such experiments as evidence of iron limitation due to the container adsorption phenomenon.

The availability of iron for phytoplankton growth may be partly a function of its absolute concentration (e.g. MENZEL and RYTHER 1961; TRANTER and NEWELL 1963) but is probably more often a function of its availability at the alkaline pH of seawater. In regeneration experiments it was found that iron was not readily released from decaying phytoplankton but that it was readily recycled in a form suitable for uptake by phytoplankton following digestion by zooplankton (FREY and SMALL 1979). While it is difficult to show that iron limits productivity, evidence indicates that it can influence species succession. For example, off the coast of Maine, increased iron concentrations from land runoff preceded a dinoflagellate bloom (GLOVER 1978).

Other metals such as *copper* and *zinc* are unlike iron because they are biologically available when they exist in the simple ionic form (SUNDA and GUILLARD 1976; CROSS and SUNDA 1977). For these metals the primary biological role of a chelator is to detoxify the metal by complexing with it (RICE et al. 1973). It was found that the diatom, *Thalassiosira pseudonana* (ERICKSON 1972), and a dinoflagellate, *Gonyaulax tamarensis* (ANDERSON and MOREL 1978), were inhibited by normal concentrations of copper in seawater, but if the copper was chelated, these organisms thrived.

Natural copper concentrations range from 1–4 μM (SCHMIDT 1978; BOYLE 1979); however, it is not the total copper concentration that is important biologically, but the concentration of the free ion. There is no suitable method of determining the concentration of this free ion and therefore bioassays are used to determine "biologically available" copper which is equated with the free ion concentration (SUNDA and GUILLARD 1976; WHITFIELD and LEWIS 1976).

Trace amounts of three vitamins, *vitamin* B_{12}, *thiamine* and *biotin*, are required, either singly or in combination, for the auxotrophic nutrition of many phytoplankton (PROVASOLI and CARLUCCI 1974). In some cases, various precursors of vitamin B_{12} or thiamine may serve the needs of certain dinoflagellates and other phytoplankton and the intact vitamin is not in this sense an absolute requirement (DROOP et al. 1959; PROVASOLI and CARLUCCI 1974). They are present in the sea in trace amounts with thiamine being the most abundant (ca. 8 to 15 $ng\,l^{-1}$) and B_{12} the least (0.1 to 3 $ng\,l^{-1}$). They do not appear to limit primary productivity per se, but in the Sargasso Sea, a diatom-coccolithophorid succession sequence was attributed to low levels of B_{12} at the end of the diatom bloom and the replacement by coccolithophorids which require primarily thiamine (MENZEL and SPAETH 1962a). This succession may have consequences further up the food chain as the composition of the diet of herbivores changes. For example, many dinoflagellates require B_{12} (PROVASOLI and CARLUCCI 1974) and these organisms may constitute most of the phytoplankton for herbivorous grazers at some times of the year (BRUNO and STAKER 1978). These vitamins are produced largely by bacteria (BURKHOLDER and BURKHOLDER 1956) and phytoplankton (CARLUCCI and BOWES 1970).

Further modifications in the nutrient cycle of the sea can be brought about by pollutants as well as some poorly investigated natural phenomena, such as allelopathy, heterotrophic and phagocytic feeding. FISHER (1977) has suggested that diatom growth is suppressed in the presence of many pollutants giving rise instead to dominance by flagellates. This may be true in some cases but it should not be regarded as the only cause of flagellate-dominated communities (EPPLEY and WEILER 1979). Allelopathy has been demonstrated in the laboratory (SHARP et al. 1979) among two competing species of diatoms, but the extent of its ecological importance is yet to be established. Heterotrophic growth of phytoplankton involving the cycling of soluble organics in seawater has been demonstrated occasionally (e.g. KUENZLER 1965), but this appears to be much more the role of marine bacteria which are capable of removing soluble organic materials from seawater at very low concentrations. Phagocytic feeding by phytoplankton has been known to occur with the dinoflagellate, *Noctiluca,* and has been more recently reported to occur among communities of microflagellates feeding on bacteria (HAAS and WEBB 1979).

3.6 The Role of Macrophytes in Nutrient Cycles

On a global basis, the production of macrophytes is only a very small fraction of the total production by phytoplankton. However, in some coastal regions, particularly those having suitable substrates and a constant supply of nutrients, seaweed production may greatly exceed phytoplankton production, yielding 1,000–2,000 g C m^{-2} yr^{-1} (e.g. MANN 1973). Two groups of seaweeds which tend to dominate in yielding this high production are the kelps (Laminariales) and rock weeds (Fucales).

Nutrient requirements by seaweeds are generally similar to those of phytoplankton, and nitrogen is again cited as the most frequent growth-limiting nutrient (e.g. CHAPMAN and CRAIGIE 1977). However, in a more detailed study of the nutrient requirements of *Fucus,* MCLACHLAN (1977) showed that growth in the laboratory could be enhanced by additions of nitrate, phosphate, bromide, iron and boron (but not silicate, as is commonly required for diatoms). Seaweeds appear to be able to use various forms of nitrogen; *Fucus* was shown to utilize ammonium, nitrite and nitrate (TOPINKA 1978) while some red algae showed a preference for ammonium over nitrate (D'ELIA and DEBOER 1978). The Michaelis-Menten uptake constant (K_s) for inorganic nitrogen in the latter reference was in the range 1–4 μM and even higher values (8–13 μM) have been reported for the large brown alga or kelp, *Macrocystis* (HAINES and WHEELER 1978). These values are higher than those reported for phytoplankton in Table 3.3. The requirement for phosphate among coralline algae appears to be unique in that high phosphate concentrations inhibit growth while low concentrations enhance growth. This is explained by the fact that high concentrations of phosphate prevent calcification of the coral exoskeleton (BROWN et al. 1977).

There are a number of examples of heterotrophic uptake of organics by seaweeds, including photoassimilation (e.g. GEMMILL and GALLOWAY 1974). However, as in the case of phytoplankton, it is unlikely that seaweeds can outcompete bacteria for organic substrates in the sea. In general, the flow of organic matter from seaweeds, both as particulate fragments and soluble organic compounds, is towards a detrital food chain which results in recycling of nutrients and a net transport of nutrients and organic material out of the seaweed bed (MANN 1972).

In conclusion, the role of seaweeds is largely competitive with that of phytoplankton in the sea. Under equal conditions phytoplankton can outcompete seaweeds by their greater growth rates, lack of a requirement for a physical substrate (with the exception of the pelagic macrophyte, *Sargassum*) and possibly their better ability to take up nutrients at low concentrations. However, in certain areas along coastlines, including areas of considerable turbulence where phytoplankton growth is inhibited or in intertidal areas where seaweeds can take advantage of nutrient concentrations released by sediments, it is apparent that the nutrient cycle of the sea will be dominated by macrophyte production.

3.7 Conclusions: Biological Consequences of Nutrient Ratios and Flux – Phytoplankton Species Succession and Coexistence

HUTCHINSON (1961) suggested that the simultaneous existence of many species of phytoplankton in an apparently uniform body of water presented a "paradox" because the continued coexistence of closely competing species of phytoplankton appeared to violate the principle of competitive exclusion. One explanation for this has been referred to as "temporal succession" in which it was supposed that the nutrient regime of aquatic habitats was changing too rapidly for equilibrium to be established (i.e. all species were in different states of reaching or declining from their maximum abundance, depending on the exact conditions at any moment of time). Another suggestion was made by RICHERSON et al. (1970) in which it was supposed that microhabitats develop within relatively small volumes of water, providing turbulence is low. Each microhabitat then favours the growth of a distinct patch of phytoplankton. This theory is known as "contemporaneous disequilibrium" because of the temporary division of an otherwise uniform body of water into microhabitats separated in space. In discussing both of the above concepts, PETERSON (1975) advanced a third hypothesis in which it is supposed that an assemblage of co-existing phytoplankton may be limited by the availability of several nutrients, with two or more species co-existing, providing each is limited by a different nutrient. Good evidence for this concept has been provided (TITMAN 1976; TILMAN 1977) and the following discussion is taken from the two references cited.

At some point where a species is being equally limited by two nutrients, it follows that the growth rate will be equally limited by both nutrients when:

$$\mu_1 = \mu_2 \tag{3.5}$$

or from Eq. (3.2)

$$\frac{[S_1]}{[S_2]} = \frac{K_1}{K_2} \tag{3.6}$$

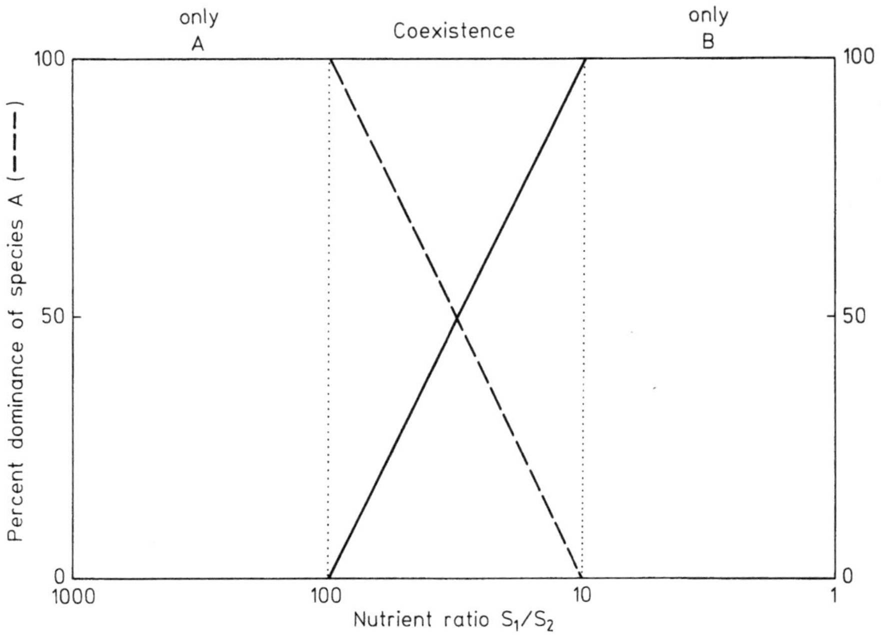

Fig. 3.4. Relative abundance of two phytoplankton species, *A* and *B*, plotted against the log of the ratio of limiting nutrients, S_1 and S_2. (Modified from TILMAN 1977)

where S_1 and S_2 are the steady-state concentrations of resources 1 and 2 and K_1 and K_2 are the half-saturation constants for growth limited by resources 1 and 2, respectively. It follows that if $[S_1]/[S_2] < K_1/K_2$, the species is limited by substrate 1. If two species are now considered, such that K_1 of species A is less than K_1 of species B, but that the K_2 of species A is greater than the K_2 of species B, then if $[S_1]$ is limiting, species A will dominate, while if $[S_2]$ is limiting, species B will dominate. Under these conditions, the nutrient ratio $[S_1]/[S_2]$ will determine which species dominates, while the difference in the K_1/K_2 ratio for each species will determine the range over which each species can co-exist in the same water mass. Thus, if K_1/K_2 for species A is 100 and for species B the ratio is 10, then coexistence of the two species, each limited by different substrates, will occur between $[S_1]/[S_2]$ ratios of 100 and 10. This is illustrated in Fig. 3.4. The model suggested here was confirmed in practice by TILMAN (1977), using two species of lake diatoms under conditions of silicate and phosphate growth limitation.

In nutrient concentration-response experiments with single species, EPPLEY et al. (1969b) suggested that the change in a single nutrient concentration over time could lead to a shift in phytoplankton species dominance, as indicated in Fig. 3.5. In an extension of this concept, the importance of specific nutrient flux (a rate measurement as opposed to nutrient concentration) in determining the outcome of competition experiments has been examined using outdoor continuous cultures containing natural phytoplankton populations (HARRISON and DAVIS 1979; TURPIN and HARRISON 1979). The results of long-term competition

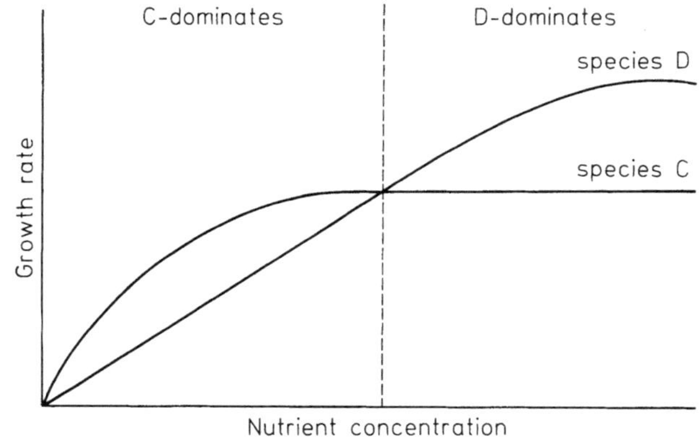

Fig. 3.5. Species domination based on differences in K_s and μ_{max} of two phytoplankton species, *C* and *D*. (Modified from EPPLEY et al. 1969)

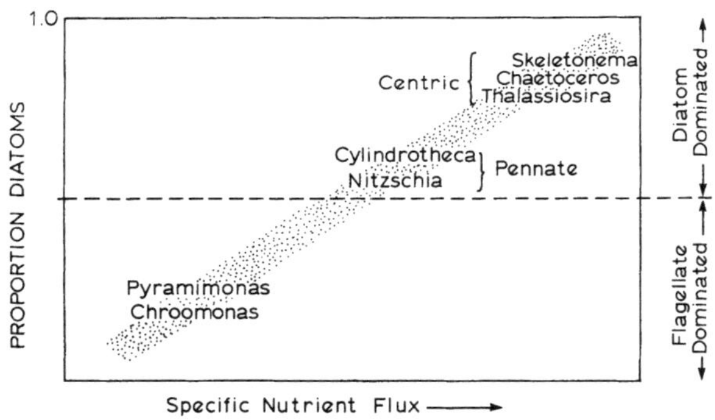

Fig. 3.6. A schematic representation of the possible relationship between specific nutrient flux (nitrogen) and community structure. (TURPIN and HARRISON 1979)

experiments run at different specific nutrient fluxes or dilution rates showed that at a high specific flux an assemblage of primarily small, fast-growing centric diatoms such as *Skeletonema costatum* and *Chaetoceros* spp. dominated with a low percentage of flagellates, while at a low specific nutrient flux, a mixture of larger, slower-growing centric diatoms, small flagellates and some pennate diatoms was obtained (Fig. 3.6). Observations from these experiments indicated that many of the small centric diatoms were high flux organisms and were successful competitors due to their high nutrient uptake and growth rates, while small flagellates were generally low nutrient flux species, exhibiting lower nutrient uptake and growth rates.

Recent results from competition experiments suggest that the patchiness of the limiting nutrient may also affect phytoplankton community structure. When the specific nutrient flux was high and the same in all experiments, different species became dominant, depending on the degree of patchiness of the limiting nutrient; under a homogeneous distribution of the limiting nutrient *Chaetoceros* spp. dominated, but when the limiting nutrient was added daily (patchy distribution), *Skeletonema costatum* dominated due to its high maximal uptake rate. Intermediate patchiness gave rise to an assemblage dominated by both *Chaetoceros* and *Skeletonema*.

Generalizations from this limited number of experiments indicate that the specific nutrient flux is a coarse-tuning variable altering the ratio of broad phytoplankton groups present in a natural assemblage, and it may be partially responsible for dominance of diatoms in upwelling areas (high nutrient flux environments) and small flagellates in oligotrophic areas or low nutrient flux environments (TURPIN and HARRISON 1979; PARSONS and TAKAHASHI 1973; PARSONS 1979 and references cited therein). The degree of patchiness of the limiting nutrient at a constant specific flux appears to be more of a fine-tuning variable determining which species within the broad phytoplankton group that was selected by the specific flux, will dominate.

In conclusion, it appears that various theories on the uptake of nutrients by phytoplankton, including the earlier concepts of "temporal succession" and "contemporaneous disequilibrium" are sufficient to illustrate the important role of nutrients in maintaining phytoplankton species diversity. In addition, other factors such as selective zooplankton grazing, temperature and salinity tolerance and inhibition of certain phytoplankton by antimetabolites (allelopathy) would serve to further broaden the number of species which could co-exist at any one time.

References

Ahmed SI, Kenner RA, Packard TT (1977) A comparative study of glutamate dehydrogenase activity in several species of marine phytoplankton. Mar Biol 39:93–101

Allen HE, Kramer JR (eds) (1972) Nutrients in natural waters. Wiley and Sons, New York

Anderson DM, Morel FMM (1978) Copper sensitivity of *Gonyaulax tamarensis*. Limnol Oceanogr 23:283–295

Antia NJ, McAllister CD, Parsons TR, Stephens K, Strickland JDH (1963) Further measurements of primary production using a large volume plastic sphere. Limnol Oceanogr 8:166–183

Ayers GP, Gras JL (1980) Ammonia gas concentrations over the Southern Ocean. Nature (London) 284:539–540

Banse K (1974a) On the interpretation of data for the carbon-to-nitrogen ratio in phytoplankton. Limnol Oceanogr 19:695–699

Banse K (1974b) The nitrogen to phosphorus ratio in the photic zone of the sea and the elemental composition of the plankton. Deep-Sea Res 21:767–771

Barbaree JM, Payne WJ (1967) Products of denitrification by a marine bacterium as revealed by gas chromatography. Mar Biol 1:136–139

Barber RT (1973) Organic ligands and phytoplankton growth in nutrient-rich seawater. In: Singer PC (ed) Trace metals and metal-organic interactions in natural waters. Ann Arbor Science, Ann Arbor, Mich, pp 321–381

Barber RT, Dugdale RC, MacIsaac JJ, Smith RL (1971) Variations in phytoplankton growth associated with the source and conditioning of upwelling water. Invest Pesq 35:171–193

Barnes H (1957) Nutrient Elements. In: Hedgpeth JW (ed) Treatise on marine ecology and paleoecology, vol I. Ecology. pp 297–344. Geol Soc Am (Mem Geol Soc Am 67)

Bowman MJ, Esaias WE (1977) Oceanic fronts in coastal processes. Springer, Berlin Heidelberg New York, pp 2–13

Boyle EA (1979) Copper in natural waters. In: Nriagu JO (ed) Copper in the environment, part 1. Ecol Cycl. Wiley and Sons, New York, pp 77–85

Bray JT, Bricker OP, Troup BM (1973) Phosphate and interstitial waters of anoxic sediments: oxidation effects during sampling procedure. Science 180:1362–1364

Brooks RH, Brezonik PL, Putman HD, Keirn MA (1971) Nitrogen fixation in an estuarine environment: the Waccasassa on the Florida Gulf coast. Limnol Oceanogr 16:701–710

Brown V, Ducker SC, Rowan KS (1977) The effect of orthophosphate concentration on the growth of articulated coralline algae (Rhodophyta). Phycologia 16:125–131

Bruno SF, Staker RD (1978) Seasonal vitamin B_{12} and phytoplankton distribution near Napeague Bay, New York (Block Island Sound). Limnol Oceanogr 23:1045–1051

Burkholder PR, Burkholder LM (1956) Vitamin B_{12} in suspended solids and marsh muds collected along the coast of Georgia. Limnol Oceanogr 1:202–208

Burton JD, Liss PS (1968) Oceanic budget of dissolved silicon. Nature (London) 220:905–906

Butler EI, Corner EDS, Marshall SM (1970) On the nutrition and metabolism of zooplankton. VII. Seasonal survey on nitrogen and phosphorus excretion by *Calanus* in the Clyde sea area. J Mar Biol Assoc UK 50:525–560

Butler EI, Knox S, Liddicoat MI (1979) The relationship between organic and inorganic nutrients in seawater. J Mar Biol Assoc UK 59:239–250

Byrne RH, Kester DR (1976) Solubility of hydrous ferric oxide and iron speciation in seawater. Mar Chem 4:255–274

Calvert SE (1968) Silica balance in the ocean and diagenesis. Nature (London) 219:919–920

Calvert SE, Price NB (1971) Upwelling and nutrient regeneration in the Benguela Current, October, 1968. Deep-Sea Res 18:505–523

Caperon J (1967) Population growth of microorganisms limited by food supply. Ecology 48:715–722

Caperon J (1968) Population growth response of *Isochrysis galbana* to nitrate variation at limiting concentrations. Ecology 49:866–872

Carlucci AF, Bowes PM (1970) Vitamin production and utilization by phytoplankton in mixed cultures. J Phycol 6:393–400

Carlucci AF, Strickland JDH (1968) The isolation, purification and some kinetic studies of marine nitrifying bacteria. J Exp Mar Biol Ecol 2:156–166

Carlucci AF, Hartwig EO, Bowes PM (1970) Biological production of nitrite in seawater. Mar Biol 7:161–166

Carpenter EJ, Guillard RRL (1971) Intraspecific differences in nitrate half-saturation constants for three species of marine phytoplankton. Ecology 52:183–185

Carpenter JH, Pritchard DW, Whaley RC (1969) Observations of eutrophication and nutrient cycles in some coastal plain estuaries. In: Eutrophication: causes, consequences and correctives. Natl Acad Sci, Washington, pp 210–221

Chapman ARO, Craigie JS (1977) Seasonal growth in *Laminaria longicruris*: relations with dissolved inorganic nutrients and internal reserves of nitrogen. Mar Biol 40:197–205

Conway HL (1977) Interactions of inorganic nitrogen in the uptake and assimilation by marine phytoplankton. Mar Biol 39:221–232

Conway HL, Harrison PJ (1977) Marine diatoms grown in chemostats under silicate or ammonium limitation. IV. Transient response of *Chaetoceros debilis, Skeletonema costatum* and *Thalassiosira gravida* to a single addition of the limiting nutrient. Mar Biol 43:33–43

Conway HL, Harrison PJ, Davis CO (1976) Marine diatoms grown in chemostats under silicate or ammonium limitation. II. Transient response of *Skeletonema costatum* to a single addition of the limiting nutrient. Mar Biol 35:187–199

Corner EDS, Davies AG (1971) Plankton as a factor in the nitrogen and phosphorus cycles in the sea. Adv Mar Biol 9:101–204

Cross FA, Sunda WG (1977) Relationship between bioavailability of trace metals and geochemical processes in estuaries. In: Wiley ML (ed) Estuarine interactions. Academic Press, London New York, pp 429–442

Cushing DH (1964) The work of grazing in the sea. In: Crisp DJ (ed) Grazing in terrestrial and marine environments. Blackwell, Oxford, pp 207–225

Davis CO, Breitner NF, Harrison PJ (1978) Continuous culture of marine diatoms under silicon limitation. 3. A model of Si-limited growth. Limnol Oceanogr 23:41–52
D'Elia CF, DeBoer JA (1978) Nutritional studies of two red algae. II. Kinetics of ammonium and nitrate uptake. J Phycol 14:266–272
Droop MR (1970) Vitamin B_{12} and marine ecology. V. Continuous culture as an approach to nutritional kinetics. Helgol Wiss Meeresunters 220:629–636
Droop, MR, McLaughlin JJA, Pinter IJ, Provasoli L (1959) Specificity of some protophytes towards vitamin B-like compounds. Prepr Int Oceanogr Congr, New York, pp 916–918
Dugdale RC (1967) Nutrient limitation in the sea: dynamics, identification and significance. Limnol Oceanogr 12:685–695
Dugdale RC (1972) Chemical oceanography and primary productivity in upwelling regions. Geoforum 7:47–61
Dugdale RC (1976) Nutrient cycles. In: Cushing DH, Walsh JJ (eds) The ecology of the seas. Saunders, Toronto, pp 141–172
Dugdale RC (1977) Modelling. In: Goldberg ED, McCave TN, O'Brien JJ, Steele JH (eds) The sea, vol VI. Wiley Interscience, New York
Dugdale RC, Goering JJ (1967) Uptake of new and regenerated forms of nitrogen in primary productivity. Limnol Oceanogr 12:196–206
Dugdale RC, Goering JJ (1970) Nutrient limitation and the path of nitrogen in Peru Current production. Anton Brunn Rep 4. Texas A and M Press, College Station, pp 5.3–5.8
Dugdale RC, Menzel DW, Ryther JH (1961) Nitrogen fixation in the Sargasso Sea. Deep-Sea Res 7:297–300
Emery KO, Orr WK, Rittenberg SC (1955) Nutrient budgets in the ocean. In: Essays in the natural sciences in honor of Captain Allan Hancock. Univ South Calif Press, Los Angeles, pp 299–310
Eppley RW, Coatsworth JL (1968) Uptake of nitrate and nitrite by *Ditylum brightwellii*. J Phycol 4:151–156
Eppley RW, Peterson BJ (1979) Particulate organic matter flux and planktonic new production in the deep ocean. Nature (London) 282:677–680
Eppley RW, Strickland JDH (1968) Kinetics of marine phytoplankton growth. Adv Microbiol Sea I:23–62
Eppley RW, Thomas WH (1969) Comparison of half-saturation constants for growth and nitrate uptake of marine phytoplankton. J Phycol 5:375–379
Eppley RW, Weiler CS (1979) The dominance of nanoplankton as an indicator of marine pollution: a critique. Oceanol Acta 2:241–245
Eppley RW, Coatsworth JL, Solórzano Lucia (1969a) Studies of nitrate reductase in marine phytoplankton. Limnol Oceanogr 14:194–205
Eppley RW, Rogers JN, McCarthy JJ (1969b) Half-saturation constants for uptake of nitrate and ammonium by marine phytoplankton. Limnol Oceanogr 14:912–920
Eppley RW, Renger EH, Harrison WG (1979a). Nitrate and phytoplankton production in southern California coastal waters. Limnol Oceanogr 24:483–494
Eppley RW, Renger EH, Harrison WG, Cullen JJ (1979b) Ammonium distribution in southern California coastal waters and its role in the growth of phytoplankton. Limnol Oceanogr 24:495–509
Erickson SJ (1972) Toxicity of copper to *Thalassiosira pseudonana* in unenriched inshore seawater. J Phycol 8:318–323
Estep M, Armstrong JE, Baalen van C (1975) Evidence for the occurrence of specific iron (III)-binding compounds in near-shore marine ecosystems. Appl Microbiol 30:186–188
Falkowski PG (1975) Nitrate uptake in marine phytoplankton: (Nitrate, Chloride)-activated adenosine triphosphatase from *Skeletonema costatum* (Bacillariophyceae). J Phycol 11:323–326
Falkowski PG, Rivkin RB (1976) The role of glutamine synthetase in the incorporation of ammonium in *Skeletonema costatum* (Bacillariophyceae). J Phycol 12:448–450

Falkowski PG, Stone DP (1975) Nitrate uptake in marine phytoplankton: Energy sources and interaction with carbon fixation. Mar Biol 32:77–84
Finenko ZZ (1978) Production in plant populations. Mar Ecol 4:13–78
Fisher NS (1977) On the differential sensitivity of estuarine and open ocean diatoms to exotic chemical stress. Am Nat 111:871–895
Fitzgerald GP (1968) Detection of limiting or surplus nitrogen in algae and aquatic weeds. J Phycol 4:121–126
Frey BE, Small LF (1979) Recycling of metabolized iron by the marine dinoflagellate *Amphidinium carterae*. J Phycol 15:405–409
Garside C, Malone TC, Roels OA, Shartstein BA (1976) An evaluation of sewage-derived nutrients and their influence on the Hudson estuary and New York Bight. Estuarine Coastal Mar Sci 4:281–289
Gemmill ER, Galloway RA (1974) Photoassimilation of ^{14}C-acetate by *Ulva lactuca*. J Phycol 10:359
Glover HE (1978) Iron in Maine coastal waters; seasonal variation and its apparent correlation with a dinoflagellate bloom. Limnol Oceanogr 23:534–537
Goering JJ (1974) Uptake of silicic acid by diatoms. Tethys 6:143–148
Goering JJ, Dugdale RC (1966) Denitrification rates in an island bay in the equatorial Pacific Ocean. Science 154:505–506
Goering JJ, Pamatmat MM (1971) Denitrification in sediments of the sea off Peru. Invest Pesq 35:233–242
Goldman JC (1979) Temperature effects on steady-state growth, phosphorus uptake, and the chemical composition of a marine phytoplankter. Microb Ecol 5:153–166
Goldman JC, McCarthy JJ, Peavey DG (1979) Growth rate influence on the chemical composition of phytoplankton in oceanic waters. Nature (London) 279:210–215
Gonye ER, Carpenter EJ (1974) Production of iron-binding compounds by marine microorganisms. Limnol Oceanogr 19:840–842
Grill EV (1970) A mathematical model for the marine dissolved silicate cycle. Deep-Sea Res 17:245–266
Grill EV, Richards FA (1964) Nutrient regeneration from phytoplankton decomposing in seawater. J Mar Res 22:51–69
Guillard RRL, Kilham P, Jackson TA (1973) Kinetics of silicon-limited growth in the marine diatom *Thalassiosira pseudonana* Hasle and Heimdal (= *Cylindrotheca nana* Hustedt). J Phycol 9:233–237
Haas LW, Webb KL (1979) Nutritional mode of several nonpigmented microflagellates from the York river estuary, Virginia. J Exp Mar Biol Ecol 39:125–134
Haines KC, Wheeler PA (1978) Ammonium and nitrate uptake by the marine macrophytes *Hypnea musciformis* (Rhodophyta) and *Macrocystis pyrifera* (Phaeophyta). J Phycol 14:319–324
Halim Y, Morcos SA (1965) Le rôle des particules en suspension dans l'eau du Nil en coue dans la répartition des sels nutritifs au lange de ses embouchunes. Rapp P-V Reun Cons Int Explor Mer 18:733–736
Harrison PJ, Davis CO (1979) The use of outdoor phytoplankton continuous cultures to analyse factors influencing species selection. J Exp Mar Biol Ecol 41:9–23
Harrison PJ, Conway HL, Dugdale RC (1976) Marine diatoms grown in chemostats under silicate or ammonium limitation. 1 Cellular chemical composition and steady-state growth kinetics of *Skeletonema costatum*. Mar Biol 35:177–186
Harrison WG (1973) Nitrate metabolism of the red tide dinoflagellate *Gonyaulax polyedra* Stein. J Exp Mar Biol Ecol 21:199–209
Harrison WG (1978) Experimental measurements of nitrogen remineralization in coastal waters. Limnol Oceanogr 23:684–694
Harrison WG, Azam F, Renger EH, Eppley RW (1977) Some experiments on phosphate assimilation by coastal marine plankton. Mar Biol 40:9–18
Hartwig EO (1976) The impact of nitrogen and phosphorus release from a siliceous sediment on the overlying water. Estuarine Proc 1:103–117
Hattori A, Wada E (1971) Nitrite distribution and its regulating processes in equatorial Pacific Ocean. Deep-Sea Res 18:557–568

Healey FP (1973) Inorganic nutrient uptake and deficiency in algae. Crit Rev Microbiol 3:69–113
Healey FP (1980) Slope of the Monod equation as an indicator of advantage in nutrient competition. Microb Ecol 5:281–286
Hollibaugh JT (1978) Nitrogen regeneration during the degradation of several amino acids by plankton communities collected near Halifax, Nova Scotia, Canada. Mar Biol 45:191–201
Hurd DC (1977) The effects of glacial weathering on the silica budget of Antarctic waters. Geochim Cosmochim Acta 41:1213–1222
Hutchinson GE (1961) The paradox of the plankton. Am Nat 95:137–145
Jackson GA (1980) Phytoplankton growth and zooplankton grazing in oligotrophic oceans. Nature (London) 284:439–441
Jansson M (1976) Phosphatases in lake water: characterization of enzymes from phytoplankton and zooplankton by gel filtration. Science 194:320–321
Ketchum BH (1939) The absorption of phosphate and nitrate by illuminated cultures of *Nitzschia closterium*. Am J Bot 26:399–407
Ketchum BH (1969) Eutrophication of estuaries. In: Eutrophication causes, consequences and correctives. Natl Acad Sci Washington, pp 197–209
Koblentz-Mishke OJ, Volkovinsky VV, Kabanova JG (1970) Plankton primary production of the world. In: Wooster WS (ed) Scientific exploration of the South Pacific. Standard Book 390-01755-d. Natl Acad Sci Washington, pp 183–193
Kuenzler EJ (1965) Glucose-6-phosphate utilization by marine algae. J Phycol 1:156–164
Kuenzler EJ (1970) Dissolved organic phosphorus excretion by marine phytoplankton. J Phycol 6:7–13
Kuenzler EJ, Ketchum BH (1962) Rate of phosphorus uptake by *Phaeodactylum tricornutum*. Biol Bull 123:134–145
Kuenzler EJ, Perras JP (1965) Phosphatase of marine algae. Biol Bull 128:271–284
Laws EA, Wong DCL (1978) Studies of carbon and nitrogen metabolism by three marine phytoplankton species in nitrate-limited continuous culture. J Phycol 14:406–416
Lean DRS (1976) Phosphorus kinetics in lake water: Influence of membrane filter pore size and low-pressure filtration. J Fish Res Board Can 33:2800–2804
Lewin J, Chen C (1973) Changes in the concentration of soluble and particulate iron in seawater enclosed in containers. Limnol Oceanogr 18:590–596
Lewin JC (1961) The dissolution of silica from diatom walls. Geochim Cosmochim Acta 21:182–198
Lui NST, Roels OA (1972) Nitrogen metabolism of aquatic organisms. II The assimilation of nitrate, nitrite and ammonia by *Biddulphia aurita*. J Phycol 8:259–264
MacIsaac JJ, Dugdale RC (1969) The kinetics of nitrate and ammonia uptake by natural populations of marine phytoplankton. Deep-Sea Res 16:45–57
Mague TM, Weane NM, Holm-Hansen O (1974) Nitrogen fixation in the North Pacific Ocean. Mar Biol 24:109–119
Malone PLG, Towe KM (1970) Microbial carbonate and phosphate precipitates from sea water cultures. Mar Geol 9:301–309
Mann KH (1972) Macrophyte production and detritus food chains in coastal waters. Mem Ist Ital Idrobio Suppl 29:353–383
Mann KH (1973) Seaweeds, their productivity and strategy for growth. Science 182:975–981
Martens CS, Berner RA, Rosenfield JK (1978) Interstitial water chemistry of anoxic Long Island Sound sediments. 2. Nutrient regeneration and phosphate removal. Limnol Oceanogr 23:605–617
McCarthy JJ (1972) The uptake of urea by marine phytoplankton. J Phycol 8:216–222
McCarthy JJ, Goldman JC (1979) Nitrogenous nutrition of marine phytoplankton in nutrient depleted waters. Science 203:670–672
McCaffney RJ, Myers AC, Davey E, Morrison G, Bender M, Luedtke N, Cullen D, Froelich P, Klinkhammer G (1980) The relation between pore water chemistry and benthic fluxes of nutrients and manganese in Narragansett Bay, Rhode Island. Limnol Oceanogr 25:31–44

McLachlan J (1977) Effects of nutrients on growth and development of embryos of *Fucus edentatus*. Pyl. (Phaeophyceae, Fucales). Phycologia 16:329–338
Menzel DW, Ryther JH (1961) Nutrients limiting the production of phytoplankton in the Sargasso Sea, with special reference to iron. Deep-Sea Res 7:276–281
Menzel DW, Spaeth JP (1962a) Occurrence of vitamin B_{12} in the Sargasso Sea. Limnol Oceanogr 7:151–154
Menzel DW, Spaeth JP (1962b) Occurrence of ammonia in Sargasso Sea waters and in rain water at Bermuda. Limnol Oceanogr 7:159–162
Murphy TP, Lean DRS, Nalewajko C (1976) Blue-green algae: their excretion of iron-selective chelators enables them to dominate other algae. Science 192:900–902
Naiman RJ, Sibert RJ (1978) Transport of nutrients and carbon from the Nanaimo River to its estuary. Limnol Oceanogr 23:1183–1193
Nelson DM, Goering JJ (1978) Assimilation of silicic acid by phytoplankton in the Baja California and northwest Africa upwelling systems. Limnol Oceanogr 23:508–517
Nelson DM, Goering JJ, Kilham SS, Guillard RRL (1976) Kinetics of silicic acid uptake and rates of silica dissolution in the marine diatom *Thalassiosira pseudonana*. J Phycol 12:246–252
Nixon WS, Oviatt CA, Hale SS (1976) Nitrogen regeneration and the metabolism of coastal marine bottom communities. In: Anderson JM, MacKadyen A (eds) The role of aquatic organisms in decomposition processes. Blackwell, Oxford, pp 269–283
Paasche E (1973) Silicon and the ecology of marine plankton diatoms. II. Silicate-uptake kinetics in five diatom species. Mar Biol 19:262–269
Paasche E (1977) Growth of three plankton diatom species in Oslofjord water in the absence of artificial chelators. J Exp Mar Biol Ecol 29:91–106
Packard TT (1979) Half-saturation constants for nitrate reductase and nitrate translocation in marine phytoplankton. Deep-Sea Res 26A:321–326
Parsons TR (1979) Some ecological, experimental and evolutionary aspects of the upwelling ecosystem. S Afr J Sci 75:536–540
Parsons TR, Takahashi M (1973) Biological oceanographic processes. Pergamon Press, New York, 186 p
Parsons TR, Brockel von K, Koeller P, Takahashi M, Reeve MR, Holm-Hansen O (1977) The distribution of organic carbon in a marine planktonic food web following nutrient enrichment. J Exp Mar Biol Ecol 26:235–247
Pasciak WJ, Gavis J (1974) Transport limitation of nutrient uptake in phytoplankton. Limnol Oceanogr 19:881–888
Perry MJ (1976) Phosphate utilization by an oceanic diatom in phosphorus-limited chemostat culture and in the oligotrophic waters of the Central North Pacific. Limnol Oceanogr 21:88–107
Peters RH (1977) Availability of atmospheric orthophosphate. J Fish Res Board Can 34:918–924
Peterson R (1975) The paradox of the plankton: An equilibrium hypothesis. Am Nat 109:35–48
Pierrou U (1976) The global phosphorus cycle. pp 75–90. In: Nitrogen Phosphorus and Sulphur – Global Cycles. SCOPE Report 7. Ecol Bull (Stockholm) 22:75–88
Pomeroy LH, Mathews HM, Min HS (1963) Excretion of phosphate and soluble organic phosphorus compounds by zooplankton. Limnol Oceanogr 8:50–55
Postma H (1971) Distribution of nutrients in the sea and the oceanic nutrient cycle. In: Costlow JD (ed) Fertility of the sea. Gordon Breach, New York, pp 336–349
Provasoli L, Carlucci AF (1974) Vitamins and growth regulators. In: Stewart WDP (ed) Algal physiology and biochemistry. Blackwell, Oxford, pp 741–787
Rakestraw MW, Brand von T (1947) Decomposition and regeneration of nitrogenous organic matter in sea water. VI. The effect of enzyme poisons. Biol Bull Woods Hole 92:110–114
Redfield AC (1934) On the proportions of organic derivatives in sea water and their relation to the composition of plankton. In: Daniel RJ (ed) James Johnstone memorial volume. Liverpool Univ Press, Liverpool

Redfield AC, Ketchum BH, Richards FA (1963) The influence of organisms on the composition of seawater. In: Hill MN (ed) The sea, vol II. Wiley-Interscience, New York, pp 26–77
Reshkin SJ, Knauer GA (1979) Light stimulation of phosphate uptake in natural assemblages of phytoplankton. Limnol Oceanogr 24:1121–1124
Rice HV, Leighty DA, McLeod GC (1973) The effects of some trace metals on marine phytoplankton. Crit Rev Microbiol 3:27–49
Richerson P, Armstrong R, Goldman CR (1970) Contemporaneous disequilibrium, a new hypothesis to explain the "paradox of the plankton." Proc Natl Acad Sci USA 67:1710–1714
Riley JP, Chester R (1971) Introduction to marine chemistry. Academic Press, London New York
Riley JP, Skirrow G (ed) (1965) Chemical oceanography vol I. Academic Press, London New York
Rowe GT, Clifford CH, Smith Jr KL (1977) Nutrient regeneration in sediments off Cape Blanc, Spanish Sahara. Deep-Sea Res 24:57–63
Schmidt RL (1978) Copper in the marine environment – Part 1. Crit Rev Environ Control 8:101–152
Senft WH (1978) Dependence of light-saturated rates of algal photosynthesis on intracellular concentrations of phosphorus. Limnol Oceanogr 23:709–718
Serra JL, Llama MJ, Cadenas E (1978) Nitrate utilization by the diatom *Skeletonema costatum*. I. Kinetics of nitrate uptake. Plant Physiol 62:987–990
Sharp JH, Underhill PA, Hughes DJ (1979) Interaction (allelopathy) between marine diatoms: *Thalassiosira pseudonana* and *Phaeodactylum tricornutum*. J Phycol 15:353–362
Simpson FB, Neilands JB (1976) Siderochromes in cyanophyceae: Isolation and characterization of schizokinen from *Anabaena* sp. J Phycol 12:44–48
Smayda TJ (1969) Some measurements of the sinking rate of fecal pellets. Limnol Oceanogr 14:621–625
Smayda TJ (1970) The suspension of sinking of phytoplankton in the sea. Oceanogr Mar Biol Annu Rev 8:353–414
Smetacek V, Bodungen von B, Brödsel von K, Zeitzschel B (1976) The plankton tower. II. Release of nutrients from sediments due to changes in the density of bottom water. Mar Biol 34:373–378
Söderlund R, Svensson BH (1976) The global nitrogen cycle. In: Nitrogen, phosphorus and sulphur – global cycles. SCOPE Report 7. Ecol Bull (Stockholm) 22:23–73
Sournia A (1970) Les Cyanophycées dans le plancton marin. Ann Biol Ser 4 9:63–76
Spencer CP (1954) Studies on the culture of a marine diatom. J Mar Biol Assoc UK 33:265–290
Spencer CP (1975) The micronutrient elements. In: Riley JP, Skirrow G (eds) Chemical oceanography, vol II (2nd edn). Academic Press, London New York, pp 245–300
Steele JH, Yentsch CS (1960) The vertical distribution of chlorophyll. J Mar Biol Assoc UK 39:217–226
Stewart WDP (1964) Nitrogen fixation by Myxophyceae from marine environments. J Gen Microbiol 36:415–422
Strickland JDH, Austin KH (1960) On the forms, balance and cycle of phosphorus observed in the coastal oceanic waters of the northwestern Pacific. J Fish Res Board Can 17:337–345
Sugimura Y, Suzuki Y, Miyake Y (1978) The dissolved organic iron in seawater. Deep-Sea Res 25:309–314
Sunda W, Guillard RRL (1976) The relationship between cupric ion activity and the toxicity of copper to phytoplankton. J Mar Res 34:511–529
Takahashi M, Ikeda T (1975) Excretion of ammonia and inorganic phosphorus by *Euphausia pacifica* and *Metridia pacifica* at different concentrations of phytoplankton. J Fish Res Board Can 32:2189–2195
Thomas WH (1969) Phytoplankton nutrient enrichment experiments off Baja California and the eastern equatorial Pacific Ocean. J Fish Res Board Can 26:1133–1145

Thomas WH, Dodson AN (1968) Effects of phosphate concentration on cell division rates and yield of a tropical oceanic diatom. Biol Bull 134:199–208
Thomas WH, Dodson AN (1975) On silicic acid limitation of diatoms in near-surface waters of the eastern tropical Pacific Ocean. Deep-Sea Res 22:671–677
Thomas WH, Seibert DLRS, Dodson AN (1974) Phytoplankton enrichment experiments and bioassays in natural coastal seawater and in sewage outfall-receiving waters off Southern California. Estuarine Coastal Mar Sci 2:191–206
Tilman D (1977) Resource competition between planktonic algae: An experimental and theoretical approach. Ecology 58:338–348
Titman D (1976) Ecological competition between algae: Experimental confirmation of resource-based competition theory. Science 192:4463–465
Topinka JA (1978) Nitrogen uptake by *Fucus spiralis* (Phaeophyceae). J Phycol 14:241–247
Tranter DJ, Newell BS (1963) Enrichment experiments in the Indian Ocean. Deep-Sea Res 10:1–9
Turpin DH (1980) Processes in nutrient based phytoplankton ecology. Ph D thesis. Univ Br Columbia, Vancouver, BC Canada
Turpin DH, Harrison PJ (1978) Fluctuations in free amino acid pools of *Gymnodinium simplex* (Dinophyceae) in response to ammonia perturbation: Evidence for glutamine synthetase pathway. J Phycol 14:461–464
Turpin DH, Harrison PJ (1979) Limiting nutrient patchiness and its role in phytoplankton ecology. J Exp Mar Biol Ecol 39:151–166
Vaccaro RF, Ryther JH (1960) Marine phytoplankton and the distribution of nitrite in the sea. J Cons Cons Int Explor Mer 25:260–271
Veeh MM, Burnett WC, Soutar A (1973) Contemporary phosphorites on the continental margin of Peru. Science 181:844–845
Walsh JJ, Whitledge TE, Kelley JC, Huntsman SA, Pillsbury RD (1977) Further transition states of the Baja California upwelling ecosystem. Limnol Oceanogr 22:264–280
Walsh JJ, Whitledge TE, Barvenik FW, Wirick CD, Howe SO (1978) Wind events and food chain dynamics within the New York Bight. Limnol Oceanogr 23:659–683
Water quality data, British Columbia, 1961–1971 (1974) Environment Canada inland waters directorate, water quality branch, Ottawa, Canada
Webb KL, Johannes RL (1967) Studies on the release of dissolved free amino acids by marine zooplankton. Limnol Oceanogr 12:376–382
Wheeler PA, North BB, Stephens GC (1974) Amino acid uptake by marine phytoplankters. Limnol Oceanogr 19:249–259
Whitfield PH, Lewis AG (1976) Control of the biological availability of trace metals to a calanoid copepod in a coastal fjord. Estuarine Coastal Mar Sci 4:255–266
Whitledge TE, Packard TT (1971) Nutrient excretion by anchovies and zooplankton in Pacific upwelling regions. Invest Pesq 35:243–250
Williams PJLeB, Berman T, Holm-Hansen O (1976) Amino acid uptake and respiration by marine heterotrophs. Mar Biol 35:41–47
Williams PM (1968) Organic and inorganic constituents of the Amazon River. Nature (London) 218:937–938

4 Modeling of Growth and Production

F.W.T. Penning de Vries

CONTENTS

4.1 Introduction 118

4.2 Concepts in Modeling 119
4.2.1 Real World Systems 119
4.2.2 Models 121
4.2.3 Use and Evaluation of Models 123
4.2.4 References to Modeling Techniques 125

4.3 Current Models in Different Phases of Development 125
4.3.1 Introduction 125
4.3.2 Regresson Models of Productivity 126
4.3.3 Preliminary Models at the Lowest Level of Crop Productivity 128
4.3.4 Comprehensive Models of Growth and Productivity 130
4.3.5 Summary Models at the Highest Production Levels 133
4.3.6 Results of Modeling Other Than Models 135

4.4 Physiological Processes in Dynamic Models 136
4.4.1 The Carbon Balance 136
4.4.1.1 Introduction 136
4.4.1.2 Distribution of Biomass 136
4.4.1.3 Efficiency of Conversion Processes 138
4.4.1.4 Maintenance Processes 139
4.4.1.5 Root Growth and Root Respiration 139
4.4.2 The Water Balance 140
4.4.2.1 Introduction 140
4.4.2.2 Water Uptake and Root Growth 141
4.4.2.3 Water Stress, Dry Matter Distribution and Death 141
4.4.3 Nutrient Balances 142
4.4.3.1 Introduction 142
4.4.3.2 Growth and Availability of N and P 142
4.4.3.3 Redistribution 143
4.4.4 Morphogenesis 144
4.4.4.1 Introduction 144
4.4.4.2 Leaf Surface Area 144
4.4.4.3 Root Surface Area and Root Profile 144

4.5 Conclusions 145

References 146

4.1 Introduction

Building models is one way of integrating knowledge. It has as important features that it

- helps to define and categorize the state of knowledge of the subject;
- helps to locate gaps in knowledge and to make hypotheses explicit, and thus helps to set priorities for research;
- provides a tool to make the integrated information operational;
- provides an effective link between scientists from adjacent disciplines, between researchers studying different levels of biological organization, and between fundamental and applied scientists.

A developed and evaluated model

- provides a key to determine progress in science;
- provides a means for disseminating knowledge;
- can be used for prediction.

These points are demonstrated in this chapter. Quite impressive crop growth models have indeed been constructed. Still, evolution of modeling has been slow in the field of crop growth, in spite of such important assets. Lack of factual knowledge, particularly of kinetics of physiological processes, has been and is still slowing down progress. In fact, some of the recent advances in crop physiology have been stimulated by modeling, as is shown in Sect. 4.4.

Knowledge of crop physiology and growth models have partly been developed in parallel and this poses a problem of the order in which they should be treated. Presuming that most readers have more background in physiology than in modeling, theory of systems and modeling has been emphasized first. The main aspects of crop physiology and the way in which they are modeled are discussed later.

Models of production and growth usually include a computation of photosynthesis and transpiration. Those processes are reviewed separately in Chaps. 8 and 16 of Vol. 12B. Growth and production models can be of great help in the study and further development of agricultural systems, as is pointed out in Chap. 5, this Volume.

In 1978, about 3% of the agricultural research community in the Netherlands used or developed models. This was estimated from the number of publications concerning simulation in agricultural research. Ten years before, this figure was still below 1%. EDMINSTER (1978) estimated that about 2% of the current scientific effort of the Agricultural Research Service of the United States of America goes into modeling. He advocated a strong further increase, for reasons listed above. This demonstrates that modeling is slowly taking an important place among research tools in agricultural sciences. A similar development is occurring in the biological sciences. It is interesting to note that most scientists involved in dynamic modeling are concentrated in small groups. Successful modeling of plant growth requires apparently, a "critical mass" of from three to five scientists from different, but related disciplines, who collaborate closely for a long period.

4.2 Concepts in Modeling

4.2.1 Real World Systems

One way of considering the real world is to divide it into systems. A functional description of a system is "a part of reality with strongly interacting elements, but with little influence on its environment". What part of the real world is singled out as a system depends first of all upon objectives. However, some elements of a system interact more than others, co-determining the shape of a system. Upon delimiting a system, one should thus take into account such natural contours, and consider all essential parts. (This is meant by the term "whole system approach".) Plant production on fields with fertile and irrigated soils, e.g., may be visualized as a system in which processes like photosynthesis, growth, maintenance and development interact intensively. The rates of these physiological processes depend strongly on weather conditions; weather is not modified noticeably by plant growth. One can thus delimit this plant production system by drawing a line between physiological and meteorological processes. As photosynthesis influences growth and vice versa, it makes little sense to exclude photosynthesis from this system even when the study of assimilation is not an objective. But when growth occurs in a greenhouse, air humidity and temperature may be modified by crop growth and the "weather" has become part of the system. The boundaries of a system can move apparently, with changes that may seem irrelevant for the objective of the study. Ideally, boundaries are chosen such that the environment influences processes of the system, but the system itself does not influence its environment.

In a typical situation of plant growth in temperate climates the growth–weather relation is of major importance at the beginning of the growing season. Nutrient shortage reduces growth at a later stage. At any time, a brief dry spell may cause water shortage and reduce growth. Plant production can thus be seen as one large system in which the relative importance of elements changes. For production and growth models it is practical to consider the structure of the system as invariable during the time span of interest, and to keep the parts of the system that are temporarily unimportant as harmless ballast. One might also argue, however, that the structure of the system changes in such a case. This conception may become effective in phases of model development still to come.

An elegant and practical delimitation of systems of growing vegetations or crops is proposed by De Wit (in: De Wit and Penning de Vries 1982). They distinguish four levels of plant production. The systems of plant growth and crop productivity at each of these levels can be considered as belonging to one broad class. Those levels are:

Production Level 1. Growth in conditions with ample plant nutrients and soil water. The growth rate of the vegetation is then limited by weather conditions and amounts to 100–350 kg dry matter ha^{-1} d^{-1}. This situation is sometimes realized in agronomic field experiments and in glasshouses. Major elements of this type of system are the dry weight of leaves, stems, reproductive or storage organs; major processes are photosynthesis, growth and maintenance, biomass distribution and leaf area development.

Production Level 2. Plant growth becomes limited by water shortage part of the time. This occurs on well-fertilized soils in semi-arid regions and in temperate climates. The situation is not very common, either in agriculture or in natural ecosystems. The extra elements in this class of systems are the plant and soil water balances; crucial processes are transpiration and other processes of loss or gain of water from the soil.

Production Level 3. Plant production is limited by shortage of nitrogen (N) and sometimes by water shortage. This is quite a common situation in agricultural systems using little fertilizer, and is also normal in nature. Important elements of this class of systems are the N in the soil and in the plant; important processes are the transformation of nitrogenous compounds in the soil and other processes of the N balance, absorption by roots, growth-availability interactions and redistribution within the plant.

Production Level 4. Plant production is limited by the availability of phosphorus (P). Growth rates are 10–50 kg dry matter $ha^{-1} d^{-1}$ over a growing season of about 100 days. This situation occurs in heavily exploited areas where no fertilizer is used, such as in the poorest areas of the world. Important elements of this class of systems are the P and N contents of the soil and plants, and important processes are their transformations in the soil, absorption by roots and the response of plant growth to their absolute and relative availabilities.

Further in this review reference will be made to these levels of production. Though it is rare to find cases that fit exactly into one of these production levels, and as also diseases, pests and weeds may interfere with growth, it is a very practical simplification in the beginning of a study to reduce specific cases to one of them. Only in more detailed studies does the complex situation in which different limitations intertwine during growth, earn consideration. The approach emphasizes dry matter production, while morphogenetic developments are of secondary importance.

De Wit's definition of systems at different levels of production is rooted in the analysis of agricultural crops. But as his delimitation is based on the effect of external factors on physiological processes, it is not restricted to agronomic situations, and applies to plant growth and production generally. Moreover, cultivation of crops has changed little or nothing in the basic physiology of species, and the similarities in physiological and biochemical functioning, e.g. between cultivated species and their ancestral forms, are often remarkably large (e.g. KHAN and TSUNODA 1970). De Wit's analysis is therefore applicable to all situations of plant growth in agricultural and in "natural" environments. But the usual heterogeneity of the vegetation of "natural areas" is a considerable handicap to the development of simulation models of their growth and production. In spite of the generality of the production level approach, this review emphasizes therefore the models of crops, forests and rangelands that are already relatively advanced.

It is characteristic for all of these systems that major elements (like plant biomass) change only gradually in value in time or in response to changing

external conditions such as weather or fertilization. Such systems are called "*continuous*", in contrast to "*discrete*" systems.

4.2.2 Models

A *model* is a schematic representation of our conception of a system. In the real world, the appearance of biological systems changes, and their models are often called "dynamic". The most obvious change in a plant production system is the growth of plants. Analysis of dynamic systems is based on the assumption that the state of a biological system at any particular time can be expressed quantitatively and that changes in the system can be described in mathematical terms. This assumption (according to De Wit and Goudriaan 1978) leads to the formulation of state-determined models in which state variables, driving variables, rate variables and auxiliary variables are distinguished. An example of a flow diagram of such a model is given in Fig. 4.1. It represents the carbon balance, the central part of many simulation models, and shows that photosynthesis, growth and respiration processes are integrated to be part of one system. State variables characterize and quantify all properties that describe the current state of the system. Examples of state variables are the amount of biomass, the leaf surface area, the N content of a part of the system etc. Driving variables are those that are not affected by elements within the system, and characterize the influence from outside. These may be meteorological quantities. It should be realized that, depending upon the choice of the boundaries of the system, the same variables may be classified as state or driving variables. Rate variables quantify the rate of change of the state variables. Their values

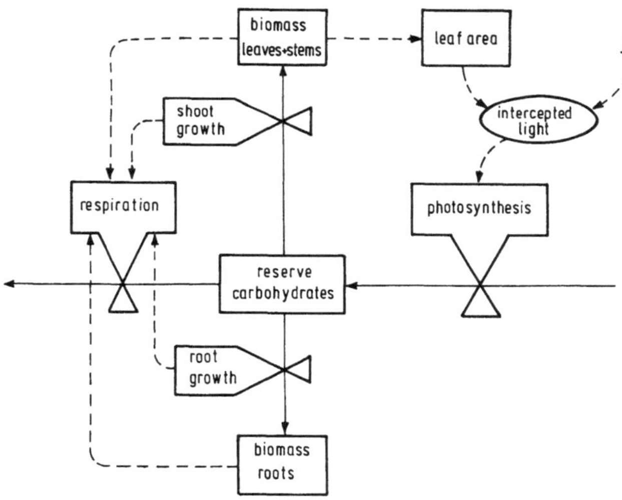

Fig. 4.1. A simple diagram of the carbon balance. *Rectangles* represent state variables, *rectangles with valve symbols* represent processes (rate variables); the *circle* represents an auxiliary variable and the *underlined variable* is the driving variable. *Full lines* represent flow of matter, *dotted lines* flow of information. (Symbols according to Forrester 1961)

are determined by the state variables and driving variables according to rules formulated from the knowledge of the underlying physiological, physical and biochemical processes. It is a basic assumption that rate variables are mutually independent, and related only through state variables. The calculation of rate variables may become so complicated that properly chosen auxiliary variables can be elucidated. The formulation of the relations between all variables is called the "structure" of a model. The "behaviour" of a model is the total of the changes in the numerical values of variables during simulation.

Given the structure of the model, the initial conditions and a certain light source, the model of Fig. 4.1 allows computation of how leaves and roots grow. To effectuate this, the growth rates are calculated and then multiplied by a short time interval. The resulting quantities are added to the appropriate state variables: weight of leaves and roots. The process of calculating rates and updating state variables is repeated until the desired time span has elapsed. The time step must be relatively small, i.e. the quantity added to a state variable should change its value by not more than 10% at a time.

Simulation is the scientific activity of building and utilizing *explanatory models*. Those models are called "explanatory", as their behaviour is based on the underlying knowledge integrated in the model. This type contrasts with *demonstrative models*, of which only the behaviour is like that of the real system, but the rules that make it behave in such a way are quite different. For example, the images of a film look like the real world, but are only a demonstrative model of it. Explanatory and demonstrative models come in many forms: dynamic mathematical models, scale models, graphical models, electric analog models. Each form has its particular advantages and disadvantages. Except for the first type, their versatility is small, they are little used and not discussed here. The general word "model" will be applied throughout this review for explanatory dynamic mathematical models. One form of demonstrative models, regression models, will be discussed below (Sect. 4.3.1) as they are often the precursors of explanatory models.

The more one analyzes processes, the more details become unraveled. A single process may then be considered separately, and treated as a subsystem. Each subsystem, characteristically, considers faster processes and shorter distances. The relaxation times of the systems and subsystems (i.e. the time to return to an equilibrium state after a sudden disturbance) differ considerably. By simulating the behaviour of a system, one links the explanatory level(s) of basic information with the higher, explainable level. Models currently comprise two or three hierarchical levels, the lower one(s) being the explanatory level(s). Relaxation times between explanatory and explainable levels differ about 10–100 fold. Time steps for simulations are typically 2–10 times smaller than the shortest relaxation time of a model.

The biological sciences together span several levels of biological organization with relaxation times as small as a fraction of a second on the enzyme level to several years on the ecosystem level, i.e. a factor 10^7–10^9 apart (Table 4.1). In theory, models that link four or more levels of organization could be constructed. However, simulation of organs, tissues or the individual cells of whole plants would require programs that are excessive for any computer system. Moreover, the development of such models would elude our comprehension, and even if it were possible to construct such giant models, their results would be too detailed and too specific to be of any use (cf. LOOMIS et al.

Table 4.1. Levels of processes in biological and agricultural systems with characteristic relaxation times

Relaxation time	Level of process	Example	Scientific discipline
Seconds	Enzymatic	TCA-cycle, protein synthesis	Biochemistry
Minutes	Cellular Organismal	Stomatal closure, cell division Water uptake, sugar transport	Plant physiology
Hours	Whole plant	Growth Development	Crop physiology Agronomy
Days	Vegetation	Competition processes among plants	Ecology and agronomy
Years	Ecosystem	Vegetation succession in plant communities	Ecology

1979). From a more pragmatic point of view, one has to establish that in many areas not even sufficient knowledge has yet been gathered to make models of plant growth connecting two levels of knowledge.

Dynamic models have been developed to simulate plant growth at each of the four levels of plant production. The degrees of understanding of underlying processes at those levels are unequal. As a result, their models are developed to different degrees. Section 4.3 deals with explanatory models in three stages of evolution: preliminary, comprehensive and summary models.

4.2.3 Use and Evaluation of Models

Models can be useful for development of science, for prediction and for instruction, but not to the same extent: scientifically interesting models are often too detailed for those who want to apply them, while models used for predictive or management purposes are often too trivial or too crude to challenge scientific interest. Table 4.2 characterizes models in different stages of development in this respect.

A *scientifically interesting model* contributes to our understanding of the real world because it helps to integrate the relevant processes of the system and to bridge areas and levels of knowledge. It helps also to test hypotheses,

Table 4.2. The relative values of certain aspects of models in different phases of development

	Predictive value	Scientific value	Instructive value	Simplicity
Regression model	++	+	+	+++
Preliminary model	+	+++	++	++
Comprehensive model	++	+++	+	+
Summary model	+++	+	+++	++

to generate alternative ones and to suggest experiments to falsify them (although data from experiments may not have sufficiently discriminatory power for this purpose, cf. FICK 1977; HUNT and LOOMIS 1979).

A *predictive model* should simulate accurately the behaviour of a part of the real world. It is therefore a good instrument to apply scientific knowledge in practice. It should predict reasonably well over a range of boundary conditions to provide its users with alternative solutions of a problem. The less detailed the desired results are, the simpler the predictive model can be. The *instructive value* of a model is its use for disseminating knowledge.

The size of a model may increase because its objectives are broadened, or because it is elaborated. In the first case, the number of parameters usually increases and the sensitivity of model behaviour to each parameter decreases. Elaborating the model of a system implies the formulation of more structure. A thorough knowledge of a complex real world system, and thus a large model of it, is always required before the model can be summarized reliably for use by others. The simpler a dynamic model that still accomplishes its purpose, the better it is for instruction and for those who want to apply it in other fields or higher up in the model hierarchy. Hence, the model attains its maximal scientific value while it is being elaborated, while its value for application increases during subsequent summarization.

Evaluation is a broad term to describe the action of judging the value of a model with a strong emphasis of its predictive value (cf. BAKER and CURRY 1976). "Evaluation" may be used for checking internal consistency and units used in a computer program for comparison of model output with real world data and for judgement of practical utility. The first thorough test of a model is often the comparison of the behavior of its output with that observed of the real system in an analogous situation. This behaviour includes, for instance, the general shape of the time course of variables, the presence of discontinuities and the sensitivity of output to parameter values.

If the behaviour of a model matches qualitatively that of the real world system, a quantitative comparison and an evaluation of the predictive success of the model should be made. At this stage, statistical tools can be useful. But even when sufficient and accurate data are available, a model cannot be proven to be correct. Sometimes model behaviour can be falsified, and thus one or more model components may be shown to be in error. Behaviour analysis is a useful form of sensitivity analysis. It is done by increasing or decreasing the value of a single parameter or group of parameters over a reasonable range and comparing direction of change and shape of the output(s) with the known or expected direction and shape. Illustrations are given by INNIS (1978), LOOMIS and NG (1978) and SCHREIBER et al. (1978).

Evaluation of models remains often limited in depth as a result of too small a data base. Some models are only "evaluated" by establishing a good correspondance between "predicted" and "observed" results, while these observed results were used to derive constants in the model. That this is a dangerous procedure needs no further emphasis. Strong experimentation is indispensable in parallel with modeling: experimentation at the explainable level for evaluation, and at the explanatory level for further improvement.

Sources of increasing concern are errors in models and in their documentation. The fundamental and most difficult errors are conceptual mistakes. Apart from these, even carefully screened simulation programs often contain simple technical errors, such as key-punching errors, dimensionally incorrect parameters and deleted variables in expressions, or deleted equations. Some of these appear when the model is used to simulate new situations, or when someone else studies the simulation program. By vigorous evaluation, modelers should eliminate as many errors as possible before releasing the model. No guarantee, however, can be given that a model is free of errors.

Another source of increasing concern is the briefness or inaccessibility of descriptions of most models and their programs. Too often, one is forced to believe, or disbelieve, an author without being able to check it. Some of the advantages of modeling (Sect. 4.1) are then lost.

4.2.4 References to Modeling Techniques

Papers and textbooks are available to introduce and accompany researchers into modeling. Instead of citing from these texts, I will refer to a selection of them.

De Wit (1970), Waggoner (1970), Hesketh and Jones (1976) and Baker and Curry (1976) express views on modeling in biology and agronomy. De Wit and Goudriaan's (1978) textbook introduces university students into analysis of dynamic ecological systems and shows ways to model them in C.S.M.P., a suitable simulation language (manuals are available from IBM 1975). A textbook on simulation of growth of crops, with examples of summary models in C.S.M.P., was published recently (Penning de Vries and Van Laar 1982). Thornley (1976) presents mathematical analyses and simulations of simple processes and small systems in plants. Brockington's (1979) textbook for beginning modelers in agricultural sciences avoids mathematical complications. Hall and Day (1977) wrote a textbook with introductions into ecological models. Innis (1978) and Rose and Harmsen (1978) give illustrations and applications of sensitivity analysis. Loomis et al. (1979) cite journals in which many models are published and discussed. Ören (1977) reviewed the available simulation languages.

4.3 Current Models in Different Phases of Development

4.3.1 Introduction

For a decade models have been used to simulate plant growth and crop productivity. The processes of the carbon and the water balance have been strongly emphasized. As a result, many aspects of models at the production levels 1 and 2 (Sect. 4.2.1) are now well developed. This will be demonstrated in Sect. 4.4. Over the last few years, some simulation studies have been directed towards relationships of plant growth and availability of nutrients from the soil. However, knowledge of the underlying processes is as yet poorly developed. As a consequence, their models are still less advanced. Phases of evolution of models will be discussed in this section. For this purpose, models of plant growth and production are divided into four groups. The *regression type model* provides the simplest characterization of a system. It relates crop yield to one or more environmental variables in a single, static equation. Next are three classes of dynamic models: the preliminary model, the comprehensive model and the summary model. During development, a model moves gradually from one phase into the next. *Preliminary models* are defined as models with structure and data that reflect current scientific knowledge, and that are simple because insight at the explanatory level is still vague and imprecise. A *comprehensive model* is a model of a system of which essential elements are thoroughly understood, and in which much of this knowledge is incorporated. *Summary models* are abstracts of comprehensive models: essential aspects are formulated in less detail than is possible for reasons of simplicity and accessibility. Summary and comprehensive models are found at the levels of production where soil water or weather limits growth, whereas models for the lowest production levels are predominantly of the preliminary and regression type. The models of these developmental phases differ considerably in their value for instruction, for prediction, for scientific progress, and in simplicity. Table 4.2 rates them on an arbitrary scale.

"Production" is defined as increase in biomass of a crop or a vegetation, and is usually considered on a time scale of weeks or a growing season; "growth" is the process of formation of biomass and its distribution over plant organs, and is typically expressed in hours or days.

4.3.2 Regression Models of Productivity

Regression models are attractive because of their simple and straightforward relation between yield and one or more environmental variables, but they are never accurate, and cannot be generalized a priori to other areas, other crops or other years. They are briefly discussed because they are often applied by those for whom agricultural productivity is basic information, such as economists and food industries. A regression model can be the best predictive tool available if the mechanism by which the environmental variable leads to a certain yield is obscure. This situation was quite common and is still not rare. For example, quantification of P-fertilization for increasing farm yields in Australia is performed successfully with the regression type model DECIDE (BENNETT et al. 1978) and is as yet impossible with dynamic models. Many other examples could be quoted. As our operational knowledge of plant growth increases, models in this group will be outperformed by comprehensive and summary models.

Regression models are also still useful for prediction of the yield of a single crop species over a large region with a variety of soil conditions, cultivation practices and disease problems: a combination of such factors is still beyond dynamic simulation. An example of a successful study on the scale of large countries is that of STEYEART et al. (1978), who related the average monthly barometric pressure (which is correlated with cloudiness and radiation) to yield of wheat to estimate productivity in inaccessible areas. How to deal with historic

Table 4.3. Some regression models

Description	Production level	Regression output	Regression input	Remark	Reference
Productivity corn and sorghum: Nebraska and Israel	3	Dry matter and grain yield	Precipitation actual and potential evaporation	Simple soil moisture balance included	HANKS 1974
Potato leaf growth	2, 3	Morphological development	Air temp., solar radiation and rainfall	Similar model used for other crops	HAUN 1975
Vegetable crops: England	1	Plant weight	Seed weight, sowing density		GREENWOOD et al. 1977
Tree height: Arctic regions	1	Daily increment tree height	Air temp., development stage		HARI and LEIKOLA 1974
Quantification P fertilization: Australia	3, 4	Economic yield	Soil type and fertilizer history, economic environment	Applied for farmers advice since 1975	BENNETT et al. 1978
Methodology of analysis of trends in yield	2–4	Yield	Historical trend, precipitation		TALPAZ et al. 1977

trends in yield increase is demonstrated by TALPAZ et al. (1977). A short selection of regression models is listed in Table 4.3.

One of the pitfalls of regression models, oversimplification, is demonstrated by results of a simulation study: VAN KEULEN et al. (1976) simulated biomass production in fertilized pastures with natural rainfall in a dry zone of Israel and concluded that in 5 out of 13 consecutive years, water was the primary limiting factor of plant growth (production level 2). In the other, wetter, years, production was determined by soil fertility (production level 3), which is not related to precipitation in the same way. This shift does not depend on the amount of precipitation only, but also on its distribution in time. A regression model cannot cope with such variation. Indeed, it is a gross underestimation of the inherent complexity of biological and agricultural systems and their environments to assume that from any (multiple) regression of yield to the many factors that exert an influence on it, a clear and accurate picture will emerge that is applicable to other crops, other geographic areas and even other years. Attempts to do so have failed, as MURATA (1975) and NELSON and DALE (1978) concluded, except when the environment is relatively uniform (REINHARDT 1975). Instead of pursuing further statistical analysis of yield and weather data, a better route is to analyze systematically such production systems and to devise explanatory models. This route is more promising and many have embarked on it, but one should realize that it is often a long way.

All categories of dynamic models may predict yield better than a regression model does in a specific case. However, the more accurate the dynamic model is, the more information is required for initialization and about driving variables. If these are unavailable, then the regression model may still be our best option. The barometric pressure-production model is a point in case. It is obvious that the quality of the prediction of crop yield with all models will remain restricted for a long time to come by our inability to predict the actual course of variables, such as precipitation.

4.3.3 Preliminary Models at the Lowest Levels of Crop Productivity

At the frontiers of knowledge, preliminary models are very common. They enable the quantification and evaluation of hypothesis and are useful as such, but they seldom survive long. The category of preliminary models shows the largest diversity of hypotheses on processes and their relationships on the explanatory level, making these models highly interesting for scientists and stimulative for experimental research. Their predictive value is generally fairly low. If preliminary crop growth models are presented unreservedly, one often finds that potential users are actually discouraged as a result.

A typical preliminary model is that of tulip growth by REES and THORNLEY (1973). It is a very small dynamic model of growth of individual plants, that describes in an unrefined way all essential processes. It consists of only 18 simple computer statements, and describes the carbon flow from a mother bulb to the developing top and daughter bulb from top emergence to leaf death.

The top is supposed to grow heterotrophically, while the daughter bulb utilizes mother bulb reserves and monopolizes also all photosynthetic products. Plant net photosynthesis equals top weight times the incident light intensity times a constant. Emergence and death occur at fixed dates. This model is labeled preliminary because of the very simple description of the photosynthesis, respiration and growth processes, and because of the absence of any consideration of environmental conditions on the date of emergence and on the rate of plant development.

A short and very incomplete list of other preliminary models is given in Table 4.4. Some models are small, others are complex. Some large models consist of many submodels, with various degrees of complexity. Such models are characterized "preliminary" on the basis of their most important section.

Most of the models on growth under nutrient stress fall into this group, as essential physiological processes are still little understood (PENNING DE VRIES 1981), the effects of extreme shortage of some micro- and macroelements being an exception (cf. WRIGHT 1977). The fact that many growth studies are performed in climate rooms, where nutritional problems are easily masked or are absent, may have contributed to the slow development. The production-limiting role of nutrients has been recognized by farmers and agronomists for over a century, and much information has been collected on details of agronomical, plant physiological and microbial processes. However, the link between them by crop physiology is only emerging (Sect. 4.4.3). Modeling seems the best tool to perform the integration, meanwhile improving our capacity to simulate plant growth at the lowest levels of plant production.

Models to simulate the development of form of plants and of its organs are still also of a preliminary nature, or even purely descriptive. Section 4.4.4 discusses some of their features.

Table 4.4. A list of preliminary models. The knowledge used to program certain aspects is characterized by R if it is as simple as a regression of output to input, with P and C if the knowledge is preliminary or comprehensive, respectively, and with S if comprehensive knowledge has been summarized. E indicates that experimental data were used. The most important sections are outlined. For further explanation, see text

Name	–	SORGF	–	–	MEDECS	ELM
Description	Grass or cereal crops	Sorghum crop	Plant growth	Cabbage growth	Growth mediterranean shrubs	Submodels including primary production N flow and P cycling
Production level	2	2	2	3, 4	3, 4	3, 4
Reference	BROCKINGTON 1978	MAAS and ARKIN 1978	RUSSO and KNAPP 1976	BARNES et al. 1976	MILLER et al. 1978	SAUER 1978; REUSS and INNIS 1977; COLE et al. 1977
Carbon balance:						
compartments	Roots, stems, old and young leaves, reserves	Leaves, roots, culms, grain	Total biomass	Roots, leaves	Old and new roots, leaves, stems, reserves	Live and dead shoot, roots, seeds, crowns, litter, soil organic material
photosynthesis	C	C			S, from transp.	R
maintenance proc.	C	P	R	R	S	R
rate and efficiency of growth	P	P			S	R
Water balance:						
compartments	Soil	Soil	Soil	Soil layers	Soil layers	Soil layers
transpiration	C	C	C	?	S	R
effect of water stress on growth	?	P	Growth proportional to transpiration	Through N-uptake	Through photosynthesis	Through photosynthesis
Nutrient balance						
compartments	–	–	–	Organic matter, NO_3^-, NH_4^+ in soil, roots, leaves	For N and P: soil layers, all C-compartments	For N and P: soil layers, all C-compartments
uptake	–	–	–	P	R	P
redistribution	–	–	–	–	P	–
effect of N level on growth	–	–	–	Reduction growth potential	P	[N] and [P] affect photosynthesis
Morphology						
leaf area	R	C	–	–	R	R
number of organs considered	–	–	–	–	Leaves, stems	–
Other important features	Germination	Germination	–	Growth response to K	Mixtures of species can be simulated	Phenology
Time step	Hour	Day	Week	Day	Day	Day
Evaluation	Some	Extensive	None	Some	Little	Extensive

Explanatory models are also not far developed for the simulation of effects of diseases and pests on crop growth. GUTIERREZ and WANG (1977) summarize how they and others have tried to link models of population growth of insects, mites and other pests to crop-growth models to give "pest management" a more solid basis. Such interdisciplinary models are essentially preliminary models at most, as the interface of micrometeorological and physiological conditions is not yet simulated in sufficient detail by any crop-growth model, and entomologists and plant pathologists hesitate to extrapolate knowledge obtained in laboratories to field situations (RABBINGE, personal communication).

Climate rooms are valuable research tools, but are also a possible source of problems: rate constants and parameters for physiological processes that are determined indoors do not always apply to outdoor conditions as a result of acclimatization of field plants to harsher conditions. The maximum rate of photosynthesis of leaves, for example, has often been underestimated as a result of this. Careful reflection is therefore needed on the extrapolability of data obtained in laboratories to field conditions.

4.3.4 Comprehensive Models of Growth and Productivity

Reasons for developing comprehensive models are scientific curiosity, and the expectation that such models may finally become predictive tools. From good comprehensive models, summary models should be made, as large and complex models are almost impossible to communicate in full to potential users. Comprehensive models should therefore not be considered as a final stage, but should necessarily be followed by a summary. Some comprehensive models are described in Table 4.5.

Though the summary model of a system may be the most utilized version, in some cases it may be necessary to employ the full complex model, e.g. when a high accuracy of results is required. Simulation of the development of diseases may be such a case, where information is required in great detail in time and space on air temperature and humidity in the canopy, and on physiological conditions of plant organs.

The model by DE WIT et al. (1978) may serve as an example of a comprehensive model. It simulates vegetative growth of crops at non-limiting levels of soil water and soil nutrients on the basis of standard meteorological observations and many physical, biochemical and physiological characteristics. It considers neither germination nor the reproductive growth phase.

Growth in this model is defined as "increase in dry weight of the structural plant material, i.e. total dry weight exclusive of those organic substances that are classified as reserves. ... Micro-weather is calculated from the weather measured at screen height, the extinction of radiant energy from sun and sky within the crop being taken into account. The infrared radiation from the canopy is also computed. A calculation of the distribution of radiation over the leaves is necessary for computation of assimilation and transpiration. The architecture of the crop determines this distribution of radiation and has to be defined. The extinction of turbulence in the canopy is also considered, so that transfer of heat, vapour and carbon dioxide can be computed. The ratio of latent and sensible heat exchange regulates to a large extent the micro-weather and this ratio is determined largely by stomatal behaviour. ... The assimilation of carbon dioxide by the canopy is calculated by adding the assimilation rates of the variously exposed leaves in successive leaf layers. These rates are dependent on light intensity,

Table 4.5. A list of comprehensive models. For further explanation, see Table 4.4 and text

Name	SIMED	SUBGOL	BACROS/PHOTON	SOYMOD	ARID CROP	GOSSYM + RHIZOS
Description	Alfalfa crop	Sugar beet crop	Vegetative growth grain crops	Soybean crop	Herbage growth and water use in arid regions	Growth and development cotton crop
Production level	1	1	1	1, 2	2	2, 3
Reference	Holt et al. 1975; Schreiber et al. 1978	Loomis and Ng 1978; Loomis et al. 1979	De Wit et al. 1978	Curry et al. 1975; Meyer et al. 1979	Van Keulen et al. 1980a	Baker et al. 1976; Lambert et al. 1976
Carbon balance:						
compartments	Roots, leaves stems	Roots, storage root, leaves, reserves	Root, shoot, reserves	Roots, leaves, stems, reprod. organs, reserves	Roots, leaves, stems, seeds	Leaves, roots, stems, fruits
photosynthesis	C	S	C	C	S, from transp.	C
maintenance proc.	R	C	C	S	S	P
rate and efficiency of growth	C	C	C	C	P	C
Water balance						
compartments	–	–	–/plant	Soil	Soil layers	Soil compartments
transpiration	–	C	C/C	C	C	C
effect of water stress on growth	–	Direct and through photosynthesis	Direct and through photosynthesis	Through photosynthesis	Direct and through photosynthesis	Through photosynthesis and N uptake
Nutrient balance						
compartments	–	–	–	Roots, stems, leaves, reprod. organs	–	Soil compartments, leaves, roots, stems, fruits
uptake	–	–	–	–	–	P
redistribution	–	–	–	R	–	?
effect of N level on growth	–	–	–	R	–	P
Morphology						
leaf area	R	R	E	R	R	R
number of organs considered	–	Leaves	–	Fruits	–	Fruits
Other important features	Carbohydrate translocation	Leaf ageing	Micrometeorology	N assimilation	Germination, water stress death	Germination, root and inflorescence morphogenesis
Time step	Hour	Hour	Hour/Second	Hour	Day	Day
Evaluation	Extensive	Extensive	Extensive	Extensive	Extensive	Extensive

CO_2 concentration in the ambient air and resistance to CO_2 diffusion from the atmosphere towards the active sites. Transpiration and CO_2 assimilation interact strongly" (see Sect. 4.4.2.1). Respiration in this model "is the sum of maintenance respiration and growth respiration. The latter is caused by the conversion of reserves into structural material and is therefore proportional to the rate of growth. The intensity of growth respiration is affected by the chemical composition of the new material. This intensity is independent of temperature, but growth respiration is indirectly influenced by temperature through the temperature dependence of the growth rate. ... The rate of maintenance respiration depends on the turnover rates of proteins and the resynthesis of other degraded compounds and the maintenance of ionic gradients. This respiration process therefore depends largely on the chemical composition of the plant. The rate of maintenance respiration is sensitive to temperature. The growth rate of the organs is dependent on the amount of reserves and temperature. Under internal water stress, growth of shoots is retarded by making a larger proportion of reserves available for growth of roots. By this mechanism a functional balance is maintained between root and shoot. ... The water status of the plant is determined by the balance between transpiration and water uptake from the soil. The transpiration rate of the crop is found by adding the transpiration rates of the variously exposed leaves in successive leaf layers of the crop. These rates are calculated from the radiation absorbed, resistance of the laminar layer, humidity and temperature of the ambient air and stomatal resistance. Stomatal resistance is either controlled through CO_2 concentration and assimilation, or through the water status of the plant. The calculation also provides leaf temperatures, which are used in the photosynthesis section and averaged to give the crop temperature that affects growth and respiration. Water uptake is determined by the conductivity of the root system, the water status of the plant and that of the soil. The water status of the soil is assumed to be optimal (field capacity), so that the transport of water within the soil can be ignored. The conductivity of the root system is dependent on the amount of roots, their degree of suberization and soil temperature" (DE WIT et al. 1978).

The model has been developed over more than a decade by DE WIT and a team of co-workers. Laboratory research, literature study and frequent evaluations led to a model that simulates growth, yield and water use quite reliably over a wide range of environmental conditions for C_3 and C_4 crops. Its structure reflects cereal and grass crops, and small but specific sets of parameters and functional relationships specify the actual species under consideration. The model is adaptable to other types of species, but this has not yet been done. Like all models in this group (Sect. 4.2.2), De Wit's model is particularly weak in the section regulation of distribution of biomass, and in the development of leaf surface area. The latter limitation is a handicap in the early stages of growth, and to its transferability to other geographic areas. The simulation of the growth of an irrigated rice crop in a tropical area, shown in Fig. 4.2, is a typical example of an application of the model BACROS to other environmental conditions than those of temporal zones in which it was developed.

De Wit's model, programmed in CSMP, has two versions of which essentially only the time steps differ: BACROS simulates growth and productivity of a field crop for a growing season with time steps of 1 h; PHOTON simulates gas exchange of a crop, usually in an enclosure, for a few days with variable time steps of tens of seconds. Two versions were developed so that the model could be evaluated with periodic harvests of the field crop (BACROS), while canopy photosynthesis and transpiration at the first lower hierarchical level could be evaluated with crop enclosures (PHOTON). At least one other group developed an enclosure for use in the field crop to allow accurate model evaluation (PHENE et al. 1978).

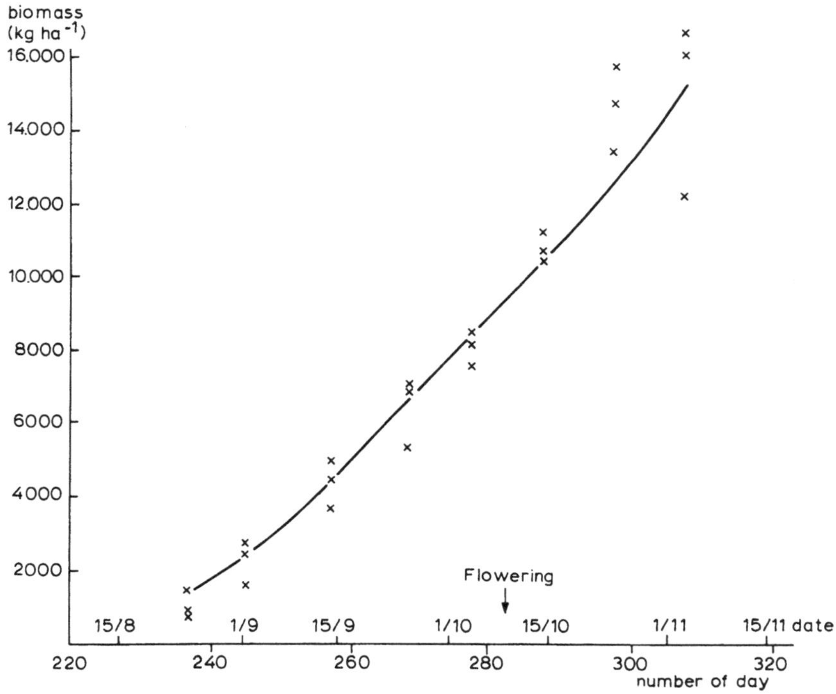

Fig. 4.2. The simulated and the actual course of the dry weight of the above ground biomass of a rice crop in Mali, West Africa. Water and nutrient shortages at any time were excluded. The individual values of periodic harvests in triplicate of an experiment in 1978 are indicated (×); the *continuous line* represents the results obtained with the simulation model BACROS (run 21/4/80)

It is not by accident that the comprehensive models currently found are all at the productivity levels 1 and 2. Processes of the carbon and water balance received much attention from crop physiologists. Only the model GOSSYM (BAKER et al. 1976) for cotton, a crop that suffers necessarily from some nutrient stress, includes also a N balance. However, as most of the farming is done under nutrient stress, the utility of the current comprehensive models is still restricted to setting maxima for yield potentials and to establishment of the contribution of the individual processes and factors to it (DE WIT and PENNING DE VRIES 1981).

4.3.5 Summary Models at the Highest Production Levels

Summarizing a comprehensive model can (and should) be done to make it (more) accessible to others in an intellectual and in a practical sense. The extent to which summarizing is useful depends on many factors, including future use of the model and its inherent complexity, but simplification should achieve a level at which the model becomes really accessible to non-specialists. In the process of summarizing, it is essential to indicate specifically the limits within which the model is valid. Construction of summary models should be done by scientists who know the comprehensive model by heart, and who are in contact with potential users for suggestions in which direction and to what extent to summarize it. Unfortunately, modelers may not always be motivated to do so, as the process provides little scientific challenge. Summary models can be made by shedding all excessive detail, using sensitivity analysis (as ROSE and HARMSEN 1978 show for ecosystem models) and by regression of model results to the main driving variable of the system.

Some static summary models have been made of aspects of the carbon and water balance. Current dynamic summary models of crop productivity simulate growth only in very good conditions of soil water and soil nutrients. In technical sciences, the results of simplification of well-developed models are called metamodels (MESAROVIC et al. 1970) or repromodels (MEISEL and COLLINS 1973). Such summary models are instructive, but the procedure to arrive at them emphasizes too greatly the reproduction of behaviour and too little the reproduction of the basic mechanisms to be advisable for agricultural and biological models.

An example of a summary model is that by VAN KEULEN (1976) for potential rice production. Its basic equation for the rate of daily growth can be given as:

$$\text{growth} = (\text{GP} \times 0.68 - \text{MC} \times \text{DW}) \times \text{CE}$$

GP stands for gross photosynthesis (in kg CO_2 m^{-2} d^{-1}; 0.68 converts it to kg glucose m^{-2} d^{-1}), DW for total dry weight (kg dry matter m^{-2}), MC for the maintenance coefficient (kg kg^{-1} d^{-1}) and CE for the conversion efficiency of the growth process (kg kg^{-1}). Van Keulen distributes biomass formed in one time-step over roots, leaves, stems and, after flowering, to inflorescences plus seeds in predetermined proportions and related to the physiological age of the crop. GP is calculated from standardized data. The leaf surface area, required in the photosynthesis calculation, is found by dividing the leaf weight by 0.1 kg m^{-2}. MC is 0.02–0.015; the effect of temperature on MC could be neglected as this model was applied in a fairly constant environment. CE is only a function of the chemical composition of the biomass formed, and a value of about 0.7 is common (Sect. 4.4.1.3). Final yield is calculated by proceeding with time steps of 10 days. The summary model is of an elegant simplicity, and is programmable on a small hand calculator. Unfortunately, the exact program was not included in the publication. For a justification of the summaries made, see the original paper.

The structure of the above summary model is correct for many other crops with a determinate growth pattern; parameters and functions might need adjustment. The model is also principally correct at lower levels of productivity. However, the equation focuses on photosynthesis as the limiting factor for growth, which is not correct. Only very cautiously may the above summary growth model thus be applied to other productivity levels. At the level where availability of water limits productivity, one could use the transpiration coefficient to calculate gross photosynthesis from the actual transpiration rate (Sect. 4.4.2.1). This principle was used in the summary model for maximum world food production by BURINGH et al. (1975).

The respiration equation of MCCREE (1970), a regression model, can be transformed into:

$$\text{growth} = 0.75 \times (\text{GP} \times 0.68) - 0.015 \times \text{DW}$$

which shows a great similarity to Van Keulen's summary model. Progress in crop physiology can be shown by examining the differences between both: (1) the summary model gives correctly a higher priority to maintenance processes than to growth processes, which the McCree equation does not, (2) additional precision for the summary model can be added at will, and (3) the limitations of the new growth model are better known.

4.3.6 Results of Modeling Other Than Models

Apart from instruction and application in other models there are areas where comprehensive and summary models have been shown, or promise, to be useful:

1. Simulation can help to suggest *new areas for research* for modelers and non-modelers: (a) During construction of models, one often finds that crucial information is lacking. Such points may be simple to resolve by experimentation, like the determination of the sensitivity of a process for temperature; or may initiate a large study, such as on the efficiency of growth processes. (b) Sensitivity analysis can help to establish which data and which structures are the most important for model behavior and hence steer research into more efficient directions. It does not regard the question whether essential processes were neglected in the model. (c) Application of a comprehensive or summary model often needs the association with models in related fields. Hence it stimulates interdisciplinary cooperation and research.

2. Prediction. This point needs little explanation. Comprehensive models have already been used to predict growth rate and water use in conditions that elude experimentation, such as increasing atmospheric CO_2 levels (Van Keulen et al. 1980b). Food, chemical and fertilizer industries are becoming interested in dynamic modeling (cf. Barrett and Peart 1979), which signals their positive expectations.

3. Genetic Engineering. To plant breeders, deliberate suggestions might be given as to what plant characteristics should receive attention (cf. Loomis et al. 1979). Landivar (1979) suggests how to use the GOSSYM model to predict the possible results of genetic manipulation in cotton crops. De Wit et al. (1979) discussed improvements that might be considered, based on their earlier work with simulation models and systems analysis. The scope for such improvements may be great. Tomato production, for example, is much enhanced as a result of the breeding of a determinate variety suitable for mechanic harvesting from an indeterminate variety, difficult to harvest (Rick 1978). However, the breeding program started long before reliable plant growth models were available. Such a success compels modelers to modesty, and recalls Passioura's (1973) statement that "there is a much quicker way of getting ... a framework in which to hang one's research ... than spending a year to create a comprehensive simulation model".

4. Optimization of Plant Environment. An important problem is often the formulation of proper criteria for optimization. Results of attempts to optimize are still not exciting, and trying to achieve such goals is probably still a long shot, but seems basically a sound idea.

Considerable progress has been made in climate control in greenhouses. Optimization is directed towards reducing the economic cost of crop production by shortening the growing period (CHALLA and VAN DE VOOREN 1980) and by using air conditioning more efficiently (SEGINER 1980). GOSSYM (BAKER et al. 1976) was employed to evaluate management options, such as soil tillage, for the cotton field crop. Optimization of cotton yield by irrigation with simulations was suggested by STAPLETON et al. (1974) and MAPP et al. (1975). Optimization of productivity by means of pest management is currently receiving attention (WELCH and CROFT 1979).

4.4 Physiological Processes in Dynamic Models

4.4.1 The Carbon Balance

4.4.1.1 Introduction

The carbon balance is part of the system at all production levels, although it controls the growth rate only at the highest level. Intensive research has been done on it, in particular on photosynthesis. Current models consider photosynthesis as a starting point for growth, and a biomass increase as its result. However, photosynthesis is also often supposed to become reduced in a reversible way at high levels of reserve carbohydrates. This situation may occur at temperatures or nutrient levels that limit carbohydrate consumption more than photosynthesis, and also if the size or number of the growing organs has become too small (cf. TOLLENAAR 1977). In almost all models of plant growth, it is assumed that photorespiration reduces the flux of carbohydrates into the plant and that this is its only effect on growth. In high light, more ATP can be formed than is absorbed in CO_2 assimilation. NO_3^- reduction and other processes can then occur at the expense of the extra ATP, and can absorb an additional 10% or more energy (PENNING DE VRIES 1975b). DE WIT et al.'s model (1978) accounts for this detail.

Photosynthetic processes and their modeling has been dealt with in Chap. 16 of Vol. 12B. This section will deal with heterotrophic processes. A conceptual framework of the heterotrophic carbon balance processes for models with short time steps has geen given in Fig. 4.1. For such models, it is essential to distinguish at least one pool of reserve carbohydrates. In models with time steps of one day or more, production and consumption of carbohydrates are roughly equal, and reserves can be neglected.

As hormones are not yet linked quantitatively to carbon balance processes, they are ignored in all growth models. WAGGONER (1977) and LOOMIS et al. (1979) reviewed recently carbon balance modeling.

4.4.1.2 Distribution of Biomass

Description of Phenomena. The growth and appearance of plant organs occurs in a fairly fixed rhythm, particularly in species with a determined development pattern. It is customary to express the increase in weight of leaves, of roots, of stems and of storage and reproductive organs as a fraction of the total weight increase of the plant. These fractions change in time. The pattern of changes becomes fairly independent of external and internal conditions if expressed on a "physiological time" basis. Physiological time is calculated simply as degree-days above a threshold temperature, or as the integrated value of the development rate which depends non-linearly on the ambient temperature (e.g. DE WIT et al. 1970). The development rate can be altered by daylength in photoperiod-sensitive species. BAKER et al. (1976) reported satisfactory results for cotton growth simulation with 1-day time steps, using a fixed distribution pattern, on which the effects of carbohydrate and nutrient stress are superimposed. They treated internodes separately.

BARNES (1979), following a different approach, expressed weight of the storage root of sugar beet as a function of shoot weight and time. This, and other very direct relationships explain little, and are basically regression models at this level of detail. They are used to circumvent dynamic simulation.

The description of biomass distribution versus physiological time is often sufficient for models with time steps of 1 day. But it is insufficient if one looks more closely at model behavior, and is unreliable in unusual conditions. Moreover, the biomass distribution pattern has to be established experimentally for each individual species. The description of biomass distribution pattern is even less generalizable for indeterminate plants, as the distribution pattern depends also on growth conditions.

Mechanisms. At the explanatory level the growth rate of an organ is generally calculated as its potential growth rate multiplied with the current effects of temperature, water stress and concentration of carbohydrate reserves, and sometimes also with those of nutrient stresses (see LOOMIS et al. 1979). The potential growth rate diminishes as tissue ages, and a complex treatment of aging is essential when individual leaves are to be considered (LOOMIS et al. 1979; cf. DE WIT et al. 1970; DE WIT et al. 1978).

The effects of water stress, temperature and concentration of reserve carbohydrates on the growth rate have assumed values in all models, as they are difficult to measure. However, their exact values are also not very important for the overall effect, when the relative effects for organs are well chosen. A fairly direct method to measure growth rates has been proposed and utilized (PENNING DE VRIES et al. 1979). They confirmed the general form of the assumed relations of carbohydrates and of temperature to growth, but the response to water stress appeared to be indirect rather than instantaneous.

The extent to which internal and external factors influence the growth rate may differ between species and might be the result of adaptation processes as well. It is questionable whether these factors exert a numerically equal effect on growth of the different organs. BROUWER and DE WIT's model (1969) with a functional balance between roots and shoots, stated that roots are more sensitive than shoots for carbohydrate stress, while leaves suffer more from water stress. Their model simulates only one level of carbohydrates and one of water stress in the plant. Many models use the same concept. THORNLEY (1976) proposed that organs have similar sensitivities to such factors, but that the sugar concentration and water stress are unequal in shoot and roots. MCKINION and WEAVER (1979) provided an intermediate proposition. On the basis of a model for the Münch hypothesis for phloem transport, they suggest that the phloem flow equals about 5% of the transpiration stream. This implies that carbohydrate translocation is proportional to the transpiration rate, and some function of the carbohydrate concentration in the leaves and of the activity of phloem loading by leaf cells. This hypothesis may help to understand consequences of temperature changes and of water stress for carbohydrate transport.

Regrowth of plant organs or tillers after cutting was modeled successfully by DAYAN and DOVRAT (1977) and by FICK (1977), basically as a normal growth process that starts from dormant buds and occurs at the expense of reserve carbohydrates in the stems or roots.

Current simulation models treat carbohydrate reserves as a homogeneous pool. However, there are indications that only soluble carbohydrates are "sensed", whereas starch is a reserve carbohydrate that only replenishes the soluble carbohydrates when their level drops too low (cf. CHALLA 1976). A pool of nitrogenous compounds must also exist, as NO_3^- reduction occurs predominantly parallel with photosynthesis, but growth is continuous. This aspect of "reserves" is probably related to protein turnover (Sect. 4.4.1.4) and redistribution of N (Sect. 4.4.3.3). Simulation of the daily course of growth of individual plant organs will require a better knowledge of the carbohydrate and protein metabolism of mature cells (THORNLEY 1977; PENNING DE VRIES et al. 1979), and of the response of individual organs to internal and external factors.

4.4.1.3 Efficiency of Conversion Processes

Assuming that the intricate conversions that occur during growth can be approached by a series of biochemical reactions of which the stoichiometry, rather than enzyme kinetics, is crucial, PENNING DE VRIES et al. (1974) computed how much plant biomass can be formed from a certain amount of carbohydrates as sole substrate, and what the concomitant synthesis respiration is. They assumed maximal efficiency for all reactions. (Note that the efficiency of growth of the process is discussed, not its rate.) It is a fortunate situation that most of the growth processes proceed in heterotrophic conditions and that cells only rarely grow with their own supply of photosynthate. This considerably facilitates computations and modeling. For their computations, biomass is divided into six chemical fractions: proteins (i.e. all nitrogeneous compounds), carbohydrates, lignin, fats plus oils, organic acids and minerals. It was derived that about 0.7 kg of vegetative plant tissue can be formed from 1 kg of photosynthate. But this ratio is a variable, and is as low as 0.3 for seeds rich in oils and proteins (soybean) or as high as 0.85 for organs that store only starch. These conversion factors are probably similar for all species of higher plants, do not change as a result of stresses and are similar at all temperatures. DE WIT et al.'s model (1978) includes a detailed calculation of conversion efficiency and plant respiration.

The cost of translocation of components was estimated to be small although the only set of experimental data shows a high value (HO and THORNLEY 1978).

A direct evaluation of the computations of the conversion efficiency has not been performed. Indirect evaluations in rapidly growing systems were made by MOLDAU and KAROLIN (1977), PENNING DE VRIES and VAN LAAR (1977), YAMAGUCHI (1978) and HUNT and LOOMIS (1979). All experimental data support the validity of the approach, and the assumption that the conversion process is quite efficient. A conversion efficiency based only on chemical constituents of the biomass formed is generally used in comprehensive models, and the subject does not need a high priority for further research.

4.4.1.4 Maintenance Processes

Maintenance processes comprise the turnover of proteins and lipids in enzymes and membranes, the maintenance of concentration gradients across membranes, plus the generation of ATP to drive these reactions. Too little is known as yet about the rates of these processes and their regulation, particularly in plants, to calculate its intensity from basic data. Protein turnover, the most important fraction of maintenance, counteracts the spontaneous degradation of enzymes, and may give the plant a capacity to adapt to new conditions where other enzymes function better (PENNING DE VRIES 1975a). In addition it enables redistribution of N through the plant (MILLER et al. 1978). Basic research is still needed to elaborate this matter.

Measurements of maintenance respiration are difficult to obtain as it is often confounded with other respiratory processes. Moreover its rate is probably not a constant but related to the overall metabolic activity of the plant. Measurements indicate "normal values" of 0.015–0.06 kg CO_2 kg^{-1} dry matter d^{-1} in vegetative tissues (MCCREE 1974; PENNING DE VRIES 1975a) but substantially lower values in fruits and in storage organs. The substrate requirement for maintenance may seem negligible, but as it is a continuous process, the integrated value of maintenance respiration over a growing season is 0.5–2 times that of growth respiration. Maintenance respiration rates at one temperature differ considerably between species, but they are always considerably stimulated by increasing temperatures.

Currently, maintenance processes are usually programmed as a constant respiration rate per unit of biomass, sometimes it is related to the quantity of N. DE WIT et al.'s (1978) model assumes also a relation to the overall metabolic activity. The effect of temperature is generally included. Further measurements and analysis of maintenance processes, their rates and regulations, will considerably improve the quality of the carbon balance models (cf. THORNLEY 1977; BARNES and HOLE 1978).

4.4.1.5 Root Growth and Root Respiration

Root growth has received little attention in comprehensive growth models, RHIZOS (LAMBERT et al. 1976) being an exception. There is little knowledge about it on the explainable and on the explanatory level. Exudation may sometimes be an important process.

Root respiration is more difficult to measure and to interpret than respiration of any other organ. A thorough analysis would be very welcome.

Root respiration consists of 3–5 components: (a) maintenance respiration; (b) growth respiration; (c) ion uptake. Uptake of a single ion requires little energy, but as ion absorption for the whole plant is concentrated in the roots, respiration linked to uptake can become a large fraction of total root respiration (VEEN, personal communication). It is unknown how many membranes each ion crosses actively, and this makes a theoretical approach unreliable; (d) in some species, part of the uptake of NO_3^- occurs in exchange with a hydrocarbonate ion (HANSEN and JENSEN 1977; DE WIT et al. 1978). This non-respiratory process causes evolution of CO_2. (e) LAMBERS (1979) observed that roots of some species utilize intensively the cyanide-resistant pathway of respiration, a pathway that yields ATP 3 times less efficiently than the "normal" pathway. He suggested that this occurs to eliminate carbohydrates supplied to roots in excess to the current demand. This explanation does not fit easily into the functional balance concept except when it occurs exclusively at low levels of plant productivity.

4.4.2 The Water Balance

4.4.2.1 Introduction

Transpiration and soil water balance models are reviewed in Chaps. 8 and 1 of Vol 12B, VAN KEULEN (1975) and HSIAO et al. (1976).

The water balance is of particular interest when it limits the growth rate of plants or crops. This may be when an excess of soil water blocks root aeration, and hence reduces root functioning (cf. Chap. 14, Vol. 12B). No publications on simulation of this situation were found.

Shortage of water causes a reduction of photosynthesis and of transpiration, and possibly also a change in the distribution of new biomass. The effects of mild to moderate stresses for brief periods have been studied in detail, and there is a fair understanding of the mechanisms involved (cf. Chap. 9, Vol. 12B and HSIAO et al. 1976). The effects of severe stresses that result in partial death of plants has been much less investigated (cf. Chap. 10, Vol. 12B).

Models to simulate the daily course of carbon and water balances produce the effect of water stress always by stomatal closure. Some growth models that use time steps of 1 day reduce photosynthesis by a factor that depends on the average water stress of the last day. Other models with 1-day time steps calculate gross photosynthesis by dividing daily transpiration by a water use coefficient. The latter approach seems superior, due to the constancy and predictability of this coefficient, particularly when transpiration is expressed relative to potential evapotranspiration (DOORENBOS and KASSAM 1979). This constancy is brought about by (1) gross photosynthesis and transpiration are almost proportional because stomata offer the largest diffusion resistance to both, particularly under water stress, (2) the efficiency of the growth process is unaffected by water stress, and (3) the rate of maintenance respiration probably diminishes slightly as a result of the lower overall metabolic activity.

A corollary about the water use coefficient is that non-stressed plants often show a similar constancy of the coefficient, as stomata open and close in correspondence with the rate of photosynthesis (see Chap. 7, Vol. 12B). This regulating mechanism maintains a constant CO_2 concentration within stomata over a wide range of photosynthetic rates (RASCHKE 1975, GOUDRIAAN and VAN LAAR 1978, WONG et al. 1979). BACROS includes such a mechanism (VAN KEULEN et al. 1980b). Further research into this mechanism seems worthwhile, as it may lead to very efficient water use.

4.4.2.2 Water Uptake and Root Growth

Models of plant growth with a water balance calculate the flow of water through the plant by dividing the difference in water potential between soil and air by the sum of plant resistances, applying Ohm's law. One of those, the resistance of the xylem vessels, is rarely considered in dynamic models, although this may be necessary under high transpirational loads (PASSIOURA 1972). The xylem resistance of the root system was modeled by SEATON and LANDSBERG (1978). Two types of preliminary models for water uptake and root growth have been developed successfully for agricultural situations.

VAN KEULEN'S (1975) model of water uptake and root growth by annual grasses in fertilized pasture is conceptually simple: canopy transpiration equals potential evapotranspiration (a meteorological characteristic) at low transpiration rates, and water is absorbed equally from all rooted soil layers. If soil water runs low in one or more layers, only a fraction of this potential rate is extracted. Root growth is modeled as a constant increment of rooting depth per day (modified for soil moisture content and temperature only). VAN KEULEN ignores rooting density, as he supposes density to be irrelevant above a low, minimum value.

LAMBERT et al. (1976) and BAR YOSEF et al. (1978) modeled root growth in a more sophisticated way, as their model deals with row crops and simulates uptake of NO_3^-. The soil is thought to be divided into compartments horizontally and vertically. Root growth per compartment is simulated in much the same way as discussed on p. 137 and in three age classes. Once a minimum amount of roots has been formed, roots can grow into lower and into adjacent compartments. They assume a constant fraction for root sloughing per unit of time.

Both models are fairly successful in simulating root growth and water uptake in the environments for which the models were devised. As their scientific depths are still quite superficial, these models can only be used in other situations with much care. Models for root growth in ecosystems have been less successful, not least because the intensive turnover of root systems makes good observations hard to come by (INNIS 1977).

4.4.2.3 Water Stress, Dry Matter Distribution and Death

The effect of mild water stress on dry matter distribution is to promote growth of the root system relative to the shoot (BROUWER and DE WIT 1969; HSIAO et al. 1976). By the reduction of photosynthesis during stress, plants suffer from carbohydrate shortage for growth and maintenance of heterotrophic tissues.

DOORENBOS and KASSAM (1979) quantified the sensitivity of the yield of 26 crops in different phases of development to water stress on the basis of a voluminous summary of experimental results. They concluded that water shortage during flowering and, to a smaller extent, during the remainder of the reproductive period, had a larger effect on yield than stress during the vegetative period (though there are exceptions to this rule). Abortion of flowers, induced by carbohydrate shortage during flowering, is probably an important aspect of it: as fruits generally do not grow beyond a certain size, their number sets a maximum to crop yield. Most of the photosynthate formed during the reproductive phase goes into the fruit, so that reduction of photosynthesis during this phase cuts directly into the final yield. Though these effects have been described and are qualitatively understood, a dynamic model of it has not been encountered.

VAN KEULEN et al. (1980a) included the effects of severe water stress on development, photosynthesis and death rate in ARID CROP. Their simulations of growth in quite different growth seasons showed a realistic behavior, although there is still little of a physiological basis below the formulation of these effects.

4.4.3 Nutrient Balances

4.4.3.1 Introduction

Only the macronutrients N and P will be discussed here. N becomes available in the soil by a complex set of largely microbiological processes an integrated picture of which is developing. These are reviewed in Chap. 5 of Vol. 12C and by BEEK and FRISSEL (1973), REUSS and INNIS (1977) and VAN VEEN (1977). The availability of P depends on microbiological and physical processes and is less understood. Interesting papers on P availability in soils have been written by COLE et al. (1977) and BEEK (1979). For reasons largely unknown, considerable differences exist in ion uptake by crops of different species (GERLOFF 1977).

Uptake of nutrients has only been modeled in a preliminary way (Sect. 4.3.3). VAN KEULEN et al. (1975) calculated that in a moist soil, all NO_3^- can be absorbed by a root in about 24 h from a cylinder with a radius of about 10 mm, the N arriving partly in the mass flow of water, partly by diffusion. Application of the same reasoning to phosphate shows that P is available only from a cylinder of about 1 mm radius in the same period. The radii of these cylinders increase proportionally to the square root of time, but root sections may remain active in uptake for only a few days. As a consequence, a rooting density that allows only the absorption of all available N in the soil does not suffice to absorb all P. This matters particularly in young plants. SELIGMAN and VAN KEULEN (1980) calculated the demand for N of the crop. The lower value of this demand and the available N is supposed to be absorbed, assuming that all NO_3^- in the rooted layers is close enough to roots to be absorbed within 1 day after becoming available. LAMBERT et al. (1976) compute NO_3^- uptake as being equal to the NO_3^- dissolved in the water absorbed for transpiration, neglecting N supply by diffusion.

Obviously, much research remains to be done to clarify, and even to pose, questions about availability of N and P in the soil and their uptake by the root system.

4.4.3.2 Growth and Availability of N and P

The N concentration in annual grasses and cereals remains within certain limits: it diminishes with age from about 5% in very young plants to 1.5% in mature plants with ample fertilizer, while these figures go from 2.5% to 1% in plants on very poor soils (SELIGMAN and VAN KEULEN 1980). These upper values apply also in tropical grasses, but the minimum values are lower: from 1.25% to

0.5% (Penning de Vries et al. 1980). A second important observation is that in experiments on poor soils a little fertilizer stimulates crop productivity considerably but increases its N concentration only little, while the N concentration increases more rapidly than biomass production does at high fertilization rates. Seligman and Van Keulen supposed that a crop grows as long as the actual N concentration remains above the minimum N concentration (or until it dies for other reasons). The N demand by the crop at any moment equals the weight of the biomass times the difference between the actual and the maximum N concentration.

Barnes et al. (1976) model crop growth in direct response to availability of N and of K from the soil. Their model is based on analysis of experiments with vegetable crops.

Both approaches can give a fair impression of growth and productivity if one is interested in total biomass. It does not suffice if the weight of a specific fraction is to be known. In such cases, the phenomenon of redistribution of N within the plant plays a predominant role, as it determines to a large extent quality and quantity of individual organs during growth under nutrient stress.

4.4.3.3 Redistribution

An illustration of redistribution of elements during growth in a sugar beet plant is given by Van Egmond (1975). The availability of nutrients from the soil diminishes, particularly in final stages of plant growth. Availability of nutrients for individual organs depends then not only on the supply from the roots but increasingly on what becomes available by export from old organs.

Explanatory models of crop growth under conditions of nutrient limitation require much more knowledge about growth and redistribution processes. Few models describe redistribution in a preliminary way, and only for all leaves, stems or fruits together: Miller et al. (1978), simulating growth of shrubs, assume that half of the N incorporated in new tissue is stable, while the other half will be exported to support new growth. Penning de Vries et al. (1975) and Fagerström (1978) modeled a similar process for N in pine needles on an annual basis. Redistribution is included in GOSSYM (Baker et al. 1976) but it is not described how. SOYMOD (Curry et al. 1975) considers growth and N concentrations independently. Seligman and Van Keulen (1980) allow the N in excess of the minimum amount of N in leaves and stems to satisfy the N demand from seeds.

Most studies of the effect of P shortage compare the concentration of P in the tissue with a minimum and an optimum level. No dynamic model of growth and P availability is known to the author. Penning de Vries et al. (1980) suggested basing a judgement of whether P shortage limits growth or not on the ratio of P to N rather than on the P concentration as such. They argue that much of the P in the cells is structurally and functionally related to N.

There is an interesting consequence to redistribution. Export of N and P from leaves that exceeds their uptake leads to reduction of leaf photosynthesis and finally to leaf death. Hence, the more rapidly redistribution occurs, the shorter the reproductive phase and the lower seed yields will be. SINCLAIR and DE WIT (1976) treated this process quantitatively in soybeans, and labeled the fast redistribution, aging and death "self-destruction".

4.4.4 Morphogenesis

4.4.4.1 Introduction

Aspects of form are usually of little interest to the modeler of growth and production. The exceptions lie where plants interface with their environment: i.e. where form assists in the interception of light, CO_2, water and nutrients. Proper treatment of form and size of interfaces is indispensible in accurate models because of a positive feedback: at the highest production levels an overestimation of leaf area leads to an overestimation of photosynthesis and hence of biomass growth and leaf area growth, causing a further overestimation of photosynthesis. At the lowest production levels, such a positive feedback exists for root growth.

Models for the simulation of development of form of organisms and of organs are still highly descriptive and explains little, but attempts are being made to give the mathematical formulas a biological basis. ERICKSON (1976) reviewed such models. HOGEWEG (1978) described cell division characteristics, and derived a preliminary model of the development of form of cellular systems in space.

4.4.4.2 Leaf Surface Area

The ratio of area to leaf weight is in the order of 0.1 kg m^{-2} foliage, but differs considerably between species and due to growth conditions in a way that is not really predictable.

The surest approach to specify leaf area in models is to give observed values as a forcing function. The common way to simulate leaf area is the integration of its growth rate, which is set equal to the growth of leaf weight divided by a constant factor. This is useful in cases where leaf area development is not crucial to model behaviour. DE WIT et al. (1970) attempted to compute leaf area growth from leaf weight increase and leaf thickness, the latter varying according to an averaged value of gross photosynthesis. They were unsuccessful as a result of the positive feedback indicated.

HORIE et al. (1979) suggest that such approaches are fundamentally wrong because plants have a strong internal regulation of leaf area growth, independent of the carbon balance. Their concepts of leaf initiation and leaf growth are promising, but are as yet too little developed to judge their usefulness for dynamic growth models. It is one of the critical areas for further research if simulation at the highest production level is to be improved.

4.4.4.3 Root Surface Area and Root Profile

The variability of the root surface area/root weight ratio is much larger than that of leaves. In spite of this, most models that simulate root growth apply a constant factor to obtain root surface, root length or even root permeability from root weight. This is not disturbing at the production levels 1 and 2: the rooted volume is then more important for water uptake than root surface. At the lowest production levels root activity and the ratio of root length or root surface to root weight are important, for P uptake in particular. No quantitative model has been published on this aspect, nor on the effect of root architecture on effectivity of nutrient absorption. RHIZOS (LAMBERT et al. 1976) is the only dynamic model to simulate the root growth at different depths.

4.5 Conclusions

Models can be used in research as tools that enable the integration of knowledge of different processes that take place in a biological system. This allows simulation of the behaviour of the system, i.e. the calculation of the course of processes, and consequently the increase of a quantity, such as biomass. Even for the study of relatively simple agricultural and biological systems, models can be of considerable help.

Phases of development of individual models can be distinguished. If knowledge of its processes is not yet well-founded, then a model of this system is preliminary. It can be useful as an integrative, scientific tool, but predictions made with it are still of little value. Some models have been developed of systems many of the processes of which are well understood and quantified. Carbon balance models fall into this category. Such comprehensive models are generally large and complex. They are helpful as an integrative tool, and predictions made with them can be fairly good. Some such large models have been summarized. Such summary models are fairly simple to use, and can be used with some confidence to make predictions within a clearly delimited set of circumstances.

Evaluation of a model is a continuing process and takes many forms. Obtaining data, for the purpose of evaluation, about the real behaviour of the modeled system, that are independent of the data with which the model was produced, is therefore often difficult and can pose a serious problem.

As more of the basic physiological, biochemical and physical processes become understood, modeling will become even more useful as an integrative technique. An important part of models will take the form of dynamic and explanatory computer simulation models.

Acknowledgements. This review is primarily based on work by colleagues at the Department of Theoretical Production Ecology of the Agricultural University and at the Centre for Agrobiological Research in Wageningen. The familiarity of the author with their work has made a strong emphasis on it unavoidable. Mrs. Merkelijn made a critical selection of the very many titles of publications on models of plant growth or crop productivity that became available by computer literature research. The help of many colleagues in The Netherlands and elsewhere, who forwarded in 1980 their recent reprints, is much appreciated.

References

Baker CH, Curry RB (1976) Structure of agricultural simulators: a philosophic view. Agric Syst 1:201–218

Baker DN, Lambert JR, Phene CJ, McKinion JM (1976) GOSSYM: a simulator of cotton crop dynamics. In: Computers applied to large–scale agricultural enterprises. Proc US – USSR Semin, Moscow Riga Kishinev, Mississippi State Univ, Mississippi

Barnes A (1979) Vegetable plant part relationships II quantitative hypothesis for shoot-storage root development. Ann Bot 43:487–499

Barnes A, Hole CC (1978) A theoretical basis of growth and maintenance respiration. Ann Bot 42:1217–1221

Barnes A, Greenwood DJ, Cleaver TJ (1976) A dynamic model for the effects of potassium and nitrogen fertilizers on the growth and nutrient uptake of crops. J Agric Sci 86(2):225–244

Barrett JR, Peart RM (eds) (1979) Abstracts of the workshop on crop simulation, May 1979. Purdue Univ, West Lafayette, Indiana, USA

Bar-Yosef B, Lambert JR, Baker DN (1978) RHIZOS, a simulator of root growth and soil processes. Tech Contr 1559. SC Agric Exp Stn, SC, USA

Beek J (1979) Phosphate retention by soil in relation to waste disposal. PhD thesis, Agric Univ, Wageningen, The Netherlands

Beek J, Frissel MJ (1973) Simulation of nitrogen behaviour in soils, PUDOC, Wageningen, The Netherlands

Bennett D, Downes PA, Janes AVJ (1978) The 1978 phosphate fertilizer rates. Manage Rep 3 (CSIRO, Australia)

Brockington NR (1978) Simulation models in crop production research. Acta Agric Scand 28:33–44

Brockington NR (1979) Computer modeling in agriculture. Oxford Univ Press, London

Brouwer R, Wit de CT (1969) A simulation model of plant growth with special attention to root growth and its consequences. In: Whittington WJ (ed) Root growth. Butterworth, London

Buringh P, Heemst van HDJ, Staring GJ (1975) Computation of the maximum food production of the world. Publ 558 Dept Trop Soil Sci. Agric Univ, Wageningen

Challa H (1977) An analysis of the diurnal course of growth carbon dioxide exchange and carbohydrate reserve content of cucumber. Agric Res Rep 861, PUDOC, Wageningen

Challa H, Vooren van de J (1980) A strategy for climate control in greenhouses in early winter production. Acta Hortic 106:159–164

Cole CV, Innis GS, Steward JWB (1977) Simulation of phosphorus cycling in semi arid grasslands. Ecology 58:1–15

Curry RB, Baker CH, Streeter JG (1975) SOYMOD I, a dynamic simulation of soybean growth and development. Trans Am Soc Agric Eng 18(5):963–974

Dayan E, Dovrat A (1977) Measured and simulated herbage production of Rhodes grass. PhD thesis, Hebrew Univ, Jerusalem

Doorenbos J, Kassam AH (1979) Yield response to water, FAO irrigation and drainage paper 33, FAO, Rome

Edminster TW (1978) Concepts for using modeling as a research tool. TM520 USDA-ARS 20/1/78. US Dept Agric Washington DC 20250

Egmond van F (1975) The ionic balance of the sugar beet plant. Agric Res Rep 832. PUDOC, Wageningen

Erickson RO (1976) Modeling of plant growth. Annu Rev Plant Physiol 27:407–434

Fagerström T (1977) Growth in Scots pine (mechanism of response to N). Oecologia 26:305–315

Fick GW (1977) The mechanism of alfalfa regrowth, a computer simulation approach. Search Agric 7(3):1–26, Cornell Univ, New York

Flavell RB, Barratt DHP (1977) Mitochondrial efficiency and grain yield in wheat. Ann Bot 41:1323–1331

Forrester JW (1961) Industrial dynamics. MIT Press, Boston, USA

Gerloff GC (1977) Plant efficiencies in use of nitrogen, phosphorus and potassium. In: Wright MJ, Ferrari SA (eds) Agric Exp Stn, Cornell Univ, Ithaca
Goudriaan J, Laar van HH (1978) Relations between leaf resistance, CO_2-concentrations and CO_2-assimilation in maize, beans, Lalang grass and sunflower. Photosynthetica 12(3):241–249
Greenwood DJ, Cleaver TJ, Loquens SMH, Niendorf KB (1977) Relationships between plant weight and growing period for vegetable crops in the United Kingdom. Ann Bot 41:987–997
Gutierrez AP, Wang Y (1977) Applied population ecology: models for crop production and pest management. In: Norton GA, Holling CS (eds) Proc Conf Pest Manage. IIASA. Vienna, Austria
Hall CAS, Day JW (1977) Ecosystem modeling in theory and practice: an introduction with case histories. Wiley and Sons, New York
Hanks RJ (1974) Model for predicting plant yield as influenced by water use. Agron J 66:660–665
Hansen GK, Jensen CR (1977) Growth and maintenance respiration in whole plants, tops and roots of *Lolium multiflorum*. Physiol Plant 39:155–164
Hari P, Leikola M (1974) Further development of the dynamic growth model of plant height growth. Flora 163(5):357–370
Haun JR (1975) Potato growth environment relationships. Agric Meteorol 15(3):325–332
Hesketh JD, Jones JW (1976) Some comments on computer simulators for plant growth-1975. Ecol Model 2(3):235–247
Ho LC, Thornley JHM (1978) Energy requirements for assimilate translocation from mature tomato leaves. Ann Bot 42:481–483
Hogeweg P (1978) Simulating the growth of cellular forms. Simulation 31:90–98
Holt DA, Bula RJ, Miles GE, Schreiber GE, Peart RM (1975) Environmental physiology, modeling of alfalfa growth. I. conceptual development of SIMED. Res Bull 907. Purdue Univ, Indiana
Horie T, Wit de CT, Goudriaan J, Bensink J (1979) A formal template for the development of cucumber in its vegetative stage I, II and III. Proc K Ned Acad Wet Ser C 82(4):433–479
Hunt WF, Loomis RS (1979) Respiration modeling and hypothesis testing with a dynamic model of sugar beet growth. Ann Bot 44:75–77
Hsiao TC, Fereres E, Acevedo E, Henderson DW (1976) Water stress and dynamics of growth and yield of crop plants. In: Lange OL, Kappen L, Schulze E-D (eds) Water and plant life. Ecol Stud Vol 19. Springer, Berlin Heidelberg New York
IBM (1975) Continuous system modeling program III. general system information manual (GH19-7000) and users manual (SH19-7001-2). IBM data processing div, White Plains, NY
Innis GS (1977) Dynamic simulation of the belowground ecosystem. In: Marshal JK (ed) The belowground ecosystem. Range Sci Dept Ser 26, Colorado State Univ, Fort Collins, Colorado
Innis GS (ed) (1978) Grassland simulation model. Ecol Stud Vol 26. Springer, Berlin Heidelberg New York
Keulen van H (1975) Simulation of water use and herbage growth in arid regions. Simulation Monograph, PUDOC, Wageningen
Keulen van H (1976) A calculation method for potential rice production. Contr Centr Res Inst Agric Bogor 21. Central Research Institut for Agriculture, Bogor, Indonesia
Keulen van H, Seligman NG, Goudriaan J (1975) Availability of anions in the growth medium of roots of an actively growing plant. Neth J Agric Sci 23:131–138
Keulen van H, Wit de CT, Lof H (1976) The use of simulation models for productivity studies in arid regions. In: Lange OL, Kappen L, Schulze ED (eds) Water and plant life. Ecol Stud, Vol 19. Springer, Berlin Heidelberg New York
Keulen van H, Seligman NG, Benjamin RW (1980a) Simulation of water use and herbage growth in arid regions: a reevaluation and further development of the model ARID CROP. Agric Syst 6:159–193
Keulen van H, Laar van HH, Louwerse W, Goudriaan J (1980b) Physiological aspects of increased CO_2-concentration. Experientia 36:786–812

Khan MA, Tsunoda S (1970) Evolutionary trends in leaf photosynthesis and related leaf characteristics among cultivated wheat species and its wild relatives. Jpn J Breed 20(3):133–140
Lambers HT (1979) Energy metabolism in higher plants in different environments. PhD thesis, Univ Groningen
Lambert JR, Baker DN, Phene CJ (1976) Dynamic simulation of processes in the soil under growing row crops: RHIZOS. In: Computers applied to the management of large scale agricultural enterprises. Proc US-USSR seminar, Moscow Riga Kishinev
Landivar JA (1979) The application of cotton simulation model GOSSYM in genetic feasability studies. MSc. thesis, Mississippi State Univ, Mississippi
Loomis RS, Ng E (1978) Influence of climate on photosynthetic productivity of sugar beet. In: Hall DO, Coombs J, Goodwin FW (eds) Proc 4th Int Congr Photosynthesis. Biochem Soc (London)
Loomis RS, Rabbinge R, Ng E (1979) Explanatory models in crop physiology. Annu Rev Plant Physiol 30:339–367
Maas SJ, Arkin GF (1978) Users guide to SORGF, a dynamic sorghum growth model with feed back capacity. Program and model documentation 78-1. Texas Agric Exp Stn, Texas A & M Univ, Texas
Mapp HP, Eidman VR, Stone JF, Davidson JM (1975) Simulating soil water and atmospheric stress – crop yield relationships for economic analysis. Tech Bull T-140. Agric Exp Stn, Oklahoma State Univ, Oklahoma
McCree KJ (1970) An equation for the rate of respiration for white clover plants under controlled conditions. In: Setlik I (ed) Prediction and measurement of photosynthetic productivity. Proc IBP/RP Tech Meet Trebon, PUDOC, Wageningen
McCree KJ (1974) Equations for the rate of dark respiration of white clover and grain sorghum as functions of dry weight, photosynthetic rate and temperature. Crop. Sci 14(4):509–514
McKinion JE, Weaver REC (1979) Simulation of plant response to primary stress factors. Trans Am Soc Agric Eng 22:586–597
Meisel WS, Collins DC (1973) Repromodelling: an approach to efficient model utilisation and interpretation. IEEE Transactions on systems, man and cybernetics, vol SMC III(4):349–358
Menon KAP, Bowonder B (1978) A model for forecasting wheat production. Technol Forecast Soc Change 11:261–271
Mesarovic MD, Macko D, Takahara Y (1970) Theory of hierarchical, multilevel systems. Academic Press, New York
Meyer GE, Curry RB, Streeter JG, Baker CH (1979) Analysis and computer simulation of flowering and reproductive growth in indeterminant soybeans. Trans ASAE (in press)
Miller PC, Stoner WA, Richards SP (1978) MEDECS, a simulator for mediterranean ecosystems. Simulation 6:173–190
Moldau H, Karolin A (1977) Effect of reserve pool on the relationship between respiration and photosynthesis. Photosynthetica 11:38–47
Murata Y (1975) Estimation and simulation of rice yield from climatic factors. Agric Meteorol 15:117–131
Nelson WL, Dale RF (1978) A methodology for testing the accuracy of yield predictions from weather-yield regression models for corn. Agron J 70:734–740
Ören TI (1977) Software for simulation of combined and discrete systems, a state of the art review. Simulation 28:33–45
Passioura JB (1972) The effect of root geometry on the yield of wheat growing on stored water. Aust J Agric Sci 23:745–751
Passioura JB (1973) Sense and nonsense in crop simulation. J Aust Inst Agric Sci 39:181–183
Penning de Vries FWT (1975a) The cost of maintenance processes in plant cells. Ann Bot 39:77–92
Penning de Vries FWT (1975b) Use of assimilates in higher plants. In: Cooper (ed) Photosynthesis and productivity in different environments. Cambridge Univ Press, Cambridge, UK

Penning de Vries FWT (1981) Simulation models of growth of crops, particularly under nutrient stress. In: Künzli, Peter von A (eds) Proc. 15th Int Potash Inst Colloq, Wageningen 1980
Penning de Vries FWT, Laar van HH (1977) Substrate utilization in germinating seeds. In: Landsberg, JJ, Cutting CV (eds) Environmental effects on crop physiology. Academic Press, London New York
Penning de Vries FWT, Brunsting AB, Laar van HH (1974) Products, requirements and efficiency of biological synthesis, a quantitative approach. J Theor Biol 45:339–377
Penning de Vries FWT, Murphy CE, Wells CG, Jorgensen JR (1975) Simulation of nitrogen distribution and its effect on productivity in even aged loblolly pine plantations. In: Howell FG, Gentry JB, Smith MM (eds) Mineral cycling in south eastern ecosystems. Natl Techn Info Serv US Dept Commerce, Springfield, VA
Penning de Vries FWT, Witlage JM, Kremer DJ (1979) Rates of respiration and of increase in structural dry matter in young wheat, ryegrass and maize plants in relation to temperature, to water stress and to their sugar content. Ann Bot 44:591–609
Penning de Vries FWT, Laar van HH (1982) Simulation of plant growth and crop production. Simulation Monograph, PUDOC, Wageningen
Penning de Vries FWT, Krul J, Keulen van H (1980) Productivity of sahelian grasslands and the availability of nitrogen and phosphorus from the soil. In: Rosswall T (ed) Nitrogen cycling in west African ecosystems. Proc SCOPE-UNEP Nitrogen Unit Workshop, Ibadan R Swed Acad Sci
Phene CJ, Baker DN, Lambert JR, Pearsons JE, McKinion JM (1978) SPAR, a soil-plant-atmosphere research system. Trans Am Soc Agric Eng 21:924–930
Raschke K (1975) Stomatal action. Annu Rev Plant Physiol 26:309–340
Rees AR, Thornley JHM (1973) A simulation model for tulip growth in the field. Ann Bot 37(149):121–131
Reinhardt H (1975) Möglichkeiten zur Vorschätzung von Ertragshöhe und Ertragsschwankungen im Körnermaisbau. Ber Landwirtsch 53:386–397
Reuss JO, Innis GS (1977) A grassland nitrogen flow simulation model. Ecology 58:379–388
Rick JR (1978) The tomato. Sci Am 8:66–76
Rose MR, Harmsen R (1978) Using sensitivity analysis to simplify ecosystem models: a case study. Simulation 31:15–26
Russo JM, Knapp WW (1976) A numerical simulation of plant growth. Int J Biometeorol 20(3):276–285
Sauer RH (1978) Simulation model for grassland primary producer phenology and biomass dynamics. In: Innis GS (ed) Grassland simulation model. Ecol Stud Vol 26. Springer, Berlin Heidelberg New York
Schreiber MM, Miles GE, Holt DA, Bula RJ (1978) Sensitivity analysis of SIMED (a basic crop growth model for alfalfa). Agron J 70(1):105–108
Seaton KA, Landsberg JJ (1978) Resistance to water movement through wheat root systems. Austr J Agricul Sci 29:913–924
Seginer I (1980) Optimizing greenhouse operations for best aerial environment. Acta Hortic 106:169–178
Seligman N, Keulen van H (1980) PAPRAN, a simulation model for annual pasture production limited by rainfall and nitrogen supply. In: Frissel MJ, van Veen JA (eds) Models for behaviour of nitrogen in soils and uptake by plants. PUDOC, Wageningen
Setlik I (1970) Prediction and measurement of photosynthetic productivity. Proc IBP/RP Tech Meet Trebon, PUDOC, Wageningen
Sinclair TR, Wit de CT (1976) Analysis of the carbon and nitrogen limitations to soybean yield. Agron J 68(2):319–324
Stapleton HN, Buxton DR, Watson FL, Nolting DJ, Baker DN (1974) COTTON, a computer simulation of cotton growth. Tech Bull 206. Agric Exp Stn, Arizona Univ, Tucson, Arizona
Steyeart LT, Duc le SK, McQuigg JD (1978) Atmospheric pressure and wheat yield modeling. Agric Meteorol 18:23–34

Talpaz H, Chun ChH, Tailor HR (1977) Prediction of dry land crop yields using rainfall. South J Agric Econ july, 115–121
Thornley JHM (1976) Mathematical models in plant physiology. Academic Press, London New York
Thornley JHM (1977) Growth, maintenance and respiration, a reinterpretation. Ann Bot 41:1191–1203
Tollenaar M (1977) Sink-source relationships during reproductive development in maize. A review. Maydica 22:49–75
Veen van JA (1977) The behaviour of nitrogen in soil. PhD thesis. Agric Univ, Wageningen
Waggoner PE (1970) Consultation on how models are made, how they are tested and what they tell us of experiments to be done. In: Setlik I (ed) Prediction and measurement of photosynthetic productivity. Proc IBP/RP Tech Meet Trebon, PUDOC, Wageningen
Waggoner PE (1977) Simulation of modeling of plant physiological processes to predict crop yields. In: Landsberg JJ, Cutting CV (eds) Environmental effects on crop physiology. Academic Press, London New York
Welch SM, Croft BA (1979) The design of biological monitoring systems for pest management. Simulation Monographs PUDOC, Wageningen
Wit de CT (1970) Dynamic concepts in biology. In: Setlik I (ed) Prediction and measurement of photosynthetic productivity. Proc IBP/RP Tech Meet Trebon, PUDOC, Wageningen
Wit de CT, Goudriaan J (1978) Simulation of ecological processes. Simulation monograph. PUDOC, Wageningen
Wit de CT, Penning de Vries FWT (1982) In: Penning de Vries FWT, Djiteye MA (eds) La Productivité des Pâturages sahéliens. Agric Res Rep 918 PUDOC, Wageningen
Wit de CT, Brouwer R, Penning de Vries FWT (1970) The simulation of photosynthetic systems. In: Setlik I (ed) Prediction and measurement of photosynthetic productivity. Proc IBP/RB Tech Meet Trebon, PUDOC, Wageningen
Wit de CT, et al. (1978) Simulation of assimilation, respiration and transpiration of crops. Simulation Monograph. PUDOC, Wageningen
Wit de CT, Laar van HH, Keulen van H (1979) Physiological potential of crop production. In: Sneep J, Hendriksen AJT (eds) Plant breeding perspectives. PUDOC, Wageningen
Wong SC, Cowan IR, Farquhar GD (1979) Stomatal conductance correlate with photosynthetic capacity. Nature (London) 282:424–426
Wright MJ (1977) Plant adaptation to mineral stress in problem soils. Proc Workshop Cornell Univ Agric Exp Stn, Cornell Univ, Ithaca, New York
Yamaguchi J (1978) Respiration and the growth efficiency in relation to crop productivity. J Fac Agric Hokkaido Univ, Jpn 59:59–129

5 Productivity of Agricultural Systems

R.S. Loomis

CONTENTS

5.1 Introduction: On the Nature of Agriculture 151
5.1.1 Agricultural Ecosystems 151
5.1.2 The Plant Materials 156

5.2 Productivity in Agriculture 157
5.2.1 Annual Production 158
5.2.2 Daily Production 163
5.2.3 Seasonal Patterns 165
5.2.4 Partitioning Patterns 166

5.3 Controlling Relationships 168

5.4 Conclusions 169

References 170

5.1 Introduction: On the Nature of Agriculture

Agriculture, one of the most ancient and diverse pursuits of man, currently touches about 30% of the ice-free land area of our globe: 1.4×10^9 ha of arable land is given to cultivated crops, while another 3.0×10^9 ha is exposed to grazing by domestic animals. Our focus here will be on the arable lands and their crops and forages, beginning with an overview of agricultural systems.

5.1.1 Agricultural Ecosystems

The arable-land ecosystems operate in fundamentally the same ways as less disturbed, wild ecosystems. Production processes, trophic chains, nutrient cycling, and other aspects of natural systems are present and obey the same basic biological and physical rules. The difference is that crop lands are subject to much greater degrees of management and exploitation. Control extends to choice of site, species and cultivar, and space relations, to timing of events such as planting and harvesting, and to inputs such as nutrients. In addition, simplification of the plant community to a monoculture of uniform plants drastically alters interference among adjacent plants. Competition is then relatively equal and has no winners and losers. Genetic traits such as those for height which contribute to competitive ability may be wasteful of photosynthate and can be eliminated, while other characters, such as delayed root extension, which may be less competitive in mixed stand, can be used in monoculture to obtain increased yield. In monoculture, soil resources not used today are available tomorrow, whereas in the natural community, unused nutrients and water may be used by a more aggressive competitor. In monoculture, we can change the timing of peak growth rate or nutrient use without concern for a competitor's behavior.

Besides nutrient applications, environmental modifications in agriculture may include applications of irrigation water, changes in land form (leveling, terracing, or use of sloping beds), exclusion of grazers, control of insects, diseases and noncrop plant species, frost protection, and alterations of the soil both physically (drainage, pH, and structural changes) and biologically (residue management and bacterial inoculation). Key features of management also include the use of various types of soil tillage, the choice of cropping sequences over years, and the integration of animal production with crop culture.

Crop communities are exploited through removal of portions of the biomass along with the nutrients that they contain. When the material is fed to livestock, significant proportions of the nutrients and organic material may be returned to the land as animal manures. At the other extreme, the produce may enter international trade, and the nutrients then end up in other continents as manures on the land or as urban effluents in the seas. Such interregional transports can be rather massive. HALL (1929) speaks of the importance of such imports over years in the improved fertility of English soils. In 1979 alone, the United States exported 55×10^6 mt of maize grain (*Zea mays*) which contained more than 800,000 mt of nitrogen and 100,000 mt of phosphorus (mt in this chapter refers to metric ton).

The nutrient stock in agricultural soils varies widely according to the nature of the virgin soil, length of time under cultivation, rainfall patterns, type of farming system imposed, and rate of natural inputs of nutrients (e.g., dust, rainfall, weathering of parent materials, biological fixation of nitrogen). After the initial stocks are severely depeleted by cropping, the productivity of the system is then limited to the rate of nutrient replacement. The most general and severe limitation is due to nitrogen, which places great importance on the management of soil organic matter, the principal source of nitrogen for unfertilized crops (see also Chap. 5, Vol. 12C). As an example, we can follow JENNY's (1930, 1941) analyses of the prairie soils of North America. The equilibrium level of soil organic matter under virgin prairie in Iowa, U.S.A., ranged from 100,000 to over 200,000 kg ha^{-1} in the surface 1 m (depending upon rainfall, temperature, and drainage) with between 5,000 and 10,000 kg of nitrogen. Assuming that 1% to 2% might be mineralized per year, 50 to 100 kg N ha^{-1} was available to the crops in each of the early years of cultivation. But the organic residues by themselves were inadequate for replacement of the soil organic matter and soil nitrogen and those stocks declined until a new equilibrium was reached. This might occur in 60 to 80 years with only 30% to 50% of the original levels if grain is cultured continuously. At that point, a mineralization rate of 1% to 2% would release only 15 to 50 kg N per year, while a similar amount from residues and natural sources would accumulate as humus each year. The net amount available for grain export would lie in the range of 10 to 30 kg N ha^{-1}, depending on the amounts of other inputs and losses. Based on the normal nitrogen content of grain, that corresponds to grain yields of 650 to 2,000 kg ha^{-1}. Such yields were characteristic of medieval agriculture (SLICHER VAN BATH 1963; STANHILL 1976). They also are what one finds today in the agriculture of many developing regions and in long-term experiments that use no fertilizer additions and do not practice rotation with grasses and legumes (Broadbalk fields at Rothamsted, GARNER and DYKE 1969; Morrow Plats, University of Illinois, STAUFFER et al. 1940; WELCH 1976).

In high-rainfall regions, the supplies of base minerals also are apt to be low due to leaching, whereas soils in semi-arid regions have higher levels of bases. The semi-arid sites have been well-suited to agriculture: some Mediterranean and Asian places have been cultivated for many centuries. But annual yields are limited there by water supply, and we again find strong deficiencies of nitrogen, and grain yields of about 1,000 kg ha^{-1}. Nutrient restrictions are generally less severe in soils of flood plains and river deltas subject to annual alluvia.

Various practices other than fertilization have been employed to cope with nutrient limitations. Medieval European systems relied heavily on fallowing practices in which the land was cropped only in alternate years, thus allowing 2 years of natural inputs per crop year. In the Broadbalk plots at Rothamsted, the immediate effect of a year of fallow is a doubling of wheat (*Triticum vulgare*) yields (GARNER and DYKE 1969). That was due mostly to nutrient replenishment but fallow also has important benefits in weed control.

Fallow is still an important feature of farming in many parts of the world. The most extreme examples of fallow are found in shifting culture in the less-developed tropical countries where fallow periods of more than 10 years are common and the land returns to forest or savannah type vegetation. The soil bank program in the United States in which the government rented land on condition that the farmer left it unused as a grass–legume rotation is analogous to the tropical shifting culture. Soil nitrogen can accumulate very rapidly under vegetated fallow (JENKINSON 1971). In the Southern areas of the Great Plains of the U.S.A., the use of fallow is more related to soil water replenishment, however, in the Northern areas, with less evaporative demand, fallow serves the purpose of replenishing both water and nutrients.

The consequence of nitrogen limitations on maize grain production under two systems of cultivation on an excellent site in Illinois, U.S.A., is illustrated in Table 5.1 (WELCH 1976). In this case, rotation with a leguminous forage crop provided strong additional inputs of nitrogen and much higher yields of the subsequent crop of grain than were obtained with continuous maize. The output of grain per farm, however, will be increased or decreased depending on the proportion of hectarage in grain and legumes and the efficiency of the legumes in supplying nitrogen. In 1950, before the widespread availability of nitrogen fertilizer, nearly half of the land in the American Cornbelt was sown with clovers and companion crops. But the best yield in rotation plus barnyard manure, 4,130 kg ha^{-1} (Table 5.1), is significantly less than the current average yield for Illinois (7,000 kg ha^{-1} in 1979), which can be sustained continuously

Table 5.1. Grain yields of maize crops grown under long-term rotations on the Morrow Plats, University of Illinois, U.S.A. Means of every 6th year, 1907–1949, when all plots were in maize the same year. (After WELCH [1976], Table 2). All values converted to dry matter at 85%

Cropping system	Maize grain yield	
	0[a] (kg/ha)	MLP[b] (kg/ha)
Continuous maize	1,310	2,690
Maize–oats–clover	2,980	4,130

[a] 0 = no additives
[b] MLP = Manure equal to that produced by feeding the crop to cattle, lime, and phosphorus added

through moderate applications of fertilizer; and it is far less than the world record yield of maize grain (18,000 kg ha^{-1}) obtained in Michigan. These yield differences are confounded with variations in moisture supply, changes in varieties and use of improved farming practices, but they do demonstrate the extreme dependence of agricultural productivity on both the nutrient supply and the technological state of farming.

The application of technology and the intensity of agriculture in turn reflect the overall state of the larger society. The per capita base of agricultural land is a fundamental factor since it dictates the demand per unit land area. Thus France with 0.6 ha caput^{-1} has less intensive agriculture than the extremely high-technology agriculture of The Netherlands (0.15 ha cap^{-1}) where agricultural yields are 30% greater. One can also enumerate roles for cultural attitudes, patterns of land ownership, levels of education and research, political and economic systems, general level of affluence, availability of capital, and many other variables and complex interactions. Collectively, such factors determine which particular crops or approaches to agriculture have a comparative economic advantage over the alternatives. In an isolated "island" agricultural system that has an urban center, one would expect that all basic food crops would find some distance-dependent place within the system. VON THÜNEN (1826) found with primitive transportation that the intensity of farming varied inversely with distance from the city market: vegetable and fruit crops and dairying close by, grain production at middle distances, and extensive grazing and forestry in the hinterlands. Such patterns still exist, but are now greatly modified (in the absence of tariffs) by the low cost of long-distance transport. But regional specialization is still very marked. California produces more than 50% of the United States' fruits and vegetables (on 800,000 ha or 0.5% of the cultivated land of that country). Iowa possesses climates and soils suited to many species, but the economic advantage for most of the species currently lies in other regions; Iowa has the advantage with maize and soybean (*Glycine max*), which now occupy 90% of its arable land. The average yields of maize and soybean are greater in California than in Iowa, but in the face of low-cost production

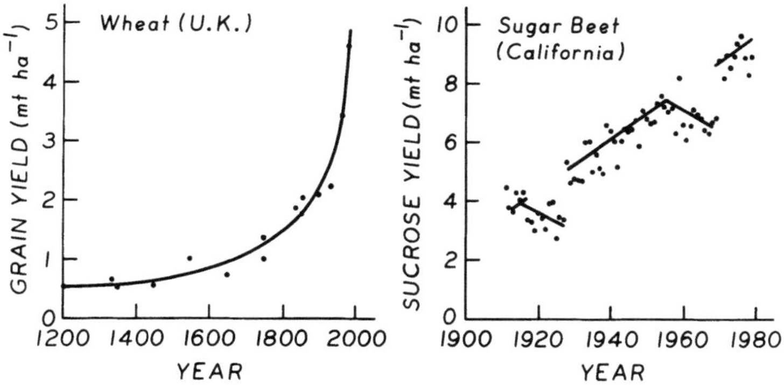

Fig. 5.1. Yield trends with changing technology. *Left* A 776-year history of wheat yields in the United Kingdom. After STANHILL (1976), with the addition of a datum point of 4.6 mt ha^{-1} as the mean yield 1974–1978. Eye-fitted trend line. *Right* A 69-year history of sucrose yields from sugar beet in California. Each datum point is a state-wide mean for a particular year. Data compiled by the California Sugar Beet Growers Association, Ltd. Eye-fitted trend line

from Iowa and higher-value alternatives in California, less than 4% of California's irrigated land is given to maize and soybean.

Dramatic changes in yield are sometimes due to changes in technology and economics. Figure 5.1 (left) illustrates the trends of wheat yields in Britain for 776 years (STANHILL 1976) and (right) of sucrose yields from sugar beet in California for 69 years. Wheat yields were relatively constant through the late middle ages, but show a doubling every 150 years since 1650; this corresponds with the shift from manorial farming, which relied heavily on cultivated fallow, to "new", yeoman farming systems centered on legume rotations (HALL 1929) and more recently to nitrogen fertilizers. On a finer scale, Stanhill found a sharp increase in yield since 1950 corresponding to fertilizer use. Also evident was a leveling-off of yield between 1870 and 1940, perhaps reflecting intense competition with low-cost grains from North America. Surprisingly, the short-term year-to-year variations averaged 7% for both the nutrient-limited medieval crops and the modern nutrient-unlimited crops.

The historic trends for sugar beet in California show behavior in a system that has always been rather advanced technologically. In an average crop, sucrose represents 75% of the dry matter of the root and perhaps 40% of the whole biomass. Improvements in varieties and cultural practices (particularly in irrigation and fertilization) account for the basic upward trend. The deviant trends, 1910–1930 and 1950–1970, represent independent "great events": epidemics, respectively, of curly-top virus and sugar beet yellows virus. Both problems were alleviated through the development of resistant varieties.

Economic conditions can have subtle influences on such trends. A decline in the comparative advantage of sugar beet causes a decline in hectarage (favoring higher yield) and a reduction in intensity (less fertilizer and insect and weed control favoring lower yield). Marginal growers (low yields) and well situated growers with attractive alternatives (high-yield group) may be the most likely to make drastic changes in their farming plans.

5.1.2 The Plant Materials

The cultivation of higher plants seems have emerged at several places and times, 10,000 to 20,000 years B.P. Although quite different plant species and cultural methods were associated with each epicenter, the same basic principles of control of genetic and nutrient resources and of simplification of the plant community to strong dominance by the economic plant were involved.

The discovery and adoption of agricultural practices must have come in rather simple steps. A change from gathering wild seeds may have occurred with the observation that a new community of the desired species arose in the middens and where gathered seeds were spilled inadvertently on disturbed land. In the case of root crops, gathering of the underground organs disturbs the environment, suppressing crowding by competitors and also promoting mineralization of nitrogen from soil organic matter and, thus, the regrowth of the harvested species from vegetative parts. Once a system of planting and harvesting was established, genetic mass selection for adaptation to monospecific, cultivated communities and for gigantism and other attributes of economic yield would be rapid.

While agricultural systems seem highly evolved from the natural state, it is surprising how little the plants differ from their wild relatives and how quickly they may revert to wild type. Related wild types generally show stronger traits for competitive ability (e.g., height growth and tillering or branching) and sometimes quite different seasonal and phasic patterns in development. The key feature of the domesticated forms is that a high proportion of their net photosynthesis flows toward economic yield and quality. Whether significant advances in photosynthetic ability have occurred with domestication is not as clear. Most crop species are capable of relatively high photosynthetic rates, but these are no greater and sometimes are even less than for their wild progenitors. Thus, high photosynthesis may be more a trait that contributed to use by man than a result of domestication.

The crop plants are mostly herbaceous annuals or, like cotton (*Gossypium* spp.), are grown as such. Longer life spans are common in forage species (also herbaceous) and in the woody perennial tree crops grown for fruit, nuts, oil, rubber, and other products. The cereal grains, fleshy root and tuber crops, and sugar crops are ubiquitous as basic food sources and occupy nearly 60% of the world's arable land. While we view these basically as energy (carbohydrate) sources, the cereals are also the major source of protein in human diets. Even rice and maize, which are the lowest in protein content (9%–10% of grain dry weight), may supply adequate amounts for adults. If enough calories are consumed to sustain life, the intakes of essential amino acids from the digestible fraction of their proteins are generally above the minimum daily requirements (CLARK 1978). These crops, along with forage species grown for animal feed and oil crops, are subject to considerable selection for yield; it is here that we find the highest annual and daily productivity rates.

Shortages of land and other resources for those extensively grown basic food crops give emphasis to production efficiency. In contrast, crops that contribute fresh fruits and vegetables to our diets seem to have been selected more for timeliness, color, shape, and flavor of the product – daily and seasonal dry matter production are generally low. The hectarages given over to fruits and vegetables and their contributions to human diets (1%–2% in each case; EVANS 1975) are so small and their economic value so high that yield has not been a key criterion. In extreme cases, essential oil crops have only 50 to 100 kg ha^{-1} of economic yield. Except for alfalfa (*Medicago* spp.), the edible-seed and forage legumes also are relatively low in photosynthetic productivity (LOOMIS and GERAKIS 1975). Whether this relates to their pattern of symbiotic nitrogen nutrition is unknown.

Much less information exists on the tree crops grown for fruit and nuts. Productivity is generally low, mainly because cultivars and cultural practices including plant spacings emphasize quality and ease of management more than high productivity. Careful analyses of oil palm (*Elaeis guineensis*) (CORLEY 1973), however, show annual production approaching 40 mt ha^{-1}. The production processes of tree crops should differ from those of annual crops in several respects. The differences relate to the perennial nature of trees and the capacity for rapid and complete foliation in early spring (in contrast to the long lag period of logarithmic growth with annuals), internal recycling of nutrients from maturing xylem (wood) to secondary vascular cambia in the stems, and presumably, to a greater maintenance respiration for the greater average amount of living biomass over the annual cycle. None of those issues has been explored critically.

5.2 Productivity in Agriculture

Our introduction has emphasized the diverse nature of agricultural systems and the constraints on productivity commonly imposed by nutrients, water, cultural practices, and genotype. Even for a relatively standard and widely occurring system such as the culture of wheat, one observes in any one year a range of yields from crop failure to near-record performances. The question arises about what statistics can serve as a measure of crop productivity. The most widely available statistics concern regional or national means in economic units like tons of air-dry hay or grain. Those numbers are available from crop-reporting services such as the U.S. Department of Agriculture and F.A.O. Along with hectarage statistics they provide the best estimates of trends and total production. Economic yields can then be related to total productivity through the use of proportionality factors (economic yield as a fraction of total crop) (LOOMIS and GERAKIS 1975). Such factors are reasonably constant for many species, providing satisfactory though only approximate estimates of the average level of production.

The quality of such statistics is variable. In some cases they are estimated by central offices, in others they are based on survey sampling techniques or, in the case of contract crops, by accurate assessment of area, yield, and quality. In addition, regional means obscure the fact that individual fields may vary widely from the mean. That is the case for sugar beet in California which was illustrated in Fig. 5.1. One reason is that beets are perishable. With limited processing capacity, a portion of the hectarage must be harvested early at low yields, while other fields can be carried to the end of the season and thus to maximum yield. The situation is complicated further by the existence of coastal and interior climates and winter and summer cropping in different areas of the state. In 1972 when the mean yield of all fields was 8,960 kg sucrose ha^{-1}, the mean of 42 high-performance fields representing 0.6% of the total hectarage was 16,300 kg ha^{-1}; two fields (21 ha) exceeded 19,000 kg ha^{-1}. But on 4% of the planted hectarage, the emerged stand was so poor that the fields were replanted with alternative crops.

5.2.1 Annual Production

Table 5.2 provides an approximate view of world-wide agriculture. The original numbers (FAO 1979) involve a series of approximations; further uncertainties are introduced with the groupings used here and with the factors selected to represent dry matter and economic yield. Cereal grains are shown as dominant, occupying 52% of the arable land now in production and supplying 64% of primary production. The estimate of 5.3×10^9 mt dry matter for world-wide net primary production corresponds to about 2.4×10^9 mt of carbon. While of the same order of magnitude, this estimate is significantly less than WHITTAKER and LIKEN's (1975) estimate of 4.1×10^9 mt carbon yr^{-1}.

A second approach to surveys of agricultural productivity is the examination of reports of individual high-performance fields or experiments. Such estimates of maximum performance can be found from reports of standard yield trials conducted annually in many agricultural areas for comparisons of cultivars, from experiment station records, and from research literature. The latter source is the most likely to give details on biomass components. Such reports are not as common as one might think. A growth-curve crop-productivity analysis is an expensive task and represents only a single time and place and an uncertain generality for the region or species. A very extensive set of growth curves along with correlations with climate and other variables was developed as part of the Japan IBP effort (MURATA 1975).

Caution and judgement are required in analyzing such data. The best data come from strip harvests of commercial fields – in the case of experimental plots, attention to adequate borders and large quadrat harvests (in contrast to "rod-rows" of cereal nurseries or single plant extrapolations). Lack of attention to such details can introduce large errors (LOOMIS and GERAKIS 1975; MONTEITH 1978). Another problem associated with small plots is that pest and disease attack may be quite different (more or less) than when the same genotype is grown over a wide area (ROBINSON 1976).

High annual biomass yields of C_3 and C_4 crops are presented in Fig. 5.2. These curves represent an envelope of maximum observations and are judged among the best present estimates of the biological potential of these species. In Fig. 5.2 (left) (revised from LOOMIS and GERAKIS 1975) the abscissa is latitude and each point represents the largest yield found for a particular latitude. Other

Table 5.2. A crude estimate of world-wide agricultural production in 1978. The principal data source for hectarage and economic production is FAO (1979). Economic yield as a fraction of biomass is estimated towards large values, thus providing a conservative estimate of annual production

Source	Area cropped (10^6 ha)	Economic production (condition) (10^6 mt)	Assumed factor: dry weight/economic yield	Comment	Primary production (10^6 mt dry)
Cereals	753	1,581 (air dry)	0.84/0.4	Factors reasonable; 75% of production accounted by wheat, rice, maize	3,360
Root and tuber crops	47	253 (wet)	0.25/0.7	95% of production accounted by potato, sweet potato, and cassava	187
Pulses	83	62 (air dry)	0.85/0.4		132
Oil crops	109	129 (air dry)	(0.85/0.3)[a]	74% of production from soybean, much of which is used for food	(366)
Sugar crops	23	91 (sucrose)	1/0.3		303
Fiber crops	28	38 (air dry)	(0.85/0.4)	63% of this production is cotton seed	(81)
Tobacco	4	6 (air dry)	0.85/0.3		17
Vegetables and melons	24	327 (wet)	(0.15/0.4)	0.4 is optimistic – may not allow sufficiently for rejected produce	(123)
Total temporary crops	1,081				4,569
Other rotation crops	293	No report		Ca. 12% is known to be in alfalfa; balance assumed to be in other forages, fallow and minor crops. Assumed average productivity at 2 mt ha^{-1} yr^{-1}	(586)
Fruits	no report	262 (wet)	(0.12/0.3)		(105)
Nuts, coffee, tea, cocoa, etc.	no report	16 (air dry)	(0.85/0.3)		(45)
Total permanent crops	88			Mean global productivity is 5,305/1,462 = 3.63 mt ha^{-1} yr^{-1}	(150)
Total farm land	1,462				5,305

[a] ( ) indicates the more uncertain assumptions and results

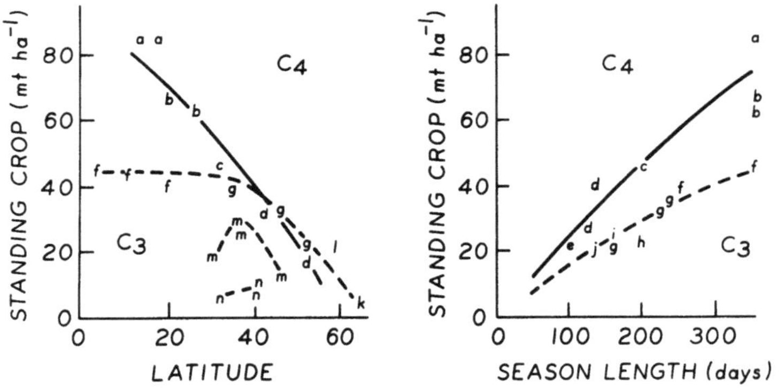

Fig. 5.2. Reports of record annual production of standing crop by various cultivated species. *Left* A distribution against latitude, revised and simplified from LOOMIS and GERAKIS (1975). New data for cassava (*f: Manihot esculenta* Crantz) are: 3 30′ N, COCK et al. (1970); 10 30′ N, HOLMES and WILSON (1977); and 20′ S (mean of several observations in Madagascar), COURS (1951). The trend lines for C_3 and C_4 species are fitted by eye, discounting somewhat the yields of napiergrass because of small size of plots. Alfalfa and soybean are included for comparison. *Right* A distribution against length of growing season. Revised from an original by MONTEITH (1978) based on compilations by COOPER (1975) and LOOMIS and GERAKIS (1975), and including the new cassava data. The eye-fitted trends are added here. *a* Napiergrass (*Pennisetum purpureum*); *b* sugar cane (*Saccharum* spp.); *c* forage sorghum (*Sorghum bicolor*); *d* maize; *e* bermudagrass (*Cynodon dactylon*); *f* cassava; *g* sugar beet; *h* rice (*Oryza sativa*); *i* potato (*Solanum tuberosum*); *j* kale (*Brassica oleracea* var. *acephala*); *k* wheat; *l* ryegrass (*Lolium perenne*); *m* alfalfa (*Medicago* spp.); *n* soybean

observations fall in a continuum from zero yield to that maximum which generally represents a "best circumstance": a long and favorable growing season matched by a late-maturing genotype, and minimum limitations due to water, nutrients, or other factors. Thus, the sugar beet data are from irrigated sites in semi-arid western United States. C_4 species are clearly superior at lower latitudes, while C_3 species are equal or superior above 40° to 45°. Lower peak irradiances and the general sensitivity of C_4 species to chilling injury contribute to their poor performance at high latitudes (functional significance of the different pathways of CO_2 fixation in photosynthesis is discussed in Chap. 15, Vol. 12B).

Figure 5.2 (right), adapted from MONTEITH (1978), in which the length of growing season is the abscissa, is perhaps more meaningful. Again, a clear superiority of C_4 species is evident. In this case the high yields of a short-season C_4 species are all from warm climates; C_3 species excel in short seasons in cool climates. In both summaries, indeterminate species (e.g., sugar cane and cassava) are more successful in long growing seasons (low latitudes) while determinate species (maize, wheat) do well in short seasons (high latitudes).

A common feature of the species illustrated here is that the composition of their biomass is generally dominated by carbohydrates (cellulose, starch, and sugar). They are basically low in protein and lipid. This observation fits well with calculations by PENNING DE VRIES et al. (1974) of the "growth yield" in plant metabolism of different organic materials from glucose (photosynthate)

Table 5.3. A calculation of the amounts of photosynthate (glucose) required for the synthesis of 100 g of maize and soybean grains using the production value approach of PENNING DE VRIES et al. (1974). Composition is from NAS (1969) and the nitrogen source is nitrate

Source	Composition					Glucose requirement	Production value[b]
	Crude[a] fiber	Ether extract	Nitrogen-free extract	Crude protein			
	(g per 100 g dry seed/g product per g glucose)					(g glucose/100 g grain)	(g grain/g glucose)
Maize grain (No. 2 yellow)	$\frac{2.2}{0.76}$ +	$\frac{4.4}{0.36}$ +	$\frac{82.2}{0.86}$ +	$\frac{10}{0.47}$	=	132	0.76
Soybean grain	$\frac{5.6}{0.76}$ +	$\frac{20.0}{0.36}$ +	$\frac{27.2}{0.86}$ +	$\frac{42.1}{0.47}$	=	184	0.54

[a] Crude fiber is taken as 85% cellulose and 15% lignin; 0.76 is a weighted average of the production values for cellulose (0.86) and lignin (0.46)
[b] Relative growth yield of maize to soybean: 0.76/0.54 = 1.4

and nitrate. Beginning with 1 g of glucose, the production yields for the major classes are (DE WIT et al. 1978): 0.86 g cellulose or starch; 0.47 g nitrogenous compounds; and 0.36 g lipid. Stated in another way, 1 g of glucose supplies sufficient carbon and metabolically available electrons for synthesis of those amounts of product.

From the proportions of these compounds in the biomass (or from the elemental analysis of the whole biomass; MCDERMITT and LOOMIS 1981) one can calculate the original photosynthate required for the synthesis, and from this, the yield of an alternate product. In Table 5.3, grains of maize and soybean are compared in this way. Because of its lower content of protein and lipid, the growth yield of maize grain is shown to be 1.4 times that of soybean for the same photosynthate supply. Applying that adjustment to the soybean yields depicted in Fig. 5.2 (i.e., expressing the grain yields in equivalent weights of corn) increases the annual yield by only 1 mt ha^{-1}.

The effect is more dramatic in oil palm. One of CORLEY's (1973) treatments averaged 32.3 mt dry matter ha^{-1} yr^{-1} production, of which 6.75 mt was oil. In original glucose equivalents, this is 6.76/0.36 = 18.8 mt, and in corn grain equivalents, 18.8 × 0.75 (from Table 5.3) = 14.1 mt. Thus, in corn equivalents, his oil palm's annual production would have been 7.3 mt greater. Growth yields for the crops depicted in Fig. 5.2 would generally exceed 0.7. With a high proportion of oxygen relative to carbon, the tissues are not highly reduced and their heats of combustion are near 17 kJ g^{-1}.

The high annual yields displayed here considerably exceed the highest values reported for natural forests and grasslands. One reason is that agriculture generally has usurped the best soil–climate situations. But the difference also reflects the very high nutrient requirements for high annual productivity. Considering that the minimum nitrogen content of herbaceous biomass is near 1.5% (ca. 9% crude protein), about 1,200 kg nitrogen ha^{-1} would be required to produce

Table 5.4. A comparison of mean grain yields of some C_3 and C_4 cereal species in 1978 for Mexico, U.S.A., and the rest of the World (FAO 1979)

Species	Photosynthetic type	Mean yields		
		Mexico (kg/ha)	U.S.A. (kg/ha)	World (kg/ha)
Wheat	C_3	3,480	2,130	1,870
Rice	C_3	3,290	4,940	2,570
Maize	C_4	1,340	6,350	2,110
Sorghum	C_4	3,000	3,460	1,020

80 mt of napiergrass biomass. This amount is much greater than might be recycled annually in an undisturbed ecosystem, indicating that ceiling photosynthetic yields can be achieved in many environments only with large external inputs. A major reason agricultural yields are normally much below these yields is that heavy fertilization is practiced in few places other than northern Europe and Japan. Because of this and the absence of irrigation, the distinction between C_3 and C_4 species is absent or much less evident in ordinary agriculture. That is illustrated by the comparisons of several C_3 and C_4 cereal species presented in Table 5.4. In these cases, mean yields are shown to depend more on location than on photosynthetic type. Mexico produces nearly twice the world average for wheat reflecting the high proportion of their acreage which is irrigated. This contrasts to dryland culture under semi-arid conditions in the major wheat regions of the U.S.A. Wheat culture in Mexico is clearly more advanced than that of maize, which is well behind maize culture in the U.S.A. and the rest of the world.

In another example, from Illinois (U.S.A.), maize and soybean are each sown to about 4×10^6 ha and at about the same technological state. The ratio of mean yields of dry grain for the entire hectarage (1977–1979) was 6,110:2,100 = 2.9, twice the ratio of 1.4 we expect from just the difference in levels of reduction. Differences in growth duration and harvest index account for some of the advantage of maize, but how much of the main effect is due to the C_4 pathway and how much to the generally poor ability of legumes (Fig. 5.2 left) is not clear.

While saturation nutrient levels are achieved in only a few agricultural ecosystems, the use of nitrogen fertilizers has increased in all parts of the world because of the wide availability and low cost of ammonia. In small grains, for example, increases in nitrogen supply coupled with lodging-resistant, stiff-straw characters have led to dramatic yield increases in many countries (DE WIT and VAN HEEMST 1976). Concerning output/input efficiency, each kg nitrogen absorbed by a crop with 1.5% nitrogen in biomass would contribute 67 kg of biomass with a heat of combustion near 17 MJ kg^{-1}. Since a modern ammonia plant can generate 1 kg of fixed nitrogen from 59 MJ of natural gas, the energy gain from using nitrogen fertilizer is 19 times the input of energy into its production ($17 \times 67/59 = 19$).

Table 5.5. Observations of maximum short-term growth rates (increase in dry biomass; compiled by MONTEITH 1978)

Source	Photosynthetic type	Growth rate ($g\ m^{-2}\ d^{-1}$)
Millet (*Pennisetum typhoides*)	C_4	54
Maize (*Zea mays*)	C_4	52
Sudangrass (*Sorghum vulgare*)	C_4	51
Potato (*Solanum tuberosum*)	C_3	37
Rice (*Oryza sativa*)	C_3	36

5.2.2 Daily Production

Table 5.5 presents a cross-section of the highest observations of maximum daily growth (net production of biomass) for C_3 and C_4 crop species. Peak rates show C_4 plants to be markedly superior. Such high rates are generally found only in warm, high-radiation environments and with only moderate photosynthetic efficiencies (combustible heat content of new biomass/total short-wave radiation received). Theoretically, the highest efficiencies would be observed at low irradiance, in which all leaves in the canopy are light-limited in their photosynthesis and thus are at a high quantum efficiency (see Chap. 3, Vol. 12A). MONTEITH (1978) noted that observations of high efficiencies for maize ranged from 3.0% at 31 $MJ\ m^{-2}\ d^{-1}$ (total solar radiation) to 4.5% at 20 $MJ\ m^{-2}\ d^{-1}$. The maximum observed efficiencies for both C_3 and C_4 species fall in the range 4.5% to 5.0%, close to the theoretical limit (5.3%) proposed by LOOMIS and WILLIAMS (1963). This suggests that the rather general assumptions made by those authors about quantum efficiency, respiration load, and reduction level of the resulting biomass are basically correct. With heats of combustion for biomass all near 17 $kJ\ g^{-1}$, it is evident from the differences in growth rates shown in Table 5.5 that the C_4 species have a higher average quantum efficiency at high irradiance. But such measurements are too crude to detect the possibility that C_3 species are superior in quantum efficiency at low irradiance (EHLERINGER and BJÖRKMAN 1977).

Early attainment of a closed canopy for maximum growth rate is a basic agricultural practice. Growth has been shown to be a saturation (diminishing returns) function of leaf area index (ha leaf ha^{-1} land area; BROUGHAM 1956; WILLIAMS et al. 1965; SHIBLES and WEBER 1965) and a linear function of PAR interception (SHIBLES and WEBER 1966; WILLIAMS et al. 1968). The curvilinear relation of leaf area index reflects the increasing degree of mutual shading that occurs as additional leaf area is added. Mutual shading is greater with horizontal leaves and clumped distributions than with erect leaves and regular distributions. Regular displays result from phototropic growth habits.

A basic problem is achieving full cover rapidly with annual plants at spacings that will later be optimal for economic yield and quality. Solutions center on selecting for speed of emergence (seedling size and vigor), branching patterns that occupy space quickly, and traits that allow for planting at high density. Many crop species show a strongly horizontal pattern of leaf display in juvenile stages and at low leaf area indices. This provides the greatest interception and growth rate per unit leaf area. The rapid expansion of leaf area and the attainment of complete cover are highly dependent upon nitrogen supply and timing. Judicious application of nitrogen often is the only way in which high leaf areas can be obtained at plant populations that are optimal for economic yield. Later in the season, a wide range of canopy architectures is evident for different crop plants. The norm seems to approximate a spatially random distribution with moderate angles of inclination (30° to 60° elevation). Such communities achieve "full cover" and nearly complete light interception at leaf area indices of 3 to 4.

Canopy architecture has been explored in detail with sophisticated calculation models that can assess the distribution of diffuse and direct radiation among leaf area elements within the canopy for any architecture and radiation geometry (LEMEUR and BLAD 1974, see also Chaps. 1 and 2, Vol. 12A). Most of the models follow either a probability point-quadrat interception approach (DUNCAN et al. 1967) or the leaf exposure astronomical approach of the Estonians (ROSS 1981). The models have been validated by light distribution under a wide variety of conditions. When coupled with a physiological function for photosynthesis, they can provide close predictions of net photosynthesis (approximating crop growth rate) for any canopy-radiation geometry.

One prediction arising from an early analysis by BOYSEN-JENSEN (1932) is that strongly erect (i.e., near 90°) leaves would give improved efficiencies and 25% to 50% greater growth rates at high irradiance if the canopies are very dense (leaf area index of 8 to 10) (e.g., LOOMIS and WILLIAMS 1967; DUNCAN 1971). This prediction has been generally confirmed experimentally (TRENBATH and ANGUS 1975). But the practical implications of the observation may be limited. Consider an example with sugar beet having 250 cm^2 of area g^{-1} lamina dry matter and a ratio of lamina weight to supporting petiole weight of 1 (CLARK and LOOMIS 1978). In this case, 2 g of dry matter equates to 250 cm^2 of photosynthetic area, and 800 kg dry matter ha^{-1} are needed per unit leaf-area index. Leaf area index 4 then requires 3,200 kg ha^{-1}, while leaf area index 10 requires 8,000 kg ha^{-1}. When the average nitrogen content of leaf material is 3%, we need 144 kg nitrogen for the additional 6 ha of leaf area. These burdens are acceptable if leaf tissue constitutes economic yield (e.g., a forage crop), but they may be unacceptable, as with sugar beet, if the photosynthate might better have gone to root yield.

Under intensive fertilization in The Netherlands, leaf area indices of wheat now reach values of 6 to 10 (SPIERTZ and ELLEN 1978). But as will be shown later, wheat, like many other determinate crops, shows a strong capability for remobilization of leaf proteins through leaf senescence during the later stages of plant growth. While the surplus leaf tissue serves a useful role as a reservoir of reduced nitrogen, production efficiency might be increased with less leafy genotypes.

5.2.3 Seasonal Patterns

Agricultural ecosystems also differ from natural systems in their distribution of production over the growing season. This results from our control of various inputs as well as from dates of planting and harvest. In temperate regions, management tactics center on use of species and cultivars that will achieve the greatest use of the growing season without undue risk of frost damage either early or late in the season. Annual crops are delayed until the soil warms sufficiently and early growth and the achievement of full cover fall behind those of perennial-dominated native vegetation. Perennial forage crops utilize the available growing season more completely, but through periodic harvest (mowing or grazing) they experience a series of regrowth cycles, beginning each time at a low leaf area index. In tropical regions, various systems of relay and overlap cropping are employed to extend the duration of green cover (DALRYMPLE 1971; WILLEY 1979) and the seasonal sequences are generally linked to rainfall patterns (HALL et al. 1979).

Figure 5.3 illustrates a seasonal pattern of potential net production in the temperate zone. This remarkable set of growth curves was obtained in The Netherlands with closed swards of perennial ryegrass (*Lolium perenne,* a C_3 species) provided with adequate nutrients and water (ALBERDA and SIBMA 1968). The crop growth rates, estimated from the slopes of these curves, increase rapidly to a maximum in June (peak radiation) and then decline steadily through October. The November–March winter period involves little or no growth. The annual cycle is asymmetric around the longest day principally because the temperatures favorable for plant growth are skewed towards late summer and fall. The maximum growth rate shown here, $20 \text{ g m}^{-2} \text{ d}^{-1}$ ($200 \text{ kg ha}^{-1} \text{ d}^{-1}$), matches that observed in the Dutch environment for other C_3 species (SIBMA 1968). Maize reaches a higher growth rate there (30 to $35 \text{ g m}^{-2} \text{ d}^{-1}$) but over a shorter season because it lacks tolerance to the low temperatures of spring and fall.

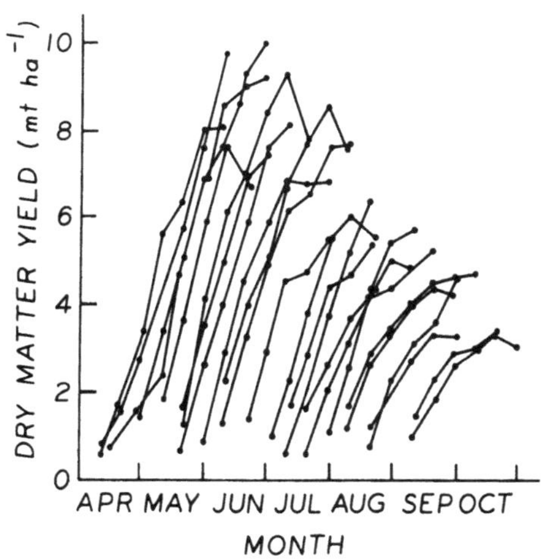

Fig. 5.3. Production of aboveground dry matter by closed swards of perennial ryegrass with different locations, soils, and years in The Netherlands; water and nutrients were non-limiting. Simplified from ALBERDA and SIBMA (1968). The slopes of the curves during May–July are near $20 \text{ g m}^{-2} \text{ d}^{-1}$, or very similar to those of other C_3 crops grown in that area (SIBMA 1968)

Where seasonal moisture supply is strongly limiting, full utilization of the frost-free period is much less important than the avoidance of drought injury. A wide array of tactics are available for such conditions. Cool-season crops permit culture during periods with the least evaporative demand, thus extending the days of transpiration possible for a given supply of water. Thus, one finds small grains (winter or spring types) and vegetables during the winter periods in Mediterranean climates, and winter wheat in the southern Great Plains of North America. Summer cropping with limited moisture may emphasize short-duration C_3 species (cucurbits and beans such as *Vigna* spp.) and C_4 cereals (e.g., *Sorghum, Panicum,* and *Pennisetum* spp.). C_4 plants are attractive options in arid climates because of their greater water-use efficiency (see Chap. 15, Vol. 12B).

Other manipulations include the culture of long-season crops but at wide spacings (giving reduced cover and a reduced rate of evapotranspiration), fallow, runoff-runon strip culture, and irrigation systems. In irrigated long-season desert regions, cropping options and seasonal patterns are very broad. Double-cropping (two crops per year) is not uncommon. Where nitrogen fertilizers are available, the grower is freed from lock-step rotation with legumes and the cropping pattern may then become very responsive to new varieties and market changes. During the past 20 years, the Davis, California environs have gone from a system of alfalfa, barley, grain sorghum, and sugar beet to one dominated by wheat, maize, and tomato (*Lycopersicon esculentum*) with cucurbits, safflower (*Carthamus tinctorius*), sugar beet and alfalfa now as secondary crops.

5.2.4 Partitioning Patterns

The most common pattern for partitioning of dry matter in annual crops is that early growth is dominated by allocations to leaf, stem, and root tissues. Following this juvenile stage, photosynthate flows primarily to reproductive organs (grains, beans, etc.) or to vegetative storage organs (potato, sugar beet). With reproductive crops, determinate flowering patterns work well with short growing seasons. While a wide range of growth durations are found in determinate species, few of them can effectively utilize the full season of tropical and subtropical environments, hence the practice of double-cropping. The indeterminate pattern is useful in fresh market crops (tomatoes, melons) and where season length is long, or uncertain, with a possibility of being long.

Figure 5.4 illustrates the partitioning pattern of spring wheat. This pattern is a fairly typical one for annual crops. The data are means for five bread wheat cultivars grown in Montana (U.S.A.) under rain-fed, semi-arid conditions (370 mm precipitation) and moderate fertility (McNEAL et al. 1966). The early dominance of leaf production followed in sequence by stem and grain is clear. Also evident is a negative growth of stem and leaves after heading and a notable decrease in their nitrogen content. The negative growth is interpreted as due mostly to mobilization and translocation of carbohydrates and nitrogenous materials to the developing grain, but respiration and the loss of dead material would also contribute. The grain yield illustrated here (3,760 kg ha^{-1}) is low

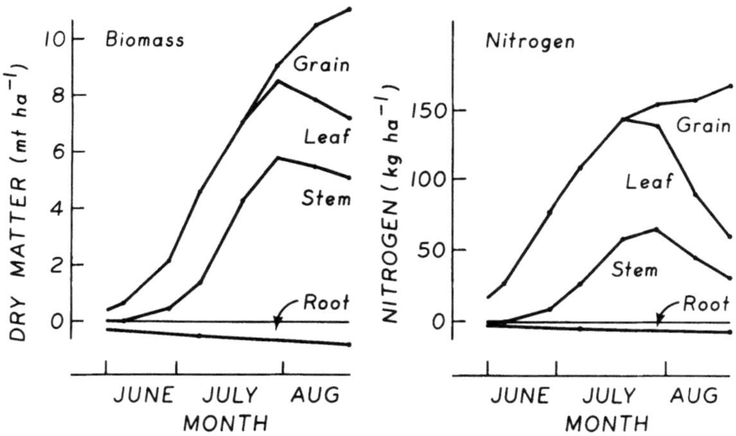

Fig. 5.4. Time course of dry matter (*left*) and nitrogen (*right*) accumulation by wheat and their distribution among various plant parts. The observations are means for five cultivars grown in a replicated trial under dryland conditions in Montana, U.S.A. (McNeal et al. 1966)

compared with the record yield of 11,900 kg ha^{-1}, yet it is nearly double the usual average for that state. The harvest index (grain weight as a fraction of aboveground biomass) was 0.34, a typical value for the older varieties and selections used in this study; 0.4 or greater is common in new varieties and varieties having a lower protein content in grain. The grain protein content for Fig. 5.4 is 16.3% (N% × 5.7) which is fairly typical for hard-red spring wheats in Montana.

Roots are shown here as only a small fraction (0.062) of the total crop at final harvest. This is similar to the 0.085–0.145 range observed by Welbank et al. (1974) at Rothamsted during the grain-filling period for wheat, but it may be an underestimation for dryland conditions where other authors have observed root fractions for wheat of 0.25 or more. Root sampling is not easy. One has difficulty distinguishing live and dead material and separating inert matter; root hair, root cap and exudate materials are not recovered.

The mobilization of nitrogen from vegetative parts was quite dramatic. On July 7, leaves and stems contained 2.9% nitrogen (129 kg ha^{-1}). Total nitrogen in leaves and stems reached a peak of 145 kg (2.0% of weight) on July 19 and then declined to 60 kg (0.8%) by the final harvest. At that time, the whole plant nitrogen content was 1.6%, but 62% of that (108 kg) was sequestered in grain. Nitrogen uptake after heading was limited. Perhaps nitrogen supply was low, perhaps drought prevented the plants from acquiring more, or perhaps drought accelerated leaf senescence and thereby reduced the late-season uptake.

Accelerated leaf senescence has been termed self-destructive (Sinclair and de Wit 1976). There is speculation that yields and perhaps harvest index could be increased through avoiding such massive redistributions of leaf nitrogen. Viewed in another way, the observed behavior has a rather high fitness. Nitrogen uptake and reduction occur principally during vegetative growth when the large cost of those processes is not competitive with the flow of current photosynthate to grain. In addition, nitrogen-use efficiency is enhanced because the yield of grain per unit of acquired nitrogen is high.

The wheat crop in Fig. 5.4 illustrates all of the basic elements of organ partitioning one sees with other crop species: a dynamic sequencing of events in morphogenesis, biomass synthesis, and production processes. Partitioning can also be considered in relation to changes in biomass composition. Forage crops, for example, are harvested at frequent intervals to obtain a high proportion of leaf tissue (high in protein) relative to stem and to avoid the increased lignin content of stems that comes with advancing age. While cell-wall materials are always a significant fraction of biomass, the proportions of protein, lipids, nonstructural carbohydrates, and other materials that accumulate within those walls can be manipulated through changes in management practices (e.g., spacing and fertilizer). The controls of both the morphological and biochemical aspects of partitioning are adaptive to the current state of the biological and environmental components of the system. An important corollary is that for much of the world's agriculture, the limits on production lie more with resources (supplies of nutrients and water) and with the dynamic aspects of growth and development than with photosynthetic capability.

A similar argument can be made that variations in adaptation to stressful environments is not a major limitation on productivity. Such adaptations are extremely important to survival, as from one rainfall to the next, but little in the way of production is accomplished under stress conditions (see Chap. 1, Vol. 12B). Over a season, a few days of high production in which water is not limiting is equal to many days under drought conditions with low stomatal conductivity (DE WIT 1958). Many species and cultivars exist with high levels of resistance to salinity, drought, low temperature and other extremes. Thus we find wheat grown in semi-arid and cold regions, barley and cotton in saline environments and so on. Lack of resistance is a problem mainly when man attempts to extend a successful crop outside its normal environmental range. Cotton, for example, suffers from chilling injury in cold environments. But even if it could be successfully bred to tolerate such conditions, its annual productivity would not be likely to exceed that of wheat except in an economic sense.

5.3 Controlling Relationships

Nutrition and partitioning have been emphasized as controlling features in agricultural productivity. Nitrogen dominates in nutrient cycle considerations because of its high concentration in herbaceous biomass and agricultural products and its relative scarcity in agricultural soils. Without external inputs, there generally is little likelihood of sustaining ceiling yields in nonleguminous crops, and the herbaceous legumes seem to have low annual productivity. Alfalfa appears as an exception. In California, alfalfa produced 25 to 30 mt forage ha^{-1} yr^{-1} for several years without inputs of nitrogen fertilizer. At 2.5% nitrogen in the harvested forage, 600 to 750 kg nitrogen ha^{-1} were removed each year. In spite of that, soil nitrogen status was higher following alfalfa than before its planting. The symbiotic nitrogen fixation system in alfalfa, like other bacterial systems, is repressed by high levels of soil and plant nitrogen. These are very high, perhaps record, rates of fixation, but it seems probable that they could be achieved only with periodic harvest (in these cases, 6 to 10 cuttings per year). In general, legume rotations do not supply amounts of rapidly cycling nitrogen sufficient for maximum yields of subsequent nonleguminous crops.

The importance of partitioning cannot be overstated. Agricultural plants must supply their own machinery for photosynthesis, nutrient and water uptake, and biosynthesis in addition to accumulating products of use to man. We saw that as much as 0.5 to 0.6 of the aboveground weight of cereal grains is in a useful form, but there are many other species in which the factor is 0.25 or less. The system is further constrained by the cellular structure of higher plants, in which wall materials account for more than 50% of the weight of vegetative organs. The wall materials are not digestible by man; this in turn leads to a lower digestibility of the cell contents. This problem accounts in part for the emphasis on seed crops (which have high ratios of digestible to indigestible substances; Table 5.3) in human diets, and for the importance of ruminant animals in agricultural systems. We may eventually come to manufacturing digestible foods from whole biomass, thus avoiding waste and by-passing ruminants, but at present, partitioning is still an important area for research. Basically, this means the study of growth and development, two areas that lag badly in plant physiological progress. Agricultural research in those areas has emphasized genetic and environmental control and, in some cases, the more basic features of morphogenesis. LOOMIS et al. (1979) have outlined methods for dealing with the dynamic and adaptive aspects of partitioning through the construction of explanatory simulation models (see Chap. 4, this Vol.). However, DE WIT et al. (1978) feel that information is still inadequate for reasonable simulations of the morphogenetic aspects of reproduction.

Such is not the case with the production processes of photosynthesis by plant stands. DE WIT et al. (1978) provide what is perhaps the state of the art in that area for a complete model (see also Chap. 16, Vol. 12B), although much more detailed models exist on component processes such as eddy transport of gaseous materials between atmosphere and crop. Work in this area emphasizes radiation exchange and the development of the crop microclimate, and their dependence upon canopy structure. Major constraints on production both diurnally and annually are shown to arise from the low ambient supply of carbon dioxide and its influence over water-use efficiency (DE WIT et al. 1978; FISCHER and TURNER 1978).

5.4 Conclusions

Arable farming is shown to occupy about 10% of ice-free land. Cereal grains, particularly wheat, rice, and maize, are the dominant crops, supplying the great bulk of energy and protein for the human population. Average economic production for those crops is about 2 mt ha^{-1}, or about 20% of the yield of those crops in intensive agriculture. The low yield is in part a limitation by water supply but nutrients seem to be a more general issue. Despite that, the cereal grain supply alone is presently equal to about 1 kg cap^{-1} d^{-1} for the world's population. That original supply is equal to about 4,000 kcal (17 MJ) and about 100 g protein for each person. The supply is modified significantly by the need for planting seed, by losses in storage, transit, and food preparation, by variations in digestibility, and by the proportion fed to livestock. But when added to other sources of food, including animal products derived by grazing of an additional 20% of the land area, it is more than adequate, if distributed properly, for the present population.

Agricultural production must be viewed as demand-dependent. Demand currently is such that agricultural potentials are pressed in only a few regions. That means that the reserve capacity of present lands is very large. In addition to the low average yield, 20% of the land is in fallow or other states of idleness. This condition, plus animal populations and the additions that might be made to the arable base, give the world a carrying capacity with present crop species and technology for several more doublings of the human population. The costs in energy and human effort, however, would be large.

Photosynthetic capacity is at present not the major issue in agriculture, and this fact is reflected by the large proportion of agricultural research effort that goes into pest management and other aspects of "maintenance" and "defect elimination." However, the argument on photosynthetic capacity does not hold in Japan, which has intensive fertilization practices; and it does not hold with legumes, in which increased capacity might also yield increases in nitrogen fixation. But with other crops, a primary limitation by nitrogen reduces photosynthetic capacity to a secondary role: with a higher leaf rate, the nitrogen supply is simply exhausted sooner. Thus, a key issue in agricultural research might be how to obtain the same production rate with less expense of nitrogen.

Acknowledgment. The advice and criticism of J. Cock, CIAT, was very helpful in the course of this work.

References

Alberda T, Sibma L (1968) Dry matter production and light interception of crop surfaces. III. Actual herbage production in different years as compared with potential values. J Br Grassl Soc 23:206–215

Boysen-Jensen P (1932) Die Stoffproduktion der Pflanzen. Fischer, Jena

Brougham RW (1956) Effect of intensity of defoliation on regrowth of pasture. Aust J Agric Sci 7:377–387

Clark EA, Loomis RS (1978) Dynamic aspects of leaf growth and development in sugarbeets. J Am Soc Sugar Beet Technol 20:97–113

Clark HE (1978) Cereal-based diets to meet protein requirements of adult man. World Rev Nutr Diet 32:27–48

Cock JH, Wholey D, Gutierrez de la Casas O (1977) Effects of spacing on cassava (*Manihot esculenta*). Exp Agric 13:289–299

Cooper JP (1975) Control of photosynthetic production in terrestrial systems. In: Cooper

JP (ed) Photosynthesis and productivity in different environments. Cambridge Univ Press, Cambridge
Corley RHV (1973) Effects of plant density on growth and yield of oil palm. Exp Agric 9:169–180
Cours G (1951) Le manioc a Madagascar. Mem Inst Sci Madagascar Ser B3:203–400
Dalrymple DG (1971) Survey of multiple cropping in less developed nations. US Dep Agric Bull FEDR-12, Washington, DC
Duncan WG (1971) Leaf angles, leaf-area, and canopy photosynthesis. Crop Sci 11:482–485
Duncan WG, Loomis RS, Williams WA, Hanau R (1967) A model for simulating photosynthesis in plant communities. Hilgardia 38:181–205
Ehleringer J, Björkman O (1977) Quantum yields for CO_2 uptake in C_3 and C_4 plants. Plant Physiol 59:86–90
Evans LT (1975) The physiological basis of crop yield. In: Evans LT (ed) Crop physiology – some case histories. Cambridge Univ Press, Cambridge
FAO (1979) Production yearbook. Food Agric Org UN, Rome
Fischer RA, Turner NC (1978) Plant productivity in the arid and semiarid zones. Annu Rev Plant Physiol 29:277–317
Garner HV, Dyke GV (1969) 3. The Broadbalk yields. Part 2, Rothamsted Exp Sta Rep 1968
Hall AD (1929) Agriculture. Encycl Br (14th edn) 1:391–397
Hall AE, Connell GH, Lawton HW (eds) (1979) Agriculture in semiarid environments. Ecol Stud Vol 34. Springer, Berlin Heidelberg New York
Holmes EB, Wilson LA (1977) Total dry matter production, tuber yield, and yield components of six local cassava cultivars in Trinidad. In: Cock J, MacIntyre R, Graham M (eds) Proc 4th Symp Int Soc Trop Root Crops. Int Dev Res Cen, Ottawa
Jenkinson DS (1971) The accumulation of organic matter in soil left uncultivated. Rothamsted Exp Stn Rep 1970
Jenny H (1930) A study on the influence of climate upon nitrogen and organic matter content of the soil. Univ Missouri Agric Exp Stn Bull 152
Jenny H (1941) Factors of soil formation. McGraw-Hill, New York
Lemeur R, Blad BL (1974) A critical review of light models for estimating the shortwave radiation regime of plant canopies. Agric Meteorol 14:255–286
Loomis RS, Gerakis PA (1975) Productivity of agricultural ecosystems. In: Cooper JP (ed) Photosynthesis and productivity in different environments. IBP3. Cambridge Univ Press, Cambridge
Loomis RS, Williams WA (1963) Maximum crop productivity: an estimate. Crop Sci 3:67–72
Loomis RS, Williams WA (1967) Productivity and the morphology of crop stands: patterns with leaves. In: Eastin JD, Haskins FA, Sullivan CY, Bavel van CHM (eds) Physiological aspects of crop yield. Am Soc. Agron, Madison
Loomis RS, Rabbinge R, Ng E (1979) Explanatory models in crop physiology. Annu Rev Plant Physiol 30:339–367
McDermitt DK, Loomis RS (1981) Elemental composition of biomass and its relation to energy content, growth efficiency, and growth yield. Ann Bot 48:275–290
McNeal FH, Berg MA, Watson CA (1966) Nitrogen and dry matter in five spring wheat varieties at successive stages of development. Agron J 58:605–608
Monteith JL (1978) Reassessment of maximum growth rates for C_3 and C_4 crops. Exp Agric 14:1–5
Murata Y (ed) (1975) Crop productivity and solar energy utilization in various climates in Japan. JIBP Synthesis, vol XI. Univ Tokyo Press, Tokyo
NAS (1969) United States – Canadian tables of feed composition. Natl Acad Sci, Washington DC
Penning de Vries FWT, Brunsting AHM, Laar van HH (1974) Products, requirements and efficiency of biosynthesis; a quantitative approach. J Theor Biol 45:339–377
Robinson RA (1976) The pathosystem concept. Plant pathosystems. Springer, Berlin Heidelberg New York

Ross J (1981) The radiation regime and architecture of plant stands. TVS 3. Junk, The Hague
Shibles RM, Weber CR (1966) Interception of solar radiation and dry matter production by various soybean planting patterns. Crop Sci 6:55–59
Sibma L (1968) Growth of closed green crop surfaces in the Netherlands. Crop Sci 11:482–485
Sinclair TR, Wit de CT (1976) Analysis of the carbon and nitrogen limitations to soybean yield. Agron J 68:319–324
Slicher van Bath BH (1963) The agrarian history of Western Europe, AD 500-1850. Olive Ordish (transl) Arnold, London
Spiertz JHJ, Ellen J (1978) Effects of nitrogen on crop development and grain growth of winter wheat in relation to assimilation and utilization of assimilates and nutrients. Neth J Agric Sci 25:210–231
Stanhill G (1976) Trends and deviations in the yield of the English wheat crop during the last 750 years. Agro-Ecosystems 3:1–10
Stauffer RS, Muckenhirn RJ, Odell RT (1940) Organic carbon, pH, and aggregation of the soil of the Morrow plats as affected by type of cropping and manurial addition. J Am Soc Agron 32:819–832
Thünen von JH (1826) Der isolierte Staat in Beziehung auf Landwirtschaft und National-Ökonomie. Perthes, Hamburg
Trenbath TR, Angus JF (1975) Leaf inclination and crop production. Field Crop Abstr 28:231–244
Welbank PJ, Gibb MJ, Taylor PJ, Williams ED (1974) Root growth of cereal crops. Part 2, Rothamsted Exp Stra Rep 1973
Welch LF (1976) The Morrow plots – hundred years of research. Ann Agron 26:881–890
Whittaker RH, Likens GE (1975) The biosphere and man. In: Lieth H, Whittaker RH (eds) Primary productivity of the biosphere. Ecol Stud Vol 14. Springer, Berlin Heidelberg New York
Willey RW (1979) Intercropping – its importance and research needs. Field Crop Abstr 32:1–10 (part 1) 73–85 (part 2)
Williams WA, Loomis RS, Lepley CR (1965) Vegetative growth of corn as affected by population density. I. Productivity in relation to interception of solar radiation. Crop Sci 5:211–215
Williams WA, Loomis RS, Duncan WG, Dovrat A, Nunez AF (1968) Canopy architecture at various population densities and the growth and grain yield of corn. Crop Sci 8:303–308
Wit de CT (1958) Transpiration and crop yields. Versl Landbouwk Onderz 64.6
Wit de CT, Heemst van HDJ (1976) Aspects of agricultural resources. In: Koetsier WT (ed) Proceedings of the plenary sessions of the first world congress on chemical engineering. Elsevier, Amsterdam
Wit de CT, et al. (1978) Simulation of assimilation, respiration and transpiration of crops. Simulation Monographs. PUDOC, Wageningen

6 Productivity of Grassland and Tundra

L.L. Tieszen and J.K. Detling

CONTENTS

6.1 Introduction 173

6.2 Factors Affecting Productivity 177
6.2.1 Photosynthetic Rates of Species 177
6.2.2 Daily and Seasonal Photosynthetic Patterns 178
6.2.3 Irradiance 179
6.2.4 Temperature 181
6.2.5 Water 186
6.2.6 Nutrients 189
6.2.7 Herbivores and Grazing 191
6.2.8 Allocation 194

6.3 Conclusions 196

References 197

6.1 Introduction

Grassland and tundra biomes are considered jointly in this chapter largely because they share a similar physiognomy and often are categorized as "grazing lands" because they support a relatively large trophic level of grazing herbivores. We will use the term "tundra" to denote treeless regions north (and south) of latitudinal treelines or above natural altitudinal treelines (see Murray 1978). Similarly, "grassland" in this chapter refers to treeless (but not desert) systems in the warmer climates of lower latitudes and lower altitudes. Savannas, systems characterized by scattered trees with open canopies, support a continuous grass understory and can be considered as a special element within grassland. Grasslands and tundras are characterized by the dominance of graminoids (grass-like monocotyledons) and the general absence of trees, but they rarely occur contiguously. They are often separated by forests from which they either can develop as a result of anthropogenic, biologic, or climatic effects.

Although tundra systems are perhaps most pragmatically defined on the basis of physiognomy and geography, they also share certain climatic features. All are characterized by generally low temperatures throughout the year, but especially during the plant growing season. Mean annual air temperature at the arctic IBP sites (Rosswall and Heal 1975) ranged from −13.0 °C at Ary-Mas, U.S.S.R., to −0.7 °C at Abisko, Sweden. In fact, tundra is sometimes delimited on the basis of a mean annual 0 °C isotherm or by the presence of permafrost (subsurface material, usually frozen, with a mean temperature ≦0 °C). These arbitrary climatic criteria, however, would exclude a number of alpine, antarctic, or maritime sites which meet our physiognomic criteria. Within "tundras", however, there is a large variation in biotic components as well as climatic factors.

On a world-wide basis, grasslands (including savanna and shrub steppe) represent the potential natural vegetation on 25% of the land surface (SHANTZ 1954), and account for about 16% (1.89×10^{10} t yr^{-1}) of the total net primary production (NNP) (1.7×10^{11} t yr^{-1}) of land-plant communities (WHITTAKER and LIKENS 1975). Although they can develop and be maintained as successional communities by burning, grazing, mowing, or other agricultural practices, the geographical distribution of most of the world's grasslands is controlled primarily by climate. They occur in areas having one or more periods during the year when water availability declines below that required for forests, but they receive sufficient precipitation to maintain graminoids as the dominant or co-dominant life form. Thus, mean annual precipitation of the 52 grassland sites reviewed by LAUENROTH (1979) generally ranged from 200 to 1,000 mm yr^{-1}, although a few higher values were reported for successional grasslands. This is a considerably narrower precipitation range than that (200 to 2,000 mm yr^{-1}) encountered in tundra ecosystems. However, grasslands occur over a much broader temperature range than do tundra systems. Mean annual temperatures of grasslands can range from near 0 °C in parts of North America and Europe to 25 °C in tropical grasslands of Africa (LAUENROTH 1979). Grasslands characterized by these low annual means, however, occupy distinctly continental climates where growing season temperatures are considerably higher. LAUENROTH (1979) indicated that while temperature and precipitation may frequently interact in controlling grassland primary production, precipitation was the single factor most closely correlated with NPP of the world's grasslands.

The large vegetation diversity has led to numerous attempts to systematically classify grassland and tundra units (see MURRAY 1978). Within the broad arctic, antarctic (see BLISS 1979), and alpine areas, these subdivisions have usually been based on the dominant vegetation. WIELGOLASKI (1972) recognizes six common groups based on the relative contributions of growth forms to the total biomass (Table 6.1). Four are dominated by woody vegetation and two by graminoids. A seventh type, consisting of patchy vegetation, the fell field, can also be identified. The vegetation often changes markedly over very short distances (WEBBER 1978) and changes in a consistent manner with increasing latitude and altitude. The flora becomes progressively reduced, and the vegetation changes as major growth forms change (BESCHEL 1969). Shrubs, especially deciduous forms, become less common and the vegetation becomes shorter and more appressed. Thus, species richness and diversity, as well as structural complexity, decrease at high latitudes and altitudes.

The recognition of vegetation types based on growth form composition suggests that growth forms reflect integrated strategies of adapting to various environmental complexes (see Chap. 18, Vol. 12B). These relationships derived from

Table 6.1. Common vegetation types identified in tundras. (WIELGOLASKI 1972)

	Total aboveground living vascular plant biomass (%)			Total aboveground living biomass (%)		
	Woody plants	Forbs	Monocotyledons	Vascular plants	Bryophytes	Lichens
Woody						
Shrub meadows	≧50	≧10–15	≦40	≧50	≦40–50	≦10
Shrub heath	≧80–90	≧10–15	≦10	Normally ≦90	≦40	Normally ≧10
Dwarf shrub bogs	≧50	≦10	≦40–50	Normally ≦50	Normally ≧45–50	≦5
Shrub-moss-lichen snowbeds	≧50	≦10–15	≦40–50	≦50	≧40–50	–5–60
Graminoid						
Monocotyledonous bogs and mires	≦50	≦10	≧40–50	Normally ≦50	Normally ≧40–50	≦5
Tundra meadows	≧50	≧10	≦90	≧50	≦40–50	≦10
Fell field	High		Low	High		

Table 6.2. Abiotic factors correlated with the presence of various growth forms in 78 circumpolar and polar tundra stands. (WEBER and FRENCH, personal communication)

Life form	Water	Temperature	Nitrogen	Calcium	Phosphorus	Aeration
1. Deciduous shrub	Mesic	Warm	High	Low	Mesic	Mesic-high
2. Evergreen shrub (> 3 cm)	Dry	Cold	Low	High	High	High
3. Single monocotyledon	Wet	Cold	High	Low	Mesic	Low
4. Caespitose monocotyledon	Dry	Warm	High	High	Mesic	High
5. Rosette dicotyledon	Dry-mesic	Mesic-warm	High	High	Mesic	Mesic
6. Erect dicotyledon	Wet	Cold	High	Low	Mesic	Low
7. Mat dicotyledon	Mesic-wet	Cold-mesic	High	High	High	Low
8. Cushion dicotyledon	Dry	Cold	Low	High	High	High
9. Bryophyte	Wet	Cold	High	Low	Low	Low
10. Lichen	Dry	Cold	Low	Mesic	Mesic	High

studies at many tundra sites (TIESZEN et al. 1981) are depicted in Table 6.2 and form a basis for the structuring of hypotheses about the physiological significance of growth forms.

Just as variability exists in the biotic components of tundra systems, so there is wide diversity across grasslands. However, no classification of grassland types based on relative contribution of various growth forms has been undertaken. In North America, for example, grassland classifications have been based upon and named according to a diversity of characteristics including dominant taxa (e.g. *Bouteloua-Buchloë*), plant height (e.g., shortgrass or tallgrass prairie), growth form of the dominants (e.g., bunchgrass steppe), perennating habit of the grasses (e.g., annual grassland), or geographical location (e.g., mountain grasslands, northern and southern mixed-grass prairies). The shrub steppe of Russia and wood grasslands (savannas) of tropical Africa are examples of grasslands with somewhat different physiognomies.

In addition to these characteristics, differences in the relative proportions of C_3, C_4 and CAM plants across grasslands can lead to differences in ecosystem function. For example, while there is variability among species with a given photosynthetic pathway, distinct differences in phenological patterns are often observed among plants with each pathway (DICKINSON and DODD 1976). As discussed in more detail later, such differences would be expected to alter seasonal patterns of photosynthesis and primary production (TIESZEN 1970; WILLIAMS and MARKLEY 1973; WILLIAMS 1974; EICKMEIER 1978; ODE et al. 1980), water use efficiency (BLACK 1971, 1973; LUDLOW 1976; BROWN and SIMMONS 1979), nutrient use efficienty (WILSON 1975; BROWN 1978; CHRISTIE 1979), above- and belowground carbon allocation patterns (CHRISTIE 1978), and quantum yield (EHLERINGER 1978); for a general discussion of the functional significance of the different pathways of CO_2 fixation in photosynthesis see Chap. 15, Vol. 12B.

In Arctic tundra, however, all species are C_3 (see TIESZEN and SIGURDSON 1973, for Northern Alaska) as TEERI and STOWE's (1976) floristic study also suggests. As far as is known, alpine tundras are also characterized by C_3 species even though some succulents do occur. A few C_4 grasses (e.g., *Andropogon* in Kenya; TIESZEN et al. 1979; and *Muhlenbergia* in North and South America) are found in low alpine sites.

This relationship with latitude is altitude-dependent, however. As Fig. 6.1 indicates, in North American grasslands (FRENCH 1979) the proportion of C_3 species increases with latitude in a manner consistent with the floristic suggestions. In tropical grasslands, C_4 grasses dominate (often exclusively) at low altitudes (TIESZEN et al. 1979) but are replaced by C_3 grasses at higher altitudes (see also HARTLEY 1973). This is probably related to temperature (VOGEL et al. 1978) and is reflected in the floristic analysis of TEERI and STOWE (1976) as well. Thus, grasslands and tundras differ with respect to the presence of some growth forms even though they share a similar overall physiognomy. In addition, the C_4 photosynthetic system becomes more important in grasslands as temperatures increase, while CAM species are sometimes important in some of the more arid grasslands (DODD and LAUENROTH 1975, 1979).

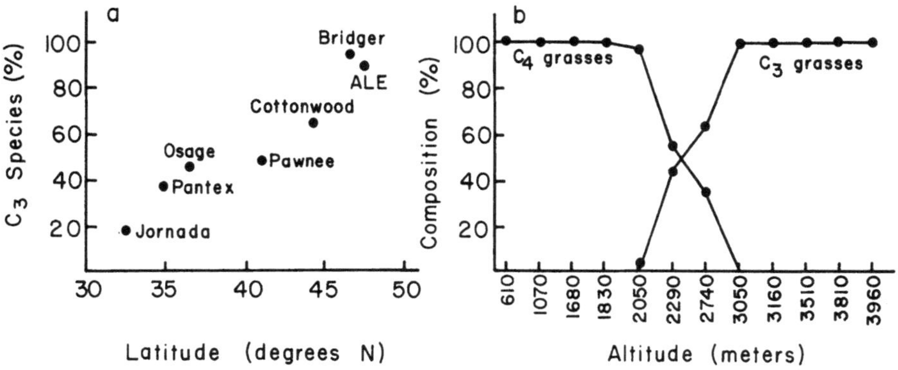

Fig. 6.1. Percent contribution of C_3 grasses to the flora of the U.S. IPB Grassland Biome sites (**a**) (redrawn from FRENCH 1979; Altitudes: Jornada, 1,350 m; Pantex, 1,075 m; Osage, 392 m; Pawnee, 1,652 m; Cottonwood, 744 m; ALE, 390 m; Bridger, 2,340 m) and **b** along an altitudinal transect in Kenya, East Africa. (TIESZEN et al. 1979)

6.2 Factors Affecting Productivity

6.2.1 Photosynthetic Rates of Species

Arctic tundra plants contrast strongly with those in alpine areas or grasslands mainly because of the short growing season, low temperatures, and continuous growing season photoperiod encountered by the former. The maximal photosynthetic rates of tundra species vary considerably from rates as high as those of temperature zone C_3 plants to substantially lower (TIESZEN et al. 1980). The closest correlation is between maximal rates and growth form (Table 6.3). Graminoids, forbs and some deciduous shrubs (e.g. *Salix*) possess the highest rates

Table 6.3. Estimates of light-saturated rates of CO_2 uptake in tundra field plants. Since in some cases only few measurements were taken, these may represent minimal rates. (Tabulated original data available from TIESZEN 1980)

Plant type	CO_2 uptake (mg CO_2 h^{-1})	
	dm^{-2} leaf area	g^{-1} dry wt.
A. Herbaceous plants		
1. Tundra graminoids	14	18
2. Tundra forbs	15	15
B. Woody plants		
1. Tundra deciduous dwarf shrubs	13	15
2. Tundra evergreen dwarf shrubs	7	5
C. Mosses		2
D. Lichens		0.4

Table 6.4. Estimates of maximum CO_2 uptake rates (mg CO_2 dm^{-2} leaf area h^{-1}) and temperature optima of North American grassland plants in the field. (Data for *Bouteloua gracilis* from DETLING et al. 1978; all other data are from A.T. HARRISON unpublished)

Plant species	Optimum temperature (°C)	Maximum CO_2 uptake
C_3 grasses		
Leucopoa kingii	15–25	16
Leersia oryzoides	20–27	16
Elymus canadensis	25	18
C_4 grasses		
Bouteloua gracilis	30	75
Panicum virgatum	38–40	48
Spartina pectinata	34	40
Andropogon hallii	30	42
Muhlenbergia filiculmis	25	17

in tundras, whereas lichens, mosses, and evergreen dwarf shrubs have lower capacities. Different longevities and maturation periods are associated with the growth forms. Species with rapid leaf turnover (JOHNSON and TIESZEN 1976) have higher photosynthetic rates; species with greater longevities take up CO_2 more slowly. Although complete costs and budgets are lacking, the data (TIESZEN et al. 1981) suggest that lifetime budgets for all growth forms may be quite comparable.

Maximum net photosynthetic (P_N) rates of vascular grassland species are closely coupled to photosynthetic pathway (Table 6.4). For example, maximum P_N rates of three C_3 grasses (16 mg CO_2 dm^{-2} h^{-1}) studied by HARRISON (personal communication) averaged about half the maximum rates of the C_4 grasses (37 mg CO_2 dm^{-2} h^{-1}). Considerable variation occurs in measured P_N rates among species with a given photosynthetic pathway. Thus, data summarized in Table 6.4 include C_4 grasses with maximum P_N rates varying from 17 mg CO_2 dm^{-2} h^{-1} in *Muhlenbergia filiculmis* (HARRISON personal communication) to 75 mg CO_2 dm^{-2} h^{-1} for *Bouteloua gracilis* (calculated by DETLING et al. 1978, from field data of BROWN and TRLICA 1974). As in the tundra, non-vascular plants are assumed to have low P_N rates but these groups constitute $<5\%$ of the standing crop in most grassland ecosystems.

6.2.2 Daily and Seasonal Photosynthetic Patterns

Daily rates under natural conditions are often as high in some tundra plants as in those from grasslands. Daily total can exceed 400 mg CO_2 dm^{-2} (e.g., in *Salix pulchra*) and commonly reach levels of 200 to 300 mg CO_2 dm^{-2} in graminoids (TIESZEN 1975). These are comparable to daily rates obtained by temperate zone grasses. Other growth forms, e.g., *Dryas* (MAYO et al. 1972, 1975) and mosses (OECHEL et al. 1975), attain lower daily levels, largely because of lower photosynthetic capacities. It is clear that the high daily totals in arctic tundra plants result from the continuous photoperiod which permits net positive CO_2 uptake 24 h a day and the ability of these plants to function well at low temperatures.

BROWN and TRLICA (1977) measured CO_2 exchange over pure swards of *Bouteloua gracilis* (C_4) at a North American shortgrass prairie site. Daily totals (DETLING 1979) ranged from 70 mg CO_2 dm^{-2} d^{-1} in August when soil water potential in the root zone was quite low (< -2.0 MPa) to 540 mg CO_2 dm^{-2} d^{-1} when soil water was less limiting (soil water potential > -1.0 MPa). By late September, maximum P_N rates had declined and photoperiods had shortened, causing daily totals to be reduced to 200 mg CO_2 dm^{-2} d^{-1} under otherwise near-optimal growing conditions. These trends are consistent with seasonal trends predicted in the grassland model of DETLING et al. (1979c).

6.2.3 Irradiance

As discussed in Chap. 3 of Vol. 12A and Chap. 15 of Vol. 12B, photosynthetic rates in individual leaves of C_4 plants do not generally become light-saturated at full sunlight, while those of C_3 plants become saturated at half or less of full sunlight (BLACK 1973). Both the photon flux density required for light saturation and CO_2 exchange rates are dependent upon past illumination (WILSON and COOPER 1969, see Chap. 2, Vol. 12A). BOARDMAN (1977) concluded that plants growing in high light environments become light-saturated at higher intensities and have higher photosynthetic rates at light saturation than do those from shady habitats. WOLEDGE (1977) performed a series of experiments on a cultivated temperate pasture grass, *Lolium perenne* (perennial ryegrass), which indicated that observed declines in photosynthetic capacity of similarly aged, newly expanded leaves over time were caused by self-shading as canopy closure progressed. Earlier simulation modeling experiments (ROBSON 1973) had suggested that total photosynthetic production of ryegrass swards was reduced by 30% by such reductions in photosynthetic capacity.

In comparing photosynthetic light responses of C_3 and C_4 species, LUDLOW (1976) noted that the marked differences which were evident in single leaves began to disappear as self- and mutual shading increased in individual plants or communities. As a result, he suggested that net photosynthesis of communities of C_3 and C_4 species responds almost linearly to solar radiation, with the slope of the line being greater in C_4 species. However, when other environmental factors limit photosynthesis, e.g., leaf temperature (WILLIAMS and KEMP 1978), light saturation does occur in single leaves of C_4 species, and their responses to fluctuations in light become more similar to those of C_3 species at the single leaf, whole plant, and community levels (LUDLOW 1976).

Photosynthesis in many tundra systems is closely coupled to photon flux density, especially in vascular plants from wet tundras (TIESZEN 1973; SHVETSOVA and VOZNESSENSKIY 1970), and highest rates are commonly observed at highest irradiances. Bryophytes (OECHEL and COLLINS 1976) and evergreen shrubs (e.g. *Dryas*), however, may show a depression at high photon flux densities associated with water stress. Thus, some growth forms have a common midday depression.

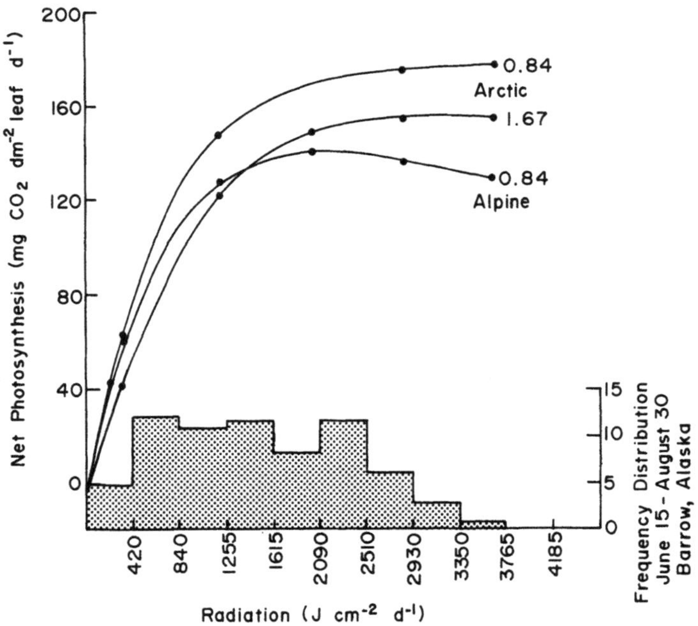

Fig. 6.2. Mean rate of CO_2 uptake of a tundra graminoid (*Dupontia*) as related to photon flux density in the Arctic and at 40 °N latitude. Canopy simulated had a leaf area index around 1.0. (MILLER et al. 1976)

Dominant species on wet meadow systems, however, do not. In an attempt to understand the ecological significance of the response curve and photon flux density we have used a refined canopy model (MILLER et al. 1976) to simulate the sensitivity of photosynthesis to daily irradiance for a wet meadow graminoid. These simulations suggest that in a natural canopy daily leaf photosynthesis is nearly linear with respect to irradiance until 1,255 J cm^{-2} d^{-1} (Fig. 6.2). It then begins to saturate and little increase occurs above 2,930 J cm^{-2} d^{-1}. If this is compared to the frequency distributions of irradiance at northern Alaska, we can see that the mean leaf is commonly light-limited in situ in the Arctic. This is in fundamental agreement with field data for the graminoid type (TIESZEN 1973; SHVETSOVA and VOZNESSENSKIY 1970) as well as with GRACE and WOOLHOUSE's (1970) data for *Calluna*. Thus, production should be positively correlated with natural variation in irradiance on a daily and seasonal basis.

If we alter the light curve to respond linearly longer and to saturate at 0.84 J cm^{-2} min^{-1}, the mean leaf attains a higher rate of uptake at all irradiances simulated. Now if we compare similar response curves and similar daily total at a 40° latitude alpine site, some interesting differences emerge. First, substantially less CO_2 is incorporated during the shortened period of illumination, indicating either poorer interception or more leaves above saturation. At high daily irradiances, daily CO_2 uptake is depressed in a plant such as *Dupontia,* primarily because of water stress around mid-day which results in low leaf water potentials and higher stomatal resistances. Thus, for comparable rates of CO_2 uptake, it appears that alpine communities require more intercepting leaf area and adaptations in the plant–water system, adaptations which would become more important in lower altitude grasslands.

6.2.4 Temperature

Tundra plants are exposed to a wide range of temperatures during the growing season (MOSER 1969, 1973; SALISBURY and SPOMER 1964; MAYO et al. 1973), ranging from −11 °C to nearly +40 °C, and they characteristically maintain positive photosynthetic rates throughout this range. It is perhaps the general ability of tundra plants to metabolize in an integrated manner at temperatures which are disruptive in many temperate zone species which provides the clearest adaptation to the Arctic. Positive net photosynthesis has been observed in *Dryas integrifolia* at +1 °C (MAYO et al. 1973) and at −2 °C in *Polytrichum alpinum* (OECHEL and COLLINS 1976), and TIESZEN (1973) reports substantial positive rates in all grasses tested at 0 °C. *Dupontia* is active until −4 °C, at which temperature cell damage occurs. Photosynthesis frequently occurs as long as the leaf is not frozen. Lower temperature compensation points for *Geum reptans* and *Ranunculus glacialis* are −3° and −5 °C respectively (MOSER 1969), and PISEK et al. (1967) report a similar compensation value of −5 °C for *Oxyria digyna. Loiseleuria* possesses a similarly low temperature compensation point (LARCHER et al. 1975). Similarly, upper temperature compensation points of 40 °C for these same species have been reported by MOSER (1969), and TIESZEN's (1973) data confirm those of ZALENSKIY et al. (1972) and SHVETSOVA and VOZNESSENSKIY (1970) for the Arctic which show positive CO_2 uptake at 35° C, a temperature about 10 °C higher than is usually attained by graminoids on days of high radiation loads and low convective losses. KALLIO and KARENLAMPI (1975) indicate a similarly wide tolerance range for lichens and mosses, and in these groups reactivation following freezing occurs more rapidly. Arctic bryophytes can maintain high rates at low temperatures (OECHEL and COLLINS 1976) and compensation points are between −5° and −10 °C (KALLIO and HEINONEN 1973). *Polytrichum,* for example, shows photosynthesis rates at 0 °C which are 60% of the maximum observed at optimal temperatures (OECHEL and COLLINS 1976). In contrast to vascular plants, however, bryophytes from temperate regions frequently share these low temperature capabilities. Bryophytes also tolerate high temperatures; appreciable rates of net photosynthesis are maintained at tissue temperatures exceeding 30 °C (OECHEL and COLLINS 1976).

Although these species photosynthesize over a wide range of temperatures, it appears that temperature optima are generally higher than the mean ambient temperatures to which plants are exposed. *Loiseleuria* has an optimum as high as 30 °C at high photon flux densities although this decreases to 15 °C at lower irradiances. The 18 °C optimum for *Calluna* (GRACE and WOOLHOUSE 1970) is high and, in part, has substantiated the suggestion that this shrub is near its northernmost extension because of its temperature responses. Optima for vascular plants on Wrangel Island (ca. 20 °C) and Barrow (15° to 20 °C) are also well above mean ambient temperatures, as apparently are tundra mosses

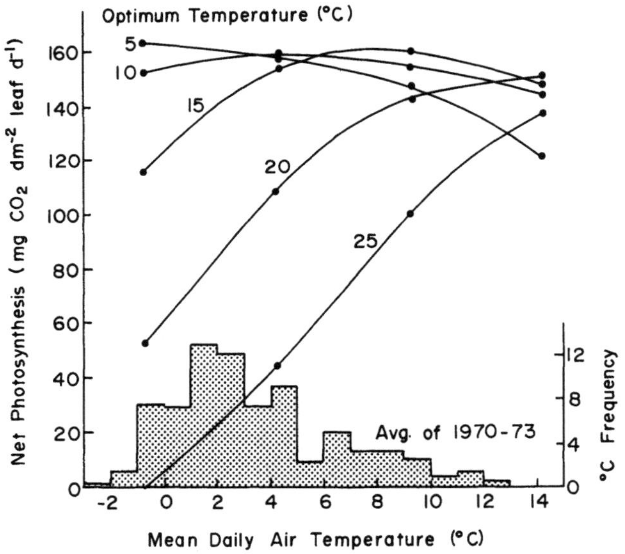

Fig. 6.3. Simulation of the daily photosynthetic response of *Dupontia fischeri* leaves as a function of temperature, and the frequency distribution of mean daily temperatures (1970 through 1973) for a tundra growing season. (Adapted from MILLER et al. 1976; TIESZEN 1978)

(RASTORFER 1970; PAKARINEN and MAYO, personal communication). Bryophytes, however, experience generally higher tissue temperatures than do vascular plants, resulting in temperature response curves which more closely reflect the actual tissue temperatures experienced (OECHEL and COLLINS 1976). *Dryas* (MAYO et al. 1973) is a seeming anomaly since its optimum is reported at 8° to 10 °C and it frequently experiences higher leaf temperatures on the Devon Island beach ridge.

The temperature response of CO_2 uptake is one factor which is important in competition among tundra plants. Thus, it is of interest to determine the significance of the shape and the position of this response curve (see Chap. 10, Vol. 12A). GRACE and MARKS (1978) have made some interesting comparisons of *Calluna,* which has a temperature optimum at 18 °C, and *Rubus,* which has a lower optimum between 12° and 18 °C. In their simulations an increase in temperature during a summer's day results in an increased rate of photosynthesis for *Calluna,* but a decrease in CO_2 uptake for *Rubus,* in which respiration exceeds photosynthesis under many summer day conditions. This indicates that *Rubus* may be near its southern limit at Moor House, performing best when temperatures are lower. In contrast, *Calluna* performs best at temperatures considerably higher than those prevailing at Moor House, as if it were best adapted to a more southerly climate. In an attempt to determine the general significance of these temperature relations, we have simulated daily courses of photosynthesis under a series of conditions. These simulations (Fig. 6.3) show a number of interesting features. A plant with an optimum of 5° to 10 °C will maintain high rates of daily CO_2 uptake across a wide range of daily temperatures. If the mean temperatures approach 10 °C, all plants with an optimum between 5° and 20 °C will have similarly high values. A plant with an optimum of 25 °C, typical of many temperate grassland species, does poorly at mean air

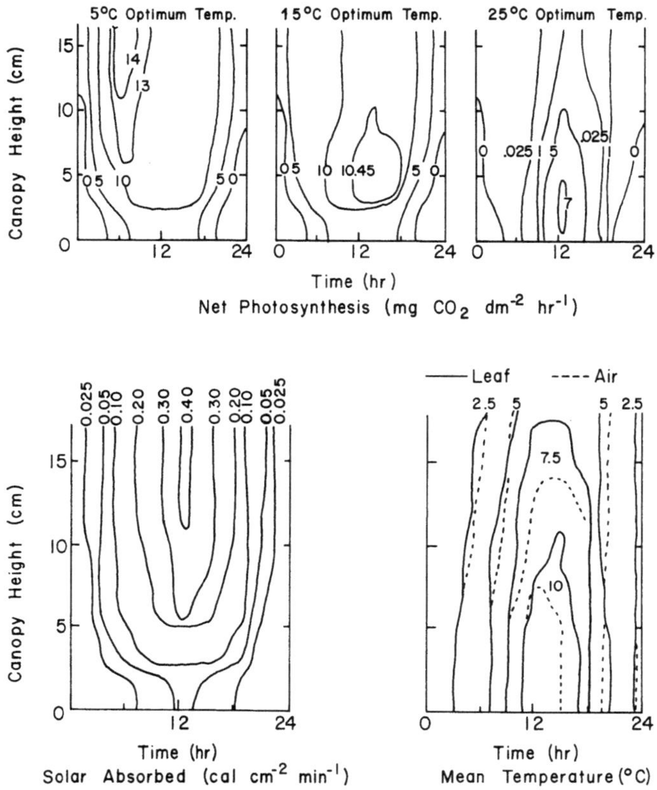

Fig. 6.4. Effect of canopy on microclimate in tundra systems and profiles of leaf CO_2 uptake. (Adapted from MILLER et al. 1976; TIESZEN 1978). *Above* Simulations of net photosynthesis using the stand photosynthesis model and photosynthesis temperature optima of 5°, 15°, and 25 °C. *Lines* represent isopleths through the canopy and through the day. Height = canopy height. *Below* Effect of canopy on absorbed solar radiation and leaf and air temperatures

temperatures below 15 °C. Again it appears that plants with optima below 15 °C should photosynthesize quite well in most arctic tundra sites.

The profiles of CO_2 uptake, light absorbed and leaf temperatures that are developed in the simulations (MILLER et al. 1976) for this "mean" day in a graminoid canopy of LAI (leaf area index) about 1 help us interpret temperature responses. Only leaves at the top of the canopy approach saturation (Fig. 6.4). These same leaves are closely coupled to air temperature. Inside the canopy, however, leaf temperatures may increase to about 10 °C even though air temperature does not exceed 7 °C. The profiles of CO_2 uptake for plants of different temperature optima are much different under these conditions. Plants with low temperature optima attain high rates at the top of the canopy where leaves are close to ambient temperatures and where they are light-saturated for at least a part of each day. On the other hand, a plant with an optimum of 25 °C receives little return from the leaves near the top and is more productive near the soil surface even though leaves are not light-saturated. These results

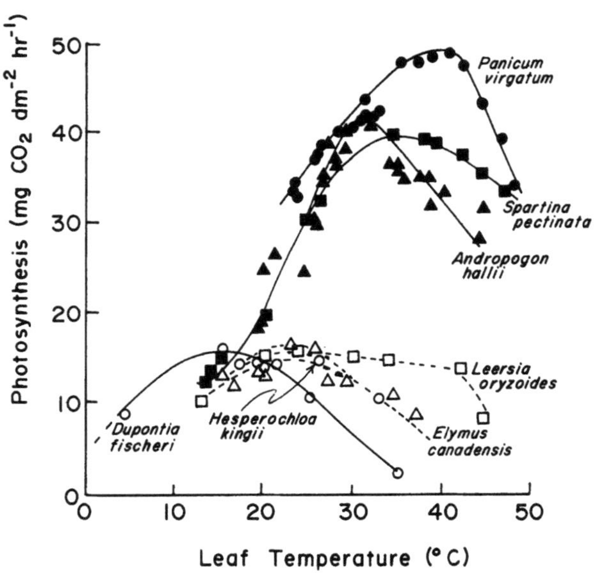

Fig. 6.5. Temperature response curves for C_3 (*open symbols*) and C_4 (*closed symbols*) grasses from North America grasslands (A.T. HARRISON, personal communication) and for an arctic grass. (TIESZEN 1973; *dashed extension* based on laboratory data)

may suggest that plants in ameliorated parts of the canopy, or certain compact life forms, should have higher temperature optima than the graminoid species with erect canopies. Furthermore, the relatively high daily totals whenever the optimum is below 15 °C suggest that adaptation to a 15 °C optimum should be sufficient to compete in the arctic tundra. In combination with only slight seasonal changes in ambient temperatures, we suggest that seasonal acclimatization (see Chap. 10, Vol. 12A) may not be as significant in arctic tundra systems as in others. This supports the arctic field and laboratory data which show a relative constancy of the temperature response (GRACE and WOOLHOUSE 1970; GERASIMENKO 1973; TIESZEN 1973; KALLIO and HEINONEN 1973). In addition, if we compare the frequency distributions of temperatures at each site, we could obtain an estimate of the effect of temperature on species with various temperature optima. In Barrow, for instance, low temperatures, per se, are not commonly limiting.

Grassland species tend to have temperature response curves with similar ranges but higher optima and upper and lower compensation points (Fig. 6.5) than tundra grasses. This clearly is true in the case of C_4 plants which characteristically have temperature optima of 30° to 45 °C (BLACK 1973; LUDLOW 1976), high temperature compensation points of 45° to 60 °C, and low temperature compensation points of 5° to 10 °C (LUDLOW 1976). In the case of C_3 grassland plants, whose temperature optima typically range from 15° to 25 °C, the differences are not always as clear-cut. As suggested above, some tundra species have optimal temperature ranges, as well as maximum light-saturated photosynthetic rates, which overlap considerably with those of many C_3 grassland species.

Another point which becomes evident when examining the photosynthetic temperature response data plotted in Fig. 6.5 is that some C_4 species may have higher absolute rates of photosynthesis at suboptimal temperatures than some

C_3 plants (either from tundra or grassland) have at their optimal temperature. For example, at 18 °C with full sunlight and adequate water, *Bouteloua gracilis* (C_4) had a net photosynthetic rate of 45 mg CO_2 dm^{-2} h^{-1}, or roughly twice that of several C_3 tundra or temperate grassland species. This does not necessarily imply that C_4 grasses are potentially better adapted or more productive at such temperatures than C_3 plants. As discussed above, tundra and grassland C_3 plants are able to maintain positive carbon balances at considerably lower temperatures than C_4 plants. Thus, as the frequency and duration of these lower temperatures increases, the C_3 plants would be expected to be progressively better adapted than the C_4 plants. Furthermore, and perhaps more importantly, literature reviewed by LUDLOW (1976) indicates that foliage of most C_4 grasses is killed at temperatures of $-2°$ to -3 °C, and most suffer chilling injury at temperatures of about 12 °C or less.

Within temperate grasslands, C_3 and C_4 species often co-exist, with their relative abundances being a function of temperature. In grasslands containing species with both photosynthetic pathways, phenological development is often offset in such a manner that C_3 species initiate spring regrowth and are most active physiologically and developmentally earlier (i.e., during periods of cooler temperatures) in the growing season than C_4 plants (DICKINSON and DODD 1976). As temperatures increase and C_4 plants become physiologically more active, the activity of C_3 plants often declines. Again in the fall, as temperatures become cooler, C_3 species often resume growth if water or other environmental factors are not limiting, while growth and physiological activity of C_4 species decline. Hence, photosynthetic production would be expected to occur predominantly via the C_3 pathway in the early and later portions of the growing season, with primary production by C_4 species becoming proportionately more important in the warmer, mid-growing season months (TIESZEN 1970; WILLIAMS and MARKLEY 1973).

This hypothesis has been addressed by laboratory growth chamber studies of WILLIAMS (1974) and KEMP and WILLIAMS (1980), who examined growth and photosynthetic temperature responses of *Bouteloua gracilis* (C_4) and *Agropyron smithii* (C_3), important grasses from North American shortgrass and northern mixed-grass prairies. In the study of KEMP and WILLIAMS (1980), plants were grown at either 35°/15 °C or 20°/15 °C day/night temperatures. The C_4 grass produced about 50% more dry matter when raised at the higher temperature regime than at the lower temperatures, while the C_3 grass exhibited the opposite trend. Interestingly, the two species produced approximately the same amount of biomass when grown at their respective optimal temperatures. In addition, *Bouteloua gracilis* had a temperature optimum around 35 °C when grown at 35°/15 °C, while *Agropyron smithii* had an optimum near 20 °C when grown at 20°/15 °C. However, when the C_4 grass was grown in the cooler temperatures and the C_3 grass was grown at higher temperatures, their respective light-saturated net photosynthetic rates were about halved. Since the daytime growth temperatures (20° or 35 °C) were comparable to daytime maxima encountered during spring or midsummer in their native habitat, these growth and photosynthetic responses were interpreted as evidence supporting the concept of offset resource utilization by species with contrasting photosynthetic

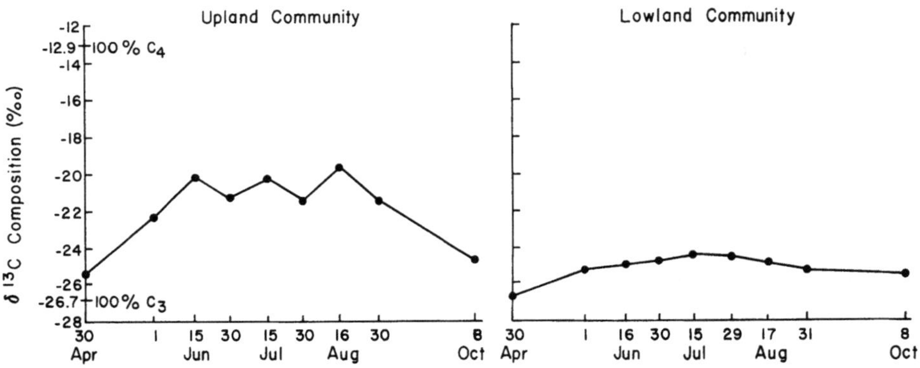

Fig. 6.6. Seasonal progression of $\delta\ ^{13}C$ values in the green biomass of a northern mixed-grass prairie in South Dakota, U.S.A. (ODE et al. 1980)

pathways. Thus, these and other similar species may be able to reduce competition for resources by being physiologically active during different parts of the season.

In addition to earlier phenological evidence (DICKINSON and DODD 1976) from the North American shortgrass prairie in support of this concept, two recent independent studies (ODE et al. 1980; COPPOCK et al. unpublished) at northern mixed-grass prairie sites in South Dakota, U.S.A., have shown that C_3 species are the predominant primary producers in the early and late seasons, while C_4 species are most important in midsummer (July–August) (Fig. 6.6). Heavy grazing by native prairie dogs and North American bison (*Bison bison*) at one of the sites (COPPOCK et al. unpublished) caused a dramatic decline in both C_3 and C_4 grasses, leaving virtually only C_3 shrubs and forbs. It is likely that this led to changes in seasonal patterns of photosynthetic production, though this was not specifically investigated.

6.2.5 Water

The availability of moisture has pronounced effects on tundra and grassland systems in two main ways. First, site moisture relationships often dictate the species which occur there and the communities which develop. Thus, plant communities in the Arctic (WEBBER 1978) as well as in North American grasslands (REDMANN 1971; LAUENROTH 1979) are frequently distributed along moisture gradients. Second, water stress, by its reduction of growth and photosynthesis, results in a direct reduction of CO_2 uptake and NPP (for a general discussion of physiological responses to water stress see Chap. 9, Vol. 12B). This direct effect is quantitatively more significant in grasslands than tundras.

A comprehensive analysis (WEBBER 1978) at Barrow, Alaska, indicated that site moisture was most highly correlated with the distribution of communities. Furthermore, ANPP (annual NNP) of vascular plants ranged from a mean of 18 g m^{-2} yr^{-1} on dry sites to 115 g m^{-2} yr^{-1} on wet sites, and ANPP was positively correlated with site moisture. The causal relationships are more difficult to establish, especially when soil moisture at all sites appears to remain high. What is clear is that monocots increase in dominance with increasing soil moisture (dicots and woody plants dominate on dry sites), and this growth form largely accounts for the differences in ANPP. This wet end of the gradient is also characterized by reduced soils, moderate to high phosphorus levels, and high decay rates. This complex of factors has selected growth forms which have high photosynthetic capabilities and yield high productivities.

The water relations of arctic and alpine tundra species appear similar and do not differ from mesic species in more temperate areas. Minimum leaf resistances and cuticular resistances are similar as are the root resistances (BROUWER 1954; JOHNSON and CALDWELL 1976; MILLER et al. 1978) which apparently are insensitive to soil temperature (MCNAUGHTON et al. 1974). In both arctic and alpine areas, the photosynthetic capacities of wet site species (CALDWELL et al. 1978) were higher than those of the more widely distributed species. Thus, under non-limiting conditions their carbon gain should be greater than the mesic species. However, they were also more sensitive to reductions in water potentials. Therefore, under conditions of developing water stress, species occupying mesic to xeric sites show a greater carbon gain. MILLER et al. (1978) concluded that along an increasing moisture gradient cell wall elasticity increases, providing a greater amount of water to be lost before stomata close. Under arctic conditions, closure is rare (TIESZEN 1975) and only infrequently affects carbon balance (see Chap. 7, Vol. 12B).

In grasslands, annual potential evapotranspiration usually exceeds precipitation, so that the ratio of actual evapotranspiration to precipitation is approximately one (LAUENROTH and SIMS 1976). As a result, water is frequently the single factor most limiting to primary production. The importance of water stress in controlling photosynthetic productivity of grassland species has been demonstrated in several studies. At a North American shortgrass prairie field site, for example, DYE (1972) examined CO_2 exchanges over relatively pure swards of *Bouteloua gracilis* during a summer with several prolonged droughts. The effect of one such month-long rainless period is illustrated in Fig. 6.7a. By 10 days after a rainfall event on June 14, the maximum daily rate of net photosynthesis had declined to about 25% of the net photosynthetic rate on the day after the rainfall. This decline of net photosynthesis closely paralleled the decline in soil water content in the upper 20 cm of soil (Fig. 6.7b).

In another study, REDMANN (1978) monitored CO_2 exchange over intact swards (28 dm^2) in a Canadian mixed-grass prairie dominated by two C_3 grass species, *Agropyron dasystachyum* and *A. smithii*. The mean net photosynthetic rates, corrected for CO_2 flux from the soil and litter, are shown together with abiotic data in Fig. 6.8. From spring through early summer, when soil moisture was relatively high, photosynthetic production generally increased as aboveground live biomass and green LAI increased. As mid-summer drought occurred in July, no further increases in aboveground biomass or LAI occurred. Rates of net photosynthesis began declining rapidly by the end of July, and from mid-August through the end of the growing season (October) net photosynthetic rates averaged only 10% or less of the maximum rates observed earlier in the summer.

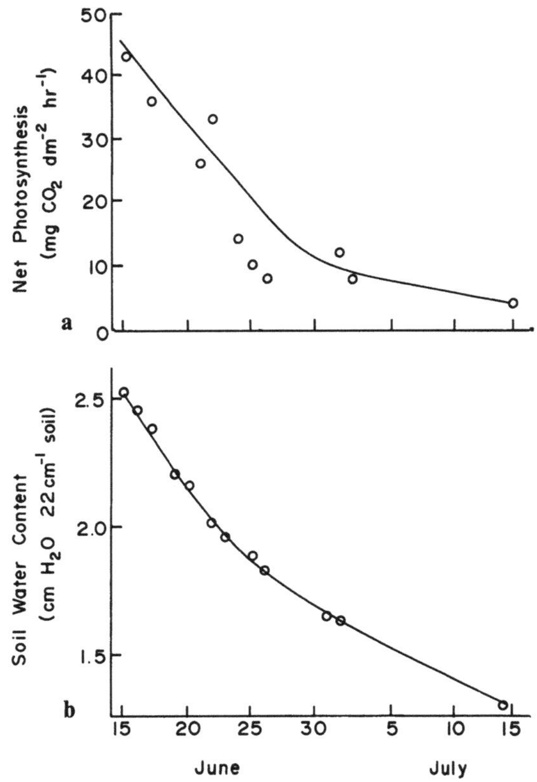

Fig. 6.7. Maximum daily net photosynthesis (**a**) and soil water content (**b**) of swards of *Bouteloua gracilis* during a 30-day drought in a North American shortgrass prairie in 1971. (Calculated from data of DYE 1972)

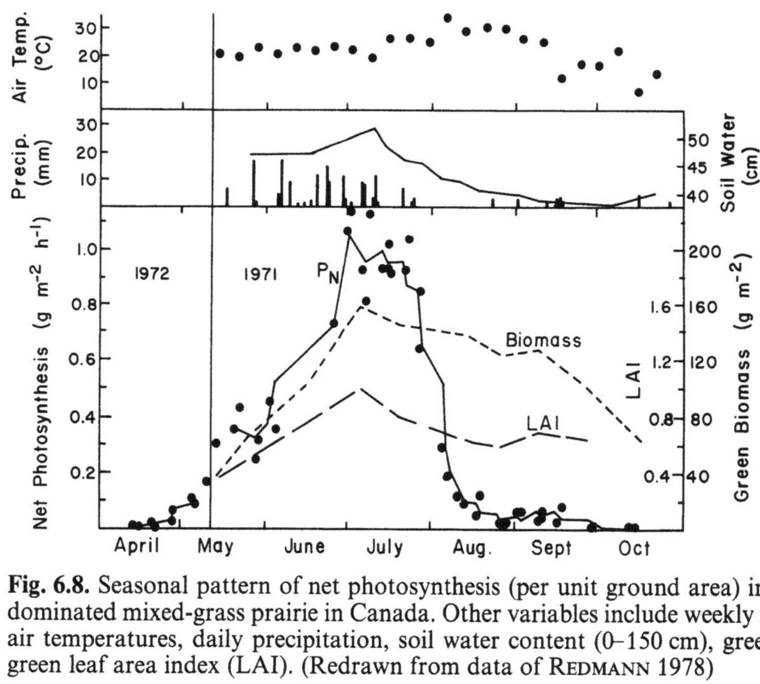

Fig. 6.8. Seasonal pattern of net photosynthesis (per unit ground area) in an *Agropyron*-dominated mixed-grass prairie in Canada. Other variables include weekly mean maximum air temperatures, daily precipitation, soil water content (0–150 cm), green biomass, and green leaf area index (LAI). (Redrawn from data of REDMANN 1978)

In temperate zone grasslands, for example, the Northern mixed-grass prairie (REDMANN 1976; MAXWELL and REDMANN 1978), there is some seasonal adjustment. During the relatively moist spring and early summer, mid-day osmotic potentials (which were closely correlated with leaf water potentials) of two C_3 grasses, *Agropyron dasystachyum* and *Koeleria cristata,* and two C_3 shrubs, *Artemisia frigida* and *Eurotia lanata,* generally ranged from -1.5 to -2.0 MPa (REDMANN 1976). As soil moisture became more limiting in July, mid-day osmotic potentials steadily declined to minimum values of -3.0 to -4.0 MPa. Following late season precipitation in September, leaf osmotic potentials again rose to values similar to those observed in the spring. Similar seasonal water stress patterns have been reported for a variety of xeric species (HICKMAN 1970; DETLING and KLIKOFF 1973; DINA et al. 1973; HALVORSON and PATTEN 1974; SALA et al. 1981).

As in tundras, grassland plant communities are frequently distributed along soil moisture gradients, suggesting differential water stress tolerances among the dominants. In the mixed-grass prairie of North Dakota (U.S.A.), REDMANN (1971) noted that the driest sites were dominated by *Bouteloua gracilis,* less dry sites were dominated by *Stipa viridula,* while *Sporobolus heterolepis* dominated the moist ravines. Although net photosynthetic rates of all three species declined with increasing water stress, *S. heterolepis* was unable to maintain positive net photosynthesis during periods of soil water stress. These results, as well as those reviewed by LUDLOW (1976) do not demonstrate any clear-cut differences between C_3 and C_4 plants with respect to their ability to photosynthesize at low water potentials. Rather, species differences in response to water stress seem to be related more closely to the environment to which the plants have adapted. This is especially interesting since it is apparent that considerably less water is used in producing a unit of C_4 plant tissue than C_3 plant tissue. While the superior water use efficiency of C_4 plants compared with C_3 plants has sometimes been attributed to lower transpiration rates in C_4 plants, these differences are largely attributable to the greater photosynthetic capacity of C_4 species (HSIAO and ACEVEDO 1974; LUDLOW 1976).

6.2.6 Nutrients

The quantitative extent to which nutrients limit NPP in tundras and grasslands is not clear; however, both systems generally respond to fertilization (especially N and P) with an increased NPP. In the Alaskan Arctic (ULRICH and GERSPER 1978) direct analyses of plant tissue coupled with determinations of critical nutrient concentrations suggest that wet tundra is a "nutritional desert" since tissues usually were on the borderline of being deficient. Nitrogen was most often limiting but was closely followed by phosphorus and then other elements. This is especially interesting because the total standing crop of nutrients (especially N) in the top 20 cm of soil (Table 6.5) is even higher than that present in temperate zone grasslands.

Most grasslands and tundras are markedly seasonal, both in productivity and nutrient status. Foliage nitrogen content is usually highest early in the

Table 6.5. Total quantity of carbon, nitrogen, and phosphorus to a depth of 20 cm present in various ecosystems

Ecosystem	Latitude	Site	Total quantity in system (g m^{-2})		
			C	N	P
Wet meadow	75°	Devon Island, Canada	17,435	1,170	81.1
Polar desert	75°	Devon Island, Canada	4,936	158	15.4
Wet meadow tundra	71°	Barrow, Alaska, U.S.A.	19,768	960	63.2
Spot tundra	71°	Agapa, U.S.S.R.	–	378	33.5
Mire	68°	Abisko, Sweden	9,446	225	10.9
Tussock tundra	65°	Eagle Creek, Alaska, U.S.A.	18,767	964	48.7
Grassland					
Cottonwood	43°	South Dakota, U.S.A.	–	275	–
Pawnee	40°	Colorado, U.S.A.	3,204	273	–
Osage	36°	Oklahoma, U.S.A.	–	261	–
Grassland	23°	Ratlam, India	–	336	–

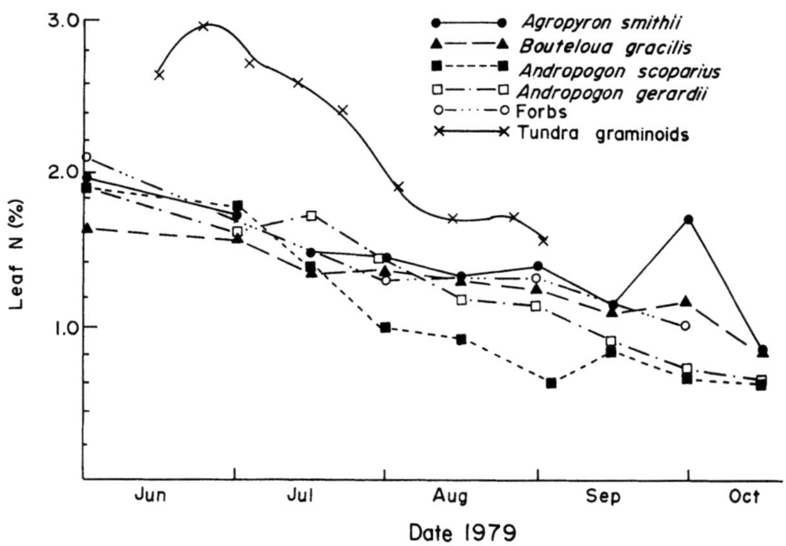

Fig. 6.9. Seasonal course of leaf N (for grasses from mixed prairie) and shoot N (for tundra grasses and sedges). Tundra leaves were 20% higher in N than other shoot components (Data from COPPOCK and DETLING, unpublished; CHAPIN et al. 1975)

growing season. Then, as aboveground growth continues, the shoot N content declines as a result of leaf senescence and dilution as shown in Fig. 6.9 (COPPOCK and DETLING, unpublished). Another frequently noted trend, though not particularly evident from the data in Fig. 6.9, is that C_3 grasses often contain higher concentrations of leaf nitrogen than C_4 grasses (COLMAN and LAZENBY 1970; WILSON and HAYDOCK 1971; WILSON 1975). It is clear in Fig. 6.9 that tundra monocots (grasses or sedges) have higher N contents. They are also C_3 plants, but they differ more from the C_3 grasses in the above temperate system than do those grasses from C_4 grasses.

Other data (see BROWN 1978; WILSON 1975) support the premise that C_4 plants require less N than C_3 grasses, probably because of a smaller investment of nitrogen in the photosynthetic carboxylation enzymes of C_4 plants than C_3 plants (BROWN 1978). Although photosynthetic rates have long been known to be influenced by nitrogen availability, very few data relating leaf N content to CO_2 exchange rates are available. TIESZEN (1978) indicated that in wet tundra leaf photosynthesis was relatively constant across a broad range of soil nutrient conditions. Productivity varied greatly, however, because allocation to leaf material was strongly dependent upon available nutrients. Net photosynthesis of both C_3 and C_4 species declines with decreasing leaf-N concentrations. However, WILSON (1975) found that net photosynthesis of C_4 grasses appears to be less limited under nitrogen-deficient conditions than net photosynthesis of C_3 grasses. As a result, C_4 species have a higher N-use efficiency than C_3 species (BROWN 1978). As discussed earlier, differences in temperature responses between C_3 and C_4 species may result in niche separation and reduced competition among the two photosynthetic pathway types in temperate grasslands. Differing nitrogen requirements of plants with these pathways might also contribute to niche separation. By initiating growth earlier in the season, C_3 species may be able to capitalize upon a relatively abundant supply of available mineral nitrogen in the spring. Growth of C_4 plants, though beginning later after some of this available nitrogen had been depleted by C_3 species, would not be as N-limited as might be expected since their nitrogen requirements are less.

6.2.7 Herbivores and Grazing

Grassland (COUPLAND 1979; FRENCH 1979) and tundra (TIESZEN and ARCHER 1979) ecosystems are believed to be determined by climate and soil; however, herbivores exert marked effects on the vegetation of both (MARK 1965; BONHAM 1972; WIELGOLASKI 1975; HARPER 1977; DYER et al. 1982). Shrubs and tussock graminoids in tundras appear sensitive to grazing and when utilized are replaced. In a northern mixed-grass prairie in South Dakota (U.S.A.), cattle grazing resulted in a shift from a predominance (78%) of cool season (C_3) grasses on ungrazed sites to a predominance (74%) of warm season (C_4) grasses on grazed sites (SIMS et al. 1978). Similarly, grazing of more xeric shortgrass prairies in Colorado and Texas caused the proportion of succulents (CAM) to nearly double, while forbs increased in importance as a function of cattle grazing in a more mesic tallgrass prairie in Oklahoma. Such grazing-induced changes in grassland community structure potentially may lead to changes in community function. Changes in the relative abundances of plants with alternative photosynthetic pathways, for instance, can be expected to alter seasonal patterns of plant phenology, photosynthesis and primary productions, water- and nutrient-use efficiency, and carbon allocation patterns.

Although grazing has a number of direct and indirect effects on net primary production, these effects are not always as extensive as might be predicted

Table 6.6. Mean annual aboveground net primary production (1970–1972) of six North American grassland types as a function of grazing during previous years by domesticated cattle (calculated from data for ten North American grassland sites examined by SIMS and SINGH 1978)

Grassland type	Ungrazed ($g\ m^{-2}$)	Grazed ($g\ m^{-2}$)
Shrub steppe	98	77
Mountain grassland	257	220
Tallgrass prairie	346	442
Mixed-grass prairie	292	243
Shortgrass prairie	198	164
Desert grassland	148	109
Mean of all grassland types	223	209

from observed consumption rates (see Chap. 17, Vol. 12C). Data from ten North American grassland sites examined by SIMS and SINGH (1978) indicated that ANPP of grazed sites was only 6% less than that of ungrazed sites (Table 6.6). Surprisingly, they also found that net root production, averaged across the ten sites, was significantly higher on grazed than on ungrazed grasslands.

In some grassland systems grazing may result in increased ANPP (JAMESON 1963). In the Serengeti Plains of Africa, for example, MCNAUGHTON (1976) showed that removal of about 85% of the plant standing crop by migratory wildebeest prevented senescence and stimulated ANPP. Thus, over approximately one month following this grazing, green biomass of grazed areas increased at a rate of 2.6 $g\ m^{-2}\ d^{-1}$, while green biomass of ungrazed areas was declining at a rate of 4.9 $g\ m^{-2}\ d^{-1}$. In a later study, MCNAUGHTON (1979) found that ANPP in this system was increased by grazing at all intensities, and depending upon soil moisture, ANPP was optimized by about 25% removal of available forage.

Although such dramatic increases in plant growth following grazing are not usually found, a certain amount of "compensatory growth" frequently occurs. In the Arctic studies (ARCHER and TIESZEN 1980), graminoids recovered from defoliation much faster than other growth forms and also showed an increase in ANPP. This, however, occurred on tillers which lost weight in the reserve (sheath and stem base) compartments. Furthermore, these plants showed a reduction in ANPP during the following year. The effect of grazing, or defoliation, on CO_2 balance in tundra forms is only now being studied. TIESZEN (1978) simulated the effects of acute grazing events on seasonal CO_2 budgets assuming no change in photosynthetic capacity, senescence, or other physiological response. These simulations illustrate the striking importance of the time of the grazing event. Grazing near mid-season (when the canopy has been exserted) had the most profound effects on CO_2 uptake even though it had only slight effects on that year's aboveground production. Obviously, reserves and root compartments were affected.

More mechanistic studies have been undertaken with grassland species. Following simulated grazing of *Agropyron smithii*, a dominant grass of the North

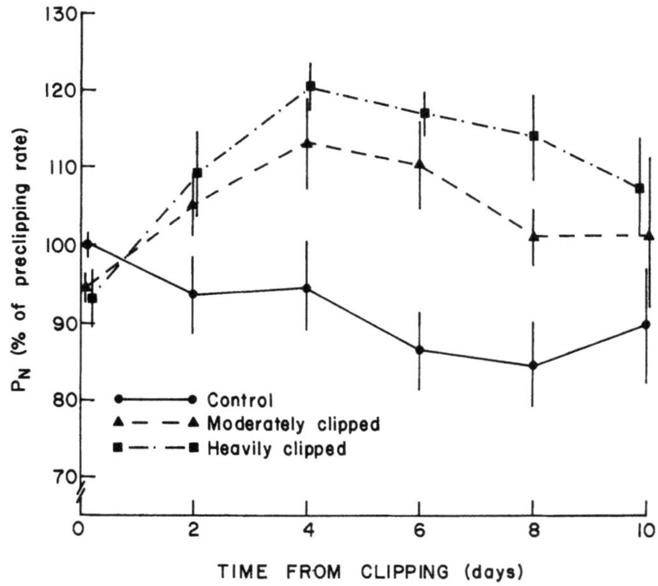

Fig. 6.10. Net photosynthesis (P_N) of a remaining undamaged leaf of *Agropyron smithii* (C_3) following removal of 50% (moderately clipped) or 75% (heavily clipped) of the tillers at a height of 5 cm. (Data from PAINTER and DETLING 1981)

American mixed-grass prairie, PAINTER and DETLING (1981) showed that net photosynthesis increased to a maximum about 4 days after defoliation (Fig. 6.10). Over the 10 days following differential tiller removal, net photosynthesis in leaves of non-defoliated plants averaged 90% of their pre-treatment rates, while net photosynthesis in remaining undamaged leaves of defoliated plants averaged 106% and 114% of preclipping rates for plants which had 50% and 75% of their tillers removed, respectively.

Similar trends have been observed in other grasses. For example, when all of the tillers except the main shoot of *Lolium multiflorum* were removed, photosynthetic rates of remaining leaves on the main shoot increased about 15% the first day and remained constant for 10 days thereafter, while net photosynthesis of similar leaves on control plants declined by about two-thirds in the same period (GIFFORD and MARSHALL 1973). The initial photosynthetic increase was attributed to a decline in stomatal resistance, while the maintenance of the enhanced photosynthetic rates over the next 10 days was attributed to the failure of both stomatal and mesophyll resistances to CO_2 diffusion to increase in defoliated plants as they did during aging of leaves from non-defoliated plants. Other studies (HODGKINSON et al. 1972) have suggested that enhanced photosynthetic rates in leaves remaining following defoliation is primarily a result of decreased mesophyll resistance.

While photosynthesis is often enhanced in remaining undamaged leaves of defoliated plants, it frequently declines in individual leaves which have been damaged. DETLING et al. (1979a) clipped notches from leaves of *Agropyron smithii* to simulate several grasshopper feeding patterns and monitored CO_2

exchange rates of the same leaves immediately before and after damage, and periodically over the next week. These treatments reduced net photosynthesis per unit of remaining leaf area by up to 30% relative to that of undamaged control leaves.

Some of the most striking responses to defoliation occur in below ground plant structures. DAVIDSON and MILTHORPE (1966) observed root elongation and respiration rates in *Dactylis glomerata* over an 8-day period following defoliation to a height of 2.5 cm. Root extension ceased almost immediately and had increased to only 5% of pre-clipping rates by 8 days after treatment. Over the same period, root respiration rates declined to 25% of pre-clipping rates. Similar decreases in root growth following defoliation have been reported for a variety of other grassland or pasture species (EVANS 1973a, b; HODGKINSON and BAAS BECKING 1977). These decreases in root activity following defoliation presumably occur because of decreased translocation of carbohydrate to belowground structures as a result of reduced photosynthetic capacity and an increase in the proportion of current photosynthate allocated to synthesis of new leaf tissue (GIFFORD and MARSHALL 1973; RYLE and POWELL 1975; DETLING et al. 1979b).

Results of the studies described above were all obtained from experiments involving clipping or other means of mechanical damage to simulate grazing. Since plants in some such studies have not responded the same as plants defoliated by natural herbivores, a few investigators have suggested that plant growth following grazing might be enhanced by growth regulators in herbivore saliva (REARDON et al. 1974; REARDON and MERRILL 1978; CAPINERA and ROLTSCH 1980; DYER and BOKHARI 1976). To test this hypothesis, DETLING et al. (1980) examined the effects of saliva from North American bison on growth of *Bouteloua gracilis*. In laboratory experiments lasting from 10–70 days following application of saliva, they found no significant effect of saliva on regrowth, net photosynthesis, or root respiration regardless of clipping intensity or nutrient status of the experimental plants. Thus, enhancement of plant regrowth by this mechanism must, at present, be considered speculative.

6.2.8 Allocation

It is clear from the preceding that the productivity of a community is a function of not only the photosynthetic capabilities of the component species and the resources available to those species, but it is also markedly dependent on plant growth rates and allocation strategies. A number of studies have indicated that plants with the greatest growth rates are those which allocate the highest proportion of their photosynthetically fixed carbon to the synthesis of additional photosynthetic tissues (MOONEY 1972; JOHNSON and TIESZEN 1976). This concept was dramatically illustrated experimentally by MCNAUGHTON (1974) who examined eight *Typha latifolia* (C_3) ecotypes with identical rates of net photosynthesis. The most productive ecotypes were those with the highest rate of canopy differentiation. He concluded biomass yield was more dependent upon rate of leaf tissue production than on the carbon balance per unit of biomass. CHRISTIE

(1975) concluded that the poorer growth of Mitchell grass relative to buffel grass (both C_4) was due to the less rapid leaf surface area development in the former. Others have found that plant growth rates are more closely related to leaf area partitioning than photosynthetic rates (e.g. POTTER and JONES 1977).

Leaf elongation is often inhibited to a greater extent than photosynthesis during the development of water stress (WARDLAW 1969; BOYER 1970; BARLOW et al. 1976), though some exceptions have been recorded (LUDLOW and NG 1976). Although the leaf water potential at which leaf elongation ceases is different among different species (cf. BOYER 1970; GANDAR and TANNER 1976; SANDS and CORRELL 1976; BUNCE 1977), there is no hard evidence among grassland plants that C_3 and C_4 species differ from one another in this regard (LUDLOW 1976). Species differences along a moisture continuum may, in part, result from differences in their relative abilities to maintain leaf growth at low soil water contents. For example, MAJERUS (1975) found that *Bouteloua gracilis,* native to the semi-arid shortgrass prairie, was able to maintain leaf growth at lower soil water potentials than *Andropogon scoparius* from more mesic locations. Higher root:shoot ratios are commonly observed in plants from drier habitats that in those from wetter ones (STRUICK and BRAY 1970; MOONEY 1972). This may result from leaf growth being inhibited by water stress to a proportionately greater extent than photosynthesis, and subsequent translocation of a higher proportion of the photosynthate to belowground structures (DETLING 1979).

Root growth is also generally favored over shoot growth in mineral-deficient soil (BENEDICT and BROWN 1944; HUNT 1975), again presumably because a smaller proportion of the current photosynthate can be utilized for shoot growth in nutrient-deficient plants. Some research indicates that the relative leaf area growth rate of C_4 grasses is greater at low leaf nitrogen contents than in C_3 grasses (WILSON and HAYDOCK 1971; WILSON 1975), thus contributing to the greater nitrogen use efficiency of C_4 species. Such a response may tend to confer an advantage upon C_4 species in nitrogen-limited habitats, especially if other factors are not severely limiting.

Optimal shoot growth of C_3 plants apparently occurs at lower temperatures than for C_4 plants. This growth response, together with differing photosynthetic rates, accounts for the differences in production of C_3 and C_4 plants grown in cool and warm environments (KEMP and WILLIAMS 1980). These differing growth responses probably contribute to reduced competition for resources between these two types of plants during different parts of the growing season (WILLIAMS 1974).

Following defoliation, allocation patterns of photosynthates within the affected plant frequently change. In addition to the reduction in photosynthetically fixed carbon produced as a result of reduced leaf area, there is often a reduction in proportion of that photosynthate which is allocated to roots and a concomitant increase in the proportion utilized in the synthesis of new leaves (GIFFORD and MARSHALL 1973; RYLE and POWELL 1975; DETLING et al. 1979b). Studies such as these led HODGKINSON and BAAS BECKING (1977) to suggest that defoliation during periods of low soil water availability could accelerate plant death because of the inability of the roots to maintain high enough growth rates to remain ahead of the drying front in the soil profile. By contrast, it has been suggested that transpirational water losses often are reduced following grazing because of reductions in leaf area, and hence that soil water would be conserved (MCNAUGHTON 1979; DETLING et al. 1979b). Thus, while the impact of such reductions in carbon allocation to roots of defoliated plants is difficult to assess, the strategy of defoliated grass plants appears to be a rapid restoration of their carbon gain capacity via increased photosynthetic rates and an increase in growth of new photosynthetic leaf surface area relative to other structures.

In tundra regions the high root to shoot biomass ratios result from the combined effects of low soil temperature and low nutrient availability (CHAPIN 1974, 1978). DENNIS (1977) observed the root:shoot ratio to change from 7:1 to 3:1 10 years after fertilization. It is clear in graminoids that the production of new tillers is positively correlated with the nitrogen content of plant tissue. It is also clear that respiration, root growth, and phosphate uptake all occur at high rates at low temperatures. The quantitative relationships between any of these factors and canopy growth rates, however, are not known.

TIESZEN (1978) and MILLER et al. (1976) have attempted to determine the significance of differing growth rates or allocation patterns by simulating community CO_2 uptake under various conditions. These simulations suggest that production in arctic tundras is strikingly sensitive to any event which stimulates canopy development or maintenance. Initiating the growing season 20 days earlier than normal, for example, resulted in a 55% increase in NPP. This increase approached 77% if the leaf area index was allowed to expand until mid-season, at which time it normally ceased expanding. Simulations during standard seasons in which the rate of leaf production was increased by 25% resulted in a 45% increase in net CO_2 uptake.

It is clear that many factors, in addition to the capability of taking up CO_2 at a high rate, are important in determining the magnitude of NPP. Foremost are those factors which stimulate an early canopy development, support a large leaf area index, and help maintain the productive tissues for long periods. It is the quantitative interaction of these factors on growth processes and the direct response of photosynthesis to environmental parameters which we must integrate as we attempt to develop a clear understanding of the controls of NPP.

6.3 Conclusions

Graminoid-dominated communities occur on 25% of the earth's vegetated land surface. Their presence is generally due to climatic control although edaphic and anthropogenic factors are also important near forest zones. These biomes span a wide range (200 to 2,000 mm yr^{-1}) of precipitation and temperature with those systems occurring where mean annual isotherms are less than 0 °C generally defined as tundras. In most systems, if not all, a seasonally dormant period is imposed by either low temperature or low moisture.

Our understanding of the magnitude and mechanistic control of primary production in these systems is imprecise at best. Future research should utilize available technology to document total carbon budgets throughout daily and yearly periods as the most feasible way of securing reliable estimates of net primary production. Furthermore, these estimates across climatic zones need to be supplemented by relating grassland and tundra data from specific sites to the range of topographic and edaphic variability present in each area. The wide range of these local effects often makes broad climatic comparisons difficult if not impossible.

Our understanding of the physiological behavior of grassland species is less well developed than that for species from other biomes, a conclusion difficult to comprehend given the importance of grassland areas for world food production. It is clear that C_4 grasses are replaced by C_3 types as either latitude or altitude increases. This is largely a consequence of changes in temperature with the result that essentially no C_4 grasses occur in tundras. When the types co-occur in temperate zones, they show seasonally displaced activity. Tundra species possess photosynthetic rates comparable to similar life forms in grasslands and their temperature response curves seem well-adapted with optima around 15 °C. Tundra C_3 grasses incorporate CO_2 on a daily basis equal to that of temperate zone C_3 grasses, largely because of the effective 24-h photoperiod at high latitudes. C_3 forms often possess a higher nutrient content (e.g. N and P) than C_4 forms; however, this basic difference is often overridden by direct climatic and edaphic effects.

In addition to the need for more systematic and comprehensive studies oriented toward the understanding of physiological and environmental controls over production, we urgently need a better understanding of allocation patterns and their control by climatic, edaphic, and biotic interactions. Short- and long-term adjustments to grazing pressure need to be understood both in terms of effects on productivity and effects on plant distributions.

Finally, sustained attention should be given to the paleoecology of graminoid dominated systems. The impact of the Pleistocene on the flora and fauna, as well as the ecology, can hardly be discerned when we know so little about the extent, complexity and dynamic distributions of past grassland and tundra systems.

References

Archer S, Tieszen LL (1980) Growth and physiological responses of tundra plants to defoliation. Arct Alp Res 12:531–552

Barlow EWR, Boersma L, Young JL (1976) Photosynthesis, transpiration and leaf elongation in corn seedlings at suboptimal soil temperatures. Agron J 69:95–100

Benedict HM, Brown GB (1944) The growth and carbohydrate responses of *Agropyron smithii* and *Bouteloua gracilis* to changes in nitrogen supply. Plant Physiol 19:481–494

Beschel RE (1969) Paper No 8. The diversity of tundra vegetation. Proc Conf Product Conserv North Circumpolar Lands 16

Black CC (1971) Ecological implications of dividing plants into groups with distinct photosynthetic production capacities. Adv Ecol Res 7:87–114

Black CC (1973) Photosynthetic carbon fixation in relation to net CO_2 uptake. Annu Rev Plant Physiol 24:253–286
Bliss LC (1979) Vascular plant vegetation of the Southern Circumpolar Region in relation to antarctic, alpine, and arctic vegetation. Can J Bot 57:2167–2178
Boardman NK (1977) Comparative photosynthesis of sun and shade plants. Annu Rev Plant Physiol 28:355–377
Bonham CD (1972) Vegetation analysis of grazed and ungrazed alpine hair-grass meadows. J Range Manage 25:276–279
Boyer JS (1970) Leaf enlargement and metabolic rates in corn, soybean, and sunflower at various leaf water potentials. Plant Physiol 46:233–235
Brouwer R (1954) Water absorption by the roots of *Vicia faba* at various transpiration strengths. II. Changes in water conductivity artificially obtained. Proc K Ned Akad Wettenschappen C57:68–80
Brown LF, Trlica MJ (1974) Photosynthesis of two important grasses of the shortgrass prairie as affected by several ecological variables. US/IBP Grassl Biomed Tech Rep 244, Colorado State Univ, Fort Collins
Brown LF, Trlica MJ (1977) Carbon dioxide exchange of blue grama swards as influenced by several ecological variables in the field. J Appl Ecol 14:205–213
Brown RH (1978) A difference in N use efficiency in C_3 and C_4 plants and its implication in adaptation and evolution. Crop Sci 18:93–98
Brown RH, Simmons RE (1979) Photosynthesis of grass species differing in CO_2 fixation pathways. I. Water-use efficiency. Crop Sci 19:375–379
Bunce JA (1977) Leaf elongation in relation to leaf water potential in soybean. J Exp Bot 28:156–161
Caldwell MM, Johnson DA, Fareed M (1978) Constraints on tundra productivity: Photosynthetic capacity in relation to solar radiation utilization and water stress in arctic and alpine tundras. In: Tieszen LL (ed) Vegetation and production ecology of an Alaskan arctic tundra. Ecol Stud Vol 29. Springer, Berlin Heidelberg New York
Capinera JL, Roltsch WJ (1980) Response of wheat seedlings to actual and simulated migratory grasshopper defoliation. J Econ Entomol 73:258–261
Chapin FS, III (1974) Morphological and physiological mechanisms of temperature compensation in phosphate absorption along a latitudinal gradient. Ecology 55:1180–1198
Chapin FS, III (1978) Phosphate uptake and nutrient utilization by Barrow tundra vegetation. In: Tieszen LL (ed) Vegetation and production ecology of an Alaskan arctic tundra. Ecol Stud Vol 29. Springer, Berlin Heidelberg New York
Chapin FS, III, Cleve van K, Tieszen LL (1975) Seasonal nutrient dynamics of tundra vegetation at Barrow, Alaska. Arct Alp Res 7:209–226
Christie EK (1975) Physiological responses of semiarid grasses. IV. Photosynthetic rates of *Thyridolepis mitchelliana* and *Cenchrus ciliaris* leaves. Aust J Agric Res 26:459–466
Christie EK (1978) Ecosystem processes in semiarid grasslands. I. Primary production and water use of two communities possessing different photosynthetic pathways. Aust J Agric Res 29:773–787
Christie EK (1979) Ecosystem processes in semiarid grasslands. II. Litter production, decomposition and nutrient dynamics. Aust J Agric Res 30:29–42
Colman RL, Lazenby A (1970) Factors affecting the response of tropical and temperate grasses to fertilizer nitrogen. In: Proc 11th Int Grassl Congr, Surfers Paradise, Australia
Coupland RT (1979) The nature of grassland. In: Coupland RT (ed) Grassland ecosystems of the world, international biological programme 18. Cambridge Univ Press, Cambridge
Davidson JL, Milthorpe FL (1966) The effect of defoliation on the carbon balance in *Dactylis glomerata*. Ann Bot 30:185–198
Dennis JG (1977) Distribution patterns of belowground standing crop in arctic tundra at Barrow, Alaska. Arct Alp Res 2:253–266
Detling JK (1979) Processes controlling blue grama production on the shortgrass prairie. In: French NR (ed) Perspectives in grassland ecology. Ecol Stud Vol 32. Springer, Berlin Heidelberg New York

Detling JK, Klikoff LG (1973) Physiological response to moisture stress as a factor in halophyte distribution. Am Midl Nat 90:307–318
Detling JK, Parton WJ, Hunt HW (1978) An empirical model for estimating CO_2 exchange of *Bouteloua gracilis* (H.B.K.) Lag. in the shortgrass prairie. Oecologia 33:137–147
Detling JK, Dyer MI, Winn DT (1979a) Effect of simulated grasshopper grazing on CO_2 exchange rates of western wheatgrass leaves. J Econ Entomol 72:403–406
Detling JK, Dyer MI, Winn DT (1979b) Net photosynthesis, root respiration, and regrowth of *Bouteloua gracilis* following simulated grazing. Oecologia 41:127–134
Detling JK, Parton WJ, Hunt HW (1979c) A simulation model of *Bouteloua gracilis* biomass dynamics on the North American shortgrass prairie. Oecologia 38:167–191
Detling JK, Dyer MI, Procter-Gregg C, Winn DT (1980) Plant–herbivore interactions: Examination of potential effects of bison saliva on regrowth of *Bouteloua gracilis* (H.B.K.) Lag. Oecologia 45:26–31
Dickinson CE, Dodd JL (1976) Phenological pattern in the shortgrass prairie. Am Midl Nat 96:367–378
Dina SJ, Klikoff LF, Keddington MB (1973) Seasonal water potential patterns in the mountain brush zone, Utah. Am Midl Nat 89:234–239
Dodd JL, Lauenroth WK (1975) Responses of *Opuntia polyacantha* to water and nitrogen perturbations in the shortgrass prairie. In: Wali MK (ed) Prairie: a multiple view. Univ North Dakota Press, Grand Forks
Dodd JL, Lauenroth WK (1979) Analysis of the response of a grassland ecosystem to stress. In: French NR (ed) Perspectives in grassland ecology. Ecol Stud Vol 32. Springer, Berlin Heidelberg New York
Dye AJ (1972) Carbon dioxide exchange of blue grama swards in the field. PhD. Dissertation. Colorado State Univ, Fort Collins
Dyer MI, Bokhari UG (1976) Plant-animal interactions: Studies of the effects of grasshopper grazing on blue grama grass. Ecology 57:762–772
Dyer MI, Detling JK, Coleman DC, Hilbert DW (1982) The role of herbivores in grasslands. In: Estes JR, Tyrl RJ, Brunken JN (eds) Grasses and grasslands: Systematics and ecology. Univ Oklahoma Press, Norman
Ehleringer JR (1978) Implications of quantum yield differences on the distributions of C_3 and C_4 grasses. Oecologia 31:255–267
Eickmeier WF (1978) Photosynthetic pathway distributions along an aridity gradient in Big Bend National Park, and implications for enhanced resource partitioning. Photosynthetica 12:290–297
Evans PS (1973a) The effect of repeated defoliation to three different levels on root growth of five pasture species. N Z J Agrid Res 16:31–34
Evans PS (1973b) Effect of seed size and defoliation at three development stages on root and shoot growth of seedlings of some common pasture species. N Z J Agric Res 16:389–399
French NR (1979) Principal subsystem interactions in grasslands. In: French NR (ed) Perspectives in grassland ecology. Ecol Stud Vol 32. Springer, Berlin Heidelberg New York
Gandar PW, Tanner CB (1976) Leaf growth, tuber growth and water potential in potatoes. Crop Sci 16:534–538
Gerasimenko TV (1973) Dependence of photosynthesis on temperature in tundra plants of Wrangel Island. Bot Zh (Leningrad) 58:493–505
Gerasimenko TV, Zalenskiy LV (1973) Diurnal and seasonal dynamics of photosynthesis in plants of Wrangel Island. Bot Zh (Leningrad) 58:1655–1666
Gifford RM, Marshall C (1973) Photosynthesis and assimilate distribution in *Lolium multiflorum* Lam. following differential tiller defoliation. Aust J Biol Sci 26:517–526
Grace J, Marks TC (1978) Physiological aspects of bog production at Moor House. In: Heal OW, Perkins DF (eds) Production ecology of British moors and montane grasslands. Ecol Stud Vol 27. Springer, Berlin Heidelberg New York
Grace J, Woolhouse HW (1970) A physiological and mathematical study of growth and productivity of a *Calluna-Sphagnum* community. I. Net photosynthesis of *Calluna vulgaris*. (L.) Hull. J Appl Ecol 7:363–381

Grace J, Woolhouse HW (1973) A physiological and mathematical study of the growth and productivity of a *Calluna-Sphagnum* community. III. Distribution of photosynthate in *Calluna vulgaris* (L.) Hull. J Appl Ecol 10:77–91
Halvorson WL, Patten DT (1974) Seasonal water potential changes in Sonoran Desert shrubs in relation to topography. Ecology 55:173–177
Harper JL (1977) Population biology of plants. Academic Press, London New York
Hartley W (1973) Studies on the origin, evolution, and distribution of the Gramineae. V. The subfamily Festucoideae. Aust J Bot 21:201–234
Hickman JC (1970) Seasonal course of xylem sap tension. Ecology 51:1052–1056
Hodgkinson KC, Baas Becking HG (1977) Effect of defoliation on root growth of some arid zone perennial plants. Aust J Agric Res 29:31–42
Hodgkinson KC, Smith NG, Miles GE (1972) The photosynthetic capacity of stubble leaves and their contribution to growth of the lucerne plant after high level cutting. Aust J Agric Res 23:225–238
Hsiao TC, Acevedo E (1974) Plant responses to water deficits, water-use efficiency, and drought resistance. Agric Meteorol 14:59–84
Hunt R (1975) Further observations on root/shoot equilibrium in perennial ryegrass (*Lolium perenne* L.). Ann Bot 39:745–755
Jameson D (1963) Responses of individual plants to harvesting. Bot Rev 29:532–594
Johnson DA, Caldwell MM (1976) Water potential components, stomatal function, and liquid phase water transport resistances of four arctic and alpine species in relation to moisture stress. Physiol Plant 36:271–278
Johnson DA, Tieszen LL (1976) Above ground biomass allocation, leaf growth, and photosynthesis patterns in tundra plant forms in Arctic Alaska. Oecologia 24:159–173
Kallio P, Heinonen S (1973) Ecology of *Rhacomitrium lanuginosum* (Hedw.) Brid. Rep Kevo Subarct Res Stn 10:43–54
Kallio P, Karenlampi K (1975) Photosynthesis in mosses and lichens. In: Cooper JP (ed) Photosynthesis and productivity in different environments. Cambridge Univ Press, London
Kemp PR, Williams GJ, III (1980) A physiological basis for niche separation between *Agropyron smithii* (C_3) and *Bouteloua gracilis* (C_4). Ecology 61:846–858
Larcher W, Cernusca A, Schmidt L, Grabherr G, Nötzel, Smeets N (1975) Mt. Patscherkofel, Austria. In: Rosswall T, Heal OW (eds) Structure and function of tundra ecosystems. Swedish natural science research council, Stockholm. Ecol Monogr 20:125–139
Lauenroth WK (1979) Grassland primary production: North American grasslands in perspective. In: French NR (ed) Perspectives in grassland ecology. Ecol Stud 32. Springer, Berlin Heidelberg New York
Lauenroth WK, Sims PL (1976) Evapotranspiration from a shortgrass prairie subjected to water and nitrogen treatments. Water Resour Res 12:437–442
Ludlow MM (1976) Ecophysiology of C_4 grasses. In: Lange OL, Kappen L, Schulze E-D (eds) Water and plant life. Ecol Stud Vol 19. Springer, Berlin Heidelberg New York
Ludlow MM, Ng TT (1976) Effect of water deficit on carbon dioxide exchange and leaf elongation rate of *Panicum maximum* var. *trichoglume*. Aust J Plant Physiol 3:401–413
Majerus ME (1975) Response of root and shoot growth of three grass species to decreases in soil water potential. J Range Manage 28:473–476
Mark AF (1965) Effects of management practices on narrow-leaved snow tussock, *Chinochloa rigida*. N Z J Bot 3:300–319
Maxwell JO, Redmann RE (1978) Leaf water potential, component potentials and relative water content in a xeric grass, *Agropyron dasystachyum* (Hook.) Scribn. Oecologia 35:277–284
Mayo JM, Despain DG, Zinderen Bakker van EM Jr (1972) CO_2 assimilation studies. In: Bliss LC (ed) Devon Island IBP project high arctic ecosystem. Project report 1970 and 1971. Univ Alberta Press, Edmonton
Mayo JM, Despain DG, Zinderen Bakker van EM Jr (1973) CO_2 assimilation by *Dryas integrifolia* on Devon Island, Northwest Territories. Can J Bot 51:581–588

Mayo JM, Despain DG, Zinderen Bakker van EM Jr (1975) CO_2 assimilation by *Dryas integrifolia* on Devon Island, Northwest Territories. Can J Bot 51:581–588
McNaughton SJ (1974) Developmental control of net productivity in *Typha latifolia* ecotypes. Ecology 55:864–869
McNaughton SJ (1976) Serengeti migratory wildebeest: Facilitation of energy flow by grazing. Science 191:92–94
McNaughton SJ (1979) Grazing as an optimization process: Grass-ungulate relationships within Serengeti. Am Nat 113:691–703
McNaughton SJ, Campbell RS, Freyer RA, Mylroie JE, Rodland KD (1974) Photosynthetic properties and root chilling responses of altitudinal ecotypes of *Typha latifolia* L. Ecology 55:168–172
Miller PC, Stoner WA, Tieszen LL (1976) A model of stand photosynthesis for the wet meadow tundra at Barrow, Alaska. Ecology 57:411–430
Miller PC, Stoner WA, Ehleringer JR (1978) Some aspects of water relations of arctic and alpine regions. In: Tieszen LL (ed) Vegetation and production ecology of an Alaskan arctic tundra. Ecol Stud Vol 29. Springer, Berlin Heidelberg New York
Mooney HA (1972) The carbon balance of plants. Annu Rev Ecol Syst 3:315–346
Moser W (1969) Die Photosyntheseleistung von Nivalpflanzen. Ber Dtsch Bot Ges 82:63–64
Moser W (1973) Licht, Temperatur und Photosynthese an der Station „Hoher Nebelkogel“ (3184 m). In: Ellenberg H (ed) Ökosystemforschung. Springer, Berlin Heidelberg New York
Murray DF (1978) Vegetation, floristics, and phytogeography of Northern Alaska. In: Tieszen LL (ed) Vegetation and production ecology of an Alaskan arctic tundra. Ecol Stud Vol 29. Springer, Berlin Heidelberg New York
Ode D, Tieszen LL, Lerman JC (1980) The seasonal contribution of C_3 and C_4 plant species to primary production in a mixed prairie. Ecology 61:1304–1311
Oechel WC, Collins NJ (1976) Comparative CO_2 exchange patterns in mosses from two tundra habitats at Barrow Alaska. Can J Bot 54:1355–1369
Oechel WC, Hicklenton P, Sveinbjörnsson, Miller PC, Stoner WL (1975) Physiological adaptations of arctic and subarctic mosses. In: Proceedings of circumpolar conference of northern ecology, Ottawa. Nat Res Counc Can
Painter EL, Detling JK (1981) Effects of simulated grazing on net photosynthesis and regrowth of western wheatgrass. J Range Manage 34:68–71
Pisek A, Larcher W, Unterholzner R (1967) Kardinale Temperaturbereiche der Photosynthese und Grenztemperaturen des Lebens der Blätter verschiedener Spermatophyten. Flora (Jena) Abt B 157:239–264
Potter JR, Jones JW (1977) Leaf area partitioning as an important factor in growth. Plant Physiol 59:10–14
Rastorfer JR (1970) Effects of light intensity and temperature on photosynthesis and respiration of two east Antarctic mosses, *Bryum argenteum* and *Bryum antarcticum*. Bryologist 73:655–556
Reardon PO, Merrill LB (1978) Response of sideoats grama grown in different soils to addition of thiamine and bovine saliva. In: Hyder (ed) Proc 1st Int Rangeland Congr, Denver, Colorado
Reardon PO, Leinweber CL, Merrill LB (1974) Response of sideoats grama to animal saliva and thiamine. J Range Manage 27:400–401
Redmann RE (1971) Carbon dioxide exchange by native Great Plains grasses. Can J Bot 49:1341–1345
Redmann RE (1976) Plant-water relationships in a mixed grassland. Oecologia 23:283–295
Redmann RE (1978) Seasonal dynamics of carbon dioxide exchange in a mixed grassland ecosystem. Can J Bot 56:1995–2005
Robson MJ (1973) The growth and development of simulated swards of perennial ryegrass. II. Carbon assimilation in a seedling sward. Ann Bot 37:501–518
Rosswall T, Heal OW (1975) The IBP tundra biome – an introduction. In: Rosswall T, Heal OW (eds) Structure and function of tundra ecosystems. Ecol Bull 20, Swed Nat Sci Res Counc Stockholm

Ryle GJA, Powell CE (1975) Defoliation and regrowth in the graminaceous plant: The role of current assimilate. Am Bot 39:297–310
Sala OE, Lauenroth WK, Parton WJ, Trlica MJ (1981) Water status and vegetation in a shortgrass steppe. Oecologia 48:327–331
Salisbury FB, Spommer GG (1964) Leaf temperatures of alpine plants in the field. Planta 60:497–505
Sands R, Correll RL (1976) Water potential and leaf elongation in radiata pine and wheat. Physiol Plant 37:293–297
Shantz HL (1954) The place of grasslands in the earth's cover of vegetation. Ecology 35:142–145
Shvetsova WM, Voznessenskiy VL (1970) Diurnal and seasonal changes of intensity of photosynthesis in some plants of western Taimyr. Bot Zh (Leningrad) 55:66–78. (Int Tundra Biome transl 2)
Sims PL, Singh JS (1978) The structure and function of ten western North American grasslands. III. Net primary production, turnover and efficiencies of energy capture and water use. J Ecol 66:573–597
Sims PL, Singh JS, Lauenroth WK (1978) The structure and function of ten western North American grasslands. I. Abiotic and vegetational characteristics. J Ecol 66:251–285
Struik GJ, Bray JR (1970) Root/shoot ratios of native forest herbs and *Zea mays* at different soil moisture levels. Ecology 51:892–893
Teeri JA, Stowe LG (1976) Climatic patterns and the distribution of C_4 grasses in North America. Oecologia 23:1–12
Tieszen LL (1970) Photosynthetic properties of some grasses in eastern South Dakota. Proc S D Acad Sci 49:78–89
Tieszen LL (1973) Photosynthesis and respiration in arctic tundra grasses: Field light intensity and temperature responses. Arct Alp Res 5:239–251
Tieszen LL (1975) CO_2 exchange in the Alaskan arctic tundra: seasonal changes in the rate of photosynthesis of four species. Photosynthetica 9:376–390
Tieszen LL (1978) Summary. In: Tieszen LL (ed) Vegetation and production ecology of an Alaskan arctic tundra. Ecol Stud Vol 29. Springer, Berlin Heidelberg New York
Tieszen LL, Archer S (1979) Physiological responses of plants in tundra grazing systems. In: Johnson D (ed) Spec Publ Am Soc Range Manage, Denver, Colorado
Tieszen LL, Sigurdson DC (1973) Temperature responses of carboxylation enzymes from three grasses in the arctic. Arct Alp Res 5:59–66
Tieszen LL, Senyimba MM, Imbamba SK, Troughton JH (1979) The distribution of C_3 and C_4 grasses and carbon isotope discrimination along an altitudinal and moisture gradient in Kenya. Oecologia 37:337–350
Tieszen LL, Miller PC, Lewis MC, Mayo JC, Oechel WC, Chapin FS (1981) Processes of primary production in tundra. In: Bliss LC, Heal OW, Moore JJ (eds) Tundra ecosystems: a comparative analysis. Cambridge Univ Press, Cambridge
Ulrich A, Gersper PL (1978) Plant nutrient limitations of tundra plant growth. In: Tieszen LL (ed) Vegetation and production ecology of an Alaskan arctic tundra. Ecol Stud Vol 29. Springer, Berlin Heidelberg New York
Vogel JC, Fuls A, Ellis RP (1978) The geographical distribution of Kranz grasses in South Africa. S Afr J Sci 74:209–215
Wardlaw IF (1969) The effect of water stress on translocation in relation to photosynthesis and growth. II. Effect during leaf development in *Lolium temulentum* L. Aust J Biol Sci 22:1–16
Webber PJ (1978) Spatial and temporal variation of the vegetation and its productivity. In: Tieszen LL (ed) Vegetation and production ecology of an Alaskan Arctic Tundra. Ecol Stud Vol 29. Springer, Berlin Heidelberg New York
Whittaker RH, Likens GE (1975) The biosphere and man. In: Lieth H, Whittaker RH (eds) Primary productivity of the biosphere. Ecol Stud Vol 14. Springer, Berlin Heidelberg New York
Wielgolaski FE (1972) Vegetation types and plant biomass in tundra. Arct Alp Res 4:291–305

Wielgolaski FE (1975) Grazing by sheep. In: Wielgolaski (ed) Fennoscandian tundra ecosystems. Part 2. Animals and systems analysis. Ecol Stud Vol 17. Springer, Berlin Heidelberg New York

Williams GJ III (1974) Photosynthetic adaptation to temperature in C_3 and C_4 grasses. A possible ecological role in the shortgrass prairie. Plant Physiol 54:709–711

Williams GJ III, Kemp RP (1978) Simultaneous measurement of leaf and root gas exchange of short-grass prairie species. Bot Gaz 139:150–157

Williams GJ III, Markley JL (1973) The photosynthetic pathway of North American shortgrass prairie species and some ecological implications. Photosynthetica 7:262–270

Wilson D, Cooper JP (1969) Effect of light intensity during growth on leaf anatomy and subsequent light-saturated photosynthesis among contrasting *Lolium* genotypes. New Phytol 68:1125–1135

Wilson JR (1975) Comparative response to nitrogen deficiency of a tropical and temperate grass in the interrelation between photosynthesis, growth, and the accumulation of non-structural carbohydrate. Neth J Agric Sci 23:104–112

Wilson JR, Haydock DP (1971) The comparative response of tropical and temperate grasses to varying levels of nitrogen and phosphorus nutrition. Aust J Agric Res 22:573–587

Woledge J (1977) The effects of shading and cutting treatments on the photosynthetic rate of ryegrass leaves. Ann Bot 41:1279–1286

Zalenskiy OV, Shvetsova VM, Voznessenskiy VL (1972) Photosynthesis in some plants of Western Taimyr. In: Wielgolaski FE, Rosswall Th (eds) Proc 4th Int Meet Biol Product Tundra. IBP Tundra Biome Steering Committee, Stockholm

7 Productivity of Desert and Mediterranean-Climate Plants

J. EHLERINGER and H.A. MOONEY

CONTENTS

7.1 Introduction 205

7.2 Biomass and Productivity of Mediterranean-Climate Plants 206

7.3 Biomass and Productivity of Desert Systems 211

7.4 Photosynthetic Rates 214
7.4.1 Photosynthetic Pathways, Rates and Life Form 215
7.4.1.1 Mediterranean-Climate Growth Forms 215
7.4.1.2 Desert Growth Forms 216
7.4.2 Temperature and Light Response 218
7.4.2.1 Mediterranean-Climate Forms 218
7.4.2.2 Desert Forms 219

7.5 Photosynthetic Adaptation to Arid Conditions 220
7.5.1 Mediterranean-Climate Forms 220
7.5.2 Desert Forms 221

7.6 Conclusions 225

References 226

7.1 Introduction

Mediterranean-climate and desert ecosystems are prominent on all continents at latitudes between approximately 15° and 30° (BROWN 1968; MCGINNIES 1969; MOONEY 1977a). Mediterranean-climate systems generally occur at these latitudes on the western or southwestern edges of the continent, whereas towards the more equatorial latitudes and also to the continental interior, there is a transition of vegetation types from mediterranean-climate types to desert forms (SHREVE 1936; MCGINNIES 1969; DI CASTRI and MOONEY 1973; MOONEY 1977a).

Both mediterranean-climate and desert ecosystems have a high diversity of life forms (SHREVE and WIGGINS 1969; DI CASTRI and MOONEY 1973; MOONEY 1977a). The dominant life form of both ecosystem types, however, is the shrub, although subshrub, herbaceous, and ephemeral life forms constitute a significant fraction of the vegetation, and under some conditions may actually become the dominant life forms. For the most part trees are not an important component of these ecosystems, except under riparian situations.

As is characteristic of arid lands, mediterranean-climate and desert ecosystems receive low precipitation. Typically, precipitation for mediterranean-

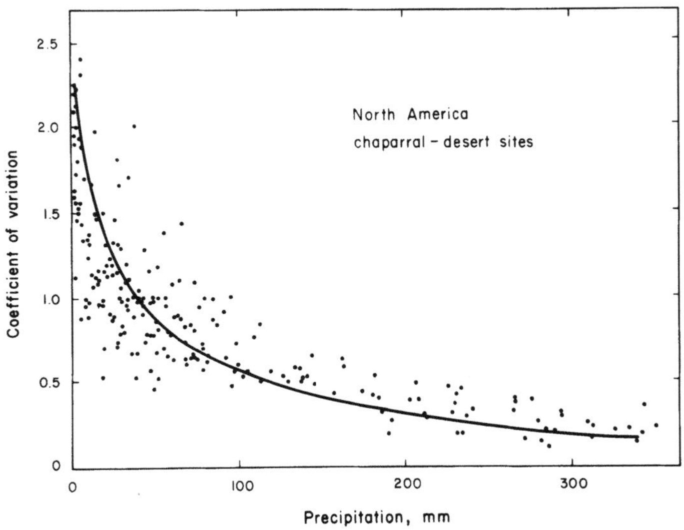

Fig. 7.1. The coefficient of variation in mean precipitation plotted as a function of mean precipitation for different locations in mediterranean-climate and desert habitats of southwestern North America. (Data from HASTINGS and HUMPHREY 1969a, b, and U.S. Weather Bureau records)

climate systems is in the range of 250–900 mm annually, with precipitation occurring almost exclusively in the cool winter months (ASCHMANN 1973). In contrast desert ecosystems receive 0–250 mm annually, and seasonally precipitation may occur predominantly during either the winter months or the summer months, or in some regions during both seasons. There is a strong positive correlation between the mean precipitation level and the predictability of that precipitation in mediterranean-climate and desert ecosystems (Fig. 7.1). This relationship indicates that desert ecosystems which have lower mean precipitation levels also have less predictable precipitation than do mediterranean-climate systems. Desert habitats which have less than 100 mm precipitation annually have a very low predictability of precipitation. SCHAFFER and GADGIL (1975) have shown that within such habitats the fraction of the flora which is ephemeral increases linearly as precipitation predictability decreases.

7.2 Biomass and Productivity of Mediterranean-Climate Plants

In mediterranean-climate regions of the world a wide range of plant growth forms may be found, ranging from evergreen trees to herbaceous annuals. Various combinations of these growth forms can be segregated in a more or less comparable manner in the different mediterranean climatic regions in response to particular climatic, soil, and disturbance gradients (NAVEH 1967; SPECHT 1969a, b; MOONEY et al. 1970; MOONEY et al. 1974a). For example, in both Chilean and Californian coastal regions, progressing from the more humid

Table 7.1. Productive characteristics of mediterranean-climate evergreen forests, evergreen scrub, and a subligneous scrub community. (From MOONEY 1980)

Vegetation type	Evergreen forest		Evergreen scrub		Subligneous scrub
Community name Locality	Pine France[a]	Evergreen oak France[b]	Chaparral California[c]	Garrigue France[b]	Phrygana Greece[d]
Stand age, yr	30	150	17–18	17	?
Height, m	10.4	11	~1.5	–	<1
Leaf area index, $m^2\ m^{-2}$	–	4.5	2.5	–	1.7
Biomass, kg ha^{-1}					
Shoot	156,800	269,000	20,390	22,500	10,950
Wood	144,100	262,000	16,720	19,500	8,860
Leaves	12,700	7,000	3,670	4,000	2,090
Roots	22,000	~50,000	~12,230	–	16,180
Litter	–	11,400	13,590	–	–
Total biomass	–	330,400	46,210	–	–
Allocation, %					
Shoot					
Wood	91.9	97.4	82.0	83.0	80.9
Leaves	8.1	2.6	18.0	17.0	19.1
Shoot	86.0	84.3	62.5	–	40.4
Root	14.0	15.7	37.5	–	59.6
Root:shoot ratio	0.16	0.19	0.60	–	1.48
Production, kg $ha^{-1}\ yr^{-1}$					
Above-ground biomass accumulation		2,600	1,300	1,100	2,020
Litter fall	7,830	3,900	2,820	2,300	2,100
Shoot, net	14,300	6,500	4,120	3,400	4,120
Root, net	1,100				
Decomposition, kg $ha^{-1}\ yr^{-1}$					
Litter	–	3,000	2,640	–	–

[a] CABANETTES 1979
[b] LOSSAINT 1973
[c] Composite data from SPECHT 1969b; KITTREDGE 1955; root biomass estimate based on KUMMEROW et al. 1977; leaf area index estimate from MOONEY et al. 1977b
[d] MARGARIS 1975 for spring

regions to the drier parts, there is a shift in community type from evergreen forests to evergreen scrub to a drought-deciduous scrub. In more interior regions or in modified sites there may be woodlands and grasslands. In regions of complex topography, substrate diversity, or land-use history, communities of widely disparate structure may co-occur within the same rainfall regime.

All of the mediterranean-climate community types have quite different productive structures and distributions of biomass, as can be seen by a comparison of the characteristics of a composite of mediterranean-climate community types (Table 7.1), as well as by a comparison of a number of community types, both natural and cultivated, found in southern France (Table 7.2).

The above-ground standing biomass of the various mediterranean woody community types varies by over an order of magnitude, from about 10,000 kg ha^{-1} for a subligneous scrub to nearly 300,000 kg ha^{-1} for an ever-

Table 7.2. Productive characteristics of four diverse Mediterranean-climate vegetation types in southern France. (ECKARDT et al. 1975)

Vegetation type	Dominant species	Growth form	Leaf specific wt. (mg cm^{-2})	Leaf area index (m^2 m^{-2})	Above-ground live biomass	Net leaf photosynthesis at 400 W m^{-2} PAR (mg CO_2 g^{-1} s^{-1})	Annual photosynthetic period (mo)	Above-ground production (kg ha^{-1} yr^{-1})
Evergreen forest	*Quercus ilex*	Phanerophyte	16	4.4	27	1.8	12	6,500
Halophytic scrub	*Salicornia fruticosa*	Chamaephyte	12.5	–	2.4	2.6	7	5,000
Wet meadow	*Arrhenatherum elatius*	Hemicryptophyte	4.2	3.6	0.4	11	10–12	11,000
Annual crop	*Helianthus annuus*	Therophyte	9.2	1.8	0.9	15.6	2	9,000

Table 7.3. Biomass accumulation in evergreen scrub communities of various mediterranean-climate regions. (Format from MOONEY 1977b)

Locality	Vegetation type	Annual biomass increment (kg ha^{-1})	Estimate period (yr)	Standing living biomass and age of mature stand (kg ha^{-1})	(yr)	Reference
California						
San Dimas	Chamise chaparral	1,200	0–5	27,260	37	SPECHT 1969b
		1,000	0–10			
Boulder Creek	Chamise chaparral	1,000	0–23	23,080	23	MOONEY et al. 1977b
Northern California (average)	Chamise chaparral	1,600	0–8	31,470	?	SAMPSON 1944
Santa Ynez Mts.	Ceanothus chaparral	1,750	0–12	48,710	21	SCHLESINGER and GILL 1980
Santa Maria	Chamise-red-shank Ceanothus chaparral	840	0–24	20,120	24	DEBANO and CONRAD 1978
France						
Montpellier	Garrigue	1,500	0–5	23,500	17	SPECHT 1969b
LePuech du Juge	Garrigue	1,380	0–17			LOSSAINT 1973
Australia						
Dark Island Soak	Heath	640	0–5			SPECHT et al. 1958
South Africa						
Jonkershoek	Fynbos – broad sclerophyll scrub	2,500	0–6			
		1,500	0–10			
		1,840	0–10			
		1,500	0–17	25,700	17	KRUGER 1977

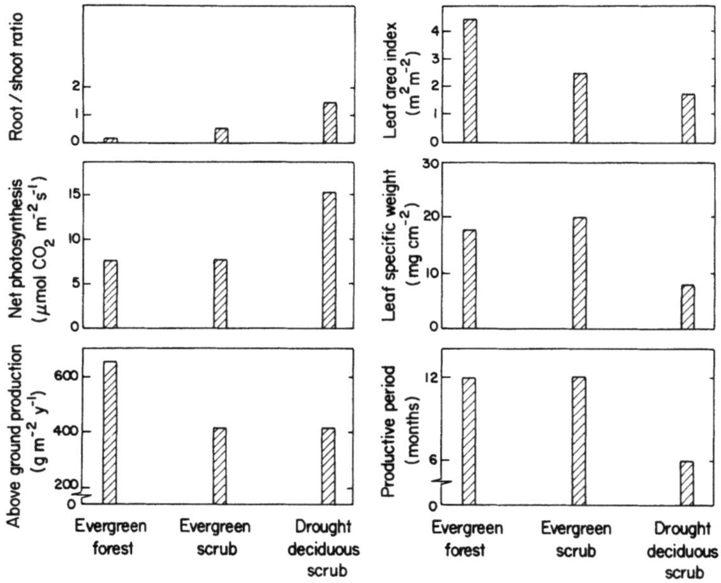

Fig. 7.2. Productive structure of three mediterranean-climate type plant communities. Specific leaf weights and photosynthetic rates are average values for plants from a wide range of communities (MOONEY 1980). Other values are from Table 7.1

green oak forest (Table 7.1). Herbaceous communities, such as Californian annual grassland, have still lower values, ranging from 500 kg ha^{-1} early in a growing season to about 5,000 kg ha^{-1} at the peak of the season (DUNCAN and WOODMANSEE 1980). These differences in biomass reflect, of course, the differences in the dominant growth forms, including their life spans, which vary from less than a year for the annuals, to tens of years for the shrubby types, to hundreds of years for the forest species. Comparable differences in biomass can be noted for the community series studied in southern France by ECKARDT et al. (1975) (Table 7.2).

The comparison of the woody community types, which is based on composite data, illustrates that, because of compensating factors in the productive structure in the community, comparable aboveground productivities can be attained (Fig. 7.2). For example, the drought-deciduous type has leaves with a high photosynthetic capacity and a low construction cost yet a short photosynthetic duration (MOONEY and DUNN 1970; MILLER and STONER 1979; see also the discussion about plant life forms in Chap. 18, Vol. 12B). In contrast, the evergreen scrub has plants with costly leaves of a low photosynthetic capacity but a higher leaf area index, and lower allocation to roots than the drought-deciduous scrub. These factors compensate each other, resulting in similar aboveground productivities between these two vegetation types. The evergreen forest has a potentially higher productivity than the evergreen scrub since, although photosynthetic capacities, duration and leaf cost are comparable, allocation to roots is less.

The French study (Table 7.2), which encompassed forest, scrub, and herbaceous communities, also indicates such internal compensation. These communities have a 30-fold difference in aboveground biomass yet only a 2-fold difference in aboveground production.

Aboveground biomass accumulation averages between 1,000 and 2,000 kg ha^{-1} y^{-1} for evergreen scrub communities in the various mediterranean-climate regions of the world (Table 7.3). SPECHT (1981) indicates that the annual increment declines in such communities with age. SCHLESINGER and GILL (1980), however, found no decline over a 22-year regrowth span for a California chaparral stand.

Annual litter fall added to annual biomass accumulation gives an index of total aboveground productivity. Average annual litter fall for a range of mediterranean-climate scrub communities is somewhat over 3,000 kg ha^{-1} y^{-1}, a little over 1,000 kg ha^{-1} y^{-1} less than an average value for mediterranean-climate evergreen forests (Table 7.4). For the evergreen scrub, then, average annual aboveground productivity (litter fall and biomass accumulation) lies in the range between 4,000 and 5,000 kg ha^{-1} y^{-1}.

There are few data available on root productivity of mediterranean-climate communities. CABANETTES (1979) indicates a value of a little over

Table 7.4. Mean annual litter fall for various evergreen forest and scrub communities

Vegetation type and dominant species	Locality	Litter fall (kg ha^{-1} yr^{-1})	Reference
Evergreen forest			
Dominant species			
Quercus agrifolia	Echo Valley, Calif.	5,130–6,460	MOONEY et al. 1977b
Quercus ilex	Montpellier, France	3,900	LOSSAINT 1973
Pinus pinea	Montpellier, France	7,830	CABENETTES 1979
Quercus ilex	Italy	4,600	SUSMEL et al. 1976
Pinus halepensis	Southern France	2,900–4,900	LOSSAINT and RAPP 1971
Cryptocarya alba	Chile	1,830–3,260	MOONEY et al. 1977b
	Overall mean =	4,760	
Evergreen scrub			
Ceanothus megacarpus	California	3,350–8,250	SCHLESINGER and GILL 1980
Adenostoma fasciculatum	California	2,730	MOONEY et al. 1977b
Mixed spp.	Chile	1,630	MOONEY et al. 1977b
Adenostoma fasciculatum	California	2,820	KITTREDGE 1955
Quercus coccifera	France	3,800	LOSSAINT 1973
	Overall mean =	3,360	
Drought-deciduous			
Phlomis fruticosa	Greece	2,100	MARGARIS 1976

1,000 kg ha^{-1} y^{-1} for a mediterranean-climate closed cone pine forest, whereas values of nearly 6,000 kg ha^{-1} y^{-1} have been measured for annual grasslands (DUNCAN and WOODMANSEE 1980). KUMMEROW et al. (1977) indicate that certain California evergreen shrub species may have root productions in excess of 10,000 kg ha^{-1} y^{-1}.

OECHEL (1980a) calculated total annual productivities for four Californian evergreen shrubs, using physiological data and simulation models. These values were lower for these plant types than indicated from the survey data given above. They do indicate, however, that a large carbon cost is required for the production and maintenance of evergreen sclerophyll tissues, a point well-documented by MILLER and STONER (1979).

7.3 Biomass and Productivity of Desert Systems

The standing biomass of desert ecosystems is low relative to that present in other ecosystems (RODIN and BAZILEVICH 1968; WHITTAKER 1975). The amount of biomass present will depend on the density and spacing of individuals in the population. For the most part the spacing of desert plants is random, although a number of studies have shown that the distribution of desert plants can be regular and/or clustered under very low precipitation conditions (BARBOUR 1969; WOODELL et al. 1969; FONTEYN and MAHALL 1978). Stand biomass will, of course, vary with shrub density, but typical values are in the range of 1,000–6,000 kg ha^{-1} (Table 7.5). These values are characteristic of desert shrub communities globally and are substantially less than values we have presented for mediterranean-climate ecosystems.

The annual productivity of the shrubs in the desert shrub ecosystems ranges widely, from a low of 300 kg ha^{-1} yr^{-1} to a high of about 1,800 kg ha^{-1} yr^{-1} (Table 7.5). The variation in annual aboveground productivity can be attributed to two factors: (1) differences in biomass between different shrub communities and (2) year-to-year variation in the amount of precipitation. When the relative productivity (annual productivity/stand biomass for aboveground components) for these desert shrub communities is calculated, it is seen to be quite high. The relative productivity reaches levels of 40%–50% for several shrub communities and falls to levels of 20%–30% for the seemingly least productive community types. It is likely that the low relative productivity of *Zygophyllum dumosum* (22%) and *Larrea-Ambrosia* (27%) shrub communities may be attributed to low precipitation during the measurement period. On the other hand, it is also likely that the high relative productivity of *Artemisia herba-alba* communities may be attributed to a high photosynthetic potential by desert plants (see Sect. 7.4) and high seasonal precipitation.

The distribution of biomass within these desert shrub communities is primarily in wood and belowground components (Table 7.6). Typically much less than one-fifth of the total biomass is found in leaves. Belowground biomass varies substantially, and as a result so does the root:shoot (weight:weight) ratio. These

Table 7.5. Stand biomass, annual above ground productivity, and relative productivity for selected desert shrub vegetation types

Stand biomass (kg ha^{-1})	Annual productivity (kg ha^{-1} yr^{-1})	Relative productivity (%)	Desert shrub type	Location	Reference
4,185	1,768	42	*Artemisia herba-alba*	Algeria	1
890	440	49	*Artemisia herba-alba*	Israel	2
6,113	2,380	39	*Artemisia sieberi*	Syria	1
1,610	470	29	*Artemisia + Zygophyllum*	Israel	2
4,600	1,440	31	*Artemisia + Atriplex*	Utah, U.S.A.	3
4,610	1,220	26	*Atriplex confertifolia*	Utah, U.S.A.	4
2,228	ca. 900	40	*Atriplex vesicaria*	Australia	5
2,810	780	28	*Ceratoides lanata*	Utah, U.S.A.	4
3,780	1,048	28	*Cercidium + Franseria*	Arizona, U.S.A.	6
2,370	795	34	*Eremophila gilesii*	Australia	7
3,320	1,430	43	*Hammada scoparia*	Israel	2
5,000	1,730	35	*Larrea divaricata*	New Mexico, U.S.A.	3
1,070	286	27	*Larrea + Ambrosia*	Nevada, U.S.A.	3
1,560	350	22	*Zygophyllum dumosum*	Israel	8

References: 1 RODIN et al. (1972); 2 EVENARI et al. (1972); 3 SZAREK (1979); 4 CALDWELL et al. (1977b); 5 CHARLEY and COWLING (1968); 6 WHITTAKER and NIERING (1975); 7 BURROWS (1972); 8 ORSHAN and DISKIN (1968)

Table 7.6. Biomass and root:shoot ratios for several dominant desert scrub communities

Community and location	Reference	Biomass (kg ha^{-1})					Root:shoot
		Above-ground	Wood	Leaves	Below-ground	Total	
Eremophila (Australia)	1	2,370	1,406	964	776	3,146	0.33
Larrea (U.S.A.)	3	4,290	3,870	420	1,550[b]	5,840[b]	0.36[a]
Atriplex (Australia)	2	2,228	1,403	825	910	3,138	0.41
Artemisia (Syria)	4	2,600	2,400	200	3,500	6,100	1.35
Hammada (Syria)	4	2,360	2,200	160	7,300	9,600	3.09
Atriplex (U.S.A.)	6	4,610	3,620	990	18,860	23,470	4.09
Ceratoides (U.S.A.)	6	2,810	2,110	700	19,010	21,820	6.77
Artemisia (U.S.S.R.)	4	530	400	130	3,800	4,300	7.17
Artemisia-Ceratoides (U.S.S.R.)	5	530	400	130	4,210	4,740	7.94

[a] Average from CHEW and CHEW (1965) and BARBOUR (1973)
[b] Calculation based on root:shoot value

References: 1 BURROWS (1972); 2 CHARLEY and COWLING (1968); 3 WHITTAKER and NIERING (1975); 4 RODIN and BAZILEVICH (1968); 5 LITVINOVA (1972); 6 CALDWELL et al. (1977b)

Table 7.7. Root:shoot ratios (weight:weight) for perennial desert shrubs from cool-winter and cold-winter desert habitats

Species	Reference	Location	Root:shoot
Cool-winter			
Atriplex vesicaria	4	Australia	0.41
Brickellia incana	6	California	0.55
Eremophila gilesii	2	Australia	0.36
Ephedra nevadensis	5	California	0.84
Franseria dumosa	6	California	0.64
Grayia spinosa	10	Nevada	0.64
Hymenoclea salsola	6	California	0.72
Krameria grayi	6	California	0.37
Larrea divaricata	6	California	0.33
Lycium andersonii	10	Nevada	0.73
Cold-winter			
Artemisia badhysi	9	U.S.S.R.	3.31
Artemisia rhodantha and *Acantholimon diapensioides*	8	U.S.S.R.	6.58
Artemisia rhodantha and *Ceratoides papposa*	8	U.S.S.R.	7.94
Atriplex confertifolia	3	Utah	4.09
Ceratoides lanata	3	Utah	6.77
Eurotia lanata	1	Utah	6.67
Eurotia ceratoides and *Artemisia terra-alba*	7	U.S.S.R.	6.30

References: 1 Bjerregaard (1971); 2 Burrows (1972); 3 Caldwell et al. (1977b); 4 Charley and Cowling (1968); 5 Evenari et al. (1972); 6 Garcia-Moya and McKell (1970); 7 Kurochkina et al. (1972); 8 Litvinova (1972); 9 Rustanov (1972); 10 Wallace et al. (1974)

root:shoot ratios vary from 0.33 in *Larrea divaricata* to as high as 7.94 in *Artemisia rhodantha-Ceratoides papposa* communities (Table 7.7). There does appear to be a systematic trend in root:shoot ratios between desert shrub communities. In those desert shrub communities with cool winters where plants can potentially be active through most of the winter season, the root:shoot ratios are less than one. However, in those desert shrub communities with long, cold winters where plants are likely to be inactive in the winter, the root:shoot ratios are generally much higher than three.

The biomass and productivity of desert annuals is quite variable and depends very much on the amount of precipitation (Beatley 1969; Evenari et al. 1976; Patten 1978). Patten (1978) reported an order of magnitude difference (952 kg ha^{-1} in 1972 and 94 kg ha^{-1} in 1973) in the biomass of desert annuals in a *Larrea-Cercidium* community in Arizona (U.S.A.) when the precipitation levels between years differed. Table 7.8 summarizes studies by Beatley (1969) and Evenari et al. (1976) in two different desert shrub communities where precipitation is strongly correlated with the seasonal productivity of annuals. The study by Evenari et al. (1976) suggests that there is a minimum precipitation level below which the annuals will not germinate. Similar observations have

Table 7.8. Biomass of desert annuals (kg ha^{-1}) in two different desert shrub community types in different years. [Data for the *Larrea* community type (USA) are from BEATLEY (1969) and for the *Hammada* community type (Israel) from EVENARI et al. (1976)]

	Peak biomass of annuals (kg ha^{-1})	Precipitation (mm)
Larrea community		
1963	67.9	102
1964	41.6	91
1965	161.7	162
Hammada community		
1970	80.0	94
1971	2,700.0	163
1972	0.0	54
1973	1,150.0	134

been made on annuals found in the Californian deserts (WENT 1949; TEVIS 1958).

The timing and phenology of desert plants plays an important role in determining rates of biomass accumulation. However, for most desert systems the factors controlling phenology and dynamics of primary production are poorly understood. Perhaps the best understood of the desert ecosystems in this respect is the Mohave Desert of North America (WENT 1949; TEVIS 1958; BEATLEY 1974; ACKERMAN and BAMBERG 1974). These studies suggest that the seasonal success of most primary producers is dependent on precipitation events occurring at particular points in time. Differences in the amounts of precipitation and temperatures during the precipitation period will have a direct effect on the germination success of different species and therefore on the potential for biomass production. Vegetative growth generally precedes flowering activities in all desert species, both woody and herbaceous.

The biomass of lower plant life forms such as lichens and mosses is quite low in desert systems and does not contribute substantially to the overall productivity within the system (KAPPEN et al. 1975; NASH et al. 1977; KAPPEN et al. 1979).

7.4 Photosynthetic Rates

There are now abundant values in the literature on the photosynthetic capacity of plants in nature from a variety of habitats. We will use many of these for our comparisons. However, rather than compiling all the available rates, we shall be somewhat selective, often choosing the most complete set by a given worker to make a particular point. We feel that this is a valid approach, considering the great amount of variability one finds in rates for a given plant dependent on measurement techniques and season. Both the light-saturated photosynthetic rates, as well as the temperature optimum of photosynthesis, can be affected by physiological, environmental, or experimental factors.

This variability will be more noticeable on some plant types than on others. For example, plants with leaves with a short life span often have very high photosynthetic rates while young, but their capacity declines rapidly with age – often without a noticeable visible sign. Then plants with particularly sensitive stomata may have quite variable rates, dependent on premeasurement handling, as well as measurement conditions.

Much of this variability can be assessed, and even compensated for if full environmental and physiological parameters such as stomatal conductance, internal CO_2 concentration, vapor pressure gradient, leaf age and some measure of carboxylating capacity are given for each photosynthetic rate reported. However, such information is rarely available to accompany any data set on photosynthetic rates.

7.4.1 Photosynthetic Pathways, Rates and Life Form

A number of recent studies have focused on determining the distributions of different photosynthetic pathways among life forms and within mediterranean-climate and desert ecosystems (MOONEY et al. 1974b; SCHULZE and SCHULZE 1976; TEERI and STOWE 1976; SYVERTSEN et al. 1976; MULROY and RUNDEL 1977; MOONEY et al. 1977a; EICKMEIER 1978; WINTER and TROUGHTON 1978; TEERI et al. 1978). In mediterranean-climate systems, the C_3 pathway predominates in all life forms; only in the coastal zones and at mediterranean-climate – desert ecotones does CAM become of any importance. Within desert ecosystems, trees and shrubs possess the C_3 pathway, except for halophytic shrubs which often possess the C_4 pathway. Succulent subshrubs are usually CAM or FCAM (facultative CAM). Among the grasses, herbs, and annuals, we find that cool season (winter) species are generally C_3, whereas warm season (summer) species are generally C_4. (For detailed discussion of the functional significance of the different pathways of CO_2 fixation in photosynthesis see Chap. 15, Vol. 12B.)

7.4.1.1 Mediterranean-Climate Growth Forms

There are now a number of studies that indicate that there are intrinsic differences in the photosynthetic capacities of different mediterranean-climate life forms. For example, evergreen species generally have lower (about $^1/_2$) photosynthetic capacities than drought-deciduous species on a leaf area basis (HARRISON et al. 1971; MOONEY et al. 1977c; GIGON 1979). In a survey of the net photosynthetic rates of mediterranean-climate growth measured in field plants, MOONEY (1981) gives an average rate of 8 µmol CO_2 m^{-2} s^{-1} for evergreen shrubs and trees and 15 µmol CO_2 m^{-2} s^{-1} for drought-deciduous shrubs.

A recent study by MARTINEZ (1979) compared the photosynthetic capacities of a whole range of Chilean mediterranean-type growth forms. Using ^{14}C techniques on field plants throughout the year, he found that the maximum photosynthetic rates observed declined in order of growth form in the following se-

Table 7.9. Maximum observed seasonal gross photosynthetic rates as determined by ^{14}C techniques on different growth forms of Mediterranean-climate Chilean plants. (MARTINEZ 1979)

Growth form	Range of photosynthetic rates ($\mu mol\ CO_2\ m^{-2}\ s^{-1}$)	Number of species observed
Drought deciduous shrubs	25–40	2
Annuals	27–35	2
Perennial herbs	10–23	2
Evergreen sclerophyll trees and shrubs	11–16	4
Chlorophyllous stemmed shrubs	11–15	2
Hygrophyllous evergreen trees and shrubs	5–12	2
Succulents	3	2

quence: drought-deciduous shrub, herbaceous annual, herbaceous perennial, sclerophyllous evergreen, chlorophyllous-stemmed shrub, hygrophyllous evergreen, and evergreen succulent (Table 7.9). Although there were only limited representations of certain forms, these results are consistent in ranking with the results of others. In a similar comparison of Californian mediterranean-climate plants, OECHEL (1980b) found results comparable to that of MARTINEZ (1979), although the spread in rates between types was less. However, OECHEL's study was conducted during a single season at the beginning of the drought (late May and early June) when the capacity of the herbaceous types in particular would be reduced.

7.4.1.2 Desert Growth Forms

The major limiting factor for the photosynthetic process in most desert plants is water availability. As noted in the previous sections, the generally low and sporadic nature of precipitation often leads to a direct correlation between amount of rainfall and magnitudes of primary productivity. Desert plants as a general rule can respond immediately to precipitation events (moving from dormant to an active state and/or seed germination) by producing leaves which have a high photosynthetic capacity. Desert plants differ considerably in the peak rates among the most common life forms. As examples of plants from the Sonoran Desert of North America, ephemerals and plants with leaves active only a brief period of the year have extremely high photosynthetic rates (Table 7.10). These high rates are the highest or among the highest photosynthetic rates measured on terrestrial plants (SESTAK et al. 1971). In contrast, relatively low photosynthetic rates are observed in succulents and ferns.

The highest photosynthetic rates and leaf conductances for desert plants are found in ephemerals (MOONEY et al. 1976; MOONEY and EHLERINGER 1978; ARMOND and MOONEY 1978; EHLERINGER et al. 1979). These ephemerals complete their entire life cycle in a period as short as 6 weeks (BEATLEY 1974). Both C_3 and C_4 photosynthetic pathways are found in desert ephemerals

Table 7.10. Examples of peak photosynthetic rates expressed on a surface area basis for different life forms found in the Sonoran Desert of North America. These data are from plants grown and measured under field conditions

Life form	Species	Photosynthetic pathway	Peak photosynthesis ($\mu mol\ CO_2\ m^{-1}\ s^{-1}$)	Maximum leaf conductance ($mm\ s^{-1}$)	Reference
Winter ephemeral	*Camissonia claviformis*	C_3	59	24	1
Summer ephemeral	*Amaranthus palmeri*	C_4	81	20	2
Perennial herb	*Tidestromia oblongifolia*	C_4	37	5	3
Deciduous leaved shrub	*Encelia farinosa*	C_3	37	17	4
Evergreen leaved shrub	*Atriplex lentiformis*	C_4	28	3	5
Evergreen leaved shrub	*Larrea divaricata*	C_3	27	11	6
Deciduous leaved tree	*Cercidium floridum*	C_3	10	–	7
Succulent leaved subshrub	*Agave deserti*	FCAM	6	1.4	8
Stem succulent subshrub	*Opuntia basilaris*	CAM	6	1.2	9
Fern	*Notholaena parryi*	C_3	5	4	10

References: 1 EHLERINGER et al. (1979); 2 EHLERINGER (1982); 3 BJÖRKMAN et al. (1972); 4 EHLERINGER and BJÖRKMAN (1978b); 5 PEARCY and HARRISON (1974); 6 MOONEY et al. (1978); 7 ADAMS and STRAIN (1968); 8 HARTSOCK and NOBEL (1976); 9 SZAREK and TING (1975); 10 NOBEL (1978)

(MULROY and RUNDEL 1977). The winter ephemerals utilize the C_3 pathway, whereas both the C_3 and C_4 pathways occur in summer ephemerals.

Perennial life forms with leaves having life expectancies similar to ephemerals (i.e., drought-deciduous leaved shrubs and herbaceous perennials) appear to have peak photosynthetic rates approaching those of the ephemerals (BJÖRKMAN et al. 1972; EHLERINGER and BJÖRKMAN 1978b). Both drought-deciduous shrubs and herbaceous perennials are active for somewhat longer periods than ephemerals. Both C_3 and C_4 pathways are found among these life forms.

Evergreen leaved shrubs have high photosynthetic rates (STRAIN 1969, 1970; OECHEL et al. 1972; PEARCY and HARRISON 1974; MOONEY et al. 1978) relative to evergreen leaved species from other habitats (SESTAK et al. 1971; DUNN 1975). Succulent subshrubs possessing crassulacean acid metabolism (CAM) and facultative CAM (FCAM, succulents which shift between C_3 photosynthesis and CAM depending on aridity conditions) and ferns have the lowest photosynthetic rates of all the life forms. These life forms also have the lowest leaf conductances.

Yet it is the CAM plants and ferns that have the greatest capacity to survive extended periods of drought (SZAREK et al. 1973; SZAREK and TING 1975; NOBEL 1978). The next most tolerant group to drought is evergreen leaved shrubs followed by drought-deciduous leaved shrubs, and the least tolerant life forms are the ephemerals (STRAIN 1970; ODENING et al. 1974; WALTER and STADELMANN 1974).

There is a direct positive correlation between maximum leaf conductance and photosynthetic rate among the C_3 plants and among C_4 plants (Table 7.10). The C_3 plants have higher leaf conductances than C_4 plants for any given photosynthetic rate, suggesting that water use efficiency should be better in C_4 plants. It is interesting to note that photosynthetic rates in C_3 plants can equal and exceed rates measured for C_4 plants. While it may be generally true that photosynthetic rates of C_4 plants exceed those of C_3 plants, there is ample evidence to suggest that this is certainly not true for plants of desert habitats. (For a general discussion of plant life forms and their carbon and water relations see Chap. 18, Vol. 12 B).

7.4.2 Temperature and Light Response

7.4.2.1 Mediterranean-Climate Forms

Just as there are considerable differences in the absolute photosynthetic capacity between leaves of various mediterranean-climate life forms, there are also differences in the season of optimum photosynthesis capacity (to a certain degree) as well as the seasonal duration of activity. In MARTINEZ' 1979 study of 16 Chilean species, encompassing seven different growth forms, he found that the period of optimum photosynthetic capacity was concentrated in the period spanning late winter to late spring (August to December), with the herbaceous plants having their highest capacity more toward the end of this period.

Herbaceous species of mediterranean climates have a photosynthetic period that coincides with the approximately 5- to 6-month winter wet season (MARTINEZ 1979). Drought-deciduous species have a photosynthetic period that is about a month or so longer than for the herbaceous plants, because of their greater rooting volume and hence greater available water stores.

The evergreen species of mediterranean-climate regions have the capacity to fix carbon year-round (DUNN 1970; ECKARDT et al. 1975; MOONEY et al. 1975; MARTINEZ 1979; OECHEL 1980b), although at somewhat reduced rates during the drought (MOONEY et al. 1975).

In a survey of the photosynthetic temperature response of a number of Californian and Chilean shrubs, it was found that most of the species could be characterized as having rather broad central regions of relative temperature insensitivity (e.g. 85% of maximum rates of certain species occurring over a 25 °C temperature range) (OECHEL 1980b). There was, however, a range of thermal optima of photosynthesis extending from near 15 °C to about 30 °C. Most of these woody species, however, had their thermal optimum near 20 °C when measured under full light in the springtime. This seems to be the characteristic optimum temperature for a wide range of mediterranean climate evergreens (MOONEY 1981).

In the study of OECHEL (1980b) noted above it was found that at optimum temperatures most species saturated at less than full sunlight. A notable exception was an evergreen with vertically oriented leaves (*Arctostaphylos*), which had a high light requirement. MOONEY (1981) proposed that the generally low light requirement of the evergreen sclerophylls was related to their low intrinsic photosynthetic capacity. One drought-deciduous species, *Encelia californica*, in contrast, has a high photosynthetic rate, which is typical for the life form, and does not saturate, even at full sunlight (EHLERINGER and BJÖRKMAN 1978b). A systematic analysis of life forms, photosynthetic capacity and light requirement of mediterranean-climate plants would be valuable.

7.4.2.2 Desert Forms

Leaves of desert plants are usually not photosynthetically light-saturated under midday irradiances (BJÖRKMAN et al. 1972; SCHULZE et al. 1972; PEARCY and HARRISON 1974; MOONEY et al. 1976; MOONEY and EHLERINGER 1978; EHLERINGER and BJÖRKMAN 1978b). Associated with the lack of light saturation are high photosynthetic capacities and high leaf conductances to water loss (Table 7.10). The initial slope of the light response curve, the quantum yield (ratio of photosynthetic rate over absorbed light quantum flux, see Chaps. 2 and 3, Vol. 12A), is 0.052 mol E^{-1} for C_3 plants and 0.053 mol E^{-1} for C_4 plants at 30 °C (EHLERINGER and BJÖRKMAN 1977). Quantum yield values measured for C_3 plants from the light response curve at 30 °C were 0.050 mol E^{-1} for the shrub *Larrea divaricata* (MOONEY et al. 1978), 0.052 mol E^{-1} for the shrub *Encelia farinosa* (EHLERINGER and BJÖRKMAN 1978b), and 0.055 mol E^{-1} for the annual *Camissonia claviformis* (MOONEY et al. 1976). A quantum yield of 0.077 mol E^{-1} has been reported for the C_3 fern *Notholaena parryi* at 15 °C (NOBEL 1978). Values of the quantum yield for C_4 plants are not generally available and are difficult to estimate based on the photosynthesis–light response curves in the literature.

The temperature response of light-saturated photosynthesis by leaves of desert plants has been the subject of intense investigations over the last decade (STRAIN and CHASE 1966; STRAIN 1969; HELLMUTH 1971; BJÖRKMAN et al. 1972; PEARCY and HARRISON 1974; LANGE et al. 1974; PEARCY et al. 1974; DEPUIT and CALDWELL 1975; LANGE et al. 1975; BJÖRKMAN et al. 1976; CALDWELL et al. 1977a; PEARCY et al. 1977; NOBEL 1978; EHLERINGER and BJÖRKMAN 1978b; BJÖRKMAN et al. 1978; LANGE et al. 1978; MOONEY et al. 1978; NOBEL et al. 1978). Generally it has been found that plants growing in desert environments show a wide variation in the thermal optima for photosynthesis, and that this optima in many species is dependent on phenology, growth temperature, and/or drought stress. Temperature optima for net photosynthesis ranging from 15° to 47 °C have been measured; these optima depend somewhat on photosynthetic pathway, with C_3 species showing lower optima than C_4 species.

7.5 Photosynthetic Adaptation to Arid Conditions

7.5.1 Mediterranean-Climate Forms

The different mediterranean-climate growth forms respond in a dissimilar manner to the seasonal drought. Annual and many perennial herbs pass through the drought in a dormant state. Many shrub species (generally subligneous, subshrubs), particularly those occupying disturbance or xeric habitats also lose all or most of their leaves during the unfavorable period. They have little stomatal control of transpiration, and they rapidly have lowered water potentials in the early drought (Poole and Miller 1975; Gigon 1979). These plants are thus drought avoiders. As noted earlier, it is plants of these types that generally have a high photosynthetic capacity.

Many mediterranean-climate subshrubs have seasonally dimorphic leaves (Orshan 1964; Margaris 1976, 1977; Montenegro et al. 1979). There is a reduction of leaf area on such plants during the drought period, with those leaves remaining differing morphologically and physiologically from the wet season leaves. On a weight basis the drought season leaves have an intrinsically lower photosynthetic capacity than the wet season leaves (Margaris 1977). Further studies on the comparative gas exchange characteristics of these two leaf types are needed.

It is the evergreen mediterranean-climate plants that have the capacity to carry on photosynthesis throughout the year. In a study of the seasonal photosynthetic activity of the Californian evergreen shrub, *Heteromeles arbutifolia*, in its native habitat, it was found that plants of these species had active photosynthesis at water potentials as low as −3.3 MPa. On a seasonal basis, however, drought is the principal limiting factor to photosynthesis. During annual drought periods these plants may have midday depressions in photosynthesis because of stomatal closure, and during the most severe drought periods photosynthesis may occur primarily in the morning (Mooney et al. 1975).

In the evergreen species morphology plays an important role in drought adaptation. For example, Morrow and Mooney (1974) found that *H. arbutifolia*, described above, had an intrinsic capacity to carry on photosynthesis at lowered water potentials comparable to those found in *Arbutus menziesii*, even though the latter species is found generally in more northerly and mesic habitats. Yet, where these species co-occur, *Arbutus* may show severe drought effects, whereas *Heteromeles* may not. This is evidently due to the smaller and more steeply angled leaves of *Heteromeles* and because of its deeper root system. Gradients of variation in leaf morphology among evergreen plants along mediterranean-region aridity gradients have been described both for within-community types (Mooney et al. 1974a) and among species of the same genus between community types (Shaver 1978). The general trend is for evergreen leaves to become smaller, more steeply angled and to have a decreased light absorptivity with increasing site aridity.

Poole and Miller (1975) found among the Californian evergreen sclerophylls that they studied that, as might be expected, the deeper-rooted species developed less water stress during the drought than did shallow-rooted species. The shallow-rooted species evidently have stomata less sensitive to water potential and are apparently thus restricted to more mesic microsites.

Although mediterranean-climate evergreen shrubs are photosynthetically active year-round, and thus during both hot and cold seasons, studies have not demonstrated any large degree of photosynthetic thermal acclimation capacity for them (HARRISON 1971). Since many of these evergreens (both Californian and Chilean) have very broad photosynthetic thermal responses (OECHEL 1980a), photosynthetic acclimation would not be very apparent.

7.5.2 Desert Forms

For many species the temperature optimum for photosynthesis will vary with the thermal environment in which the plants are growing, such that the thermal optima remain in tune with mean daytime air temperatures (MOONEY and WEST 1964; STRAIN and CHASE 1966; STRAIN 1969; LANGE et al. 1974; BJÖRKMAN et al. 1975; LANGE et al. 1975; PEARCY 1977; MOONEY et al. 1978). In their study of the C_4 shrub *Hammada scoparia* from the Negev Desert, LANGE et al. (1974) found that the temperature optimum for photosynthesis varied from 28.6 °C in late March to 40.9 °C in mid-August (Fig. 7.3). In association with the change in temperature optimum, there were also corresponding changes in the upper temperature compensation point from a minimum of 46.3 °C in late March to 60.0 °C in mid-August. MOONEY et al. (1978) have conducted similar measurements on the C_3 shrub *Larrea divaricata* of the Sonoran Desert. They found a shift in the temperature optima from about 20 °C in January

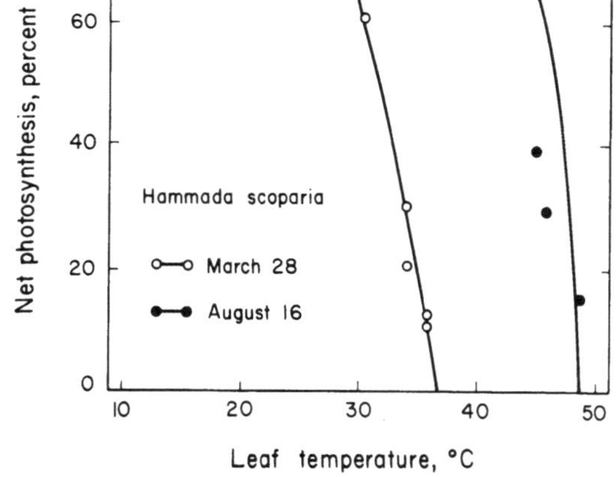

Fig. 7.3. Temperature dependence of photosynthesis at high irradiances for field-grown plants of the C_4 shrub *Hammada scoparia* in the Negev Desert, Israel, during late March (spring) and mid-August (summer) conditions. (Modified from LANGE et al. 1974)

to 32 °C in September. In laboratory studies under controlled growth regimes, identical changes in thermal optima were observed. While the upper temperature compensation point was found to vary in the same manner as the temperature optima of photosynthesis, MOONEY et al. (1978) do not present specific values for *L. divaricata*. These two studies demonstrate that, while temperature optima may be higher in C_4 plants, both C_3 and C_4 plants are equally capable of thermal acclimation. While for both *Hammada scoparia* and *Larrea divaricata* the temperature optimum for net photosynthesis varied in concert with the seasonal air temperatures, the absolute photosynthetic rate at the respective temperature optimum showed little change seasonally.

The upward shift in photosynthetic temperature optimum which results from growth or exposure to high temperatures has been interpreted as photosynthetic acclimation. It is presumed to be of adaptive value to the shrubs by allowing the plant to operate at a high photosynthetic capacity despite wide fluctuations and often extremely high air temperatures. Few calculations have been made to estimate the net benefits of thermal acclimation on the daily carbon gain of a leaf. LANGE et al. (1978) made such calculations for *Prunus armeniaca* (which shows thermal acclimation) and *Hammada scoparia*. Their conclusion was that thermal acclimation did result in a greater rate of daily carbon gain in *P. armeniaca*, but that it was not beneficial in *H. scoparia*. MOONEY (1980) made a similar analysis on the consequence of thermal acclimation in *Larrea divaricata* and concluded that thermal acclimation did result in a greater daily carbon gain by the leaf.

Although many desert species exhibit a capacity for thermal acclimation of their photosynthetic apparatus, a number of species, such as *Encelia farinosa* (Fig. 7.4), do not show changes in temperature optima when grown under differ-

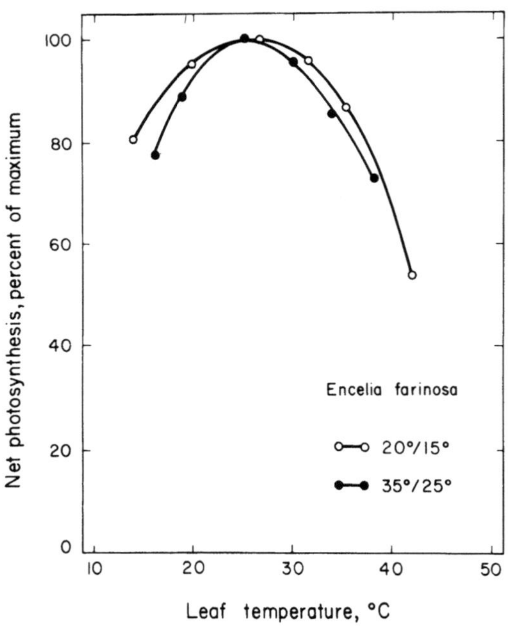

Fig. 7.4. Temperature dependence of photosynthesis of high irradiances for laboratory grown plants of the C_3 shrub *Encelia farinosa* from the Sonoran Desert. The growth temperatures presented are day–night temperatures. (Data from EHLERINGER and BJÖRKMAN 1978b)

ent thermal regimes (STRAIN 1969; PEARCY et al. 1974; BJÖRKMAN et al. 1975; EHLERINGER and BJÖRKMAN 1978b). These plants exhibit a morphological change (or acclimation) in leaf structure rather than physiological acclimation in response to changes in the thermal environment (CUNNINGHAM and STRAIN 1969; EHLERINGER et al. 1976; MOONEY et al. 1977c; EHLERINGER and BJÖRKMAN 1978a; EHLERINGER and MOONEY 1978). Plants such as the C_3 shrub *Encelia farinosa* and the C_4 shrub *Atriplex hymenelytra* show changes in leaf spectral properties under changing thermal regimes. As air temperatures and/or drought stress increase, leaves from both of these plants increase their reflectance, reducing heat load and thus leaf temperatures. The two shrubs utilize different mechanisms to achieve an increased reflectance. *Encelia farinosa* uses a thick layer of pubescence to reflect light (EHLERINGER and BJÖRKMAN 1978a), whereas the crystallization of sodium chloride in epidermal salt glands is the mechanism used by *A. hymenelytra* (MOONEY et al. 1977c). The net effect on leaf temperature is the same; temperatures are reduced and remain close to the respective thermal optima of photosynthesis (MOONEY et al. 1977c; EHLERINGER and BJÖRKMAN 1978b; EHLERINGER and MOONEY 1978).

Since precipitation events are infrequent and amounts are often low, water stress is a continual limiting factor for desert plants. Stomatal control under low soil moisture conditions is the major cause of reduced net photosynthetic activity in plants under water stress. Desert plants exhibit two fundamental adaptive patterns to adapt the photosynthetic apparatus to low soil moisture conditions: desiccation tolerance and drought avoidance.

Under moderate water stress, many desert plants respond with a partial midday closure of stomata, which results in a bimodal peak of daily photosynthetic activity (STRAIN 1969; LANGE et al. 1969; HELLMUTH 1971). Most true xerophytes and desert halophytes are able to survive extremely low leaf water potentials (WALTER and STADELMANN 1974; LANGE et al. 1976). Positive net photosynthesis, although severely depressed by low stomatal conductances, still persists for such plants under extreme drought (STRAIN 1970; KAPPEN et al. 1972; ODENING et al. 1974). Within the nonsucculent desert plants, evergreen shrubs are most desiccation-tolerant, drought-deciduous shrubs are the next most tolerant, followed by trees, and the least tolerant are ephemerals and herbaceous perennials (Fig. 7.5) (WALTER and STADELMANN 1974; ODENING et al. 1974).

An alternative to drought tolerance in photosynthetic adaptation is drought avoidance. Three life forms occur within this category: perennial shrubs with drought-deciduous leaves (ZOHARY 1961; ORSHAN and ZAND 1962; SHREVE and WIGGINS 1964; CUNNINGHAM and STRAIN 1969; EHLERINGER and BJÖRKMAN 1978a), ephemerals (SHREVE and WIGGINS 1964; WALTER and STADELMANN 1974; MULROY and RUNDEL 1977), and succulent CAM plants (NEALES et al. 1968; SZAREK et al. 1973; LANGE et al. 1976; NOBEL 1977; KLUGE and TING 1978). Drought-deciduous shrubs and ephemerals avoid drought stress by achieving high rates of photosynthesis during non-stress periods and senescing photosynthetic activity as leaf water potentials decline. CAM plants, on the other hand, avoid drought stress through internal water storage and allowing only a minimal water loss at night when stomates are open.

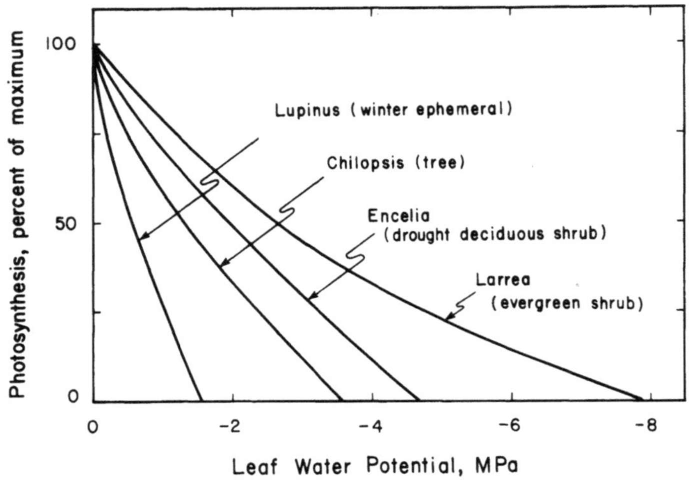

Fig. 7.5. Relationships between decreasing leaf water potential and net photosynthetic rate for different life forms found in the Sonoran Desert. (Data for *Lupinus arizonicus* calculated utilizing values from ARMOND and MOONEY 1978 and FORSETH and EHLERINGER 1980. Data for *Larrea divaricata, Encelia farinosa,* and *Chilopsis linearis* from ODENING et al. 1974)

Within drought-deciduous perennials and ephemerals, several species show adaptations which allow these plants to extend the photosynthetic activity to greater levels of drought stress. In drought-deciduous shrubs such as *Encelia farinosa* there are several changes in external and internal leaf morphology that will allow the leaves to be active to low leaf water potentials (CUNNINGHAM and STRAIN 1969; EHLERINGER and MOONEY 1978). The phenomenon of solar tracking in desert ephemerals (leaves move to become perpendicular to the sun's rays) results in leaves operating at maximum photosynthetic capacity throughout the day, greatly increasing the daily carbon gain over fixed orientation leaves (WAINWRIGHT 1977; MOONEY and EHLERINGER 1978; FORSETH and EHLERINGER 1980). Under mild drought stress, solar trackers can become solar-avoiders (leaves move to become parallel to the sun's rays) and thus extend their period of productivity (albeit severely reduced) into drought conditions (BEGG and TORSELL 1974; FORSETH and EHLERINGER 1980).

In drier habitats, especially in coastal desert zones, there is an increase in the occurrence of CAM plants (MOONEY et al. 1974b; SCHULZE and SCHULZE 1976; MOONEY et al. 1977a; WINTER and TROUGHTON 1978). Since stomates open only at night when vapor pressure deficits are lowest, CAM plants have low rates of water loss and consequently high water use efficiencies (SZAREK et al. 1973; SZAREK and TING 1975; NOBEL 1978; KLUGE and TING 1978). As leaf water potentials decline in CAM plants, leaf conductances decrease (SZAREK and TING 1975; NOBEL 1978; KLUGE and TING 1978). Total CO_2 fixation decreases as stem water potentials decline in *Opuntia basilaris,* and a point of zero net CO_2 fixation occurs at a fairly highly stem water potential of −1.5 MPa (Fig. 7.6). During periods of zero net CO_2 fixation, CAM plants are in an "idling mode" and refix internally generated CO_2. Because of their extremely

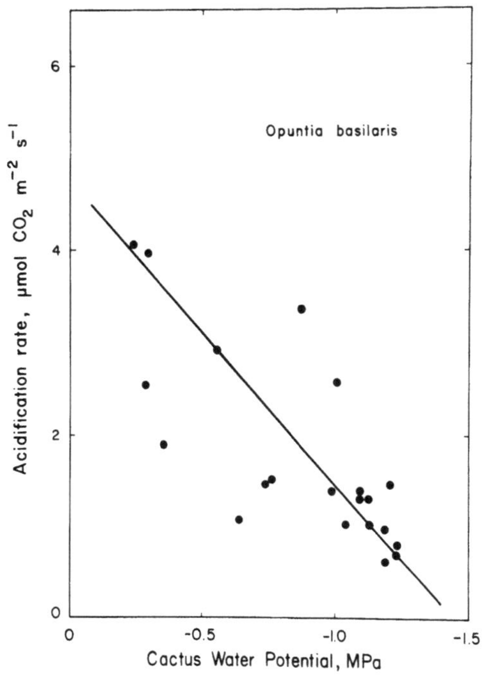

Fig. 7.6. Effects of cactus pad leaf water potential in the nocturnal acidification rate in *Opuntia basilaris* from the Sonoran Desert. (Data from SZAREK and TING 1975)

thick cuticles, which virtually eliminate any water loss, CAM plants can survive extended periods of drought in this "idling mode" (KLUGE and TING 1978).

7.6 Conclusions

Although there are many similarities in the life forms present in mediterranean-climate and desert ecosystems, rates of annual productivity and amounts of standing biomass are vastly different. Almost all of this difference can be attributed to increased precipitation in the mediterranean-climate ecosystem. One major biomass difference between these two ecosystem types is the very high root:shoot ratios that occur in cold-winter desert habitats; a high root:shoot ratio evidently does not occur in mediterranean-climate plants.

There are differences in the photosynthetic adaptations of plants in these two ecosystems. These differences seem related to the seasonality as well as the predictability of precipitation events. In mediterranean-climate systems, precipitation and plant growth occurs almost exclusively in the winter and spring months. Since both temperature and precipitation are fairly predictable, mediterranean-climate plants exhibit a broad temperature response in photosynthesis and apparently lack a thermal acclimation potential. In contrast, in desert systems precipitation and growth can generally occur over a wide range of temperatures and are less predictable. Consequently, most desert plants either show adaptations to morphologically or physiologically acclimate to changes in thermal environments or to be active during thermally favorable periods, and remain dormant or inactive during unfavorable periods.

References

Ackerman TL, Bamberg SA (1974) Phenological studies in the mojave desert at Rock Valley (Nevada test site). In: Lieth H (ed) Phenology and seasonality modeling. Ecol Stud Vol 8. Springer, Berlin Heidelberg New York

Adams MS, Strain BR (1968) Photosynthesis in stems and leaves of *Cercidium floridum:* spring and summer diurnal field response and relation to temperature. Oecol Plant 3:285–297

Armond PA, Mooney HA (1978) Correlation of photosynthetic unit size and density with photosynthetic capacity. Carnegie Inst Wash Yearb 77:234–237

Aschmann H (1973) Distribution and peculiarity of mediterranean ecosystems. In: Castri di F, Mooney HA (eds) Mediterraean type ecosystems. Ecol Stud Vol 7. Springer, Berlin Heidelberg New York

Barbour MG (1969) Age and space distribution of the desert shrub *Larrea divaricata.* Ecology 50:679–685

Barbour MG (1973) Desert dogma reexamined: root/shoot productivity and plant spacing. Am Midl Nat 89:41–57

Beatley JC (1969) Biomass of desert winter annual plant populations in southern Nevada. Oikos 20:261–273

Beatley JC (1974) Phenological events and their environmental triggers in Mojave Desert ecosystems. Ecology 55:856–863

Begg JE, Torsell BWR (1974) Diaphotonastic and parahelionastic leaf movements in *Stylosanthes humilis* H.B.K. (Townsville Stylo). R Soc N Z Bull 12:277–283

Bjerregaard RS (1971) The nitrogen budget of two salt desert shrub plant communities of western Utah. PhD Diss, Utah State Univ, Logan

Björkman O, Pearcy RW, Harrison AT, Mooney HA (1972) Photosynthetic adaptation to high temperatures: a field study in Death Valley, California. Science 175:786–789

Björkman O, Mooney HA, Ehleringer J (1975) Photosynthetic responses of plants from habitats with contrasting thermal environments. Carnegie Inst Wash Yearb 74:743–748

Björkman O, Boynton J, Berry JA (1976) Comparison of heat stability of photosynthesis, chloroplast membrane reactions, photosynthetic enzymes, and soluble protein in leaves of heat-adapted and cold-adapted C_4 species. Carnegie Inst Wash Yearb 75:400–407

Björkman O, Badger MR, Armond PA (1978) Thermal acclimation of photosynthesis: effect of growth temperature on photosynthetic characteristics and components of the photosynthetic apparatus of *Nerium oleander.* Carnegie Inst Wash Yearb 77:262–276

Brown GW (ed) (1968) Desert biology, vol I. Academic Press, London New York

Burrows WH (1972) Productivity of an arid zone shrub (*Eremophila gilesii*) community in southwestern Queensland. Aust J Bot 20:317–329

Cabanettes A (1979) Croissance, biomasse et productivité de *Pinus pinea* L. en Petite Camargue. Thesis, Univ Montpellier

Caldwell MM, Osmond CB, Nott DL (1977a) C_4 pathway photosynthesis at low temperature in cold-tolerant *Atriplex* species. Plant Physiol 60:157–164

Caldwell MM, White RS, Moore RT, Camp LB (1977b) Carbon balance, productivity, and water use of cold-winter desert shrub communities dominated by C_3 and C_4 species. Oecologia 29:275–300

Castri di F, Mooney HA (eds) (1973) Mediterranean type ecosystems: origin and structure. Ecol Stud Vol 7. Springer, Berlin Heidelberg New York

Charley JL, Cowling SW (1968) Changes in soil nutrient status resulting from overgrazing and their consequences in plant communities of semi-arid areas. Proc Ecol Soc Aust 3:28–38

Chew RM, Chew AE (1965) The primary productivity of a desert shrub (*Larrea tridentata*) community. Ecol Monogr 35:255–275

Cunningham GL, Strain BR (1969) Ecological significance of seasonal variability in a desert shrub. Ecology 50:400–408

DeBano LF, Conrad CE (1978) The effects of fire on nutrients in a chaparral ecosystem. Ecology 59:489–497

Depuit EJ, Caldwell MM (1975) Gas exchange of three cool semidesert species in relation to temperature and water stress. J Ecol 63:835–858

Duncan DA, Woodmansee RG (1980) Plant biomass, soil water dynamics, and primary production in an annual grassland ecosystem (unpublished)

Dunn EL (1970) Seasonal patterns of carbon dioxide metabolism in evergreen sclerophylls in California and Chile. PhD Diss, Univ California, Los Angeles

Dunn EL (1975) Environmental stresses and inherent limitations affecting CO_2 exchange in evergreen sclerophylls in mediterranean climates. In: Gates DM, Schmerl RB (eds) Perspectives in biophysical ecology. Ecol Stud Vol 12. Springer, Berlin Heidelberg New York

Eckardt F, Heim G, Methy M, Sauvezon R (1975) Interception de l'énergie rayonnante, échanges gazeaux et croissance dans une forêt méditerranéenne à feuillage persistant (Quercetum ilicis). Photosynthetica 9:145–156

Ehleringer J, Björkman O (1977) Quantum yields for CO_2 uptake in C_3 and C_4 plants: dependence on temperature, CO_2, and O_2 concentration. Plant Physiol 59:86–90

Ehleringer J, Björkman O (1978a) Pubescence and leaf spectral characteristics in a desert shrub, *Encelia farinosa*. Oecologia 36:137–148

Ehleringer JR, Björkman O (1978b) A comparison of photosynthetic characteristics of *Encelia* species possessing glabrous and pubescent leaves. Plant Physiol 62:185–190

Ehleringer JR, Mooney HA (1978) Leaf hairs: effects on physiological activity and adaptive value to a desert shrub. Oecologia 37:183–200

Ehleringer JR, Björkman O, Mooney HA (1976) Leaf pubescence: effects on absorptance and photosynthesis in the desert shrub. Science 192:376–377

Ehleringer JR, Mooney HA, Berry JA (1979) Photosynthesis and microclimate of *Camissonia claviformis*, a desert winter annual. Ecology 60:280–286

Eickmeier WG (1978) Photosynthetic pathway distributions along an aridity gradient in Big Bend National Park, and implications to enhanced resource partitioning. Photosynthetica 12:290–297

Evenari M, Schulze ED, Lange OL (1972) Ecophysiological investigation in the Negev Desert. Part III. The diurnal course of carbon dioxide exchange and transpiration and its balance in regard to primary production. Eco-Physiol Found Ecosyst Prod Arid Zone. Nauka, Moscow Leningrad

Evenari M, Schulze ED, Lange OL, Kappen L, Buschbom U (1976) Plant production in arid and semi-arid areas. In: Lange OL, Kappen L, Schulze E-D (eds) Water and plant life. Ecol Stud Vol 19, Springer, Berlin Heidelberg New York

Fonteyn PJ, Mahall BE (1978) Competition among desert perennials. Nature (London) 275:544–545

Forseth I, Ehleringer JR (1980) Solar tracking response to drought in a desert annual. Oecologia 44:159–163

Garcia-Moya E, McKell CM (1970) Contribution of shrubs to the nitrogen economy of a desert-wash plant community. Ecology 51:81–88

Gigon A (1979) CO_2-gas exchange, water relations and convergence of mediterranean shrub-types from California and Chile. Oecol Plant 14:129–150

Harrison AT (1971) Temperature-related effects on photosynthesis in *Heteromeles arbutifolia* M. Roem. PhD thesis, Stanford Univ

Harrison A, Small E, Mooney HA (1971) Drought relationships and distribution of two mediterranean-climate Californian plant communities. Ecology 52:869–875

Hartsock TL, Nobel PS (1976) Watering converts a CAM plant to daytime CO_2 uptake. Nature (London) 262:574–576

Hastings JR, Humphrey RR (1969a) Climatological data and statistics for Sonora and northern Sinoloa. Univ Ariz Inst Atm Phys Tech Rep 19

Hastings JR, Humphrey RR (1969b) Climatological data and statistics for Baja California. Univ Ariz Inst Atm Phys Tech Rep 18

Hellmuth EO (1971) Ecophysiological studies on plants in arid and semi-arid regions of western Australia. III. Comparative studies on photosynthesis, respiration and water relations of ten arid zone and two semi-arid zone plants under winter and late summer climatic conditions. J Ecol 59:225–260

Kappen L, Lange OL, Schulze ED, Evenari M, Buschbom U (1972) Extreme water stress and photosynthetic activity of the desert plant *Artemisia herba-alba* Asso. Oecologia 10:177–182

Kappen L, Oertli JJ, Lange OL, Schulze E-D, Evenari M, Buschbom U (1975) Seasonal and diurnal courses of water relations of the arido-active plant *Hammada scoparia* in the Negev Desert. Oecologia 21:175–192

Kappen L, Lange OL, Schulze ED, Evenari M, Buschbom U (1979) Ecophysiological investigations on lichens of the Negev Desert. IV. Annual course of the photosynthetic production of *Ramalina maciformis* (Del.) Bory. Flora 168:85–108

Kittredge J (1955) Litter and forest floor of the chaparral in parts of the San Dimas Experimental Forest, California. Hilgardia 23:563–596

Kluge M, Ting IP (1978) Crassulacean acid metabolism: analysis of an ecological adaptation. Ecol Stud Vol 30. Springer, Berlin Heidelberg New York

Kruger FJ (1977) A preliminary account of aerial plant biomass in fynbos communities of the mediterranean-type climate zone of the Cape Province. Bothalia 12:299–305

Kummerow J, Krause D, Jon W (1977) Seasonal changes of fine root density in southern California chaparral. Oecologia 37:201–212

Kurochkina LY, Osmanova LT, Borovskaya TA (1972) Bioecological characteristics and productivity of psammophilous communities in the South Balkash area. In: Ecophysiol Found Ecosyst Prod Arid Zone. Nauka, Moscow Leningrad

Lange OL, Koch W, Schulze ED (1969) CO_2-gas exchange and water relationships of plants in the Negev Desert at the end of the dry period. Ber Dtsch Bot Ges 82:39–61

Lange OL, Schulze ED, Evenari M, Kappen L, Buschbom U (1974) The temperature-related photosynthetic capacity of plants under desert conditions. I. Seasonal changes of the photosynthetic response to temperature. Oecologia 17:97–110

Lange OL, Schulze ED, Kappen L, Evenari M, Buschbom U (1975) CO_2 exchange pattern under natural conditions of *Caralluma negevensis,* a CAM plant of the Negev Desert. Photosynthetica 9:318–326

Lange OL, Kappen L, Schulze E-D (eds) (1976) Water and plant life. Ecol Stud Vol 19. Springer, Berlin Heidelberg New York

Lange OL, Schulze E-D, Evenari M, Kappen L, Buschbom U (1978) The temperature-related photosynthetic capacity of plants under desert conditions. III. Ecological significance of seasonal changes of photosynthetic response to temperature. Oecologia 34:89–100

Litvinova NP (1972) Productivity of high mountain deserts (Pamirs). In: Ecophysiological Found. Ecosyst Prod Arid Zone. Nauka, Moscow Leningrad

Lossaint P (1973) Soil vegetation relationships in mediterranean ecosystems of southern France. In: Castri di F, Mooney H (eds) Mediterranean type ecosystems. Ecol Stud Vol 7. Springer, Berlin Heidelberg New York

Lossaint P, Rapp M (1971) Le cycle du carbone dans les forêts de *Pinus halepensis*. In: Duvigneaud P (ed) Productivity of forest ecosystems. UNESCO, Paris

Margaris N (1975) Effect of photoperiod on seasonal dimorphism of some mediterranean plants. Ber Schweiz Bot Ges 85:96–102

Margaris NS (1976) Structure and dynamics in a phryganic (East Mediterranean) ecosystem. J Biogeogr 3:249–259

Margaris NS (1977) Physiological and biochemical observations in seasonal dimorphic leaves of *Sarcopoterium spinosum* and *Phlomis fruticosa*. Oecol Plant 12:343–350

Martinez P (1979) Photosynthetic responses of different life-forms in a mid-elevation mediterranean matorral of central Chile. PhD thesis, Univ Georgia

McGinnies WG (1969) Deserts of the world, an appraisal of research into their physical and biological environments. Univ Ariz Press, Tucson

Miller PC, Stoner WA (1979) Canopy structure and environmental interactions. In: Solbrig OT, Jain S, Johnson GB (eds) Topics in plant population biology. Columbia Univ Press, New York

Montenegro G, Hoffmann AJ, Aljaro N, Hoffmann AE (1979) *Satureja gilliesii*, a poikilolydric shrub from the Chilean mediterranean vegetation. Can J Bot 57:1206–1213
Mooney HA (ed) (1977a) Convergent evolution in Chile and California. Dowden Hutchinson & Ross, Stroudsburg
Mooney HA (1977b) The carbon cycle in mediterranean-climate evergreen scrub communities. In: Mooney HA, Conrad E (eds) Proc Symp Environ Consequences Fire Fuel Manag Mediterranean Ecosyst. US For Serv Gen Tech Rep 3:107–115
Mooney HA (1980) Seasonality and gradients in the study of stess adaptation. In: Turner N, Kramer P (eds) Adaptation of plants to stress. Wiley and Sons, New York
Mooney HA (1981) Primary production in mediterranean-climate regions. In: Castri di F, Goodall DW, Specht RL (eds) Ecosystems of the world, vol 11. Mediterranean type shrublands. Elsevier, Amsterdam
Mooney HA, Dunn EL (1970) Photosynthetic systems of mediterranean-climate shrubs and trees of California and Chile. Am Nat 104:447–453
Mooney HA, Ehleringer JR (1978) The carbon gain benefits of solar tracking in a desert annual. Plant Cell Environ 1:307–311
Mooney HA, West M (1964) Photosynthetic acclimation of plants of diverse origin. Am J Bot 51:825–827
Mooney HA, Dunn EL, Shropshire F, Song L (1970) Vegetation comparisons between the mediterranean-climatic areas of California and Chile. Flora 159:480–496
Mooney HA, Gulmon SL, Parsons DJ, Harrison AT (1974a) Morphological changes within the chaparral vegetation type as related to elevational gradients. Madroño 22:281–316
Mooney HA, Troughton JH Berry JA (1974b) Arid climates and photosynthetic systems. Carnegie Inst Wash Yearb 73:793–805
Mooney HA, Harrison AT, Morrow PA (1975) Environmental limitations of photosynthesis on a California evergreen shrub. Oecologia 19:293–301
Mooney HA, Ehleringer J, Berry JA (1976) High photosynthetic capacity of a winter annual in Death Valley. Science 194:322–324
Mooney HA, Troughton JH, Berry JA (1977a) Carbon isotope ratio measurements of succulent plants in southern Africa. Oecologia 30:295–305
Mooney HA, Kummerow J, Johnson AW, Parsons DJ, Keeley S, Hoffmann A, Hays RI, Giliberto J, Chu C (1977b) The producers – their resources and adaptive responses. In: Mooney H (ed) Convergent evolution in Chile and California. Dowden, Hutchinson & Ross, Stroudsburg
Mooney HA, Ehleringer J, Björkman O (1977c) The energy balance of leaves of the evergreen desert shrub, *Atriplex hymenelytra*. Oecologia 29:301–310
Mooney HA, Björkman O, Collatz GJ (1978) Photosynthetic acclimation of temperature in the desert shrub, *Larrea divaricata*. Plant Physiol 61:406–410
Morrow PA, Mooney HA (1974) Drought adaptations in two Californian evergreen sclerophylls. Oecologia 15:205–222
Mulroy TW, Rundel PW (1977) Annual plants: adaptations to desert environments. BioScience 27:109–114
Nash TH, White SL, Marsh JE (1977) Lichen and moss distribution and biomass in hot desert ecosystems. Bryologist 80:470–479
Naveh Z (1967) Mediterranean ecosystems and vegetation types in California and Israel. Ecology 48:445–459
Neales TF, Patterson AA, Hartney VJ (1968) Physiological adaptation to drought in the carbon assimilation and water loss of xerophytes. Nature (London) 219:469–472
Nobel PS (1977) Water relations and photosynthesis of a barrel cactus, *Ferrocactus acanthodes*, in the Colorado Desert. Oecologia 27:117–133
Nobel PS (1978) Microhabitat, water relations, and photosynthesis of a desert fern, *Notholaena parryi*. Oecologia 31:293–309
Nobel PS, Longstreth DL, Hartsock TL (1978) Effect of water stress on the temperature optima of net CO_2 exchange for two desert species. Physiol Plant 44:97–101
Odening W, Strain BR, Oechel WC (1974) The effects of decreasing water potential on net CO_2 exchange of intact desert shrubs. Ecology 55:1086–1095

Oechel W (1980a) Carbon allocation and utilization. In: Miller P (ed) Resource use by chaparral and matorral. Springer, Berlin Heidelberg New York
Oechel W (1980b) Energy and carbon acquisition. In: Miller P (ed) Resource use by chaparral and matorral. Ecol Stud Vol 39. Springer, Berlin Heidelberg New York
Oechel WC, Strain BR, Odening WR (1972) Tissue water potential, photosynthesis, ^{14}C-labeled photosynthate utilization, and growth in the desert shrub *Larrea divaricata*. Cav Ecol Monogr 42:127–141
Orshan G (1964) Seasonal dimorphism of desert and mediterranean chamaephytes and its significance as a factor in their water economy. In: Rutter AJ, Whitehead FH (eds) The water relations of plants. Blackwell, Oxford
Orshan G, Diskin S (1968) Seasonal changes in productivity under desert conditions. In: Eckardt E (ed) Functioning of terrestrial ecosystems at the primary production level. UNESCO, Paris
Orshan G, Zand G (1962) Seasonal body reduction of certain desert half shrubs. Bull Res Counc Isr Sect D 110:35–42
Patten DT (1978) Productivity and production efficiency of an upper Sonoran Desert ephemeral community. Am J Bot 65:891–895
Pearcy RW (1977) Acclimation of photosynthetic and respiratory CO_2 exchange rates in coastal and desert races of *Atriplex lentiformis*. Oecologia 26:245–255
Pearcy RW, Harrison AT (1974) Comparative photosynthetic and respiratory gas exchange characteristics of *Atriplex lentiformis* (Torr.) Wats. in coastal and desert habitats. Ecology 55:1104–1111
Pearcy RW, Harrison AT, Mooney HA, Björkman O (1974) Seasonal changes in net photosynthesis of *Atriplex hymenelytra* shrubs growing in Death Valley, California. Oecologia 17:111–121
Pearcy RW, Berry JA, Fork DC (1977) Effects of growth temperature on the thermal stability of the photosynthetic apparatus of *Atriplex lentiformis* (Torr.) Wats. Plant Physiol 59:873–878
Poole DK, Miller PC (1975) Water relations of selected species of chaparral and coastal sage communities. Ecology 56:1118–1128
Rodin LE, Bazilevich NI (1968) Production and mineral cycling in terrestrial vegetation. Oliver and Boyd, London
Rodin LE, Bazilevich NI, Miroshnichenko YM (1972) Productivity and biogeochemistry of *Artemisia* in the mediterranean area. In: Ecophysiol Found Ecosyst Prod Arid Zone. Nauka, Moscow Leningrad
Rustanov IG (1972) Phytomass quantitative characteristics and productivity of subshrublet communities of the Krasnovodsk plateau. In: Ecophysiol Found Ecosyst Proc Arid Zone. Nauka, Moscow Leningrad
Sampson AW (1944) Plant succession on burned chaparral lands in northern California. Univ Calif Agric Exp Stn Bull 685:144
Schaffer WM, Gadgil MD (1975) Selection for optimal life histories in plants. In: Cody ML, Diamond JM (eds) Ecology and evolution of communities. Harvard Univ Press, Cambridge
Schlesinger WH, Gill DS (1978) Demographic studies of the chaparral shrub, *Ceanothus megacarpus* in the Santa Ynez Mountains, California. Ecology 59:1256–1263
Schlesinger WH, Gill DS (1980) Biomass, production, and changes in the availability of light, water and nutrients during the development of pure stands of the chaparral shrub, *Ceanothus megacarpus*, after fire. Ecology 61:781–789
Schulze ED, Schulze I (1976) Distribution and control of photosynthetic pathways in plants growing in the Namib Desert, with special regard to *Welwitschia mirabilis* Hook F. l. Madoqua 9:5–13
Schulze ED, Lange OL, Koch W (1972) Eco-physiological investigations on wild and cultivated plants in the Negev Desert. III. Daily courses of net photosynthesis and transpiration at the end of the dry period. Oecologia 9:317–340
Sestark Z, Catsky J, Jarvis PG (1971) Plant photosynthetic production. Manual of methods. Junk, The Hague
Shaver GR (1978) Leaf cycle and light absorptance of *Arctostaphylos* species (Ericaceae) along environmental gradients. Madroño 25:133–138

Shreve F (1936) The transition from desert to chaparral in Baja California. Madroño 3:257–264
Shreve F, Wiggins IL (1964) Vegetation and flora of the Sonoran Desert. Stanford Univ Press, Stanford
Specht RL (1969a) A comparison of the sclerophyllous vegetation characteristics of mediterranean type climates in France, California and southern Australia. I. Structure, morphology, and succession. Aust J Bot 17:277–292
Specht RL (1969b) A comparison of the sclerophyllous vegetation characteristics of mediterranean type climates in France, California and southern Australia. II. Dry matter, energy, and nutrient accumulation. Aust J Bot 17:293–308
Specht RL (1981) Primary production in mediterranean-climate ecosystems regenerating after fire. In: Castri di F, Goodall DW, Specht RL (eds) Ecosystems of the world, vol II. Mediterranean type shrublands. Elsevier, Amsterdam
Specht R, Rayson P, Jackman M (1958) Dark Island Heath (ninety-mile plain, South Australia), VI. Pyric succession: changes in composition, coverage, dry weight, and mineral nutrient status. Aust J Bot 6:59–88
Strain BR (1969) Seasonal adaptations in photosynthesis and respiration in four desert shrubs growing in situ. Ecology 50:511–513
Strain BR (1970) Field measures of tissue water potential and carbon dioxide exchange in the desert shrubs *Prosopis julifera* and *Larrea divaricata*. Photosynthetica 4:118–122
Strain BR, Chase VC (1966) Effect of past and prevailing temperatures on the carbon dioxide exchange capacities of some woody desert perennials. Ecology 47:1043–1045
Susmel L, Viola F, Bassata G (1976) Ecologia della lecceta del supramonte di orgosolo. III. Contributo: Produzione primaria, produzione secondaria (erbivori), Condizioni attuali e possibilita di conservazione. CEDAM Padova
Syvertsen JP, Nickell GL, Spellenberg RW, Cunningham GL (1976) Carbon reduction pathways and standing crop in three Chihuahuan Desert plant communities. Southwest Nat 21:311–320
Szarek SR (1979) Primary production in four North American deserts: indices of efficiency. J Arid Environ 2:187–209
Szarek SR, Ting IP (1975) Physiological responses to rainfall in *Opuntia basilaris* (Cactaceae). Am J Bot 62:602–609
Szarek SR, Johnson HB, Ting IP (1973) Drought adaptation in *Opuntia basilaris*. Plant Physiol 52:539–541
Teeri JA, Stowe LG (1976) Climatic patterns and the distribution of C_4 grasses in North America. Oecologia 23:1–12
Teeri JA, Stowe LG, Murawski DA (1978) The climatology of two succulent plant families, Cactaceae and Crassulaceae. Can J Bot 56:1750–1758
Tevis L Jr (1958) A population of desert ephemerals germinated by less than an inch of rain. Ecology 39:681–687
Wainwright CM (1977) Sun-tracking and related leaf movements in a desert lupine (*Lupinus arizonicus*). Am J Bot 64:1032–1041
Wallace A, Bamberg SA, Cha JW (1974) Quantitative studies of roots of perennial plants in the Mojave Desert. Ecology 55:1160–1162
Walter H, Stadelmann E (1974) A new approach to the water relations of desert plants. In: Brown GW (ed) Desert biology, Vol 2. Academic Press, London New York
Went FW (1949) Ecology of desert plants. Part II. The effect of rain and temperature on germination and growth. Ecology 30:1–13
Whittaker RH (1975) Communities and Ecosystems. MacMillan, New York
Whittaker RH, Niering WA (1975) Vegetation of the Santa Catalina Mountains, Arizona. V. Biomass, production and diversity along the elevation gradient. Ecology 56:771–790
Winter K, Troughton JH (1978) Photosynthetic pathways in plants of coastal and inland habitats of Israel and the Sinai. Flora 167:1–34
Woodell SRJ, Mooney HA, Hill AJ (1969) The behaviour of *Larrea divaricata* (creosote bush) in response to rainfall in California. J Ecol 57:37–44
Zohary M (1961) On hydro-ecological relations of Near East desert vegetation. In: Plant-water relationships in arid and semiarid conditions. UNESCO, Paris

8 Productivity of Temperate, Deciduous and Evergreen Forests

P.G. JARVIS and J.W. LEVERENZ

CONTENTS

8.1 Introduction: Temperate Forest 234
8.1.1 Scope 234
8.1.2 Production of Forest Trees 235

8.2 A Conceptual Growth Model 237
8.2.1 Growth 237
8.2.2 Light Interception 238
8.2.3 Photosynthetic Efficiency 239
8.2.4 Losses by Respiration 240
8.2.5 Losses of Plant Parts 241
8.2.6 Potential Production 242

8.3 Radiation Interception 243
8.3.1 Direct Measurement 244
8.3.2 Canopy Properties: the Status Quo 245
8.3.3 Estimation of Canopy Interception 246
8.3.3.1 The Model Approach 247
8.3.3.2 The Empirical Approach 248
8.3.3.3 Conclusions 250

8.4 Leaf Area Index 250
8.4.1 Amount of Leaf 250
8.4.2 Vertical Distribution of Foliage 251
8.4.3 Grouping or Clumping of Foliage 252
8.4.4 Seasonal Variation 252

8.5 Photosynthetic Efficiency 255
8.5.1 Dependence on Structural Properties 256
8.5.2 Dependence on Leaf Photosynthetic Properties 258
8.5.3 Dependence on Environment 260

8.6 Respiratory Losses 262
8.6.1 Respiration and Temperature 262
8.6.2 Maintenance Respiration 263
8.6.3 Growth Respiration 265
8.6.4 Total Respiration 266

8.7 Mortality Losses 266
8.7.1 Litter Fall 267
8.7.2 Tree Mortality 267
8.7.3 Losses from Insect Attack 268
8.7.4 Fine Root Turnover 268

8.8 Conclusions 269

References 272

8.1 Introduction: Temperate Forest

8.1.1 Scope

For the purposes of this discussion, temperate forest is regarded as occurring in broad latitudinal bands between the taiga towards the poles and the mediterranean flora towards the equator. Within these bands temperate forest occurs at moderate elevations below the altitudinal extremes of climate and associated alpine vegetation and away from the continental extremes of temperature and dryness. Temperatures are moderate, usually not falling below about −15 °C in winter, and the soil freezes for less than 3 months. In summer temperatures rarely exceed 35 °C, precipitation exceeds ca. 400 mm and is fairly uniformly distributed through the year so that a dry season lasts for less than ca. 3 months. A more exact definition can be found in Walter (1970).

Temperate forest thus covers a wide climatic range and is often subdivided into cool-temperate and warm-temperate categories. Cool-temperate forest is typified by the evergreen coniferous forest of the northern hemisphere grading into the taiga to the north and into mixed deciduous forest to the south. Warm-temperate forest in the northern hemisphere is typified by the mixed coniferous-deciduous forest of southeastern North America. Warm-temperate forest in the southern hemisphere consists largely of evergreen Angiosperms, particularly *Eucalyptus* spp. and *Nothofagus* spp. and grades into dry sclerophyll forest in the drier areas and into rainforest in the more humid areas.

The seasonal periodicity in climates in middle latitudes in which temperate forest occurs is largely determined by the seasonal variation in radiation. This not only affects temperature, day-length, humidity and evaporation, but also has a direct influence on productivity. From an analysis of the production processes in temperate agricultural crops, Monteith (1977, 1981) arrived at the conclusion that "the amount of light *intercepted* by a crop is a major discriminant (of production)". We shall follow this line of argument in analysing the productivity of temperate forest. The seasonal changes in the incident solar radiation received by temperate forest are shown in Fig. 8.1, using the definition of the temperate forest region given above. The solar radiation income of the majority of temperate forests lies within the shaded area and fluctuates seasonally by a factor of three.

The annual income of solar radiation for temperate forests ranges from about 2,500 to 6,000 MJ m^{-2}. The low limit is depressed by the considerable cloudiness in some maritime temperate regions such as the British Isles. A representative figure for temperate regions is probably about 4,600 MJ m^{-2} yr^{-1} solar irradiance or 2,300 MJ m^{-2} yr^{-1} PAR (photosynthetically active radiation, see Chap. 2, Vol. 12A).

The productivity of a forest ecosystem depends primarily on the productivity of the overstorey species. In this analysis we shall limit ourselves to the environmental and physiological variables that determine the productivity of the major tree species, since we cannot take into account the whole field of ecosystem dynamics. We shall refer to information pertaining to both unmanaged and managed forests, including plantations.

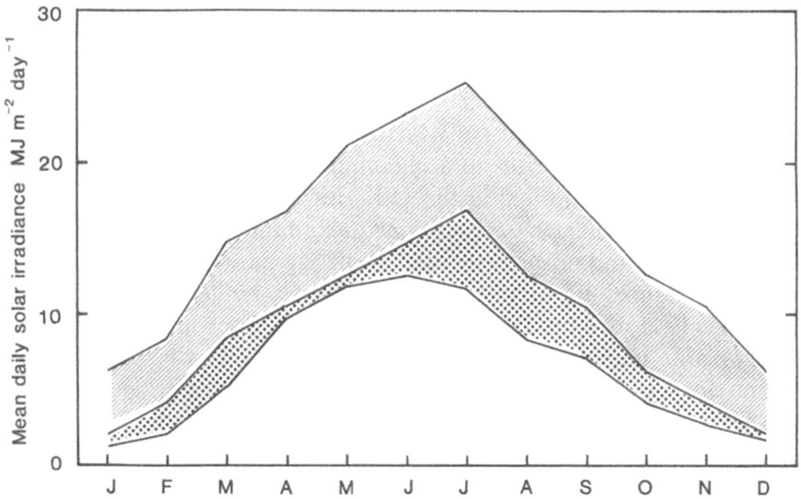

Fig. 8.1. The monthly variation in the mean daily solar irradiance (Q_0) in temperate regions of the northern hemisphere. The figures have been taken from the maps of isopleths of solar radiation from sun and sky on a horizontal surface presented by BLACK (1956). *Shaded area* shows the range of variation in Q_0 over the latitudinal range of temperate climates. *Stippled area* indicates downward depression of the lower limit in cloudy maritime-temperate areas, such as the British Isles

8.1.2 Production of Forest Trees

Measured growth rates of temperate forest range from <1 to ca. 40 Mg ha^{-1} yr^{-1} (see for example OVINGTON 1962; WHITTAKER 1966; KIRA and SHIDEI 1967; SATOO 1970; GORDON 1975; KIRA 1975, 1977; WESTMAN and WHITTAKER 1975; FUJIMORI et al. 1976; FUJIMORI 1977; MADGWICK et al. 1977, 1981; GRIER and LOGAN 1977; WARING and FRANKLIN 1979; VYSKOT 1979; JARVIS 1981; FORD 1982). Some of these data are derived from special studies designed to investigate forest productivity, but such data can be extensively supplemented from the yield tables widely used by foresters. Yield tables give the rate of volume production of harvestable product. Assuming appropriate values for the basic density (i.e. dry weight/fresh volume) and for the harvest index (HI; yield of stem/total yield), estimates of growth rate can be derived. Basic densities do, of course, depend on species, and are about 400 (300–500) kg m^{-3} for conifer stem wood and 600 (300–800) kg m^{-3} for hardwoods; HI is poorly known because few studies have included the measurement of root production, but is probably about 0.65.

A number of variables enter into the range of values cited above, not least being the period of the life cycle during which the growth rate was measured. In uneven aged stands at equilibrium the living, standing biomass may be constant with time and the growth rate zero (GRIER and LOGAN 1977). At the other extreme, in plantations the biomass is steadily increasing but at a variable rate. Consequently the growth rate depends not only on when during the growth cycle it is measured, but also on the period of time over which it is averaged.

For a particular stand, growth rates averaged over the entire life of the stand (mean annual growth rates, $\bar{G}$) are smaller than growth rates measured over short periods during the life of the stand (current annual growth rates, G). $\bar{G}$ is usually calculated as the accumulated dry matter (W) over the number of years (n) to the time (m) at which $\Sigma W/\Sigma n$ is maximum (i.e. $\bar{G} = \Sigma^m W/\Sigma^m n$). During this period G rises to a maximum $1\frac{1}{2}$ to 2 times $\bar{G}$ and then falls again. In plantations of conifers in the U.K. for example, G reaches a peak between 20 to 40 years, whereas $\bar{G}$ reaches its maximum between 50 to 80 years.

A current annual growth rate of 40 Mg ha^{-1} yr^{-1} ranks with the highest recorded figures for dry matter production by temperate agricultural crops (COOPER 1975; JARVIS 1981). However such figures must be regarded as record figures and unrealistic in general practice (Table 8.1). The national average values for G for production forests of the six most widely planted species in the U.K. lie in the range 6 to 12 Mg ha^{-1} yr^{-1} and the corresponding values for $\bar{G}$ are about $\frac{2}{3}$ of these (JARVIS 1981). These are still good figures in relation to national average values for agricultural crops, which are comparable in size although the crops are grown on better soils and with more intensive management (GORDON 1975; JARVIS 1981). Values of G for colder, more continental cool-temperate forests are about $\frac{1}{4}$ the above (i.e. <3 Mg ha^{-1} yr^{-1}), whereas

Table 8.1. Maximal rates of dry matter production of some temperate forest stands. The rates have been measured over short periods of a few years and are current or periodic dry matter increments for the whole tree including the roots. Where information on the roots was not provided, 20% has been added. Leaves and fine roots that die and are shed during the year are not included. Data from single rows, very small plots and mixed variety trials have been excluded

Species	Rate of production (Mg ha^{-1} yr^{-1})	Source
a) Coniferous		
Cryptomeria japonica	53	KIRA (1975)
Pinus radiata	34	MADGWICK (1981)
Tsuga heterophylla	30	FUJIMORI (1971)
Abies sachalinensis	29	KIRA (1975)
Pseudotsuga menziesii	28	KESTEMONT (1977)
Pinus nigra var. *nigra*	25	MINDERMAN (1967)
Picea sitchensis	24	FORD (1982)
Sequoia sempervirens	23	WESTMAN and WHITTAKER (1975)
Abies grandis	22	OVINGTON (1956)
Picea abies	22	KESTEMONT et al. (1977)
Thuja plicata	20	HANLEY (1976)
b) Broad-leaved		
Eucalyptus grandis	41	SCHÖNAU and PENNEFATHER (1975)
Eucalyptus globulus	40	FABIÃO (personal communication)
Salix purpurea	36	HATHAWAY (1980)
Salix matsudana × alba	34	HATHAWAY (1980)
Populus balsamifera × tristis	33	CRIST and DAWSON (1975)
Populus deltoides × nigra	23	ANDERSON and ZSUFFA (1975)
Eucalyptus nitens	22	MADGWICK et al. (1981)

values for warm-temperate forests may be more than twice these figures (i.e. 12–24 Mg ha^{-1} yr^{-1}) (e.g. KIRA 1975; ASHTON 1976a; MADGWICK et al. 1981).

It seems probable that $\bar{G}$ reaches only $\frac{2}{3}$ of G because the trees fail to exploit the site fully throughout the growth cycle. In oak forests, for example, the leaf area index increases steadily to a maximum of ca. 5.5 at 30 years and then starts to decline towards 3.5 at 200 years (RAUNER 1976). During the period that the leaf area index is less than maximal, the opportunity for interception of radiation is being lost in part. Consequently, during the early stages of growth before the canopy is closed, and again later as the leaf area index declines, the opportunity for growth is also less than maximal, leading to the situation that $\bar{G}$ is less than G at 30 years. The question of lost opportunity will be discussed in more detail later.

There are large differences in growth rate amongst species growing in the same environment (e.g. SATOO 1970) and within the same species on different sites (e.g. ALBREKTSON 1980; KEYES and GRIER 1981). Clearly also, there are a large number of variables, both environmental and physiological, which might influence the rate of growth in any particular situation (see for example the review by ŠESTÁK et al. 1971). It would be tedious to list all the variables known to have an influence on the growth physiology of trees. Instead we propose to follow through a simple model of growth in an attempt to pin-point some of the more important variables to concentrate on in the future. The model essentially makes a distinction between processes concerned with the interception of light energy, the efficiency with which it is used and the associated losses, both those inevitable and those subject to considerable variation. A useful start along these lines was made by KIRA and SHIDEI (1967).

8.2 A Conceptual Growth Model

8.2.1 Growth

The conservation equation for the mass of dry matter in the trees in an area of forest requires that over an interval of time (0→t) in which production processes are occurring

$$\int_0^t (A_n - R)\,dt - \Delta M = \Delta W, \tag{8.1}$$

where A_n is the net rate of assimilation by the foliage, R is the rate of respiration by the trees (excluding the foliage), ΔM is the change in tree biomass resulting from death and loss of parts through insect predation, root death, leaf and branch fall etc, and ΔW is the net change in tree biomass (kg ha^{-1}). The total dry matter production per unit time, the stand growth rate, is:

$$G = \int_0^t \frac{dW}{dt} = \frac{\Delta W}{t}. \tag{8.2}$$

For agricultural crops Eq. (8.1) is either integrated over the whole life of the crop from planting to harvest, or applied to a succession of short periods of about 10 days during the growth and development of the crop. For assessing the production efficiency of different kinds of forest, an appropriate time interval would seem to be either the life of the stand, if it is determinate as in the case of a plantation, or the complete cycle of seasons encompassed within one year.

Without going into the detail of complex growth models (e.g. LOOMIS et al. 1979, see also Chap. 4, this Vol.), Eq. (8.1) can be amplified to give insight into the factors controlling production efficiency of stands and to emphasise the areas of adequate knowledge and the areas deficient in knowledge with respect to temperate forest.

8.2.2 Light Interception

If photosynthesis by the canopy as a whole is light-limited, the assimilation rate, A, would be expected to be proportional to the absorbed flux density of photosynthetically active radiation (PAR), i.e.

$$A = \varepsilon Q_a,$$

where ε is the efficiency with which absorbed radiation is used in fixing carbon dioxide. In practice, the radiation intercepted by the canopy (Q_i) is more frequently measured, so we may write

$$A = \varepsilon Q_a = a \varepsilon Q_i, \qquad (8.3)$$

where $a \approx (1 - \alpha)$ and α is the reflectance of PAR (see GALLAGHER and BISCOE 1978). Q_i is measured as the difference between the accumulated radiation flux density measured on a horizontal surface above the canopy and that measured beneath the canopy.

In recent years, it has been shown in practice that the growth rate (G) of several agricultural crops increases *linearly* with Q_i (either as PAR or total solar radiation) (MONTEITH 1972, 1977; GALLAGHER and BISCOE 1978; LEGG et al. 1979). The slope of this relationship in temperate climates is remarkably consistent over a wide range of conditions and genotypes. For example, the average value of the slope was 3.0 g per MJ of intercepted PAR for a large number of wheat and barley cultivars ranging from the UK to Mexico to Australia (FISCHER 1983). [On an intercepted solar radiation basis the value would be approximately 1.5 g MJ^{-1} because in temperate climates PAR is on average close to 50% of the solar irradiance (SZEICZ 1974)].

Environmental stresses, such as mild drought, suboptimal temperatures or nitrogen availability, that have a primary influence on the production of leaf area influence crop growth rate through their effects on the interception of radiation.

Other variables, such as low temperature or severe drought during the

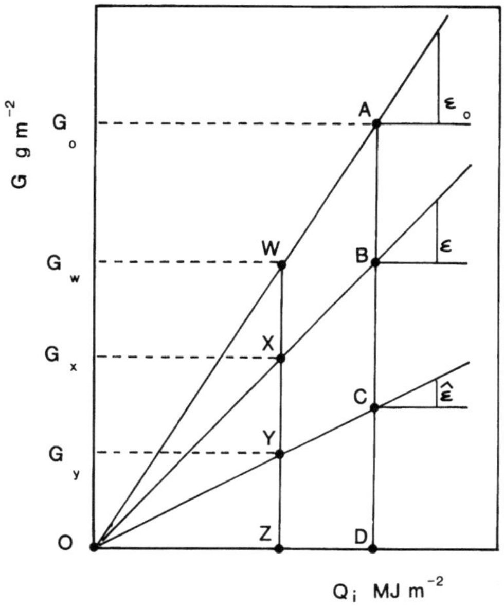

Fig. 8.2. A diagram to show the relationship between stand growth rate, G, and intercepted photosynthetically active radiation, Q_i. The line OA depicts G as a result of interception by the canopy operating at maximum efficiency and with no losses (i.e. with slope ε_0) and leading to the potential growth rate G_0. If canopy development and hence light interception are restricted in any way G_0 declines along the line AW to G_w. If photosynthetic efficiency is less than maximum because of the particular structural, physiological and biochemical properties of the canopy and leaves, or because of environmental stress, G_w falls along the line WX to G_x, leading to the line OB of slope ε. Because of losses through respiration, turnover of leaves and roots and mortality, G_x declines further along the line WZ leading to the actual response line OYC with a slope of the effective efficiency of utilisation of intercepted PAR, $\hat{\varepsilon}$. Thus OWA is the upper boundary line for G which will fall within the shaded area AWOZD at a level depending on Q_i and the extent to which $\varepsilon < \varepsilon_0$

growing season, reduce ε so that observations of G fall in the space below the linear relationship of maximum slope (Emmingham and Waring 1977; Legg et al. 1979). Thus the growth rate, G, varies up and down the straight line A0 in Fig. 8.2 as a result of variation in the amount of incident solar radiation and the intercepting properties of the foliage, particularly the area of leaf present, and up and down vertically in the direction WZ below the limiting line as ε varies with genotype, environmental history and stress.

8.2.3 Photosynthetic Efficiency

What determines the maximum slope of the relationship, dG/dQ_i? In the absence of any respiratory or other losses, this will be given by the maximum value of $\varepsilon, (\varepsilon_0)$. Let us assume as a practical value a minimum quantum requirement of photosynthesis of 20 mol quanta required per mol net CO_2 fixed in

C_3 photosynthesis (EHLERINGER and BJÖRKMAN 1977, see Chap. 3, Vol. 12A), an equivalence in temperate climates of ca. 4.2 mol quanta per MJ of PAR (personal measurement) and an equivalence of 1 mol CO_2 to 1 mol standard drymatter, i.e. a conversion factor for CO_2 to drymatter of $(\frac{30}{44} \times 0.72)$ (PENNING DE VRIES et al. 1974, 1975b, 1979). [LEDIG et al. (1976) quote an empirically determined value of this conversion factor for *seedlings* of *Pinus rigida, Pinus densiflora, Cryptomeria japonica* and *Populus tremuloides* of 0.55.] Then we would expect a dry matter assimilation of 4.5 g MJ^{-1}. If, alternatively, we assume a lower quantum requirement of 12, then the limit would be 7.6 g MJ^{-1}. These figures represent two estimates of the theoretical upper limit to the relationship between G and $Q_a(\varepsilon_0)$, and provide an appropriate standard against which to assess the growth rates realised in practice.

If all the leaves in a canopy are evenly illuminated below quantum saturation, ε should approach ε_0. In practice canopies of different properties will variously depart from this ideal, so we define

$$\varepsilon = f\varepsilon_0 . \tag{8.4}$$

ε is estimated from the linear relationship between A and Q_a at low light as

$$\varepsilon = dA/dQ_a \approx (A_n + R_d)/Q_a \tag{8.5}$$

where R_d is the dark respiration of the leaves and A_n is the net influx of CO_2. Then from Eqs. (8.3), (8.4) and (8.5)

$$\int_0^t A_n dt = \int_0^t (Q_i a f \varepsilon_0 - R_d)\, dt. \tag{8.6}$$

Combining Eq. (8.6) with Eqs. (8.1) and (8.2) and taking one year as the time interval

$$G = \Delta W = \int_0^1 (Q_i a f \varepsilon_0 - R_d - R)\, dt - \Delta M. \tag{8.7}$$

8.2.4 Losses by Respiration

The integrated product $\int_0^t (Q_i a f \varepsilon_0)\, dt$ gives the initial amount of assimilate which is subsequently reduced by respiration. In studies of the production of annual crops it has become usual to consider respiration in two parts, one associated with the maintenance of the plant structure, the maintenance respiration, and the other associated with the synthesis of new tissues, the growth respiration (MCCREE 1970, 1974; PENNING DE VRIES 1975a, b).

The proportion of the assimilate not used in maintenance respiration is

$$Y_m = \left(1 - \frac{R_m}{A}\right) \tag{8.8}$$

where R_m is the rate of maintenance respiration of a tree or part of a tree. That part of the assimilate remaining after reduction by Y_m is available (A_v) for growth and some is synthesised into new structures and some is utilised in the enhanced respiration associated with synthesis. Thus the rate of growth of new tissues is

$$Y_g \cdot A_v = Y_m \cdot Y_g \cdot A, \tag{8.9}$$

where $A_v = A - R_m$ and Y_g is the efficiency of conversion of the assimilate available for growth and is given by:

$$Y_g = \left(1 - \frac{R_g}{A_v}\right) = \left(1 - \frac{R_g}{Y_m A}\right). \tag{8.10}$$

Synthesis of new tissue does, of course, often occur at times when there is no net increase in assimilate and the mass and carbon balance of the plant is negative. This is particularly the case in trees at the time of bud break and leaf flushing. To take this into account over periods of days or weeks, a more complex model, such as that of THORNLEY (1977), that distinguishes pools of reserve carbohydrates and degradable and structural materials, is necessary. For the present purpose it is assumed that over a sufficiently long period of, say, a year the change in stored reserve carbohydrates is zero and hence that synthesis of new tissue can be related to the amount of assimilation. The overall efficiency with which assimilate is used in the production of new plant material is then

$$Y = Y_m \cdot Y_g. \tag{8.11}$$

8.2.5 Losses of Plant Parts

Regular annual losses of dry matter are most evident in the seasonal leaf fall of deciduous trees. Other regular losses which make less of a visual impact are the more uniformly distributed needle fall from many conifers (BEADLE et al. 1982) and the sometimes massive fall of leaves and branches from trees such as *Quercus* spp. and *Eucalyptus* spp. (ATTIWELL 1979; CHRISTENSEN 1977). Still less evident are the enormous seasonal losses resulting from the turnover of the fine root population (DEANS 1979; PERSSON 1980; GRIER et al. 1981; KEYES and GRIER 1981). Substantial less regular losses also occur from the chronic grazing of particular predators and very heavy irregular losses may occur from the acute attack of predators leading to massive defoliation of both deciduous (e.g. VARLEY 1970) and coniferous trees (e.g. PARRY 1969).

These losses result in a proportion of the assimilate being used to replace tissues and replenish reserves rather than being used in growth. In extreme circumstances of, say, repeated defoliation by an insect predator, coupled with a number of cycles of fine root turnover as a result of drought, and a lack of available nitrogen, ΔM may exceed A with the result that G is negative and a proper growth ring is not formed (MCLAUGHLIN and SHRINER 1980). For convenience, therefore, we can express the losses as a fraction of the assimilation and define a coefficient of loss as follows

$$d = \left(1 - \frac{\Delta M}{A}\right). \tag{8.12}$$

Substituting Eqs. (8.11) and (8.12) into Eq. (8.7) gives

$$G = \Delta W = \int_0^1 (Q_i\, a\, f\, \varepsilon_0\, Y_m\, Y_g\, d)\, dt, \tag{8.13}$$

where the differences have now been replaced by a series of factors for ease of handling. Over a year the majority of the factors can be regarded as constant and removed from the integral and replaced with average values so that

$$G = \bar{a}\, \bar{f}\, \varepsilon_0\, \bar{Y}_m\, \bar{Y}_g\, \bar{d} \int_0^1 Q_i\, dt \tag{8.14a}$$

$$= \hat{\varepsilon} \int_0^1 Q_i\, dt, \tag{8.14b}$$

where $\hat{\varepsilon}$ is an *effective* efficiency of utilisation of intercepted PAR.

8.2.6 Potential Production

The potential production of a stand of trees depends on the incident PAR and could be defined as

$$G_0 = a\, \varepsilon_0 \int_0^1 Q_0\, dt. \tag{8.15a}$$

However, it seems unreasonable to ignore completely the inevitable thermodynamic requirements of growth and maintenance so that a more realistic definition would assume their minimum values, $Y_{g,0}$ and $Y_{m,0}$ respectively, i.e.

$$G_0 = a\, \varepsilon_0\, Y_0 \int_0^1 Q_0\, dt. \tag{8.15b}$$

Thus the ratio of actual to potential production is

$$\frac{G}{G_0} = \frac{a\, f\, \varepsilon_0\, Y_m\, Y_g\, d \int_0^1 Q_i\, dt}{a\, \varepsilon_0\, Y_{m,0}\, Y_{g,0} \int_0^1 Q_0\, dt}. \tag{8.16a}$$

Since our best estimate of Y_g is derived theoretically anyway, $Y_g = Y_{g,0}$, and making other cancellations gives

$$\frac{G}{G_0} = f \cdot d \frac{Y_m}{Y_{m,0}} \frac{\int_0^1 Q_i \, dt}{\int_0^1 Q_0 \, dt}. \tag{8.16b}$$

This may also be written as

$$\frac{G}{G_0} = d \frac{A}{A_0} \frac{Y_m}{Y_{m,0}} \bar{i}, \tag{8.16c}$$

where $\bar{i}$ is the mean annual interceptance by the canopy and A_0 is the potential rate of assimilation.

Taking a as 0.8, Y_0 as 0.6, ε_0 as 4.54 g MJ^{-1} PAR, Eq. (8.15b) leads to estimates of G_0 for temperate regions ranging from 27 to 65 Mg ha^{-1} yr^{-1} depending on Q_0. The estimates by MONTEITH (1977) of 54 and DE WIT (1965) of 36 Mg ha^{-1} yr^{-1} for a 12-month or 6-month growing season in northern Europe fall within this range as does the estimate of 54 Mg ha^{-1} yr^{-1} for warm-temperate evergreen forest arrived at by KIRA and YABUKI (1978). The maximal measured rates of dry matter production of forest stands in temperate climates fall within the range 20 to 53 Mg ha^{-1} yr^{-1} (Table 8.1) and thus lie within this range of potential production.

For temperate agricultural crops the maximum value of dG/dQ_a is ca. 3.6 g MJ^{-1} over the life of a sugar beet crop (BISCOE personal communication). The difference between this figure and the figure of 4.5 g MJ^{-1} (or possibly 7.6 g MJ^{-1}) represents the fall in efficiency resulting from reduced ε, respiration, and other losses. Since the energy equivalent of photosynthesis is ca. 17 kJ g^{-1} of dry matter, a figure of 3.6 g MJ^{-1} corresponds to an efficiency of PAR conversion of ca. 6%. This is as high a conversion efficiency as is likely to be attained by forest, and in the absence of any published relationships between G and Q_i for temperate forest may be assumed to provide an upper boundary line against which to assess the production efficiency of different kinds of forest. Having made generous allowance for a and Y_0 in the above calculation, it would seem that the main limitation to the regular achievement of high productivities lies in the values of i, ε and d found in practice and we shall now consider what is known about these parameters for temperate forest.

8.3 Radiation Interception

The radiation intercepted by a canopy depends on the radiation above the canopy, the structure of the canopy and the spectral properties of the foliage and the surface beneath the canopy. Above the canopy the total irradiance, the proportions of direct beam and of diffuse radiation, the solar elevation and the distribution of diffuse radiation over the hemisphere all influence interception by the vegetation. Consequently, the interceptance by a canopy does not have a unique value but varies appreciably (ANDERSON 1964b; for a general discussion of radiation relations see Chap. 1, Vol. 12A).

8.3.1 Direct Measurement

Radiation interception by forest canopies has been measured many times to define the radiation environment for the ground flora (see ANDERSON 1964a for a review of the earlier work). However, such measurements have infrequently been accompanied by measurements of the area of leaf present in the overstorey because of the obvious practical difficulties in making such measurements. Mea-

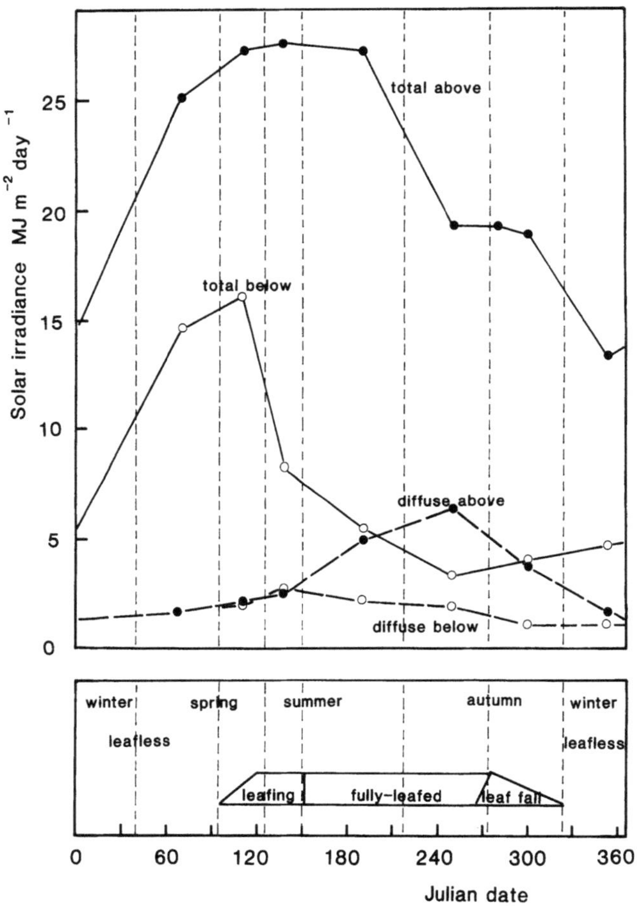

Fig. 8.3. A diagram to show the effect of the phenoseasons on radiation in a stand of tulip poplar, *Liriodendron tulipifera*, in east Tennessee, U.S.A. (HUTCHISON and MATT 1977)

surements of the canopy structure are even rarer and measurements of stand or tree growth in relation to radiation interception non-existent.

In forest particularly there are considerable sampling problems in measuring the radiation below the canopy. Depending on the spatial arrangement of the trees, a very long transect of measurements every few mm of 10–50 m or a large number of spot measurements may be necessary to arrive at an acceptable estimate of instantaneous $\bar{Q}_i$ (e.g. NORMAN and JARVIS 1974; HUTCHISON and MATT 1977). This becomes a particularly difficult problem when the trees are well spaced out, say <500 trees ha^{-1}. There are also large variations from day to day in the radiation transmitted at a point, in both the amount and in the proportions of direct beam and diffuse radiation (ANDERSON 1964b). Much of this variation results from changing sky conditions, but there is a seasonal component resulting from changes in leaf area index and canopy structure. This is particularly evident in a deciduous forest in spring and autumn (Fig. 8.3). ANDERSON (1964b) shows considerable variation from year to year in the timing and period of both leaf expansion and leaf fall in a mixed deciduous wood in southern England: the phenoseason varied over 3 years from about 190 to 240 days. In the deciduous temperate forests of central Europe and the central and eastern United States the phenoseason is about 140–190 days.

8.3.2 Canopy Properties: the Status Quo

The total amount of radiation intercepted by a canopy over a year depends on the incident irradiance, the area of green leaf present, the structure of the canopy and its phenology. Twigs, branches and stems also intercept a significant amount of radiation but this is even less well defined (NORMAN and JARVIS 1974; HUTCHISON and MATT 1977). The structural properties of a canopy which determine its interception of radiation are, in order of importance:

a) the leaf area index (L) (see Chap 1, Vol. 12A: also LAI),
b) the vertical distribution of the foliage,
c) the distribution of leaf inclination angles or, failing that, the average leaf inclination angle,
d) the leaf reflectance and transmittance for the visible wavelengths,
e) the grouping or clumpiness of the foliage (i.e. its three-dimensional distribution),
f) the distribution of leaf azimuthal angles, if there is a preferred azimuthal direction.

Such information is sufficient to define explicitly the effects of different canopy structures on the interception of radiation at a particular time (NORMAN 1978, 1980). However, even for well-studied agricultural crops detail of this kind is only very rarely available, and it is presently only available for one or two forest stands (e.g. NORMAN and JARVIS 1974, 1975).

There is an increasing number of observations on the leaf area index of forest stands and these are discussed in the next section. Arising from some of these studies some useful generalisations are beginning to emerge about the

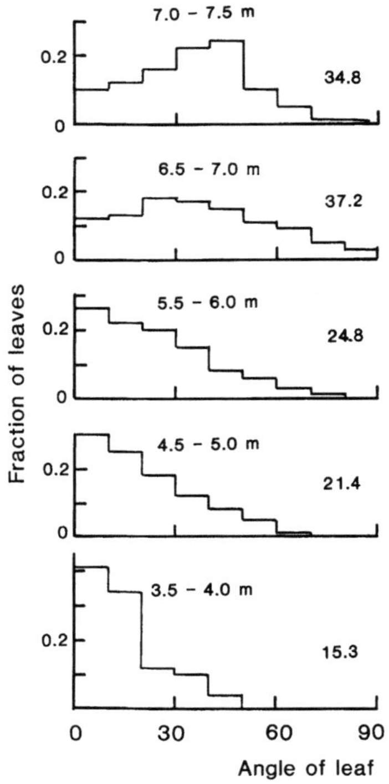

Fig. 8.4. The distribution of leaf inclination angles to the horizontal at different heights in a stand of coppiced sweet chestnut, *Castanea sativa*. The measurements were made with a protractor. The mean values are shown to the right side. (FORD and NEWBOULD 1971)

vertical distribution of leaf area density (e.g. BEADLE et al. 1982), but measurements of the three-dimensional distribution leading to an evaluation of the grouping of the foliage are rare (see NORMAN and JARVIS 1975 for one example). Similarly, the distribution of leaf angles has been rarely measured. An example for a stand of *Castanea sativa* taken from FORD and NEWBOULD (1971) is shown in Fig. 8.4. Whilst the reflectance and transmittance of tree leaves has often been measured (e.g. WOOLLEY 1971), such data have rarely been coordinated into interception models. In general, data on the leaf area index and possibly its vertical distribution are the most that can be hoped for at the present time for the purpose of defining canopy properties and radiation interception.

8.3.3 Estimation of Canopy Interception

In this situation we can either attempt to apply simplified models of radiation penetration or fall back upon the use of empirically determined extinction coefficients. In both cases significant assumptions are necessary: we must assume that the foliage is randomly distributed in space and that leaf inclination angles are spherically distributed. Measurements of radiation penetration have shown that even in canopies in which the foliage is not randomly distributed, such as in *Picea sitchensis* (NORMAN and JARVIS 1975) and sumac (NORMAN et al. 1971), the random assumption has not resulted in too great errors. However, the random assumption produces absurd results for dense crowns of widely spaced trees. In these conditions it is adequate to assume random and spherical distributions within the crowns and to define the crowns by ellipsoids that are appropriately distributed throughout the canopy space (THORPE et al. 1978; NORMAN and WELLES 1983).

8.3.3.1 The Model Approach

The transmittance of the direct beam in a canopy with a random and spherical distribution of the foliage is

$$\tau_b = \exp(-L \cos\alpha / \cos\theta) \quad (8.17)$$

where α is the mean leaf-sun angle and θ is the solar zenith angle. For a spherical angular distribution α is 60°, so that the extinction coefficient $\cos\alpha = 0.5$, but α may vary from 45° to 75° for a wide range of sun angles and different canopies. Whilst the exact definition of the distribution of leaf angles in a forest canopy seems an unlikely possibility, KUROIWA (1970) showed that different distributions of leaf angles had but little effect on total interceptance, although there were large effects on the distribution of light. CAMPBELL (personal communication) has shown that the use of a mean leaf angle is almost as satisfactory as the use of up to six leaf angle classes and is vastly superior to the spherical assumption for canopies in which the mean α is not close to 60°. Considerable hope is provided for the future by the recent use of integral inversion techniques to derive the mean leaf angle, and leaf area index, from measurements of the fraction of a transect that is in sunflecks (i.e. the direct beam transmittance, or gap fraction) (NORMAN et al. 1979). This technique has been thoroughly tested now in agricultural crops and is eminently suitable for use in forest where access below the canopy is simpler to obtain. Using a similar technique to analyse hemispherical photographs, ANDERSON (1981) measured the average inclination angle in a number of stands of *Eucalyptus* spp. to range from 41°–87° with the majority at $70 \pm 10°$: for oaks α was 20°.

Diffuse radiation in a canopy arises from the sky and from scattering of the beam by other foliage and the understore or soil. A simple adequate layer by layer treatment of the diffuse and scattered radiation in a canopy is given by NORMAN (1978). The transmittance of unintercepted diffuse radiation through a layer is obtained by integrating Eq. (8.17) over the hemisphere for an isotropic sky. Treatment of the scattered radiation originating from other leaves is, however, more difficult to handle and requires information about the upward and downward reflectance and transmittance of the leaves. Consequently, this treatment, although relatively simple, has only been applied to one or two forest canopies (e.g. NORMAN and JARVIS 1975).

In a careful but empirical simplification of his 1978 treatment, NORMAN (1981) approximated the diffuse transmittance as

$$\tau_d = \exp(-L^{0.7}) + S, \quad (8.18)$$

where S is the scattering component given by

$$S = 0.07 \frac{(1-\phi)}{\phi} (1.1 - 0.1\,L) \exp(-\cos\theta), \tag{8.19}$$

and where ϕ is the fraction of incident radiation which is diffuse and $(1-\phi)$ the direct beam fraction. The irradiance below the overstore canopy is then

$$Q_z = Q_0 \tau = Q_0 [(1-\phi)\,\tau_b + \phi\,\tau_d]. \tag{8.20}$$

Since the intercepted radiation is $Q_i = Q_0 - Q_z$, we can define the fraction of intercepted radiation as an interceptance

$$i = 1 - \tau = Q_i/Q_0 = 1 - [(1-\phi)\,\tau_b + \phi\,\tau_d]. \tag{8.21}$$

Making the assumption then of a spherical and random leaf distribution, these equations will give the total intercepted radiation within a few $W\,m^{-2}$ of more sophisticated treatments, utilising only the incident irradiance and its diffuse fraction, the solar zenith angle and the leaf area index (NORMAN 1981). In southern England, a maritime-temperate climate, ϕ averaged 65% (SZEICZ 1974); it is probably rather less on average in less cloudy temperate regions.

8.3.3.2 The Empirical Approach

When radiation profiles are measured in conjunction with profiles of leaf area density, it has become customary to express the transmittance of diffuse and beam radiation in terms of an empirical extinction coefficient (K):

$$\tau = \exp(-KL). \tag{8.22}$$

This equation is closely similar to Eq. (8.17) for the transmittance of direct beam radiation alone in a canopy with a spherical leaf angle distribution and, like Eq. (8.17), applies only when the foliage is randomly positioned (NORMAN 1975). It does not, of course, take into account explicitly the dependence of radiation penetration on solar elevation or the scattering resulting from the interactions between the beam and the foliage. The intercepted radiation is then given by:

$$Q_i = Q_0 \tau = Q_0 [1 - \exp(-KL)]. \tag{8.23}$$

Since this equation is slightly less demanding than even the simplified model presented in Eqs. (8.17) to (8.20), it has been generally used, despite its limitations.

The dependence of the interceptance (Q_i/Q_0) on (KL) is shown in Fig. 8.5. The leaf area index for "full" interception, say 95% in practice, is $\ln 0.05/(-K)$. Thus full interception occurs when the product $(KL) = 3$: for example if $K = 0.5$, full interception occurs at a leaf area index of 6. Some values of L, i, K and L_{95} for forest canopies are given in Table 8.2. Although there are only few

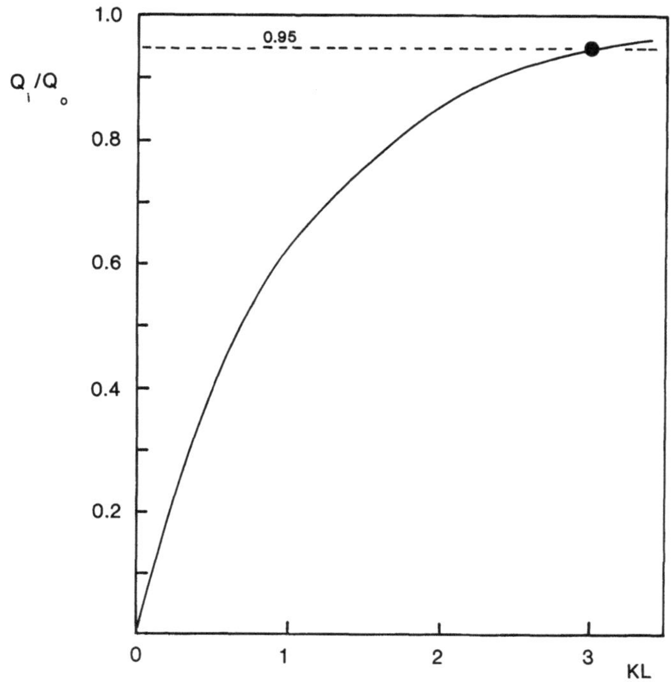

Fig. 8.5. The relationship between canopy interceptance Q_i/Q_0 ($=1-\tau$) and the product (KL) for a canopy with a random and spherical leaf angle distribution. *Dashed line* indicates 95% interception and intersects the curve at KL = 3

Table 8.2. Extinction coefficients and interceptances for solar radiation or PAR in several temperate forest canopies. K has been calculated for the maximum solar elevations given (i.e. it is the minimum value)

Species	L	K	L_{95}	i	Reference
a) Coniferous					
Pinus resinosa	2.6	0.40	7.5	0.65	Waggoner and Turner (1971)
Pinus resinosa/strobus	3.1	0.45	6.7	0.75	Mukammal (1971)
Pinus radiata	8.3	0.51	5.9	0.99	Whitehead (unpublished)
Pinus sylvestris	2.8	0.62	4.8	0.82	Whitehead and Jarvis (unpublished)
Picea sitchensis	9.8	0.53	5.7	0.99	Norman and Jarvis (1974)
Picea abies	8.4	0.28	10.7	0.90	Baumgartner (unpublished)
Pseudotsuga menziesii	11.0	0.48	6.2	0.99	Ungs (1981)
b) Broad-leaved					
Eucalyptus maculata	2.8	0.57	5.3	0.80	Dunin and Reyenga (unpublished)
Liriodendron tulipifera	6.0	0.29	10.3	0.83	Hutchison and Matt (1976)
Fagus crenata	7.0	0.65	4.6	0.99	Kira et al. (1969)
Castanopsis cuspidata	7.0	0.50	6.0	0.97	Kira et al. (1969)
Quercus robur	5.0	0.39	7.7	0.86	Rauner (1976)
Populus tremula	3.8	0.39	7.7	0.78	Rauner (1976)

data, it seems that for coniferous forest K is between about 0.4 and 0.65 and for broadleaf forest between 0.5 and 0.8.

8.3.3.3 Conclusions

When the intercepted radiation is not measured directly it can be calculated. Whether Eq. (8.21) or Eq. (8.23) is used, it is clear that the main determinants of Q_i are Q_0 and L, but that in addition, as a minimum, variation in θ and ϕ should be taken into account. Q_0 varies geographically within the temperate region and, of course, seasonally as discussed earlier in relation to Fig. 8.1 and θ and ϕ also vary geographically and seasonally: K and L also vary from stand to stand and they vary seasonally. Variation in L will be considered in the next section, here we consider an example of variation in K.

HUTCHISON and MATT (1976) made use of a model developed by Reifsnyder to calculate the diffuse and direct beam irradiances below the overstory canopy in a stand of tulip poplar (*Liriodendron tulipifera*) in east Tennessee in relation to time of day at four different times of year. Their calculations lead to estimates of K at solar noon of 0.17 and 0.29 when the canopy was fully leafed ($L=6.0$) in midsummer ($\theta=76°$) and autumn ($\theta=62°$), respectively. Their values of K are low partly because of high solar elevations, but also because of low diffuse fractions ($\phi<19\%$) (Table 8.3). As would be expected, the transmittance, particularly of the diffuse radiation, was much higher during the leafless phase than when the canopy was fully leafed. In general, however, the information available on canopy structure and light penetration is quite inadequate to show whether K varies amongst temperate forest with season as a result of changes in canopy structure as well as changes in solar elevation.

8.4 Leaf Area Index

8.4.1 Amount of Leaf

The leaf area index (projected area basis) of temperate forest stands ranges from about 1 (CARBON et al. 1979) to about 21 (WARING et al. 1978). This variation depends on the type of stand, its stage of development and the local topoclimate. As a forest stand develops, the leaf area increases to a maximum and, in many cases, subsequently falls (OVINGTON 1957; KIRA and SHIDEI 1967; RAUNER 1976; FORD 1982). In the deciduous forests of the eastern United States, for example, the maximum leaf area may be re-established 5 years after disturbance (MARKS and BORMAN 1972; MARKS 1974); in dense plantations of *Picea sitchensis* in Scotland, the maximum leaf area index was found at age 16 (FORD 1982) and in *Pinus sylvestris* in Sweden at age 25 (ALBREKTSON 1980). In other cases, however, an equilibrium leaf area is reached, e.g. in *Abies veitchii, Pinus densiflora, Pinus taeda* (TADAKI 1966; SWITZER et al. 1966; KIRA and SHIDEI 1967). On the other hand, in the Pacific Northwest of North America the leaf area index of coniferous stands continues to rise for 40 years or more (LONG and TURNER 1975), reaching values of over 15 (WESTMAN and WHITTAKER 1975; GHOLZ et al. 1976). In contrast L of broadleaf deciduous forest seldom exceeds about 6 (SATOO 1970; ANDERSON 1981). It is generally supposed that this is because the structure of broadleaf canopies results in high extinction coefficients, but the data available do not bear this out (Table 8.2). It seems more likely that L is limited in deciduous species by the annual cost of renewing an entire set of leaves (TADAKI 1977; SCHULZE et al. 1977b). The lowest values of L of ca. 1–3 are found in dry sclerophyll forests of *Eucalyptus* spp. in Australia (CARBON et al. 1979; ANDERSON 1981).

GRIER and RUNNING (1977) demonstrated a close correlation between L and precipitation and WARING et al. (1978) extended this approach to 40 stands in the mountains of Oregon to conclude that high potential evaporation and low night temperatures result in low values of L. In recent years much attention has focussed on the linear relationship between the leaf area on a tree (A_l) and the cross-sectional area of sapwood (A_{sw}) at 1.3 m (WHITEHEAD and JARVIS 1981). More xerophytic species in drier topoclimates have a smaller leaf area per unit sapwood area (i.e. lower dA_l/dA_{sw}) than species in more humid topoclimates, and in addition their sapwood area is a larger proportion of the basal area (WHITEHEAD, paper read to the IBC 1981). This is consistent with the interpretation that the slope, dA_l/dA_{sw}, is proportional to the hydraulic conductivity of the wood (k) and inversely proportional to the saturation deficit (D) and stomatal conductance of the foliage ($\bar{g}_s$) (WHITEHEAD and JARVIS 1981). Thus the leaf area carried by a tree depends on the capacity of the conducting system ($k \cdot A_{sw}$) and on the stress imposed upon it ($D \cdot \bar{g}_s$).

WARING et al. (1978) found that L was lower than anticipated from the topoclimate in sites on ultrabasic soils because of unbalanced nutrition. There is no doubt that deficiency of nutrients, especially of N, P and K, leads to a reduction in leaf area on particular sites (ALBREKTSON et al. 1977; TAMM 1979). Addition of N leads to needle retention for longer periods and to more and larger needles in conifers (e.g. MILLER and MILLER 1976; MILLER et al. 1976).

8.4.2 Vertical Distribution of Foliage

In a substantial number of closed coniferous forest canopies the leaf area of the stand is *normally* distributed in the vertical between the top of the tree and the base of the live crown (see reviews by STEPHENS 1969; BEADLE et al. 1982). The midpoint of the distribution closely matches the midpoint of the canopy and the standard deviation of the leaf area with respect to height varies between about 15% and 20%. Both positive and negative skewness and kurtosis have been shown to be significant in particular cases but the unmodified normal distribution provides an adequate fit to most data. It is noteworthy that in *Pinus sylvestris,* and probably also in other species, the leaf area distribution on individual trees in the stand may be bi- or trimodal, although the distribution for the stand as a whole fits a continuous normal function (BEADLE et al. 1982).

There is less regularity in the vertical distribution of foliage in deciduous broadleaf trees. Neutral distributions with respect to height (ABER 1979), normal distributions (SHINOZAKI and KIRA 1977) and distributions heavily skewed both upwards and downwards have been described (SATOO 1970; RAUNER 1976; KINERSON 1979; HOUSSARD and ESCARRE 1981). Consequently, the particular distribution in any situation must be evaluated.

8.4.3 Grouping or Clumping of Foliage

Although crucial to the penetration of the direct beam through the canopy, and hence to the distribution of direct and diffuse radiation in the canopy, grouping appears to have been evaluated quantitatively in a forest canopy only once. By comparison between model and measurement, NORMAN and JARVIS (1975) showed clearly that by treating the foliage of *Picea sitchensis* as grouped into whorls and into shoots much better predictions of canopy transmittance were made than on the random distribution assumption. The latter resulted in estimates of τ that were too small by a factor of nearly 10 (i.e. 0.1% as opposed to 1%). Whilst this has a small effect on the total interceptance, it does provide a more effective distribution of light and so explains why L of certain canopies is much larger than the value for 95% interception as predicted by the random model (i.e. KL=3).

8.4.4 Seasonal Variation

It is a matter of common observation that L changes markedly through the year in both coniferous evergreen forest and in broadleaf deciduous forest. An adequate description of these changes is necessary to obtain $\int_0^t Q_i \, dt$ over the year. Although the phenology of the ground vegetation has frequently been followed (LIETH 1970), the phenology and population dynamics of leaf area in forest canopies is much less well known because of the practical difficulties of quantitative measurement.

The usual procedure is to couple regular measurements of litter fall (see NEWBOULD 1967) with a single destructive harvest and to make the assumption that over a year the canopy is in a steady state, with the year's production of leaf area equal to the year's loss. Even with this assumption, however, measurements of the emergence and expansion of new leaves in the canopy are necessary to enable an estimate of L to be made for any day of the year (see for example FORD and NEWBOULD 1971; KINERSON et al. 1974). Since in-canopy measurements of leaf area are difficult and laborious, reliable seasonal estimates of L are hard to come by, particularly for species in which flushing of new leaves occurs throughout the growing season. Where the period of new leaf growth is short and well-defined, as in most temperate conifers, it is a little easier to model changes in L through the year since once the new leaves have expanded, the only changes in leaf area are caused by leaf deaths (Fig. 8.6). But even then regular in-canopy measurements may also be necessary if a sepa-

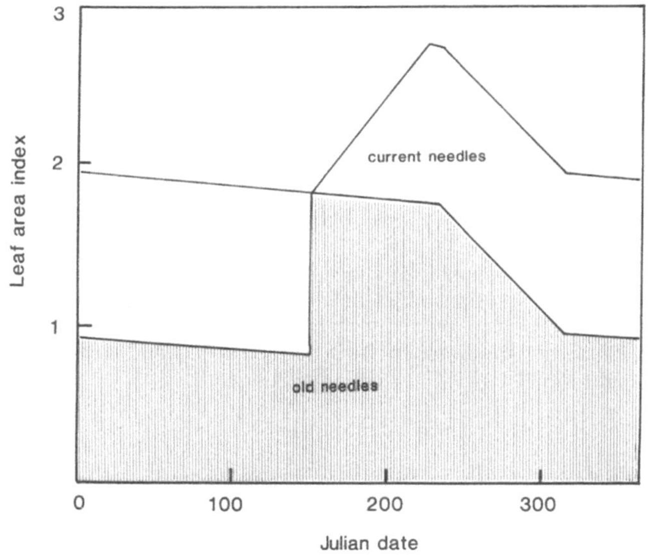

Fig. 8.6. The changes in leaf area index of current and old needles during a year in a 50-yr-old stand of Scots pine, *Pinus sylvestris,* based on the following model. The leaf area index on 1 Jan is the same in each year. New needles appear on 1 June and continue expansion until 12 August. They remain without loss until 6 Sept in the following year. They then fall at a constant rate until 21 August in the following year after which they fall at a more rapid rate until none are left on the trees by 12 Nov. A cohort of needles has therefore a life of nearly $2\frac{1}{2}$ years. (BEADLE et al. 1982)

rate account of needles of different ages is to be kept, because they differ in physiological properties (see Sect. 8.5 below) (FLOWER-ELLIS and PERSSON 1980).

There are large obvious differences between evergreen conifers and deciduous broadleaf trees in the amount and duration of green foliage through the year, and these would be expected to have major effects on light interception and hence on the annual production. Full, i.e. 95%, light interception occurs throughout the year when $(KL) > 3$ the whole time. If at any time $(KL) < 3$ interception will be less than 95%. The loss in intercepted radiation is $\int (Q_0 \Delta i)\,dt$ where Δ_i is the reduction in interceptance below the potential interceptance, i.e.

$$\int_0^{365} (Q_0 \Delta i)\,dt = \int_0^{365} [Q_0(0.95 - i)]\,dt$$

$$= \int_0^{365} [Q_0(\exp(-KL) - 0.05]\,dt. \qquad (8.24)$$

Essentially Δi represents the loss of opportunity in effectively exploiting the site. We shall consider three examples.

Firstly, we consider a coniferous forest canopy of, say, *Pinus sylvestris* in Scotland for which $K = 0.62$ and $L = 2.8$ for a year in which $Q_0 = 3{,}080$ MJ m^{-2}. The loss in potential radiation interception as a result of the low leaf area

Table 8.3. The transmission of direct and diffuse solar radiation by a canopy of tulip poplar, *Liriodendron tulipifera,* at midday on clear sunny days at four times of year. The subscripts b and d refer to the direct beam and diffuse components of radiation. The symbols are defined in the text, Q_0 is in W m^{-2}. (HUTCHISON and MATT 1976)

Time of year	Phenoseason	Q_{0b}	Q_{0d}	θ	ϕ	L	τ_b	τ_d	τ	K
Jan 20	Winter leafless	588	50	33°	0.08	0	0.48	0.93	0.51	–
March 8	Spring leafless	910	56	49°	0.06	0	0.66	0.89	0.67	–
June 8	Summer fully leafed	809	177	76°	0.18	6	0.32	0.56	0.36	0.17[a]
Sept 2	Autumn fully leafed	683	157	62°	0.19	6	0.14	0.34	0.17	0.29[a]

[a] Mean transmittance for both periods is 0.16 and mean K is therefore 0.31. (HUTCHISON and MATT 1977)

index is 389 MJ m^{-1} yr^{-1} or 13%. The loss is less than might have been expected because of the compensating effect of the value of K (>0.5). Secondly, consider an evergreen broadleaf canopy of *Eucalyptus* spp. with K=0.5 and L=1 for a year in which Q_0=6,000 MJ m^{-2}. The loss in radiation intercepted as a result of the very low leaf area index is 3,340 MJ m^{-2} yr^{-1} or 56%. Finally, consider a deciduous broadleaf canopy of, say, *Liriodendron tulipifera* in east Tennessee with K=0.31 and L=6 for the phenoseason mid-April to mid-October of 180 days (see Table 8.3). Out of an assumed total solar irradiance of 5,571 MJ m^{-2} yr^{-r}, 1,884 MJ m^{-2} are lost during the leafless period and a further 574 MJ m^{-2} during the fully leafed period as a result of the low extinction coefficient, making a total loss of potential interception of 2,458 MJ m^{-2} yr^{-1} or 44%. It is noteworthy that 77% of this loss occurs during the leafless period.

In all three examples significant quantities of radiation are not intercepted because of low values of either L or K. In particular the long winter leafless period of the deciduous tulip poplar stand results in a very substantial loss of intercepted radiation. In contrast, a spruce or fir canopy with K$\simeq$0.5 and L>9 intercepts virtually all of the radiation incident throughout the year. The relative importance of the amount or duration of leaf area will vary with the situation. SCHULZE et al. (1977b) attributed the higher annual photosynthetic production of *Picea abies* than *Fagus sylvatica* (see Table 8.7) in central Europe to the longevity of the spruce needles rather than to the difference in phenoseasons. As a result of retention of needles by the spruce for ca. 5 years, a much higher leaf area index was maintained by the spruce (11.6 vs. 5.7) but at a very similar annual replacement cost. Unfortunately, at the present time we do not have any relationships between growth rate and intercepted radiation for forest of the kind that exists for agricultural crops. If, however, we assume a figure of 3.0 g MJ^{-1} PAR for this relationship, on the lines of experience with field crops, then the losses of solar irradiance of 389, 3,340 and 2,522 MJ m^{-2} correspond to reductions in growth of 5.8, 50.1 and 37.8 Mg ha^{-1} yr^{-1}. These represent large reductions in yield as a result of incomplete radiation interception.

In agricultural crops a very substantial part of the variation in yield is attributable to variation in leaf area development and light interception caused by variation in genotype, weather and environmental stresses (MONTEITH 1972, 1977). We wish to emphasise strongly that reliable information on leaf area index and its seasonal changes is crucial to an understanding of the growth and productivity of temperate forest. Intensive physiological studies of the influence of environmental variables, especially temperature, on the development and growth of leaves in forest canopies are needed, coupled with demographic studies of the leaf populations. The lack of information in this area contrasts markedly with the substantial body of information on photosynthetic processes (see Sect. 8.5), but without one it is impossible to make proper use of the other.

8.5 Photosynthetic Efficiency

We are here concerned with the extent to which ε falls below ε_0 in practice as a result of structural, physiological and environmental limitations. For example, in a young open stand of *Pinus sylvestris* ($L=1$), ε was 1.42 g MJ^{-1} over the year and 1.80 g MJ^{-1} in the irrigated and fertilised plots ($L=2.5$), whereas in the most productive month of August ε was much higher (Fig. 8.7) reaching 3.14 g MJ^{-1} in the irrigated and fertilised plots (LINDER personal communication). Since ε is defined as dA/dQ_a for the canopy and ε_0 is dA_0/dQ_a (where A_0 is the potential rate of photosynthesis) our concern is with the extent to which $A/A_0<1$. For broad comparisons f ($=\varepsilon/\varepsilon_0=A/A_0$) can be removed from the integral in Eq. (8.13) and replaced with an average value as in Eq. (8.14a) but an appreciation of the limiting effects of impaired photosynthesis

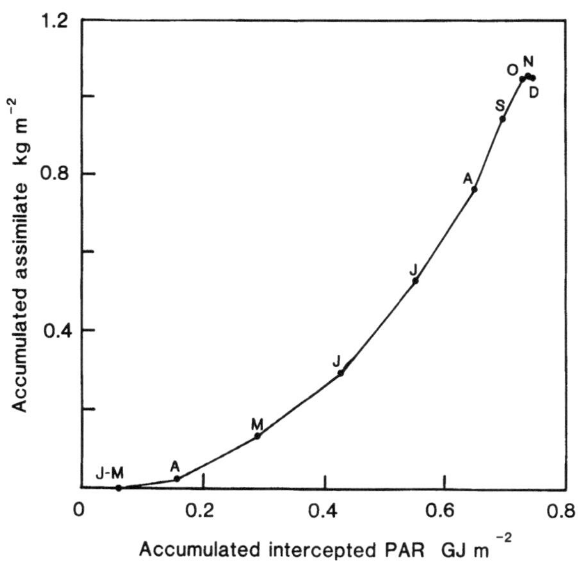

Fig. 8.7. Accumulated photosynthate (expressed as equivalent dry matter) in relation to the accumulated intercepted PAR in a natural stand of young Scots pine, *Pinus sylvestris*, over 1 year. The photosynthetic properties of needles of different age and position were taken into account: the stand was untreated with a leaf area index of ca. 1.0. (Based on LINDER and TROENG 1980; TROENG and LINDER 1982a and b; LINDER and LOHAMMER unpublished)

on growth requires an understanding of how ε responds to environmental stresses and changes in canopy structure over short periods. It is therefore, necessary to break down ε into its constituents and to consider their dependence on environmental and canopy variables.

8.5.1 Dependence on Structural Properties

The presumption that $G \propto \int (\varepsilon \cdot Q_i) \cdot dt$ implies a linear relationship between A and Q_i. At first sight, this seems unlikely because the response of *leaf* photosynthesis to quantum flux density is essentially asymptotic in form. Consequently, an increase in Q_i might be expected to lead to a progressive decrease in ε as a larger proportion of the leaves in the canopy become light-saturated. However, for canopies, as opposed to leaves, the relationship between A and Q_i is more nearly linear because of the structural properties of the canopy, as first pointed out by BOYSEN-JENSEN (1932). Figure 8.8 shows that this is certainly true for spruce forest and it is probably also true for other forest canopies with $L > 3$. Thus it is reasonable to assume a constant value for ε over the diurnal and seasonal range of Q_i, as is implied by Eq. (8.3) and the subsequent argument. Figure 8.8 shows that the value of $1/\varepsilon$ for the Sitka spruce forest canopy at low quantum flux densities is ca. 18 incident mol quanta per mol CO_2 assimilated – a surprisingly low figure – and closely similar to the apparent quantum requirement for the shoots and needles (LEVERENZ and JARVIS 1979). The *average* value for $1/\varepsilon$ over the full range of quantum fluxes is about twice this figure at low vapour pressure deficits (JARVIS et al. 1976).

What are the properties of a canopy that result in a high value of ε largely independent of quantum flux density? This would result if the radiation incident on the canopy is distributed through the canopy in such a way that all the leaves are exposed to intermediate, non-saturating quantum flux densities such that A_n/Q (cf. dA/dQ) for each leaf is maximum. We suggest that this is achieved in forest canopies in the following three ways (a) through the particular grouping of the foliage into crowns, whorls of branches and foliated shoots; (b) through the particular spatial distribution of the leaves and the distribution of their inclination angles within these groupings; and (c) through some compensation in physiological and anatomical properties for the low quantum flux densities deep in the canopy as a result of "shade" acclimation.

Grouping of the foliage results in the efficient transmission of beam radiation through the canopy and effective scattering where it is intercepted. The scattering component of diffuse radiation within the canopy depends on the spectral properties of the leaves (NORMAN 1978) but these are comparatively conservative and can be approximated as in Eq. (8.19). HUTCHISON and MATT (1976) calculated that half the diffuse radiation in the overstorey canopy of *Liriodendron tulipifera* at midsummer arises from scattering of the direct beam. Secondly, as the groups of foliage get further apart the proportion of penumbral radiation at intermediate quantum flux densities increases.

The grouped distribution of the leaves along the branches and twigs and the distribution of their inclination angles is such that the majority of leaf

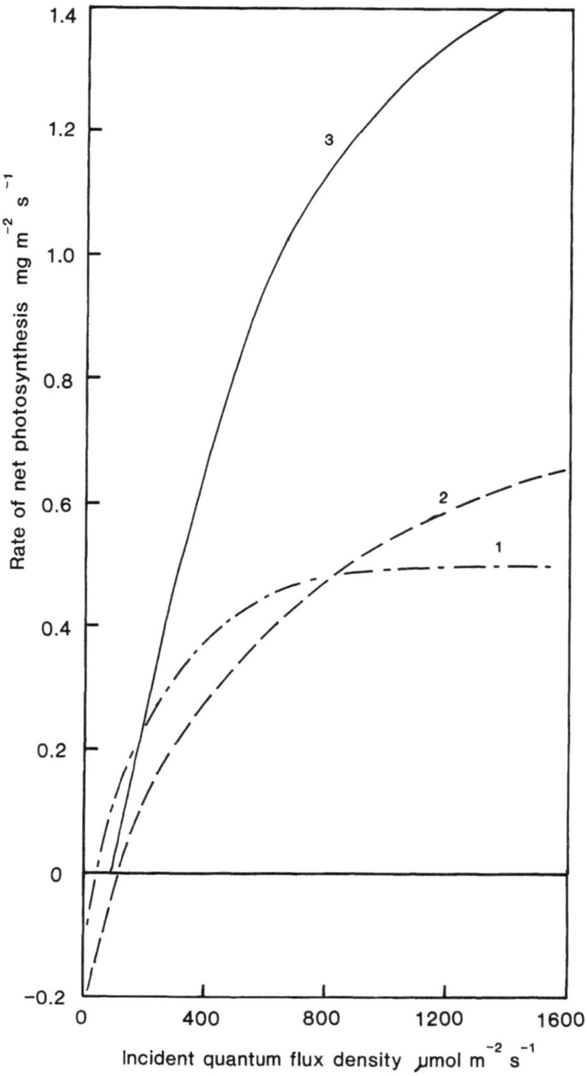

Fig. 8.8. The relation between rate of net photosynthesis, A_n and incident quantum flux density, Q_0, in Sitka spruce, *Picea sitchensis*, for *1* "sun" *needles* illuminated bilaterally, normal to the beams, without any mutual shading (LEVERENZ and JARVIS 1979), *2* a "sun" *shoot* illuminated bilaterally with some mutual shading (LEVERENZ and JARVIS 1980) and *3* a forest *canopy* illuminated naturally from above (JARVIS et al. 1976). In each case A_n and Q_0 are expressed on the same area basis, viz. projected needle area in *1*, shoot silhouette area in *2*, and ground area in *3*

surfaces are exposed to intermediate quantum flux densities so that ε tends to remain high (LEVERENZ and JARVIS 1980). If, for example, the angular distribution of the leaves is spherical, then the degree of light saturation is independent of the solar zenith angle and small because the average leaf–sun angle is 60°. In many forest canopies leaf inclination varies from predominantly vertical in the upper canopy to predominantly horizontal in the lower canopy (BÜSGEN and MÜNCH 1931; MILLER 1967; LEVERENZ and JARVIS 1980). This, coupled with an appropriate degree of clumping, leading to mutual shading with respect to the solar beam, and a substantial proportion of diffuse radiation in the canopy, leads to a fairly uniform light distribution with the majority of leaf surfaces at intermediate quantum flux densities (KUROIWA 1970).

Acclimation in the lower parts of the canopy leads to some compensation for loss of efficiency at the upper surface of the canopy. The quantum requirement of shade-acclimated leaves may be less than that of the saturation-susceptible "sun" leaves in the upper canopy (SCHULZE 1970; BOARDMAN 1977; LEVERENZ and JARVIS 1979), and the photosynthesis light-response curves more convex because of a more even distribution of light within and among shade leaves (LEVERENZ and JARVIS 1979, 1980).

Values of $L >$ ca. 6 are not likely to change the interception of radiation, but may affect the linearity of the A/Q_i relationship through associated changes in canopy structure. This may be another reason why some forest canopies considerably exceed the leaf area index required for 95% interception (e.g. NORMAN and JARVIS 1974). Differences between species in the structure of shoot, branch and crown that affect the distribution of radiation in the canopy and thereby lead to *higher* values of K would be expected to reduce ε and enhance the non-linearity of the A/Q_i relationship. However, as far as we know at present, this relationship is close to linear for forest canopies, and this is also borne out by simulations of canopy photosynthesis as a function of Q_i built up from the hyperbolic response curves of individual leaves (MIRANDA and JARVIS unpublished).

8.5.2 Dependence on Leaf Photosynthetic Properties

The efficiency with which the intercepted light is used in photosynthesis depends on the rate of photosynthesis of all the leaves in the canopy. The rate of photosynthesis of a leaf depends on its potential rate which is determined by a number of internal variables (JARVIS 1981; VON CAEMMERER and FARQUHAR 1981) and on the extent to which this potential is realised in the prevailing environmental conditions (BERRY and DOWNTON 1982). We can use a model of leaf photosynthesis to evaluate the relative importance of these different variables in determining ε (e.g. REED et al. 1976; THORNLEY 1976; WATSON et al. 1978, see also Chap. 16, Vol. 12B).

At an appropriate level of complexity, the rate of photosynthesis by a leaf depends on the stomatal conductance for CO_2 transfer (g_s), the mesophyll or internal conductance (g_m), the quantum yield (ϕ) and the CO_2 compensation concentration (Γ). The boundary layer is included with the stomatal conductance here because in a temperate forest canopy the boundary layer resistance of the leaves is small and usually non-limiting because of the extreme aerodynamic roughness of the canopy (GRACE 1980; JARVIS 1980). For conifers and probably also deciduous broadleaf trees, it is appropriate to describe the response of photosynthesis to light by a non-rectangular hyperbola since the convexity coefficient (θ) is frequently 0.7 to 0.9 (LEVERENZ and JARVIS 1979; MARSHALL and BISCOE 1980). Thus

$$\theta A^2 - A(\alpha Q + A_m) - \alpha Q A_m = 0 \tag{8.25}$$

where α is the initial slope (dA/dQ) and A_m is the asymptotic value of A.

A is related to the net CO_2 flux through the stomata (A_n) by

$$A = A_n + R_d \quad \text{and similarly } A_m = A_{nm} + R_d . \tag{8.26}$$

The mean intercellular space CO_2 concentration is

$$C_i = C_a - A_n/g_s . \tag{8.27}$$

Assuming for simplicity a linear dependence of A_n on C_i over the range of interest

$$A_{nm} = (C_i - \Gamma)\, g_m . \tag{8.28}$$

Combining Eqs. (8.25) to (8.28) with the elimination of C_i gives a quadratic expression for A_n in terms of the leaf variables g_s, g_m, α, θ, R_d and Γ and the environmental variables Q and C_a. The leaf variables do, of course, depend on a number of environmental variables and their effects on ε are discussed in Sect. 8.5.3 below. To apply this model to calculate the photosynthesis by every leaf in the forest canopy, submodels of radiation distribution and stomatal conductance are required. We calculate the upward and downward fluxes of direct beam and diffuse radiation in thin canopy layers each with L of not more than 0.5 using NORMAN's (1978) model. Stomatal conductance is calculated using an empirical description of its dependence on environmental variables (JARVIS 1976; THORPE et al. 1980).

The sensitivity of photosynthesis by a Sitka spruce forest canopy to the leaf variables is indicated in Table 8.4. It is clear that A_n is extremely sensitive to both increases and decreases in g_m, to decreases in α and to increases in θ. REED et al. (1976) also showed that in *Liriodendron tulipifera* A_n was very sensitive to g_m and α. Values of these leaf variables for several temperate forest species are listed in Table 8.5; unfortunately it is extremely difficult to make

Table 8.4. The sensitivity of photosynthesis by Sitka spruce, *Picea sitchensis*, to leaf variables. The rate of net photosynthesis, A_n, at a quantum flux density of 400 μmol m^{-2} s^{-1}, is expressed relative to the rate with the standard values of the variables A_s, when the variable being considered, x, is taken as half or twice its standard value, x_s. (MIRANDA unpublished)

Leaf variable	x_s	A_n/A_s for x/x_s equal to	
		0.5	2.0
α	2.44 g mol^{-1}	0.77	1.10
g_s	10.0 mm s^{-1}	0.90	1.08
g_m	1.2 mm s^{-1}	0.64	1.54
Γ	75.0 mg m^{-3}	1.05	0.85
θ	0.68	0.90	1.23[a]

[a] Increased to $\theta = 1$, its maximum value, rather than doubled

Table 8.5. Some examples of the photosynthetic parameters for leaves of temperate trees. The symbols are defined in the text

	Parameter						Reference
	A_{nm} mg m^{-2} s^{-1}	Γ (20 °C) mg m^{-3}	α g CO_2 mol^{-1}	g_m mm s^{-1}	g_s mm s^{-1}	θ	
Abies balsamea	0.51	68	–	3.3	1.5	–	1
Picea sitchensis	0.50	75	2.40	1.6	2.6	0.86	2
Pinus radiata	0.41	104	–	0.9	4.8	–	3
Pinus contorta	0.28	–	–	1.3	2.5	–	4
Pinus nigra	0.22	–	3.52	–	–	–	5
Eucalyptus pauciflora	0.81	100	–	4.0	4.2	–	6
Liriodendron tulipifera	0.70	–	2.25	0.7	2.5	–	7
Betula pendula	0.65	–	3.65	3.1	2.5	–	8
Quercus alba	0.41	80	–	1.8	1.0	–	9
Populus tremula	0.39	–	3.61	0.9	3.3	–	8

References: 1 LITTLE and LOACH (1975); 2 LEVERENZ and JARVIS (1979); 3 BENNETT and ROOK (1978); 4 DYKSTRA (1974); 5 EL AOUNI and MOUSSEAU (1974); 6 SLATYER and FERRAR (1977); 7 REED et al. (1976); 8 TSEL'NIKER (1979); 9 DOUGHERTY et al. (1979).

a good comparative compilation of all five variables for a number of species because of lack of data. Gas-exchange data indicate that some or all of the leaf variables change seasonally and with position in the canopy (e.g. KIRA et al. 1969; HELMS 1970; WOODMAN 1971; LUDLOW and JARVIS 1971; NEILSON et al. 1972; KÜNSTLE and MITSCHERLICH 1975; WATTS et al. 1976; SCHULZE et al. 1977a; KUSUMOTO 1978; LEVERENZ and JARVIS 1979; LEVERENZ 1981; SCHULZE 1981; TROENG and LINDER 1982a, b). However, in only a few cases have the changes in the leaf variables been quantified.

8.5.3 Dependence on Environment

Photosynthesis by all temperate forest trees is, as far as is known, of the C_3 type, and responds to environmental variables in a typical C_3 manner (JARVIS 1981; BERRY and DOWNTON 1982). The leaf variables discussed above are variously dependent on extremes of temperature, vapour pressure deficit and leaf water potential. In temperate trees much of the reduction in g_s during the summer can be accounted for by the effects of high vapour pressure deficits in closing the stomata; low leaf water potentials can also have a significant influence on g_s in lengthy periods without rain, and low temperatures may reduce g_s at other times of the year (see reviews by WHITEHEAD and JARVIS 1981; KRIEDEMANN and BARRS 1981; LANDSBERG and JONES 1981; HINCKLEY et al. 1981, see also Chap. 7, Vol. 12B); g_m is particularly sensitive to both low and high temperatures (e.g. NEILSON et al. 1972) and to low leaf water potentials (e.g. BEADLE et al. 1981); α declines markedly at higher temperatures in many species, although this has not been demonstrated particularly in trees

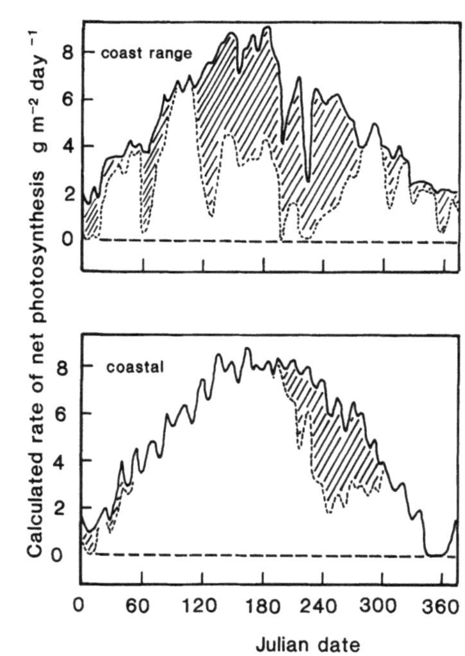

Fig. 8.9. The seasonal course of potential and actual photosynthesis of Douglas-fir, *Pseudotsuga menziesii,* at two sites in western Oregon. Photosynthesis was calculated from leaf and environmental variables using a model which included effects of temperature on g_m, and effects of water stress, low temperature and frost on g_s (a modification of that given by REED et al. 1976). Potential photosynthesis was based on average daily temperatures, daylength and Q_0, with no water stress: actual photosynthesis was based on minimum night temperatures and included water stress. Calculated annual actual photosynthesis correlated well with wood production. (EMMINGHAM and WARING 1977)

(e.g. EHLERINGER and BJÖRKMAN 1977); and Γ increases with both increasing temperature and decreasing leaf water potential (e.g. BRIX 1968; BEADLE et al. 1981).

It is to be expected that the effects of such environmental stresses on the leaf variables would effectively reduce ε. Although there have been many investigations of the influence of environmental stresses on photosynthesis in trees (see for example reviews by LINDER 1979; KRAMER and KOZLOWSKI 1979; LARCHER 1969, 1980), there are only a few cases in which the effects of stress on the annual carbon budget have been evaluated (*Pinus cembra,* TRANQUILLINI 1959; *Pinus aristata,* SCHULZE et al. 1967; *Fagus sylvatica,* SCHULZE 1970; *Pseudotsuga menziesii,* EMMINGHAM and WARING 1977; *Picea abies,* FUCHS et al. 1977; *Pinus sylvestris,* ÅGREN and AXELSSON 1980; TROENG and LINDER 1982a). In all cases the amount of carbon assimilated annually was reduced below the potential by low temperatures in spring and autumn or by summer drought, or by both (see the example in Fig. 8.9), in one case by as much as 50%. These cases are nearly all from alpine, montane or continental climates where extremes are commonplace. In the coastal example in Fig. 8.9, the reduction below the potential was less and in other cool maritime-temperate climates, dry matter has been shown to increase in every month and even to double over winter (HAGEM 1962; BRADBURY and MALCOLM 1978). Whilst it seems very probable that variation in ε contributes to the enormous range of growth rates that occur, there is little evidence to show that differences in the annual growth rate result from differences in the efficiency with which intercepted radiation is used. Furthermore, many of the environmental variables that would be expected to reduce ε are also likely to have at least as large effects on leaf development and expansion (HSIAO et al. 1976) and hence on light interception.

8.6 Respiratory Losses

Respiration by trees varies diurnally (e.g. NEGISI 1975, 1978), seasonally (e.g. SCHULZE 1970; WOODWELL and BOTKIN 1970; LINDER and TROENG 1981) and with the age of the tree (RAUNER 1976). The rate of tree respiration also varies strongly with temperature (e.g. BUTLER and LANDSBERG 1981) and it would be expected to vary in relation to the availability of stored reserves and readily degradable structural materials (THORNLEY 1976).

In herbaceous plants it has proved useful to distinguish between respiration associated with the maintenance of existing tissues and respiration associated with the growth of new tissues (e.g. MCCREE 1970; HORIE 1977; BARNES and HOLE 1978). Whilst these concepts are helpful in explaining some of the phenomena listed above, there are some problems in their application to large woody perennials. However, conceptually more satisfying models such as THORNLEY's (1976) three-compartment model with storage, degradable and non-degradable structures require more information and are therefore at present even less practical (see also BARNES and HOLE 1978).

8.6.1 Respiration and Temperature

In general, respiration increases exponentially with temperature and this is also true of the woody organs of trees for which

$$R(T) = R_0 e^{\beta T} \tag{8.29}$$

where R_0 is the respiration rate at T = 0 °C. The parameter β is relatively conservative. For the aerial parts of whole apple trees β was 0.084 with a range of ± 0.005 for the tree with or without leaves, blossom or fruits over 8 months of the year (BUTLER and LANDSBERG 1981); in stems and roots of *Pinus sylvestris* β was ca. 0.070 throughout the year (LINDER and TROENG 1981) and the respiration rate of the stems of several other conifers increased exponentially with approximately the same value of β (TRANQUILLINI and SCHÜTZ 1970; NEGISI 1975; ROOK and CORSON 1978).

R_0 is, however, highly variable. In apple R_0 was higher for the whole tree than for the bare branches; in *Picea abies* R_0 was higher in the small twigs than in the branches (ZIEGLER 1957); and in *Pinus sylvestris* R_0 was higher in the coarse roots than in the stem (LINDER and TROENG 1981). In the whole apple trees, too, R_0 increased nearly three times when leaf flushing occurred and in a number of studies on stem sections R_0 doubled during periods of rapid growth (e.g. TRANQUILLINI and SCHÜTZ 1970; LINDER and TROENG 1981).

Diurnal variation in R is probably largely attributable to temperature variations; the variation in R_0 between organs may reflect the availability of substrate at the time of the measurements; the seasonal variation in R_0 is probably largely the result of a shift in emphasis from maintenance to growth respiration. However, we have little information on these points for trees.

8.6.2 Maintenance Respiration

In herbaceous plants the rate of maintenance respiration is usually taken as proportional to the weight of the plant or organ (McCREE 1970) viz.

$$R_m = m\,W. \tag{8.30}$$

Thus the "maintenance coefficient" (m) is the specific rate of the maintenance respiration and is expected to be a constant for a particular tissue. There have been very few attempts to measure the maintenance respiration of woody plants as distinct from the total respiration. LEDIG et al. (1976) estimated m in the stems and roots of seedlings of *Pinus rigida* as 28 and 7 g kg^{-1} d^{-1} (i.e. CO_2 efflux per unit dry matter per day), respectively, at 29 to 32 °C. Values for herbaceous plants at 20 to 25 °C generally lie in the range 5 to 70 g kg^{-1} d^{-1} at 20 to 25 °C (PENNING DE VRIES 1975a; HORIE 1977; SILSBURY 1979). There is some biochemical evidence to suggest that rates between 20 to 35 g kg^{-1} d^{-1} (13 to 23 g CH_2O kg^{-1} d^{-1}) might be expected (PENNING DE VRIES 1975a).

If we assume a figure of 25 g kg^{-1} d^{-1} for R_m and apply this to a forest with a standing biomass of, say, 400 Mg ha^{-1}, we arrive at an estimate of the daily maintenance respiration of 1,000 g m^{-2} on a ground area basis. This exceeds the maximum possible photosynthesis by the canopy of 250 g m^{-2} day by four times (average $A_n = 0.5$ mg m^{-2} s^{-1} on a leaf area basis over a 14-h day, $L = 10$)! HOLE and BARNES (1980) have pointed out that in bulky tissues such as fruits m is inversely proportional to the weight of the organ (i.e. $m = a + b/W$). Such a decline in m with size of organ may result from progressive increase in the proportion of tissue with very low maintenance costs or from increasing internal diffusion resistance to the efflux of CO_2. The respiration rate of the stems and branches of trees is more closely related to the diameter or surface area than to the weight, because of the largely peripheral distribution of the living tissues and the large proportion of internal non-living xylem tissue (JOHANSSON 1933; MÜLLER and NIELSEN 1965; NEGISI 1977; YODA 1978). Thus it is more appropriate to define R_m as

$$R_m = m'\,S, \tag{8.31}$$

where m′ is the rate of maintenance respiration per unit surface area (S) of stem and branches. Since for a segment of stem of length l and radius r, $S = 2\pi r l$ and $W = \pi r^2 l \rho$, where ρ is the density of the segment of stem

$$m = 2m'/(\rho r).$$

Thus in stems, coarse roots and branches m must certainly decline with increasing size in inverse proportion to the radius, as YODA found in tropical trees (see KIRA and SHIDEI 1967). Equation (8.30) is therefore appropriate for leaves, twigs and fine roots, whereas Eq. (8.31) should be used for the more bulky tissues. WHITTAKER and WOODWELL (1967) specify the relationships between surface area and weight for a number of tree species. Using this approach,

GRIER and LOGAN (1977) calculated R_m for five stands of old-growth *Pseudotsuga menziesii* in Oregon to be as high as 150–300 Mg ha^{-1} yr^{-1} (CO_2 efflux). These stands were of exceptionally large biomass (590–1,170 Mg ha^{-1}) and other estimates of the respiration rates of forest stands are an order of magnitude smaller (e.g. WOODWELL and BOTKIN 1970; KINERSON et al. 1977; YODA 1978). Since the number of living cells in wood is proportional to the volume of the sapwood, this may turn out to be an even better variable to use than either weight or surface area of wood (HAVRANEK 1981).

In the old-growth Douglas-fir stands and in stands of *Abies sachalinensis* (YODA et al. 1965), two-thirds of the respiration was by the leaves; in mixed stands of evergreen oak (*Castanopsis cuspidata, Cyclobalanopsis gilva*) about one-half the respiration was by the leaves (YODA 1978); whereas in the young stands of *Pinus taeda* and the deciduous *Liriodendron tulipifera* in southeastern USA only one-third of the annual respiration was by the foliage (KINERSON et al. 1977).

Probably the most precise estimates of R_m for trees at the present time are measurements made in darkness on the whole tree, or parts of it, at times of the year when growth of cambium, roots and leaves is not in evidence, i.e. early spring and autumn. On that basis, m′ of the aerial parts of apple trees was ca. 1.7 g m^{-2} d^{-1} at 20 °C (BUTLER and LANDSBERG 1981) and m′ of stem and branches of conifers and of *Quercus coccinea* 3 to 5 g m^{-2} d^{-1} (TRANQUILLINI and SCHÜTZ 1970; WOODWELL and BOTKIN 1970; LINDER and TROENG 1981). Two points should be borne in mind in making use of these figures. First some reassimilation may occur when the trunk is exposed to light (WIEBE 1975; SHEPARD 1975; FOOTE and SCHAEDLE 1976; PARKER 1978; DOUGHERTY et al. 1979). LINDER and TROENG (1980) showed that in young trees of *Pinus sylvestris* reassimilation by chlorophyll in the outer bark reduced CO_2 efflux from the trunk and branches by about 25%. Secondly, the respiration rate of the small twigs and roots may be up to twice that of the trunk when expressed on a surface area basis (ZIEGLER 1957; ROOK and CORSON 1978; LINDER and TROENG 1981). These effects are difficult to quantify, so that the best estimate is probably given by the whole apple trees. Taking R_m = 2 g m^{-2} d^{-1} on a stem area basis together with a stem + branch area index of 5 appropriate for pole-stage *Picea sitchensis* (NORMAN and JARVIS 1975), and a measured rate of canopy photosynthesis of, e.g. 40 g m^{-2} d^{-1} (average $A_n = 0.8$ mg m^{-2} s^{-1} on a ground area basis over 14 h, JARVIS et al. 1976), leads to an estimate of

$$Y_m = 1 - \frac{R_m}{R_m + A_n} = 0.8.$$

For older, more open stands of deciduous trees, the stem and branch area index is <1.5 (WHITTAKER and WOODWELL 1967) leading to lower estimates of R_m and hence $Y_m > 0.9$ (see also ÅGREN et al. 1980). On the other hand, for the massive stands of old-growth *Pseudotsuga menziesii* analysed by GRIER and LOGAN (1977), $Y_m < 0.1$. Clearly these figures can be regarded only as a general indication of the likely magnitude of Y_m.

8.6.3 Growth Respiration

High rates of respiration in the range 5 to 20 g m^{-2} d^{-1} at 20 °C have been measured on shoots, branches and trees at times of high meristematic activity (e.g. ZIEGLER 1957; SCHULZE 1970; TRANQUILLINI and SCHÜTZ 1970; KINERSON 1975; ROOK and CORSON 1978; LINDER and TROENG 1981; BUTLER and LANDSBERG 1981). In herbaceous plants the rate of growth respiration has been related both to recent CO_2 assimilation (e.g. MCCREE 1970; MCCREE and SILSBURY 1978) and to the current rate of growth (dW/dt) (e.g. HORIE 1977; BARNES and HOLE 1978). Neither approach seem wholly appropriate for large woody perennials which carry large reserves of degradable carbohydrates and amino acids for much of the year. Since the growth of new tissues is usually at the expense of these reserves (DOUGHERTY et al. 1979), some measure of their mobilisation and turnover, irrespective of the rate of recent assimilation or current change in weight, would be preferable. At the time of rapid shoot extension and leaf development, for example, the growth of new tissues with its associated respiration may occur while the growth rate of the organ or tree is negative.

However, over a long enough period of time for the short-term changes in reserves to be unimportant, the rate of growth respiration of organ or tree can be defined as

$$R_g = \bar{g}\, dW/dt. \tag{8.32}$$

The "growth coefficient" $\bar{g}$ is, therefore, the rate of respiration per unit of dry matter increase averaged over a sufficient period to smooth out the short-term fluctuations in dW/dt. None the less this approach still fails to take into account the costs associated with the growth of tissues such as the fine roots which may turnover several times during the growing season but with no net change in weight at the end (see Sect. 8.7).

The value of g for a particular tissue depends on the composition of that tissue and on the biosynthetic pathways. Biochemical calculations and a large number of measurements indicate that g varies between about 0.25 and 0.5 depending on the growing tissue (excluding seeds) (PENNING DE VRIES et al. 1974; PENNING DE VRIES 1975b). For example in the production of 1 g of leaf of typical composition from sucrose (1.055 g) and amino acids (0.305 g), 0.33 g of CO_2 are produced as a result of the concurrent respiration. Similar figures for the production of 1 g of non-woody stem, woody stem and standard dry matter (SDM) referred to earlier are 0.28, 0.43 and 0.47 g of CO_2, respectively (PENNING DE VRIES et al. 1974). In seedlings of *Pinus rigida* 0.47 and 0.19 g of CO_2 were respired in the production of 1 g of shoot and root, respectively (LEDIG et al. 1976).

The conversion efficiency (Y_g) defined earlier can be expressed as $1/(1+g)$ (THORNLEY 1976) and consequently varies between about 0.7 and 0.8 depending on the tissue being produced. A complete analysis of the growth costs of leaves, roots, branches and trunk of trees along these lines has not been made, but in view of the substantial amounts of wood produced Y_g is likely to be at the low end of this range.

8.6.4 Total Respiration

From the foregoing it is clear that we are unable to distinguish quantitatively between growth and maintenance respiration in trees at present because the coefficients m′ and $\bar{g}$ are not known reliably in any particular case.

Qualitatively, however, the concepts are very helpful in looking at the changing balance of respiration and assimilation during the life of a tree or stand of trees. Over a year the total respiration is

$$\int_0^1 R \cdot dt = \int_0^1 R_m \cdot dt + \int_0^1 R_g \cdot dt$$
$$= \int_0^1 m' S \cdot dt + \int_0^1 \bar{g} \frac{dW}{dt} \cdot dt. \tag{8.33}$$

The size of R_m relative to R_g clearly depends on the stage of growth of the tree. During the early years when S is small but dW/dt significant, it is probable that $R_g \cdot dt > R_m \cdot dt$. However as the tree grows larger and the maintenance requirement with it, this position will be reversed. In the small 6-year-old apple tree of ca. 1 kg dry weight used by BUTLER and LANDSBERG (1981), it appears that $\int R_m \cdot dt$ was about three times $\int R_g \cdot dt$, although maximum R_g was about three times R_m. RAUNER (1976) described stands of *Quercus robur* in which dW/dt is maximum at about 30 years when the leaf area index reaches a maximum of 4 to 6. At that time ε is ca 1.2 g MJ^{-1} and the ratio of S to $L < 0.5$. As the stand ages L falls towards 2.0 and S increases until S/L reaches 1.5 to 2.0. As a result, ε falls to about 1/4 of its earlier maximum. Thus increasing age and size of tree increase R_m and pull the growth rate down into the variable space below the boundary line AO in Fig. 8.2.

8.7 Mortality Losses

Assimilate is used to produce parts of the tree which are short-lived and some entire trees which quickly die. The shedding of parts of the tree and the death of entire individuals in a stand can substantially reduce the accumulation of dry matter by living trees in the stand. Whilst the turnover of dead dry matter and the associated recycling of nutrients contribute to the overall functioning of the stand, the growth rate is none the less reduced below the potential by the amount of shed material and dead individuals. Thus shedding and death contribute in varying degrees to filling the variable space below the boundary line BO in Fig. 8.2.

The shedding of leaves, fruits, bark and branches is generally referred to as litter fall and the amounts are well-documented (e.g. BRAY and GORHAM 1964) but the amounts involved in root turnover or the death of individual trees are much less well-known. It will be clear from the sections that follow that these losses are very large, very variable and very poorly quantified or understood.

In addition to the losses of dry matter, account must also be taken of the respiratory costs of growing a tissue which then dies during the course of the year without adding to the permanent dry matter income of the tree. If this happens several times the respiratory costs are considerable and are not included in the growth respiration efficiency discussed earlier, since that is related to the net dry matter income over the period. Consequently the losses of material discussed below should be increased by at least 30% to take this into account.

These losses of material and the associated respiratory costs are the largest single factor reducing growth rate and consequently deserve to be far better known. At present we can only guess that d is of the order of 0.2 to 0.4.

8.7.1 Litter Fall

Average rates of annual litter fall for a large number of temperate forest stands range between ca. 1 and 9 Mg ha^{-1} yr^{-1} (BRAY and GORHAM 1964; KIRA and SHIDEI 1967; SAITO 1977; ATTIWILL et al. 1978; NISHIOKA and KIRITA 1978). On average this is made up of about 70% leaves and 7% each of branches, bark and fruits. However, variation between species and localities is large: leaves 53%–75%, branches 12%–29%, bark <1%–14%, and fruits, flowers, bud scales etc. <1%–15%. *Wood* litter fall in temperate deciduous forests is highly variable but averages about 3 Mg ha^{-1} yr^{-1} (CHRISTENSEN 1977; NISHIOKA and KIRITA 1978; SOLLINS 1982). Annual variation in litter fall is related to the occurrence of extremes in the weather and, particularly, to the incidence of insect attacks. The analysis by BRAY and GORHAM (1964) of their tabulated data gave average values for cool and warm temperate forests of 3.5 (2.5) and 5.5 (3.6) Mg ha^{-1} yr^{-1}, respectively (leaf figures in brackets). The amount of leaf litter increased with a number of variables associated with increased leaf production and was maximal on fertile, moist sites at low latitudes and intermediate altitudes where the basal area index was high. Average litter production by evergreen Gymnosperms and deciduous Angiosperms was very similar, viz. 3.7 (2.6) and 3.2 (2.4) Mg ha^{-1} yr^{-1}, respectively.

8.7.2 Tree Mortality

BRAY and GORHAM (1964) emphasised the problems in measuring the losses resulting from tree mortality from sampling tree fall and were not able to arrive at quantitative estimates from the data they reviewed. Studies on competition in young stands (e.g. FORD 1975; ASHTON 1976b) provide some indication of the losses, which clearly must be very large as numbers of seedlings self-thin from, for example, 20,000 ha^{-1} to 1,000 ha^{-1} over say 25 years. In such a situation mortality may more than halve the stand growth rate. ATTIWILL (1979) describes the tree mortality in a forest of *Eucalyptus obliqua* over a 22-year period. When the average age was 44 years, there were 914 trees ha^{-1} with a biomass above ground of 242 Mg ha^{-1}: 22 years later there were 568 trees ha^{-1} with a biomass of 371 Mg ha^{-1}. During the 22 years an average of 19 trees ha^{-1} died each year causing an average loss of biomass of 1.9 Mg ha^{-1} yr^{-1}. Over the same period litter fall averaged 3.7 Mg ha^{-1} yr^{-1} and the actual average growth rate of the living trees was 5.7 Mg ha^{-1} yr^{-1}. Thus mortality and litter fall together were responsible for halving the stand growth rate.

In areas with endemic wind-throw, such as the British Isles, a large and variable amount of mortality occurs annually. At high elevations up to 5% of the forest may be wind-thrown annually and the growth rate of individual stands suddenly halved. Catastrophic wind-throw strikes irregularly and can reduce the growth rate of large areas to zero (see also Chap. 15, Vol. 12A).

8.7.3 Losses from Insect Attack

Grazing and defoliation of the canopy by insects reduces both light interception and, if a new generation of leaves is produced, the availability of reserves for growth (see Chap. 17, Vol. 12C). Hence growth rates would be expected to move down both the line BO and the line BC in Fig. 8.2. FRANKLIN (1970), RAFES (1970) and VARLEY et al. (1973) reviewed the effects of particular species of insects on forest canopies and tree growth. Moderate grazing leads to the loss of 5%–10% of leaf area; irregular defoliation of up to ca. 30% leads to a reduced canopy in the following year and evident reduced growth; repeated defoliation of 30% or more leads to death which occurs in one or two years if the defoliation repeatedly exceeds 70%. Depending, therefore, on the severity of the attack, growth rates will be found throughout the variable space below the boundary line BO in Fig. 8.2. The absence of the native insect pests is one reason why the growth rate of exotic species in frequently very high and exceeds the growth rate of native species, e.g. *Pinus radiata* in New Zealand, *Eucalyptus globulus* in Portugal, *Picea sitchensis* in Scotland. A number of physiological and biological changes also occur as the result of insect attack. For example, stressed trees are reported to mobilise a large part of their stored carbohydrates into sugars and to convert protein and other nitrogenous compounds into more soluble forms such as amino acids (MCLAUGHLIN and SHRINER 1980). Clearly these changes will have more subtle effects on productivity.

8.7.4 Fine Root Turnover

The population of fine roots increases and decreases during the growing season (ATKINSON 1980). Root growth tends to increase with increasing temperature during spring and summer but may be inhibited by shortage of reserves and water stress. During periods of rapid shoot growth, fine root growth may cease and the existing fine roots die (LYR and HOFFMANN 1967; HEAD 1973; DEANS 1979). Small decreases in soil matric potential may also lead to cessation of growth and substantial death of fine roots (SANANTONIO 1979; HINCKLEY et al. 1979; DEANS 1979). Fine roots rank a close second behind leaf and shoot growth in precedence for reserves for growth, and regrowth occurs after rewetting of the surface soil horizons where they are concentrated. It has been shown in stands of *Pinus sylvestris* (PERSSON 1978, 1979, 1980), *Pseudotsuga menziesii* (SANANTONIO et al. 1977; SANANTONIO 1979), *Picea sitchensis* (FORD and DEANS 1977; DEANS 1979 and personal communication), in *Abies amabilis* (GRIER et al. 1981), *Quercus alba* (HINCKLEY et al. 1979) and *Liriodendron* forest (HARRIS et al. 1977) that the fine root populations turnover 5–10 times per season involving the production of 2 to 9 Mg ha^{-1} yr^{-1} of new tissue. This represents a direct loss of carbon in both substance and respiration which might otherwise have contributed to the annual growth rate. Estimates of the annual carbon balance of a stand of young *Pinus sylvestris* in central Sweden showed that 65%–75% of the annual photosynthetic production was used below ground largely in the production of fine roots. The regular application of low concentration liquid fertiliser during the growing season over 5 years reduced this to about 40% of the annual carbon income (LINDER and ROOK 1983).

8.8 Conclusions

We have identified six properties of the forest all of which can have substantial effects on growth. We have attempted to evaluate the six properties quantitatively as factors which diminish the rate of growth below the maximum set by a quantum requirement of 20 mol per mol of CO_2 fixed or 4.5 g of dry matter per MJ of PAR. Annual growth rate is reduced below this potential in proportion to the combined value of the parameters as defined by Eq. (8.16). Crude estimates of the factors for cool and warm evergreen and deciduous temperate forest are proposed in Table 8.6. When combined they lead to estimates of ε of 0.2 to 0.8 g MJ^{-1} and of G of 4.6 to 11.7 Mg ha^{-1} yr^{-1}. Table 8.6 indicates that the major discriminants of growth rate are the interception of quanta by the canopy, and the losses resulting from root, leaf and tree mortality.

Clearly each stand will have a unique combination of the six factors. The estimates in Table 8.6 are really only guesses. To understand the reasons under-

Table 8.6 a. Suggested values of the six factors that determine growth rate [see Eq. (14a)] for cool and warm, evergreen and deciduous temperate forest

Factor	Cool		Warm	
	Evergreen	Deciduous	Evergreen	Deciduous
$\bar{a}$	0.9	0.8	0.8	0.9
$\bar{i}$	0.9	0.5	0.55	0.45
$\bar{f}$	0.8	0.9	0.9	0.5
$\bar{Y}_g$	0.7	0.7	0.7	0.7
$\bar{Y}_m$	0.95	0.95	0.8	0.8
$\bar{d}$	0.4	0.4	0.3	0.3
Combined effect	0.17	0.10	0.07	0.03

Table 8.6b. The effect of the combined factors in leading to the effective $\hat{\varepsilon}$ and in predicting growth rates, $\hat{G}$, in relation to the potential G_0

Factor	Cool		Warm	
	Evergreen	Deciduous	Evergreen	Deciduous
$\hat{\varepsilon}$ (g MJ^{-1})	0.78	0.45	0.30	0.15
$\hat{G}(=\hat{\varepsilon}\int Q_0 \cdot dt)$[a] (Mg ha^{-1} yr^{-1})	11.7	6.8	9.1	4.6
$\hat{G}/G_0$ [from Eq. (16)]	0.27	0.17	0.12	0.05

[a] $Q_0 = 1{,}500$ and 3,000 MJ m^{-2} yr^{-1} for PAR in cool and warm temperate regions, respectively

Table 8.7. Annual dry matter balances for three stands for which both assimilation and dry matter data are available. The gains and losses are expressed in Mg ha^{-1} yr^{-1} of dry matter. Net photosynthetic production is the net production of assimilate by the leaves after leaf respiration has been deducted, i.e. it is the integral of what is measured in a cuvette over 24 h

Species	*Pinus sylvestris*		*Fagus sylvatica*		*Picea abies*	
Source	ÅGREN et al. (1980)		SCHULZE (1970)		SCHULZE et al. (1977a)	
Age (yr)	14		120		89	
Height (m)	2.8		27		26	
Basal area (m^2 ha^{-1})	2.9		—		~35	
Leaf area index	1.0		5.7		11.6	
Gains (Mg ha^{-1} yr^{-1})						
Net photosynthetic production		3.44		14.6		29.7
Losses (Mg ha^{-1} yr^{-1})						
Respiration of						
Stems	0.10		6.6[a] (respiration and fine root turnover combined)		17.9[a] (respiration and fine root turnover combined)	
Branches	0.03					
Roots	0.22					
Buds, fruits etc.	[b]					
Fine root turnover	1.84					
Litter fall	0.57		3.8		3.5	
Tree mortality	[b]		[b]		[b]	
Other losses	0.05		[b]		[b]	
Total loss		2.80		10.4		21.3
Balance (Mg ha^{-1} yr^{-1}) (=growth rate of stem, branches and coarse roots)		0.64		4.2		8.4

[a] Estimated as a residual
[b] Data not given

lying the growth rate of a particular stand, it is necessary to evaluate these factors quantitatively. A recent attempt to do this for a young stand of *Pinus sylvestris* is presented by ÅGREN et al. (1980) (Table 8.7). Their carbon budget is unique in that they attempted to evaluate all the terms from measurements and to balance the budget without the necessity of assigning the remainder

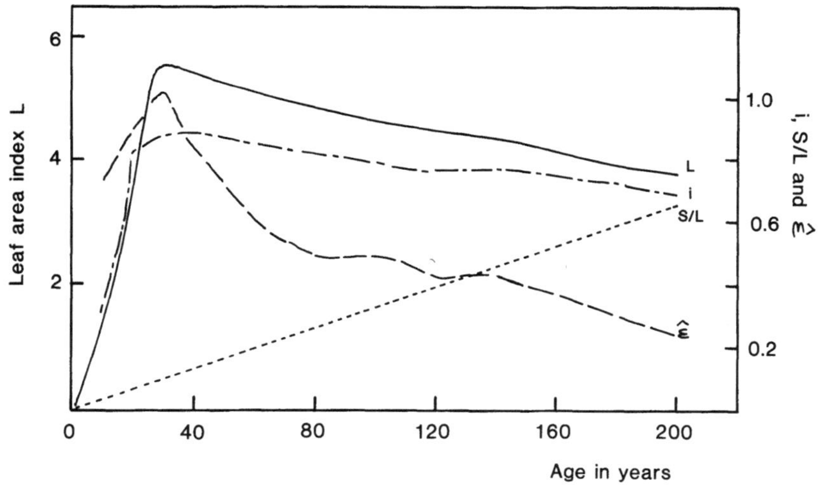

Fig. 8.10. The changes with age of leaf area index (L), interceptance of solar radiation (i), the ratio of non-leaf surface area to leaf surface area (S/L) and the effective efficiency of PAR utilisation ($\hat{\varepsilon}$, g MJ^{-1}) in forests of pedunculate oak, *Quercus robur*. (Adapted from RAUNER 1976)

to one of the processes. There is a large discrepancy between their estimates of respiration loss and those derived here, emphasising the difficulties and uncertainties in both approaches. Much larger estimates of respiration (plus fine root turnover) are included in the budgets for *Fagus sylvatica* and *Picea abies* but these have been obtained as the residual and are, therefore, to some extent suspect. In all three cases, some uncertainty lies in the calculation of net photosynthetic production for the stand because of the impossibility of adequately sampling the variation in the canopy over the year with a few assimilation chambers, and because none of the budgets includes an allowance for mortality. Criteria for determining the CO_2-exchange components of the carbon balance of coniferous trees are discussed by LINDER and LOHAMMAR (1981). [There is a number of other fairly complete dry matter or carbon budgets in the literature calculated solely from harvest data (e.g. GRIER and LOGAN 1977; KINERSON et al. 1977; KIRA and YABUKI 1978; GRIER et al. 1981) but these suffer from the serious limitation that the net photosynthetic production was obtained by summation rather than by measurement, and they have not therefore been included in this discussion].

At present then, we lack a complete carbon budget for any one stand, yet changes in the budget are certain to occur with age. Extensive work in oak forests in the USSR, summarised by RAUNER (1976), shows a decline in leaf area index and an increase in the proportion of non-leaf, respiring surface as the stands get older and these changes lead to reductions in both the radiation interceptance and $\hat{\varepsilon}$ (Fig. 8.10), and consequently must result in a substantial fall in G. That G falls with age is, of course, well known from forest mensuration data, but the factors responsible are largely unknown. Without doubt more and better data relating to the factors discussed here are required to elucidate the factors controlling forest production. It is our hope that the framework adopted here will help to direct efforts where they will be most rewarding. It is evident from the deficiencies encountered in making this analysis that a disproportionate amount of work has been done and is continuing to be done on leaf photosynthesis without regard to the crucial role of other processes such as leaf population dynamics, leaf growth and canopy development, maintenance respiration, mortality and shedding, especially of the fine roots.

Acknowledgement. This work was supported by National Science Foundation, Grant No. DEB 8100348 to J.W. LEVERENZ.

References

Aber JD (1979) Foliage-height profiles and succession in northern hardwood forests. Ecology 60:18–23

Ågren GI, Axelsson B (1980) PT – a tree growth model. In: Persson T (ed) Structure and function of northern coniferous forests – an ecosystem study. Ecol Bull (Stockholm) 32:525–536

Ågren GI, Axelsson B, Flower-Ellis JGK, Linder S, Persson H, Staaf H, Troeng E (1980) Annual carbon budget for a young Scots pine. In: Persson T (ed) Structure and function of northern coniferous forests – an ecosystem study. Ecol Bull (Stockholm) 32:307–313

Albrektson A (1980) Relations between tree biomass fractions and conventional silvicultural measurements. In: Persson T (ed) Structure and function of northern coniferous forests – an ecosystem study. Ecol Bull (Stockholm) 32:315–327

Albrektson A, Aronsson A, Tamm CO (1977) The effect of forest fertilisation on primary production and nutrient cycling in forest ecosystems. Silva Fenn 11:233–239

Anderson HW, Zsuffa L (1975) Yield and wood quality of hybrid cottonwood grown in a two-year rotation. Ont Div For Min Nat Res Rep 101:1–35

Anderson MC (1964a) Light relations of terrestrial plant communities and their measurement. Biol Rev 39:425–486

Anderson MC (1964b) Studies of the woodland light climate II. Seasonal variation in the light climate. J Ecol 52:643–663

Anderson MC (1981) The geometry of leaf distribution in some south-eastern Australian forests. Agric Meteorol 25:195–205

Aouni El MH, Mousseau M (1974) Relation d'échange de CO_2 chez les aiguilles du pin noir d'Autriche (*Pinus nigra* Arn.) avec l'age, la teneur en chlorophylle et réassimilation. Photosynthetica 8:78–86

Ashton DH (1976a) Phosphorous in forest ecosystems at Beenak, Victoria. J Ecol 64:171–186

Ashton DH (1976b) The development of even-aged stands of *Eucalyptus regnans* F. Muell in central Victoria. Aust J Bot 24:397–414

Atkinson D (1980) The distribution and effectiveness of the roots of tree crops. Hortic Rev 2:424–490

Attiwill PM (1979) Nutrient cycling in a *Eucalyptus obliqua* (L'Herit.) forest. III Growth, biomass and net primary production. Aust J Bot 27:439–458

Attiwill PM, Guthrie HB, Leuning R (1978) Nutrient cycling in a *Eucalyptus obliqua* (L'Herit.) forest. I Litter production and nutrient return. Aust J Bot 26:79–91

Barnes A, Hole CC (1978) A theoretical basis of growth and maintenance respiration. Ann Bot (London) 42:1217–1221

Beadle CL, Neilson RE, Jarvis PG, Talbot H (1981) Photosynthesis as related to xylem water potential and carbon dioxide concentration in Sitka spruce. Physiol Plant 52:391–400

Beadle CL, Talbot H, Jarvis PG (1982) Canopy structure and leaf area index in a mature Scots pine forest. Forestry 55:2–20
Bennett KJ, Rook DA (1978) Stomatal and mesophyll resistance in two clones of *Pinus radiata* D. Don known to differ in transpiration and survival rate. Aust J Plant Physiol 5:201–230
Berry JA, Downton WJS (1982) Environmental regulation of photosynthesis. In: Govindjee (ed) Photosynthesis II. Carbon metabolism and plant productivity. Academic Press, London New York, pp 263–343
Black JN (1956) The distribution of solar radiation over the earth's surface. Arch Meteorol Geophys Bioklimatol Ser B7:165–189
Boardman NK (1977) Comparative photosynthesis of sun and shade plants. Annu Rev Plant Physiol 28:355–377
Boysen-Jensen P (1932) Die Stoffproduktion der Pflanzen. Fischer, Jena
Bradbury IK, Malcolm DC (1978) Dry matter accumulation by *Picea sitchensis* seedlings during winter. Can J For Res 8:207–213
Bray JR, Gorham E (1964) Litter production in forests of the world. Adv Ecol Res 2:101–157
Brix H (1968) Influence of light intensity at different temperatures on rate of respiration of Douglas-fir seedlings. Plant Physiol 43:389–393
Büsgen M, Münch E (1931) The structure and life of forest trees. Translated by T. Thomson. Chapman and Hall Ltd., London
Butler DR, Landsberg JJ (1981) Respiration rates of apple trees, estimated by CO_2-efflux measurements. Plant Cell Environ 4:153–159
Caemmerer von S, Farquhar GD (1981) Some relationships between the biochemistry of photosynthesis and the gas exchange of leaves. Planta 153:376–387
Carbon BA, Bartle GA, Murray AM (1979) Leaf area index of some eucalypt forests in south-west Australia. Aust For Res 9:323–326
Christensen O (1977) Estimation of standing crop and turnover of dead wood in a Danish oak forest. Oikos 28:177–186
Cooper JP (1975) Control of photosynthetic production in terrestrial systems. In: Cooper JP (ed) Photosynthesis and productivity in different environments. CUP, London, pp 593–621
Crist JB, Dawson DH (1975) Anatomy and dry weight yields of two *Populus* clones grown under intensive culture. USDA For Serv Res Pap NC-224:1–15
Deans JD (1979) Fluctuations of the soil environment and fine root growth in a young Sitka spruce plantation. Plant Soil 52:195–208
Dougherty PM, Teskey RO, Phelps JE, Hinckley TM (1979) Net photosynthesis and early growth trends of a dominant white oak (*Quercus alba* L.) Plant Physiol 64:930–935
Dykstra GF (1974) Photosynthesis and carbon dioxide transfer resistance of lodgepole pine seedlings in relation to irradiance, temperature and water potential. Can J For Res 4:201–206
Ehleringer JR, Björkman O (1977) Quantum yields for CO_2 uptake in C_3 and C_4 plants. Dependence on temperature, CO_2 and O_2 concentration. Plant Physiol 59:86–90
Emmingham WH, Waring RH (1977) An index of photosynthesis for comparing forest sites in western Oregon. Can J For Res 7:165–174
Fischer RA (1983) Growth and yield of wheat. In: Potential productivity of field crops under different environments. Int Rice Res Inst, Manila (in press)
Flower-Ellis JGK, Persson H (1980) Investigation of structural properties and dynamics of Scots pine stands. In: Persson T (ed) Structure and function of northern coniferous forests – an ecosystem study. Ecol Bull (Stockholm) 32:125–138
Foote KC, Schaedle M (1976) Diurnal and seasonal patterns of photosynthesis and respiration by stems of *Populus tremuloides* Michx. Plant Physiol 58:651–655
Ford ED (1975) Competition and stand structure in some even-aged plant monocultures. J Ecol 63:311–333
Ford ED (1982) High productivity by a pole stage Sitka spruce stand and its relation to canopy structure. Forestry 55:1–17

Ford ED, Deans JD (1977) Growth of a Sitka spruce plantation: spatial distribution and seasonal fluctuations of lengths, weights and carbohydrate concentrations of fine roots. Plant Soil 47:463–485
Ford ED, Newbould PJ (1971) The leaf canopy of coppiced deciduous woodland. I Development and structure. J Ecol 59:843–862
Franklin RT (1970) Insect influences of the forest canopy. In: Reichle DE (ed) Analysis of temperate forest ecosystems. Ecol Stud Vol I. Springer, Berlin Heidelberg New York, pp 86–99
Fuchs M, Schulze ED, Fuchs MI (1977) Spacial distribution of photosynthetic capacity and performance in a mountain spruce forest in northern Germany II. Climatic control of carbon dioxide uptake. Oecologia 29:329–340
Fujimori J, Kawanabe S, Saito H, Grier CC, Shidei T (1976) Biomass and primary production in forests of three major vegetation zones of the northwestern United States. J Jpn For Soc 58:360–373
Fujimori T (1971) Primary productivity of a young *Tsuga heterophylla* stand and some speculations about biomass of forest communities on the Oregon coast. USDA For Serv Res Pap PNW-123:1–11
Fujimori T (1977) Stem biomass and structure of a mature *Sequoia sempervirens* stand on the Pacific coast of northern California. J Jpn For Soc 59:435–441
Gallagher JN, Biscoe PV (1978) Radiation absorption, growth and yield of cereals. J Agric Sci 91:47–60
Gholz HL, Fitz FK, Waring RH (1976) Leaf area difference associated with old-growth forest communities in the western Oregon Cascades. Can J For Res 6:49–57
Gordon JC (1975) The productive potential of woody plants. Iowa State J Res 49:267–274
Grace J (1980) Some effects of wind on plants. In: Grace J, Ford ED, Jarvis PG (eds) Plants and their atmospheric environment. Blackwell, Oxford, pp 31–56
Grier CC, Logan RS (1977) Old-growth *Pseudotsuga menziesii* communities of a western Oregon watershed: biomass distribution and production budgets. Ecol Monogr 47:373–400
Grier CC, Running SW (1977) Leaf area of mature north-western coniferous forests: relation to a site water balance. Ecology 58:893–899
Grier CC, Vogt KH, Keyes MR, Edmonds RL (1981) Biomass distribution and above- and below-ground production in young and mature *Abies amabilis* zone ecosystems of the Washington Cascades. Can J For Res 11:155–166
Hagem O (1962) Additional observations on the rate of dry matter increase of coniferous seedlings in winter. Investigations in an oceanic climate. Medd Vestl Forstl Forsoeksstn 37:1–253
Hanley DP (1976) Tree biomass and productivity estimated for three habitat types of northern Idaho. Univ Idaho Coll For Wildl Range Sci Bull 14:1–5
Harris WF, Kinerson RS, Edwards NT (1977) Comparison of belowground biomass of natural deciduous forest and loblolly pine plantations. Pedobiologia 17:369–381
Hathaway RL (1980) Effect of planting density and harvesting cycle on biomass production of willows. Min Works Devel NZ Aokautere Sci Cent Rep 16:1–6
Havranek WM (1981) Stammatmung, Dickenwachstum und Photosynthese einer Zirbe (*Pinus cembra* L.) an der Waldgrenze. In: Dickenwachstum der Bäume, Mitteil Forst. Bundesversuchsanstalt, Wien 142:443–468
Head GC (1973) Shedding of roots. In: Kozlowski TT (ed) Shedding of plant parts. Academic Press, London New York, pp 237–293
Helms JA (1970) Summer net photosynthesis of ponderosa pine in its natural environment. Photosynthetica 4:243–253
Hinckley TM, Dougherty PM, Lassoie JP, Roberts JE, Teskey RO (1979) A severe drought: impact on tree growth, phenology, net photosynthetic rate and water relations. Am Midl Nat 102:307–316
Hinckley TM, Teskey RO, Duhme F, Richter H (1981) Temperate hardwood forests. In: Kozlowski TT (ed) Water deficits and plant growth, vol VI. Academic Press, London New York, pp 153–208
Hole CC, Barnes A (1980) Maintenance and growth components of carbon dioxide efflux from growing pea fruits. Ann Bot 45:295–307

Horie T (1977) Simulation of sunflower growth I. Formulation and parametrisation of dry matter production, leaf photosynthesis, respiration and partitioning of photosynthates. Bull Natl Inst Agric Sci Ser A 24:45–70

Houssard C, Escarre J (1981) Mesures de la quantité relative de feuillage par strates dans des taillis de chêne pubescent. Ann Sci For 38:449–468

Hsiao TC, Acevedo E, Fereres E, Henderson DW (1976) Stress metabolism – water stress, growth, and osmotic adjustment. Philos Trans R Soc London Ser B 273:479–500

Hutchison BA, Matt DR (1976) Beam enrichment of diffuse radiation in a deciduous forest. Agric Meteorol 17:93–110

Hutchison BA, Matt DR (1977) The distribution of solar radiation within a deciduous forest. Ecol Monogr 47:185–207

Jarvis PG (1976) The interpretation of the variations in leaf water potential and stomatal conductance found in canopies in the field. Philos Trans R Soc London Ser B 273:593–610

Jarvis PG (1980) Stomatal conductance, gaseous exchange and transpiration. In: Grace J, Ford ED, Jarvis PG (eds) Plants and their atmospheric environment. Blackwell, Oxford, pp 175–204

Jarvis PG (1981) Production efficiency of coniferous forest in the U.K. In: Johnson CB (ed) Physiological processes limiting plant productivity. Butterworth, London, pp 81–107

Jarvis PG, James GB, Landsberg JJ (1976) Coniferous forest. In: Monteith JL (ed) Vegetation and the atmosphere 2, Case studies. Academic Press, London New York, pp 171–240

Johansson N (1933) The relation between the tree-stem's respiration and its growth. Sven Skogsvardsfoer Tidskr 10:53–134 (in Swedish with English summary)

Kestemont P (1977) Biomasse et productivité primaire de la douglasière de Mirwart. (Plantation de *Pseudotsuga menziesii*). In: Duvigneaud P, Kestemont P (eds) Productivité biologique en Belgique. Duculot, Paris, pp 177–189

Kestemont P, Duvigneaud P, Paulet E (1977) Biomasse et productivité primaire d'une pessière à Mirwart. (Plantation de *Picea abies*). In: Duvigneaud P, Kestemont P (eds) Productivité biologique en Belgique. Duculot, Paris, pp 161–176

Keyes MR, Grier CC (1981) Above- and below-ground net production in 40-year-old Douglas fir stands on low and high productivity sites. Can J For Res 11:599–605

Kinerson RS (1975) Relationship between plant surface area and respiration in loblolly pine. J Appl Ecol 12:965–971

Kinerson RS (1979) Studies of photosynthesis and diffusion resistance in paper birch (*Betula papyrifera* Marsh.) with synthesis through computer simulation. Oecologia (Berlin) 39:37–49

Kinerson RS, Higginbotham KO, Chapman RC (1974) The dynamics of foliage distribution within a forest canopy. J Appl Ecol 11:347–353

Kinerson RS, Ralston CW, Wells CG (1977) Carbon cycling in a loblolly pine plantation. Oecologia 29:1–10

Kira T (1975) Primary production of forests. In: Cooper JP (ed) Photosynthesis and productivity in different environments. CUP, London

Kira T (1977) Production rates. In: Shidei T, Kira T (eds) Primary productivity of Japanese forests. JIBP Synth, Tokyo 16:101–114

Kira T, Shidei T (1967) Primary production and turnover of organic matter in different forest ecosystems of the western Pacific. Jpn J Ecol 17:70–87

Kira T, Yabuki K (1978) Primary production rates in the minamata forest. In: Kira T, Ono Y, Hosokawa T (eds) Biological production in a warm-temperate evergreen oak forest of Japan. JIBP Synth, Tokyo 18:131–139

Kira T, Shinozaki K, Hozumi K (1969) Structure of forest canopies as related to their primary productivity. Plant Cell Physiol 10:129–142

Kramer PJ, Kozlowski TT (1979) Physiology of woody plants, 2nd edn. Academic Press, London New York

Kriedemann PE, Barrs HD (1981) Citrus orchards. In: Kozlowski TT (ed) Water deficits and plant growth Vol VI. Academic Press, London New York, pp 325–417

Künstle E, Mitscherlich G (1975) Photosynthese, Transpiration und Atmung in einem Mischbestand im Schwarzwald. I. Photosynthese. Allg Forst Jagdztg 146:45–62
Kuruoiwa S (1970) Total photosynthesis of foliage in relation to inclination of leaves. In: Šetlik I (ed) Prediction and measurement of photosynthetic productivity. Proc Trebon Symp. PUDOC, Wageningen, pp 79–89
Kusumoto T (1978) Photosynthesis and respiration of leaves of main component species. In: Kira T, Ono Y, Hosokawa T (eds) Biological production in a warm-temperate evergreen oak forest of Japan. JIBP Synth, Tokyo 18:88–98
Landsberg JJ, Jones HG (1981) Apple orchards. In: Kozlowski TT (ed) Water deficits and plant growth Vol VI. Academic Press, London New York, pp 419–469
Larcher W (1969) The effect of environmental and physiological variables on the carbon dioxide gas exchange of trees. Photosynthetica 3:167–198
Larcher W (1980) Physiological plant ecology, 2nd edn. Springer, Berlin Heidelberg New York
Ledig FT, Drew AP and Clark JG (1976) Maintenance and constructive respiration, photosynthesis and net assimilation rate in seedlings of pitch pine (*Pinus rigida* Mill.). Ann Bot 40:289–300
Legg BJ, Day W, Lawlor DW, Parkinson KJ (1979) The effects of drought on barley growth: models and measurements showing the relative importance of leaf area and photosynthetic rate. J Agric Sci 92:703–716
Leverenz JW (1981) Photosynthesis and transpiration in large forest-grown Douglas-fir: diurnal variation. Can J Bot 59:349–356
Leverenz JW, Jarvis PG (1979) Photosynthesis in Sitka spruce VIII. The effects of light flux density and direction on the rate of net photosynthesis and the stomatal conductance of needles. J Appl Ecol 16:919–932
Leverenz JW, Jarvis PG (1980) Photosynthesis in Sitka spruce (*Picea sitchensis* (Bong.) Carr.) IX. The relative contribution made by needles at various positions on the shoot. J Appl Ecol 17:59–68
Lieth H (1970) Phenology in productivity studies. In: Reichle DE (ed) Analysis of temperate forest ecosystems. Ecol Stud, vol I. Springer, Berlin Heidelberg New York, pp 29–46
Linder S (1979) Photosynthesis and respiration in conifers: a classified reference list 1891–1977. Stud For Suec 149:1–71
Linder S, Lohammar T (1981) Amount and quality of information on CO_2-exchange required for estimating annual carbon balance of coniferous trees. In: Linder S (ed) Understanding and predicting tree growth. Stud For Suec 160:73–87
Linder S, Rook DA (1983) Effects of mineral nutrition on carbon dioxide exchange and partitioning of carbon in trees. In: Bowen GD, Nambier EKS (eds) Nutrition of forest trees in plantations. Academic Press (in press)
Linder S, Troeng E (1980) Photosynthesis and transpiration of 20-yr-old Scots pine. In: Persson T (ed) Structure and function of northern coniferous forests – an ecosystem study. Ecol Bull 32:165–181
Linder S, Troeng E (1981) The seasonal variation in stem and coarse root respiration of a 20-yr-old Scots pine (*Pinus sylvestris* L.) In: Tranquillini W (ed) Radial growth in trees. Mitt Forstl Bundes-Versuchsanst Wien 142:125–139
Little CMA, Loach K (1975) Effect of gibberellic acid on growth and photosynthesis in *Abies balsamea*. Can J Bot 53:1805–1810
Long JN, Turner J (1975) Aboveground biomass of understory and overstory in an age sequence of four Douglas-fir stands. J Appl Ecol 12:179–188
Loomis RS, Rabbinge R, Ng E (1979) Explanatory models in crop physiology. Annu Rev Plant Physiol 30:339–367
Ludlow MM, Jarvis PG (1971) Photosynthesis in Sitka spruce (*Picea sitchensis* (Bong.) Carr.). I. General characteristics. J Appl Ecol 8:925–953
Lyr H, Hoffmann G (1967) Growth rates and growth periodicity of tree roots. Int Rev For Res 2:181–236
Madgwick HAI (1981) Above-ground dry-matter content of a young close-spaced *Pinus radiata* stand. N Z J For Sci 11:203–209
Madgwick HAI, Jackson DS, Knight PJ (1977) Above-ground dry matter, energy and nutrient contents of trees in an age series of *Pinus radiata* plantations. N Z J For Sci 7:445–468

Madgwick HAI, Beets P, Gallagher S (1981) Dry matter accumulation, nutrient and energy content of the above-ground portions of 4-yr-old stands of *Eucalyptus nitens* and *E. fastigata*. N Z J For Sci 11:53–59
Marks PL (1974) The role of pin cherry (*Prunus pennsylvanica* L.) in the maintenance of stability in northern hardwood ecosystems. Ecol Monogr 44:73–88
Marks PL, Borman FH (1972) Revegetation following forest cutting: mechanisms for return to steady-state nutrient cycling. Science 172:914–915
Marshall B, Biscoe PV (1980) A model for C_3 leaves describing the dependence of net photosynthesis on irradiance I. Derivation. J Exp Bot 31:29–39
McCree KJ (1970) An equation for the rate of respiration of white clover plants grown under controlled conditions. In: Šetlik I (ed) Production and measurement of photosynthetic productivity. Proc Trebon Symp. PUDOC, Wageningen, pp 221–229
McCree KJ (1974) Equations for the rate of dark respiration of white clover and grain sorghum, as functions of dry weight, photosynthetic rate and temperature. Crop Sci 14:509–514
McCree KJ, Silsbury JM (1978) Growth and maintenance requirements for subterranean clover. Crop Sci 25:53–58
McLaughlin SB, Shriner DS (1980) Allocation of resources to defense and repair. In: Horsfall JG, Cowling EB (eds) Plant disease Vol V. Academic Press, London New York, pp 407–431
Miller HG, Miller JD (1976) Effect of nitrogen supply on net primary production in Corsican pine. J Appl Ecol 13:249–256
Miller HG, Cooper JM, Miller JD (1976) Effect of nitrogen supply on nutrients in litter fall and crown leaching in a stand of Corsican pine. J Appl Ecol 13:233–248
Miller PC (1967) Leaf temperatures, leaf orientation and energy exchange in quaking aspen (*Populus tremuloides*) and Gambell's oak (*Quercus gambellii*) in central Colorado. Oecol Plant 2:241–270
Minderman G (1967) The production of organic matter and the utilisation of solar energy by a forest plantation of *Pinus nigra* var. *austriaca*. Pedobiologia 7:11–22
Monteith JL (1972) Solar radiation and productivity in tropical ecosystems. J Appl Ecol 9:747–766
Monteith JL (1977) Climate and efficiency of crop production in Britain. Philos Trans R Soc London Ser B 281:277–294
Monteith JL (1981) Does light limit crop production? In: Johnson CB (ed) Physiological processes limiting plant productivity. Butterworth, London, pp 23–38
Mukammal EI (1971) Some aspects of radiant energy in a red pine forest. Arch Met Geophys Bioklimatol Ser B 19:29–52
Müller D, Nielsen J (1965) Production brutes pertés par respiration et production nette dans la forêt ombrophile tropicale. Forstl Forsogsvaes Dan 29:69–160
Negisi K (1975) Diurnal fluctuations of CO_2 release from the stem bark of standing young *Pinus densiflora* trees. J Jpn For Soc 57:375–383
Negisi K (1977) Respiration in forest trees. In: Shidei T, Kira T (eds) Primary productivity of Japanese forest. JIBP Synth, Tokyo 16:86–99
Negisi K (1978) Daytime depression in bark respiration and radial shrinkage in stem of a standing young *Pinus densiflora* tree. J Jpn For Soc 60:380–382
Neilson RE, Ludlow MM, Jarvis PG (1972) Photosynthesis in Sitka spruce (*Picea sitchensis* (Bong.) Carr.) II. Response to temperature. J Appl Ecol 9:721–745
Newbould PJ (1967) Methods for estimating the primary production of forests. Blackwell, Oxford
Nishioka M, Kirita H (1978) Litterfall. In: Kira T, Ono Y, Hosokawa T (eds) Biological production in a warm-temperate evergreen oak forest of Japan. JIBP Synth, Tokyo 18:231–238
Norman JM (1975) Radiative transfer in vegetation. In: De Vries DA, Afgan NH (eds) Heat and mass transfer in the biosphere 1 Transfer processes in plant environment. Scripta Book, Washington, DC, pp 187–205
Norman JM (1978) Modelling the complete crop canopy. In: Barfield BJ, Gerber JF (eds) Modification of the aerial environment of plants. Am Soc Agric Eng, St Joseph, Michigan, pp 249–277

Norman JM (1980) Interfacing leaf and canopy light interception models. In: Hesketh JD, Jones JW (eds) Predicting photosynthesis for ecosystem models II. CRC, Boca Raton, pp 49–67
Norman JM (1981) Simulation of microclimates. In: Hatfield J (ed) Application of biometeorology to integrated pest management. Academic Press, London New York, pp 65–99
Norman JM, Jarvis PG (1974) Photosynthesis in Sitka spruce (*Picea sitchensis* (Bong.) Carr.) III. Measurement of canopy structure and interception of radiation. J Appl Ecol 11:375–398
Norman JM, Jarvis PG (1975) Photosynthesis in Sitka spruce (*Picea sitchensis* (Bong.) Carr.) V. Radiation penetration theory and a test case. J Appl Ecol 12:839–878
Norman JM, Welles J (1983) Radiative transfer in an array of canopies. Agron J (in press)
Norman JM, Miller EE, Tanner CB (1971) Light intensity and sunfleck size distributions in plant canopies. Agron J 63:743–748
Norman JM, Perry SG, Fraser AB, Mach W (1979) In: 14th conference on agriculture and forest meteorology. Am Meteorol Soc, Boston, Mass, pp 184–185
Ovington JD (1956) The form, weights and productivity of tree species in close stands. New Phytol 55:289–304
Ovington JD (1957) Dry matter production in *Pinus sylvestris*. Ann Bot (London) 21:257–314
Ovington JD (1962) Quantitative ecology and the woodland ecosystem concept. In: Cragg JB (ed) Adv Ecol Res 1:103–192
Parker J (1978) Seasonal variation in photosynthesis in black oak twigs. Photosynthetica 12:423–427
Parry WH (1969) A study of the relationship between defoliation of Sitka spruce and population levels of *Elatobium abietinum* (Walker). Forestry 42:69–82
Penning de Vries FWT (1975a) The cost of maintenance processes in plant cells. Ann Bot 39:77–92
Penning de Vries FWT (1975b) Use of assimilates in higher plants. In: Cooper JP (ed) Photosynthesis and productivity in different environments. CUP, London, pp 459–507
Penning de Vries FWT, Brunsting AHM, Laar van AH (1974) Products requirements and efficiency of biosynthesis, a quantitative approach. J Theor Biol 45:339–377
Penning de Vries FWT, Witlage JM, Kremer D (1979) Rates of respiration and of increase in structural dry matter in young wheat, ryegrass and maize plants in relation to temperature, to water stress and to sugar content. Ann Bot 44:595–609
Persson H (1978) Root dynamics in a young Scots pine stand in central Sweden. Oikos 30:508–519
Persson H (1979) Fine root production, mortality and decomposition in forest ecosystems. Vegetatio 41:101–109
Persson H (1980) Death and replacement of fine roots in a mature Scots pine stand. In: Persson T (ed) Structure and function of northern coniferous forests – an ecosystem study. Ecol Bull (Stockholm) 32:251–260
Rafes PM (1970) Estimation of the effects of phytophagous insects on forest production. In: Reichle DE (ed) Analysis of temperate forest ecosystems. Ecol Stud Vol 1. Springer, Berlin Heidelberg New York, pp 100–106
Rauner JL (1976) Deciduous forests. In: Monteith JL (ed) Vegetation and the atmosphere Vol II. Case studies. Academic Press, London New York, pp 241–264
Reed KL, Hamerly ER, Dinger BE, Jarvis PG (1976) An analytical model for field measurements of photosynthesis. J Appl Ecol 13:925–942
Rook DA, Corson MJ (1978) Temperature and irradiance and the total daily photosynthetic production of a *Pinus radiata* tree. Oecologia 36:371–382
Saito H (1977) Litterfall. In: Shidei T, Kira T (eds) Primary productivity of Japanese forests. JIBP Synth, Tokyo 16:65–75
Sanantonio D (1979) Seasonal dynamics of fine roots in mature stands of Douglas fir of different water regimes – a preliminary report. In: Riedacker A, Gagnaire-Michard J (eds) Physiologie des racines et symbioses. C R Reunions Groupe d'Etude Racines, Nancy, pp 190–203

Sanantonio D, Hermann RK, Overton WS (1977) Root biomass studies in forest ecosystems. Pedobiologia 17:1–31
Satoo T (1970) A synthesis of studies by the harvest method: primary production relations in the temperate deciduous forests of Japan. In: Reichle DE (ed) Analysis of temperate forest ecosystems. Ecol Stud Vol I. Springer, Berlin Heidelberg New York, pp 55–72
Schönau APG, Pennefather M (1975) A first account of profits at harvesting as a result of fertilising *Eucalyptus grandis* at time of planting in Southern Africa. S Afr For J 94:29–35
Schulze E-D (1970) Der CO_2-Gaswechsel der Buche (*Fagus silvatica* L.) in Abhängigkeit von den Klimafaktoren im Freiland. Flora 159:177–232
Schulze E-D (1981) Carbon gain and wood production in trees of deciduous beech (*Fagus sylvatica*) and trees of evergreen spruce (*Picea excelsa*). In: Tranquillini W (ed) Radial growth in trees. Mitt Forstl Bundes-Versuchsanst Wien 142:105–123
Schulze E-D, Mooney MA, Dunn EL (1967) Wintertime photosynthesis of bristlecone pine (*Pinus aristata*) in the White Mountains of California. Ecology 48:1044–1047
Schulze E-D, Fuchs MI, Fuchs M (1977a) Spacial distribution of photosynthetic capacity and performance in a mountain spruce forest of northern Germany I. Biomass distribution and daily CO_2 uptake in different crown layers. Oecologia 29:43–61
Schulze E-D, Fuchs M, Fuchs MI (1977b) Spacial distribution of photosynthetic capacity and performance in a mountain spruce forest of northern Germany III. The significance of the evergreen habit. Oecologia 30:239–248
Šesták Z, Jarvis PG and Čatský J (1971) Criteria for the selection of suitable methods. In: Šesták Z, Čatský J, Jarvis PG (eds) Plant photosynthetic production: Manual of methods. Junk, The Hague
Shepard RK (1975) Radial distribution of corticular photosynthate in stems of big tooth and trembling aspen. For Sci 21:370–372
Shinozaki K, Kira T (1977) Canopy structure and light utilization. In: Shidei T, Kira T (eds) Primary productivity of Japanese forests. JIBP Synth, Tokyo 16:75–86
Silsbury JH (1979) Growth, maintenance and nitrogen fixation of nodulated plants of subterranean clover (*Trifolium subterraneum* L.) Aust J Plant Physiol 6:165–176
Slatyer RO, Ferrar PJ (1977) Altitudinal variation in the photosynthetic characteristics of snow gum, *Eucalyptus pauciflora* Sieb. ex Spreng. II Effects of growth temperature under controlled conditions. Aust J Plant Physiol 4:289–299
Sollins P (1982) Input and decay of coarse woody debris in coniferous forest stands in western Oregon and Washington. Can J For Res 12:18–28
Stephens GR (1969) Productivity of red pine I. Foliage distribution in tree crown and stand canopy. Agric Meteorol 6:275–282
Switzer GL, Nelson LE, Smith WH (1966) The characterisation of dry matter and nitrogen accumulation by loblolly pine (*Pinus taeda* L.). Soil Sci Soc Am Proc 30:114–119
Szeicz G (1974) Solar radiation for plant growth. J Appl Ecol 11:617–636
Tadaki Y (1966) Some discussions on the leaf biomass of forest stands and trees. Bull Gov For Exp Stn Tokyo 184:1–35
Tadaki Y (1977) Leaf biomass. In: Shidei T, Kira T (eds) Productivity of terrestrial communities. JIBP Synth Tokyo 16:39–44
Tamm CO (1979) Nutrient cycling and productivity of forest ecosystems. In: Leaf AL (ed) Impact of intensive harvesting on forest nutrient cycling. State Univ New York, Coll Environ Sci For. Syracuse, New York, pp 2–21
Thornley JHM (1976) Mathematical models in plant physiology. Academic Press, London New York
Thornley JHM (1977) Growth, maintenance and respiration: a reinterpretation. Ann Bot 41:1191–1203
Thorpe MR, Saugier B, Auger S, Berger A, Methy M (1978) Photosynthesis and transpiration of an isolated tree: model and validation. Plant Cell Environ 1:269–277
Thorpe MR, Warrit B, Landsberg JJ (1980) Responses of apple leaf stomata: a model for single leaves and a whole tree. Plant Cell Environ 3:23–27
Tranquillini W (1959) Die Stoffproduktion der Zirbe (*Pinus cembra* L.) an der Waldgrenze während eines Jahres. Planta 54:107–151

Tranquillini W, Schütz W (1970) Über die Rindenatmung einiger Bäume an der Waldgrenze. Zentralbl Ges Forstwes 87:42–60
Troeng E, Linder S (1982a) Gas exchange of a 20-yr-old stand of Scots pine I. Net photosynthesis of current and one-yr-old shoots between and within seasons. Physiol Plant 54:7–14
Troeng E, Linder S (1982b) Gas-exchange in a twenty-yr-old stand of Scots pine II. Variation in net photosynthesis and transpiration within and between trees. Physiol Plant 54:15–23
Tsel'niker YL (1979) Resistances to CO_2 uptake at light saturation in forest tree seedlings at different adaptation to shade. Photosynthetica 13:124–129
Ungs MJ (1981) Distribution of light within the crown of an open-grown Douglas-fir. PhD Thesis, Oregon State Univ, Corvallis, Oregon
Varley GC (1970) The concept of energy flow applied to a woodland community. In: Watson A (ed) Animal populations in relation to their food resources. Blackwell, Oxford
Varley GC, Gradwell GR, Hassell MP (1973) Insect population ecology, an analytical approach. Blackwell, Oxford
Vyskot M (1979) Biomass of the tree layer of a spruce forest in the Bohemian uplands. In: Stability of spruce forest ecosystems MAB. UNESCO, Brno, pp 135–167
Waggoner PE, Turner NC (1971) Transpiration and its control by stomata in a pine forest. Bull Conn Agric Exp Stn 726:1–87
Walter H (1970) Vegetationszonen und Klima. Ulmer, Stuttgart
Waring RH, Franklin JF (1979) Evergreen coniferous forests of the Pacific northwest. Science 204:1380–1386
Waring RH, Emmingham WH, Gholz HL, Grier CC (1978) Variation in maximum leaf area of coniferous forests in Oregon and its ecological significance. For Sci 24:131–140
Watson RL, Landsberg JJ, Thorpe MR (1978) Photosynthetic characteristics of the leaves of "Golden Delicious" apple trees. Plant Cell Environ 1:51–58
Watts WR, Neilson RE, Jarvis PG (1976) Photosynthesis in Sitka spruce (*Picea sitchensis* (Bong.) Carr.) VII. Measurement of stomatal conductance and $^{14}CO_2$ uptake in forest canopy. J Appl Ecol 13:623–638
Westman WE, Whittaker RH (1975) The pygmy forest region of northern California: studies on biomass and primary productivity. J Ecol 63:493–520
Whitehead D, Jarvis PG (1981) Coniferous forests and plantations. In: Kozlowski TT (ed) Water deficits and plant growth, VI. Academic Press, London New York, pp 49–152
Whittaker RH (1966) Forest dimensions and production in the Great Smokey Mountains. Ecology 47:103–121
Whittaker RM, Woodwell GM (1967) Surface area relations of woody plants and forest communities. Am J Bot 54:931–939
Wiebe H (1975) Photosynthesis in wood. Physiol Plant 33:245–246
Wit CT de (1965) Photosynthesis of leaf canopies. Wageningen Inst Biochem Chem Res Field Crops Herbage, Agric Res Rep 663
Woodman JN (1971) Variation of net photosynthesis within the crown of a large forest-grown conifer. Photosynthetica 5:50–54
Woodwell GM, Botkin DB (1970) Metabolism of terrestrial ecosystems: the Brookhaven approach. In: Reichle DE (ed) Analysis of temperate forest ecosystems. Ecol Stud Vol 1. Springer, Berlin Heidelberg New York, pp 73–85
Woolley JT (1971) Reflectance and transmittance of light by leaves. Plant Physiol 47:656–662
Yoda K (1978) Estimation of community respiration. In: Kira T, Ono Y, Hosokawa T (eds) Biological production in a warm-temperate evergreen oak forest of Japan. JIBP Synth, Tokyo 18:112–131
Yoda K, Shinozaki K, Ogawa H, Hozumi K, Kira T (1965) Estimation of the total amount of respiration in woody organs of tree and forest communities. J Biol Osaka City Univ 16:15–26
Ziegler H (1957) Über den Gaswechsel verholzter Achsen. Flora 144:229–250

9 Productivity of Tropical Forests and Tropical Woodlands

E. MEDINA and H. KLINGE

CONTENTS

9.1 Introduction 281

9.2 Biomass Distribution 282

9.3 Structural and Functional Relationships 284
9.3.1 Vertical Distribution of Leaf Area and Light Penetration 284
9.3.2 Leaf Size and Specific Leaf Area in Different Tropical Ecosystems . . . 286
9.3.3 Development of Leaf Area Index and Transpiration 287
9.3.4 Photosynthetic Characteristics of Rain Forest Plants 287
9.3.4.1 Photosynthesis and Respiration of Tropical Plants 287
9.3.4.2 Photosynthetic Pathways and Life Forms 291

9.4 Productivity of Tropical Forests 292
9.4.1 Net and Gross Productivity 292
9.4.1.1 Litter Production 293
9.4.1.2 Gross and Net Production Data for Selected Tropical Forests . 297
9.4.2 $\delta^{13}C$ Values of Rain Forest Woody Plants 298

9.5 Conclusions 299

References 300

9.1 Introduction

The forests to be considered in this chapter are roughly located between 23° on either side of the equator and are climatically characterized by high total annual insolation $>4.5 \cdot 10^5$ J cm^{-2}, and thermal regimes in which daily temperature oscillations are frequently more pronounced than seasonal average monthly differences. The magnitude of the seasonal differences in temperature increases towards higher latitudes from a climatic equator, by definition the belt with more homogeneous climate in which weather is most predictable (UNESCO 1978). Near the equator, rainfall is generally high (>2,000 mm per year) and evenly distributed during the year, with seasonality increasing with one or two annual dry periods towards higher latitudes. Total amount of rainfall varies greatly within the tropics, and because of predominantly high potential evapotranspiration, there exists a great diversity of forest ecosystems regulated by water availability during the year. Altitude modifies average temperature and water availability, because mountain ranges change incidence of insolation and rainfall regime.

Tropical lowland soils are frequently relatively poor. Under similar climatic conditions, soil nutrient availability determines a series of structurally and functionally well differentiated forest ecosystems. The diversity of the physiognomy and structure of tropical forest ecosystems has been documented in a comprehensive state-of-knowledge report recently published by UNESCO (1978).

Existing data on organic matter production in tropical forests have been reviewed recently (MURPHY 1975, 1977; GOLLEY 1975; UNESCO 1978; JORDAN 1982). We will limit ourselves to the possible generalizations at hand of the most reliable and illustrative data, and emphasis will be given to the eco-physiological properties of the forest components as related to organic matter production.

There is no general classification of tropical forests, and occasionally, comparisons between continents may be confused by the use of different denominations for similar forest types. Hence, the descriptions given by the respective authors, paying special attention to degree of seasonality, will be followed.

9.2 Biomass Distribution

Biomass accumulation by itself is not an indication of organic matter production. However, allocation of biomass or energy may reveal patterns of ecosystem function.

Data presented in Table 9.1 show that total biomass tends to decrease along altitudinal and water availability gradients as could be predicted from physiognomic and structural studies. In lowland rain forests with slight or no seasonal water stress biomass values average 425 t ha^{-1}. With increasing water stress during the year tree density and biomass accumulation are correspondingly reduced. This decrease may be observed along the series rain forests, deciduous forests and dry deciduous forests. Tropical woodlands occupy habitats characterized not only by strongly seasonal water stress, but also by low soil fertility and frequently a regular fire impact during the dry season. Because of their open canopy, tropical woodlands have the lowest biomass values for tropical forest communities. Woody biomass values range from 150 t ha^{-1} in a Miombo Woodland in Zaire (MALAISSE 1978) to 3.5 t ha^{-1} in a shrubby savanna in Senegal (BILLE and POUPON 1972).

The amount of leaves in lowland rain forests is relatively uniform, averaging 8.5 ± 0.9 t ha^{-1}. Montane rain forests also have a similar leaf biomass in spite of lower total biomass accumulation.

Root biomass values are less reliable owing to the difficulty of measurement, but observed differences are large enough to reveal clear patterns of biomass allocation related to environmental constraints.

The proportion of leaves from total biomass in lowland rain forests is near 2%, but root percentage varies from a low of about 4% in Pasoh and Panama forests to a high of 33% in an Amazonian rain forest on podsolized sandy soils. The shoot: root ratios decrease from more than 20 in Pasoh and Panama forests, to 10 in Khao Chong and Banco, 5–6 in Chékô and Amazonian forests

Table 9.1. Selected biomass data of tropical forests and woodlands
Reference: 1 Kira (1978); 2 Jordan and Uhl (1978); Jordan and Escalante; 3 Klinge and Herrera (in press); 4 Ogawa et al. (1961); 5 Hozumi et al. (1969); 6 Huttel and Bernhard-Reversat (1975); 7 Golley et al. (1977); 8 Klinge (1976); 9 Brun (1976); 10 Edwards and Grubb (1977); 11 Odum (1970); 12 Tanner, from Grubb (1977); 13 Misra (1972); 14 Menaut (1977)

Forest type	Location	Rainfall (mm) and temperature (°C)	Dry season (months)	Biomass t ha^{-1}				LAI (ha ha^{-1})	Reference
				Stem and branches	Roots	Leaves	Total		
Evergreen lowland rain forests	Pasoh, Malaysia	1,807; 27	0	433	20	7.8	461	6.9	1
	On laterite, San Carlos, Venezuela	3,500; 26	0	328	56	8.0	392	5.2	2
	On podsol, San Carlos, Venezuela	3,500; 26	0	263	135	8.0	406	5.1	3
	Seasonal, Khao Chong, Thailand	2,696; 27	1	325	32	8.4	365	–	4
	Seasonal, Cheko, Cambodia	~3,700; ~26	1	338	70	7.3	415	7.4	5
	Seasonal, Banco, Ivory Coast	2,095; 26	1–2	504	49	9.0	562	–	6
	Seasonal, Panama	~2,000; ~26	1–2	306	11	10.0	327	–	7
	Seasonal, Manaus, Brazil	~2,000; 27	2–3	397	67	9.5	473	–	8
Montane rain forests	La Carbonera, 2,300 m, Venezuela	1,500; 12	0	340	73	7.7	421	–	9
	Mt. Kerigomma, 2,400–2,600 m, New Guinea	3,960–4,010; 8–18	0	301	40	8.9	350	5.5	10
	El Verde, 500 m, Puerto Rico	3,760; 22	0	255	71	8.1	334	6.4	11
	Mor Ridge, 1,500 m, Jamaica	>4,000; 13	0	209	54	7.5	269	5.3	12
Deciduous forests	Monsoon, Ping Kong, Thailand	1,300; ~25	3–4	263	25	4.7	293	6.6	4
	Vadanasi, India	1,040; 25	4–6	199	34	6.2	239	–	13
Savanna (Woodland) forests	Ping Kong, Thailand	1,300; ~25	3–4	66	10	2.7	79	3.0	4
	Lamto, Ivory Coast	1,260; 27	2	33	14	1.7	49	0.4	14

on lateritic soils, to 2 in the podsol forest in Venezuela. In the few montane rainforests listed, there is an increase in leaf biomass (2%–3%) and in the proportion of roots (11%–21%). Shoot:root ratios vary accordingly between 4%–8%. The higher root proportion in forest communities has been associated with low soil fertility in Amazonian forests (HERRERA et al. 1978); this factor may also be relevant for low shoot:root ratios in montane forests (GRUBB 1977).

9.3 Structural and Functional Relationships

The productive capacity of an ecosystem depends on the photosynthetic efficiency of the leaf area, and the total amount of leaf area built per unit area of soil occupied (Leaf Area Index, LAI). LAI of different forests may vary considerably during the year, therefore information on LAI should be supplemented with Leaf Area Duration (LAD) in order to obtain more precise indication of their potential productive capacity (EVANS 1972; KIRA and SHIDEI 1967). In evergreen tropical forests LAD is around 11–13 months, while deciduous forests have LAD values always below 9 months. LAD in savanna woodlands may vary depending on the predominant tree flora. Deciduous trees are the most common in African savannas, while evergreen trees are dominant in the American savannas.

The amount of biomass that is invested in the development of a certain leaf area varies significantly among ecosystems. This relationship is clearly expressed as the specific leaf area (SLA = leaf area/leaf dry weight). Mean SLA calculated from Table 9.1 shows a variation from 65 $cm^2\ g^{-1}$ in evergreen rain forests in the Amazon basin to about 140 $cm^2\ g^{-1}$ in a monsoon forest in Thailand. There seems to be a tendency toward lower SLA in montane forests (GRUBB 1977) and in forests growing in oligotrophic soils (MEDINA 1978).

9.3.1 Vertical Distribution of Leaf Area and Light Penetration

Tropical rain forests present two to four distinguishable strata. The major fraction of LAI accumulates in the upper and middle canopy, but there exists a substantial amount of leaf area spread in the lower strata of the forest. The vertical distribution of leaves can be conveniently expressed as leaf area density ($ha\ ha^{-1}\ m^{-1}$), or as leaf biomass density ($t\ ha^{-1}\ m^{-1}$) (KIRA et al. 1969). Leaf area density is much smaller in forests as compared with herbaceous canopies. Nevertheless, the extinction of light seems to follow a similar pattern in both community types.

MONSI and SAEKI (1953) expressed the relationship between leaf area index (LAI) and relative illuminance as analogous to the Lambert-Beer relation to light extinction

$$\frac{I}{Io} = e^{-K(LAI)}.$$

The equation defines the light intensity at a certain level within the forest (I) as a function of the prevailing light intensity outside the forest (Io) and the amount of intercepting leaf area above this level (LAI). K is named, by analogy, the light extinction coefficient of the plant community. This equation has been modified to introduce the fraction transmitted by the intercepting leaf area (m) (SAEKI 1960)

$$\frac{I}{Io} = \frac{e^{-K(LAI)}}{(1-m)}.$$

The logarithmic extinction of light has been measured in widely different evergreen forests (KIRA et al. 1969; HOZUMI et al. 1969; ODUM 1970; YODA 1974; HEUVELDOP 1978), and K can be considered to be a reliable indicator of forest canopy structure.

YODA (1974) studied the vertical light distribution in a tropical rain forest in Pasoh, West Malaysia, and found that the frequency distribution of relative light intensity at various levels fits a lognormal distribution. The geometric mean is therefore more representative of the relative illuminance value at a certain forest level than the arithmetic mean.

The size of leaves increases with depth in the canopy and can be related to the relative illuminance in the forest. CAIN et al. (1956) and KLINGE and HERRERA (1983), among others, have observed that average leaf size increases from microphyll to mesophyll (after Raunkiaer, see Chap. 18, Vol. 12B) in the upper canopy and from the mesophyll to macrophyll in the lower canopy (Table 9.2). This size increase is probably related to the energy balance of the leaves and the cost relationships between leaf area development for photosynthesis and the metabolic cost for support and water supply. GIVNISH and VERMEIJ (1976) proposed a model to explain known patterns of leaf size distribution in tropical rain forests. It assumes that net carbon assimilation equals the net photosynthetic gains minus the night leaf respiration and the possible metabolic cost for supplying a certain water flow rate to the leaves. This model predicts bigger leaves at intermediate levels in the canopy and smaller, though more variable sizes in the lower canopy. Predictions of this model have been specifically tested for the leaf size of lianas but it seems to be of general applicability.

SLA depends on the relative illuminance under which the leaf has developed, and increases with depth in the forest canopy. An almost linear relationship

Table 9.2. Distribution of leaf area among different strata in a tropical rain forest in Brazil (values in %). (CAIN et al. 1956)

Leaf area (cm^2)	Macrophanerophytes (>30 m)	Mesophanerophytes (9–30 m)	Microphanerophytes + nanophanerophytes (0.25–8 m)	Lianes
0–80	84	67	48	58
80–160	14	18	40	23
>160	2	14	12	19

between SLA and depth in the canopy was shown by YODA (1974) in several species of the Pasoh forest. SLA increased from an average of 86 $cm^2\ g^{-1}$ in the upper canopy to 120 $cm^2\ g^{-1}$ in the middle and lower canopies, and 164 $cm^2\ g^{-1}$ near the forest floor. Similarly, in a low montane forest in Puerto Rico, average SLA in dominant tree species increased from 43 $cm^2\ g^{-1}$ in the upper canopy to 68 $cm^2\ g^{-1}$ in the middle canopy and 93 $cm^2\ g^{-1}$ in the lower canopy (LIETH in ODUM 1970).

9.3.2 Leaf Size and Specific Leaf Area in Different Tropical Ecosystems

Dominant leaf size decreases both with increasing aridity and altitude. A comparison between rain forests and dry tropical forest shows that in the rain forest the mesophyll size (after Raunkiaer) predominates, while microphyll to nanophyll sizes tend toward higher frequencies in drier forests (GIVNISH 1978; HAMANN 1979). A similar effect has been documented for montane rain forests, in which leaf size decreases from mesophyll in lowland and lower montane rain forests, to microphyll and nanophyll in upper montane and subalpine rain forests (GRUBB 1977). Here also the decrease in average leaf size is paralleled by a decrease in specific leaf area.

The pronounced variation of SLA in dominant, upper canopy trees in different forest ecosystems is given in Table 9.3. Evergreen species have lower SLA values than deciduous species. The higher biomass investment per unit area in the former can be compensated by a longer leaf duration.

The lower values of SLA are correlated with lower nitrogen content per unit leaf dry weight, but, in general, values of N per unit area are similar in leaves of both deciduous and evergreen species.

Table 9.3. Average specific leaf area (SLA) of canopy trees of tropical forests in relation to their nitrogen (N) content. (Partially from MEDINA 1978)

	SLA ($cm^2\ g^{-1}$)	N content ($mg\ g^{-1}$)
1. Evergreen rain forests (Amazonas)		
On laterite	93	15.8
On podsol Tall Caatinga	78	13.2
Low Caatinga	54	9.2
2. Deciduous tropical forests		
Evergreens	58	9.6
Deciduous	195	15.7
3. Dry deciduous forests		
Evergreens	35	11.8
Deciduous	123	21.0
4. Montane tropical forests		
El Verde, Puerto Rico	83	13.6
Elfin Forest, Puerto Rico	47	10.3
Cloud Forest, Venezuela	68	9.0

9.3.3 Development of Leaf Area Index and Transpiration

There is a clear relationship between organic matter production and actual evapotranspiration (ODUM 1970). This is due to the fact that the amount of actual evapotranspiration is related to incident light energy and availability of water; both factors are strongly correlated with plant productivity (LIETH 1975). Under a certain set of environmental conditions actual volume of water transpired by a plant community is proportional to the amount of leaf area exposed, providing that there is no restriction in water supply to the plant. Moreover, the development of a certain leaf area is accompanied by a proportional development of the hydraulic system for water transport from the soil to the upper canopy (ZIMMERMANN and BROWN 1971). JORDAN and KLINE (1977) found a linear relationship between the amount of water transpired ($l\,d^{-1}$) and the area of sapwood available for conduction in a forest on laterite in the Amazon basin (JORDAN and UHL 1978). In *Micranderas sprucei,* a dominant species of podsol forest in the same region, KLINGE and HERRERA (1983) also found a linear relationship between sapwood area at 1.5 m height and the total leaf area per tree. These relationships suggest that the measurement of sapwood area is a feasible way to evaluate potential water use in the community.

9.3.4 Photosynthetic Characteristics of Rain Forest Plants

The study of photosynthetic characteristics of tropical forest trees is still fragmentary and in most cases the techniques used for measurements of leaf gas exchange are inadequate (MOONEY et al. 1980). Simultaneous measurements of water vapor and CO_2 exchange have generally not been performed, therefore calculation of leaf diffusion resistance cannot be made. Stomatal effects are obvious in published curves of CO_2 exchange, indicating that lack of leaf temperature and humidity control has led to variable results. In many cases, detached leaves or branches have been used in the measurements without accounting for the effect of severance on stomatal behavior. Methods employed for CO_2 exchange measurements are also variable in the degree of accuracy (alkali CO_2 absorption techniques; infrared gas analyzer in absolute or differential modes).

The few data available in the literature do not allow a general synthesis of photosynthetic performance of tropical plants, but certain patterns appear to be present and worth exploring.

The older literature on CO_2 exchange of tropical trees was summarized by LARCHER (1969). Subsequently, a few studies have appeared (HOZUMI et al. 1969; STEPHENS and WAGGONER 1970; LUGO 1970; ODUM et al. 1970; KOYAMA 1974; SAN JOSE 1978; LUGO et al. 1978). Measurements of respiration of tropical plants have been reported in the above-cited papers as well as by YODA (1967, 1974) in Malaysia and Thailand, and LÖHR (1969) in Ivory Coast.

9.3.4.1 Photosynthesis and Respiration of Tropical Plants

The vertical extinction of light intensity regulates both photosynthetic capacity and respiration rate of the forest plants. STOCKER (1935), referring to the relative

Table 9.4. Photosynthesis and respiration of tropical plants

	Max. Photo-synthesis	Respiration	Temp. Range	Refer-ence
	(mg CO_2 $dm^{-2}h^{-1}$)		(°C)	
Pioneer species				
Cecropia peltata	26	0.8	24	1
Mallotus paniculatus	22	1.9	30	2
Antocephalus cadamba	22	–	–	3
Macaranga hypoleuca	21	0.9	28	2
Rain forest trees				
Shorea leprosula	24	4.0	29	2
Mangletia glauca	22	1.2	–	4
Erythrophleum fordii	18	0.9	–	4
Dipterocarpus crinitus	15	2.2	29	2
Shorea pauciflora	12	1.7	29	2
S. macroptera	12	0.5	28	2
Dipterocarpus sublamellatus	9	1.0	28	2
Cinnamomum parthenoxylon	9	1.0	29	2
Parashorea densiflora	7	0.6	26	2
Alangium ridleri	7	1.2	20	2
Cordia alliodora	8	1.2	28–23	1
Goethalsia meiantha	7	0.7	28–23	1
Virola koschnyi	6	0.8	24–24	1
Shade plants of undergrowth				
Woody plants				
Protium glabrum	3.4	0.3	24–20	1
Croton glabellus	2.6	0.3	24–23	1
Melastomataceae ssp.	1.8	0.1	26–20	1
Herbs				
Cordyline rubra	4.4	0.06	17	5
Alocasia macrorhiza	4.3	0.05	20	5
Heliconia latispatha	2.4	0.2	24–20	1

Reference: 1 STEPHENS and WAGGONER (1970); 2 KOYAMA (1974); 3 LUGO (1970); 4 after LARCHER (1969); 5 BJÖRKMAN et al. (1972)

illuminance under which leaves developed, differentiated between sun and shade leaves. These leaf types differ both in physiological and morphological characteristics. The stratification of the rain forest points to a continuum between sun and shade leaves (KOYAMA 1974). The morphogenetic influence of light and its results on leaf morphology, ultrastructure, and gas exchange patterns have been discussed by BJÖRKMAN et al. (1973) when analyzing the photosynthetic properties of rain forest shade plants (see also Chap. 3, Vol. 12A).

Table 9.4 gives maximal values of photosynthetic rates reported for various rain forest plants. There is a pattern indicating higher photosynthetic rates in the dominant species in the canopy and decreasing maximal rates in the woody plants belonging to the middle or lower canopy. Woody plants and

Fig. 9.1. Net photosynthesis as dependent on light intensity for species of a secondary lower montane rain forest (Costa Rica) (STEPHENS and WAGGONER 1970)

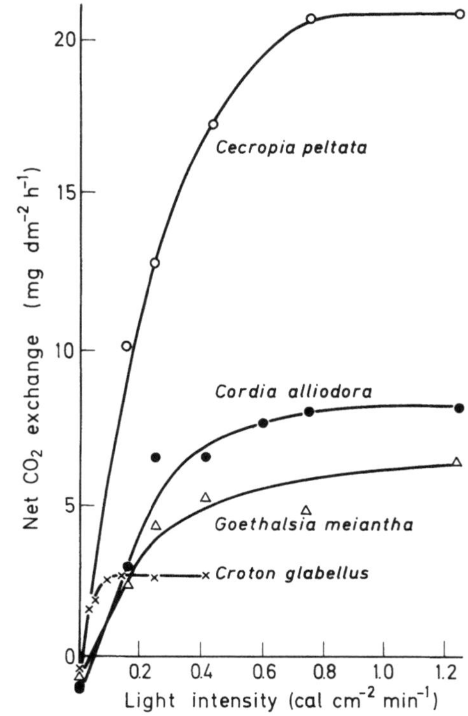

herbs of the forest floor are also different, herbs having higher photosynthetic rates under comparable light conditions.

Pioneer species are characterized by relatively high photosynthetic rates probably associated with their high growth rates. Light curves of photosynthesis for species of a secondary lower montane rain forest in Costa Rica are depicted in Fig. 9.1. *Cecropia peltata,* a well-known pioneer species of Central America, shows the highest rates, while the dominant trees in the upper canopy, such as *Goethalsia meiantha* and *Cordia alliodora,* have relatively modest rates. *Croton glabellus,* a shrub from the understory, presents the lower respiration rate and low light intensity for photosynthesis saturation typical for a shade plant.

There are no comparable measurements of photosynthesis in trees of drier tropical forest ecosystems, where deciduous and evergreen species coexist in different proportions along the gradient of water availability. However, there are indications that photosynthetic capacity is lower in evergreen trees than in deciduous trees. *Curatella americana* and *Byrsonima crassifolia,* evergreen trees of the savannas in Central Venezuela, have rates of photosynthesis between 10–12 mg CO_2 dm^{-2} h^{-1} (SAN JOSÉ 1977; MEDINA in press), while deciduous trees, such as *Cochlospermum vitifolium,* present significantly higher rates (MEDINA unpublished). Similar patterns are known in evergreen broadleaf and deciduous trees of Mediterranean regions (MOONEY 1972).

The reason for this difference is not clear and it has been hypothesized that deciduous trees, with leaves of higher SLA and higher N content per unit leaf dry weight, allocate a higher proportion of this nitrogen into photosynthetic enzymes, while evergreen trees, with more sclerophyllous leaves, are less photosynthetically efficient but better adapted to grow in environments with lower capacity for nutrient supply, mainly N and P (MEDINA 1981).

Adaptation to a shady environment can be shown by means of the light compensation point (LIETH 1960). For extreme shade-adapted plants of a rain forest floor in Queensland, Australia, BJÖRKMAN et al. (1972) reported light compensation points of 1.7–2.0 $\mu E\ m^{-2}\ s^{-1}$ PAR. An extensive inventory of light compensation points of Venezuelan cloud forest (montane rain forest) plants by HUBER (1978) showed that the lower values are between 1–2 $\mu E\ m^{-2}\ s^{-1}$. Lowest relative illuminance values reported by the authors were 0.5% for the Queensland rainforest and 1%–3% for the cloud forest in Venezuela.

The photosynthesis-respiration ratios in rain forest seedlings have scarcely been investigated. The seedlings of the climax, dominant trees, frequently developing from seeds with substantial reserves, grow readily in the forest floor to a certain stage, with their roots normally distributed in the litter or upper soil layers. After that stage, the seedlings grow very slowly and many die. LUGO (1970) measured the percentage of seedling survival in El Verde montane forest in Puerto Rico. Potted seedlings of sun species, *Cecropia peltata* and *Anthocephalus cadamba,* did not tolerate the shade conditions of the forest floor, while seedlings of climax forest species, *Dacryodes excelsa* and *Sloanea berteriana,* survived although their growth was very slow. In natural populations of these species, LUGO (1970) calculated a seedling half-life of 8 months.

The survival of seedlings in the forest floor is a function of the degree of adaptation of their photosynthetic system to low light intensity. In fact, the low tolerance of seedlings of pioneer species to shady conditions in the forest floor may be explained on the basis of their high respiration rates.

The effect of root competition in the upper soil layers could be more important for survival than light intensity itself, as has been demonstrated by KARPOV for a temperate *Picea* forest (after WALTER 1973). Tropical rain forests, mainly those growing in very oligotrophic soils as in the Amazon basin, have a very high density of active small roots in the upper layers of the soil (HERRERA et al. 1978; STARK and SPRATT 1978; KLINGE and HERRERA 1978), with root competition being presumably very intense. Observations in rain forests of the Rio Negro basin in southern Venezuela indicate that survival of potted seedlings and saplings of *Eperua leucantha, Aspidosperma cruentum,* and *Licania heteromorpha* is 100% after 2 years of transplanting, while natural populations of the same species are almost eliminated within one year.

The adaptability of the photosynthetic system of the seedlings to different light intensities has to be investigated in order to understand the underlying mechanisms of forest succession after disturbance. The hypothesis of KOYAMA (1974) stating the sequence in succession of pioneer species, followed by invasion of climax sun-type species, and finally shade-type canopy species and shade-adapted undercanopy plants, is worth testing on the basis of photosynthetic behavior of seedlings.

There are only a few data on temperature dependence of photosynthesis and respiration of tropical plants. Reported values indicate that lowland rain forest plants have temperature optima well above 20 °C (STOCKER 1935; BJÖRKMAN et al. 1972). There is evidence that rates of respiratory CO_2 production during the growing season in tropical plants is similar to that of plants in temperate ecosystems. STOCKER (1935) states that high night temperatures in the tropics do not result in increased CO_2 losses, since respiration rates of tropical trees at 30 °C are similar to respiration rates of arctic plants at 10 °C. The same conclusions were reached by YODA (1974) when comparing respiration activity of tropical trees in Pasoh forest, Malaysia, with the results obtained in a broadleaf warm-temperate forest in Minamata, Japan. This adaptation is a logical consequence of similar energy requirements for the maintenance and growth of trees in different latitudinal regions during the growing season.

9.3.4.2 Photosynthetic Pathways and Life Forms

The diversity of life forms in tropical forests is very high. Lianas, vascular epiphytes, and the presence of large-sized leaves give the tropical rain forests their peculiar character.

Lianas are characterized by their climbing habit, and therefore allocate much more photosynthates for the development of leaf area than for building support organs. This characteristic gives them a very high competitive capacity in comparison with forest trees (GIVNISH and VERMEIJ 1976).

The specific contribution of lianes and epiphytes to organic matter production of tropical forests has not been quantitatively assessed, but it can be evaluated by their proportion of leaf area and leaf weight. In the study of CAIN et al. (1956), near Belem, Brazil, 14% of the 216 plant species recorded in a 2-ha block were lianes. The proportion of lianes to total biomass in Pasoh, Malaysia, has been shown to be about 2% stems + branches, 5% of leaf biomass, and about 7% of total leaf area (KIRA 1978). In Ivory Coast, Africa, wood biomass of lianes is about 5% of total aboveground biomass (HUTTEL and BERNHARD-REVERSAT 1975).

Tropical montane forests are characterized by a heavy development of epiphytes (GRUBB 1977; HUBER 1978). In a montane rain forest in the Andes of Venezuela, BRUN (1976) reports that about 36% of total leaf biomass belongs to vascular epiphytes.

Tropical forest plants predominantly present the C_3 (Calvin) pathway of photosynthesis. Only in the relatively low and open rain forests of Hawaii, arborescent woody C_4 species of the genus *Euphorbia* (up to 8 m tall) have been reported (PEARCY and TROUGHTON 1975). The authors interpret this finding as another indication that the C_4 pathway of photosynthesis is of relatively recent evolutionary origin.

Plants with crassulacean acid metabolism, CAM plants, are heavily represented, at least in American and Australasian rain forests. CAM plants in American rain forests are epiphytes belonging to the Bromeliaceae, Orchidaceae, and Cactaceae families (COUTINHO 1963; MEDINA 1975). In the rain forests of north eastern Australia an increasing number of epiphytic CAM species are being found. Ferns and many orchid species have been added to the list of known CAM species (WINTER personal communication).

The distribution of CAM species within the forest follows a clear pattern of light intensity requirement and resistance to drought. The upper canopy layers of the tropical rain forest are relatively dry because there is no water accumulating substrate and, therefore, a few days without rain may represent a considerable drought. In the Bromeliaceae, PITTENDRIGH (1948) described a variety of adaptations to cope with the problem of water availability. The evolution of tank bromeliads (the leaves can form a water-tight container where rainwater is accumulated) and succulence are important morphological adaptations in this respect. In general, the more drought-tolerant species occupy the upper canopy and present CAM metabolism, while the ones in the lower forest layers have the C_3 photosynthetic pathway (see also Chap. 15, Vol. 12B).

9.4 Productivity of Tropical Forests

9.4.1 Net and Gross Productivity

In a plant community, the net primary productivity (NPP) may be defined as the sum of: (a) biomass increase; (b) replacement of dead plant and plant parts (total litter production); and (c) herbivory (amount of plant biomass consumed by non-autotrophic organisms). All terms are rates expressed as biomass units per unit soil area and per unit time. This NPP plus the amount of organic matter respired during the energy-releasing processes for growth and structure maintenance is called the gross primary production (GPP).

During succession in forest ecosystems the ratio NPP/GPP decreases as a consequence of the increasing proportion of non-photosynthetic biomass. In the process, the stage is eventually reached in which there is no biomass increase during reasonably long periods, then the forest may be considered to be in steady state. At this point, the NPP/GPP ratio would be the theoretical minimum and NPP would be represented by the amount of total litter production plus the amount of organic matter consumed by herbivores. In most tropical forests herbivory is ecologically important but quantitatively small; therefore, the amount of total litter production has been considered to be a reliable index of the net production capacity of forest ecosystems (BRAY and GORHAM 1964).

Several approaches have been employed to assess the productivity of tropical forests. The most reliable up to now is the summation method (NEWBOULD 1967; KIRA et al. 1967; OGAWA 1978). Essentially, the method consists of measuring biomass increment, rate of shedding of plant parts, and herbivory over relatively long time periods (at least a year). For the accurate estimation of the forest biomass, very elaborate allometric relationships have been developed, based on tree diameter and height (KIRA and OGAWA 1971; HOZUMI et al. 1969; JORDAN and UHL 1978).

The measurement of respiration of plants is required to estimate gross production of the forest. Techniques for the measurement of tree respiration in tropical forests were first applied by MÜLLER and NIELSEN (1965) in a disturbed lowland rain forest in Ivory Coast, Africa, and YODA (1967, 1974) in forests of Thailand, Malaysia. Respiration rates under field conditions have been frequently measured by using CO_2 absorption methods and severed organs. This technique presents many problems, mainly those related to the efficiency of CO_2 absorption by the alkali solution after detachment. However, the general design for calculating total plant respiration has been proven worthy of development and has been applied also in the extensive studies of the Japanese IBP primary production program (YODA 1978).

A second approach for measuring plant productivity is based on the analysis of CO_2 exchange of plant parts during the growing season. This approach also requires an accurate estimation of the biomass of the forest components. In the simplest case, a summation procedure is adopted in which leaf photosynthesis and respiration rates of all organs are measured. To obtain the fluxes of carbon per unit area and time, the rates are multiplied by the corresponding biomass values (ODUM 1970; LUGO et al. 1978).

SAEKI (1960) developed a model for determining gross production based on light curves of photosynthesis and the logarithmic extinction of light with depth in the forest. The model is based upon the photosynthetic properties of the leaves at different heights in the forest, as well as an accurate determination of the extinction coefficient of the forest canopy and the light-transmission properties of the leaves. Several versions of this model have been used in tropical forests (HOZUMI et al. 1969; KOYAMA 1974).

There are two main problems with this method for the assessment of gross primary productivity. The first is the accurate measurement of the light regime prevailing at different levels in the forest, and the second is the assumption that dark respiration of leaves is identical to respiration in the light. This last assumption has to be considered as an approximation, because the process of photorespiration is quite distinct from dark respiration, and is strongly dependent both on light intensity and temperature. The same observation is valid for the measurement of total leaf respiration in the summation method previously described.

A third approach for assessment of forest production is based on micrometeorological techniques. They have been used in several tropical forests with limited success (LEMON et al. 1970; ODUM et al. 1970; YABUKI and AOKI 1974). Micrometeorological methods allow the estimation of the total carbon balance of the ecosystem by measuring the in- and outgoing fluxes of CO_2 and requires the accurate measurement of several properties in a forest vertical profile such as CO_2 concentration, wind velocity, humidity, and temperature. The application of this methodology to tropical forests still requires further development.

9.4.1.1 Litter Production

Litter fall refers generally to fine litter production. That is, the fall of the big branches and trunks is not included (KLINGE 1977). In many cases, the amount of leaf fall recorded has been used to approximate the average life span of leaves and to determine seasonality of forest productive capacity. Table 9.5 presents selected data of litter fall for forests differentiated as slightly seasonal, strongly seasonal, and montane. A more comprehensive review of litter fall

Table 9.5. Selected data of fine litter fall in tropical forests

	Leaf fall	Total litter fall	Reference
	(t ha^{-1} yr^{-1})		
Lowland, slight seasonal rain forests			
Kade, Ghana	7.9	9.7	JOHN (1973)
Yangambi, Zaire	–	13.7	LAUDELOUT and MEYER (1954)
Omo forest, Nigeria	7.2	–	HOPKINS 1966
Banco, Ivory Coast	8.2	12.4	HUTTEL and BERNHARD-REVERSAT (1975)
Yapo, Ivory Coast	7.1	9.6	ibid (1975)
Mucambo, Brazil	8.0	9.9	KLINGE (1977)
Tierra Firme, Manaus Brazil	5.6	7.3	KLINGE and RODRIGUES (1968)
Laterite, San Carlos, Venezuela	–	6.2	JORDAN (unpublished)
Podsol, San Carlos, Venezuela	–	5.7	JORDAN (unpublished)
Mora forest, Trinidad	5.9	–	CORNFORTH (1970)
Darien, Panama	–	11.3	GOLLEY et al. (1977)
Barro Colorado, Panama	5.8	11.1	HAINES and FOSTER (1977)
Pasoh forest, Malaysia	5.3	10.6	OGAWA (1974)
Lowland, strongly seasonal forests			
Ibadan, Nigeria	3.4	5.6	MADGE (1965)
Olokemeji, Nigeria	4.7	–	HOPKINS (1966)
Muhulu, Zaire	4.7	9.2	MALAISSE (1978)
Calabozo, Venezuela	–	8.2	MEDINA and ZELWER (1972)
Varanasi, India	–	7.7	MISRA (1972)
Udaipur, India	4.4	–	RANAWAT and VYAS (1975)
Tropical woodlands			
Miombo, Zaire	2.9	4.3	MALAISSE (1978)
Guinea Savanna, Nigeria	2.4	3.3	COLLINS (1977)
Tropical montane forests			
Jamaica, Blue Mountains	5.7	6.5	TANNER, cited in GRUBB (1977)
Lower montane, Puerto Rico	5.5	–	ODUM (1970)
Cloud forest, Venezuela	–	7.8	MEDINA and ZELWER (1972)
Mt. Kerigomma, New Guinea	6.4	7.6	EDWARDS (1977)
Montane, Gundar, India	3.9	5.5	BLASCO and TASSY (1975)

in the tropics is given by BRAY and GORHAM (1964), JORDAN and MURPHY (1978), and JORDAN (1982).

The amount of leaf fall decreases from evergreen rain forests to seasonal forests and woodlands in correspondence with the development of leaf biomass. Average leaf life span in the lowland rain forests is generally approximately 12 months (RICHARDS 1952; HAINES and FOSTER 1977), although a diversity of patterns of leaf renewal can be observed in different species (e.g., MEDWAY 1972). Toward drier climates, average life span decreases.

Leaf litter production is associated with leaf flushing, and both are dependent on patterns of rainfall distribution in lowland forests. In evergreen rain forests, rates of litter fall fluctuate widely with no obvious relation to major climatic parameters. The community as a whole, however, has an annual pattern of leaf replacement as observed by MEDWAY (1972) in Malaysia.

Rates of litter production in Pasoh, Malaysia, fluctuate between 8 and 40 g m^{-1} wk^{-1} (OGAWA 1974), while in San Carlos, Venezuela, JORDAN (unpublished) reports values between 5 and 26 g m^{-2} wk^{-1}. Seasonal humid forests in Africa have rates between 5 and 25 g m^{-2} wk^{-1} in Ivory Coast (HUTTEL and BERNHARD-REVERSAT 1975), and about 9 and 50 g m^{-2} wk^{-1} in Kade, Nigeria (JOHN 1973). In a dry deciduous forest in Puerto Rico, LUGO et al. (1978) reported rates between and 1 and 9 g m^{-2} wk^{-1}.

It seems that in spite of approximately homogeneous environmental conditions in the humid tropics, the population as a whole presents a seasonal pattern both in flowering and litter production. Triggering factors for the synchronization of physiological events are difficult to assess, but frequency of dry spells, slight variations in photoperiods and thermoperiods, or rain after prolonged dry periods, might be relevant (OPLER et al. 1976). More homogeneous environments, mainly in regard to humidity, may prevail in tropical montane forests, where a constant supply of humidity is provided through rainfall and frequent fog. For a cloud forest in northern Venezuela, an almost constant rate of litter production has been reported, varying between 10 and 13 g m^{-2} wk^{-1} (MEDINA and ZELWER 1972).

It has been established that the amount of litter production is higher in the humid tropics than in ecosystems elsewhere (BRAY and GORHAM 1964). An analysis of world data for leaf and wood fall lead to the hypothesis that, while leaf production is higher in tropical than in temperate regions, wood production remains constant. Therefore, considering the gradient of available light energy, the efficiency of leaf production is higher in tropical than in temperate regions, the reverse being true for wood production (JORDAN 1971; JORDAN and MURPHY 1978; JORDAN 1982). Table 9.6 gives average values of litter and wood production calculated by JORDAN (1982) for mesic forests along a latitudinal gradient expressed by the radiation balance. JORDAN and MURPHY (1978) hypothesize that trees with high wood production rates during the growing season have been selected in temperate environments.

The causal relationship behind this pattern may be the cost of maintenance of non-photosynthetic organs in both the tropics and in temperate regions.

Table 9.6. Litter and wood production of mesic forests along a latitudinal gradient characterized by the radiation balance. (JORDAN 1982)

Radiation balance ($J\ cm^{-2}\ yr^{-1} \cdot 10^{-5}$)	Litter production	Wood production
	($g\ m^{-2}\ yr^{-1}$)	
1.0–1.7	2.8 ± 1.3	9.7 ± 4.0
1.7–2.1	3.3 ± 1.1	6.6 ± 2.5
2.1–2.5	3.7 ± 1.2	6.1 ± 5.4
2.5–2.9	6.4 ± 2.0	7.6 ± 2.9
> 2.9	9.6 ± 3.6	7.3 ± 2.8

Table 9.7. Aboveground biomass, respiration losses, and maintenance costs in tropical and temperate forests. (With data from MÜLLER and NIELSEN 1965; KIRA and YABUKI 1978; KIRA 1978)

Forest	Aboveground biomass ($t\ ha^{-1}$)		Respiration losses ($t\ ha^{-1}\ yr^{-1}$)		Maintenance costs ($t\ ton^{-1}\ yr^{-1}$)	
	Wood	Leaves	Wood	Leaves	Wood	Leaves
Fagus forest, 46 years old, Denmark	129	2.7	4.5	4.6	0.035	1.7
Warm-temperate, evergreen, Minamata, Japan	351	7.5	12.1	19.1	0.035	2.6
Tropical rain forest, Pasoh, Malaysia	414	7.6	18.8	29.1	0.045	3.8

In tropical regions, the maintenance costs for support of the photosynthetic system are more or less constant during the whole year, while in temperate regions the respiratory losses during the winter season are reduced to very low levels due to low temperatures. Pertinent data to substantiate this assumption are those reported by KIRA and YABUKI (1978) for a warm-temperate evergreen oak forest in Minamata, Japan, and KIRA (1978) for a tropical rain forest in Pasoh, Malaysia. Table 9.7 gives the estimated respiratory losses during the year for the above ground biomass of both forests. The values for a 46-year old *Fagus* forest in Denmark as given by MÜLLER and NIELSEN (1965) are added for purposes of comparison.

Maintenance costs are clearly higher in the tropical rain forest and this increase has to be compensated for with a relative decrease in wood production in favor of a more pronounced development of the photosynthetic biomass. This difference in maintenance costs can be assessed by studying the fluctuations of carbohydrate reserves in deciduous forests in both temperate and tropical regions.

JANZEN and WILSON (1974) followed during one year the total carbohydrate content of *Jacquinia pungens*, an understory shrub in deciduous forests in Costa Rica, which remains dormant during the rainy season and becomes photosynthetically active during the dry season, when the canopy trees are leafless (JANZEN 1970). This shrub does not produce flowers or fruits during this leafless period. The authors showed that carbohydrate reserves were reduced for about 70% during the dormant period (about 6 months). In contrast to that, SIMINOVITCH et al. (1953) found that trees of *Robinia pseudoacacia* in central USA showed reduction of total reserve carbohydrates in the living bark of about 1% between the winter months of December and April.

The reduction of carbohydrate reserves of deciduous trees in the tropics can be very drastic due to the reproductive activity which occurs during the dry season, when they are leafless (OPLER et al. 1976). Under the high temperatures characteristic of the lowland tropics, this metabolic activity has to be reflected in a reduction of total organic matter production. OLOFIMBOBA (1969)

Table 9.8. Production data of selected tropical rain forests (data in t ha^{-1} yr^{-1})

	Aboveground biomass increment	Root production	Leaf litter	Other litter	Wood fall	Net production	Reference
Asia							
Evergreen forest							
Pasoh	5.3[a]	(5.5)[c]	7.0	1.4	7.1	27[b]	1
Africa							
Moist seasonal							
Kade	11.9	(2.6)[c]	7.0	3.5		25	2
Banco	4.6	(0.7)[c]	8.2	1.1	2.6	17	3
America							
Evergreen forests							
Lowland							
San Carlos	–[d]	2.0	6.2		3.4	12	4
Montane							
El Verde	0.8	(2.1)[c]	5.0	6.6		15	5

[a] Wood + leaf mass measurement
[b] Grazing included
[c] Estimated
[d] Biomass increase is approximately equalled by wood fall (JORDAN personal communication)

Reference 1 KIRA (1978); 2 GREENLAND and KOWAL (1960); 3 HUTTEL and BERNHARD-REVERSAT (1975); 4 JORDAN and ESCALANTE (1980); 5 ODUM (1970)

measured in *Antiaris africana* a reduction of about 20% of the sucrose content in the xylem during the short leafless period of that tree species. In 2 months its leaves fall, and flowers and fruits are set. When the leafless period is longer, as in dry and very dry deciduous tropical forests, the losses may be expected to be higher.

A carbohydrate cost analysis along water and temperature gradients should be carried out in tropical forest ecosystems in order to understand patterns of biomass allocation and productivity (see also Chap. 18, Vol. 12B).

9.4.1.2 Gross and Net Production Data for Selected Tropical Forests

Production data of tropical forests have been summarized in the recent state-of-knowledge report by UNESCO (1978). Reliable data is very scarce and in most cases root production has only been estimated. As far as we know, the most reliable data is given in Table 9.8. It includes forests of Asia, Africa, and America.

The accuracy of biomass determination and measurement of litter production rates is not the same for all forest samples listed. The most precise data are those of Pasoh forest, in which grazing and decomposition of dead material before collection was accounted for. The best root production value is perhaps the one for the San Carlos forest on laterite reported by JORDAN and ESCALANTE (1980). These authors were able to measure root growth directly by taking advantage of the presence of a root mat on the mineral soil, which represents 36.5% of the total root biomass (STARK and SPRATT 1978).

The highest net production value corresponds to Pasoh forest in Malaysia. This value approximates previous reports for other Asiatic tropical forests, mainly the Khao Chong forest in Thailand, which was not included in Table 9.8 because of over-estimation of litter fall (KIRA et al. 1967; KIRA 1978). The high value reported for the Kade forest may be related to the secondary character of the plot studied. The Amazon forest on laterite has a very low net productivity, a fact which has been associated with low soil fertility.

The proportion of total litter and wood fall from total net production varies from 42% to 80%. This proportion seems to increase from secondary to mature forests.

MÜLLER and NIELSEN (1965) pioneered the measurement of gross productivity in tropical forests. Similar efforts were developed in Asia (KIRA 1978), and by an integrated ecological project in Puerto Rico (ODUM 1970). The few available data for gross production indicate values up to 120 t ha^{-1} yr^{-1}. As expected, a substantial fraction of gross production is consumed in respiration of the productive structure of the forest. The respiratory losses accounted for 75% of total gross production in Ivory Coast, 67% in Pasoh forest, and 87% in a Puerto Rican low montane rain forest. Estimates of gross production based on canopy photosynthesis or micrometeorological methods are still in the preliminary stages.

Based on estimations of gross production in tropical and temperate forests, KIRA and SHIDEI (1967) and KIRA (1975) showed that there is a linear relationship between gross production and leaf area duration (leaf area index × leaf life span). The correlation also holds for needle-leaved forests. This relationship implies that the gross photosynthetic capacity of a certain amount of leaf canopy is similar regardless of the latitude location, when water supply during the growing season is not limiting. This conclusion would be a very important generalization and requires further study.

9.4.2 $\delta^{13}C$ Values of Rain Forest Woody Plants

As mentioned before, the overwhelming majority of rainforest woody plants have the Calvin photosynthetic pathway (C_3 plants). As such, they are characterized by $\delta^{13}C$ values below $-25^0/_{00}$ (among others TROUGHTON et al. 1974; see also Chap. 15, Vol. 12B).

Within a closed forest, there are strong diurnal variations of the CO_2 concentration of the air as a result of the rhythm of canopy photosynthesis and respiration, and also because of the substantial amounts of CO_2 released from the forest floor (the amount of CO_2 derived from litter decomposition and root respiration). The organic matter oxidized by soil respiration has a $\delta^{13}C$ value well below $-25^0/_{00}$. Under conditions of low turbulence a gradient of $\delta^{13}C$

Table 9.9. Height distribution of $\delta\,^{13}C$ values in a tropical rain forest in San Carlos de Rio Negro. (MEDINA and MINCHIN 1980)

	$\delta\,^{13}C$ ‰
Upper canopy leaves (>20 m)	-29.5 ± 1.8
Lower canopy leaves (2–10 m)	-33.7 ± 1.5
Undergrowth leaves	-35.2 ± 1.2
Seedlings	-28.7
Composite leaf litter	-29.4

of CO_2 available for photosynthesis could develop, with the $\delta\,^{13}C$ values being more negative near the forest floor. If this is true, a similar gradient should be detected in the living biomass.

Such a profile of $\delta\,^{13}C$ was demonstrated by VOGEL (1978) for a *Fagus* forest in Central Europe. A similar, though more pronounced pattern has been reported for an Amazonian forest (MEDINA and MINCHIN 1980). Table 9.9 shows that upper canopy trees in a rain forest in San Carlos, Venezuela, have an average $\delta\,^{13}C$ value of -29.5‰, while those of the undergrowth have an average of -35.2‰. Seedlings of dominant trees, which develop from seeds with bulky reserves, present $\delta\,^{13}C$ values similar to the upper canopy trees, indicating that they are still dependent on reserve substances. Several-years-old saplings present characteristic $\delta\,^{13}C$ values of the shade flora. Considering that the composite litter sample from the forest floor has a value of -29.4‰, it can be presumed that the CO_2 evolving from soil respiration has a similar $\delta\,^{13}C$ value. The pattern described indicates that understory plants accomplish a substantial direct recycling of CO_2 from soil respiration.

9.5 Conclusions

Sound interpretation of ecosystem stability, succession, and evolution requires the accurate knowledge of underlying productive properties of the whole system. The basic concepts of productivity of tropical ecosystems have been laid in the recent past, but the further improvement and application of field techniques are required before a definitive picture of these complex systems can be established.

Acknowledgements. We would like to acknowledge the valuable assistance of Mrs. JoAnne Feheley, Chief Librarian of the USDA Institute of Tropical Forestry in Puerto Rico, in the literature search of the Asiatic regions; to Ms. Elvira Cuevas for her cooperation in the editing of the manuscript; and to our colleagues from the San Carlos project, mainly Dr. Carl Jordan (Institute of Ecology, University of Georgia, Athens, GA., U.S.A.) who provided valuable background data to this work.

References

Bernhard F (1970) Étude de la litière et de sa contribution au cycle des éléments minéraux en forêt ombrophile de Côte-d'Ivoire. Oecol Plant 5:247–266
Bille J-C, Poupon H (1972) Biomasse végétale et production primaire nette. In: Recherches écologiques sur une savane sahelienne du Ferlo Septentrional, Sénégal. Terre Vie 26:366–382
Björkman O, Ludlow MM, Morrow PH (1972) Photosynthetic performance of two rain forest species in their native habitat and analysis of their gas exchange. Carnegie Inst Yearb 71:94–102
Blasco F, Tassy B (1975) Étude d'un écosystème forestiér montagnard du sud de l'Inde. Bull Ecol 6:525–539
Bray JR, Gorham E (1964) Litter production in forests of the world. Adv Ecol Res 2:101–157
Brun R (1976) Methodik und Ergebnisse zur Biomassenbestimmung eines Nebelwaldökosystemes in den venezolanischen Anden. IVI IUFRO World Congr, Oslo, Div I
Cain SA, Oliveira Castro de GM, Murca Pires J, Tomas de Silva N (1956) Application of some phytosociological techniques to Brazilian rainforest. Am J Bot 43:911–946
Collins NM (1977) Vegetation and litter production in Southern Guinea Savanna, Nigeria. Oecologia 28:163–175
Cornforth JS (1970) Leaf fall in a tropical rain forest. J Appl Ecol 7:603–608
Coutinho LM (1963) Algumas informacoes sobre a occorrencia de "Efeito de De Saussure" em epifitas e herbaceas terrestres da mata pluvial. Bol Fac Filos Cienc Let USP Bot 20:81–98
Edwards PJ (1977) Studies on mineral cycling in a montane rain forest in New Guinea. II. The production and disappearance of litter. J Ecol 65:971–992
Edwards PJ, Grubb PJ (1977) Studies of mineral cycling in a montane rain forest in New Guinea. I. The distribution of organic matter in the vegetation and soil. J Ecol 65:975–992
Evans GC (1972) The quantitative analysis of plant growth. Univ Press, Berkeley, Calif
Givnish TJ (1978) Adaptive significance of compound leaves with particular reference to tropical trees. In: Tomlinson PG, Zimmermann MH (eds) Tropical trees as living systems. Cambridge Univ Press, Cambridge
Givnish TJ, Vermeij GJ (1976) Size and shape of liane leaves. Am Nat 110:743–778
Golley FB (1975) Productivity and mineral cycling in tropical forests. In: Reichle DE, Franklin JF, Goodall DW (eds) Productivity of world ecosystems. Natl Acad Sci, Washington
Golley FB, McGinnis JT, Clements RG, Child GI, Duever MJ (1975) Mineral cycling in a tropical moist forest ecosystem. Univ Georgia Press, Athens
Greenland DJ, Kowal JL (1960) Nutrient content of the moist tropical forest of Ghana. Plant Soil 12:154–174
Grubb PJ (1977) Control of forest growth and distribution on wet tropical mountains: with special reference to mineral nutrition. Annu Rev Ecol Syst 8:83–107
Haines B, Foster RB (1977) Energy flow through litter in a Panamanian forest. J Ecol 65:147–155
Hamann O (1979) On climatic conditions, vegetation types and leaf size in the Galapagos Islands. Biotropica 11:101–122
Herrera R, Jordan CF, Klinge H, Medina E (1978) Amazon ecosystems. Their structure and function with particular emphasis on nutrients. Interciencia 3:223–232
Heuveldop J (1978) The International Amazon MAP rain forest ecosystem pilot project of San Carlos de Rio Negro: Micrometeorological studies. 2nd Int MAB-IUFRO Workshop on tropical rain forest ecosystem research, Jakarta
Hopkins B (1966) Vegetation of the Olokemeji Forest reserve, Nigeria. IV. The litter and soil with special reference to their seasonal changes. J Ecol 54:687–703
Hozumi K, Yoda K, Kira T (1969) Production ecology of tropical rain forests in southwestern Cambodia. II. Photosynthetic production in an evergreen seasonal forest. Nat Life SE Asia 6:57–81

Huber O (1978) Light compensation point of vascular plants of a tropical cloud forest and an ecological interpretation. Photosynthetica 12:382–390
Huettel Ch, Bernhard-Reversat F (1975) Recherches sur l'écosysteme de la forêt subéquatoriale de basse à Côte d'Ivoire. V. Biomasse végétale et productivité primaire. Cycle de la matière organique. Terre Vie 29:203–228
Janzen DH (1970) *Jacquinia pungens*, a heliophile from the understory of tropical deciduous forest. Biotropica 2:112–119
Janzen DH, Wilson DE (1974) The cost of being dormant in the tropics. Biotropica 6:260–262
John DM (1973) Accumulation and decay of litter and net production of forest in tropical West Africa. Oikos 24:430–435
Jordan CF (1971) Productivity of a tropical forest and its relation to a world pattern of energy storage. J Ecol 59:127–142
Jordan CF (1982) Productivity of tropical rain forest ecosystems and implications for their use as future wood and energy sources. In: Golley FB (ed) Tropical rain forest, ecosystems structure and function. Elsevier, Amsterdam pp. 117–136
Jordan CF, Escalante G (1980) Root productivity in an Amazonian rain forest. Ecology 61:14–18
Jordan CF, Kline J (1977) Transpiration of trees in a tropical rain forest. J Appl Ecol 14:853–860
Jordan CF, Murphy PG (1978) A latitudinal gradient of wood and litterproduction, and its implications regarding competition and species diversity in trees. Am Midl Nat 99:415–432
Jordan CF, Uhl Ch (1978) Biomass of a "tierra firme" forest of the Amazon basin. Oecol Plant 13:387–400
Kira T (1975) Primary production of forests. In: Cooper JP (ed) Photosynthesis and productivity in different environments. Cambridge Univ Press, Cambridge
Kira T (1978) Community architecture and organic matter dynamics in tropical lowland rain forests of Southeast Asia with special reference to Pasoh forest, West Malaysia. In: Tomlinson PB, Zimmermann MH (eds) Tropical trees as living systems. Cambridge Univ Press, Cambridge
Kira T, Ogawa H (1971) Assessment of primary production in tropical and equatorial forests. In: Duvigneaud P (ed) Productivity of forest ecosystems. Proc Brussels Symp Unesco, Paris
Kira T, Shidei T (1967) Primary production and turnover of organic matter in different forest ecosystems of the Western Pacific. Jpn J Ecol 17:70–87
Kira T, Yabuki K (1978) Primary production rates in the Minamata forest. In: Kira T, Onu Y, Hosokawa T (eds) Biological production in a warm-temperate evergreen oak forest of Japan. JIBP-Synthesis Vol 18. Univ Tokyo Press, Tokyo
Kira T, Ogawa H, Yoda K, Ogino K (1967) Comparative ecological studies on three main types of forest vegetation in Thailand. IV. Dry matter production, with special reference to the Khao Chong rain forest. Nat Life SE Asia 5:149–174
Kira T, Shinozaki K, Hozumi K (1969) Structure of forest canopies as related to their primary productivity. Plant Cell Physiol 10:129–142
Klinge H (1976) Bilanzierung von Hauptnährstoffen im Ökosystem tropischer Regenwälder (Manaus) – Vorläufige Daten: Trop. Regenwald Biographica 7:59–76
Klinge H (1977) Fine litter production and nutrient return to the soil in three natural forest stands of Eastern Amazonia. Geo Eco Trop 1:159–167
Klinge H, Herrera R (1978) Biomass studies in Amazon Caatinga forest in Southern Venezuela. 1. Standing crop of composite root mass in selected stands. Trop Ecol 19:93–110
Klinge H, Herrera R (1983) Phytomass structure of the Amazon Caatinga ecosystem in Southern Venezuela. 1. Tall Amazon Caatinga. Vegetatio (in press)
Klinge H, Rodrigues WA (1968) Litter production in an area of Amazon terra firme forest. Amazoniana 1:287–310
Koyama H (1974) Photosynthesis studies in Pasoh forest. IBP-Synthesis meeting, Kuala Lumpur, Mimeograph

Larcher W (1969) The effect of environmental and physiological variables on the carbon dioxide gas exchange of trees. Photosynthetica 3:167–198
Laudelout H, Meyer J (1954) Les cycles d'éléments minéraux et de matière organique en forêt équatoriale congolaise. 5th Congr Int Sci Sol 2:267–272
Lemon E, Allen LH, Muller L (1970) Carbon dioxide exchange of a tropical rain forest. Part II. BioScience 20:1054–1059
Lieth H (1960) Über den Lichtkompensationspunkt der Landpflanzen. Planta 54:530–576
Lieth H (1975) Modeling the primary productivity of the world. In: Lieth H, Whittaker RH (eds) Primary productivity of the biosphere. Ecol Stud Vol 14. Springer, Berlin Heidelberg New York
Löhr E (1969) Respirationsintensität in Stämmen, Zweigen und Blättern von Laubbäumen im tropischen Regenwald und in temperierten Wäldern. Physiol Plant 22:86–93
Lugo A (1970) Photosynthetic studies on four species of rain forest seedlings. In: Odum HT, Pigeon RF (eds) A tropical rain forest. Div Tech Inf US At Energ Comm
Lugo A, Gonzalez-Liboy JA, Cintrón B, Dugger K (1978) Structure, productivity and transpiration of a sub-tropical dry forest. Biotropica 10:278–291
Madge DS (1965) Leaf fall and litter disappearance in a tropical forest. Pedobiologia 5:272–288
Malaisse F (1978) The Miombo ecosystem. In: Tropical forest ecosystems, Unesco. Nat Resour Res 14, Unesco, Paris
Malaisse F, Fresson R, Goffinet G, Malaisse-Mousset M (1975) Litter fall and litter breakdown in Miombo. In: Golley FB, Medina E (eds) Tropical ecological systems. Ecol Stud Vol 11. Springer, Berlin Heidelberg New York
Medina E (1975) Dark CO_2 fixation, habitat preference and evolution within the Bromeliaceae. Evolution 28:677–686
Medina E (1978) Significación eco-fisiológica del contenido foliar de nutrientes y el área foliar específica en ecosistemas tropicales. 2nd Congr Lat Am Bot, Univ Brasilia
Medina E (1981) Physiological ecology of neotropical savanna plants. In: Huntley B, Walker B (eds) Ecology of tropical savannas. Ecol Stud Vol 42. Springer, Berlin Heidelberg New York
Medina E, Minchin P (1980) Stratification of $\delta^{13}C$ values in amazonian forests. Oecologia (Berl.) 45:377–378
Medina E, Zelwer M (1972) Soil respiration in tropical plant communities. Golley P, Golley FB (eds) Tropical ecology, with emphasis on organic matter production. Proc IInd Int Symp Trop Ecol. Univ Georgia, Athens
Medway Lord (1972) Phenology of a tropical rain forest in Malaya. Biol J Linn Soc 4:117–146
Menaut J-C (1977) Analyse quantitative des ligneux dans une savane arbustive préforestière de Côte d'Ivoire. Geo Eco Trop 1:77–94
Misra R (1972) A comparative study of net primary productivity of dry deciduous forest and grassland of Varanasi, India. In: Golley PM, Golley FB (eds) Tropical ecology with an emphasis on organic productivity. Univ Georgia Press, Athens
Monsi M, Saeki T (1953) Über den Lichtfaktor in den Pflanzengesellschaften und seine Bedeutung für die Stoffproduktion. Jpn J Bot 14:22–52
Mooney H (1972) The carbon balance of plants. Annu Rev Ecol Syst 3:315–346
Mooney H, Björkman O, Hall A, Medina E, Tomlinson PB (1980) The study of the physiological ecology of tropical plants – Current status and needs. BioScience 30:22–26
Müller D, Nielsen J (1965) Production brute, pertes par respiration et production nette dans la forêt ombrophile tropicale. Forstl Forsogsvaes Dan 29:69–160
Murphy PG (1975) Net primary productivity of tropical terrestrial ecosystems. In: Lieth H, Whittaker RH (eds) Primary productivity of the biophere. Ecol Stud Vol 14. Springer, Berlin Heidelberg New York
Murphy PG (1977) Rates of primary productivity in tropical grassland, savanna and forest. Geo Eco Trop 1:95–102
Newbould PJ (1967) Methods for estimating the primary production of forests. IBP Handbook No 2. Blackwell, Oxford

Odum HT (1970) Summary: an emerging view of the ecological system at El Verde. In: Odum HT, Pigeon RF (eds) A tropical rain forest. Div Tech Inf US At Energ Comm
Odum HT, Lugo A, Cintrón G, Jordan CF (1970) Metabolism and evapotranspiration of some rain forest plants and soil. In: Odum HT, Pigeon RF (eds) A tropical rain forest. Div Tech Inf US At Energ Comm
Ogawa H (1974) Litter production and carbon cycling in Pasoh forest. IBP-Synthesis meeting, Kuala Lumpur, Mimeograph
Ogawa H (1978) Principles and methods of estimating primary production in forests. In: Kira T, Ono Y, Hosokawa T (eds) Biological production in a warm-temperate evergreen oak forest of Japan. JIBP, Univ Tokyo Press, Tokyo
Ogawa H, Yoda K, Ogino K, Kira T (1961) Comparative ecological studies on three main types of forest vegetation in Thailand. II. Plant biomass. Nat Life SE Asia 4:49–86
Olofimboba MU (1969) Seasonal variations in the carbohydrates in the xylem of *Antiaris africana*. Ann Bot 33:339–349
Opler PA, Frankie GW, Baker HG (1976) Rainfall as a factor in the release, timing, and synchronization of anthesis by tropical trees and shrubs. J Biogeogr 3:231–236
Pearcy RW, Troughton J (1975) C_4 photosynthesis in tree form *Euphorbia* species from Hawaiian rain forest sites. Plant Physiol 55:1054–1056
Pittendrigh CS (1948) The Bromeliad-Anopheles-Malaria complex in Trinidad. I. The Bromeliad flora. Evolution 2:58–89
Ranawat MPS, Vyas LN (1975) Litter production in deciduous forests of Koriyat, Udaipur (South Rajasthan), India. Biologia (Bratislava) 30:41–47
Richards PW (1952) The tropical rain forest: an ecological study. Cambridge Univ Press, Cambridge
Saeki T (1960) Interrelationships between leaf amount, light distribution and total photosynthesis in a plant community. Bot Mag 73:55–63
San José JJ (1977) Potencial hídrico e intercambio gaseoso de *Curatella americana* L. en la temporada seca de las sabanas de *Trachypogon*. Acta Cient Venez 28:272–279
Siminovitch D, Wilson CM, Briggs DR (1953) Studies on the chemistry of the living bark of the black locust in relation to its frost hardiness. V. Seasonal transformations and variations in the carbohydrates: starch-sucrose interconversions. Plant Physiol 28:383–400
Stark N, Spratt M (1978) Root biomass and nutrient estimation in Venezuelan tropical soils. Trop Ecol 18:1–9
Stephens GR, Waggoner PE (1970) Carbon dioxide exchange of a tropical rain forest Part I. BioScience 20:1050–1053
Stocker O (1935) Assimilation und Atmung westjavanischer Tropenbäume. Planta 24:402–445
Troughton JH, Card KA, Hendy CH (1974) Photosynthetic pathways and carbon isotope discrimination by plants. Carnegie Inst Yearb 73:768–780
UNESCO (1978) Tropical forest ecosystems. A State-of-Knowledge report prepared by UNESCO-UNEP-FAO. Nat Resour Res 14, Paris
Vogel JC (1978) Recycling of carbon in a forest environment. Oecol Plant 13:89–94
Walter H (1973) Die Vegetation der Erde Bd 1. Die tropischen und subtropischen Zonen, 3rd edn. Fischer, Jena
Yabuki M, Aoki M (1974) Micrometeorological assessment of primary production rate of Pasoh forest. IBP-Synthesis meeting, Kuala Lumpur, Mimeograph
Yoda K (1967) Comparative ecological studies on three main types of forest vegetation in Thailand. III. Community respiration. Nat Life SE Asia 5:83–148
Yoda K (1974) Three-dimensional distribution of light intensity in a tropical rain forest of West Malaysia. Jpn J Ecol 24:247–254
Yoda K (1978) Community respiration. In: Kira T, Ono Y, Hosokawa T (eds) Biological production in a warm-temperate evergreen oak forest of Japan. JIBP Univ Tokyo Press, Tokyo
Zimmermann MH, Brown LL (1971) Trees structure and function. Springer, Berlin Heidelberg New York

10 Phytoplankton Productivity in Aquatic Ecosystems

W.R. BOYNTON, C.A. HALL, P.G. FALKOWSKI, C.W. KEEFE, and W.M. KEMP

CONTENTS

10.1 Introduction 305

10.2 Annual Rates and Seasonal Patterns of Primary Production 306
10.2.1 Lakes and Rivers 308
10.2.2 Estuarine Systems 309
10.2.3 Marine Systems 309

10.3 Regulation of Primary Productivity 311
10.3.1 Macroscopic Factors at the Ecological Scale 312
10.3.1.1 Light and Latitude 313
10.3.1.2 Nutrients 313
10.3.1.3 Temperature 315
10.3.1.4 Herbivory 315
10.3.2 Physiological Adaptations 316
10.3.2.1 Responses to Light 316
10.3.2.2 Temperature Responses 317
10.3.2.3 Nutrient Uptake 317

10.4 Synthesis 318

10.5 Conclusions 320
10.5.1 Physiological Studies 320
10.5.2 Measurement Techniques 321

References 323

10.1 Introduction

In recent years substantial efforts have been expended in developing and synthesizing data concerned with primary production in aquatic ecosystems. Salt marsh interactions with aquatic systems have been critically reviewed by NIXON (1980), eutrophication issues have received great attention (e.g. NATIONAL ACADEMY OF SCIENCES 1969; SCHINDLER 1977; NEILSON and CRONIN 1981; see also Chap. 13, this Vol.) and notable progress has been made concerning factors regulating primary production rates in lakes (e.g. BRYLINSKY and MANN 1973; SCHINDLER 1978; FEE 1979). RILEY (1972) and KOBLENTZ-MISHKE et al. (1970) have reviewed aspects of marine productivity. Such reviews are helpful in active research areas as mechanisms to confirm or rephrase basic research hypotheses and to provide those involved in managing aquatic systems with current (although not necessarily generally accepted) ecological information.

Phytoplankton production is of central importance in most aquatic environments because of its role in supporting many, if not most, food webs. Mass balance models of carbon fluxes and more recent results of stable carbon isotope studies support this concept (HAINES 1976; THAYER et al. 1978). The late STRICKLAND (1960) emphasized this in his suggestion that the effects of a disinterest in algal production would "not be unlike that which would prevail if livestock raisers lacked knowledge of, or interest in pasture conditions." This practical viewpoint can be construed easily to include more theoretical concepts concerning energy transfer in complex food webs (e.g. ULANOWICZ and KEMP 1979), benthic-pelagic coupling of nutrients and primary production (ROWE et al. 1975; BOYNTON et al. 1980; NIXON 1981) as well as other issues.

Studies of primary production in aquatic systems have been made at an accelerating pace over the past 50 years, particularly since the radioactive carbon method of STEEMANN-NIELSEN (1952) came into general use. Despite the extensive number of productivity measurements available, the degree to which the factors regulating this rate are understood varies considerably between different types of aquatic systems. In general, rates of phytoplankton production appear to be regulated by a complex of interacting physical, chemical and biological factors; however, the relative role of each of these factors has yet to be resolved into a manageable paradigm. Clearly, there is yet much to be accomplished.

An unsettling but perhaps overdue result of the accelerated pace of productivity studies has been the recent questions raised concerning the validity and comparability of productivity methods for measuring plankton primary production. A number of recent studies have questioned the validity of results obtained from the standard radioactive carbon methods, which support reservations expressed by RILEY (1953) nearly three decades ago, and argue that many ^{14}C-based estimates of oceanic production are four to ten times too low. The potential importance of this to those involved in all aspects of carbon flux is obvious.

With this brief background it is important to state our present objectives clearly. These are modest, as it is not possible to review in detail all aspects of this research area. We present here (1) a representative review of the seasonal patterns and annual magnitude of phytoplankton primary production in freshwater, estuarine and marine systems, (2) a general discussion of factors regulating primary production at the macroscopic and physiological levels, (3) a preliminary synthesis of physiological and ecological aspects of phytoplankton production and (4) critical comments concerning several areas of research and debate.

10.2 Annual Rates and Seasonal Patterns of Primary Production

A summary of primary production rates in aquatic environments is presented in Table 10.1. Although the results are useful for comparative purposes it should be noted that each contains some generally unevaluated inconsistency attributable to differences in methodology, as well as environmental conditions at the time of measurement (HALL and MOLL 1975; PETERSON 1980); as yet satisfactory and versatile intercalibration techniques between or within methods are not generally available. With this caveat in mind, we provide some comments on the magnitudes and patterns of photosynthesis for three major categories of aquatic habitat.

Table 10.1. Summary of mean daily rates and annual ranges in phytoplankton productivity ($g\ C\ m^{-2}\ d^{-1}$) in selected aquatic ecosystems

Ecosystem type	Primary production	
	Mean	Range
Lakes[a]		
Oligotrophic		
Meretta, NWT, Can.	0.01	0.00–0.17
Castle, Calif.	0.01	0.00–0.32
Lawrence, Mich.	0.11	0.00–0.49
Borax, Calif.	0.25	0.01–0.52
Lake Michigan	–	0.07–1.03
Mesotrophic		
Erken, Sweden	0.20	0.04–2.21
Clear, Calif.	0.44	0.00–2.44
Estrom, Denmark	0.37	0.02–0.42
Furesø, Denmark	0.46	0.60–1.38
Walter, Indiana	0.33	0.06–1.01
Martin, Indiana	0.56	0.03–1.07
Eutrophic		
Victoria, Africa	1.75	1.70–3.80
Lanao, Phillipines	1.70	0.40–5.00
Sylvan, Indiana	1.56	0.00–4.95
Sollerød, Denmark	1.43	0.00–3.80
Fredericksborg Slotssø, Denmark	1.03	0.01–4.16
Estuarine Systems[b]		
Non-stratified Systems		
Upper San Francisco Bay, Calif.	0.16	0.07–0.25
Upper Patuxent, Md.	0.23	0.04–0.52
W. Wadden Sea, Neth.	0.40	0.03–0.70
Venice Lagoon, Italy	0.48	0.10–1.10
Barataria Bay, La.	0.58	0.10–1.60
Upper Chesapeake, Md.	0.72	0.10–1.50
Raritan Bay, NJ	0.78	0.10–1.50
Apalachicola Bay, Fla.	1.00	0.70–1.10
Ems Estuary, Neth.	0.08	0.01–0.20
E. Wadden Sea, Neth.	0.31	0.03–1.00
Narragansett, RI	0.85	0.30–2.10
Beaufort Sound, NC	0.26	0.13–0.50
Peconic Bay, NY	0.35	0.20–0.70
Chinocoteague Bay, Md.	0.55	0.20–1.10
Port Hacking, Australia	0.90	0.30–1.70

[a] Data concerning lakes adapted from WETZEL (1975)
[b] Estuarine data from BOYNTON et al. (1981)

Table 10.1 (continued)

Ecosystem type	Primary production	
	Mean	Range
Moderately–Strongly Stratified Systems		
Lower San Francisco Bay, Calif.	0.63	0.40–0.80
Hudson River Mouth, NY	0.64	0.05–2.20
Long Island Sound, NY	0.54	0.35–0.70
Mid-Patuxent, Md.	1.10	0.50–1.70
Mid-Chesapeake, Md.	1.15	0.50–1.50
Baltic Sea	0.10	0.01–0.16
Bedford Basin, NS	0.62	0.07–0.90
Frazer R., BC	0.24	0.14–0.34
Str. of Georgia, BC	0.29	0.20–0.40
Byfjord, Sweden	0.55	0.10–1.00
St. Margarets Bay, NS	0.75	0.17–1.90
Indian Arm, BC	1.70	0.20–4.00
Marine Systems[c]		
Oligotrophic waters		
Central subtropical halistatic areas	0.07	0.03–0.10
Transition between subtropical and subpolar zones; extremity of equatorial divergences	0.14	0.10–0.15
Equatorial divergence and subpolar zones	0.20	0.15–0.25
Coastal marine waters	0.34	0.25–0.50

[c] Marine data adapted from KOBLENTZ-MISHKE et al. (1970)

10.2.1 Lakes and Rivers

In lakes there are sharp differences in primary production between those in different trophic categories. The representative data given in Table 10.1 suggest that rates in oligotrophic lakes are comparable to those observed in the open ocean, while rates observed in mesotrophic and eutrophic lakes are somewhat lower and higher, respectively, than those observed in estuarine systems. The annual pattern of primary production in temperate fresh waters often consists of low, sunlight-limited winter production, a spring bloom, a summer period of low, nutrient-limited production (in stratified lakes and ponds), sometimes an autumn bloom followed again by low winter production. While there are various studies of annual cycles of production in freshwater environments that show the above cycle, there are many other patterns observed; for example a unimodal seasonal pattern with high rates in summer and low rates in winter is common in shallow lakes and a few lakes even have maximum productivity under the ice (WETZEL 1975). Those lakes that show a seasonal cycle with a summer maximum tend to be characterized by a lack of summer nutrient depletion in the euphotic zone.

Streams do not have a thermocline-induced summer nutrient depletion, and the annual pattern of nutrient availability is not well studied for many streams. Obviously, human and edaphic factors are important, but are often irregular.

Sunlight availability clearly determines the annual pattern for at least some streams. One study that may characterize some types of woodland streams is that of HALL (1972), who used the free-water oxygen method to determine photosynthesis and respiration over the annual cycle in New Hope Creek, North Carolina. The pattern for New Hope Creek was not unlike the "classic" view of lakes discussed above in that photosynthesis showed peaks in both spring and fall. There was not, however, a clear pattern of nutrients becoming less abundant during the summer. The pattern of photosynthesis did correspond rather closely with the amount of sunlight available under the forest canopy, suggesting that the stream was sunlight-limited over most of the year.

10.2.2 Estuarine Systems

Mean daily rates of primary production for selected estuarine systems are also given in Table 10.1. Rates ranged from near zero to 4.0 g C m^{-2} d^{-1}. The average for all estuarine systems was about 0.59 g C m^{-2} d^{-1} (216 g C m^{-2} yr^{-1}), a value substantially higher than the 100 g C m^{-2} yr^{-1} reported by RYTHER (1963) for coastal areas and on the same order as the 300 g C m^{-2} yr^{-1} estimated for upwelling areas. In the majority of non-stratified estuaries, maximum rates appear to occur during the summer. Despite the large range in freshwater input, physical morphology, insolation and other factors, maximum rates tended to occur when water temperatures were at annual maxima. This pattern has been interpreted to indicate that temperature-regulated metabolism strongly influences nutrient recycling processes and planktonic growth rates which are, in turn, important factors maintaining high photosynthetic rates (e.g. WILLIAMS 1966; THAYER 1971; FLEMER 1970). While high production rates often occur in summer in moderately and highly stratified systems, the pattern is not nearly as consistent. Factors such as nutrient exhaustion in euphotic waters, deep mixing of plankton (RILEY 1946), dispersion of phytoplankton stocks (WELCH 1968) and others have been invoked to explain these patterns. A final observation concerns the magnitude of primary production between the ecosystem types as defined here. While there are considerable differences in the physical structure of these systems, the range of values observed within one group overlap values in the other group. In short, estuaries with different physical characteristics commonly have comparable rates of primary production. This suggests that there are system-specific biotic and physical mechanisms which interact to different degrees to produce comparable photosynthetic rates in substantially different estuaries.

10.2.3 Marine Systems

On the average phytoplankton production in marine systems (Table 10.1) is about 60 g C m^{-2} yr^{-1} (range 25–500 g C m^{-2} yr^{-1}; DE VOOYS 1979) compared with about 370 g C m^{-2} yr^{-1} by plants in the terrestrial biosphere (LIETH 1975) and 220 g C m^{-2} yr^{-1} for estuarine systems (BOYNTON et al. 1981). Large-scale

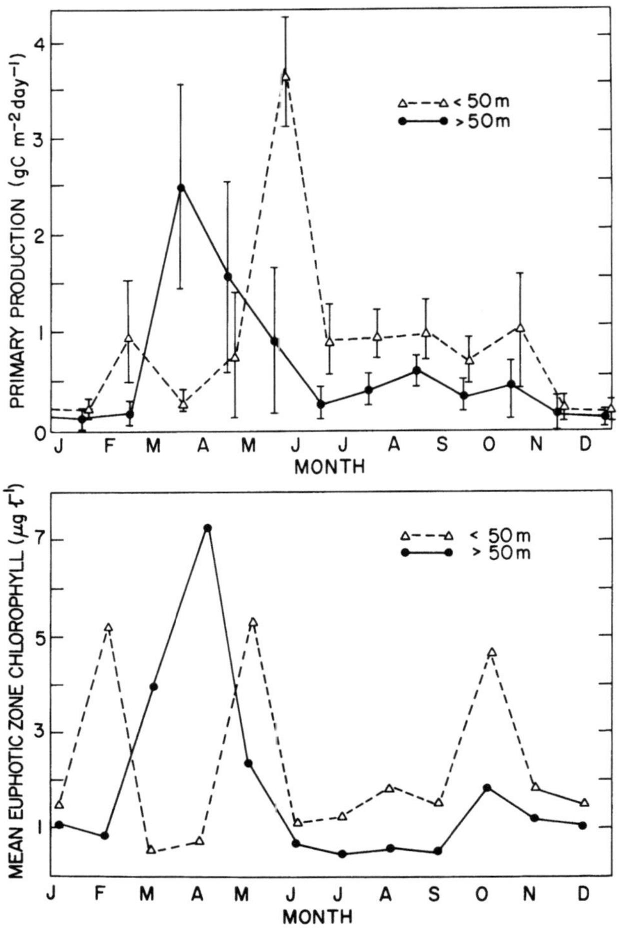

Fig. 10.1. Average seasonal distribution of primary production (*top*) and chlorophyll (*bottom*) in the Middle Atlantic Bight off the Northeast Coast of the U.S. Values are for nearshore (< 50 m) and offshore (> 50 m) regions

geographic patterns in biomass parallel variations in primary productivity (MALONE 1980), and the first-order influences on both are the rates at which nutrients are supplied to the euphotic zone and the depth of the mixed layer relative to the euphotic zone. These two factors are seasonally influenced in turn by latitude and the total depth of the water column. For example, primary productivity in coastal waters is higher on average than in the open ocean because nutrients are supplied more directly to coastal waters from terrestrial runoff, upwelling, benthic regeneration and coastal boundary layer turbulence. Offshore, in the open ocean, the rate of primary production is controlled primarily by the rate at which nutrients diffuse from the deep, nutrient-rich waters across the pycnocline into the euphotic zone. The seasonal dependence of this process is markedly influenced by latitude and open ocean phytoplankton blooms are primarily confined to temperate latitudes. Seasonal variations in primary productivity are most apparent in boreal and temperate coastal waters where large-scale variations in water temperature, coupled with large kinetic energy inputs from air–sea interactions and tidal mixing can cause seasonal turnovers or vertical mixing of the water column and the subsequent injection of large amounts of nutrients into the euphotic zone.

There are numerous seasonal studies of the distribution of chlorophyll, primary productivity, nutrients, grazing pressure, density, and light regimes in temperate coastal waters (e.g. RILEY 1946; RYTHER and YENTSCH 1958; MANDELLI et al. 1970; FALKOWSKI and STONE 1975; MALONE 1976; O'REILLY et al. 1976). On temperate continental shelves, phytoplankton productivity typically follows a bimodal bloom pattern, with a dominant spring bloom prior to the onset of thermal stratification and a secondary fall bloom corresponding to the erosion of thermal stratification (Fig. 10.1). The distributions of phytoplankton communities, relative to sites of nutrient input, are thus strongly influenced by the presence or development and dissipation of horizontal and vertical density gradients. The bloom may be terminated by either an increased grazing pressure from zooplankton or by sinking. For example, primary productivity in the Middle Atlantic Bight varies seasonally by over an order of magnitude and is highly correlated with the distribution of chlorophyll biomass (Fig. 10.1). Offshore productivity is highest in March–April, corresponding with the offshore phytoplankton bloom. Productivity during these two months amounts to ca. 40% of the annual productivity in waters >50 m. Productivity as high as 5.4 g C m^{-2} d^{-1} has been measured, and values of 2.5 g C m^{-2} d^{-1} are typical of the area during the spring bloom. The onshore phytoplankton bloom in April–May is greater than the offshore late spring bloom; the average productivity for waters <50 m during this period is ca. 3.7 g C m^{-2} d^{-1}, and values as high as 4.9 g C m^{-2} d^{-1} have been measured. The summer inshore primary production is as high as 1.5 g C m^{-2} d^{-1} within 10 km of the coast. The secondary phytoplankton bloom across the shelf in October–November accounts for ca. 1.3 g C m^{-2} d^{-1}. In the early winter, productivity offshore is low, averaging ca. 0.2 g C m^{-2} d^{-1}, and is only slightly higher nearshore.

10.3 Regulation of Primary Productivity

There are a variety of factors which regulate the magnitude, seasonal pattern and species composition of phytoplankton photosynthesis across the biosphere. These include light, temperature, nutrients, physical transport processes, herbivory and others, and are operative on both the ecological (macroscopic) and physiological scales. As indicated in Fig. 10.2, at the ecological scale these factors exert influence through general environmental modifications which affects species selection, yielding relatively broad ranges of "optimal response" to a given factor. On the other hand, at the physiological scale, adaptation to changing environmental conditions is regulated by biochemical mechanisms within cells which involve shifts in pigment and enzyme composition and which pro-

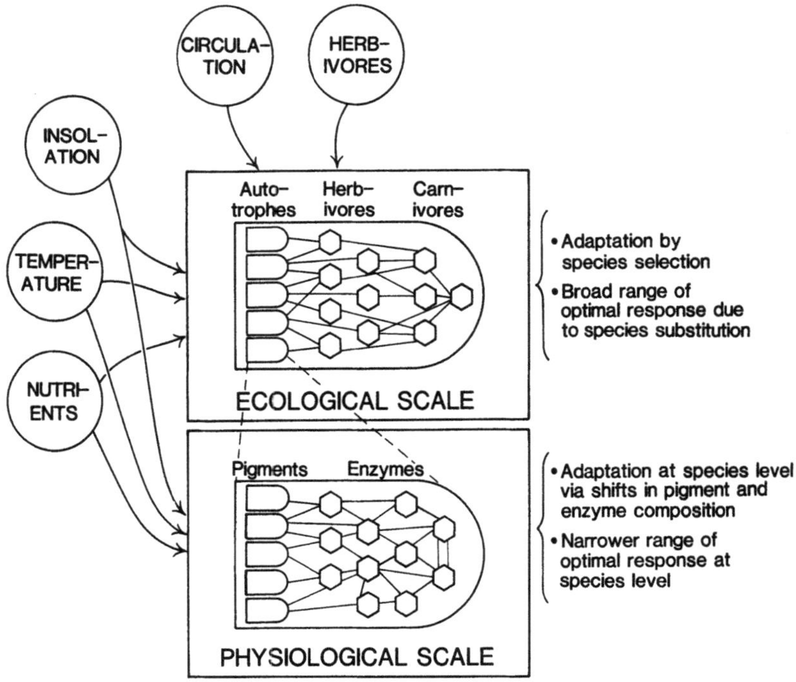

Fig. 10.2. Schematic representation of factors regulating primary production at both ecological and physiological scales

vides a capability for specific species to maintain near-optimal responses. The relative significance of these factors apparently varies from one system to the next and it is our purpose here to compare and contrast the importance of these factors. (The aquatic environment and responses to light in aquatic plants are discussed in Chaps. 8 and 9 of Vol. 12A; for nutrient cycling in freshwater and marine ecosystems see Chaps. 2 and 3, this Vol.)

10.3.1 Macroscopic Factors at the Ecologial Scale

In one of the first papers to consider the full gamut of factors controlling phytoplankton productivity, RILEY (1946) utilized a combination of statistical regressions and mechanistic equations to explain the magnitude and temporal importance of regulating factors. He concluded that the seasonal pattern of gross productivity of George's Bank was generally controlled by light and temperature, but that vertical mixing and nutrient shortages limited actual rates by about 20% of their potential in the spring and summer, respectively. Net productivity of phytoplankton was further reduced by respiration and zooplankton grazing to about 25% of the gross rates, with respiration largely a function of temperature and grazing most significant in late spring. Since Riley's work, considerable effort has been expended in the study of large-scale events thought to regulate photosynthetic rates.

10.3.1.1 Light and Latitude

Of obvious importance to phytoplankton production is incident solar radiation which provides the predominant energy source for autotrophic organisms. BRYLINSKY and MANN (1973) analyzed global data from 43 lakes and 12 reservoirs and found significant correlations between annual plankton production and various measures of solar energy (including latitude and temperature). SCHINDLER (1978) demonstrated that a significant relation could be obtained between latitude and phosphorus loading for a different but overlapping set of lake ecosystems. He concluded that at least a portion of the Brylinsky/Mann correlation between photosynthesis and latitude could be explained in terms of phosphorus input. WALSH (1974) has reported a striking negative relationship between potential yield of upwelling fisheries and latitude, which he interpreted to indicate the importance of light in the productivity of these coastal ecosystems.

The significance of light in the initiation of winter–spring diatom blooms has been demonstrated for Long Island Sound and Narragansett Bay (RILEY 1967; HITCHCOCK and SMAYDA 1977; NIXON et al. 1979). In these temperate areas peak annual concentrations of chlorophyll often occur during this winter–spring phenomenon; however, maximum photosynthesis typically occurs later in the year (June–July) when solar insolation reaches its annual peak (RILEY 1967; NIXON personal communication). Thus, annual cycles of sunlight appear to play a strong role in observed seasonal patterns of plankton photosynthesis in many temperate areas.

Vertical mixing in coastal and oceanic water-columns can strongly modulate this relationship between light and plankton photosynthesis by transporting cells below the depth at which there is sufficient light to maintain their growth. The early work of RILEY (1942) showed a positive relationship between water column stability (as indicated by the inverse of mixed-layer depth) and plankton production at George's Bank. SVERDRUP (1953) later derived an elegant formulation to estimate the "critical depth" below which vertical mixing of phytoplankton surpresses their growth. In most instances this critical depth can be approximated as the ratio of insolation at the surface to that at compensation depth divided by the light attenuation coefficient. Several investigators (e.g. PARSONS et al. 1966) have compared critical depth with depth of the mixed layer to predict or reconstruct geographic patterns of phytoplankton productivity in the ocean. Surprisingly little attempt has been made to use this notion of critical depth in estuarine environments; however, FISHER (personal communication) has shown that differences in plankton productivity between the Delaware and Chesapeake Bays can be explained in terms of vertical mixing and extinction coefficient. A possible manifestation of this relationship is embodied in the data of WETZEL (1975) and LORENZEN (1976), who report negative relationships between productivity and depth of the euphotic zone for lakes and seas, respectively.

10.3.1.2 Nutrients

In apparent contradiction to the findings of BRYLINSKY and MANN (1973), SCHINDLER (1978) reported strong correlations between annual phytoplankton production and phosphorus loading for some 50–90 lakes. Unfortunately, because of inaccessibility of certain data, Schindler's lakes were not the same as those of BRYLINSKY and MANN (though there was some overlap), so that it is impossible to thoroughly compare their respective findings. More recently, SMITH (1979) has reported that volumetric productivity rates (g C m^{-3} d^{-1}) correlate more strongly with nutrient conditions than do integral rates (g C m^{-2} d^{-1}) due to effects of transparency. Her lake data exhibit good regressions between productivity and both phosphorus and nitrogen concentrations. FEE (1979) has derived an index of nutrient recycling for lakes which relates directly to morphometric variables of area, volume, and depth and which correlates well with production in the Canadian experimental lakes. Hence, we see that nutrients are important whether they are derived from watershed inputs or recycled from internal sources.

The actual source of nutrient inputs to aquatic productivity is an issue of some concern for understanding the magnitude of production which can be supported in a given environment. EPPLEY and PETERSON (1979) have calculated that about 20% of oceanic production is driven by inputs of "new" nitrogen (nitrate), of which the vast majority is from upwelling, since rivers and rain account for less than 0.1% of total new nitrogen (DUGDALE 1976). In coastal and estuarine waters, both water-column recycling (e.g. CARPENTER and MCCARTHY 1978) and benthic remineralization (e.g. ROWE et al. 1975) can contribute substantially to the nutrient input budget for phytoplankton production (NIXON 1980). HARRISON (personal communication) has suggested that a general trend of decreasing rates of plankton production observed from the estuary across the continental shelf to the ocean corresponds to the decreasing importance of benthic remineralization. Similarly, moderately to non-stratified estuaries appear to be generally more productive over the year than highly stratified systems, implying the importance of maintaining some connection between the benthic and euphotic zones (Table 10.1).

In many aquatic systems periodic wind-driven circulation provides the necessary physical mechanism to bring nutrient-rich demersal waters to the euphotic zone (e.g., PLATT 1971; IVERSON et al. 1974). WALSH et al. (1978) estimate that about one-third of the nutrients incorporated in phytoplankton production over the New York Bight are derived from storm generated winds. Water-column stability as influenced by wind also may affect size-selection of phytoplankton in these regions (MALONE 1976; MALONE and CHERVIN 1979). Moreover, the periodicity of wind events and resulting turbulent spectra may have profound influence on the concentration and spatial dimensions of plankton blooms (WALSH et al. 1977, 1978). Several investigators have demonstrated reasonably close correspondence between power spectra of turbulence and phytoplankton chlorophyll in lakes (POWELL et al. 1975) and coastal embayments (PLATT 1972; THERRIAULT 1978), suggesting, again, the relationship between wind (and tidal) circulation and phytoplankton abundance. Tidal circulation in estuaries can also influence phytoplankton production both directly via dilution and flushing of cells (KETCHUM 1954, 1967) and indirectly by inducing water-column destratification and hence nutrient upwelling (WEBB and D'ELIA 1980).

While river input of nutrients is of little consequence to annual oceanic phytoplankton production, it is (as with lakes) extremely important for estuarine and coastal environments. For instance, SUTCLIFFE (1972) has demonstrated a significant regression between land drainage and planktonic chlorophyll a over an annual cycle in the Canadian maritime. Furthermore, a strong relationship was observed between annual discharge of the St. Lawrence River and yield of five major fisheries (adjusted for age at maturity). COPELAND (1966) has reported a similar relation between rainfall and harvest of white shrimp (2-year lag) in the Texas bays, and MIHURSKY et al. (1981) find a consistent correlation between spring river flow and striped bass year-class success in the Potomac River estuary. The negative relationship between distance-from-shore and plankton productivity, inferred by HARRISON (personal communication) to suggest the importance of benthic remineralization, has been interpreted by TURNER et al. (1979) to suggest the importance of particulate and dissolved nutrient inputs from terrigenous and estuarine sources. Finally, WELCH et al. (1972) have shown that the timing, location and magnitude of plankton blooms in the Duwamish Estuary are strongly related to river flow.

10.3.1.3 Temperature

In his thorough review article, EPPLEY (1972) has shown that the general shape of curves for phytoplankton photosynthetic response to temperature are quite similar for a wide variety of species. Temperature optima, however, span a wide range from about 10°–40 °C for different species, and thermal adaptation is available for a given population. Still, it is accepted as the general rule among many that in nature temperature, per se, is not a major factor controlling productivity in the oceans, since comparable rates have been observed over a broad geographic range of latitudes (e.g. WALSH 1974; BUNT 1975). On the other hand, the fact that many influences other than temperature bear on phytoplankton photosynthesis (EPPLEY 1972) should not obviate the potential role of thermal regimes. For example, it is not surprising that nutrient-rich upwelling waters in the arctic may be more productive in the summer than oligotrophic tropical seas despite relatively cold temperatures. Furthermore, numerous studies of power plant effects of phytoplankton have demonstrated significant enhancement of productivity during cold months but an inhibition at peak summer temperatures (e.g. see review in KEMP 1977).

YENTSCH and his colleagues have provided extensive data to further examine this issue (YENTSCH 1974; YENTSCH et al. 1974). They found that from a given temperate coastal environment, mixed assemblages of phytoplankton exhibit temperature response curves very similar to those for pure cultures, regardless of the ambient temperature from which they were sampled. Apparently, little temperature acclimatization occurs for a given assemblage, but rather temperature exerts a selective force for populations having temperature optima coincidental with the conditions occurring in the given locale.

10.3.1.4 Herbivory

The importance of zooplankton grazing on phytoplankton stocks as a mechanism regulating productivity has been an issue of some debate for several decades past. STEEMANN-NIELSEN (1958b), for instance, argued that the commonly observed seasonal patterns of coincidental phytoplankton and zooplankton abundance evidenced the fact that grazing maintained algal populations in a steady-state balance suited to the limitations of other environmetal conditions (i.e. light, nutrients, temperature). In contrast, CUSHING (1959) utilized a simple predator–prey model to conclude that grazing did, indeed, affect the magnitude and timing of phytoplankton stocks, and that a lag between peak abundances of phytoplankton and zooplankton populations could be observed consistently, especially in colder climates.

Extreme population densities of zooplankton in both lakes and estuaries are a relatively common occurrence, and, in many cases the energy needed to support associated herbivory must be explained either in terms of tightly coupled turnover rates (CUSHING 1976) or organic carbon sources other than phytoplankton (HEINLE and FLEMER 1975). In estuaries such as Chesapeake and Narragansett Bays, the remarkably high non-specific filter-feeding rates of ctenophores and medusae on plankton may serve to keep the zooplankton grazing in balance (HEINLE 1974). WALSH (1976) has speculated that, in the sea, the relatively low productivity in offshore divergence zones compared to coastal upwellings may be attributable to differential grazing pressures in the two areas, since nutrient regimes are similar in both. Nevertheless, we must conclude that, in general, while herbivory may be an important control on phytoplankton productivity for certain seasons in certain environments (e.g. MARTIN 1970), it is not likely to be a severe limitation overall (e.g. OVIATT et al. 1979).

10.3.2 Physiological Adaptations

Two principal features of aquatic environments dominate the physiological responses of algae, namely the high density and the dynamic fluidity of the medium. While terrestrial plants necessarily evolved extensive support structures, algae, differing little in density from water, are suspended in the matrix. The support structures of higher plants are multi-purpose, however, and serve as pipelines for nutrients and water as well as an anchor. Algae, having no vascular tissue, are evolutionally confined to size scales where molecular diffusion is a viable means of exchanging gases, nutrients, and water. Thus, even the largest seaweeds are only a few millimeters thick, and the overwhelming majority of aquatic plant biomass is unicellular.

10.3.2.1 Responses to Light

While diffusion and boundary layer effects may present physiological and anatomical limitations for algae, the requirement for survival, that a cell must divide before it leaves the euphotic zone, places it under severe restrictions. For the most part algae are at the mercy of the motion of the water column to remain within the euphotic zone. Aquatic environments characteristically have steep vertical gradients in light, nutrients and temperature; consequently, mixing processes potentially expose algae to a wide range of environmental regimes. Many algae, especially the phytoplankton, have evolved a remarkable degree of physiological plasticity allowing them to adapt (physiologically) to changes in many environmental conditions on time scales of minutes to hours; i.e. similar to the time scales of meso-scale turbulence. Turbulence can affect productivity by altering the vertical distribution of the phytoplankton cells during the day, thereby influencing the light and temperature regimes experienced by the cells, as well as by mixing nutrients both within and into the mixed layer (see also Chaps. 8 and 9, Vol. 12A).

Phytoplankton cells experience large variations in light intensity over the course of a day. Because a cell's photosynthetic response to light is non-linear, the amount of carbon it fixes might be influenced by temporal variations in light intensity as well as the total light it receives (e.g. MARRA 1978a). The first-order variation is clearly created by the sun as it travels across the sky. The path of the sun gives the total daily radiation, the length of the day, and the maximum intensity at the sea surface. Local atmospheric conditions, clouds, and aerosols modulate the light reaching the sea surface. The light received by a cell is perhaps best predicted from the surface intensity and depth of the cell. In a turbulent ocean, cells are displaced vertically about a daily mean depth and these displacements also create variations in light. MARRA (1978b) suggested that vertical turbulence may increase water column productivity and inferred that the effect is due to variations in light experienced by the cells. Many species of marine phytoplankton maximize their photosynthetic capacity by acclimating to variation in ambient light intensity through changes in photosynthetic pigment content. As the cells are mixed by physical processes through the euphotic zone, photo-adaptation on the same time scales of vertical mixing is greatly facilitated by rapid pigment metabolism. Changes in intracellular photosynthetic pigments such as chlorophylls (light–shade adaptation) or accessory pigments (chromatic adaptation) are sometimes observed within 15 min after exposure to a new light regime (FALKOWSKI 1980).

10.3.2.2 Temperature Responses

Temperature has a well-known modifying influence on enzymatic processes and Arrhenius plots of phytoplanktonic light-saturated photosynthetic rates are not uncommon but have most often been obtained for a given species at one growth temperature. In cases where a species has been cultured at various temperatures prior to assay it is apparent that the maximum and optimum temperatures of photosynthesis, as well as the photosynthetic rates, can be physiologically modified. This adaptive strategy primarily occurs on time scales of 1 or 2 days, and has been suggested to increase the fitness of non-motile organisms in temperate environments and of diurnally migrating motile organisms exposed to vertical temperature gradients (LI 1980).

10.3.2.3 Nutrient Uptake

An adaptive strategy, similar in its time-dependence to photoadaptation, has been proposed for the uptake of inorganic nutrients (MCCARTHY and GOLDMAN

1979). They postulate small-scale patchiness (either spatial or temporal) in nutrient supplies, as might be expected if zooplankton excretion is a source of inorganic nutrients in situ. The variability in the small-scale nutrient patches is suggested to be coupled with an enhanced uptake capacity by the phytoplankton, induced by nutrient limitation, making it possible for phytoplankton to maintain near-maximum growth rates under apparently nutrient-deficient conditions. The short-term enhanced uptake rates for nitrogen or phosphorus are temporally uncoupled from growth, resulting in an increase in the internal pool of the nutrient immediately following the rapid uptake process. The concept of uncoupling between nutrient uptake and growth is not new, and was suggested to account for the observations of "luxury consumption" of nutrients by phytoplankton, advanced in the 1930's (e.g. KETCHUM 1939); however, the observation of the induction of enhanced short-term nutrient uptake capability increases the complexity of interactions that can be expected in phytoplankton, and further accents the degree of physiological influence in regulating photosynthetic and growth processes.

10.4 Synthesis

The variation in annual phytoplankton production among most freshwater, estuarine and marine systems and across broad geographic regimes appears to be surprisingly small, despite widely varying environmental conditions (e.g. Table 10.1). Maximum productivity for fertile estuaries and coastal upwelling systems may be only 10–40 times greater than values for the oligotrophic Sargasso Sea (RYTHER 1963; WALSH 1974). Moreover, there is some tentative reason to believe that part of this apparent difference is a function of methodological problems. For example, RILEY (1972) has estimated (based on carbon budgeting) that only a three fold difference separates annual production in an estuarine system, Long Island Sound, and the oligotrophic Sargasso Sea (see also following section). Hence, it may be that the overall effects of various environmental and physiological factors on productivity are more a matter of fine-tuning than gross adjustment.

In most aquatic systems the seasonal distribution in photosynthetic rates and biomass appears to be a consequence of multiple factors operating together or in a time sequence. As WALSH (1976) has said, "simple consideration of single control factors of... primary production such as light, nutrients, mixing... can lead to rather naive and perhaps inaccurate conceptualizations of the dynamics of non-linear aquatic ecosystems." With this cautionary note in mind, what generalizations can we make? It would appear that the most important variables regulating aquatic photosynthetic rates are light and nutrients, and in the case of nutrients, most often nitrogen or phosphorus. Moreover, nutrient supply appears to be a central feature regulating the level of plankton biomass.

Obviously, the mechanisms which regulate light and nutrient availability vary considerably between different system types. In the case of many lakes,

SCHINDLER (1978) found strong correlations between annual plankton production and phosphorus loading; SMITH (1979) reported good correlations between volumetric productivity and nitrogen and phosphorus concentrations, and FEE (1979) reported good correlations between annual production in lakes and an index of nutrient recycling potential of the benthos. In estuarine and marine systems there are numerous examples where the rate of nutrient supply appears to regulate photosynthetic rates as well as levels of plankton biomass. RILEY (1972) concluded that of the factors regulating productivity, the rate of supply of nutrients to surface waters was of overwhelming importance. In this regard then, the available stock of nutrients in deep stratified systems and vertical transfer rates become critical features regulating production rates (RILEY 1946; IVERSON et al. 1974; EPPLEY and PETERSON 1979). Thus, it appears that nutrient supply is an essential feature whether the source is from land drainage, in situ recycling or upwelling from deep water reserves.

An additional consistent feature of systems where nutrients are supplied primarily from deep water storages is juxtapositioning in time of very high and considerably lower productivity rates. RILEY (1972) concluded that a major reason for higher production in African upwelling waters than in Long Island Sound was that productivity in the Sound relied on biological recycling of nutrients rather than accessible deep water supplies and that recycling proceeded at a slower rate than physical upwelling. BOYNTON et al. (1981) found highest rates (5 g C m^{-2} d^{-1}) reported for estuarine-coastal ecosystems to be primarily related to those where substantial upwelling or riverine inputs occurred. Data reported by IVERSON et al. (1974) are particularly striking in this regard.

As indicated earlier, light availability is also of central importance in regulating productivity rates. Here again, the mechanisms regulating the amount of insolation vary considerably both due to climatic differences as well as internal system features. For example, in some estuaries, particularly in oligohaline regions, it appears that rapid light attenuation due to suspended particles severely limits production rates (e.g. STROSS and STOTTLEMEYER 1966). In coastal, oceanic and deeper estuarine areas vertical mixing obviously modulates nutrient regimes as well as the light available to plankton cells. Mixing of cells below the "critical depth" (SVERDRUP 1953) has generally been invoked as a primary factor regulating production. When stable vertical structure is established such that plankton cells are not actively mixed below the critical depth, plankton populations rapidly increase. In recent years the role of nutrient supply has overshadowed the importance of light in many lake studies even though BRYLINSKY and MANN (1973) earlier reported good correlations between annual production and insolation from a global sampling of lakes. It would be helpful if the latter data set could be reanalyzed in view of results reported by SCHINDLER (1978), FEE (1979) and SMITH (1979).

Thus, in our view, nutrient supply and insolation are key variables regulating photosynthetic rates in aquatic systems. There is, however, considerable variability between systems in the mechanisms regulating nutrient supply and light and these are presumably responsible for the differences in magnitude and seasonal patterns observed. We tend to view direct effects of temperature and herbivory, and physiological adaptations as important, and critical at the species level, but appropriately viewed as second-order regulatory factors at the ecosystem level of organization.

10.5 Conclusions

From the preceding sections it is clear that while much has been achieved concerning understanding of phytoplankton production in aquatic environments, conflicts in data interpretation remain unresolved and new research themes have developed. In this section we provide comment on two such areas.

10.5.1 Physiological Studies

Studies on the physiology and biochemistry of phytoplankton and macrophytes increased enormously during the 1960's and 1970's. The vast majority of these research efforts were directed toward the elucidation of one or more species' responses to a single environmentally relevant variable, such as light, nutrients, temperature or pollutant. It was hoped that these studies would provide a mechanistic basis for understanding and improving ecologically derived models of phytoplankton production, distribution and basic floristic composition. In fact, in many cases the accuracy of ecological predictions was significantly enhanced by experimental physiological studies. This is especially true in essentially single variable problems such as point-source eutrophication (Caperon et al. 1971), or interannual variations in light extinction (Falkowski et al. 1980). The early modeling efforts of more complex systems suggested, however, that the environmental outcome of integrated physiological processes, such as production, growth or floristic composition, cannot be accurately inferred from additive or multiplicative constructions of single variable responses (i.e. physiological responses to multivariable perturbations often appear to be synergistic).

In an effort to understand synergistic behavior more fully in physiological processes and its importance in ecological theory, physiological studies of algae have centered on two central themes: (1) multiple factor analysis, such as nutrient–light relationships, or nutrient–temperature relationships, as exemplified by the studies of Goldman (1979) and Laws and Bannister (1980), and (2) time-course studies of kinetic responses and physiological adaptations such as those of Hitchcock (1979) and Prézelin and Matlick (1980). For the most part these approaches are a natural outgrowth and necessary extension of the earlier single variable experiments. It is hoped that these types of approach will lead to the formulation of ecological models at the species level although this premise has been challenged on philosophical grounds (Medawar 1967; Dunbar 1980). Perhaps, somehow, the species and ecosystem-level perspectives can be blended into an operative hierarchial paradigm.

10.5.2 Measurement Techniques

One of the most fundamental problems to be addressed in any review of phytoplankton productivity must concern the comparison and interpretation of various methodologies for measuring photosynthesis. It is a sobering thought that the controversy which was raised among oceanographers in the early 1950's regarding the comparability of such methodologies remains today far from resolution. Recently, PETERSON (1980) has provided an excellent review which takes us through the controversial 30-year history of the ^{14}C technique and its interpretation. We shall not attempt to reconstruct this fascinating story here; however, we will highlight the major issues and add a few current research findings not included in PETERSON's review.

There are a wide variety of techniques for measuring phytoplankton photosynthesis, including: O_2 production and consumption; CO_2 change as indicated by pH and total inorganic carbon; changes in concentrations of total particles or particulate carbon; and incorporation of ^{14}C-bicarbonate. The O_2 technique was among the first finding wide usage (GAARDER and GRAN 1927), and it enjoyed a prominent position among limnological and oceanographic studies for some 20 years until the introduction of the ^{14}C technique (STEEMANN-NIELSEN 1952), which promised a simple, highly sensitive approach to photosynthetic research. A wide variety of technical problems with the ^{14}C method have been elaborated over the past years, and these are summarized by PETERSON (1980) under 13 categorical headings.

STEEMANN-NIELSEN employed his ^{14}C measurements to update the estimate of total annual oceanic production. His calculations yielded a value which was an order-of-magnitude less than that provided earlier by RABINOWICH (1945) based on O_2 measurements of SEIWELL (1935) and RILEY (1939). The ensuing debate, which lasted more than a decade, led to a flurry of laboratory and field experimental comparisons of the two methods. These experiments addressed such questions as: bactericidal effects of light and algal antibiotics; effects of incubation time; growth phase, nutrient status and ratio of photosynthesis to respiration; and magnitude and effects of light on respiration. PETERSON (1980) describes the results of some 19 such comparative studies which, taken as a group, are amazingly contradictory. The most consistent conclusion throughout was that O_2 tended to give higher estimates of photosynthesis than the ^{14}C method (e.g. RYTHER and VACCARO 1954; STEEMANN-NIELSEN 1958a; MCALLISTER et al. 1961; ANTIA et al. 1963; THOMAS 1963, 1964; EPPLEY and SLOAN 1965).

In several of these experiments, as well as in subsequent studies, direct comparisons were also made between O_2 and/or ^{14}C methods and other techniques such as pH/CO_2 and particle concentration changes. The results of this work were also generally inconsistent, but they did lead to some tentative insights. Some experiments showed a good correspondence between particulate carbon production and ^{14}C uptake (e.g. RYTHER and MENZEL 1965); others found that ^{14}C underestimated direct observations on carbon production (e.g. TUNZI and PORCELLA 1974), while still other studies concluded that ^{14}C measurements exceeded carbon production (e.g., PETERSON 1978). Efforts to budget carbon fluxes have demonstrated the potential importance of excretion of dissolved organic carbon (much of which is ^{14}C-labelled), and this observation has led to still another controversy (e.g. SHARP 1977). Since the early work of BEYERS and ODUM (1960), perhaps the strongest proponent of the pH/CO_2 method for photosynthetic measurements has been JACOB VERDUIN, and his results have consistently shown that ^{14}C uptake is far exceeded by CO_2 fixation inferred through pH change (e.g., VERDUIN 1960, 1975). VERDUIN has emphasized relentlessly the potential problems of bottle effects (HALL and MOLL 1975), stressing the advantage of O_2 and pH/CO_2 techniques over ^{14}C in that the former can be used readily in non-contained, open water columns.

It has been assumed generally that the highly sensitive ^{14}C technique would be most appropriate for estimating productivity in the oligotrophic waters of the open ocean. Yet most attempts to develop carbon budgets for an oceanic water column have led to the conclusion that ^{14}C estimates were low by a factor of about 10 (POMEROY and JOHANNES 1968; RILEY 1972). These calculations are based largely on measurements of zooplankton standing stocks and respiration rates, since benthic carbon demand appears to be very small (SMITH 1978; HINGA et al. 1979). Early attempts to construct oxygen and phosphorus budgets from oceanographic data led to similar estimates of photosynthetic production on the order of 1.0 g C m^{-2} d^{-1} (RILEY 1951). Finally, data provided by GIESKES et al. (1979) strongly suggest (once again) that "bottle effects" and pigment degradation associated with standard ^{14}C incubation techniques can lead to serious underestimates of productivity. A companion study reported by TIJSSEN (1979) used a sensitive photometric titration technique to generate diurnal O_2 balances in the mixed layer of the mid-Atlantic Ocean. Productivity estimates from both of these studies indicated values an order-of-magnitude higher than those measured using typical ^{14}C techniques. The recent work of MCCARTHY and GOLDMAN (1979) and GOLDMAN et al. (1979) suggests that such high photosynthetic rates may indeed be possible in the face of vanishingly small nutrient concentrations due to tight coupling between nutrient uptake and remineralization.

In reading these papers on comparative methodologies, one is struck by the interpretive tone of various authors, which often seems to suggest a bias toward one technique or another. Many of these authors interpret their results to conclude that one method over- or underestimates another, implying that the latter is a touchstone to which the former must be compared. PETERSON's (1980) review calls for a more balanced perspective on these methodological problems; however, his own point of view fails to emphasize the potential utility of O_2 techniques in many aquatic environments. He suggests a suite of continuous-culture experiments which could provide the necessary impetus toward clarifying this troublesome problem.

Acknowledgments. We thank Chris D'Elia, John Barlow, and Bruce Peterson for criticisms of the manuscript. Discussions with C. Wirick, T.E. Malone, S. Nixon and T. Fisher were particularly helpful. Contribution No. 1342, Center for Environmental and Estuarine Studies of the University of Maryland.

References

Antia NJ, McAllister CD, Parsons TR, Stephens K, Strickland JDH (1963) Further measurements of primary production using a large volume plastic sphere. Limnol Oceanogr 8:166–183

Beyers RJ, Odum HT (1960) Differential filtration with strong acids or bases vs. CO_2 water for productivity studies. Limnol Oceanogr 5:228–230

Boynton WR, Kemp WM, Osborne CG (1980) Nutrient fluxes across the sediment–water interface in the turbid zone of a coastal plain estuary. In: Kennedy VS (ed) Estuarine processes. Academic Press, London New York

Boynton WR, Keefe C, Kemp WM (1981) A review of phytoplankton processes in estuarine environments. Chesapeake Biological Laboratory, Solomons, Maryland

Brylinsky M, Mann KH (1973) An analysis of factors governing productivity in lakes and reservoirs. Limnol Oceanogr 18(1):1–14

Bunt JS (1975) Primary productivity of marine ecosystems. In: Lieth H, Whittaker RH (eds) Primary productivity in the biosphere. Ecol Stud Vol 14. Springer, Berlin Heidelberg New York

Caperon J, Cattell SA, Krasnick G (1971) Phytoplankton kinetics in a subtropical estuary: eutrophication. Limnol Oceanogr 16:599–607

Carpenter EJ, McCarthy JJ (1978) Benthic nutrient regeneration and high rate of primary production in continental shelf waters. Nature (London) 274:188–189

Copeland BJ (1966) Effects of decreased river flow on estuarine ecology. J Water Pollut Contr Fed 38:1831–1839

Cushing DH (1959) The seasonal variation in oceanic production as a problem in population dynamics. J Cons Int Explor Mer 24:455–464

Cushing DH (1976) Grazing in Lake Erken. Limnol Oceanogr 21:349–356

Dugdale RC (1976) Nutrient cycles. In: Cushing DH, Walsh JJ (eds) The ecology of the seas. Saunders, Philadelphia

Dunbar MJ (1980) The blunting of Occam's razor, or to hell with parsimony. Can J Zool 58:123–128

Eppley RW (1972) Temperature and phytoplankton growth in the sea. Fish Bull 70:1063–1085

Eppley RW, Peterson BJ (1979) Particulate organic matter flux and planktonic new production in the deep ocean. Nature (London) 282:677–680

Eppley RW, Sloan PR (1965) Carbon balance experiments with marine phytoplankton. J Fish Res Board Can 22:1083–1097

Falkowski PG (1980) Light–shade adaptation in marine phytoplankton. In: Falkowski PG (ed) Primary productivity in the sea. Plenum, New York

Falkowski PG, Stone DP (1975) Nitrate uptake in marine phytoplankton: energy sources and the interaction with carbon fixation. Mar Biol 32:77–84

Falkowski PG, Hopkins TS, Walsh JJ (1980) An analysis of factors affecting oxygen depletion in the New York Bight. J Mar Res 38:479–506

Fee EJ (1979) A relation between lake morphometry and primary productivity and its use in interpreting whole-lake eutrophication experiments. Limnol Oceanogr 24(3):401–416

Flemer DA (1970) Primary production in the Chesapeake Bay. Chesapeake Sci 11:117–129

Gaarder T, Gran HH (1927) Investigation of the production of plankton in the Oslo Fjord. Rapp et Proc Verb. Cons Int Explor Mer 42:1–48

Gieskes WWC, Kraay GW, Baars MA (1979) Current ^{14}C methods for measuring primary production: gross underestimates in oceanic waters. Neth J Sea Res 13(1):58–78

Goldman JC (1979) Temperature effects on steady-state growth, phosphorus uptake, and the chemical composition of a marine phytoplankton. Microb Ecol 5:153–166

Goldman JC, McCarthy JJ, Peavey DG (1979) Growth rate influence on chemical composition of phytoplankton in oceanic waters. Nature (London) 279:210–215

Haines EB (1976) Relation between the stable carbon isotope composition of fiddler crabs, plants, and soils in a salt marsh. Limnol Oceanogr 21(6):880–883

Hall CAS (1972) Migration and metabolism in a temperate stream ecosystem. Ecology 53:585–604

Hall CAS, Moll R (1975) Methods of assessing aquatic primary productivity. In: Lieth H, Whittaker RH (eds) Primary productivity in the biosphere. Ecol Stud Vol 14. Springer, Berlin Heidelberg New York

Heinle DR (1974) An alternative grazing hypothesis for the Patuxent Estuary. Chesapeake Sci 15:146–150

Heinle DR, Flemer DA (1975) Carbon requirements of a population of the estuarine copepod, *Eurytemora affinis*. Mar Biol 31:235–247

Hinga KR, Sieburth JMcN, Heath GR (1979) The supply and use of organic material at the deep-sea floor. J Mar Res 37:557–579

Hitchcock GL (1977) The time course of photosynthetic adaptation, the growth rate response, and variations in pigment, carbohydrate, and protein content of *Skeletonema costatum* and *Detonula confervacea* to changes in light intensity. PhD thesis, Univ Rhode Island, Narragansett, Rhode Island

Hitchcock GL, Smayda TJ (1977) The importance of light in the initiation of the 1972–1973 winter-spring diatom bloom in Narragansett Bay. Limnol Oceanogr 22:126–131

Iverson RL, Curl HC Jr, O'Connors HB Jr, Kirk D, Zakar K (1974) Summer phytoplankton blooms in Auke Bay, Alaska, driven by wind mixing of the water column. Limnol Oceanogr 19(2):271–278

Kemp WM (1977) Energy analysis and ecological evaluation of a coastal power plant. PhD dissertation Dept Environmental Engineering. Univ Fla, Gainesville

Ketchum BH (1939) The absorption of phosphate and nitrate by illuminated cultures of *Nitzschia closterium*. Am J Bot 26:399–407

Ketchum BH (1954) Relation between circulation and planktonic populations in estuaries. Ecology 35:191–200

Ketchum BH (1967) Phytoplankton nutrients in estuaries. In: Lauff G (ed) Estuaries. Am Assoc Adv Sci Publ No 83, Washington, DC

Koblentz-Mishke OJ, Volkovinsky VV, Kabanova JG (1970) Plankton primary production of the world ocean. In: Wooster WS (ed) Scientific exploration of the South Pacific. Wash DC Natl Acad Sci

Laws EA, Bannister TT (1980) Nutrient- and light-limited growth of *Thalassiosira fluviatilis* in continuous culture, with implications for phytoplankton growth in the ocean. Limnol Oceanogr 25:457–473

Li WKW (1980) Temperature adaptation in phytoplankton: cellular and photosynthetic characteristics. In: Falkowski PG (ed) Primary productivity in the sea. Plenum, New York

Lieth H (1975) Primary production of the major vegetation units of the world. In: Lieth H, Whittaker RH (eds) Primary productivity of the biosphere. Ecol Stud Vol 14. Springer, Berlin Heidelberg New York

Lorenzen CH (1976) Primary production in the sea. In: Cushing DH, Walsh JJ (eds) The ecology of the seas. Saunders, Philadelphia

Malone TC (1976) Phytoplankton productivity in the Apex of the New York Bight: environmental regulation of productivity/chlorophyll *a*. In: Gross MG (ed) The Middle Atlantic Continental Shelf and New York Bight. Limnol Oceanogr Spec Symp 2

Malone TC (1980) Size-fractionated primary productivity of marine phytoplankton. In: Falkowski PG (ed) Primary productivity in the sea. Plenum, New York

Malone TC, Chervin MB (1979) The production and fate of phytoplankton size fractions in the plume of the Hudson River, New York Bight. Limnol Oceanogr 24:683–696

Mandelli EF, Burkholder PR, Doheny TE, Brody R (1970) Studies of primary productivity in coastal waters of southern Long Island, New York. Mar Biol 7:153–160

Marra J (1978a) Effect of short-term variations in light intensity on photosynthesis of a marine phytoplankter: a laboratory simulation study. Mar Biol 46:191–202

Marra J (1978b) Phytoplankton photosynthetic response to vertical movement in a mixed layer. Mar Biol 46:203–208

Martin JH (1970) Phytoplankton-zooplankton relationships in Narragansett Bay. IV. The seasonal importance of grazing. Limnol Oceanogr 15:413–418

McAllister CD, Parsons TR, Stephens K, Strickland JDH (1961) Measurements of primary production in coastal seawater using a large-volume plastic sphere. Limnol Oceanogr 6:237–258

McCarthy JJ, Goldman JC (1979) Nitrogenous nutrition of marine phytoplankton in nutrient-depleted waters. Science 203:670–672

Medawar PB (1967) The art of the soluble. Penguin Books, Harmondsworth, England

Mihursky JA, Boynton WR, Setzler-Hamilton EM, Wood KV, Polgar TT (1981) Freshwater influences on striped bass population dynamics. In: Cross RD, Williams DL (eds) Proc Natl Symp Freshwater Inflow to Estuaries. US Dep Inter Fish Wildl Serv. FWS/OBS 81/04:140–167

National Academy of Sciences (1969) Eutrophication: causes, consequences, correctives. Proc Int Symp Eutrophication. Natl Acad Sci, Washington, DC

Neilson BJ, Cronin LE (1981) Estuaries and nutrients. Humana, Clifton

Nixon SW (1980) Between coastal marshes and coastal waters – A review of twenty years of speculation and research on the role of salt marshes in estuarine productivity and water chemistry. In: Hamilton P, MacDonald KB (eds) Estuarine and wetland processes. Plenum, New York

Nixon SW (1981) Remineralization and nutrient cycling in coastal marine ecosystems. In: Neilson BJ, Cronin LE (eds) Estuaries and nutrients. Humana, Clifton

Nixon SW, Oviatt CA, Kremer JN, Perez K (1979) The use of numerical models and laboratory microcosms in estuarine ecosystem analysis: simulations of a winter phytoplankton bloom. In: Dame RF (ed) Marsh-estuarine systems simulation. Univ SC Press, Columbia

O'Reilly JE, Thomas JP, Evans C (1976) Annual primary production (nannoplankton, netplankton, dissolved organic matter) in the Lower New York Bay. In: McKeon, Lauer GJ (eds) 4th Symposium on Hudson River ecology. Hudson River Environ Soc Inc. New York, Paper No 19

Oviatt CA, Nixon SW, Perez KT, Bucklay B (1979) On the season and nature of perturbations in microcosm experiments. In: Dame RF (ed) Marsh-estuarine systems simulation. Univ SC Press, Columbia

Parsons TR, Giovando LF, LeBrasseur RJ (1966) The advent of the spring bloom in the eastern subarctic Pacific Ocean. J Fish Res Board Can 23:539–546

Peterson BJ (1978) Radiocarbon uptake: Its relation to the net particulate carbon production. Limnol Oceanogr 23:179–184

Peterson BJ (1980) Aquatic primary productivity and the ^{14}C-CO_2 method: A history of the productivity problem. Annu Rev Ecol Syst 11:359–385

Platt T (1971) The annual production of phytoplankton in St. Margaret's Bay, Nova Scotia. J Cons Perm Int Explor Mer 33:324–333

Platt T (1972) Local phytoplankton abundance and turbulence. Deep-Sea Res 19:183–188

Pomeroy LR, Johannes RE (1968) Occurrence and respiration of ultraplankton in the upper 500 meters of the ocean. Deep-Sea Res 15:381–391

Powell TM, Richerson PJ, Dillon TM, Agee BA, Dozier BJ, Godden DA, Myrup LO (1975) Spatial scales of current speed and phytoplankton biomass fluctuations in Lake Tahoe. Science 189:1088–1089

Prézelin BB, Matlick HA (1980) Time-course of photoadaptation in the photosynthesis–irradiance relationship of a dinoflagellate exhibiting photosynthetic periodicity. Mar Biol 58:85–96

Rabinowich EI (1945) Photosynthesis and related processes vol I. Wiley-Interscience, New York

Riley GA (1939) Plankton studies II. The Western North Atlantic, May–June 1939. J Mar Res 2:145–162

Riley GA (1942) The relationship of vertical turbulence and spring diatom flowerings. J Mar Res 5:67–87

Riley GA (1946) Factors controlling phytoplankton population on George's Bank. J Mar Res 6:54–73

Riley GA (1951) Oxygen, phosphate, and nitrate in the Atlantic Ocean. Bull Bingham Oceanogr Coll 13:1–26
Riley GA (1953) Letter to the Editor. J Cons Int Explor Mer 19:85–89
Riley GA (1967) The plankton of estuaries, In: G. Lauff (ed) Estuaries. Am Assoc Adv Sci Publ No 83, Washington, DC
Riley GA (1972) Patterns of production in marine ecosystems. In: Wiens JA (ed) Ecosystem structure and function. Univ Oregon Press, Corvallis
Rowe GT, Clifford CH, Smith KL Jr, Hamilton PL (1975) Benthic nutrient regeneration and its coupling to primary productivity in coastal waters. Nature (London) 255:215–217
Ryther JH (1963) Geographical variations in productivity. In: Hill MN (ed) The sea Vol II. Wiley-Interscience, New York
Ryther JH, Menzel DW (1965) Comparison of the ^{14}C technique with direct measurement of carbon fixation. Limnol Oceanogr 10:490–492
Ryther JH, Vaccaro RF (1954) A comparison of O_2 and ^{14}C methods of measuring marine photosynthesis. J Cons Perm Int Explor Mer 20:25–37
Ryther JH, Yentsch CS (1958) Primary production of continental shelf waters off New York. Limnol Oceanogr 3:327–335
Schindler DW (1977) Evolution of phosphorus limitation in lakes. Science 195:260–262
Schindler DW (1978) Factors regulating phytoplankton production and standing crop in the world's freshwaters. Limnol Oceanogr 23(3):478–486
Seiwell HR (1935) The annual organic production and nutrient phosphorus requirement in the tropical western North Atlantic. J Cons Int Explor Mer 10:20–32
Sharp JH (1977) Excretion of organic matter by marine phytoplankton: Do healthy cells do it? Limnol Oceanogr 22:381–399
Smith KL (1978) Benthic community respiration in the N.W. Atlantic Ocean: in situ measurements from 40 to 5200 m. Mar Biol 47:337–347
Smith VH (1979) Nutrient dependence of primary productivity in lakes. Limnol Oceanogr 24:1051–1064
Steemann-Nielsen E (1952) The use of radio-active carbon for measuring organic production in the sea. J Cons Explor Mer 18:117–140
Steemann-Nielsen E (1958a) Experimental methods for measuring organic production in the Sea. Rapp Proc Verb, Cons Perm Int Explor Mer 144:38–46
Steemann-Nielsen E (1958b) The balance between phytoplankton and zooplankton in the sea. J Cons Explor Mer 23:178–188
Strickland JDH (1960) Measuring the production of marine phytoplankton. Bull No 122. Fish Res Board Can, Ottawa
Stross RG, Stottlemeyer JR (1966) Primary production in the Patuxent River. Ches apeake Sci 6:125–140
Sutcliffe WH (1972) Some relations of land drainage, nutrients, particulate material, and fish catch in two eastern Canadian bays. J Fish Res Board Can 29:357–362
Sverdrup HU (1953) On conditions for the vernal blooming of phytoplankton. J Cons Explor Mer 18:287–295
Thayer GW (1971) Phytoplankton production and the distribution of nutrients in a shallow unstratified estuarine system near Beaufort, NC. Chesapeake Sci 12:240–253
Thayer GW, Parker PL, LaCroix MW, Fry B (1978) The stable carbon isotope ratio of some components of an eelgrass, *Zostera marina* bed. Oecologia 35(1):1–12
Therriault JC (1978) Spatial variability of phytoplankton turnover in relation to physical processes in a coastal environment. Limnol Oceanogr 23:900–911
Thomas WH (1963) Physiological factors affecting the interpretation of phytoplankton production measurements. In: Doty MS (ed) Primary productivity measurement, marine and fresh water, Conf Proc USAEC. TID-7633. Washington, DC
Thomas WH (1964) An experimental evaluation of the C^{14} method for measuring phytoplankton production, using cultures of *Dunaliella primolecta* Butcher. Fish Bull 63:273–292
Tijssen SB (1979) Diurnal oxygen rhythm and primary production in the mixed layer of the Atlantic Ocean at 20° N. Neth J Sea Res 13:79–84

Tunzi MG, Porcella DB (1974) Carbon-14 assimilation, chlorophyll and particulate organic matter in steady state systems at Lake Tahoe. Limnol Oceanogr 19:420–428
Turner RE, Woo SW, Jitts HR (1979) Estuarine influences on a continental shelf plankton community. Science 206:218–220
Ulanowicz RE, Kemp WM (1979) Toward canonical trophic aggregations. Am Nat 114:871–883
Verduin J (1960) Phytoplankton communities of western Lake Erie and the CO_2 and O_2 changes associated with them. Limnol Oceanogr 5:372–380
Verduin J (1975) Photosynthetic rates in Lake Superior. Verh Int Ver Limnol 19:689–693
Vooys CGN de (1979) Primary production in aquatic environment. In: Bolin B, Degens ET, Kempe S, Ketner P (eds) The global carbon cycle. Wiley and Sons, New York
Walsh JJ (1974) Primary production in the sea. In: 1st Int Congr Ecol, Proc Center Agric Publ Doc, Wageningen
Walsh JJ (1976) Herbivory as a factor in patterns of nutrient utilization in the sea. Limnol Oceanogr 21:1–13
Walsh JJ, Kelley JC, Whitledge TE, Huntsman SA, Pillsbury RD (1977) Further transition states of the Baja California upwelling ecosystem. Limnol Oceanogr 22:264–280
Walsh JJ, Whitledge TE, Barvenik FW, Wirick CD, Howe SO, Esaias WE, Scott JT (1978) Wind events and food chain dynamics within the New York Bight. Limnol Oceanogr 23:659–683
Webb KL, D'Elia CF (1980) Nutrient and oxygen redistribution during a spring-neap tidal cycle in a temperate estuary. Science 207:983–985
Welch E (1968) Phytoplankton and related water-quality conditions in an enriched estuary. J Water Pollut Contr Fed 40:1711–1727
Welch EB, Emery RM, Matsuda RI, Dawson WA (1972) The relationship of algal growth in an estuary to hydrographic factors. Limnol Oceanogr 17:734–737
Wetzel RG (1975) Limnology. Saunders, Philadelphia
Williams RB (1966) Annual phytoplanktonic production in a system of shallow temperate estuaries. In: Barnes H (ed) Some contemporary studies in marine sciences. Allen and Unwin, London
Yentsch CS (1974) Some aspects of the environmental physiology of marine phytoplankton: A second look. Oceanogr Mar Biol Annu Rev 12:41–75
Yentsch CS, Yentsch CM, Strube LR, Morris I (1974) Influence of temperature on the photosynthetic efficiency in natural populations of marine phytoplankton. In: Gibbons JW, Sharitz RR (eds) Thermal ecology. USAEC Conference Series CONF-730505. NTIS, Springfield, Virginia

11 Effects of Biocides and Growth Regulators: Physiological Basis

K. PFISTER and W. URBACH

CONTENTS

11.1 Introduction 329

11.2 History and Development 330

11.3 Selectivity of Herbicides 333

11.4 Mode of Action 335
11.4.1 General Aspects 335
11.4.2 Herbicides 336
11.4.2.1 Herbicides Affecting Photosynthesis 336
11.4.2.2 Herbicides Affecting Respiration 348
11.4.2.3 Herbicides Affecting Pigment Metabolism 349
11.4.2.4 Herbicides Affecting Lipid Metabolism and Membrane Structure and Function 352
11.4.2.5 Herbicides Affecting Protein and Nucleic Acid Synthesis . . 354
11.4.2.6 Herbicides Affecting Cell Division 355
11.4.2.7 Herbicides Affecting Growth 356
11.4.3 Growth Regulators 357
11.4.4 Insecticide Effects on Plants 363
11.4.5 Herbicide-Safeners (Antidotes) 366

11.5 Herbicide Resistance 368

11.6 Integrated Pest Management 371

11.7 Conclusions 372

References 372

11.1 Introduction

Improvement of food production and minimization of crop yield losses due to pests have been a major concern in agriculture. Until the last century the only major progress in agriculture had been the improvement of mechanical equipment for cultivation and irrigation. Since the mid 19th century the introduction of artificial fertilizers has helped increasing crop production substantially. A tremendous boost in productivity occurred after World War II when high-yield varieties became prevalent and large-scale chemical pest control developed. The term "green revolution" was coined to characterize this development.

Parallel to our growing understanding of plant and animal physiology and of biochemistry, more and more chemicals became available which allow the control of undesired plants and animal species. These chemicals are called in general pesticides or biocides. Depending on their target organism, one distinguishes between insecticides, herbicides, and fungicides. As minor groups, acaricides, molluscicides, nematicides and rodenticides should be mentioned for the sake of completeness. An important addition to the biocides used in agriculture are the plant growth regulators. Some of these compounds interfere with the action of endogenous plant hormones and offer possibilities to influence plant growth in a desirable way.

Plant and animal ecosystems are strongly affected by environmental pollutants and by the application of biocides (see also Chaps. 13, 14 and 16, this Vol.). Also, agricultural practices are influenced by chemical pest control measures. This underlines the need for extensive studies on the physiological and ecological impact of biocides. The present chapter tries to characterize how biocides and growth regulators interfere with physiological reactions of plants. The following contribution (Chap. 12, this Vol.) considers the ecological consequences of these chemicals in terrestrial ecosystems; pollution of fresh water ecosystems with pesticides is discussed in Chap. 13, this Vol. We want to focus our survey mainly onto the interference of commercial herbicides and growth regulators with physiological reactions of plants. Also, the action of selected experimental chemicals will be discussed. Where possible, special emphasis will be given on the molecular mechanism of action. In addition, some toxic effects of insecticides on plant metabolism will be discussed.

Due to the abundant amount of the literature published during the last few years, it is not possible to pay adequate attention to every related aspect. For this reason, at the beginning of our sections we have listed more comprehensive review articles or books, which cover certain areas of interest in much more detail: WEAVER (1972), CORBETT (1974), KURTH (1975), KEARNEY and KAUFMAN (1976), AUDUS (1972, 1976), BÜCHEL (1977), GRÄSER (1977), WEGLER, Vol. 1–7 (1970–1981), LETHAM et al. (1978), GEISSBÜHLER (1978), MORELAND (1980), ASHTON and CRAFTS (1981), SANDERS (1981), FEDTKE (1982).

Fungicides share quite an important part in the biocide market. They are applied to protect plants from diseases in agriculture, horticulture and forest management. This topic is not covered in this review. For general information see MARTIN and WORTHING (1976), KRÄMER (1977), GREWE (1977), SCHEINPFLUG et al. (1977), SIEGEL and SISSLER (1977).

The nomenclature in this chapter follows the recommendations of the Weed Science Society of America (WSSA) given in the latest edition of their Herbicide Handbook (1979). Further references are given in the Pesticide Manual of the British Crop Protection Council (WORTHING 1979). Chemicals not listed there or not having been assigned a common name are named in the text or in Table 11.2 and 11.4.

11.2 History and Development

Until the end of the last century, manual and mechanical pest control was dominant in agriculture. Chemical weed control began with the use of simple, inorganic compounds as iron or copper sulfate, cyanates, chlorates, borates, arsenates and even sulfuric acid. Most of these compounds were applied in large quantities and mainly for total vegetation control because of the lack of reasonable selectivity (for more details see CORBETT 1974; AUDUS 1976; WEGLER 1977; BÜCHEL 1977). The area of modern organic herbicides started in 1935 when the first of a series of nitrophenol derivatives, DNOC, was described as a selective herbicide (TRUFFAUT and PASTAC 1935). The next important development occurred in the early 1950's with the discovery of the herbicidal properties of substituted phenylureas like monuron or diuron (BUCHA and TODD 1951; WESSELS and VAN DER VEEN 1956). Later, the group of s-triazines was found to contain powerful and selective herbicides such as simazine and atrazine (GAST et al. 1955). Even today, ureas and triazines are among the most widely used groups of herbicides, whose mode of action is considered to be inhibition of photosynthesis. Later, herbicides from several more chemical families were discovered, which are also thought to act as inhibitors of photosynthesis. Examples of these herbicides are the uracils, pyridazinones, some carbamates, amides, triazinones, anilides and diphenylethers. Bipyridylium salts like diquat and paraquat were introduced in 1955 as another group of herbicides which interfere with photosynthetic reactions (BRIAN et al. 1958).

The first commercial compounds described as acting as growth regulators were NAA (naphthylacetic acid) and MCPA, followed by the phytotoxic phenoxyacetic acid derivatives 2.4-D and 2.4.5-T, which have been shown to interfere with the function of the endogenous plant hormones. The development of these compounds started towards the end of World War II (HAMNER and TUKEY 1944). In the meantime, several growth regulators have become available. Many of these compounds have been shown to perturb or enhance the action of endogenous plant hormones such as cytokinins, gibberellins or ethylene.

The development of the present large scale use of biocides and growth regulators was caused by the increasing need to control undesired pests which limit potential crop yield. Plant diseases, fungi, insects and weeds are considered as pests, the latter of which are of special interest for the present chapter. Whereas diseases, fungi and insects induce direct damage to crop plants, weeds influence crop yield indirectly by competing with the crop for light, water, and nutrients. In some cases, weeds serve as hosts for fungi and insects. There are several studies which demonstrate that an already surprisingly low number of weeds per acreage can produce substantial food losses (summarized by MCEWEN 1978 and KURTH 1975). On a global basis food losses due to pests are tremendous. Estimations from different sources claim that food losses for the most important crops range between 20% and 45% of the potential harvest despite use of biocides (see CRAMER 1967; BÜCHEL 1977; KURTH 1975; MCEWEN 1978; PIMENTEL et al. 1978; PIMENTEL 1981). In total, world food losses for the five most important crops were estimated to represent a value of 200 billion U.S.$ (FAO yearbook 1980). This is one aspect underlining the economic need for efficient pest control practices. Today pest control relies mainly on the use of chemicals. However, we have to realize that in the long run an increasing and extensive use of biocides may be harmful to our ecosystems. Alternative methods and an intensive development and employment of integrated pest management (IPM, Sect. 11.6) will help to reduce food losses without similar

Table 11.1 World population, agricultural production, fertilizer and biocides

	1951–55	1960–70	1971–75	1985	2000
World population (billion)[a]	3		4		6.35
mal-nourished people[a] (mill.)			400–600		1,300
Agricultural area[a] (bill. ha)	1.27	1.38	1.47	1.51	1.54
Total cereal production[a]					
(mill. tons)		1.1	1.2		2.14
($t\ ha^{-1}$)		1.5	1.75	2.2	2.8
Fertilizer production[a]					
(mill. t)	18.3	37.8	79.9	134.9	220.3
($kg\ ha^{-1}$)	15	30	55	90	145
Biocide sales[b]					
(mill. US$)			5,138	7,500	

Source: [a] BARNEY (1980)
[b] PIMENTEL (1981)

hazards. From the ecological point of view the use of every chemical which harmfully interferes with ecosystems and disturbs their ecological balance should be restrained as far as possible.

In the following, some basic data concerning food demand and production will be given (Table 11.1). This information was taken from recent, very comprehensive studies, which try to predict future developments based on a careful analysis of our present situation [Global 2000 Report to the President (1980) and Global Future (1981): BARNEY (1980, 1981); CRC Handbook of Pest Management in Agriculture: PIMENTEL (1981)].

Most estimates indicate that the total world population will rise from presently 4 billion to 6.3 billion people in the year 2000. Whereas today 400–600 million people are considered to be starving or at least mal-nourished, this number will increase to 1.3 billion in the year 2000. It has to be recognized, however, that a more responsible utilization of the natural resources on earth and a more even distribution of agricultural products in the world, can – even today – considerably diminish the number of starving people. The area available for agricultural use will increase very slightly, but a steady decline of the agricultural area on a per capita basis is expected. Thus not much more food will be available by the end of the century (Table 11.1). All these estimates are based on rather optimistic models, assuming no significant changes in politics, world economy or energy prices. The latter very probably will influence future agricultural development and practices by increasing costs for chemical pest control, fertilizers, and use of mechanical equipment. It is estimated that total world production of fertilizers and of biocides will double by the year 2000, whereas considerably higher growth rates of biocide application are expected for the underdeveloped countries.

Increases in agricultural productivity since the end of World War II have relied heavily on widespread use of fertilizers, high-yield crop varieties and biocides. Today, extensive use of biocides should be considered with great concern. The various ecological problems created by a continuously growing use of biocides might exceed the benefits and advantages of biocide application, as the following examples illustrate:

- Accumulation of persistent biocides in diverse food chains will lead to increasing amounts of residues in living organisms. Although in few cases toxicity of biocide residues has already been observed, there is an almost complete lack of knowledge about long-term effects.
- Development of resistant species of noxious insects, weeds and fungi reduces the effectiveness of chemical pest control and will increase the costs for repeated and more extensive biocide application. This forces the development of more and more new chemicals.
- In the near future, due to further environmental pollution, a considerable number of all existing plant and animal species may be lost. However, the variety of living species represents an important genetic resource for generating new crop species and food.

Intact ecosystems are the basic requirements for the healthy development and survival of mankind. Therefore it is necessary to use refined methods for agricultural and medical pest control. Today, the development of concepts for Integrated Pest Management (IPM) is regarded as representing a realistic new method for pest control. IPM combines several techniques, including chemical, biological and agricultural means for pest control, as will be discussed below. Since 1972 IPM programs have been established in the United States [Apple and Smith 1976; Global 2000 Report to the President, Barney 1980].

11.3 Selectivity of Herbicides

In nearly all cases, the agricultural use of herbicides requires selective action of herbicides against weeds, whereas the crop plant should tolerate the treatment without injury. In contrast to this, non-selective herbicides are widely used for complete plant eradication, e.g. in industrial areas, railroads, highways or airfields. Also in certain non-crop applications (e.g. cotton) non-selective herbicides are used to facilitate harvesting. Typical representatives of herbicides used for total vegetation control are the bipyridyls paraquat and diquat and phenylureas such as diuron and monuron. Selectivity cannot be considered as an inherent feature of a particular chemical. It depends strongly on the amount of herbicide applied under a given condition (plant species, soil, climatic conditions). Usually, selective herbicides have a sufficient safety margin for recommended application purposes. If this safety margin is exceeded by an incorrect application (wrong concentration, wrong adjustment of spray equipment, changing of environmental conditions) injury to crop plants is likely to occur. It is very difficult, if not impossible, to correlate the factors controlling herbicide selectivity with the chemical structure or with the mode of action of a particular group of herbicides. For this reason, determining the selectivity of a herbicide is mostly an empirical task. For a comprehensive coverage and for a discussion of factors determining the selectivity of individual herbicides, the reader should refer to HOLLY (1976), WAIN and SMITH (1976), SARGENT (1976), VAN OORSCHOT (1979). Several factors, however, can be listed which confer selectivity:

By careful choice of the time for an herbicide application selective action against weeds can be obtained. Pre-emergence applications will affect weeds before germination or emergence of the crop plant.

Leaf surface properties of crops and weeds establish another simple mechanism for selectivity. It is obvious that much smaller amounts of sprayed herbicide will be retained and absorbed by a small, sometimes wax-coated surface of a grassy cereal crop plant as compared to a broadleaf weed. The limited selectivity of the early inorganic herbicides was based mostly on this property. Wetting agents or detergents abolish nearly all selective herbicide action, if selectivity is based on a lower wettability of the crop leaves. On the contrary, wetting agents are added to prepare a formulation of a herbicide, in order to facilitate absorption by the weeds, if wettability of the weed leaves previously limited the herbicide uptake.

Soil adsorption is a key factor in controlling selectivity of several herbicides. A strong adsorption in the upper zones of the soil may prevent herbicides from reaching deep-rooting crop plants and allows action on shallow-rooting weeds or vice versa. Also a rapidly leaching herbicide can have differential effects on weeds and crop plants, depending on the sizes of their root systems. A strong soil adsorption accompanied by a fast degradation may prevent a foliar applied contact herbicide from reaching the roots of a crop plant. This latter is the case with several bipyridylium herbicides such as diquat or paraquat.

After uptake of herbicide by the leaves or roots, a differential translocation in the xylem or phloem of crops or weeds will lead to a selective action (SARGENT 1976). Translocation is severely influenced by environmental conditions as wind, humidity, light or CO_2 levels.

Metabolism of a compound with different rates is a common and very important mechanism for selective action (WAIN and SMITH 1976; NAYLOR 1976). A more rapid enzymatic detoxification by the crop plant leads to an herbicide tolerant crop as compared to the susceptible weed. Such a different rate of metabolism is the basis for tolerance of corn (*Zea mays*) to triazine herbicides such as atrazine or simazine. After application of a triazine herbicide to corn, photosynthesis initially becomes inhibited but recovers during the following hours due to a progressive detoxification of the herbicide. For a more detailed coverage of the triazine metabolism see KEARNEY and KAUFMAN (1976), ASHTON and CRAFTS (1981). Also the selectivity of thiocarbamate herbicides is based on differential rates of metabolism in crops and weeds (CASIDA et al. 1975). This enzymatic detoxification can even be stimulated by externally applied chemicals (antidotes, safeners) to yield improved thiocarbamate tolerance in crops as corn, rice or sorghum (see Sect. 11.4.5). In contrast to these examples, where metabolism led to non-toxic compounds, in some cases inactive chemicals are applied which are in turn converted to toxic herbicides inside the plant. Ioxynil-octanoate and dinoseb-acetate are relatively ineffective as inhibitors of photosynthesis in a cell free system compared to their parent compounds ioxynil and dinoseb. Phytotoxicity of ioxynil-octanoate and dinoseb-acetate in vivo is believed to be based on their parental compounds formed after cleavage of the octanoate or acetate substituent, respectively.

11.4 Mode of Action

11.4.1 General Aspects

In the last decade great progress has been made in understanding the molecular mechanisms of action of several groups of biocides and growth regulators which are presently used for chemical pest control in agriculture. Intensive investigations have been made on the effect of biocides on plants, especially on those herbicides which interfere with specific metabolic reactions. The enormous increase in research work has been caused by more sensitive experimental techniques at the cellular and molecular level and by attempts to coordinate the results obtained in the different disciplines. This has led to a better understanding of the physiological and biochemical mechanism of herbicidal action. Moreover, increasing efforts in studies on the effects of growth regulators and insecticides on plant metabolism have been made. Therefore, this chapter is primarily confined to the interference of herbicides and growth regulators with plant metabolic reactions. Some phytotoxic effects of insecticides will be summarized.

Generally a distinction can be made between a primary and secondary action of biocides (CORBETT 1974; MORELAND 1980). Primary mechanisms of action are mainly related to processes in which biocides interfere biochemically or biophysically with molecules or structures of cells such as proteins, DNA, RNA, membranes or enzymes. Primary mechanisms of action, for instance, are inhibition of photosynthesis at a specific endogenous electron carrier (see Sect. 11.4.2.1), formation of toxic radicals which in turn oxidize pigments and lipids (see Fig. 11.3), inhibition of enzymes synthesizing carotenoids or chlorophylls (see Sect. 11.4.2.3) or interference with membranes influencing the ion transport (see Sect. 11.4.2.4). Since these interactions, which are the first observable effects at lowest herbicide concentrations, can be measured very quickly after herbicide application, they permit an insight into the mechanism of biocidal actions. Studies with isolated cells, cell organells, particles, membranes and artificial membranes will help to elucidate the primary actions of biocides on plants.

Secondary actions of biocides indicate the sequence of reactions following the primary response and leading to physiological alterations of plant metabolism and cell structures and finally to the death of a plant. They include the interference with photosynthetic reactions, destruction of chlorophyll (bleaching), lipid peroxidation, loss of membrane functions, suppression of cell division and disorganization of the plant cell ultrastructure and can only be described by physiological responses.

In this connection often a distinction between direct and indirect effects of biocides has to be made. Herbicides, for instance, which directly affect the function of the thylakoid membrane by blocking the photosynthetic electron transport also indirectly affect this membrane by inhibiting the synthesis of ATP or by inducing photooxidations of the inhibited electron transport system. Depending on the concentration applied, in many cases a combination of several primary effects of commercially used biocides will result in a severe disturbance of the metabolism followed by death of the plant. Besides their primary and secondary actions, lethal or sublethal applications of biocides to ecosystems in combination with other environmental conditions have important ecological implications, which will be reviewed in the following chapter by SCHUBERT (Chap. 12, this Vol.).

11.4.2 Herbicides

The herbicidal action of chemical compounds depends on their interaction with a multitude of major biochemical reactions in the plant. The majority of commercial herbicides attacks the photosynthetic apparatus. Others interfere with respiratory and phosphorylating processes, pigments and lipid metabolism, protein synthesis, nucleic acid metabolism, cell division, metabolic regulation and growth reactions and destroy structures and functions of membranes. Usually a herbicide affects more than one metabolic reaction in plants. Studies to identify the single reaction which is primarily affected by a given herbicide have significantly extended our knowledge on herbicide actions. However, regarding the total herbicidal effect, the other metabolic sites of action have to be considered. Since the most sensitive site of action may not be the process which mainly contributes to phytotoxicity, an understanding of the different metabolic sites of action is as important in herbicide research as in studies on the primary mechanism of herbicidal action. It is especially important that initial research on the mode of action of a given herbicide includes several metabolic reactions (ASHTON et al. 1977). There are a number of screening methods and algal bioassays which are recommended for the detection and investigation of herbicidal activity (KRATKY and WARREN 1971; SUMIDA and UEDA 1974; HOROWITZ 1976; HESS 1980; BÖHME et al. 1981).

In the following some effects of herbicides and their representation in the recent literature will be reviewed briefly. For more detailed information the reader is referred to recent reviews on these topics as mentioned at the beginning of the different sections.

11.4.2.1 Herbicides Affecting Photosynthesis

Photosynthesis, as the central series of metabolic reactions in higher plants offers a preferential target site for a large number of herbicides. It is estimated that more than half of all commercial herbicides act primarily by interfering with photosynthetic reactions. For a detailed description of fundamental properties of photosynthetic reactions, the reader should refer to the following general reviews. Most aspects of photosynthesis are compiled in several volumes of this series, i.e. Volume 5 and 6 [see also WITT (1971), TREBST (1974), GOVINDJEE (1975), BARBER (1976, 1977) and AKOYUNOGLOU (1981)]. Organization of photosynthetic light reactions is indicated schematically in Fig. 11.1. All photosynthetic reactions take place inside a cell organelle – the chloroplast. Electron carriers involved in the primary reactions are located in the chloroplast lamellar system, the thylakoids. Most of the enzymes catalyzing CO_2 fixation are contained in the chloroplast's stroma. No commercial product has yet been developed, which directly affects the Calvin cycle.

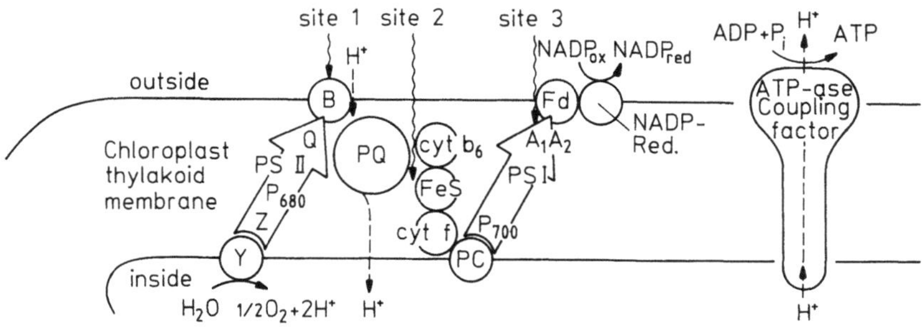

Fig. 11.1. Schematic description of the organization of the light reactions of photosynthesis. Two photosystems are imbedded into the thylakoid membrane and are connected by a chain of electron carriers. Light absorption by the chlorophyll molecules of photosystem II (*PS II*) leads to a charge separation at the reaction center of PS II by oxidizing the PS II donor *Z* and in turn by oxidizing water, accompanied by the reduction of the primary acceptor, *Q*. The electron then passes from the reduced semiquinone Q^- to the quinoid 2-electron carrier *B*, which is considered to be a protein-bound plastoquinone molecule. From B electron pairs are transfered into the bulk plastoquinone pool (*PQ*). Oxidation of PQ takes place at a protein complex comprising cytochrome b_6, a non-heme iron sulfur center (Rieske FeS center) and a f-type cytochrome. Plastocyanin (PC), located at the inside of the membrane, serves as an electron donor for the PS I light reaction (P_{700}). In this reaction, electrons are passed to the acceptors A_1 and A_2 (non-heme iron sulfur), to ferredoxin (*Fd*) and finally via the Fd-NADP-reductase NADP becomes reduced. Synthesis of ATP requires a proton motive force (proton gradient and membrane potential) across the thylakoid membrane (Mitchell hypothesis). Formation of a proton gradient is supported by (1) proton uptake during plastohydroquinone formation near the outside of the thylakoids (2) by proton liberation into the inner space of the membrane during oxidation of the bulk PQH_2 and during splitting of water. The proton gradient in connection with a membrane potential serves as driving force for the enzymatic synthesis of ATP, which takes place at an enzyme complex called ATPase or coupling factor CF_1 (right). Three sites are indicated in the model, where photosynthetic electron transport is affected by commercial herbicides. Acting at *site 1*, electron flow is interrupted by compounds termed PS II- or diuron (DCMU)-type herbicides. At *site 2*, inhibition by DBMIB-type herbicides is located. Bipyridylium-type herbicides act at *site 3*. The model of the electron transport chain is based on the current understanding of photosynthetic light reactions. Electron transport as related to photosynthesis inhibitors is more thoroughly discussed by IZAWA and GOOD (1977), WRIGHT and CORBETT (1979) and TREBST (1980)

Functional chloroplasts can be easily isolated from nearly all plant material using standard procedures ("intact chloroplast": WALKER 1980; "broken chloroplasts": REEVES and HALL 1980). Isolated chloroplasts are a preferential material for studying effects of chemical agents on the light reactions of photosynthesis. Also unicellular algae are frequently used in studies of photosynthesis (cf. URBACH 1977).

As indicated schematically in Fig. 11.1, three major sites can be distinguished whereby herbicides affect photosynthesis. The majority of all commercial products, termed PS II herbicides, acts at site 1, which is the reducing site of PS II. A few compounds of the DBMIB-type are found to act at site 2. The group of bipyridylium herbicides (viologens) acts at the reducing site of PS I (site 3).

Herbicides Acting at Site 1 (PS II Herbicides, Diuron-Type)

A wide variety of herbicides has been described in the past to "inhibit photosynthesis". This statement was based mainly on the observation that CO_2 fixation in intact leaves or O_2 evolution of isolated chloroplasts (Hill reaction) could be blocked efficiently by application of the agents. During the past few years numerous studies have been performed using biochemical and biophysical techniques in order to obtain a deeper understanding of the molecular mechanisms of action of the PS II herbicides. It should be emphasized that herbicide research, as well as basic photosynthetic research, has gained mutual benefit from these studies. From all this work, more and more evidence has emerged which supports the view that all the PS II herbicides – although chemically very diverse – follow the same molecular mechanism of action and have the same molecular target in common.

The most important chemical classes of PS II herbicides together with representative members are listed in Table 11.2. Properties of these classes in relation to uptake, metabolism, molecular fate, mode of action etc. are extensively discussed in the books of AUDUS (1976), KEARNEY and KAUFMAN (1976), ASHTON and CRAFTS (1981), FEDTKE (1982). Several chemical classes of PS II inhibitors which have not yet proved successful as commercial herbicides are of particular interest in photosynthesis research. These chemicals are discussed by IZAWA and GOOD (1977), VAN ASSCHE (1978), WRIGHT and CORBETT (1979) and by TREBST (1980, 1981).

A localization of an inhibition site inside the photosynthetic electron transport chain can be obtained by comparing herbicide action on several partial reactions of the electron transport chain. Using artificial electron donor and acceptor couples, the activity of most segments of the electron transport system can be studied separately (TREBST 1974). From the studies on the effects of PS II herbicides on electron transport reactions a common pattern emerges: activity of PS I is not affected by the PS II herbicides, whereas PS II reactions can be inhibited. From measurements of several different electron transport systems, the inhibition site of the PS II herbicides was localized to be at the reducing side of PS II in the Q-B-PQ region.

Further detailed insight into the mechanism of herbicide action is based on a second highly important research technique. Using chlorophyll fluorescence as a probe, photochemical reactions of PS II can specifically be monitored. For further information on this topic see GOVINDJEE and PAPAGEORGIOU (1981), LAVOREL and ETIENNE (1977). From the characteristics of the chlorophyll fluorescence rise curve – together with electron transport measurements – it is concluded that PS II herbicides affect the function of the secondary acceptor quinone B. The mechanism of inactivation of the acceptor B was first discussed by VELTHUYS and AMESZ (1974). Further evidence for this mechanism is summarized by PFISTER and ARNTZEN (1979), TREBST (1980), and PFISTER and SCHREIBER (1983). It is proposed that the PS II herbicides shift the redox potential of B towards more negative values, thus making B thermodynamically inaccessible for a reduction by Q. Alternatively, it has been proposed that a PS II herbicide molecule displaces the quinone B from the membrane (VELTHUYS 1981). B is considered to be a protein-bound plastoquinone-9 molecule (BOWES

Table 11.2

Chemical group	Common name (chemical name) of representatives	Representative references concerning mechanism of action
1. Herbicides affecting photosynthesis		
a) Herbicides acting at site 1		reviewed by TREBST (1980) WRIGHT and CORBETT (1979)
Acylanilides	Propanil (3′,4′-dichloro-propionanilide)	MORELAND (1969)
Buthidazoles	Buthidazole (3[5-(1,1-dimethylethyl)-1,3,4-thiadiazol-2-yl]-4-hydroxy-1-methyl-2-imidazolidinone)	YORK et al. (1981)
Benzothiadiazinones	Bentazon (3-isopropyl-1H-2,1,3-benzothiadiazin-4(3H)-one-2,2-dioxide)	PFISTER et al. (1974) MINE and MATSUNAKA (1975)
Carbamates	Phenmedipham (methyl m-hydroxycarbanilate m-methylcarbanilate); SWEP	SCHULZ (1969); TREBST et al. (1968)
Phenols:		
nitriles	Ioxynil (4-hydroxy-3,5-diiodobenzonitrile); bromoxynil	GROMET-ELHANAN (1968); TREBST (1979); KERR and WAIN (1964b)
nitrophenols	Dinoseb (2-sec-butyl-4,6-dinitrophenol); dinoterb DNOC (2,4-dinitro-6-methylphenol)	MORELAND and HILTON (1976); LAASCH et al. (1981) VAN RENSEN (1977, 1978)
Pyridazinones	Pyrazon (5-amino-4-chloro-2-phenyl-3(2H)-pyridazinone)	HILTON et al. (1969); HERCZEG et al. (1979)
Triazines	Atrazine (2-chloro-4-(ethylamino)-6-(iso-propylamino)-s-triazine); simazine; terbutryn	EBERT and DUMFORD (1976)
Triazinones	Metribuzin (4-amino-6-tert-butyl-3-(methylthio)-as-triazin-5(4H)-one)	TREBST and WIETOSKA (1975); DRABER et al. (1969)
Uracils	Bromacil (5-bromo-3-sec-butyl-6-methyluracil); lenacil; terbacil	HILTON et al. (1964)
Ureas	Diuron (=DCMU, 3-(3,4-dichlorophenyl)-1,1-dimethylurea); monuron; linuron; (metha)benzthiazuron	WESSELS and VAN DER VEEN (1956); Most widely studied PS II-herbicide, see reviews by TREBST (1980, 1981) and this chapter for more references
b) Herbicides acting at site 2		reviewed by TREBST (1980, 1981); TREBST and DRABER (1979)
Benzoquinones	DMBIB (dibromomethylisopropylbenzoquinone)	Note: experimental inhibitor. BÖHME et al. (1971); OETTMEIER et al. (1978)
Dinitroanilides	Trifluralin (α,α,α-trifluoro-2,6-dinitro-N,N-dipropyl-p-toluidine);	ROBINSON et al. (1977)
	fluchloralin (N-(2-chloroethyl)-2,6-dinitro-N-propyl-4-(trifluoromethyl)aniline)	TREBST and DRABER (1979)
Diphenylethers	Nitrofluorfen (2-chloro-1-(4-nitrophenoxy)-4-(trifluoromethyl)benzene)	BUGG et al. (1980)
c) Herbicides acting at site 3		reviewed by SUMMERS (1980)
Bipyridyls	Paraquat (1,1′-dimethyl-4,4-bipyridinium ion); diquat	BRIAN et al. (1958)

Table 11.2 (continued)

Chemical group	Common name (chemical name) of representatives	Representative references concerning mechanism of action
2. Herbicides interfering with the synthesis of ATP (phosphorylation)		reviewed by MORELAND (1980)
a) Uncouplers:		
–	Perfluidone (1,1,1-trifluoro-N-[2-methyl-4-(phenylsulfonyl)phenyl]methanesulfonamide	MORELAND (1981)
b) Energy-transfer-inhibitors:		
Nitrophenyl-ethers	Nitrofen (2,4-dichlorophenyl-p-nitrophenyl ether)	LAMBERT et al. (1979)
Thiadiazoyl-phenylureas	–	HAUSKA et al. (1975)
c) Inhibitory uncouplers:		reviewed by MORELAND (1980)
Phenols:		
nitrophenols	DNOC (4,6-dinitro-o-cresol) dinoseb	VAN RENSEN (1977, 1978) MORELAND and HILTON (1976)
nitriles	Ioxynil (4-hydroxy-3,5-diiodobenzonitrile); bromoxynil	GROMET-ELHANAN (1968)
3. Herbicides affecting pigments (bleaching)		reviewed by SANDMANN and BÖGER (1981)
Pyridazinones	Norflurazon (4-chloro-5-(methylamino)-2-(α,α,α-trifluoro-m-tolyl)-3-(2H)-pyridazinone); metflurazon	HILTON et al. (1969); BARTELS and MCCULLOUGH (1972); URBACH et al. (1976)
Triazole	Amitrole (3-amino-s-triazole)	BURNS et al. (1971)
Bipyridyls	Paraquat (1,1′-dimethyl-4,4′-bipyridinium ion); diquat	VAN RENSEN (1975) ELSTNER and OSSWALD (1980)
Diphenylethers	Oxyfluorfen (2-chloro-1-(3-ethoxy-4-nitro-phenoxy)-4-(trifluoromethyl)benzene; nitrofen	KUNERT and BÖGER (1981)
4. Herbicides affecting lipids and membranes		reviewed by MORROD (1976); RIVERA and PENNER (1979)
Bipyridyls	Paraquat (1,1′-dimethyl-4,4′-bipyridinium ion)	HARRIS and DODGE (1972); ELSTNER (1980); BOEHLER-KOHLER al. (1981)
Diphenylethers	Oxyfluorfen (2-chloro-1-(3-ethoxy-4-nitro-phenoxy)-4-(trifluoromethyl)benzene; nitrofen	SANDMANN and BÖGER (1981)
Pyridazinones	Norflurazon (4-chloro-5-(methylamino)-2-(α,α,α-trifluoro-m-tolyl)-3-(2H)-pyridazinone); metflurazon	ST. JOHN and HILTON (1976)

Table 11.2 (continued)

Chemical group	Common name (chemical name) of representatives	Representative references concerning mechanism of action
Thiocarbamates	EPTC (S-ethyl dipropylthiocarbamate); diallate; cycloate	WILKINSON (1978)
Chloroacetamides	Metolachlor (2-chloro-N-(2-ethyl-6-methylphenyl)-N-(2-methoxy-1-methylethyl) acetamide); alochlor	MELLIS et al. (1982)
5. Herbicides affecting protein and nucleic acid synthesis		reviewed by CHERRY (1976)
Phenoxyacetic acids	2,4-D (2,4-dichlorophenoxy acetic acid); 2,4,5-T	CHEN et al. (1972)
Miscellaneous compounds	Amitrole (3-amino-s-triazole);	MANN et al. (1965)
	propachlor (2-chloro-N-isopropylacetanilide); alachlor	MORELAND et al. (1969)
	trifluralin (α,α,α-trifluoro-2,6-dinitro-N,N-dipropyl-p-toluidine)	DUKE et al. (1975)
	chlorsulfuron (2-chloro-N-(4-methoxy-6-methyl-1,3,5-triazin-2yl)-benzene sulfonamide)	RAY (1980)
	glyphosate (N-(phosphonomethyl)glycine)	JAWORSKI (1972) AMRHEIN et al. (1980)
6. Herbicides affecting cell division (microtubules)		reviewed by CARTWRIGHT (1976); DRABER and FEDTKE (1979)
Dinitroanilines	Trifluralin (α,α,α-trifluoro-2,6-dinitro-N,N-dipropyl-p-toluidine); oryzalin	PARKA and SOPER (1977); UPADHYAYA and NOODEN (1977)
Carbamates	Propham (isopropyl carbanilate); chlorpropham; barban	BARTELS and HILTON (1973)
Miscellaneous compounds	Pronamide (3,5-dichloro(N-1,1-dimethyl-2-propynyl)-benzamide; terbutol (2,6-di-tert-butyl-p-tolyl methylcarbamate)	DRABER and FEDTKE (1979)
7. Auxin herbicides		reviewed by WEAVER (1972); LOOS (1975); SCHNEIDER and WIGHTMAN (1978)
Chlorophenoxy acetic acids	2,4-D (2,4-dichlorophenoxy acetic acid); 2,4,5-T; MCPA	PENNER and ASHTON (1966); CHEN et al. (1972)
Chlorobenzoic acids	Chloramben (3-amino-2,5-dichlorobenzoic acid); dicamba	FREAR (1976)
Miscellaneous compounds	Picloram (4-amino-3,5,6-trichloropicolinic acid); naptalam (NPA, N-1-naphthylphthalamic acid)	FOY (1976)

et al. 1980) serving as a gating mechanism which accepts two electrons sequentially and donates an electron-pair into the PQ pool (reviewed by Velthuys 1980). Further evidence for the proposal of B binding to a chloroplast membrane protein (B-protein) representing the target site for PS II herbicides came from studies in which the proteolytic enzyme trypsin was used to selectively modify this surface exposed protein. It has been shown in several studies that trypsin digestion of chloroplast membranes interrupts the photosynthetic electron transport at the acceptor B (Regitz and Ohad 1976; several studies by Renger 1976, Renger et al. 1981; Tischer and Strotmann 1979; Trebst 1979; Böger and Kunert 1979; van Rensen and Kramer 1979; Steinback et al. 1981a, 1982; Oettmeier and Masson 1980) and gradually eliminates the binding of PS II herbicides (Tischer and Strotmann 1979; Trebst 1979; Steinback et al. 1981a, 1982). These studies served as support for the general conclusion that the B protein creates the binding site for the PS II herbicides. Using covalent labeling techniques (photoaffinity label) the B protein has recently been identified as a 32,000 molecular weight protein of thylakoid membranes (Pfister et al. 1981a, b). Photoaffinity labeling studies by Oettmeier et al. (1980) indicate that binding of phenol-type herbicides in addition requires a 41,000 molecular weight protein. The conclusion that the 32,000 molecular weight protein is the PS II herbicide receptor is further supported by several indirect approaches (Croze et al. 1979; Mattoo et al. 1981).

Inactivation of the secondary acceptor quinone B apparently is the common mechanism of action of PS II herbicides, which comprise of a wide variety of chemically different substances. A comparison of the chemical structures of PS II herbicides made apparent it that most of them contain a structural element which is considered to be a $-\overset{\|}{C}-\bar{N}-$, or more specifically, a sp_2-hybrid carbon adjacent to a nitrogen, carrying a lone electron pair (Hansch 1969; Moreland 1969; Trebst and Harth 1974). This element seems to be essential for PS II inhibitors. As it can be seen from a number of studies of quantitative structure-activity relationships (QSAR), herbicide efficiency in vivo and in vitro is greatly influenced by lipophilic substituents surrounding the essential element. These studies have been useful for designing more efficient herbicides within a class of PS II inhibitors (Büchel and Draber 1969; Hansch 1969; Moreland 1969; Gabbott 1969; Draber et al. 1969; Büchel 1972; Oettmeier et al. 1978; Trebst and Draber 1979; Oettmeier 1979). It should be emphasized that the "essential element" cannot be found in the structure of phenol-type herbicides, i.e. ioxynil, dinoseb, DNOC. Probably our view of the element essential for the inhibition has to be even more generalized to fit the structural properties of the phenol-type herbicides. Alternatively, a second "essential element" would have to be proposed. There are several observations which distinguish the phenol herbicides from those carrying the "essential element" (Pfister and Schreiber 1983; van Assche 1981). Current research in several laboratories is concerned with differential inhibitory and binding properties of these two groups of PS II herbicides.

From the fact that a common functional mechanism of action and a common structural element is found for the PS II herbicides, it seems reasonable – although not conclusive – that also the molecular target or the binding site should be the same for all PS II herbicides. Studies on the binding site of the PS II herbicides have been greatly influenced by the pioneering work of TISCHER and STROTMANN (1977), who first quantitatively analyzed herbicide binding to chloroplast membranes using radioactively labeled compounds. Based on this work several research groups have contributed to a more detailed understanding of herbicide binding. Several findings should be mentioned:

1. Analysis of herbicide binding revealed that the concentration of binding sites on a chlorophyll basis was to be 300–500 chlorophyll per one binding site (2–3 nMol sites mg^{-1} chl), depending on the plant material and on the herbicide used (LEE and FANG 1972; TISCHER and STROTMANN 1977; PFISTER et al. 1979, 1981b; TREBST 1979; LAASCH et al. 1981; OETTMEIER and MASSON 1980; OETTMEIER et al. 1982). This stoichiometry suggests a 1:1 ratio between the concentration of binding sites and the number of electron transport chains, which means that one herbicide molecule is sufficient to inhibit one electron transport chain. Occupation of herbicide-binding sites has been shown to be proportional to the degree of inhibition of PS II reactions and is called "specific binding". In some cases, low affinity binding processes have been observed at concentrations when the electron transport was already fully inhibited (TISCHER and STROTMANN 1977; LAASCH et al. 1981, 1982). This process, called "unspecific" binding, represents binding to different sites, where additional reactions may be affected. Whether this observation can account for secondary herbicide effects – which are described for nearly all herbicides – has not yet been investigated.

2. Binding studies allow the quantitative determination of herbicide affinity towards its receptor site. Binding affinities expressed as equilibrium binding constants K_b are described now for several PS II herbicides and are listed together with I_{50}-concentrations in Table 11.3. Generally, the binding affinity is corre-

Table 11.3. Inhibitory efficiency, binding affinity and field application rate of several groups of PS II inhibitors. I_{50}-concentrations represent the concentrations of herbicides required for 50% inhibition of a PS II activity assay in a cell-free system. K_b is the binding constant describing herbicide affinity towards isolated thylakoid membranes. Field application rates are given for selective use (Herbicide Handbook WSSA 1979).

	I_{50}-concentration for inhibition of PS II electron transport	K_b, binding constant for isolated thylakoid membranes	Field application rate in kg ha^{-1}
Diuron	3.0×10^{-8} M[a]	1.9×10^{-8} M[a]	0.7–7.2
Atrazin	5.0×10^{-8} M[a]	6.7×10^{-8} M[a]	2.0–4.5
Metribuzin	1.1×10^{-7} M[b]	6.9×10^{-8} M[b]	0.3–1.1
Phenmedipham	7.2×10^{-8} M[b]	2.1×10^{-8} M[b]	1.0–2.0
Ioxynil	3.0×10^{-7} M[a]	2.4×10^{-8} M[a]	0.5–1.1
Dinoseb	1.0×10^{-6} M[a]	3.6×10^{-7} M[a]	0.8–13.0
Bentazon	2.0×10^{-4} M[c]	2.3×10^{-5} M[d]	0.8–2.2

[a] LAASCH et al. 1981
[b] TISCHER and STROTMANN 1977
[c] PFISTER et al. 1974
[d] PFISTER et al. 1982

lated with the efficiency of a herbicide. The higher the affinity, the better is usually the inhibitory efficiency of a compound. The K_b-values given in Table 11.3 for some representatives of important groups of herbicides range mostly between 10^{-8} M and 5×10^{-7} M. The data obtained with isolated thylakoid membranes are similar to those found on intact algal cells (LAASCH et al. 1981). This demonstrates the high degree of binding specificity found for most of the compounds listed in Table 11.3. For agricultural use, these herbicides are usually applied at a rate of approximately 1–5 kg ha^{-1}. The correlation of herbicide efficiency in a cell free system with the field application rate (phytotoxicity) does not hold true in every case (DRABER 1979). Bentazon, for example, is considered to be a relatively poor inhibitor of PS II reactions in a cell free system (I_{50}: 2×10^{-4} M, PFISTER et al. 1974). The field application rate of bentazone, however, (1–3 kg ha^{-1}) is comparable to herbicides having I_{50}-concentrations up to 10,000-fold lower than bentazon (see Table 11.3).

3. A very central result of PS II herbicide-binding studies is the observation that all compounds tested so far are able to displace another, previously bound PS II herbicide from its binding site. This is interpreted that all PS II herbicides compete for the same common binding site at the chloroplast membrane (TISCHER and STROTMANN 1974; PFISTER and ARNTZEN 1979; TREBST 1980; PFISTER et al. 1982). Competition between herbicides regardless from which chemical class has been shown to be due to "competitive binding" (LAASCH et al. 1982). Studies on herbicide binding, together with observations on herbicide resistant algae and especially by using chloroplasts from triazine-resistant weeds (see Sect. 11.5), have led to a model of the herbicide-binding site, which describes very schematically the interaction of chemically diverse groups of herbicides with a receptor site at the chloroplast membrane. The model is mainly based on the notion that in algal mutants and in herbicide resistant weeds the efficiency of one PS II herbicide can be entirely lost, whereas the sensitivity towards chemically different classes of PS II herbicides is maintained (PFISTER and ARNTZEN 1979), as will be discussed more thoroughly in Sect. 11.5. According to the model given in Fig. 11.2, the herbicide-binding site consists of two "domains". The central part is formed by an receptor for the "essential element" (Fig. 11.2a, receptor 1). This domain is common to all PS II herbicides

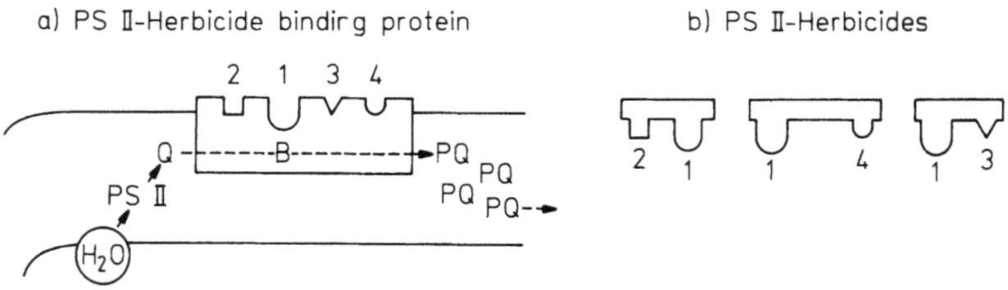

Fig. 11.2a, b. Model of the PS II herbicide binding site. **a** The binding site is located at the chloroplast membrane protein containing the functional quinone B (see Fig. 11.1). After binding of a herbicide involving several of the subreceptors (*2–4*) the receptor for the "essential element" (*1*) becomes occupied thus inducing an inactivation of *B*. All herbicides symbolized in **b** share a common portion of the binding site and are therefore competitive

(Fig. 11.2b) and has to be occupied for inhibition. Surrounding the "essential element domain" subreceptors (Fig. 11.2a, receptors 2–4) for specific structures differ between each chemical class of herbicides (Fig. 11.2b). Similar models are described by TREBST and DRABER (1979), PFISTER et al. (1979), PFISTER and ARNTZEN (1979) and by TREBST (1980). All models have in common the concept of subreceptors creating overlapping herbicide binding sites. A molecular model based on the geometry of triazinone-type herbicides is given by DRABER and FEDTKE (1979). Two recent theoretical approaches to characterize the molecular architecture of the binding site should be mentioned (SHIPMAN 1981; CARLES and VAN ASSCHE 1980).

Herbicides Acting at Site 2 (DBMIB-Type)

The experimentally extremely useful inhibitor DBMIB (dibromomethylisopropylbenzoquinone) interrupts the photosynthetic electron flow after the plastoquinone pool. This inhibitor has been introduced and characterized by the group of TREBST (BÖHME et al. 1971). DBMIB is thought to prevent the oxidation of the reduced plastoquinone (plastoquinole) by competing with the plastoquinole for a binding site at the FeS-Rieske center (Fig. 11.1). Properties of DBMIB have recently been reviewed by TREBST (1980) and OETTMEIER et al. (1982). Although many DBMIB analogs tested so far show good inhibitory activity (OETTMEIER et al. 1978; TREBST et al. 1979), none of them has yet been commercially developed, mainly because of the lack of chemical stability.

Several herbicides – although structurally unrelated to DBMIB – are thought to have inhibitory properties like DBMIB (Table 11.2). ROBINSON et al. (1977) have studied the herbicides trifluralin and diallate with respect to their inhibition of partial reactions of photosynthetic electron flow. The authors assume for both compounds an inhibition site between the two pigment systems, as for DBMIB. According to the I_{50}-concentrations for electron transport inhibition (10–30 μM), inhibition of photosynthesis may at least be partially responsible for phytotoxicity of these compounds. Several diphenylethers have been described by TREBST (1979) as experimental inhibitors acting at the DMBIB site. The herbicides fluchloralin and RHL 1288 (4[2,6-dichloro-4trifluoromethylphenoxy]benzonitril) are considered to act in analogous fashion to DBMIB (TREBST and DRABER 1979). Also the herbicides HOE 29152 (methyl-2[4-(4-trifluoromethoxy)phenoxy]propanoate) and nitrofluorfen seem to act at site 2 (BUGG et al. 1980). In all cases, however, it is difficult to distinguish from the present information which of the mechanisms of action claimed for a particular herbicide contributes most to the overall phytotoxicity.

Herbicides Acting at Site 3 (Bipyridylium-Type, Viologens)

Several compounds from the chemical class of bipyridylium salts, namely diquat and paraquat possess high herbicidal activity (Table 11.2; BRIAN et al. 1958). There are several reviews about the mode and mechanism of action of the bipyridyls, e.g. CALDERBANK and SLADE (1976), DODGE et al. (1970), SUMMERS (1979). The most recent and complete coverage of the bipyridyls is the book of SUMMERS (1980). Compared to herbicides acting as inhibitors of the photosynthetic electron flow at site 1 and site 2 (Fig. 11.1), the mechanism of action of the bipyridyls is substantially different: they do not act as inhibitors but instead as electron acceptors, thus becoming reduced upon illumination of the chloroplasts.

In the light reactions of photosynthesis a strong reductant is formed at the reducing site of PS I (A_1 and A_2 in Fig. 11.1), which has a very negative normal potential. Thus paraquat ($E_0 = -450$ mV) can be reduced by the strong reductant of PS I. For a survey of the redox potentials of the bipyridyls see SUMMERS (1980). In extensive studies structural requirements and chemical properties have been defined to yield bipyridylium salts with high herbicidal activity (SUMMERS 1979, 1980). Requirements for herbicidal activity are a normal potential between -300 mV and -500 mV, the ability for a one-electron transfer and to form a stable cation radical in aqueous environment, autooxidability in presence of oxygen and finally a planar arrangement of the molecule (Fig. 11.3). As a consequence of these properties the mechanism of action of the bipyridyls can be described as their reduction by PS I, which diverts the photosynthetic electron flow from the endogenous acceptor NADP towards artificial ones. Prevention of CO_2-fixation alone is not considered as the principal mode of action of the bipyridyls. Instead, the formation of toxic products during reoxidation of the reduced compound is thought to be responsible for phytotoxicity. During reoxidation of reduced bipyridylium superoxide anion (O_2^-), peroxide (H_2O_2) and presumably singlet oxygen ($^1O_2$) and free hydroxyl radicals (OH^-) are generated (Fig. 11.3; FARRINGTON et al. 1973; ELSTNER 1982). Although H_2O_2 as an oxidizing agent and cell poison may contribute to the herbicide efficiency (DODGE et al. 1970), main phytotoxicity is believed to originate from the production of highly reactive radicals.

At cell level, destruction of cell membranes (plasmalemma, tonoplast) seems to be the first detectable herbicidal effect, followed by chlorophyll bleaching and oxidation of pigments and lipids (HARRIS and DODGE 1972; YOUNGMAN and DODGE 1979). Peroxidation of unsaturated lipids is indicated by the formation of malondialdehyde (BÖGER and KUNERT 1978; ELSTNER and OSSWALD 1980). It becomes obvious from the mechanism of action – as discussed above – that light and oxygen are required for best herbicidal activity (VAN OORSHOT 1981; ELSTNER et al. 1980a).

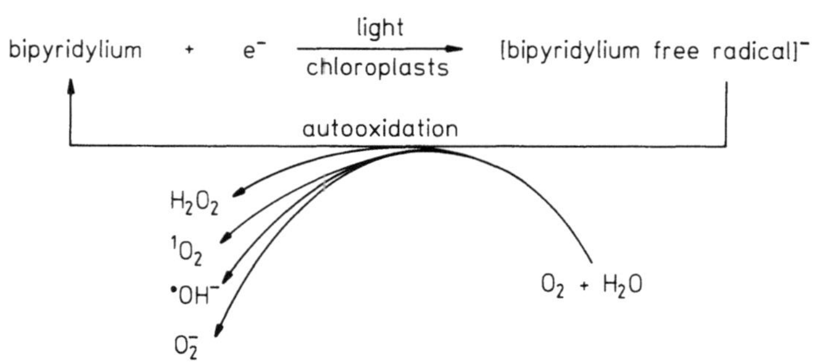

Fig. 11.3. Schematic mechanism of action of the bipyridylium herbicides. The herbicides are reduced in the light by uptake of one electron forming stable free radicals. Under consumption of oxygen the autooxidizable radicals form peroxide, O_2^- and presumable $^1O_2$ and $^{\bullet}OH^-$

The bipyridyls were applied as fast-acting contact herbicides. Due to their extremely strong soil adsorption and rapid breakdown uptake by roots plays a minor role. Diquat and paraquat possess limited selectivity and hence are applied usually for total vegetation control or as desiccants and defoliants to facilitate mechanical harvesting.

Herbicides Affecting the Synthesis of ATP (Phosphorylation)
ATP is generated in the photosynthetic light reactions in a process called photophosphorylation (Fig. 11.1). Synthesis of ATP occurs at an enzyme complex (ATPase, coupling factor CF_1) embedded in the thylakoid membrane. The current understanding of the energy conservation reactions of chloroplasts have been extensively reviewed (see Vol. 5, this Series).

There are four basic ways by which herbicides can interfere with the process of photophosphorylation (see scheme given by McCarty 1977):

Inhibition of Electron Flow. No transmembrane pH-gradient is formed when the linear electron flow is blocked by an inhibitor. Depending on the type of inhibitor, ATP may still be generated in a cyclic system involving light reaction I only. Inhibitors of photosynthesis have been discussed in a previous section.

Uncoupling of Photophosphorylation. The term "uncoupling" is used for compounds or treatments which dissipate the energy stored temporarily in the electrochemical gradient. The proton gradient is dissipated either by increasing the leakiness of the thylakoid membrane for protons or by internally buffering the accumulated protons. Thus uncouplers do not interfere with the enzymes involved in the ATP synthesis nor do they inhibit electron transport reactions. On the contrary, the electron flow is stimulated in presence of uncouplers. Many chemicals are known which serve as experimental uncouplers by dissipating the pH by various mechanisms, e.g. amines (NH_4Cl), atebrin, phenylhydrazones as FCCP and CCCP, the ionophores nigericin and gramicidin and others (reviewed by Good 1977). Perfluidone seems to be the only herbicide so far, which is considered to act as a pure uncoupler (Moreland 1981). Recently, a series of aryloxylalkylamines has been described to combine uncoupler activity and inhibition of carotenoid biosynthesis. These compounds are currently under test for field application (Wright et al. 1980).

Energy Transfer Inhibitors. This term applies to compounds which inhibit ATP synthesis by preventing the transduction of the membrane-stored energy at the coupling factor. Several chemically unrelated compounds, e.g. phlorozidin, dio-9, DCCD or ethylmaleimide are thought to prevent the terminal step in the synthesis of ATP (reviewed by McCarty 1977). A series of thiadiazoyl-phenyl-ureas seem to carry potential herbicidal properties of this kind (Hauska et al. 1975). The only commercial product so far described as an energy transfer inhibitor is the nitrophenylether nitrofen, which is shown to compete with ADP for a binding site on the chloroplast ATP-ase (Lambert et al. 1979).

Inhibitory Uncouplers. This term was coined for herbicides which inhibit electron flow and also act as uncouplers. The group comprises a number of chemicals which can be shown to act as inhibitors of photosynthetic electron flow at site 1 and as well as uncouplers of photophosphorylation. Inhibitory uncouplers are listed and discussed in detail by MORELAND and HILTON (1976) and MORELAND (1980). In respect to commercial herbicides phenol-type compounds seem to be the most important group of inhibitory uncouplers. Nitrophenols such as DNOC, dinoseb, dinoterb, and nitrils such as ioxynil and bromoxynil belong to this group. In chloroplast preparations phenolic type herbicides are more effective as inhibitors of electron flow than in their action as uncouplers of photophosphorylation (KERR and WAIN 1964a; GROMET-ELHANAN 1968; VAN RENSEN et al. 1977, 1978; TREBST 1979; OETTMEIER and MASSON 1980; LAASCH et al. 1981). Uncoupling of oxidative phosphorylation of mitochondria, as it will be discussed below, is assumed to be a major mechanism of action of the phenols. The molecular mechanism of action by which the phenols uncouple photophosphorylation has not yet been fully understood. Also some benzimidazols (e.g. chlorfluorazol; BÜCHEL et al. 1966), and diphenylamines (OETTMEIER 1979) seem to carry properties of inhibitory uncouplers.

11.4.2.2 Herbicides Affecting Respiration

A survey of mechanisms and modes of action of herbicides reveals that only few compounds are thought to interfere strongly with respiration (CORBETT 1974; KIRKWOOD 1976). One reason for this fact may be the potential animal toxicity of respiration-affecting agents. In contrast to this, the target sites of herbicides and growth regulators are preferably the photosynthetic apparatus, plant pigment synthesis or phytohormones, which all have no counterpart in animal physiology. This might explain the generally low animal toxicity of herbicides and of growth regulators in comparison to the toxicity of other biocides. Interference with respiration can occur – by analogy with interference to photosynthesis – by two general mechanisms: by inhibiting the enzyme activity of electron transport reactions or by affecting synthesis of ATP.

There are no commercial herbicides which are thought to act primarily by inhibiting respiration. However, rotenone (an insecticide), azides, cyanides and antimycin have to be mentioned as effective inhibitors of respiratory electron transport systems. Uncoupling of mitochondrial oxidative phosphorylation is the major mechanism by which several herbicides affect respiration. Uncoupling prevents the formation of ATP in the mitochondrial electron transport system by very similar mechanisms as discussed above for the uncoupling of photophosphorylation. As a consequence of uncoupling, the respiratory electron flow is stimulated, thus leading to a much faster consumption of substrates. The toxic properties of uncouplers are considered to be due to the lack of ATP production coupled with the stimulated respiration.

The classical uncoupler used in numerous studies of mitochondrial phosphorylation is dinitrophenol (DNP). The most efficient uncoupling herbicides are derivatives of the parent compound DNP, e.g. the dinitrophenols dinoseb, dinoterb and DNOC (KERR and WAIN 1964a, b). The benzonitriles ioxynil and bromoxynil have been shown as uncouplers (KERR and WAIN 1964a). All phenols mentioned above are also described as "inhibitory uncouplers" in chloroplasts (MORELAND 1980). CORBETT (1974) lists uncoupling activities of several phenol-type biocides. Concentration for 50% uncoupling of plant mitochondrial ATP-formation is in a range between 10 to 50 μM, which is comparable to the I_{50}-concentrations for inhibition of photosynthetic electron flow (ioxynil and dinoseb: see Table 11.3; DNOC: VAN RENSEN et al. 1977, 1978). From this it is difficult to estimate which mechanism of action contributes most to the overall toxicity of the phenol-herbicides. In vivo toxicity very probably is a combination of the effects on photosynthesis and on respiration. Also for several diphenylether herbicides interference with respiratory electron transport and phosphorylation has been described (MORELAND et al. 1970).

It should be noted that there are reports about interference with respiration for nearly every chemical group of herbicides and growth regulators (reviewed by KIRKWOOD 1976 and VAN OORSCHOT 1976). Most of these results are obtained either at relatively high herbicide concentrations or might be the consequence of other primary effects. The significance of these results for the in vivo toxicity seems rather doubtful.

11.4.2.3 Herbicides Affecting Pigment Metabolism

Another rather important and wide-spread mode of action of herbicides is the bleaching effect. The decrease in the pigment content (carotenoids and chlorophylls) finally leads to chlorosis of leaves. Chlorosis can be induced by a number of chemically unrelated herbicides. Since with most of these herbicides the bleaching effect is caused by interference with carotenoid biosynthesis or by destruction of carotenoids, a distinction can be made between (1) herbicides inducing chlorosis indirectly by photooxidation of chlorophylls, which is due to the loss of carotenoids, and (2) herbicides interfering directly with steps of chlorophyll biosynthesis or with chloroplast biogenesis (Fig. 11.4). The mode of action of herbicidal bleaching has been reviewed recently by SANDMANN and BOEGER (1981) and by RIDLEY (1982).

1. Chlorosis caused by the destruction of chlorophyll is induced (a) by herbicides inhibiting the carotenoid biosynthesis, (b) by PS II herbicides such as diuron inhibiting the photosynthetic electron transport, (c) by herbicides such

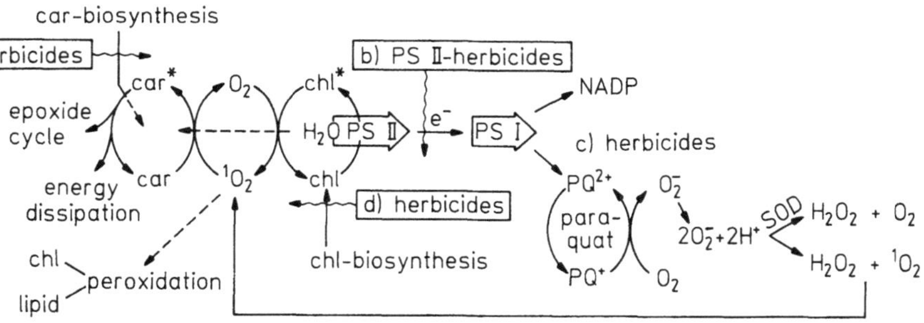

Fig. 11.4. Proposed scheme for the sites of action of chlorosis inducing herbicides (bleaching effect). *site a* inhibition of carotenoid biosynthesis; *site b* inhibition of photosynthetic electron flow at PS II; *site c* photoreduction of autooxidizable bipyridylium herbicides (i.e. paraquat PQ^{2+}) followed by formation of superoxide (O_2^-), peroxide (H_2O_2), singlet oxygen ($^1O_2$), or presumably OH^- radicals; *SOD* superoxide dismutase; *site d* inhibition of chlorophyll biosynthesis

as diquat and paraquat, acting as electron acceptors at PS I or by diphenylethers like oxyfluorfen (Fig. 11.4).

a) Various herbicides with unrelated chemical structures are known to inhibit carotenoid biosynthesis (Fig. 11.4). This effect has been demonstrated for amitrole, dichlormate, pyriclor (BURNS et al. 1971), for the pyridazinones metflurazone (SAN 6706), norflurazone (SAN 9789) and other derivatives (HILTON et al. 1969; BARTELS and MCCULLOUGH 1972; BEN-AZIZ and KOREN 1974; KÜMMEL and GRIMME 1975; URBACH et al. 1976; VAISBERG and SCHIFF 1976; BARTELS and WATSON 1978; SANDMANN et al. 1979; KUNERT and BÖGER 1979; RIDLEY and RIDLEY 1979; FROSCH et al. 1979; BUSCHMANN et al. 1980; GRUMBACH and BACH 1979), for the experimental herbicide difunone (EMD-IT 5914) (URBACH et al. 1976; KUNERT and BÖGER 1978) and for some other chemical compounds such as fluridone, fluometuron, R-4024 and haloxydine (DODGE and LAWES 1972; BARTELS and WATSON 1978; DEVLIN et al. 1979). Most of these herbicides are thought to affect the biosynthetic pathway of carotenoids at the same site(s), before α- and β-carotene formation, since in the presence of these herbicides, the content of these pigments is strongly reduced, while the colorless precursors phytoene and phytofluene are accumulated. It has been proposed that these herbicides interfere with the mechanism of desaturation by blocking the formation of the desaturases and stepwise dehydrogenation of precursors of colored carotenes (DRABER and FEDTKE 1979; EDER 1979). In a cell free enzyme system from carotenogenic fungi it was shown that norflurazon acts primarily at a reaction which is catalysed by phytoene synthetase (SANDMANN et al. 1980). A number of potential aryloxyalkylamine herbicides was shown to inhibit carotenoid biosynthesis between lycopene and β-carotene in addition to their action as uncouplers (WRIGHT et al. 1980). Structure-activity studies with several pyridazinone derivatives demonstrate a correlation of the bleaching activities with electronic parameters of the various substituents (SANDMANN et al. 1981a). Inhibition of carotenoid biosynthesis is not only followed by loss of colored carotenoids and chlorophylls, but also by degradation of 70 S ribosomes and chloroplast ultrastructures (BARTELS and HYDE 1970a, b; BURNS et al. 1971; DODGE and LAWES 1972; ST. JOHN and HILTON 1976; BARTELS and WATSON 1978).

Carotenoid biosynthesis seems to be an important and ideal target for herbicide action since carotenoids are essential for the functioning of the photosynthetic apparatus (BRITTON 1979). Loss of carotenoids means loss of the protective function against photooxidation of chlorophylls, followed by bleaching and death of the plant cell (KRINSKY 1966, 1978; WITT 1971).

b) Bleaching of plant pigments can indirectly be caused by PS II herbicides (Fig. 11.4). Studies with monuron, diuron and other PS II herbicides revealed that these herbicides induce a loss of carotenes and chlorophylls in chloroplasts and plant cells during illumination (STANGER and APPLEBY 1972; RIDLEY 1977; PALLET and DODGE 1979; GIANNOPOLITIS and AYERS 1978; ELSTNER and PILS 1979; MUSCHINEK et al. 1979; ELSTNER and OSSWALD 1980; URBACH et al. 1983). There are several hypotheses concerning the protection of chlorophylls and the photosynthetic apparatus against photooxidation by carotenoids (KRINSKY 1966, 1978; WITT 1971; RIDLEY 1977). If the electron transport chain is interrupted by PS II herbicides, the photoexcited chlorophyll molecules might be able to pass the energy directly on to molecular oxygen (Fig. 11.4). Carotenoids might quench the excess excitation energy of the chlorophylls or – if produced – remove singlet oxygen, which is possibly leading to an overloading of the carotenoids and in consequence to a breakdown of the protection mechanism. If singlet oxygen cannot be removed, it is able to catalyze peroxidation reactions in chlorophylls, membranes and lipids, which consequently causes the destruction of the photosynthetic apparatus (ELSTNER and PILS 1979; PALLET and DODGE 1979; TAKAHAMA 1978; ELSTNER 1980; URBACH et al. 1983). In absence of oxygen or after addition of the antioxidant BHT (butylated hydroxytoluene) or singlet oxygen quenchers like DABCO (1,4-diazobicyclo(2,2,2)octane) (ANDERSON and KRINSKY 1973), the pigments in chloroplasts and in intact plant cells are to some degree protected against the effects of the tested herbicides (PALLET and DODGE 1979; ELSTNER 1980; URBACH et al. 1983), indicating the participation of oxygen radicals like singlet oxygen in the bleaching activity of PS II herbicides (Fig. 11.4). The proof of NADH or NADPH participation in the protective system (STANGER and APPLEBY 1972; GIANNOPOLITIS and AYERS 1978) gave rise to the assumption that carotenoids remove excess exitation energy from chlorophyll through the epoxide cycle (KRINSKY 1966, 1978; SIEFERMANN-HARMS 1977). Regardless the protection mechanism in the plant cell, bleaching of photosynthetic pigments is always correlated with peroxidation of lipids, which can be measured by the formation of malondialdehyde or ethane, and is followed by degradation of membraneous cell structures (TAKAHAMA 1978; PALLET and DODGE 1979; ELSTNER and PILS 1979; ELSTNER 1980, 1982). Therefore death of the plant after treatment with PS II herbicides is probably not due to "starvation", but rather to damage by failure of an effective protection of the photosynthetic apparatus in light (WRIGHT and CORBETT 1979).

c) Loss of carotenoids and subsequently of chlorophylls is also induced by bipyridylium herbicides (diquat, paraquat) acting at site 3 of the photosynthetic electron transport chain (Fig. 11.1) (HARRIS and DODGE 1972; VAN RENSEN 1975; BÖGER and KUNERT 1978; YOUNGMAN and DODGE 1979; ELSTNER and OSSWALD 1980; SUMMERS 1980; URBACH et al. 1983). The mechanism of action of these herbicides in photosynthesis has already been discussed in Sect. 11.4.2.1. These compounds – acting as acceptors after PS I – are reduced and subsequently reoxidized by molecular oxygen. This in turn leads to the formation of superoxide anion, hydrogen peroxide and especially of singlet oxygen (Fig. 11.4) (FARRINGTON et al. 1973; ELSTNER et al. 1980a, b; ELSTNER 1982). As already mentioned, singlet oxygen seems to be the reactive species involved in the carotenoid and chlorophyll breakdown caused by these herbicides in the light. The loss of pigments is accompanied by lipid peroxidation and consequently by a destruction of cell membranes, as it has been observed after treatment with PS II herbicides. Despite the similar effects caused by bipyridylium and by PS II herbicides in plant cells, the mechanism of action leading to photooxidation seems to be different (ELSTNER 1980).

In the past few years also studies on the bleaching effect of diphenylether herbicides like nitrofen and oxyfluorfen have been made (MATSUNAKA 1976; LAMBERT et al. 1979; KUNERT and BÖGER 1981; SANDMANN et al. 1981b). Also some diphenylethers are reported as energy transfer and electron transport inhibitors (LAMBERT et al. 1979; BUGG et al. 1980; PRITCHARD et al. 1980). Although the exact mechanism of the bleaching has not been identified yet, destruction of carotenoids and chlorophylls is correlated with lipid peroxidation and membrane degradation. Presumably, however, bleaching is caused by a mechanism different from that of bipyridylium and PSII herbicides, as singlet oxygen might not be involved (KUNERT and BÖGER 1981; SANDMANN and BÖGER 1981).

2. Evidence has been given that some chlorosis-inducing herbicides have direct effects on chlorophyll synthesis and chloroplast biogenesis (Fig. 11.4). Besides an inhibition of carotenoid biosynthesis by pyridazinone herbicides, chlorophyll synthesis has been found to be suppressed simultaneously (LICHTENTHALER and KLEUDGEN 1977; KLEUDGEN 1979; KLOCKARE et al. 1981). These findings might indicate an interference of these herbicides with the formation of chloroplast ribosomes, which leads to a reduced synthesis of enzymes involved in carotene and chlorophyll biosynthesis. Several authors discuss a reduction of plastidic ribosomes as a secondary effect of chlorosis-inducing herbicides, causing changes in the chloroplastic ultrastructure (BARTELS and HYDE 1970a; BURNS et al. 1971; ST. JOHN and HILTON 1976; BARTELS and WATSON 1978), and an inactivation of enzymes in chloroplasts and leaf peroxisomes (FEIERABEND and SCHUBERT 1978; FEIERABEND et al. 1979). Interference with biosynthesis of chlorophyll has been described for amitrol, glyphosate and other compounds (RÜDIGER et al. 1976; RÜDIGER and BENZ 1979; FEIERABEND et al. 1979; KITCHEN et al. 1981).

Finally in this connection several studies on physiological adaptation of plants treated with sublethal doses of herbicides should be mentioned (FEDTKE 1973, 1974, 1979; BÖGER and SCHLUE 1976; KLEUDGEN 1978; LICHTENTHALER 1979; LICHTENTHALER et al. 1980; MEIER et al. 1980). Under conditions where PS II herbicides (e.g. diuron, bentazon, methabenzthiazuron, s-triazines) do not inhibit photosynthetic electron transport, there are additional physiological effects on plants which induce a shade-type growth response. These physiological effects of herbicides – described as shade adaption and greening process – are observed during the development of the photosynthetic apparatus. They include changes in chloroplast ultrastructure (broader and higher grana stacks), pigment composition (lower chlorophyll a:b ratios and carotenoid content:dry weight) as well as enzyme (RuBP-carboxylase) and metabolite contents of chloroplasts. Besides these alterations in chloroplasts also other metabolic reactions (respiration, nitrogen metabolism) are affected (FEDTKE 1973, 1979).

11.4.2.4 Herbicides Affecting Lipid Metabolism and Membrane Structure and Function

Herbicide-induced changes in the membrane lipid composition affect membrane structures and functions of the plant cell. As already dicussed in the previous section, bleaching of pigments caused by several groups of herbicides (Fig. 11.4) is mostly accompanied by an increase in the lipid peroxidation. As indicated before, this lipid breakdown in the light, studied by measuring the formation of malondialdehyde or ethane, may be either caused by PS II herbicides (TAKAHAMA and NISHIMURA 1975, 1976; KRINSKY 1977; TAKAHAMA 1978; LAI et al.

1978; FOOTE 1979; ELSTNER and PILS 1979; PALLET and DODGE 1979), by the bipyridylium herbicides diquat and paraquat (HARRIS and DODGE 1972; VAN RENSEN 1975; BÖGER and KUNERT 1978; YOUNGMAN and DODGE 1979; BÖHLER-KOHLER et al. 1981; ELSTNER and OSSWALD 1980; URBACH et al. 1983) or by diphenylethers like oxyfluorfen (SANDMANN and BÖGER 1981). Oxygen radicals such as singlet oxygen and hydroxyl radicals (OH^-) as the active species for starting lipid peroxidation are currently under discussion (FRIDOVICH 1975; MEAD 1976; HARBOUR and BOLTON 1978; ELSTNER 1980, 1982). These radicals probably start a peroxidative chain reaction resulting in the production of α-linolenate peroxide and the gradual degradation of the fatty acids. Paraquat, for instance, causes a decrease in the level of linoleic and linolenic acid and changes the fatty acid composition of different lipid classes (YOUNGMAN and DODGE 1979; BOEHLER-KOHLER et al. 1980). In case of oxyfluorfen unsaturated fatty acids are thought to be directly peroxidized by nitrophenylether radicals (SANDMANN and BÖGER 1981). Structural organization and function of membranes, however, cannot only be impaired by the biochemical breakdown of the lipids but also by the interference of herbicides with steps of the lipid biosynthesis or the lipid turnover rates within the membranes. Herbicides of different chemical structure such as ioxynil, dinoben and endothal are known to inhibit the incorporation of ^{14}C-malonic acid into lipids, whereas other compounds (e.g. 2,4-D, 2,4,5-T, picloram) stimulate the lipid synthesis (MANN and PU 1968). Studies with substituted pyridazinones revealed a direct effect of these compounds on the lipid synthesis (ST. JOHN 1976; ST. JOHN and HILTON 1976; ST. JOHN et al. 1979; EDER 1979; KHAN et al. 1979). This interference with lipid biosynthesis is reflected by an alteration in the lipid composition of membranes especially of polar lipids (galactolipids) and in the ratio of linolenic:linoleic acid. Such changes in the linolenic acid content in the membrane may impair cellular function and modify plant sensitivity to environmental stress. They can be related to plant responses to low and high temperature and may increase in the resistance to temperature stress (ST. JOHN and CHRISTIANSEN 1976; ST. JOHN et al. 1979). Moreover, reduction of polar lipids is expected to substantially alter the membrane structure and function, as polar phospholipids and glycolipids are the main lipid constituents of the cell membranes – especially of chloroplasts and mitochondria. Thiocarbamates (e.g. EPTC) were found to inhibit chloroplast fatty acid and epicuticular wax synthesis and quantitatively and qualitatively altering the fatty acid contents in plant organelles (WILKINSON 1978). Similar effects are reported for chloroacetamides, e.g. metolachlor (MELLIS et al. 1982). Inhibitory and occasionally stimulatory effects on lipid synthesis are reported for many other herbicides, e.g. phenols, phenyl ureas, triazines, ureas, dinitroanilines, benzoic acids and phenoxy acetic acids (SUMIDA and UEDA 1973; ST. JOHN and HILTON 1974; SUMIDA et al. 1975; Muslih and LINSCOTT 1977; ASHTON et al. 1977; RIVERA and PENNER 1979). These effects on the lipid metabolism are presumably due to interactions of herbicides with specific enzyme systems of the lipid synthesis or indirectly caused by limiting the chloroplast and mitochondrial ATP production. Consequences of herbicidal interference with the lipid metabolism of plant cells, modifications of cellular membranes, e.g. loss of membrane integrity accompanied by ultrastructural changes, and alterations and impairments of essential membrane functions (permeability, ion transport, membrane potential) have been observed and comprehensively described (ST. JOHN and HILTON 1974; MORROD 1976; LINCK 1976; RIVERA and PENNER 1979).

11.4.2.5 Herbicides Affecting Protein and Nucleic Acid Synthesis

As already discussed in the preceding sections, application of herbicides very often causes changes in plant constituents by impairment of metabolic pathways. In particular this is true for the protein and nucleic acid metabolism as the basic part of the entire plant metabolism. Distinct changes in the amino acid, protein, RNA and DNA content in tissues and cells after herbicide treatment have been observed (ASHTON and BAYER 1976; CHERRY 1976). The decrease or even increase in the level of a compound may be indirectly caused by herbicides inhibiting synthesis of ATP required for the biosynthetic reactions of these substances. Most of the herbicides which reduce the level of ATP (e.g. dinoseb, ioxynil, propanil, chlorpropham, pyriclor, 2,4,5-T) also show strong inhibition of RNA and protein synthesis (GRUENHAGEN and MORELAND 1971). However, also direct herbicidal effects on protein, DNA and RNA synthesis have been demonstrated (WEEKS and BAXTER 1972; PENNER and EARLY 1972a), although many studies did not precisely differentiate between direct and indirect herbicidal action on the biosynthetic reactions. Inhibition of thymidine incorporation into DNA has been demonstrated for the recently introduced herbicide chlorsulfuron (DPX 4189, substituted sulfonyl urea, see Table 11.2). A remarkable property of chlorsulfuron is that this compound becomes already effective at rates of few grams per ha, which is far better than any other herbicide known so far (PALM et al. 1980; RAY 1980).

Evidence has been given that protein synthesis can be affected at different levels by herbicides. At the transcription level DNA dependent RNA synthesis is thought to be influenced by many auxin herbicides. Protein synthesis at the translation level may be affected by interaction of herbicides with specific RNAs and ribosomes. Among all herbicides tested so far the effects of 2,4-D on nucleic acids and protein synthesis have been most precisely investigated (VENIS 1971; CHEN et al. 1972; CHERRY 1976). In addition there is quite a number of other herbicides (dinitroanilines, phenylcarbamates, nitrophenols) showing effects similar to those of the phenoxyacetic acids. These studies with chemically unrelated herbicides including dinoseb, ioxynil, propanil, pyriclor, chlorpropham and paraquat revealed a more or less strong inhibition of the incorporation of labelled precursors into protein, DNA and RNA (MANN et al. 1965; MORELAND et al. 1969; KERR and AVERY 1972; ASHTON et al. 1977; BELL et al. 1976). Especially, the ^{14}C-leucine incorporation has been widely used to investigate herbicidal effects on protein synthesis (MORELAND et al. 1969; ASHTON et al. 1977).

Biosynthesis of proteins can also be prevented by inhibition of amino acid synthesis. It has been shown for the herbicide glyphosate that it blocks the synthesis of aromatic amino acids (JAWORSKI 1972). There is evidence accumulating that glyphosate inhibits a step in the biosynthetic sequence from shikimate to chorismate (AMRHEIN et al. 1980).

In general, various influences of different herbicides (e.g. phenoxy acetic acids, thiocarbamates, triazines and amitrole) on nucleic acid synthesis are described in the literature (BARTELS and HYDE 1970b; VENIS 1971; BESTE and SCHREIBER 1972; CHEN et al. 1972). Numerous structural changes in treated cells are caused by the blocking of some steps in the nucleic acid metabolism (ARNTZEN et al. 1970; ANDERSON and THOMSON 1972). Moreover, some herbicides (e.g. trifluralin, atrazine) have been shown to affect the DNA and RNA synthesis directly, as they are bound to the chromatin, thus changing its integrity or derepressing transcription (PENNER and EARLY 1972a, b; BURROWS 1978). Although protein synthesis is often affected by herbicides, a specific inhibition of this process, for instance of peptide chain initiation or of other partial reactions, has only rarely been documented (WEEKS and BAXTER 1972; DUKE et al. 1975). A detailed survey of the literature on herbicidal effects on protein and nucleic acid synthesis has been presented in some previous reviews (CHERRY 1976; MORELAND 1980). As already mentioned, inhibition of nucleic acid and protein synthesis is often indirectly caused by the interaction of herbicides with phosphorylating ractions which reduce the ATP content in cells that is necessary for various biosynthetic processes (MORELAND et al. 1969; GRUENHAGEN and MORELAND 1971; FEDTKE 1974). These studies, however, revealed a great variety of results, which was probably due to the plant species or experimental conditions. By treatment of plants with a number of different herbicides (e.g. amitrole, methabenzthiazuron, phenoxy acids, triazines) changes have been repeatedly observed in the nitrogen metabolism (ASHTON and BAYER 1976) and in the contents of storage components (ANDERSON and THOMSON 1972), in an increase in the nitrate, protein and amino acid level and in the nitrate reductase activity similar to those observed in shade-adapted plants (RIES et al. 1967; SINGH et al. 1972; FEDTKE 1973, 1974, 1979). The latter phenomenon is widely known, for example the increase in plant constituents, after subtoxic applications of herbicides (especially phenoxy acids, triazines, substituted ureas and uracils). This promotes alterations in plant composition, particularly in proteins, and induces growth stimulation (WIEDMAN and APPLEBY 1972; RIES 1976). So far, however, no practical use has been made of the stimulatory effect of subtoxic treatment with herbicides.

11.4.2.6 Herbicides Affecting Cell Division

In many cases the action of herbicides on cell division is related to the effect of these chemicals on the synthesis of DNA or its precursors, as well as to a change in the DNA or chromatin properties. In a more direct way, herbicides interact with the spindle formation, the microtubule system and the formation of the cell wall (LIGNOWSKI and SCOTT 1972; BARTELS and HILTON 1973; HESS and BAYER 1974; ROST and BAYER 1976; SUMIDA et al. 1977). Since the mitotic activity depends on the availability of ATP, the cell division again is indirectly influenced by herbicidal substances which inhibit ATP formation (see Table 11.2, WEBSTER and VAN'T HOF 1969; ROST 1977). Earlier literature on the effects of several herbicides (e.g. carbamates, thiocarbamates, dinitroanilines, uracils, pronamides, phenoxy acids) on cell division and mitotic activity has been summarized in some reviews (CARTWRIGHT 1976; LINCK 1976), which document in detail the long-known fact that quite a number of these herbicides act as mitotic poisons.

In the past few years our knowledge about the correlation between mitosis and the function of the microtubule system has been considerably improved by the use of antimitotic substances (DUSTIN 1978). Probably many of the substances which block the cell division or mitosis interfere with the formation of microtubules. A survey on some compounds (e.g. trifluralin, chlorpropham, pronamide, amiprophos-methyl) which interact with the microtubule formation like colchicine has been presented by DRABER and FEDTKE (1979). However, the exact mechanism of action of these herbicides in the very complex process of microtubuli formation and organization is still unknown. There is evidence that either the synthesis of microtubule subunits or the microtubule organizing center is affected (HEPLER and JACKSON 1969; BARTELS and HILTON 1973; JACKSON and STETLER 1973; PARKA and SOPER 1977; UPADHYAYA and NOODEN 1977). It was shown that dinitroaniline herbicides like trifluralin, which induce similar effects to colchicine, are bound to a tubulin-like protein from *Chlamydomonas* flagella and interact with microtubule sub-unit protein (HESS and BAYER 1977). These findings, however, could not be substantiated yet by binding experiments with roots of higher plants (UPADHYAYA and NOODEN 1980). While the most important function of microtubules is the distribution of daughter chromosomes during mitosis, they are also involved in the formation and structure of cilia and flagellae such as in algae. This leads to the assumption that flagellae formation is suppressed by herbicides like trifluralin or amiprophos-methyl. Further antimicrotubule effects are the inhibition of the postmitotic migration of the nucleus in the alga *Micrasterias* (KIERMAYER and FEDTKE 1977). As cortical microtubules are essential for the orientation of cellulose fibrils during growth of the cell wall, antitubuline herbicides cause morphological alterations in the tissues by swelling of their cells. Many of the herbicidal effects on the microtubule system have been extensively described in the past few years (PARKA and SOPER 1977; DRABER and FEDTKE 1979; MORELAND 1980).

11.4.2.7 Herbicides Affecting Growth

One of the main purposes of herbicidal use in agriculture is to prevent growth of undesirable plants in crop fields. Plant growth and development can be affected by interference with many essential metabolic reactions and physiological functions of plant cells and tissues (see preceding sections). Also many general growth reactions like seed germination, cell enlargement and differentiation are influenced by herbicide treatment (cf. CARTWRIGHT 1976). The better the understanding of plant growth at the molecular level, the more the general suppression of growth by herbicides will be recognized as a result of specific effects on one or more different basic processes of plant metabolism. However, very often the specific herbicidal effects are still unknown and only become evident generally as growth inhibition. Moreover, an attempt will be made to distinguish between the herbicidal action of growth inhibitors and the effects of growth regulators including the natural plant hormones. The latter will be discussed below.

The auxin-type herbicides have been known as growth inhibitors and weed killers for a long time and they are still of great importance and widely used for weed control in agriculture and forest. These herbicides can be grouped into several general classes: (1) the chlorophenoxy compounds (e.g. 2,4-D, 2,4,5-T, MCPA, silvex); (2) the chlorobenzoic acids (e.g. 2,3,6-TBA, dicamba, chloramben) and (3) a heterogenous group of compounds (e.g. picloram) (Table 11.2). The chlorophenoxy acetic acids were the first active herbicides with selectivity and a widespread and successful application. These herbicides have been extensively investigated and show a multitude of rather different actions depending on the concentrations used. Trace amounts of 2,4-D stimulate extension growth and enzyme activity as does IAA, whereas higher concentrations show inhibitory effects on synthesis of nucleic acids and proteins and other reactions (PENNER and ASHTON 1966; COBLE and SLIFE 1971; CHEN et al. 1972; WEAVER 1972; LOOS 1975; GRÄSER 1977; FEDTKE 1982). One advantage of these chemicals is their selectivity in suppressing growth of broad-leaved weeds in grain fields. Similar modes of actions are observed with benzoic acid herbicides and with picloram and related compounds (FREAR 1976; FOY 1976). The manifold and complex effects of auxin herbicides on growth, cell division, cell elongation, metabolic reactions and cell structures have been extensively described in several reviews (WEAVER 1972; LOOS 1975; CARTWRIGHT 1976; LINCK 1976). Also CHERRY (1976) has summarized attempts to determine the molecular interactions of 2,4-D and other phenoxyacetic acids with the cell and their effects on nucleic acid and protein synthesis.

11.4.3 Growth Regulators

In comparison to herbicides the plant growth regulators have a smaller share of the market, although their applied use has been known for a long time (WITTWER 1971). In trying to minimize the rate of biocide use, growth regulator technology, by chemical manipulation of plant growth and development might have the advantage of contributing to improved food production and reduced environment pollution.

Differentiation between plant growth regulators and herbicides is quite a question of definition. Whereas herbicides are mainly used to suppress the growth of undesirable plants in agricultural and horticultural crop production and therefore allow an optimal weed control, plant growth regulators primarily induce specific modifications in the growth of crops, fruits, vegetables and ornamentals with the aim of increased yield, higher quality, ease of harvesting, uniformity and further improvements. Due to their various effects at different conditions, however, several chemicals can be denoted as herbicides as well as growth regulators (e.g. maleic hydrazide, cycloheximide, endothal). Depending on the concentrations used, typical growth regulators in high concentrations may kill plants, whereas herbicides in sub-lethal concentrations sometimes stimulate growth and modify development and metabolism of plants (RIES 1976). Therefore a clear separation between these two groups of chemicals seems rather difficult. Also in the literature on this subject growth inhibitors and hormone-type (auxin-type) herbicides very often are characterized as growth regulators

(CARTWRIGHT 1976; WITTWER 1978). In the group of plant growth regulators the natural phytohormones have to be included. Therefore we have to subdivide these chemicals into natural and artificial or synthetic growth regulators. The mode of action of both, artificial and natural growth regulators in respect to their applied uses in agriculture will be shortly summarized. For detailed information the reader is referred to several books and reviews on hormone physiology (ABELES 1973; JONES 1973; MILBORROW 1978; LETHAM et al. 1978; LIEBERMAN 1979; STODDART and VENIS 1980; ZERONI and HALL 1980; WALTON 1980; BEARDER 1980) and on other plant growth substances and their commercial use (AUDUS 1972, 1976; WEAVER 1972; THOMAS 1974; WAIN 1975, 1979; WAREING 1976; DRABER 1977; WITTWER 1978; SCOTT 1979; SKOOG 1980; MORGAN 1980; NICKELL 1982).

For more than 40 years plant physiologists have worked on chemical regulation of plant growth. The studies on auxins led to the discovery of NAA, 2,4-D and other phenoxyacetic acids with highly active herbicidal effects for the control of plant growth (HAMNER and TUKEY 1944). After the introduction and availability of gibberellins, cytokinins and abscisins, the interest in plant hormones was stimulated by the search for applied uses of chemical growth regulators in agriculture and horticulture (AUDUS 1972; WITTWER 1978). In addition to the natural phytohormones numerous synthetic compounds with similar or antagonistic effects are now available (DRABER 1977; MORGAN 1980). The development of the ethylene generator ethephon or ethrel (2-chloroethyl phosphonic acid), for instance, which can control release of ethylene, opened new ways for the practical use of chemical regulators in agriculture. However, the effects and functions of a single plant hormone are very variable and dependent on interrelation with other growth regulators. Their behaviour rarely can be explained by one specific reaction within the plant. When supplied exogenously to plants, the different types of phytohormones and related compounds exhibit a number of similar effects and may act synergistically or antagonistically. This fact complicates the development of commercially used growth regulators and may be one reason for the limited number of such chemicals available for use in agriculture compared to herbicides or insecticides. Many aspects of hormonal regulation of plant growth and development and the complexity of interaction between the different growth regulators is extensively discussed by LETHAM et al. (1978) and SKOOG (1980).

In addition to the discovery of plant hormones and their synthetic analogues a great number of chemical growth regulators has been developed, which are chemically not related to the phytohormones and mostly show antagonistic effects (Table 11.4). Indeed, most of the plant growth regulators currently used are growth inhibitors rather than growth promotors. However, crop yields are very often increased indirectly by preventing losses, accelerating the production cycle, changing the ratio between roots and shoots of root crops or facilitating mechanical operations during harvesting. Only in few cases is the actual productivity increased by growth regulators when the plant growth is stimulated or the photosynthates are diverted to the harvested products (THOMAS 1974). Besides increases in yield, however, other aspects such as quality and resistance

Table 11.4. Classification of plant growth regulators

Classes	Common name (chemical name)	Commercial uses
Auxins Analogues	IAA (Indolyl acetic acid) IBA (Indol-3-ylbutyric acid) NAA (Naphthalene acetic acid)	Rooting of cutting; fruit thinning; increasing fruit set; preventing fruit drop
Antiauxins	TIBA (2,3,5-triiodobenzoic acid) NPA (Naptalam, naphthylphthalamic acid) Morphactins (e.g. 2-chloro-9-hydroxy-fluorene-9-carboxylic acid)	Inducing flowering; fruit thinning; growth control of grasses; leaf defoliation; stimulating branching
Auxin-herbicides	2,4-D; 2,4,5-T; MCPA and others (see Table 11.2)	
Gibberellins Analogues	GA (gibberellic acid, GA_3, GA_4, GA_7) Pro-Gibb	Germination; potatoes (tuber dormancy); introducing parthenocarps; delay fruit ripening; quality improvement
Anti-gibberellins	Chloromequat (CCC, 2-chloroethyltrimethylammonium chloride) Phosphon D (dichlorobenzyl-n-butyl-phosphonium chloride) Daminozide (Alar, B-9, SADH, N-(dimethylamino)-succinamic acid) PIX (1,1-dimethyl-piperidinium-chloride)	Inhibition of shoot growth; prevention of lodging; controlled growth of flower crops; increasing fruit sets (grapes); control flowering; fruit quality; maturity control; promotion of ripening
Cytokinins Analogues	Zeatin 6-BAP (6-Benzylaminopurine)	Prevention of chlorophyll degradation; increasing fruit set; senescence inhibition; breaking dormancy
Anti-cytokinins	Pyrazalopyrimidine Pyrrolopyrimidine	(experimental compounds)
Abscisins Analogues	ABA (abscisic acid) Xanthoxine ABA-esters Pydanon (4-hydroxy-2,6-dioxy-hexahydro-pyridazinyl-4-acetic acid)	Growth inhibitor; influencing dormancy and flowering; fruit ripening; flower and leaf abscission; stomatal closure; antitranspirants; promotion of senescence
Ethylene Ethylene generators	Ethylene Ethephon (Ethrel)	Stimulation of fruit ripening and abscission; inducing flowering; promotion of latex flow; fruit thinning; prevention of wheat lodging
Unrelated compounds	MH (maleic hydrazide) Cycloheximide Glyphosine (bis(phosphonomethyl)-glycine)	Suppression of growth (grasses); control of storage sprouting; control of fruit abscission; promotion of ethylene synthesis; promotion in ripening (sugar cane)

to disease and physiological disorders, uniformity, speedy harvesting operation, short-term storage etc. can be improved by using these chemical growth regulators (HUDSON 1976; ANDERSON 1979; HARDY 1979). For these reasons a significant increase in their commercial use can be anticipated. It has been estimated that between 1980 and 1984 the market of plant growth regulators will grow by 12% per year, which is three times more than that predicted for herbicides and insecticides (MORGAN 1980).

Plant growth regulators can be classified into various groups based on their chemical structure and in relation to the known classes of phytohormones and their activities (Table 11.4). This classification – following the suggestions of DRABER (1977) – seems rather arbitrary, but has the advantage of arranging the non-uniform group of synthetic growth regulators as chemical compounds acting as analogues or antagonists in relation to natural plant hormones.

The auxins and their synthetic analogues as well as the auxin-type herbicides are chemicals most widely applied for growth regulation in agriculture and weed control (Table 11.4). Indol-3yl-acetic acid (IAA), the parent compound and some analogues like indol-3yl-butyric acid (IBA) and naphth-1yl-acetic acid (NAA), for instance, were applied as rooting compounds, for thinning of fruits, for increased flowering and fruit set. These basic activities and interactions of the auxins within plant metabolisms are reviewed in detail by SCHNEIDER and WIGHTMAN (1978).

Based on its antagonism to the auxins, another group of chemicals can be placed together. They include the classical antiauxins, primarily compounds that compete with IAA for a specific receptor, as well as substances which affect auxin activity, e.g., by inhibiting the auxin transport (Table 11.4). Of special interest in this connection are the synthetic compounds 2,3,5-triiodobenzoic acid (TIBA), N-(naphthyl)-phthalamic acid (NPA = naptalam) and the morphactins (derivatives of fluorene-9-carboxylic acid). A new class of anti-auxins, aryl-pyrazolyl-benzoic acid derivatives have been described (BEYER et al. 1976). All of them mainly inhibit the transport of auxin (KATEKAR and GEISSLER 1977; SCHNEIDER and WIGHTMAN 1978). TIBA seems to act at the auxin binding site at the plasma membrane (HERTEL et al. 1972) and causes several secondary effects, among others induction of flowering, reduction of lodging and in some cases an increase in the yield of some crops. Specific binding of naptalam has been demonstrated at a different site than IAA, since this auxin does not compete with naptalam binding (LEMBI et al. 1971; THOMSON et al. 1973). On the other hand it shows a stronger effect on basipetal auxin transport than TIBA.

The morphactins form a group of synthetic plant regulators, which act on a wide variety of processes within the plant (for review see SCHNEIDER 1970). Their inhibitory effects on the auxin transport seems to evoke inhibition of cell division, seedling growth, stem elongation, stimulation of branching and others. The agricultural uses of morphactins at present are still limited, but they are applied to control growth of grasses in rough areas, along railways, etc., often in combination with maleic hydrazide (MH).

The gibberellins are phytohormones with activities in regulating dormancy, flowering, fruit setting and stimulating germination of seeds and extending growth of shoots. Over 50 naturally occurring compounds have been chemically characterized with a multitude of responses (WITTWER 1971; JONES 1973; GRAEBE and ROPERS 1978), but only few (e.g., GA_3, GA_4, GA_7) are available for commercial use (RAPPAPORT 1980). Well known applications of the gibberellins include growing seedless grapes and delaying of citrus fruit ripening on the tree.

The important growth retardants chloromequat (CCC), phosphon D and AMO 1618 may act as anti-gibberellins, since they inhibit gibberellin biosynthesis (GRAEBE and ROPERS 1978). In western Europe chloromequat and analogues are used commercially mainly to increase wheat production, as they cause stiffer and shorter stems that resist lodging. They also have some importance as chemical ripeners. Recently 1,1-dimethyl-piperidinium-chloride (DPC, PIX) has been introduced, whose mode of action resembles that of chloromequat. It has been used successfully as growth regulator in cotton and barley (WALLER et al. 1980; KNITTEL et al. 1981). Daminozide (Alar, B-9, SADH) is also known as gibberellin antagonist, although inhibition of gibberellin biosynthesis in vitro was not found. It has been used successfully with several horticultural crops bringing about improved storage quality, fruit setting and maturity control.

A further group of plant hormones, the cytokinins (Table 11.4) show some effects similar those of gibberellins in breaking the dormancy of a wide range of seeds and in an increased fruit set (ADDICOTT 1976; PARKER 1976; LETHAM 1978a). These hormones mainly stimulate cell division and prevent chlorophyll degradation. Some synthetic analogues, in particular 6-benzylaminopurine (6-BAP) were used as senescence inhibitors (WITTWER 1978). Although low solubility, poor transport and fast degradation within the plant limit the commercial use of cytokinins, new synthetic compounds with antagonistic effects have been developed (LETHAM 1978a).

The abscisic acid (ABA) and derivatives (e.g. esters, ABA metabolites, xanthoxin) are also naturally occurring phytohormones that regulate plant growth and metabolism in various ways and have been detected in nearly all plants (MILBORROW 1978). They are involved in the abscission of plant organs, induction and promotion of senescence, breaking dormancy of seeds and vegetative buds, in regulation of fruit ripening and generally in retardation of growth (WALTON 1980). No practical uses of these growth regulators have yet been realized. Recently, however, a commercial interest has been noted on the effect of abscisins on stomatal closure as anti-transpirants, which could permit a control of water loss through transpiration and reduce water stress of crops by hormone application (WITTWER 1978; WRIGHT 1978).

Ethylene and its role in the growth and development of plants, especially in the ripening of fruits, has been known for a long time and attempts have been made to utilize these properties for regulating and increasing food production (ABELES 1973; OSBORNE 1978; LIEBERMAN 1979). Since the introduction of ethylene generators like 2-chloroethylphosphonic acid (Ethephon, ethrel) and many derivatives, these growth regulators have been commonly applied in agriculture and horticulture. Ethephon can selectivily promote fruit ripening and abscission and therefore shorten the harvesting time of fruits. Induction of flowering and stimulation of latex flow in the natural production of rubber in *Hevea brasiliensis* are further important commercial applications (for further details compare reviews by Wittwer 1971, 1978; Morgan 1980). The biosynthesis of ethylene can also be influenced by phytohormones (auxins) and by auxin-herbicides (Morgan 1976). Ethylene production is also influenced when plants are grown under chemical and environmental stress. By modifying the ethylene physiology (Wright 1978; Wittwer 1978) it should be possible to alleviate undesirable effects of stress.

In addition to these groups of plant hormones and their analogues and antagonists a number of synthetic growth regulators, chemically unrelated to the phytohormones, has been developed and successfully used. Maleic hydrazide (MH) for instance (Audus 1972; Nooden 1972) blocks nucleic acid and protein synthesis, prevents cell division and causes chromosome breakage. Maleic hydrazide is one of the most widely used synthetic growth regulators in agriculture, especially for the control of storage, sprouting of onions and potatoes, inhibition of suckering in tobacco and retardation of the growth of grass along streets, railways etc. (Wittwer 1978). Very often maleic hydrazide is applied in a mixture together with herbicides. Binding studies using labelled maleic hydrazide in plant tissues, however, did not allow conclusions on the actual mechanism by which this chemical affects synthesis of nucleic acids (Nooden 1970).

The protein inhibitor cycloheximide shows growth-regulatory effects by suppressing the fruit abscission and promoting ethylene synthesis, but unfortunately it is highly toxic for mammals. The glycine derivative glyphosine promotes the ripening in sugar-cane and sugar-beets and also endothal has been used to increase yield in sugar-cane production (Draber 1977).

At least it should be mentioned that until now a great number of naturally occurring plant growth regulators other than the principal hormones of higher plants have been isolated (Kefeli 1978; Letham 1978b; Steffens et al. 1979) and are used as models for the synthesis of new commercial growth substances (Schreiber 1979). Also growth regulators produced by microorganisms (e.g. helminthosporol, fusicoccin) are of interest for applied as well as for basic research on the mechanism of action and in the detection of specific receptors in plant tissues (Dohrmann et al. 1977; Tamura 1979; Ballio 1979).

In the past few years numerous papers have been published with the final aim of elucidating the primary mechanisms of actions of phytohormones and other plant growth regulators. Although until 1975 – in contrast to the progress in animal hormone physiology – no plant hormone receptor has yet been isolated (Kende and Gardner 1976), there have been great efforts in the meantime to characterize similar systems in plants. Studies on the binding of growth regulators like NPA, TIBA and morphactins to corn coleptile membranes stimulated very much the research on plant hormone receptors (Lembi et al. 1971; Hertel et al. 1972; Thomson et al. 1973; Thomson and Leopold 1974). The characterization of the binding sites of natural and synthetic growth regulators by measuring binding constants, competition and displacement by unlabelled compounds and other methods led to the isolation of binding proteins and to a better understanding of the molecular requirements for the biological activities of growth regulators (Batt and Venis 1976; Dohrmann et al. 1978; Cross and Briggs 1978; Katekar 1979; Rubery 1981). Recently, covalent labelling techniques have been employed for identification of the auxin receptor (MELHADO et al. 1981). Most of this work has been done with auxins, their analogues and auxin transport inhibitors, but also binding proteins for cytokinins (FOX and ERION 1977; TAKEGAMI and YOSHIDA 1977; PRATT and FOX 1978; POLYA and DAVIS 1978), gibberellins (STODDARD 1979) and further growth regulators have been thoroughly investigated (VENIS 1977, 1979). The new findings on molecular and subcellular aspects of growth regulator action have been recently reviewed in Vol. 9, this Series (STODDARD and VENIS 1980; ZERONI and HALL 1980).

In summarizing the discussion in this section, it can be said that in comparison to herbicides the effects of plant growth regulators seem to be more complex and manifold and the mechanisms of action are generally less understood. It became evident that natural and synthetic growth regulators primarily affect the synthesis and metabolism of nucleic acids and proteins, the synthesis of enzymes and the regulation of enzyme activities, the general metabolic processes (respiration, photosynthesis, ATP-dependent reactions), the membrane organization and ion transport through membranes (see reviews by JABOBSEN 1977; PENNY and PENNY 1978; HIGGINS and JACOBSEN 1978; ZERONI and HALL 1980). The progress in the last 5 years in the basic understanding of hormone physiology gives us hope to discover and to develop better growth regulators for agricultural uses with advantages compared to biocides (HUDSON 1976; WAIN 1979; ANDERSON 1979; HARDY 1979). Future uses of plant growth regulators might include the modification of carbon metabolism and processes which influence crop quality (protein and carbohydrate content, amino acid balance etc.), the regulation of flowering, fruiting, harvest and storage, the improvement of nitrogen and mineral uptake and of plant responses to environmental stress, the allocation of the assimilates in the plant and the manipulation of the plant development (MORGAN 1980). If antitranspirants like abscisins could be used in areas with high transpiration rates and high temperatures, problems related to water shortage might be diminished. Thus the plant growth regulators represent an interesting group of chemicals besides biocides and are generally less harmful to plants, to the ecosystems and finally to the total environment. Plant growth regulators will play a growing role in Integrated Pest Management (see Sect. 11.6).

11.4.4 Insecticide Effects on Plants

The insecticides represent an important group of biocide chemicals that are used for agricultural and medical pest control. Among them the organochlorine insecticides (e.g. DDT, lindane, dieldrin and endosulfan) are known to be highly effective and very persistent. Residues in soils, aquatic organisms, plants and animals reaching up to the upper levels of food chains may produce long-term ecological hazards. It can, however, be noted that the use of persistent insecticides has decreased during the past 20 years, while the use of less persistent, but more immediately toxic chemicals, e.g. organophosphorus and carbamate insecticides, has increased.

Whereas in western Europe the consumption of insecticides has reached about 21% of the total biocide market, its portion increased to 28% in the United States and Canada. On the other hand, the herbicides hold a market share of 47% in western Europe and even 65% in the United States and Canada (BÜCHEL 1977). The use of insecticides for pest control in agriculture and diverse plant ecosystems – in addition to the contamination of the total environment by persistent biocides (EDWARDS 1973; BEVENUE 1976; PERRING and MELLANBY 1977; BROWN 1978; MCEWEN and STEPHENSON 1979) – gives rise to the question, whether a particular biocide accumulates in the soil, whether such residues are phytotoxic and how this becomes apparent. Very often both, insecticides and herbicides, are provided to crop and non-crop plants in such a way that biocide interactions release synergistic effects in these higher plants (PUTMAN and PENNER 1973; HAYES et al. 1979).

In this review some aspects of phytotoxicity by insecticides will briefly be mentioned. Detailed information on the general mode of action of insecticides is given by WEGLER (1970), EDWARDS (1973), BROOKS (1974), CORBETT (1974), HAQUE and FREED (1975), COULSTON and KORTE (1976) and BÜCHEL (1977). The insecticides can be classified into four main groups:

1. Chlorinated hydrocarbons (organochlorines, e.g. DDT, lindane, aldrin, dieldrin, endrin, chlordane, dicofol, endosulfan, heptachlor, toxaphene).
2. Organophosphates (e.g. malathion, parathion, phosphamidon).
3. Carbamates (e.g. dimetan, pyrolan, isolan, carbaryl, promecarb, aldicarb).
4. Naturally occurring insecticides (pyrethrine, allethrine, rotenone) and their synthetic analogues.

At present the latter group is attracting increasing interest (CASIDA 1973; BÜCHEL 1977; NAUMANN 1981). Finally the polychlorinated biphenyls (PCB's) should be mentioned, as they are similar in structure and properties to the organochlorine insecticides, widely distributed, very persistent and with similar toxic effects (HUTZINGER et al. 1974; HARDING and PHILLIPS 1978; KALMAZ and KALMAZ 1979; FALKNER and SIMONIS 1982).

The amount of insecticide residues in plants depends on the uptake and penetration of these chemicals by the roots and leaves, as well as on the accumulation and metabolism within the plant (FINLAYSON and MACCARTHY 1973; VOERMAN and BESEMER 1975; STEINWANDTER 1976; BUTLER 1977). Especially plants and crops which contain lipophilic constituents (e.g. carrots, rape and mustard) are able to absorb unusually large amounts of insecticides and to accumulate them in the roots or other tissues.

Insecticides are of greater hazard to aquatic ecosystems and phytoplankton than to most terrestrial higher plants, because they are more susceptible to them and accumulate these compounds (WARE and ROAN 1970; HURLBERT 1975; NEUDORF and KHAN 1975; SCHAUBERGER and WILDMAN 1977; GREICHUS et al. 1978). Since the water in rivers and seas, which ultimately receives the biocides applied to the environment, is generally more contaminated than the soil, the biocides are primarily concentrated in the phytoplankton. Therefore, our attention should be more directed to possible effects of insecticides on algae as primary members of the aquatic food chains. The reduction of the algal biomass production will have a multitude of aggravating consequences. Most of the investigations on effects of insecticides on plant growth and other metabolic reactions have been carried out with algae (SÖDERGREEN 1971; KOPECEK et al. 1975; POWERS et al. 1977; BIRMINGHAM and COLMAN 1977; SIMONIS and LEE-KADEN 1979). Chlorinated hydrocarbons are the best-studied insecticides in this respect. Even with low concentrations of DDT and other insecticides (e.g. lindane, dieldrin, aldrin) growth of algae in the phytoplankton is generally reduced (MENZEL et al. 1970; BATTERTON 1971; BOWES 1972; MACFARLANE et al. 1972; POWERS et al. 1977; KOPECEK et al. 1975; SCHAUBERGER and WILDMAN 1977; KRISHNAKUMARI 1977; BIGGS et al. 1979). Although there are few data on effects of organophosphorus insecticides on algae, several papers have been published on various inhibitory effects on algal growth (DERBY and RUBER 1970; BIRMINGHAM and COLMAN 1977). Influence of polychlorinated biphenyls on algal growth and photosynthesis was studied by several authors (FISCHER and WURSTER 1973; COLE and PLAPP 1974; EWALD et al. 1976; LARSSON and TILLBERG 1975; HARDING and PHILLIPS 1978).

The effect of insecticides on photosynthesis was investigated more directly by measuring photosynthetic partial reactions of chloroplasts isolated from algae (BOWES 1972) or higher plants (ROGERS et al. 1972; OWEN et al. 1975, 1977; WILD et al. 1977; DELANEY et al. 1977, 1978). It was shown that DDT inhibits photosynthetic electron transport at two sites, before PS II and between the photosystems (see Fig. 11.1), affecting both non-cyclic and cyclic photophosphorylation (ROGERS et al. 1972; OWEN et al. 1977; DELANEY et al. 1978). Additionally – as a consequence of the use in agriculture and by comparison with herbicides – it became evident that some crop species are adversely affected by DDT. While certain varieties of barley became resistant to DDT, others remained susceptible (OWEN et al. 1975). Experiments on binding of DDT to isolated chloroplasts of susceptible and resistant plants might help elucidation of the mechanism of this insecticide action. Also lindane has been found to partially inhibit basal, coupled and uncoupled electron transport from water to ferricyanide of chloroplasts isolated from leaves of *Pisum sativum*, whereas the activity of PS I was stimulated (WILD et al. 1977). However, it should be noted that our current understanding of the interaction of insecticides with photosynthetic reactions is not as advanced as our knowledge of the mechanism of herbicide action on photosynthetic electron transport (see Sect. 11.4.2.1).

Besides their effects on growth and photosynthesis of algae and higher plants, various insecticides (lindane, DDT) interfere with cell division, especially with microtubuli function (KIERMAYER and MEINDL 1979; JEANNE 1979), and cause ultrastructural changes in cells (BORGHI et al. 1973; PUISEUX-DAO et al. 1977; SKULT and PAATERO 1979). Also the synthesis of nucleic acids and protein is affected by insecticides (ANDEREGG et al. 1977; SIMONIS and LEE-KADEN 1979) and polychlorinated biphenyls (EWALD et al. 1976). In various algae the incorporation of labelled amino acid leucine is inhibited by lindane, which points to an interaction of insecticides with transport processes in the membrane (BORGHI et al. 1973).

The fact that the toxicity of insecticides on insects is mainly based on alterations of properties of nerve membranes gives rise to the assumption that membranes are also a favoured target in plant cells. Numerous studies with synthetic membranes (e.g. liposomes or various bilayers) reveal that insecticides interact with membrane lipids (HILTON and O'BRIEN 1970; HILTON et al. 1973; O'BRIEN et al. 1977; REINERT and TINSLEY 1977; LAKOWICZ et al. 1977; ANTUNES-MADEIRA and MADEIRA 1979; LAKOWICZ and HOGEN 1980). Recently, the first studies on plant membranes and their interference with insecticides by using electrophysiological techniques have been reported (SCHEFCZIK and SIMONIS 1979, 1980; SCHEFCZIK 1981). In cells of the aquatic plant *Elodea densa* sublethal doses of lindane (5×10^{-5} M) inhibit light induced membrane potential changes, decrease the level of the diffusion potential and increase membrane resistance. The loss of the potassium selectivity points to a direct effect of lindane at the plasmalemma of *Elodea* cells (SCHEFCZIK and SIMONIS 1980). The accumulation of lindane in the lipid phases of membranes disturbs lipid-protein interactions followed by alterations of cell membrane properties with serious consequences on ion uptake, cell metabolism and development (SCHEFCZIK 1981).

Summing up, it may be said that insecticides do not only affect insects, but also interfere with growth and development of plants. As a result of environmental pollution they are accumulated in plants, especially in algae and subsequently in the entire food web. Besides the exact monitoring of insecticide residues in plant material, crops and food, further studies on the modes of action of these chemicals in plants are of fundamental importance. As shown in this review, a multitude of phytotoxic effects has been observed. However, only in few cases could the primary mechanisms of action of insecticides within plants be explained.

11.4.5 Herbicide-Safeners (Antidotes)

Although today a wide variety of selective herbicides is available, there are still conditions where a satisfying weed control in crops is either difficult to obtain or is even impossible. This can be the control of some broadleaf weed species in broadleaf crop plants or the control of weed strains of the same species as the crop (wild oat in oat, for example). In some cases, the safety margin for herbicide use in a particular crop is relatively small and unusual environmental conditions can narrow down this margin until injury to crop plants occurs. From this the need for more selective herbicides becomes obvious. The discovery of a large number of selective herbicides in the near future does not seem very probable, because in the last 10 years only a few new herbicides have been introduced and costs for development of a single new herbicide are growing rapidly. Two approaches to this problem of limited herbicide selectivity are possible: breeding of herbicide-tolerant crops or the protection of crop plants against herbicide injury by chemical means.

Introduction of herbicide tolerance by genetic means has been reviewed by several authors (GRESSEL et al. 1982; FAULKNER 1982; MEREDITH and CARLSON 1982). Examples where first successes were obtained are discussed by these authors. There are recent attempts to transfer the genetic trait conferring triazine resistance from resistant weeds to crop plants (reviewed by SOUZA-MACHADO 1982).

Chemical protection of crop plants against herbicide injury is relatively new in weed science. Since the early 1960's there has been a growing number of reports on this subject and until today there are several commercial products available. A compound which protects a crop plant against herbicide injury is called a safener, an antidote or a protectant. These terms are used interchangeably. The properties of some herbicide antidotes and their mechanism of action – as far as is known already – are reviewed by BLAIR et al. (1976) and are compiled by PALLOS and CASIDA (1978).

HOFFMANN of Gulf Oil Co. introduced 1,8-naphthalic anhydride (NA) as the first commercial antidote (HOFFMANN 1978b). NA acts as seed applied antidote against several thiocarbamate (i.e. EPTC) and chloroacetamide (i.e. metolachlor) herbicides in corn, rice, grain sorghum and oats. Several dichloroacetamides were developed by Stauffer Chemical Co. as protectants against thiocarbamate herbicides (CASIDA et al. 1975) and also against acetamide herbicides (LEAVITT and PENNER 1978). The antidotes from the group of dichloroacetamides not only act after seed treatment (as NA does), but also when the herbicide is applied simultaneously to the soil. The most active compound of this series was R-25788 (N,N-diallyl-2,2-dichloroacetamide). In an extensive study of structure-activity relationships STEPHENSON and CHANG (1978) established that compounds which have structural similarities with thiocarbamate herbicides are potentially good antidotes against these herbicides. The results of their study indicate that the antidotes may act as competitors to the herbicides, while root uptake of EPTC was not affected. As an alternative mechanism, a stimulation of herbicide metabolism by an antidote has been proposed by CASIDA et al. (1975). These authors observed that thiocarbamates – after transformation into toxic sulfoxides inside the cell – are metabolized to a non-toxic glutathione conjugate. LAY and CASIDA (1976, 1978) and CARRINGER et al. (1978) demonstrated that the antidote R-25788 acts by increasing the glutathione (GSH) content and the activity of the glutathione-S-transferase. EZRA and GRESSEL (1981) questioned that enhanced GSH levels and stimulated GSH-S-transferase activity play a primary role of the mechanism of action of NA or R-25788. According to their observations the antidote effects become apparent long before the cellular GSH level has increased. Also HATILOS and PENNER (1980) concluded that the chloracetamide safener effect is not based on the GSH-mechanism. They found that NA and CDAA (N,N-diallyl-2-chloroacetamide) are potential safeners for the PS II herbicide buthidazole (YORK et al. 1981), the mechanism of action and chemical structure of which is entirely different to the herbicides from thiocarbamates or chloroacetamides.

Recently CGA-43089 (α-cyanomethoximino-benzacetonitrile) has been introduced by CIBA-GEIGY Co. as an antidote which increases the tolerance of sorghum to the potent grass herbicide metolachlor (ELLIS et al. 1980; NYFFELER et al. 1980; PARADIES et al. 1981) and other acetamide herbicides (SIMKINS et al. 1980). CGA-43089 prevents the herbicide-induced depletion of waxes which cover the leaf surface (EBERT 1982) and is thought to interfere with the biosynthesis of long-chain lipids (EBERT and RAMSTEINER in preparation).

Several diazosulfonates were described by PHILLIPS and BHAGSARI (1978) as potential antidotes against atrazine injury to soybeans. The protection of the highly triazine susceptible soybeans is of particular interest, since under special environmental conditions triazine residues ("carry-over") induce losses in soybean crop yield after crop rotation (usually corn–soybean). The most active compound in the studies of PHILLIPS and BHAGSARI (1978) was p-(dimethylamino)-benzenediazosulfonate, originally known as the fungicide dexon. The authors suspect that the diazosulfonates may either limit atrazine uptake or induce rapid detoxification of the herbicide.

It should be noted that until now no antidote has been found against the primary mechanism of action of the PS II herbicides. These commercially very important herbicides have all the same target site at the chloroplast membrane, as was discussed above. An antidote acting at this common target site would be useful against a large number of PS II herbicides. According to HOFFMANN (1978a, b) more than 6,000 compounds have been tested to no avail. Recently, a new theoretical approach was proposed for designing antidotes against PS II herbicides (PFISTER and ARNTZEN 1979; PFISTER et al. 1982). Based on models of the binding mechanism of PS II herbicides (Fig. 11.2) it has been argued that a potential safener could be a compound which lacks the "essential element" (a chemical structure common to most of the PS II herbicides; see p. 344), but which still contains the lipophilic substituents responsible for binding to the herbicide target site. A compound modified in that way should be able to bind to its target site at the chloroplast membrane, but would be unable to produce an inhibitory effect. Thus, a safener would compete with a PS II herbicide for the same binding site and therefore decrease or even abolish herbicide efficiency. This new concept is presently under investigation.

11.5 Herbicide Resistance

Resistance to biocides, especially to insecticides and fungicides, and also the microbial resistance to antibiotics are well known problems. As the classical example, the rapid development of DDT resistance by houseflies should be mentioned. In herbicide applications, resistance was almost unknown until about 10 years ago, when resistance of *Senecio vulgaris* against triazine herbicides was first reported (RYAN 1970). Until today, triazine resistance has remained the only case, where – in certain areas – resistance had created a major, but temporary agricultural problem. Most aspects of herbicide resistance are comprehensively covered in the book edited by LE BARON and GRESSEL (1982).

Selectivity, i.e. a different susceptibility of crop plants and weeds, is an essential requirement in most herbicide applications (see Sect. 11.3). Differential susceptibility, however, is not only found between different species of plants, but also inside a population of a single species. This indicates that repeated use of herbicides will tend to eliminate the more susceptible biotypes and will increase the percentage of more herbicide-tolerant biotypes. Several examples where weed populations became more tolerant to biocides are compiled by BANDEEN et al. (1982) and by GRESSEL et al. (1982). Consequences for shifts in the composition of weed populations are mentioned in the following chapter by SCHUBERT (this volume, Chap. 12). The term "resistance" is used to characterize a biotype which is of extreme herbicide tolerance so that – under realistic agricultural conditions – weeds of this biotype can no longer be killed by a particular chemical.

Following the first report by RYAN (1970), appearance of triazine resistance was subsequently confirmed and studied by several researchers. For references and case histories see BANDEEN et al. (1982). In the meantime there are a large number of reports describing triazine resistance for more than 15 weed species in Canada, United States (BANDEEN et al. 1982) and some western European countries (GRESSEL 1982; several studies by BARRALIS and GASQUEZ 1979). In almost all cases, resistance originated in areas were triazine herbicides have been used for many years, mainly in monocultures of corn. Several hundred thousand hectares are reported to be heavily infested by resistant weed biotypes, which illustrates the local importance of this problem. *Chenopodium album, Amaranthus hybridus, Senecio vulgaris* and *Brassica campestris* are among the most important resistant species. Without anticipating details from the following paragraphs, it should be mentioned that the molecular basis for the expression of resistance is the same for every triazine-resistant weed biotype investigated so far – regardless of its origin. For this reason, most of the results from studies of a particular weed species can be generalized.

Several lines of evidence indicate that triazine resistance has not developed due to a spontaneous mutation but due to selection: Triazine resistance originated independently in about 15 weed species at different locations in the western hemisphere. In each case, the same modification of the herbicide receptor protein in chloroplast membranes conferred resistance (for details see below). Selection of resistant biotypes is further indicated by the observation that weed species originating in Europe and brought to the United States became resistant in Europe only. In contrast, weeds originating from the American continent, which were brought to Europe, became resistant in the United States, but not in Europe. This is probably due to the lack of genetic variability for the few plants transferred from one continent to the other (GRESSEL et al. 1982). In the past, the frequency of resistant plants of a particular weed species was kept low by a natural mechanism. It is documented by several studies that the resistant (R-)biotype is less "fit" than the susceptible (S-)biotype. This means that an individual R-plant is inferior to the S-plant in competition for light, water and nutrients (CONRAD and RADOSEVICH 1979; RADOSEVICH and HOLT 1982; SIMS-HOLT et al. 1981). The removal of this natural selection pressure by repeated herbicide use favored the spreading of the R-biotype.

Differential ecological fitness of the two biotypes is one factor which determines the rate of appearance of triazine resistance. GRESSEL developed a quantitative model for studying the dependence of the development of triazine resistance on several parameters (GRESSEL and SEGEL 1978, 1982). Key factors in this model are the "fitness", persistance of a herbicide, soil seed reservoir and the "kill rate" of a herbicide application. The model shows that under given conditions resistance develops exponentially, but without being noticed in the field for the first 12–15 years. Within the last 1–2 generations an almost completely resistant population emerges. His analysis is in accordance with the actually observed situation, and underlines the fact that there are no easily recognizable warning signs for a current selection of resistance under field conditions.

The fact that triazine resistance having once occurred has not yet infested waste areas is partially due to the lack of competitive fitness of the R-biotype. More important in this respect seems to be the genetics of triazine resistance. Several studies confirm that the resistant trait is cytoplasmatically, maternally inherited (Souza-Machado et al. 1978a; Darr et al. 1981; Scott and Putwain 1981; Souza-Machado 1982). This indicates the possible involvement of the chloroplast genome. Further support for this notion comes from the independent observation that the herbicide binding protein, a 32,000 molecular weight chloroplast membrane protein (B-protein, see p. 342f.) which is modified in the R-chloroplasts, is actually coded for by the plastid DNA (Steinback et al. 1981b). Therefore, resistance is not spread by pollen transfer.

The physiological mechanism of triazine resistance has become subject of a series of studies which first tried to explain the extreme lack of triazine susceptibility on the basis of alterations of triazine uptake, translocation, accumulation or metabolism. In this respect no differences between the S- and R-biotype were found which were large enough to influence herbicide susceptibility (Radosevich and Devilliers 1976; West et al. 1976; Radosevich 1977, 1982; Jensen et al. 1977). It was first noted by Radosevich (1977) that triazine resistance is not only a property of the whole plant but is also observed in a cell-free preparation of broken thylakoid membranes, isolated from the R-biotype of *Senecio vulgaris*. This finding was unique, because never before had selectivity, susceptibility or even resistance been related to the molecular target of a herbicide or of a growth regulator. In the following years, inhibition of photosynthetic reactions was extensively studied mainly by the group of Arntzen (Souza-Machado et al. 1977, 1978b; Jensen et al. 1977; Arntzen et al. 1979; Radosevich et al. 1979; Pfister and Arntzen 1979; Pfister et al. 1982; Arntzen et al. 1982). In these studies the sensitivity of the R-chloroplasts towards triazine herbicides was shown to be more than 1,000 times lower than that of the S-chloroplasts. In addition to the extreme cross resistance between the S-triazine herbicides, for pyridazinone, triazinone and uracil herbicides an approximately 100 times lower susceptibility was found, whereas the urea herbicides were almost as effective in the R-chloroplasts as in the S-chloroplasts. Some herbicides, e.g. phenols such as dinoseb, DNOC, ioxynil and the structurally unrelated bentazon were even more effective in the chloroplasts from the R-biotype. Differential susceptibility of chloroplasts from the two biotypes towards several groups of PS II herbicides is explained by alterations in the binding affinity of these herbicides. By studying binding of radiolabelled atrazine and diuron, it was shown that resistance is based on a complete loss of triazine binding affinity, while binding of diuron was only slightly diminished (Pfister et al. 1979). It is proposed that a minor modification in the triazine specific subreceptor at the PS II herbicide binding protein (Fig. 11.2) has established resistance in the R-biotype.

From studies mentioned above and from field work several ways have been proposed to overcome the problem of herbicide resistance in agriculture. Obviously, the R-biotype can be controlled by several non-triazine herbicides, e.g. by ureas or by compounds not affecting primarily photosynthesis. Changing between different herbicides may also require crop rotation or application of some additional, mechanical weed control measures. By applying these practices, appearance of resistance may be considerably delayed or even prevented.

Besides herbicide resistance which has developed in the field, several laboratories have reported selection of herbicide resistant algal strains with or without previous mutagenic treatment. Some strains of *Chlamydomonas, Euglena, Bumilleriopsis* are reported to be urea, triazine or triazinone resistant, respectively (LIEN et al. 1977; several reports of LAVAL-MARTIN et al. 1977; CALVAYRAC et al. 1979; ASTIER et al. 1979; BÖGER et al. 1981; JANATKOVA and WILDNER 1982). In *Ankistrodesmus* and other green algae, a temporary tolerance to phenol herbicides can be induced (URBACH et al. 1979). Studies of herbicide resistance utilizing either weeds or algae have been and will be extremely useful in elucidating details of the interaction between a herbicide and its receptor site.

11.6 Integrated Pest Management

While discussing physiological effects of biocides and growth regulators it seems appropriate to point briefly to present developments which influence agricultural crop production. More and more it becomes evident that the exclusive use of chemical methods for pest control will not be sufficient to solve the increasing problems which appear with the growing world population, growing food demand and the rise of total environmental pollution. The overuse, accumulation and distribution of persistent biocides all over the world, the development of biocide-resistant species of noxious organisms and reduction of natural enemies require the intensified employment of different and integrated methods for pest control. Not least, the spiraling costs of chemical protection in the last ten years stimulated the development of the Integrated Pest Management (IPM). IPM is a conceptual approach, which involves a multiplicity of pest control strategies adapted to each specific situation (APPLE and SMITH 1976). It includes the combination of:

1. Agricultural practices, e.g. crop rotation or selection of special crop varieties.
2. Biological control, e.g. using natural enemies or microorganisms.
3. Biotechnological methods, e.g. breeding of pest resistant crops or using attractants and repellents.
4. Chemical pest control, e.g. by using more selective and efficient compounds.

By utilizing these techniques, IPM tries to keep pest populations below an economic injury level and to establish an ecologically more acceptable pest control. It has to be realized that an extended employment of IPM programs will be a slow and gradual process, which requires a profound knowledge on pest-control interactions.

General information on IPM is given by APPLE and SMITH (1976), FRYER and MATSUNAKA (1977), FLINT and VAN DER BOSCH (1981) and PIMENTEL (1981). Aspects of biological control are discussed by DEBACH (1974), HUFFAKER and

MESSENGER (1976), FRANZ and KRIEG (1976) and BATRA (1982). A comprehensive coverage of the present stage of IPM research is reported in Weed Science Suppl. 1, Vol. 30 (1982).

11.7 Conclusions

In our review we have tried to summarize some physiological aspects of the mode and mechanism of actions of biocides and growth regulators in plants. We focused our attention on specific biochemical reactions which are considered to represent the primary targets of biocides. Some recent topics have been discussed in more detail, e.g. the analysis of herbicide binding to membrane proteins, the improvement of selectivity of herbicides by safeners, properties of herbicide-resistant weeds and the development of IPM programs. The apparent progress in the last decade in basic studies of plant metabolic processes (e.g. photosynthesis, respiration, phytohormone actions) helped significantly in understanding the mechanisms of biocide action. On the other hand the use of biocides as tools strongly stimulated research in plant physiology. Obviously both disciplines, basic as well as applied research, gained mutual benefit. A better understanding of plant growth, membrane functions and bioenergetic regulation at the molecular level will provide a more fundamental interpretation of phytotoxicity. The present knowledge of the mode of action of biocides lets us generally recognize that a multitude of apparently unrelated effects of biocides on plants is based on the interference with a few common biochemical mechanisms. It should be realized, however, that in addition to the insight into biocide action at the molecular level, knowledge of physiological properties of the whole plant will help to evaluate the ecological consequences of biocide application.

Acknowledgements. We thank our colleagues Dr. Simonis, Dr. Ebert, Dr. Hartung, Dr. Kaiser and especially Dr. Rienits for their criticism and valuable suggestions. Several colleagues provided us with preprints of their current work, which is gratefully acknowledged. We wish to express our special thanks to C. and H.J. Geisweid, E. Reisberg, S. Franzmann and C. Dorschky for their valuable help during preparation of the manuscript. Own research cited in this paper was supported in part by a grant to W. Urbach from the Deutsche Forschungsgemeinschaft (DFG).

Trademarks or proprietary products mentioned in this chapter are used for the benefit of the reader and do not exclude other products.

References

Abeles FB (1973) Ethylene in plant biology. Academic Press, London New York
Addicott FT (1976) Actions on abscission, defoliation and related responses. In: Audus LJ (ed) Herbicides Vol I. Academic Press, London New York, pp 191–219
Akoyunoglou GA (ed) (1981) Photosynthesis I–VI, Balaban International Science Services, Philadelphia
Amrhein N, Deus B, Gehrke P, Steinrücken HC (1980) The site of the inhibition of

the shikimate pathway by glyphosate II. Interference of glyphosate with chorismate formation in vivo and in vitro. Plant Physiol 66:830–834
Anderegg BN, Lichtenstein EP, Kemp JD (1977) Effects of lindane on DNA, RNA, and protein synthesis in corn roots. J Agric Food Chem 25:923–928
Anderson AS (1979) Plant growth retardants: present and future use in food production. In: Scott TK (ed) Plant regulation and world agriculture. Plenum, New York, pp 251–277
Anderson JL, Thomson WW (1972) The effects of herbicides on the ultrastructure of plant cells. Residue Rev 47:167–189
Anderson SM, Krinsky NJ (1973) Protective action of carotenoid pigments against photodynamic damage to liposomes. Photochem Photobiol 18:403–408
Antunes-Madeira MC, Madeira VC (1979) Interaction of insecticides with lipid membranes. Biochim Biophys Acta 550:384–392
Apple JL, Smith RF (1976) Integrated pest management. Plenum, New York
Arntzen CJ, Linck AJ, Cunningham WP (1970) The effect of 3-amino-s-triazole on flax root apices. Bot Gaz 131:14–23
Arntzen CJ, Ditto C, Brewer P (1979) Chloroplast membrane alteration in triazine-resistant *Amaranthus retroflexus* biotypes. Proc Natl Acad Sci USA 76:278–282
Arntzen CJ, Pfister K, Steinback KE (1982) The mechanism of chloroplast triazine resistance: Alterations in the herbicide site of action. In: LeBaron H, Gressel J (eds) Herbicide resistance in plants. Wiley and Sons, New York
Ashton FM, Bayer DE (1976) Effects on solute transport and plant constituents. In: Audus LJ (ed) Herbicides, Vol I. Academic Press, London New York, pp 219–253
Ashton FM, Crafts AS (eds) (1981) Mode of action of herbicides. Wiley-Interscience, New York
Ashton FM, Villiers de OT, Glenn RK, Duke WB (1977) Localization of metabolic sites of action of herbicides. Pestic Biochem Physiol 7:122–141
Assche van CJ (1978) Characterization of a common molecular target for selected structures of photosynthesis inhibiting herbicide. In: Geissbühler H (ed) Advances in pesticide science, Vol III. Pergamon, Oxford, pp 494–498
Assche van CJ (1981) Chemical-triggered luminescence in dark-adapted *Chlorella:* A probe for DCMU-like inhibitors. In: Akoyunoglou GA (ed) Photosynthesis II, Balaban International Science Services, Philadelphia, pp 227–236
Astier C, Vernotte C, Der-Vartanian M, Joset-Espardellier F (1979) Isolation and characterization of two DCMU-resistant mutants of the blue-green alga *Aphanocapsa 6714.* Plant Cell Physiol 20:1501–1510
Audus LJ (1972) Plant growth substances. Chemistry and physiology, Vol I. Hill, London
Audus LJ (ed) (1976) Herbicides. Physiology, biochemistry, ecology, Vols I, II. Academic Press, London New York
Ballio A (1979) Chemistry and plant growth regulating activity of fusicoccin derivatives and analogues. In: Geissbühler H (ed) Advances in pesticide science, part II. Pergamon, Oxford, pp 366–372
Bandeen JD, Stephenson GR, Cowett ER (1982) Discovery and distribution of herbicide resistant weeds in North America. In: LeBaron H, Gressel J (eds) Herbicide resistance in plants. Wiley and Sons, New York
Barber J (ed) (1976) The intact chloroplast, Vol I. Elsevier, Amsterdam
Barber J (ed) (1977) Primary processes of photosynthesis, Vol II. Elsevier, Amsterdam
Barney GO (ed) (1980) The Global 2000 Report to the President. Counc Environ Qual State Dep. US Government Printing Office, Washington DC
Barney GO (ed) (1981) Global future: Time to act. Counc Environ Qual State Dep. US Government Printing Office, Washington DC
Barralis G, Gasquez J (1979) Comportement ecologique et physiologique des dicotyledones resistantes a l'atrazine en France. Proc EWRS Symp 1979:217–224
Bartels PG, Hilton JL (1973) Comparison of trifluralin, oryzalin, pronamide, propham and colchicine treatments on microtubules. Pestic Biochem Physiol 3:462–472
Bartels PG, Hyde A (1970a) Chloroplast development in 4-chloro-5-(dimethylamino)-2-(α,α,α-trifluoro-m-tolyl)-3(2H)-pyridazinone (San 6706) treated wheat seedlings. Plant Physiol 45:807–810

Bartels PG, Hyde A (1970b) Buoyant density studies of chloroplast and nuclear DNA from control and 3-amino-1,2,4-triazole-treated wheat seedlings, *Triticum vulgare*. Plant Physiol 46:825–830
Bartels PG, McCullough C (1972) A new inhibitor of carotenoid synthesis in higher plants: 4-chloro-5-(dimethylamino)-2-(α,α,α-trifluoro-m-tolyl)-3(2H)-pyridazinone (Sandoz 6706). Biochem Biophys Res Commun 48:16–22
Bartels PG, Watson CW (1978) Inhibition of carotenoid synthesis by fluridone and norflurazone. Weed Sci 26:198–203
Batra SWT (1982) Biological control in agroecosystems. Science 215:134–139
Batt S, Venis MA (1976) Separation and localization of two classes of auxin binding sites in corn coleoptile membranes. Planta 130:15–21
Batterton JC (1971) Growth response of blue-green algae to aldrin, dieldrin and their metabolites. Bull Environ Contam Toxicol 6:589–594
Bearder JR (1980) Plant hormones and other growth substances – their background, structures and occurrence. In: MacMillan J (ed) Encyclopedia of plant physiology new Ser Vol IX. Hormonal regulation of development I. Springer, Berlin Heidelberg New York, pp 9–112
Bell SL, Schwarz OJ, Hughes KW (1976) Studies of the herbicide paraquat I. Effects on the cell cycle and DNA synthesis in *Vicia faba*. Can J Genet Cytol 18:93–99
Ben-Aziz A, Koren E (1974) Interference in carotenogenesis as a mechanism of action of the pyridazinone herbicide Sandoz 6706: Accumulation of C-40 carotenoid precursors, inhibition of β-carotene synthesis and enhancement of phytoene epoxidation. Plant Physiol 54:916–920
Beste CE, Schreiber MM (1972) RNA synthesis as a basis for EPTC and 2,4-D antagonism. Weed Sci 20:8–11
Bevenue A (1976) The "bioconcentration" aspects of DDT in the environment. Residue Rev 61:37–112
Beyer EM, Johnson AL, Sweetser PB (1976) A new class of synthetic auxin transport inhibitors. Plant Physiol 57:839–841
Biggs DC, Rowland RG, Wurster CF (1979) Effects of trichloroethylene hexachlorobenzene and polychlorinated biphenyls on the growth and cell size of marine phytoplankton. Bull Environ Contam Toxicol 21:196–201
Birmingham BC, Colman B (1977) The effect of two organophosphate insecticides on the growth of freshwater algae. Can J Bot 55:1453–1456
Blair AM, Parker C, Kassasian L (1976) Herbicide protectants and antidotes – a review. PANS 22:65–74
Böger P, Kunert KJ (1978) Phytotoxic action of paraquat on the photosynthetic apparatus. Z. Naturforsch 33c:688–694
Böger P, Kunert KJ (1979) Differential effects of herbicides upon trypsin-treated chloroplasts. Z Naturforsch 34c:1015–1025
Böger P, Schlue U (1976) Long-term effects of herbicides on the photosynthetic apparatus. I. Influence of diuron, triazines and pyridazinones. Weed Res 16:149–154
Böger P, Sandmann G, Miller R (1981) Herbicide resistance in a mutant of the microalga *Bumilleriopsis filiformis*. Photos Res 2:61–74
Boehler-Kohler BA, Schopf M, Böger P (1981) The effect of paraquat on membrane lipids. In: Akoyunoglou GA (ed) Photosynthesis VI, Balaban International Science Services, Philadelphia
Böhme H, Reimer S, Trebst A (1971) The effect of dibromothymoquinone (DBMIB) an antagonist of plastoquinone on non cyclic and cyclic electron flow systems in isolated chloroplasts. Z Naturforsch 26b:341–352
Böhme H, Kunert KJ, Böger P (1981) Sites of herbicidal action on photosynthesis: A fluorescence assay study. Weed Sci 29:371–375
Borghi H, Puiseux-Dao S, Bonotto S, Hoursiangou-Neubrunn D (1973) The effects of lindane on *Acetabularia mediterranea*. Protoplasma 78:99–112
Bowes GW (1972) Uptake and metabolism of 2,2-bis(p-chlorophenyl)-1,1-trichlorethane (DDT) by marine phytoplankton and its effect on growth and chloroplast electron transport. Plant Physiol 49:172–176

Bowes J, Crofts AR, Arntzen CJ (1980) Redox reactions on the reducing site of photosystem II in chloroplasts with altered herbicide binding properties. Arch Biochem Biophys 200:303–308

Brian RC, Homer RF, Stubbs J, Jones RL (1958) A new herbicide. 1:1-ethylene-2:2-dipyridylium dibromide. Nature (London) 181:446–447

Britton G (1979) Carotenoid biosynthesis – a target for herbicide activity. Z. Naturforsch 34c:979–985

Brooks GT (1974) Chlorinated insecticides Vols I, II. CRC Press, Cleveland

Brown AWA (1978) Ecology of pesticides. Wiley and Sons, New York

Bucha HC, Todd CW (1951) 3-(p-chlorophenyl)-1,1-dimethylurea – a new herbicide. Science 114:493–494

Büchel KH (1972) Mechanisms of action and structure activity relations of herbicides that inhibit photosynthesis. Pestic Sci 3:89–110

Büchel KH (ed) (1977) Pflanzenschutz und Schädlingsbekämpfung. Thieme, Stuttgart

Büchel KH, Draber W (1969) Structure-activity relationship of photosynthesis-inhibition by NH-acidic π-excessive heteroaromatica. In: Metzner H (ed) Progress in photosynthesis research Vol III. Tübingen, pp 1777–1788

Büchel KH, Draber W, Trebst A, Pistorius E (1966) Zur Hemmung photosynthetischer Reaktionen in isolierten Chloroplasten durch Herbicide des Benzimidazol-Typs und deren Struktur-Aktivitäts-Beziehung unter Berücksichtigung des Verteilungskoeffizienten und des Pk_A-Wertes. Z. Naturforsch 21b:234–254

Bugg MW, Whitmarsh J, Rieck CE, Cohen WS (1980) Inhibition of photosynthetic electron transport by diphenylether herbicides. Plant Physiol 65:47–50

Burns ER, Buchanan GA, Carter MC (1971) Inhibition of carotenoid synthesis as a mechanism of action of amitrole, dichlormate and pyriclor. Plant Physiol 47:144–148

Burrows WJ (1978) Effects of glycerol ethers on monocotyledon nucleic acid and protein biosynthesis. Pestic Biochem Physiol 8:137–145

Buschmann C, Grumbach KH, Bach TJ (1980) Herbicides which inhibit photosystem II, or produce chlorosis and their effect on production and transformation of pigments in etiolated radish seedlings (*Raphanus sativus*). Physiol Plant 49:455–458

Butler GL (1977) Algae and pesticides. Residue Rev 66:19–22

Calderbank A, Slade P (1976) Diquat and paraquat. In: Kearney PC, Kaufman DD (eds) Herbicides Vol II. Dekker, New York, pp 501–540

Calvayrac R, Ledoigt G, Laval-Martin D (1979) Analysis and characterization of DCMU-resistant *Euglena gracilis*. III. Thylakoid modifications and dark "recovery" of photosynthesis. Planta 145:259–267

Carles PM, Assche van CJ (1980) Interactions between PS II inhibitors and a simulated target: A molecular orbital approach. In: Akoyunoglou GA (ed) Photosynthesis II. Balaban International Science Services, Philadelphia

Carringer RD, Riek CE, Bush LP (1978) Effect of R-25788 on EPTC metabolism in corn (*Zea mays*). Weed Sci 26:167–171

Cartwright PM (1976) General growth responses of plants. In: Audus LJ (ed) Herbicides Vol I. Academic Press, London New York, pp 55–82

Casida JE (1973) Pyrethrum, the natural insecticide. Academic Press, London New York

Casida JE, Kimmel EC, Lay MM, Ohkawa H, Rodebush JE, Gray RA, Teseng CK, Tilles H (1975) Thiocarbamate sulfoxide herbicides. In: Coulston F, Korte F (eds) Environmental quality and safety Vol III. Pesticides. Thieme, Stuttgart, pp 675–679

Chen LG, Switzer CM, Fletcher RA (1972) Nucleic acid and protein changes induced by auxin-like herbicides. Weed Sci 20:53–55

Cherry JH (1976) Actions on nucleic acid and protein metabolism. In: Audus LJ (ed) Herbicides Vol I. Academic Press, London New York, pp 525–546

Coble HD, Slife FW (1971) Root disfunction in honeyvine milkweed caused by 2,4 D. Weed Sci 19:1–3

Cole DR, Plapp FW (1974) Inhibition of growth and photosynthesis in *Chlorella pyrenoidosa* by a polychlorinated biphenyl and several insecticides. Environ Entomol 3:217–220

Conrad SG, Radosevich SR (1979) Ecological fitness of *Senecio vulgaris* and *Amaranthus retroflexus* biotypes susceptible or resistant to atrazine. J Appl Ecol 16:171–177
Corbett JR (1974) The biochemical mode of action of pesticides. Academic Press, London New York
Coulston F, Korte F (eds) (1976) Environmental quality and safety Vol V. Global aspects of chemistry, toxicology and technology as applied to the environment. Thieme, Stuttgart; Academic Press, London New York
Cramer HH (1967) Pflanzenschutz und Welternte. Pflanzenschutz-Nachr 20:1–523
Cross JW, Briggs WR (1978) Properties of a solubilized microsomal auxin-binding protein from coleoptiles and primary leaves of *Zea mays*. Plant Physiol 62:152–157
Croze E, Kelly M, Horton P (1979) Loss of sensitivity to diuron after trypsin digestion of chloroplast photosystem II particles. FEBS Lett 103:22–26
Darr S, Souza Machado V, Arntzen CJ (1981) Uniparental inheritance of a chloroplast photosystem II polypeptide controlling herbicide binding. Biochim Biophys Acta 634:219–228
DeBach P (1974) Biological control by natural enemies. Cambridge Univ Press, Cambridge
Delaney ME, Owen WJ, Rogers LJ (1977) Accumulation of sugars and polysaccharide accompanying an inhibition of the light reaction in photosynthesis. J Exp Bot 28:1153–1162
Delaney ME, Jones M, Rogers LJ (1978) Effect of 1,1,1-trichloro-2,2-bis(p-chlorophenyl)ethane on cytochromes of the photosynthetic electron transport system. J Exp Bot 29:25–30
Derby SB, Ruber E (1970) Primary production: Depression of oxygen evolution in algae cultures by organophosphorus insecticides. Bull Environ Contam Toxicol 5:553–558
Devlin RM, Kisiel MJ, Kostusiak AS (1979) Influence of R-40244 on pigment content of wheat and corn. Weed Res 19:59–61
Dodge AD, Lawes GB (1972) Some effects of the herbicide haloxydine on the structure and development of chloroplasts. Acta Bot 36:315–323
Dodge AD, Harris N, Baldwin BC (1970) The mode of action of paraquat and diquat. Biochem J 118:43–44P
Dohrmann U, Hertel R, Pesci P, Cocucci SM, Marré E, Randazzo G, Ballio A (1977) Localization of in vitro binding of the fungal toxin fusicoccin to plasma-membrane rich fractions from corn coleoptiles. Plant Sci Lett 9:291–299
Dohrmann U, Hertel R, Kowalik H (1978) Properties of auxin binding sites in different subcellular fractions from maize coleoptiles. Planta 140:97–106
Draber W (1977) Natürliche und synthetische Pflanzenwachstumsregulatoren. In: Wegler R (ed) Chemie der Pflanzenschutz- und Schädlingsbekämpfungsmittel Vol IV. Springer, Berlin Heidelberg New York, pp 1–63
Draber W (1979) Neuere Ergebnisse zur Struktur-Aktivitäts-Beziehung. Z Naturforsch 34c:973–978
Draber W, Fedtke C (1979) Herbicide interaction with plant biochemical systems. In: Geissbühler H (ed) Advances in pesticide sciences, part 3. Pergamon, Oxford, pp 475–486
Draber W, Büchel KH, Dickoré K, Trebst A, Pistorius E (1969) Structure-activity correlations of 1,2,4-triazinones, a new group of photosynthesis inhibitors. In: Metzner H (ed) Progress in photosynthesis research. Tübingen, pp 1789–1795
Duke WB, Slife FW, Hanson JB, Butler HS (1975) An investigation on the mechanism of action of propachlor. Weed Sci 23:142–147
Dustin P (1978) Microtubules, Springer, Berlin Heidelberg New York
Ebert E (1982) The role of plant surface structures on the uptake and herbicidal action of metolachlor in *Sorghum* and the activity of metolachlor safener CGA 43089. Weed Res 22:305–311
Ebert E, Dumford SW (1976) Effects of the triazine herbicides on the physiology of plants. In: Gunther FA, Gunther JA (eds) Residue reviews Vol 65. Springer, Berlin New York Heidelberg
Eder FA (1979) Pyridazinones, their influence on the biosynthesis of carotenoids and the metabolism of lipids in plants. Z Naturforsch 34c:1052–1054

Edwards CA (ed) (1973) Evironmental pollution by pesticides. Plenum, New York London
Ellis JF, Peek JW, Boehle J, Müller G (1980) Effectiveness of a new safener for protecting Sorghum (*Sorghum bicolor*) from metolachlor injury. Weed Sci 28:1–5
Elstner EF (1980) Mechanisms and products of photosynthetic oxygen activation. In: Bannister JV, Hill HAO (eds) Chemical and biochemical aspects of superoxide and superoxide dismutase. Developments in biochemistry Vol XIA. Elsevier/North-Holland, Amsterdam, pp 390–401
Elstner EF (1982) Oxygen activation and oxygen toxicity. Annu Rev Plant Physiol 33:73–96
Elstner EF, Osswald W (1980) Chlorophyll photobleaching and ethane production in dichlorophenyldimethylurea- (DCMU) or paraquat-treated *Euglena gracilis* cells. Z Naturforsch 35c:129–135
Elstner EF, Pils J (1979) Ethane formation and chlorophyll bleaching in DCMU treated *Euglena gracilis* cells and isolated spinach cloroplast lamellae. Z Naturforsch 34c:1040–1043
Elstner EF, Lengfelder E, Kwiatkowski G (1980a) Paraquat-catalyzed photodestructions in subchloroplast particles are independent on photosynthetic electron transport. Z Naturforsch 35c:303–307
Elstner EF, Fischer HP, Osswald W, Kwiatkowski G (1980b) Superoxide- and ethane-formation in subchloroplast particles: Catalysis by pyridinium derivatives. Z Naturforsch 35c:770–775
Ewald WG, French JE, Champ MA (1976) Toxicity of polychlorinated biphenyls to *Euglena gracilis*: Cell population growth, carbon fixation, chlorophyll level, oxygen consumption and protein and nucleic acid synthesis. Bull Environ Contam Toxicol 16:71–80
Ezra G, Gressel J (1982) Rapid interactions between a thiocarbamate herbicide and dichloroacetamide protectant in maize cultures. Pestic Biochem Physiol 17:48–58
Falkner R, Simonis W (1982) Polychlorierte Biphenyle im Lebensraum Wasser. In: Elster HJ, Ohle W (eds) Ergebn. Limnol. Beih. 17. Schweizerbart, Stuttgart
FAO Agricultural Series 10 (1980) The state of food and agriculture 1979, Rome
Farrington JA, Ebert M, Land EJ, Fletcher K (1973) Bipyridylium quaternary salts and related compounds. V. Pulse radiolysis studies of the reaction of paraquat radical with oxygen. Implications for the mode of action of bipyridyl herbicides. Biochim Biophys Acta 314:372–381
Faulkner JS (1982) Breeding herbicide tolerant crop cultivars by conventional methods. In: LeBaron H, Gressel J (eds) Herbicide resistance in plants. Wiley and Sons, New York, pp 235–255
Fedtke C (1973) Effects of the herbicide methabenzthiazuron on the physiology of wheat plants. Pestic Sci 4:653–664
Fedtke C (1974) Influence of methabenzthiazuron on ATP-level and protein synthesis in wheat. Pestic Biochem Physiol 4:386–392
Fedtke C (1979) Plant physiological adaptation induced by low rates of photosynthesis. Z Naturforsch 34c:932–935
Fedtke C (1982) Biochemistry and physiology of herbicide action. Springer, Berlin Heidelberg New York
Feierabend J, Schubert B (1978) Comparative investigation of the action of several chlorosis-inducing herbicides on the biogenesis of chloroplasts and leaf microbodies. Plant Physiol 61:1017–1022
Feierabend J, Schulz U, Kemmerich P, Lowitz T (1979) On the action of chlorosis-inducing herbicides in leaves. Z Naturforsch 34c:1036–1039
Finlayson DG, MacCarthy HR (1973) Pesticide residues in plants. In: Edwards CA (ed) Environmental pollution by pesticides. Plenum, New York, pp 57–86
Fischer NS, Wurster CF (1973) Individual and combined effects of temperature and polychlorinated biphenyls on the growth of three species of phytoplankton. Environ Pollut 5:205–212
Flint ML, Bosch van der R (1981) Introduction to integrated pest management. Plenum, New York

Foote CS (1979) Detection of singulet oxygen in complex systems. In: Caughey WS (ed) Biochemical and clinical aspects of oxygen. Academic Press, London New York, pp 603–626

Fox JE, Erion JL (1977) Cytokinin binding proteins in higher plants. In: Pilet PE (ed) Plant growth regulation. Springer, Berlin Heidelberg New York, pp 139–146

Foy CL (1976) Picloram and related compounds. In: Kearney PC, Kaufman DD (eds) Herbicides Vols I, II. Dekker, New York, pp 777–813

Franz JM, Krieg A (1976) Biologische Schädlingsbekämpfung, 2nd edn. Parey, Berlin

Frear DS (1976) The benzoic acid herbicides. In: Kearney PC, Kaufman DD (eds) Herbicides Vol II. Dekker, New York, pp 541–607

Fridovich I (1975) Superoxide dismutase. Annu Rev Biochem 44:147–159

Frosch S, Jabben M, Bergfeld R, Kleinig H, Mohr H (1979) Inhibition of carotenoid biosynthesis by the herbicide SAN 9789 and its consequences for the action of phytochrome on plastogenesis. Planta 145:497–505

Fryer JD, Matsunaka S (eds) (1977) Integrated control of weeds. Univ Tokyo Press, Tokyo

Gabbott PA (1969) Inhibition of photoreactions in isolated chloroplasts by 2-azido-4-alkylamino-6-alkylamino-s-triazines. In: Metzner H (ed) Progress in photosynthesis research Vol III. Tübingen, pp 1712–1727

Gast A, Knuesli E, Gysin H (1955) Über Pflanzenwachstumsregulatoren: Chlorazin, eine phytotoxisch wirksame Substanz. Experientia 11:107–108

Geissbühler H (ed) (1978) Advances in pesticide science Vols I–III. Pergamon, New York

Giannopolitis GN, Ayers GS (1978) Enhancement of chloroplast photooxidations with photosynthesis-inhibiting herbicides and protection with NADH or NADPH. Weed Sci 26:440–443

Good NE (1977) Uncoupling of electron transport from phosphorylation in chloroplasts. In: Trebst A, Avron M (eds) Photosynthesis I. Encyclopedia of plant physiology new ser Vol 5. Springer, Berlin Heidelberg New York, pp 426–456

Govindjee (1975) Bioenergetics of photosynthesis. Academic Press, London New York

Govindjee, Papageorgiou G (1971) Chlorophyll fluorescence and photosynthesis: fluorescence transients. In: Giese AC (ed) Photobiology Vol VI. Academic Press, London New York, pp 1–46

Graebe JE, Ropers HJ (1978) Gibberellins. In: Letham DS, Goodwin PB, Higgings TJV (eds) Phytohormones and related compounds. A comprehensive treatise Vol I. Elsevier/North-Holland, Amsterdam, pp 107–204

Gräser H (1977) Biochemie und Physiologie der Phytoeffektoren. Chemie, Weinheim

Greichus YA, Greichus A, Draayder HA, Marshall B (1978) Insecticides, polychlorinated biphenyls and metals in african lake ecosystems. II. Lake McIlwaine, Rhodesia. Bull Environ Contam Toxicol 19:444–453

Gressel J, Segel LA (1978) The paucity of plants evolving genetic resistance to herbicides: possible reasons and implications. J Theor Biol 75:1–23

Gressel J, Segel LA (1982) Interrelating factors controlling the rate of appearance of resistance: The outlook for the future. In: LeBaron H, Gressel J (eds) Herbicide resistance in plants. Wiley and Sons, New York, pp 325–347

Gressel J, Ammon HU, Fogelfors H, Gasquez J, Kay QON, Kees H (1982) Discovery and distribution of herbicide resistant weeds outside North America. In: LeBaron H, Gressel J (eds) Herbicide resistance in plants. Wiley and Sons, New York, pp 31–55

Grewe F (1977) Fungi und Fungizide. In: Wegler R (ed) Chemie der Pflanzenschutz- und Schädlingsbekämpfungsmittel Vol. 4. Springer, Berlin Heidelberg New York, pp 68–115

Gromet-Elhanan Z (1968) The inhibition of photoreactions of chloroplasts by ioxynil. Biochem Biophys Res Commun 30:28–31

Gruenhagen RD, Moreland DE (1971) Effects of herbicides on ATP levels in excised soybean hypocotyls. Weed Sci 19:319–323

Grumbach KH, Bach TJ (1979) The effect of PS II herbicides, amitrol and SAN 6706 on the activity of 3-hydroxy-3-methylglutaryl-coenzyme-A-reductase and the incorporation of (2-^{14}C) Acetate and (2-^{3}H) Mevalonate into chloroplast pigments of radish seedlings. Z Naturforsch 34c:941–943

Hamner CL, Tukey HB (1944) The herbicidal action of 2,4-dichlorophenoxyacetic acid and 2,4,5-trichlorophenoxyacetic acid on bindweed. Science 100:154–155
Hansch C (1969) Theoretical considerations of the structure-activity relationship in photosynthesis inhibitors. In: Metzner H (ed) Progress in photosynthesis research Vol III. Tübingen, pp 1685–1692
Haque R, Freed VH (eds) (1975) Environmental dynamics of pesticides Vol VI. Plenum, New York London
Harbour JR, Bolton JR (1978) The involvement of the hydroxyl radical in the destructive photooxidation of chlorophylls in vivo and in vitro. Photochem Photobiol 28:231–234
Harding LW, Phillips JH (1978) Polychorinated biphenyls: Transfer from microparticulates to marine phytoplankton and the effects on photosynthesis. Science 202:1189–1192
Hardy RWF (1979) Chemical plant growth regulation in world agriculture. In: Scott TK (ed) Plant regulation and world agriculture. Plenum, New York London, pp 165–206
Harris N, Dodge AD (1972) The effect of paraquat on flax cotyledon leaves: Physiological and biochemical changes. Planta 104:210–219
Hatilos KK, Penner D (1980) Potential antidotes against buthidazole injury to corn. Weed Sci 28:273–276
Hauska G, Trebst A, Kötter C, Schulz H (1975) 1,2,3-Thiadiazoylphenylureas, new inhibitors of photosynthetic and respiratory energy conservation. Z Naturforsch 30c:505–510
Hayes RM, Yeargan KV, Witt WW, Raney HG (1979) Interaction of selected insecticide herbicide combinations on soybeans (Glycine max). Weed Sci 27:51–54
Hepler PK, Jackson WT (1969) Isopropyl N-phenylcarbamate affects spindle microtubule orientation in dividing endosperm cells of *Haemanthus katherinae* Baker. J Cell Sci 5:727–743
Herbicide Handbook of the Weed Science Society of America (1979) 4th edn. Champaign, Illinois
Herczeg T, Lehoczki E, Szalay L (1979) The prompt effect of pyridazinone herbicides on the primary processes of photosynthesis. FEBS Lett 108:226–228
Hertel R, Thomson KS, Russo, VEA (1972) In-vitro auxin binding to particulate cell fractions from corn coleoptiles. Planta 107:325–340
Hess FD (1980) A *Chlamydomonas* algal bioassay for detecting growth inhibitor herbicides. Weed Sci 28:515–520
Hess FD, Bayer DE (1974) The effect of trifluralin on the ultrastructure of dividing cells of the root meristem of cotton (*Gossypium hirsutum* L. 'acata 4–42') J Cell Sci 15:429–441
Hess FD, Bayer DE (1977) Binding of the herbicide trifluralin to *Chlamydomonas* flagellar tubulin. J Cell Sci 24:351–360
Higgins TJV, Jacobsen JV (1978) Phytohormones and subcellular structural modification. In: Letham DS, Goodwin PB, Higgins TJV (eds) Phytohormones and related compounds. A comprehensive treatise Vol I. Elsevier/North-Holland, Amsterdam, pp 419–465
Hilton BD, O'Brien RD (1970) Antagonism by DDT of the effect of valinomycin on a synthetic membrane. Science 168:841–843
Hilton BD, Bradkowski TA, Yamada M, Narahashi T, O'Brien RD (1973) The effects of DDT-analogues upon potassium conductance in synthetic membranes. Pestic Biochem Physiol 3:14–19
Hilton JL, Monaco TJ, Moreland DE, Gentner WA (1964) Mode of action of substituted uracil herbicides. Weeds 12:129–131
Hilton JL, Scharen AL, St. John JB, Moreland DE, Norris KH (1969) Modes of action of pyridazinone herbicides. Weed Sci 17:541–547
Hoffmann OL (1978a) Herbicide antidotes: concept to crop. Chemtech 8:488–492
Hoffmann OL (1978b) Herbicide antidotes: from concept to practice. In: Casida JE, Pallos FM (eds) Chemistry and action of herbicide antidotes. Academic Press, London New York, pp 1–13

Holly K (1976) Selectivity in relation to formulation and application methods. In: Audus LJ (ed) Herbicides Vol II. Academic Press, London New York, pp 249–277
Horowitz M (1976) Application of bioassay techniques to herbicide investigations. Weed Res 16:209–215
Hudson JP (1976) Future roles for growth regulators. Outlook Agric 9:95–98
Huffaker CB, Messenger PS (1976) Theory and practice of biological control. Academic Press, London New York
Hurlbert SH (1975) Secondary effects of pesticides on aquatic ecosystems. Residue Rev 57:81–148
Hutzinger O, Safe ASW, Zitko V (1974) The chemistry of PCBs. CRC-Press Cleveland, Ohio
Izawa S, Good NE (1977) Inhibition of photosynthetic electron transport and photophosphorylation. In: Trebst A, Avron M (eds) Photosynthesis I. Encyclopedia of plant physiology new ser Vol 5. Springer, Berlin Heidelberg New York
Jackson WT, Stetler DA (1973) Regulation of mitosis. IV. An in vitro and ultrastructural study of effects of trifluralin. Can J Bot 51:1513–1518
Jacobsen JV (1977) Regulation of ribonucleic acid metabolism by plant hormones. Annu Rev Plant Physiol 28:537–564
Janatkova H, Wildner GF (1982) Isolation and characterization of metribuzin resistant *Chlamydomonas rheinhardii* cells. Biochim Biophys Acta 682:227–233
Jaworski EG (1962) Mode of action of N-phosphonomethylglycine: Inhibition of aromatic amino acid biosynthesis. J Agric Food Chem 20:1195–1198
Jeanne N (1979) Effect of lindane on division cell cycle and biosynthesis in 2 unicellular algae. Can J Bot 57:1464–1472
Jensen KIN, Bandeen JD, Souza-Machado V (1977) Studies on the differential tolerance of two lambsquarters selections to S-triazine herbicides. Can J Plant Sci 57:1169–1177
Jones RL (1973) Gibberellins: Their physiological role. Annu Rev Plant Physiol 24:571–598
Kalmaz EV, Kalmaz GD (1979) Transport distribution and toxic effects of polychlorinated biphenyls in ecosystems: a review. Ecol Model 6:223–251
Katekar GF (1979) Auxins: On the nature of the receptor site and molecular requirements for auxin activity. Phytochemistry 18:223–233
Katekar GF, Geissler AE (1977) Auxin transport inhibitors, III. Chemical requirements of a class of auxin transport inhibitors. Plant Physiol 60:826–829
Kearney PC, Kaufman DD (eds) (1976) Herbicides. Chemistry, degradation, and mode of action Vols I, II. Dekker, New York
Kefeli VJ (1978) Natural plant growth inhibitors and phytohormones. Junk, The Hague
Kende H, Gardner G (1976) Hormone binding in plants. Annu Rev Plant Physiol 27:267–290
Kerr MW, Avery RJ (1972) Studies on the mode of action of anilinopropionamide herbicides. Biochem J 128:132–138
Kerr MW, Wain RL (1964a) The uncoupling of oxidative phosphorylation in pea shoot mitochondria by 3,5-diiodo-4-hydroxybenzonitrile (ioxynil) and related compounds. Ann Appl Bot 54:441–446
Kerr MW, Wain RL (1964b) Inhibition of the ferricyanide-Hill reaction of isolated bean leaf chloroplasts by 3,5-diiodo-4-hydroxybenzonitrile (ioxynil) and related compounds. Ann Appl Bot 54:447–450
Khan MU, Lem NW, Chandorkar KR, Williams JP (1979) Effects of substituted pyridazinones SAN 6706, SAN 9774 on glycero lipids and their associated fatty acids in the leaves of *Vicia faba* and *Hordeum vulgare*. Plant Physiol 64:300–305
Kiermayer O, Fedtke C (1977) Strong anti-microtubule action of amiprophosmethyl (AMP) in *Micrasterias*. Protoplasma 92:163–166
Kiermayer O, Meindl U (1979) Antimicrotubule action of various substances in *Micrasterias denticulata*. Eur J Cell Biol 20:130–131
Kirkwood RC (1976) Action on respiration and intermediary metabolism. In: Audus LJ (ed) Herbicides Vol I. Academic Press, London New York, pp 443–492
Kitchen LM, Witt WW, Rieck CE (1981) Inhibition of δ-aminolevulinic acid synthesis by glyphosate. Weed Sci 29:571–577

Kleudgen HK (1978) The influence of the herbicide methabenzthiazuron on the synthesis of plastidic lipids in *Hordeum vulgare* seedlings. Pestic Biochem Physiol 9:57–60
Kleudgen HK (1979) Changes in composition of chlorophylls, carotenoids, and prenylquinones in green seedlings of *Hordeum* and *Raphanus* induced by the herbicide San 6706 – an effect possibly antagonistic to phytochrome action. Pestic Biochem Physiol 12:231–238
Klockare B, Axelsson L, Ryberg H, Sandelius AS, Widell KO (1981) The function of carotenoids during chloroplast development. I. Effects of the herbicide SAN-9789 on chlorophyll synthesis and plastid ultrastructure. In: Akoyunoglou GA (ed) Photosynthesis V, Balaban International Science Services, Philadelphia, pp 277–284
Knittel H, Behrendt S, Schott PE (1981) Einfluß eines Wachstumsregulators auf Wachstum und Ertrag von Wintergerste bei unterschiedlichem Lagerdruck. Z Acker Pflanzenbau 150:50–61
Kopecek K, Füller F, Ratzmann W, Simonis W (1975) Lichtabhängige Insektizidwirkung auf einzellige Algen. Ber Dtsch Bot Ges 88:269–281
Krämer W (1977) Wirkstoffe gegen Pflanzenkrankheiten (Fungizide und Bakterizide). In: Büchel KH (ed) Pflanzenschutz und Schädlingsbekämpfung. Thieme, Stuttgart, pp 111–154
Kratky BA, Warren GF (1971) A rapid bioassay for photosynthetic and respiratory inhibitors. Weed Sci 19:658–661
Krinsky NJ (1966) The role of carotenoid pigments as protective agents against photosensitized oxidations in chloroplasts. In: Goodwin TW (ed) Biochemistry of chloroplasts Vol I. Academic Press, London New York, pp 423–430
Krinsky NJ (1977) Singlet oxygen in biological systems. Trends Biochem Sci 2:35–38
Krinsky NJ (1978) Non-photosynthetic functions of carotenoids. Philos Trans R Soc London Ser B 284:581–590
Krishnakumari MK (1977) Sensitivity of the alga *Scenedesmus acutus* to some pesticides. Life Sci 20:1525–1532
Kümmel HW, Grimme LH (1975) The inhibition of carotenoid biosynthesis in green algae bei SAN 6706: accumulation of phytoene and phytofluene in *Chlorella fusca*. Z Naturforsch 30c:333–336
Kunert KJ, Böger P (1978) Action of EMD-IT 5914 on chloroplasts. Weed Sci 26:292–296
Kunert KJ, Böger P (1979) Influence of bleaching herbicides on chlorophyll and carotenoids. Z Naturforsch 34c:1047–1051
Kunert KJ, Böger P (1981) The bleaching effect of the diphenyl ether oxyfluorfen. Weed Sci 29:169–173
Kurth H (1975) Chemische Unkrautbekämpfung. VEB Fischer, Jena
Laasch H, Pfister K, Urbach W (1981) Comparative binding of photosystem II-herbicides to isolated thylakoid membranes and intact algae. Z Naturforsch 36c:1041–1049
Laasch H, Pfister K, Urbach W (1982) High and low affinity binding of photosystem II-herbicides to isolated chloroplast membranes and intact algal cells. Z Naturforsch 37c:620–631
Lai EK, Fong KL, McCay PB (1978) Studies on the properties of the singlet oxygen like factor produced during lipid peroxidation. Biochim Biophys Acta 528:497–506
Lakowicz JR, Hogen D (1980) Chlorinated hydrocarbon cell membrane interactions studied by the fluorescence quenching of carbazole labeled phospholipids: Probe synthesis and characterization of the quenching methodology. Chem Phys Lipids 26:1–40
Lakowicz JR, Hogen D, Omann G (1977) Diffusion and partitioning of a pesticide lindane into phosphatidylcholine bilayers. A new fluorescence quenching method to study chlorinated hydrocarbon membrane interactions. Biochim Biophys Acta 471:401–411
Lambert R, Kunert KJ, Böger P (1979) On the phytotoxic mode of action of nitrofen. Pest Biochem Physiol 11:267–274
Larsson CM, Tillberg JE (1975) Effects of the commercial polychlorinated biphenyl mixture Aroclor 1242 on growth, variability, phosphate uptake and oxygen evolution in *Scenedesmus*. Plant Physiol 33:256–260

Laval-Martin D, Dubertret G, Calvayrac R (1977) Photosynthetic properties of a DCMU resistant strain of *Euglena gracilis*. Plant Sci Lett 10:185–195
Lavorel J, Etienne AL (1977) In vivo chlorophyll fluorescence. In: Barber J (ed) Primary processes of photosynthesis. Elsevier, Amsterdam, pp 203–268
Lay MM, Casida JE (1976) Dichloroacetamide antidotes enhance thiocarbamate sulfoxide detoxification by elevating corn root glutathione content and glutathione S-transferase activity. Pestic Biochem Physiol 6:442–456
Lay MM, Casida JE (1978) Involvement of glutathione and glutathione S-transferase in the action of dichloroacetamide antidotes for thiocarbamate herbicides. In: Pallos FM, Casida JE (eds) Chemistry and action of herbicide antidotes. Academic Press, London New York, pp 151–160
Leavitt JRC, Penner D (1978) Protection of corn (*Zea mays*) from acetanilide herbicide injury with the antidote R-25788. Weed Sci 26:653–659
LeBaron H, Gressel J (eds) (1982) Herbicide resistance in plants. Wiley and Sons, New York
Lee SS, Fang SC (1972) Association of monuron in chloroplasts in relation to inhibition of Hill Reaction. Phytochemistry 11:2693–2699
Lembi CA, Morre DJ, Thomson KS, Hertel R (1971) N-1-Naphthylphthalamic-acid-binding activity of a plasma membrane-rich fraction from maize coleoptiles. Planta 99:37–45
Letham DS (1978a) Cytokinins. In: Letham DS, Goodwin PB, Higgins TJV (eds) Phytohormones and related compounds. A comprehensive treatise Vol I. Elsevier/North-Holland, Amsterdam, pp 205–263
Letham DS (1978b) Naturally-occurring plant growth regulators other than the principal hormones of higher plants. In: Letham DS, Goodwin PB, Higgins TJV (eds) Phytohormones and related compounds. A comprehensive treatise Vol I. Elsevier/North-Holland, Amsterdam, pp 349–417
Letham DS, Goodwin PB, Higgins TJV (eds) (1978) Phytohormones and related compounds. A comprehensive treatise Vols I, II. Elsevier/North-Holland, Amsterdam
Lichtenthaler HK (1979) Effect of biocides on the development of the photosynthetic apparatus of radish seedlings grown under strong and weak light conditions. Z Naturforsch 34c:936–940
Lichtenthaler HK, Kleudgen HK (1977) Effect of the herbicide SAN 6706 on biosynthesis pigments and prenylquinones in *Raphanus* and *Hordeum* seedlings. Z Naturforsch 32c:236–240
Lichtenthaler HK, Burkard G, Grumbach KH, Meier D (1980) Physiological effects of photosystem II-herbicides on the development of the photosynthetic apparatus. Photos Res 1:29–43
Lieberman M (1979) Biosynthesis and action of ethylene. Annu Rev Plant Physiol 30:533–591
Lien S, McBridge JC, Togasaki RK, San Pietro A (1977) A comparative study of photosystem II specific inhibitors: the differential action on a DCMU resistant mutant strain of *Chlamydomonas reinhardii*. Special issue. Plant Cell Physiol 3:243–256
Lignowski EM, Scott EG (1972) Effect of trifluralin on mitosis. Weed Sci 20:267–270
Linck AJ (1976) Effects on the cytology and fine structure of plant cells. In: Audus LJ (ed) Herbicides Vol I. Academic Press, London New York, pp 83–125
Loos MA (1975) Phenoxyalkanoic acids. In: Kearney PC, Kaufman DD (eds) Herbicides Vol I. Dekker, New York, pp 1–128
MacFarlane RB, Glooschenko WA, Harriss RC (1972) The interaction of light intensity and DDT concentration upon the marine diatom *Nitzschia delicatissima* Cleve. Hydrobiologia 39:373–382
Mann JD, Pu M (1968) Inhibition of lipid synthesis by certain herbicides. Weed Sci 16:197–198
Mann JD, Jordan LS, Day BE (1965) A survey of herbicides for their effect upon protein synthesis. Plant Physiol 40:840–843
Martin H, Worthing CR (1976) Insecticide and fungicide handbook. Blackwell, Oxford
Matsunaka S (1976) Diphenylethers. In: Kearney PC, Kaufman DD (eds) Herbicides. Dekker, New York, pp 709–739

Mattoo AK, Pick U, Hoffmann-Falk H, Edelmann M (1981) The rapidly-metabolized, 32,000-Dalton polypeptide of the chloroplast is the proteinaceous shield regulating photosystem II electron transport and mediating diuron herbicide sensitivity. Proc Natl Acad Sci USA 78:1572–1576

McCarty RE (1977) Energy transfer inhibitors of photophosphorylation in chloroplasts. In: Trebst A, Avron M (eds) Photosynthesis I. Encyclopedia of plant physiology new ser Vol 5. Springer, Berlin Heidelberg New York, pp 437–447

McEwen FL (1978) Food production – the challenge for pesticides. BioScience 28:773–777

McEwen FL, Stephenson GR (1979) The use and significance of pesticides in the environment. Wiley and Sons, New York

Mead JF (1976) Free radical mechanisms of lipid damage and consequences for cellular membranes. In: Pryar WA (ed) Free radicals in biology Vol I. Academic Press, London New York, pp 51–68

Meier D, Lichtenthaler HK, Burkard G (1980) Change of chloroplast ultrastructure in radish seedlings under the influence of the photosystem II-herbicide bentazon. Z Naturforsch 35c:656–664

Melhado LL, Jones AM, Leonard NJ, Vanderhoef LN (1981) Azido auxins: Synthesis and biological activity of fluorescent photoaffinity labeling agents. Plant Physiol 68:469–475

Mellis JM, Pillai P, Davis DE, Truelove B (1982) Metolachlor and alachlor effects on membrane permeability and lipid synthesis. Weed Sci 30:399–404

Menzel DW, Anderson J, Randtke A (1970) Marine phytoplankton vary in their response to chlorinated hydrocarbons. Science 167:1724–1726

Meredith CP, Carlson PS (1982) Herbicide resistance in plant cell cultures. In: LeBaron H, Gressel J (eds) Herbicide resistance in plants. Wiley and Sons, New York, pp 275–291

Milborrow BV (1978) Abscisic acid. In: Letham DS, Goodwin PB, Higgins TJV (eds) Phytohormones and related compounds. A comprehensive treatise Vol I. Elsevier/North-Holland, Amsterdam, pp 295–347

Mine A, Matsunaka S (1975) Mode of action of bentazon: Effect on photosynthesis. Pestic Biochem Physiol 5:444–450

Moreland DE (1969) Inhibitors of chloroplast electron transport: Structure-activity relations. In: Metzner H (ed) Progress in photosynthesis research Vol III. Tübingen, pp 1693–1711

Moreland DE (1980) Mechanism of action of herbicides. Annu Rev Plant Physiol 31:597–638

Moreland DE (1981) Interaction of perfluidone with mitochondrial, thylakoid and liposome membranes. Pestic Physiol Biochem 15:21–31

Moreland DE, Hilton JL (1976) Actions on photosynthetic systems. In: Audus LJ (ed) Herbicides Vol I. Academic Press, London New York, pp 493–523

Moreland DE, Malhotra SS, Gruenhagen RD, Shokraii EH (1969) Effects of herbicides on RNA and protein synthesis. Weed Sci 17:556–563

Moreland DE, Blackmon WJ, Todd HG, Farmer FS (1970) Effects of diphenylether herbicides on reactions of mitochondria and chloroplasts. Weed Sci 18:636–642

Morgan PW (1976) Effects on ethylene physiology. In: Audus LJ (ed) Herbicides Vol I. Academic Press, London New York, pp 255–280

Morgan PW (1980) Synthetic growth regulators: Potential for development. Bot Gaz 141:337–346

Morrod RS (1976) Effects on plant cell membrane structure and function. In: Audus LJ (ed) Herbicides Vol I. Academic Press, London New York, pp 281–340

Muschinek Gy, Garab GyI, Mustardy LA, Faludi-Daniel A (1979) The mechanism of linuron phytotoxicity in maize. Weed Res 19:101–107

Muslih RK, Linscott DL (1977) Regulation of lipid synthesis in soybeans by two benzoic acid herbicides. Plant Physiol 60:730–735

Naumann K (1981) Chemie der synthetischen Pyrethroid-Insektizide. In: Wegler R (ed) Chemie der Pflanzenschutz- und Schädlingsbekämpfungsmittel Vol VII. Springer, Berlin Heidelberg New York

Naylor AW (1976) Herbicide metabolism in plants. In: Audus LJ (ed) Herbicides Vol I. Academic Press, London New York, pp 397–426
Neudorf S, Khan MAQ (1975) Pick-up and metabolism of DDT, dieldrin and photodieldrin by a fresh water alga (*Ankistrodesmus amalloides*) and a microcrustacean (*Daphnia pulex*). Bull Environ Contam Toxicol 13:443–450
Nickel LG (1982) Plant growth regulators. Agricultural uses. Springer, Berlin Heidelberg New York
Nooden LD (1970) Metabolism and binding of ^{14}C-maleic hydrazide. Plant Physiol 45:46–52
Nooden LD (1972) Inhibition of nucleic acid synthesis by maleic hydrazide. Plant Cell Physiol 13:609–621
Nyffeler A, Gerber HR, Hensley JR (1980) Laboratory studies on the behavior of the herbicide safener CGA-43089. Weed Sci 28:6–10
O'Brien RD, Herbert TY, Hilton BD (1977) The effect of DDT upon flat and vesicular bilayers by a fluorescent probe. Pestic Biochem Physiol 7:416–425
Oettmeier W (1979) Quantitative structure activity relationship of diphenylamines as inhibitors of photosynthetic electron transport and photophosphorylation. Z Naturforsch 34c:1024–1027
Oettmeier W, Masson K (1980) Synthesis and thylakoid membrane binding of the radioactively labeled herbicide dinoseb. Pestic Biochem Physiol 14:86–97
Oettmeier W, Reimer S, Link K (1978) Quantitative structure activity relationship of substituted benzoquinones as inhibitors of photosynthetic electron transport. Z Naturforsch 33c:695–703
Oettmeier W, Masson K, Johanningmeier U (1980) Photoaffinity labelling of the photosystem II herbicide binding protein. FEBS Lett 118:267–270
Oettmeier W, Johanningmeier U, Trebst A (1982) Inhibitors of plastquinone function as tool for identification of its binding proteins in chloroplasts. In: Trumpower BL (ed) Function of quinones in energy conserving systems. Academic Press, London New York, pp 425–441
Oorshot van JLP (1976) Effects in relation to water and carbon dioxide exchange of plants. In: Audus LJ (ed) Herbicides Vol I. Academic Press, London New York, pp 305–333
Oorshot van JLP (1979) Types of selective action by herbicides which inhibit photosynthesis. Z Naturforsch 34c:900–904
Oorshot van JLP (1981) Some effects on photosynthesis and photorespiration in bean plants by different herbicides. In: Akoyunoglou GA (ed) Photosynthesis VI. Balaban International Science Services, Philadelphia, pp 607–613
Osborne DJ (1978) Ethylene. In: Letham DS, Goodwin PB, Higgins TJV (eds) Phytohormones and related compounds. A comprehensive treatise Vol I. Elsevier/North-Holland, Amsterdam, pp 265–294
Owen WJ, Rogers LJ, Hayes JD (1975) Inhibition of Hill-reaction and photophosphorylation by 1,1,1-trichloro-2,2-bis(p-chlorophenyl) ethane. J Exp Bot 26:692–701
Owen WJ, Delaney ME, Rogers LJ (1977) Studies on the sites of inhibition of photosynthetic electron transport in barley by DDT (1,1,1-trichloro-2,2-bis(p-chlorophenyl)ethane). J Exp Bot 28:986–998
Pallet KE, Dodge AD (1979) The role of light and oxygen in the action of the photosynthetic inhibitor herbicide monuron. Z Naturforsch 34c:1058–1061
Pallos FM, Casida JE (eds) (1978) Chemistry and action of herbicide antidotes. Academic Press, London New York
Palm HL, Riggleman JD, Allison DA (1980) Worldwide review of the new cereal herbicide DPX 4189. Proc Br Crop Protect Conf, pp 1–6
Paradies I, Ebert E, Elstner EF (1981) Metolachlor (2-chloro-N-(2-ethyl-6-methylphenyl)-N-(2-methoxy-1-methylethyl)-acetamide) and the metolachlor safener CGA 43089 (α-(cyanomethoximino)-benzacetonitrile) in *Sorghum* seedlings. Correlations between morphological effects and ethylene formation. Pestic Biochem Physiol 15:209–212
Parka SJ, Soper OF (1977) The physiology and mode of action of the dinitroaniline herbicides. Weed Sci 25:79–87

Parker C (1976) Effects on the dormancy of plant organs. In: Audus LJ (ed) Herbicides Vol I. Academic Press, London New York, pp 165–190
Penner D, Ashton FM (1966) Biochemical and metabolic changes in plants induced by chlorophenoxy herbicides. Residue Rev 14:39–113
Penner D, Early RW (1972a) Action of trifluralin on chromatin activity in corn and soybean. Weed Sci 20:364–366
Penner D, Early RW (1972b) Effect of atrazine on chromatin activity in corn and soybean. Weed Sci 20:367–370
Penny P, Penny D (1978) Rapid responses to phytohormones. In: Letham DS, Goodwin PB, Higgins TJV (eds) Phytohormones and related compounds Vol II. Elsevier/North-Holland, Amsterdam, pp 537–597
Perring FH, Mellanby K (eds) (1977) Ecological effects of pesticides. Academic Press, London New York
Pfister K, Arntzen CJ (1979) The mode of action of photosystem II specific inhibitors in herbicide-resistant weed biotypes. Z Naturforsch 34c:996–1009
Pfister K, Schreiber U (1983) Different inhibitory actions of diuron- and phenol-type herbicides on the donor and acceptor site of photosystem II (in prep.)
Pfister K, Buschmann C, Lichtenthaler HK (1974) Inhibition of the photosynthetic electron transport by bentazon. In: Avron M (ed) Proc 3rd Int Congr Photosynth Res Elsevier, Amsterdam, pp 675–681
Pfister K, Radosevich SR, Arntzen CJ (1979) Modification of herbicide binding to photosystem II in two biotypes of *Senecio vulgaris* L. Plant Physiol 64:995–999
Pfister K, Steinback KE, Gardner G, Arntzen CJ (1981a) Photoaffinity labeling of an herbicide receptor protein in chloroplast membranes. Proc Natl Acad Sci USA 78:981–985
Pfister K, Steinback KE, Arntzen CJ (1981b) Characterization of the thylakoid membrane protein which binds photosystem II inhibitors. In: Akoyunoglou GA (ed) Photosynthesis VI, Balaban International Sciences Services, Philadelphia, pp 595–606
Pfister K, Watson J, Arntzen CJ (1982) Analysis of the PS II-inhibitor binding site. Weed Sci (in press)
Phillips BA, Bhagsari AS (1978) Diazosulfonates as protectants against atrazine toxicity to soybeans. In: Pallos FM, Casida JE (eds) Chemistry and action of herbicide antidotes. Academic Press, London New York, pp 21–34
Pimentel D (ed) (1981) CRC handbook of pest management in agriculture Vols I–III. CRC Press, Boca Raton
Pimentel D, Krummel J, Gallahan D, Hough J, Merrill A, Schreiner I, Vittum P, Koziol F, Back E, Yen D, Fiance S (1978) Benefits and costs of pesticide use in US food production. BioScience 28:772–784
Polya GM, Davis AW (1978) Properties of a high-affinity cytokinin-binding protein from wheat germ. Planta 139:139–147
Powers CD, Rowland RG, Wurster CF (1977) Dieldrin-induced destruction of marine algal cells with concomitant decrease in size of survivors and their progeny. Environ Pollut 12:17–25
Pratt HM, Fox JE (1978) Subunit structure and amino acid composition of a cytokinin-binding protein. Plant Physiol Suppl 61:10
Pritchard MK, Warren GF, Dilley RA (1980) Site of action of oxyfluorfen. Weed Sci 28:640–645
Puiseux-Dao S, Jeanne-Levain N, Roux F, Ribier J, Borghi H, Brun C (1977) Analyse des effects du lindane, insecticide organochloré au niveau cellulaire. Protoplasma 91:325–341
Putman AR, Penner D (1973) Pesticide interactions in higher plants. Residue Rev 50:73–110
Radosevich SR (1977) Mechanisms of atrazine resistance in Lambsquarters and Pigweed. Weed Sci 25:316–318
Radosevich SR, Holt JS (1982) Physiological responses to triazine herbicides in susceptible and resistant weed biotypes. In: LeBaron H, Gressel J (eds) Herbicide resistance in plants. Wiley and Sons, New York, pp 163–183

Radosevich SR, Devilliers OT (1976) Studies on the mechanism of s-triazine resistance in common groundsel. Weed Sci 24:229–232
Radosevich SR, Steinback KE, Arntzen CJ (1979) Effect of photosystem II inhibitors on thylakoid membranes of two common groundsel (*Senecio vulgaris*) biotypes. Weed Sci 27:216–218
Rappaport L (1980) Applications of gibberellins in agriculture. In: Skoog F (ed) Plant growth substances. Springer, Berlin Heidelberg New York, pp 377–391
Ray TB (1980) Studies on the mode of action of DPX 4189. Proc Br Crop Conf Weeds, pp 7–14
Reeves SG, Hall DO (1980) Higher plant chloroplasts and grana: General preparative procedures (excluding high carbon dioxide fixation ability chloroplasts). Methods Enzymol 69:85–94
Regitz G, Ohad I (1976) Trypsin sensitive photosynthetic activities in chloroplast membranes from *Chlamydomonas reinhardii* y-1. J Biol Chem 251:247–252
Reinert JC, Tinsley IJ (1977) The incorporation of DDT and related compounds into lecithin bilayers. Pestic Biochem Physiol 7:578–588
Renger G (1976) Studies on the structural and functional organization of system II of photosynthesis. The use of trypsin as a structurally selective inhibitor at the outer surface of the thylakoid membrane. Biochim Biophys Acta 440:287–300
Renger G, Hagemann R, Dohnt G (1981) Properties of the proteinaceous component acting as apoenzyme for the functional plastoquinone groups on the acceptor side of the system II. Biochim Biophys Acta 636:17–26
Rensen van JJS (1975) Lipid peroxidation and chlorophyll destruction caused by diquat during photosynthesis in *Scenedesmus*. Physiol Plant 33:42–46
Rensen van JJS, Kramer HJM (1979) Short circuit electron transport insensitive to diuron type herbicides induced by treatment of isolated chloroplasts with trypsin. Plant Sci Lett 17:21–28
Rensen van JJS, Vet van der E, Vliet van WPA (1977) Inhibition and uncoupling of electron transport in isolated chloroplasts by the herbicide 4,6-dinitro-o-cresol. Photochem Photobiol 25:579–583
Rensen van JJS, Wong D, Govindjee (1978) Characterization of the inhibition of photosynthetic electron transport in pea chloroplasts by the herbicide 4,6-dinitro-o-cresol by comparative studies with 3-(3,4-dichlorophenyl)-1,1-dimethylurea. Z Naturforsch 33c:413–420
Ridley SM (1977) Interaction of chloroplasts with inhibitors. Induction of chlorosis by diuron during prolonged illumination in vitro. Plant Physiol 59:724–732
Ridley SM (1982) Carotenoids and herbicide action. In: Britton G, Goodwin TW (eds) Carotenoid chemistry and biochemistry. Pergamon, Oxford New York, pp 353–369
Ridley SM, Ridley J (1979) Interaction of chloroplasts with inhibitors. Location of carotenoid synthesis and inhibition during chloroplast development. Plant Physiol 63:392–398
Ries SK (1976) Subtoxic effects on plants. In: Audus LJ (ed) Herbicides Vol II. Academic Press, London New York, pp 313–344
Ries SK, Chmiel H, Dilley DR, Filner P (1967) The increase in nitrate reductase activity and protein content of plants treated with simazine. Proc Natl Acad Sci USA 58:526–532
Rivera CM, Penner D (1979) Effects of herbicides on plant cell membrane lipids. Residue Rev 70:45–76
Robinson SJ, Yocum CF, Ikuma H, Hayashi F (1977) Inhibition of chloroplast electron transport reactions by trifluralin and diallate. Plant Physiol 60:840–844
Rogers LJ, Owen WJ, Delaney ME (1972) Sites of inhibition of photosynthetic electron transport by 1,1,1-trichloro-2,2-bis-(p-chlorophenyl) ethane (DDT). In: Forti G, Avron M, Melandri A (eds) Proc 2nd Int Congr Photos Res 1:689–699
Rost TL (1977) Responses of the plant cell cycle to stress. In: Rost TL, Gifford EM Jr (eds) Mechanisms and control of cell division. Dowden Hutchinson and Ross, Stroudsburg, pp 111–143
Rost TL, Bayer DE (1976) Cell cycle population kinetics of pea root tip meristems treated with propham. Weed Sci 24:81–87

Rubery PH (1981) Auxin Receptors. Annu Rev Plant Physiol 32:569–596
Rüdiger W, Benz J (1979) Influence of aminotriazol on the biosynthesis of chlorophyll and phytol. Z Naturforsch 34c:1055–1057
Rüdiger W, Benz J, Lempert U, Schoch S, Steffens D (1976) Hemmung der Phytol-Akkumulation mit Herbiziden. Geranylgeraniol und Dihydrogeranylgeraniol-haltiges Chlorophyll aus Weizenkeimlingen. Z Pflanzenphysiol 80:131–143
Ryan GF (1970) Resistance of common groundsel to simazine and atrazine. Weed Sci 18:614–616
Sanders HJ (1981) Herbicides. Chem Eng News 3:20–35
Sandmann G, Böger P (1982) Mode of action of herbicidal bleaching. In: Moreland DE, St John JD, Hess FD (eds) Biochemical responses induced by herbicides. ACS Symposium, Series 181, Am Chem Soc. Washington DC, pp 111–130
Sandmann G, Kunert KJ, Böger P (1979) Biological systems to assay herbicidal bleaching. Z Naturforsch 34c:1044–1046
Sandmann G, Bramley PM, Böger P (1980) The inhibitory mode of action of the pyridazinone herbicide norflurazon on a cell-free carotenogenic enzyme system. Pestic Biochem Physiol 14:185–191
Sandmann G, Kunert KJ, Böger P (1981a) Bleaching activity and chemical constitution of phenylpyridazinones. Pestic Biochem Physiol 15:288–293
Sandmann G, Lambert R, Böger P (1981b) Multifunctional mode of action of substituted nitrodiphenylethers in *Scenedesmus* cells. Z Naturforsch 36c:633–637
Sargent JA (1976) Relationship of selectivity to uptake and movement. In: Audus LJ (ed) Herbicides Vol II. Academic Press, London New York, pp 303–312
Schauberger CW, Wildman RB (1977) Accumulation of aldrin and diedrin by blue-green algae and related effects on photosynthetic pigments. Bull Environ Contam Toxicol 17:534–541
Schefczik KW (1981) Elektrische Eigenschaften von Blattgewebezellen höherer Wasserpflanzen und ihre Beeinflussung durch das Insektizid Lindan und seine Stereoisomeren. PhD Thesis, Univ Würzburg
Schefczik K, Simonis W (1979) Action of lindane on the current-voltage relationship in the plasma membrane of *Elodea* under passive conditions. Z Naturforsch 34c:1072–1074
Schefczik K, Simonis W (1980) Side effects of chlorinated hydrocarbon insecticides on membranes of plant cells. I. The influence of lindane on the membrane potential of *Elodea densa* leaf cells. Pestic Biochem Physiol 13:13–19
Scheinpflug H, Schlör H, Widdig A (1977) Chemie der Fungizide. In: Wegler R (ed) Chemie der Pflanzenschutz- und Schädlingsbekämpfungsmittel Vol IV. Springer, Berlin Heidelberg New York, pp 120–238
Schneider EA, Wightman F (1978) Auxins. In: Letham DS, Goodwin PB, Higgins TJV (eds) Phytohormones and related compounds. A comprehensive treatise Vol I. Elsevier, Amsterdam, pp 29–105
Schneider G (1970) Morphactins: physiology and performance. Annu Rev Plant Physiol 21:499–520
Schreiber K (1979) Naturally-occurring plant growth inhibitors as models for the synthesis of new growth retardants. In: Geissbühler H (ed) Advances in pesticide science. Pergamon, Oxford, pp 336–342
Schulz H (1969) Structure-activity relationships of the inhibition of photosynthesis by biscarbamate herbicides. In: Metzner H (ed) Progress in photosynthesis research Vol III. Tübingen, pp 1752–1760
Scott KR, Putwain PD (1981) Maternal inheritance of simazine resistance in a population of *Senecio vulgaris*. Weed Res 21:137–140
Scott TK (ed) (1979) Plant regulation and world agriculture. Plenum, New York London
Shipman LL (1981) Theoretical study of the binding site and mode of action for photosystem II herbicides. J Theor Biol 90:123–148
Siefermann-Harms D (1977) The xanthophyll cycle in higher plants. In: Tevini M, Lichtenthaler MK (eds) Lipids and lipid polymers in higher plants. Springer, Berlin Heidelberg New York, pp 218–230
Siegel MR, Sissler HD (1977) Antifungal compounds Vols I, II. Dekker, New York

Simkins GS, Moshier LJ, Russ OG (1980) Influence of acetamide herbicide applications on the efficacy of the protectant CGA-43089 in grain Sorghum (*Sorghum bicolor*). Weed Sci 28:646–649

Simonis W, Lee-Kaden J (1979) Selektive Wirkungen von Lindan (γ-1,2,3,4,5,6-Hexachlorcyclohexan) auf Photosynthese, Aminosäure-Membrantransport und Proteinsynthese bei *Anacystis nidulans*. Z Naturforsch 34c:1062–1065

Sims Holt JD, Stemler AJ, Radosevich SR (1981) Differential light responses of photosynthesis by triazine resistant and triazine susceptible *Senecio vulgaris* biotypes. Plant Physiol 67:744–748

Singh B, Campbell WF, Salunkhe DD (1972) Effects of s-triazines on protein and fine structure of cotyledons of bush beans. Am J Bot 59:568–572

Skoog F (ed) (1980) Plant growth substances. Springer, Berlin Heidelberg New York

Skult H, Paatero G (1979) Notes on the ultrastructure of root cells of barley grown in media containing DDT. Ann Bot Fenn 16:367–373

Södergren A (1971) Accumulation and distribution of chlorinated hydrocarbons in cultures of *Chlorella pyrenoidosa* (Chlorophyceae). Oikos 22:215–220

Souza Machado V (1982) Inheritance and breeding potential of triazine tolerance and resistance in plants. In: LeBaron H, Gressel J (eds) Herbicide resistance in plants. Wiley and Sons, New York, pp 257–273

Souza Machado V, Bandeen JD, Stephenson GR, Jensen KIN (1977) Differential atrazine interference with the Hill reaction of isolated chloroplasts from *Chenopodium album* L. biotypes. Weed Res 17:407–413

Souza-Machado V, Bandeen JD, Stephenson GR, Lavigine P (1978a) Uniparental inheritance of chloroplast atrazine tolerance in *Brassica campestris*. Can J Plant Sci 58:977–981

Souza Machado V, Arntzen CJ, Bandeen JD, Stephenson GR (1978b) Comparative triazine effects upon system II photochemistry in chloroplasts of two common lambsquarters (*Chenopodium album*) biotypes. Weed Sci 26:318–322

Stanger CE, Appleby AP (1972) A proposed mechanism for diuron-induced phytotoxicity. Weed Sci 20:357–363

Steffens GL, Buta JG, Gregory LE, Mandava NB, Mendt WJ, Worley JF (1979) New plant growth regulators isolated from higher plants. In: Geissbühler H (ed) Advances in pesticide science. Pergamon, Oxford, pp 343–346

Steinback KE, Pfister K, Arntzen CJ (1981a) Trypsin-mediated removal of herbicide binding sites within the photosystem II complex. Z Naturforsch 36c:98–108

Steinback KE, McIntosh L, Arntzen CJ (1981b) Identification of the 32 Kdalton triazine binding polypeptide of thylakoid membranes as a chloroplast gen product. Proc Natl Acad Sci USA 78:7463–7467

Steinback KE, Pfister K, Arntzen CJ (1982) Identification of the receptor site for triazine herbicides in chloroplast thylakoid membranes. In: Moreland DE, St John JB, Hess FD (eds) Biochemical responses induced by herbicides. ACS Symp Ser 181, Am Chem Soc, Washington DC, pp 38–55

Steinwandter H (1976) Zum Lindanmetabolismus in Pflanzen. II. Bildung von α-HCH. Chemosphere 5:221–225

Stephenson GR, Chang FY (1978) Comparative activity and selectivity of herbicide antidotes. In: Pallos FM, Casida JE (eds) Chemistry and action of herbicide antidotes. Academic Press, London New York, pp 35–61

St John JB (1976) Manipulation of galactolipid fatty acid composition with substituted pyridazinones. Plant Physiol 57:38–40

St John JB, Christiansen MN (1976) Inhibition of linolenic acid synthesis and modification of chilling resistance on cotton seedlings. Plant Physiol 57:257–259

St John JB, Hilton JL (1974) Herbicide-lipid interaction. In: Kolm GK (ed) Mechanism of pesticide action. ACS Symposium, Ser 2, Washington, pp 69–79

St John JB, Hilton JL (1976) Structure versus activity of substituted pyridazinones as related to mechanism of action. Weed Sci 24:579–582

St John JB, Rittig FR, Ashworth EN, Christiansen MN (1979) Structure-activity correlations in control of linolenic acid content of plant membranes by substituted pyridazin-

ones. In: Geissbühler H (ed) Advances in pesticide science. Pergamon, Oxford, pp 271–273
Stoddart JL (1979) Interaction of (^{3}H)gibberellin A_1 with a sub-cellular fraction from lettuce (*Lactuca sativa* L.) hypocotyls. II. Stability and properties of the association. Planta 146:363–368
Stoddart JL, Venis MA (1980) Molecular and subcellular aspects of hormone action. In: MacMillan J (ed) Hormonal regulation of development I. Encyclopedia of plant physiology new ser Vol IX. Springer, Berlin Heidelberg New York, pp 445–510
Sumida S, Ueda M (1973) Studies on pesticide effects on *Chlorella* metabolism. I. Effect of herbicides on complex lipid biosynthesis. Plant Cell Physiol 14:781–785
Sumida S, Ueda M (1974) *Chlorella* as a model system to study herbicidal mode of action and its application to a new herbicide O-ethyl-O-(3-methyl-6-nitrophenyl)-N-sec-butyl phosphorothioamidate (S-2846). In: Kolm GK (ed) Mechanism of pesticide action. ACS Symposium, Ser 2, Washington, pp 156–168
Sumida S, Yoshida R, Ueda M (1975) Studies of pesticides effects on *Chlorella* metabolism. II. Effect of DCMU on galactolipid metabolism. Plant Cell Physiol 16:257–264
Sumida S, Yoshida R, Ueda M (1977) Studies of pesticides effects on *Chlorella* metabolism. III. Effects of isopropyl 3-chlorocarbanilate (chlorpropham) on cell cycle and biosynthesis. Plant Cell Physiol 18:9–16
Summers LA (1979) Chemical constitution and activity of bipyridylium herbicides. In: Geissbühler H (ed) Advances in pesticide science. Pergamon, Oxford, pp 244–247
Summers LA (1980) The bipyridylium herbicides. Academic Press, London New York
Takahama U (1978) Suppression of lipid peroxidation by β-carotene in illuminated chloroplast fragments: Evidence for β-carotene as a quencher of singlet molecular oxygen in chloroplasts. Plant Cell Physiol 19:1565–1569
Takahama U, Nishimura M (1975) Formation of singlet molecular oxygen in illuminated chloroplasts. Effects on photoinactivation and lipid peroxydation. Plant Cell Physiol 16:737–748
Takahama U, Nishimura M (1976) Effects of electron donor and acceptors, electron transfer mediators, and superoxide dismutase on lipid peroxidation in illuminated chloroplast fragments. Plant Cell Physiol 17:111–118
Takegami T, Yoshida K (1977) Specific interaction of cytokinin binding protein with 40 S ribosomal subunits in the presence of cytokinin in vitro. Plant Cell Physiol 18:337–345
Tamura S (1979) Plant growth regulators produced by microorganisms. In: Geissbühler H (ed) Advances in pesticide science. Pergamon Press, Oxford, pp 356–365
Thomas TH (1974) An assessment of the value of plant growth regulators and herbicides for crops grown for the processed food industries. Pestic Sci 5:87–100
Thomson KS, Leopold AC (1974) In vitro binding of morphactins and 1-N-naphthylphthalamic acid in corn coleoptiles and their effect on auxin transport. Planta 115:259–270
Thomson KS, Hertel R, Muller S, Tavares JE (1973) 1-N-naphthylphthalamic acid and 2,3,5-triiodobenzoic acid. In vitro binding to particulate cell fractions and action on auxin transport in corn coleoptiles. Planta 109:337–352
Tischer W, Strotmann H (1977) Relationship between inhibitor binding by chloroplasts and inhibition of photosynthetic electron transport. Biochim Biophys Acta 460:113–125
Tischer W, Strotmann H (1979) Some properties of the DCMU-binding site in chloroplasts. Z Naturforsch 34c:992–995
Trebst A (1974) Energy conservation in photosynthetic electron transport of chloroplasts. Annu Rev Plant Physiol 25:423–458
Trebst A (1979) Inhibition of photosynthetic electron flow by phenol and diphenylether herbicides in control and trypsin-treated chloroplasts. Z Naturforsch 34c:986–991
Trebst A (1980) Inhibitors in electron flow: Tools for the functional and structural localization of carriers and energy conserving sites. In: San Pietro A (ed) Methods in enzymology Vol 69. Academic Press, London New York, pp 675–715
Trebst A (1981) Action mechanism of herbicides in photosynthetic electron transport.

In: Akoyunoglou GA (ed) Photosynthesis VI, Balaban International Sciences Services, Philadelphia, pp 507–520
Trebst A, Draber W (1979) Structure-activity correlations of recent herbicides in photosynthesis. In: Geissbühler H (ed) Advances in pesticide science Vol II. Pergamon, New York London, pp 223–235
Trebst A, Harth E (1974) Herbicidal N-alkylated-ureas and ringclosed N-acylamides as inhibitors of photosystem II. Z Naturforsch 29c:232–235
Trebst A, Wietoska H (1975). Hemmung des photosynthetischen Elektronentransports von Chloroplasten durch Metribuzin. Z Naturforsch 30c:499–504
Trebst A, Pistorius E, Boroschewski G, Schulz H (1968) Die Hemmung photosynthetischer Reaktionen durch Herbizide des Biscarbamat-Typs. Z Naturforsch 23b:342–348
Trebst A, Reimer S, Draber W, Knops HJ (1979) The effect of analogues of dibromothymoquinone and of bromonitrothymol on photosynthetic electron flow. Z Naturforsch 34c:831–840
Truffaut G, Pastac I ((1935) Br Patent 424, 295, May 29th
Upadhyaya MK, Nooden LD (1977) Mode of dinitroaniline herbicide action. I. Analysis of the colchicine-like effects of dinitroaniline herbicides. Plant Cell Physiol 18:1319–1330
Upadhyaya MK, Nooden LD (1980) Mode of dinitroaniline herbicide action. II. Characterization of (^{14}C)oryzalin uptake and binding. Plant Physiol 66:1048–1052
Urbach D, Suchanka M, Urbach W (1976) Effect of substituted pyridazinone herbicides and of difunone (EMD-IT 5914) on carotenoid biosynthesis in green algae. Z Naturforsch 31c:652–655
Urbach W (1977) Algal and bacterial photosynthesis. I. Eucaryotic algae. in: Trebst A, Avron M (eds) Photosynthesis I. Encyclopedia of plant physiology new ser Vol V. Springer, Berlin Heidelberg New York, pp 603–624
Urbach W, Lurz G, Hartmeyer H, Urbach D (1979) Induction of reversible tolerance of algal cells to various herbicides. I. Inhibition of photosynthesis by phenol herbicides and dibromothymoquinone, its reversal and development of insensitivity to different herbicides. Z Naturforsch 34c:951–956
Urbach W, Urbach D, Reisberg E (1983) Pigment destruction and lipid peroxidation caused by phenyl ureas and paraquat in isolated chloroplasts and in intact algal cells. Z Naturforsch (in prep)
Vaisberg AJ, Schiff JA (1976) Events surrounding the early development of *Euglena* chloroplasts. 7. Inhibition of carotenoid biosynthesis by the herbicide SAN 9789 (4-chloro-5-(methylamino)-2-(α,α,α-trifluoro-m-tolyl)-3(2H)pyridazinone) and its developmental consequences. Plant Physiol 57:260–269
Velthuys BR (1980) Mechanisms of electron flow in photosystem II and toward photosystem I. Annu Rev Plant Physiol 31:545–567
Velthuys BR (1981) Electron-dependent competition between plastoquinone and inhibitors for binding to photosystem II. FEBS Lett 126:277–281
Velthuys BR, Amesz J (1974) Charge accumulation at the reducing side of system 2 of photosynthesis. Biochim Biophys Acta 333:85–94
Venis MA (1971) Stimulation of RNA transcription from pea and corn DNA by protein retained on sepharose coupled to 2,4-dichlorophenoxyacetic acid. Proc Natl Acad Sci USA 68:1824–1827
Venis MA (1977) Receptors for plant homones. Adv Bot Res 5:53–88
Venis MA (1979) Receptors for plant growth regulators. In: Geissbühler H (ed) Advances in pesticide science, part II. Pergamon, Oxford, pp 487–493
Voerman S, Besemer AFH (1975) Persistence of dieldrin, lindane and DDT in a light sandy soil and their uptake by grass. Bull Environ Contam Toxicol 13:501–515
Wain RL (1975) Some developments in research on plant growth inhibitors. Proc R Soc London Ser B 191:335
Wain RL (1979) Potential for regulation of plant growth and development. In: Scott TK (ed) Plant regulation and world agriculture. Plenum, New York London, pp 155–164

Wain RL, Smith MS (1976) Selectivity in relation to metabolism. In: Audus LJ (ed) Herbicides Vol II. Academic Press, London New York, pp 279–302
Walker DA (1980) Preparation of higher plant chloroplasts. In: San Pietro A (ed) Methods in enzymology Vol 69. Academic Press, London New York, pp 94–104
Waller H, Gausman HW, Rittig FR, Namken LN, Escobar DE, Rodriguez RR (1980) Effect of mepiquat chloride on cotton plant leaf and canopy structure and dry weights of its components. Pwc Beltwide Cotton Prod Res Conf, St Louis
Walton DC (1980) Biochemistry and physiology of abscisic acid. Annu Rev Plant Physiol 31:453–489
Ware GW, Roan CC (1970) Interaction of pesticides with aquatic microorganisms and plankton. Residue Rev 33:15–47
Wareing PF (1976) Modification of plant growth by hormones and other growth regulators. Outlook Agric 9:42–45
Weaver RJ (1972) Plant growth substances in agriculture. Freeman, San Francisco
Webster PL, Hof van't J (1969) Dependence on energy and aerobic metabolism of initiation of DNA synthesis and mitosis by G1 and G2 cells. Exp Cell Res 55:88–94
Weeks DP, Baxter R (1972) Specific inhibition of peptide chain initiation by 2-(4-methyl-2,6-dinitroanilino)-N-methylpropionamide. Biochemistry 11:3060–3064
Wegler R (ed) (1970–1981) Chemie der Pflanzenschutz- und Schädlingsbekämpfungsmittel, 6 Vols. Springer, Berlin Heidelberg New York
Wessels JSC, Veen van der R (1956) The action of some derivatives of phenylurethan and of 3-phenyl-1,1-dimethylurea on the Hill-reaction. Biochim Biophys Acta 19:548–549
West LD, Muzik TJ, Witters RE (1976) Differential gas exchange responses of two biotypes of redroot pigweed to atrazine. Weed Sci 24:68–72
Wiedman SJ, Appleby AP (1972) Plant growth stimulation by sublethal concentrations of herbicides. Weed Res 12:65–74
Wild A, Oberweiss AL, Ruehle W (1977) The effect of the insecticides allethrin, lindane and jacutin fogetten sublimate on the photosynthetic electron transport. Z Pflanzenphysiol 82:161–172
Wilkinson RE (1978) Physiological response of lipid components to thiocarbamates and antidotes. In: Casida JE, Pallos FM (eds) Chemistry and action of herbicide antidotes. Academic Press, London New York, pp 85–108
Witt HT (1971) Coupling of quanta, electrons, fields, ions and phosphorylation in the functional membrane of photosynthesis. Results by pulse spectroscopic methods. Q Rev Biophys 4:365–477
Wittwer SH (1971) Growth regulants in agriculture. Outlook Agric 6:205–217
Wittwer SH (1978) Phytohormones and chemical regulators in agriculture. In: Letham DS, Goodwin PB, Higgins TJV (eds) Phytohormones and related compounds. A comprehensive treatise Vol II. Elsevier/North-Holland, Amsterdam, pp 599–615
Worthing CR (ed) (1979) Pesticide manual. Br Crop Protect Counc, 6th edn. Boots, Nottingham
Wright BJ, Baillie AC, Wright K, Dowsett JR, Sharpe TM (1980) Synthesis of potential herbicides designed to uncouple photophosphorylation. Phytochemistry 19:61–66
Wright K, Corbett JR (1979) Biochemistry of herbicides affecting photosynthesis. Z Naturforsch 34c:966–972
Wright STC (1978) Phytohormones and stress phenomena. In: Letham DS, Goodwin PB, Higgins TJV (eds) Phytohormones and related compounds. A comprehensive treatise Vol II. Elsevier/North-Holland, Amsterdam, pp 495–536
York AC, Arntzen CJ, Slife FW (1981) Photosynthetic electron transport inhibition by buthidazole. Weed Sci 29:59–64
Youngman RJ, Dodge AD (1979) Mechanism of paraquat action: Inhibition of the herbicidal effect by a copper chelate with superoxide dismutating activity. Z. Naturforsch 34c:1032–1035
Zeroni M, Hall M (1980) Molecular effects of hormone treatment on tissue. In: MacMillan J (ed) Hormonal regulation of development I. Encyclopedia of plant physiology new ser Vol IX. Springer, Berlin Heidelberg New York, pp 511–586

12 Effects of Biocides and Growth Regulators: Ecological Implications

R. SCHUBERT

CONTENTS

12.1 Introduction 393

12.2 Effects of Biocides and Growth Regulators on Plant Populations 394
    12.2.1 External Environmental Factors 394
    12.2.2 Internal Factors 396

12.3 Effects of Biocides and Growth Regulators on Soil Microbial Populations . 398

12.4 Effects of Biocides and Growth Regulators on Soil Faunal Populations . . 399

12.5 Effects of Biocides and Growth Regulators on Terrestrial Faunal Populations 400

12.6 Effects of Biocides and Growth Regulators on Agro-Ecosystems 401

12.7 Effects of Biocides and Growth Regulators on Grassland Ecosystems . . . 404

12.8 Effects of Biocides and Growth Regulators on Forest Ecosystems 405

12.9 Conclusions 406

References 406

12.1 Introduction

The use of agrochemicals has become the prevalent method of pest control in most areas of the world during the last 30 years. Annual pesticide use is estimated to be 2×10^9 kg (PIMENTEL 1981). Given this massive present-day use of growth regulators and biocides, it is extremely important to investigate their complex effects on plants, on entire agro-ecosystems, and on the landscape. It is necessary to pay more attention in future to the investigation of the distribution and ecological impact of these man-made compounds within ecosystems (DIERCKS 1976; SCHUBERT 1980). We should, however, remember that even ecosystems which are not influenced by man, contain allelopathic agents produced by various plants and microorganisms. These include colins, marasmines, phytocides, toxins, antibiotics, bacteriotoxins, and telergones (STUGREN 1978). Our present knowledge of cycling of these compounds is still rather poor.

The physiological basis for effects of biocides and growth regulators on plant metabolism and plant performance is treated in Chapter 11, this Volume. In this contribution, impact of these agrochemicals on plants in agro-ecosystems will be discussed (pesticides in freshwater ecosystems will be treated in Chap.

13, this Vol.). We will begin by examining in some detail the effects of growth regulators and herbicides at the level of populations, including plant, soil microbial as well as animal populations. Afterwards, we will discuss the overall effects on different agro-ecosystems, i.e. cultivated fields and grasslands, but also on forests.

12.2 Effects of Biocides and Growth Regulators on Plant Populations

12.2.1 External Environmental Factors

Environmental factors have significant influence on toxicity and selectivity of biocides with respect to plant populations (KURTH 1968). Several factors influencing herbicide selectivity are discussed in Chapter 11, this Volume (see also ASHTON and CRAFTS 1981).

The particular pathway by which herbicides and growth regulators enter plants will vary with the mode of application. When chemicals are applied directly to the above-ground organs of a crop and its weed population (post-emergence application), the leaves and stems will be the primary intercepting and absorbing organs. If the herbicides and growth regulators are applied to or incorporated into the soil before most of the weeds appear (pre-emergence application), then the roots and emerging shoots are the important pathways of entry. Penetration into the leaves and roots of plants can be described in simplified fashion as follows: (a) absorption into the cuticle, if present, (b) movement across the cuticular membrane, if present, (c) absorption into the apoplast and (d) uptake by the underlying cells (cf. BUKOVAC 1976).

These processes can be influenced by a number of physicochemical and biological factors which ultimately determine the selective phytotoxic effect of the chemicals (KURTH 1968).

To be considered, for example, are (a) the chemical and physical properties of the herbicide, (b) the dosage level of herbicides (quantity per ha and concentration per plant), (c) the morphological, anatomical and histological differences in the structure of the plants, (d) herbicide solubility and transport of the herbicide within the plant, (e) the physiological differences due to biological rhythms, (f) herbicide metabolism, (g) the developmental stage of crops and weeds at the time of application, (h) the physical and chemical qualities of the soils, and (i) the climatic conditions, especially prevailing light, temperature and rainfall.

When chemicals are sprayed on the leaves, generally both surfaces are penetrated, but there is often a greater permeability of the lower leaf surface, due to the greater number of stomata and trichomes. The absorption of foliar-applied chemicals from spray droplets is initially rapid. With increasing time after treatment, penetration decreases (LUCKWILL and LLOYD-JONES 1962). When penetration is measured directly it is generally linearly related to external herbicide concentration (SARGENT and BLACKMAN 1972).

The stage of plant development may markedly influence spray retention (DAVIES et al. 1967). Mature, fully expanded leaves generally absorb less than immature expanding leaves (BUKOVAC 1965). Leaves damaged mechanically or by insects are more permeable than non-damaged leaves. Moisture which keeps the cuticle in a highly hydrated state favors absorption (VAN OVERBECK 1956). Moisture stress and low soil and air temperature, on the other hand, reduce absorption (HULL 1970).

Dry environmental conditions often produce stunted, slowly growing plants with small leaves and a thick cuticle. These will be more tolerant to herbicides than rapidly growing plants of the same age with large thin leaves. Plants which have recently undergone severe water stress to the point of wilting will also be more resistant. The occurrence of wind and rain before treatment enhances phytotoxicity, but may reduce ability to obtain a species-specific response (MUZIK 1976).

The effects of other environmental factors such as light, temperature and humidity on foliar absorption are very complex. They indirectly affect absorption because they alter the quantity, fine structure and chemistry of epi-cuticular wax, which in turn leads to alterations in wettability, water retention, and permeability. However, they also have a direct influence on penetration. Because light, temperature and humidity are highly correlated with each other under field conditions, this direct effect on uptake of herbicides is extremely difficult to analyze (BAKER 1972; WHITECROSS and ARMSTRONG 1972).

For the best results, most susceptible species have to be treated during a period of active growth and high humidity. For most plants of temperate regions high temperatures up to 30 °C, and a light intensity near the saturation point for the particular species will lead to increased sensitivity towards phytotoxic chemicals. Low temperatures (5° to 10 °C) following treatment often delay the expression of phytotoxic symptoms. As an exception, amitrole shows a drastic increase in effects when low temperatures follow application. Drought conditions or temperatures below 0 °C are implicated in some serious grain losses in wheat treated with 2,4-D, bromoxynil and MCPA, terbutryne and diuron. It seems that these herbicides place a stress on the treated plants which makes them more susceptible to frost and drought conditions (MUZIK and MAULDIN 1964).

When herbicides are applied to the soil, roots are the primary absorbing organs. They present a large absorbing surface with zones preferentially permeable to water and solutes. In general, the time course of rate of uptake of biocides into the roots consists of a rapid phase during the first 30 min to 2 h, followed by a second phase of slower but constant uptake (ISENSEE et al. 1971). Biocide-uptake by roots is decreased by increase in pH, or by lowering temperature, the latter having more effect during the second phase. Uptake is an approximately linear function of biocide concentration in the external solution. Transpiration is thought to play an important role in uptake of water-soluble herbicides by facilitating the movement of the absorbed chemical to the aerial plant parts, thereby reducing root concentration. Thus a steeper concentration gradient is maintained between the external medium and the root tissue. However, herbicide absorption and transpiration are not always directly correlated (BRUINSMA 1967). There are several reports of root exudation of organic chemicals, including herbicides, following their application to the foliage of plants. Many of the compounds are exuded unaltered, and in quantities up to more than 10% of the original dose applied (SHARMA et al. 1971; LINDER et al. 1964; REID and HURTT 1970).

The effective use of biocides and growth regulators requires a knowledge of their translocation within plants. Many weeds regrow from buds on the aboveground organs or from the roots or rhizomes below the ground. If these are to be killed, sufficient herbicides must move to the growing points (ASHTON and CRAFTS 1981; HAY 1976). Translocation of herbicides is possible through the symplast, the apoplast or both (amphiplast, CRAFTS and CRISP 1971). Free carboxyl groups appear to be necessary for symplast transport (JACOB et al. 1973). The rate of degradation of herbicides, as well as their ability to combine with plant constituents, may alter distribution patterns.

A supply of carbohydrate is necessary for the movement of herbicides from leaves to provide energy for phloem loading. Downward translocation to the roots has been reported in response to increased relative humidity. High humidity probably reduces the transpiration rate. Dry soil conditions reduce the translocation of symplast transported herbicides. Lack of soil moisture may reduce the supply of carbohydrates for phloem loading (BASLER et al. 1961). All these facts demonstrate the influence of the external factors, such as weather, soil, type of application, concentration of biocides, crop rotation and agrotechnical treatment, on the development of weed populations. The elimination of the more sensitive species is therefore often followed by the formation of new populations. Repeated application of similar products can lead on this way to the development of "resistant" forms also of originally sensitive species (HANF 1979; ROLA 1979; SCHMIDT 1979; ZEMÁNEK 1979).

The influence of the stage of development on plant sensitivity to herbicides varies with plant species and type of herbicide used. Most of the phanerogamic perennials and winter annuals have a period of maximum susceptibility after fruit set. Generally sensitivity decreases as plants get older (ÅBERG and STECKO 1976). Sometimes the symptoms in treated plants, e.g. potatoes treated with mecoprop, become apparent only in the next year (ARLT and FEYERABEND 1972; HANF 1957). Interactions of biocides applied in combination may result in an increased biocide persistence in soils and plants and sometimes in toxic effects to plants different from the single application (WALLNÖFER et al. 1979).

12.2.2 Internal Factors

Ever since biocides and growth regulators came into wide use, differences in response of varieties or cultivars of the same species have been discussed. Such differences can depend on the genotype of the plants and on the morphological and physiological variability related to it (ÅBERG and STECKO 1976). The differential toxicity of the applied chemicals, resulting from these internal factors, may lead to selection for certain individuals of a population. In the context of a crop species, resistance to herbicides can be one of the objectives of plant

breeding. Differences in responses are often apparent between cultivars of crops, after they have been sprayed with chemicals. Triploid sugar beets are less sensitive to delapon than diploids and tetraploids. Differences in susceptibility to treatment with simazine are also found in cultivars of apples. Among several hundred cultivars of soybean (WAX et al. 1974), eleven varieties were highly sensitive to bentazon, 2,4-DB, bromoxynil and chloroxuron.

Differences in sensitivity to chemicals are also found in weed populations. Herbicide applications may allow invasion by those weed species which are more tolerant. *Cirsium arvense* includes many regional races that are distinct, genetically fixed ecotypes, varying in phenology, morphology and anatomy, and which respond differently to 2,4-D and amitrole (HODGSON and MOORE 1972). The lipid content of the leaves of the thistles and the resistance to 2,4-D are highly positively correlated (HODGSON 1973). HARVEY and MUZIK (1973) found that noticeable differences in sensitivity to 2,4-D of two clones of *Convolvulus arvensis* were most likely explained by differential binding of 2,4-D within the cells.

Also for related ecotypes of monocotyledons, differences are found in sensitivity to biocides. ZÉMANEK (1970) reports differences in the susceptibility of winter wheat cultivars treated with soil herbicides. Out of six cultivars of oats, two were more susceptible to treatment with amitrole (BENGTSSON 1970). Wild oats have a large number of different ecotypes. Eighteen ecotypes which differed markedly from others in herbicide response were isolated in Sweden. Wild oat plants originating from 214 wild plants in North America differed very much in reaction to diallate (JACOBSOHN and ANDERSEN 1968). SCHOOLER et al. (1972) showed that three dominant factors control the different tolerance to siduron in 36 collections of *Hordeum jubatum*. Differences in susceptibility for ecotypes are also reported for other monocots and dicots such as *Sorghum halepense* (MCWHORTER 1971), *Agropyron repens, Cynodon dactylon* (ROCHECOUSTE 1962), *Echinochloa crus-galli* (ROCHE and MUZIK 1964) and *Centaurea cyanus* (HOJDEN 1973). Increased resistance to herbicides due to selection is probably responsible for reported population increases in *Avena fatua, Apera spica-venti, Chenopodium album, Galium aparine, Agropyron repens, Cynodon dactylon* and *Digitaria sanguinalis* (SCHUBERT 1971).

In intensive agriculture with a high use of biocides or growth regulators there is an evolution of specific resistant weed species and in consequence the formation of new weed populations (STRYCKERS 1979). Sometimes the application of herbicides together with the simplification of crop rotations caused the development of resistant weed populations. At sites where maize has been cultivated for several years, a biotype of *Stellaria media* resistant to atrazine was found. The same is suspected in biotypes of *Chenopodium album* and *Chenopodium serotinum* (KEES 1979). BARRALIS et al. (1979) found resistant populations of susceptible species appearing recently in monocultures weeded with triazines, e.g. *Amaranthus retroflexus, Chenopodium album, Polygonum lapathifolium, P. persicaria* and *Solanum nigrum*. Several different weed species in the U.S.A., in Canada and Western Europe have been shown to be of extreme resistance to triazine herbicides and partially also to other herbicides which inhibit photosynthesis. In all documented cases, resistance has appeared in areas where triazine herbicides were used in monocultures with only little additional weed control. Weeds from the resistant biotype are not considered to represent spontaneous mutants but rather to be selected from the heterogeneous population. Current knowledge on herbicide resistance is comprehensively covered by LE BARON and GRESSEL (1982). Physiological properties of weeds resistant to herbicides have been discussed in detail in Chapter 11, this Volume.

12.3 Effects of Biocides and Growth Regulators on Soil Microbial Populations

Soil microbial populations are influenced by soil organic matter content, soil moisture content, temperature, pH, cation exchange capacity, and aeration as well as by numerous other parameters. Many reports show that the soil microorganisms are responsible for the degradation of most herbicides (cf. KAUFMAN and KEARNEY 1976). While most herbicides are degraded by microbes producing special enzymes (HURLE and RADEMACHER 1970; SANDMANN 1974; FRYER and KIRKLAND 1970), other herbicides are degraded in normal microbe metabolic processes (BLAKE and KAUFMAN 1975). Microbial metabolism of herbicides often results in substances which are non-toxic to the ecosystem.

On the other hand, biocides affect the soil microflora and microfauna both in pure cultures and in the soil (KOCH and HURLE 1977). It is remarkable that inhibitory or stimulatory effects on populations can be detected in the soil, since relatively low concentrations of biocides are usually applied, because microorganisms are able to adjust to biocide stress, and since population changes due to biocides must be distinguished from the differences caused by environmental conditions and overall fluctuations with time (GROSSBARD 1976; CRAMER 1977). Direct stimulating effects of 2,4-D and simazine on the population density of bacteria, actinomycetes and fungi were found by HELMECKE et al. (1977). After stimulation an inhibitory phase was observed. Chlorophyll content can be used to measure density of soil algal populations (BECK 1970; LEVIS et al. 1978). EPTC, monuron and atrazine at 5 kg ha^{-1} reduced the total algal population in a light and in a humus soil (MICHAJLOVA and KRUGLOV 1973). To assess the influence of biocides on the overall activity of all microorganisms, carbon dioxide evolution and oxygen uptake are frequently used. These non-specific measures give no indication of the suppression of sensitive species. Resistant survivors may multiply profusely and produce as much CO_2 as did the population extant before biocide application (GROSSBARD 1970; BEZUGLOV et al. 1976; PAESCHKE and HEITEFUSS 1978; MARSH et al. 1978).

Biocides may act to decrease or increase the concentration of enzymes in the soil, though some may exert only a negligible effect. Trifluralin even at the high concentration of 15 kg ha^{-1} did not lead to change in the concentration of hydrolytic soil enzymes or catalase (GRUZDEV et al. 1973). After an application of simazine (4 kg or 8 kg ha^{-1}) a depression of dehydrogenase occurred, followed by an increase but then again a decrease; 2,4-D applied in the next year after simazine treatment, caused decline in soil dehydrogenase content as well as MCPA (HELMECKE et al. 1977). In some cases there is a connection between the application of herbicides and plant diseases mostly caused by fungi living in the soil. KATAN and ESHEL (1973) found increasing damage of cotton by *Rhizoctonia solani* when trifluralin was applied. When wheat was treated with triazine herbicides a reduced damage by *Cercosporella herpotrichoides* was observed.

12.4 Effects of Biocides and Growth Regulators on Soil Faunal Populations

Soil faunal organisms are attacked by biocides and growth regulators in three ways: (1) through direct contact on the body surface, (2) by ingestion into the digestive tract, and (3) by volatile substances, through diffusion mainly via the respiratory organs (EIJSACKERS and VAN DER DRIFT 1976). The animals' degree of mobility and their reaction to repellent or attractant stimuli have a strong influence on their degree of contact with the compounds (MÜLLER et al. 1980). DNOC caused a high mortality of the earthworm, *Allopophora caliginosa* during the spring, especially when the weather was wet, whereas during the summer, when the species is in diapause, the mortality was almost nil (VAN DER DRIFT 1963).

The impact of phenoxy acids on soil fauna is rather mild, whereas triazines, ureas and especially phenolic-herbicides often cause a distinct decline in populations (MÜLLER 1972). In nematodes a differential sensitivity has been found between free-living saprophagous and plant-parasitizing species. RAO and PRASAD (1968) found that low doses of EPTC were not toxic for the former but were for the latter.

The influence of biocides on the soil fauna is not caused exclusively by direct means. Indirect effects of herbicides via their influence on plants may also contribute to complex soil faunal population changes (CURRY 1970; PRASSE 1975) found that long-term herbicide treatments influenced soil fauna of a loess-black soil, depending on the time of application and the type of herbicide. The effect of 2,4-D on *Collembola* and *Acari* populations disclosed no statistically significant differences, however, the general trend pointed toward overpopulation. Two weeks after each separate herbicide application, populations increased in the soil layer at 5–10 cm. An overpopulation of mites occurred in November in the uppermost soil layer. A decrease in population size which began for the *Collembola* in autumn, and for the mites evidently later, had not yet been redressed in April of the following year. Simazine, however, significantly decreased the abundance of *Collembola* in almost all soil layers throughout the year of the investigation. Its effect was greatest in the uppermost soil layer. The *Acari* population was also reduced immediately after treatment, recovering slightly in April. Data obtained in April of the following year suggested that the depression of the populations had persisted during the second winter.

12.5 Effects of Biocides and Growth Regulators on Terrestrial Faunal Populations

There is no doubt that different biocides influence faunal populations in a different manner (KOEMAN 1978). Undesirable environmental effects of biocides have been reported from many countries. The effects include excessive mortality and a reduced reproductive potential in birds, fishes and other organisms, changes in the abundance of species and the diversity of ecosystems, a reduction in the productive potential of natural resources and the development of biocide resistance in target and non-target species.

At present efforts are being made in many places to increase the predictive capability with respect to the assessment of the possible hazards of chemicals in the environment. However, the prediction of safe or acceptable levels is still complicated because of the many uncertainties which implicitly accompany extrapolations from relatively simple tests to generally complex ecosystems (MÜLLER 1974).

Attempts have been made to classify biocides into categories of danger solely on the basis of their LD_{50} (median lethal dose) for laboratory rats (BARNES 1976). Such data only provide a preliminary guide. It is interesting to note that, whereas 102 out of 200 insecticides tested had an LD_{50} below 250 mg kg^{-1}, out of 150 herbicides tested only 17 fell in this category (BEN-DYKE et al. 1970). The low toxicity of many herbicides in the experiments with rats is also seen in tests with birds and fishes. While some pesticides have caused serious or fatal poisoning to animals or man (DDT), only a few herbicides cause poisoning in animals (LAUDIN 1972). It can be generally stated that herbicides and growth regulators are less toxic to vertebrates as compared to insecticides. As discussed in Chapter 11, this Volume, the mechanism of action of the majority of herbicides and growth regulators is the interference with plant-specific biochemical reactions, i.e. pigment and hormone metabolism or especially in photosynthesis. For this reason, most herbicides do not pose toxicological problems to animals. The few compounds which are known to be toxic in this respect interfere with the mechanism of phosphorylation which is basically the same in animals and plants. These herbicides are mostly substituted phenols such as DNOC or dinoseb. Today, application of this kind of herbicides is greatly reduced or restricted.

Changes within a biotope caused by biocide application may be favorable or unfavorable for the existing fauna. Often a decrease in diversity and number of insectivorous birds is due to a decrease in insects, which in turn is the consequence of a destruction of the weed vegetation. The butterflies *Pieris brassicae* and *Pieris rapae* disappear when the weeds *Rhaphanus raphanistrum* and *Sinapis arvensis* are controlled by chemicals (LAUDIN 1972). A correlation was also found between the decrease in pheasant (*Phaseanus colchicus*) and partridge (*Perdix perdix*) populations and the decrease in arthropods caused by weedkilling (POTTS 1970; FRANK 1970). In other cases, it has been possible to increase bird populations by deforestation with arboricides (BLASZYK 1970).

ADAMS and DREW (1969) found in oats and barley treated with 2,4-D that the number of aphids (*Rhopalosiphum padi* and *Macrosiphum avenae*) was higher than in untreated stands. MAXWELL and HARWOOD 1960 (in MÜLLER 1974) pointed out that after treatment of young *Vicia faba* plants with 2,4-D there was a highly increased number of the aphid *Acyrthosiphon pisum* on the top of the twigs. A higher amount of alanine, aspartate and glutamate is considered to be the reason for this increase. EBLE (1977) demonstrated a depressing influence of 2,4-D and simazine on the epigaeic entomofauna. In other cases, growth regulators decrease the populations of sucking insects. Thus cycocel and CCC cause higher mortality and less fertility in *Brevicoryne brassicae* (VAN EMDEN 1969 in MÜLLER 1974). STECHE (1972) found that direct spraying of biocides may harm bees. Flowering weeds should be sprayed when bees are not flying. It is unlikely that bees will forage for nectar containing herbicides and thus contaminate honey. The main influence of biocides especially herbicides and growth regulators on wildliving animals appears to arise from the effects on vegetation which provides niches or food supplies (WAY 1969).

12.6 Effects of Biocides and Growth Regulators on Agro-Ecosystems

Because the cultivated field is a continuously disturbed area, it is inhabited by those weed species which are successful during the early stages of a plant succession. Due to fluctuating environmental conditions and the influence of man, rapid changes in species composition or dominance occur. Dominance of a particular species rarely exceeds 1 year and generally continues for less than 6 months. Chemicals are one of the most important factors affecting these ephemeral plant communities (BURNSIDE et al. 1980).

The influence of growth regulators and biocides on structure and composition of weed communities is very severe. Certain species can be totally eradicated at a specific site, but no weed species is known which has disappeared completely as a result of biocide use. However, the latter might be the case due to liming, seed cleaning, and amelioration (SCHUBERT and HILBIG 1969; MITTNACHT et al. 1979). On the other hand, records exist of a considerable increase in weeds following eutrophication or liming of arable land, or as a result of long-term, uninterrupted use of herbicides. BACHTHALER (1966) recorded an increase of *Avena fatua* and ROLA (1973) an increase of *Galium aparine, Stellaria media* and *Apera spica-venti* as a result of adjustment to changing competitive pressure. Other species, e.g. *Sinapis arvensis, Papaver rhoeas* and *Ranunculus arvensis,* declined *Scleranthus annuus* and *Rumex acetosella* (HAAS and STREIBIG 1982).

In sugar cane plantations in Trinidad, continuous and extensive use of 2,4-D resulted in a predominance of grassy weeds, which were then controlled with newer herbicides. In response, the grasses were subsequently replaced by sedges.

Similar changes occurred in Hawaiian sugar cane fields treated for over 50 years with chemicals (HANSON 1972).

MAHN and HELMECKE (1979) investigated the effects of continued herbicide application over several years on the structure of an agro-ecosystem on black soil. Differences in the weed community, which became more pronounced in the course of the experiment, could already be observed after a single application of herbicide. Due to the repeated application of herbicides, a reduction in weed density occurred. Weed density in the non-treated control area increased. In this case, changes in the structure of the weed community were primarily quantitative. Most species originally present continued to persist, although changes were observed in dominance (HELMECKE et al. 1977; SCHUBERT 1977). Species dominance was apparently determined by their ability to withstand direct or indirect effects of herbicide application at annual intervals. Only a few of the species, which were dominant before application remained dominant after application, whereas certain previously suppressed populations increased dramatically, e.g. *Polygonum convolvulus, Veronica polita* and *Euphorbia exigua*. When the density of weeds was low, the application of herbicides caused no significant increase in crop biomass and grain yield (MAHN and HELMECKE 1979). It is therefore necessary to check whether the existing or expected weed density has reached a threshold value, which from the economic standpoint necessitates and ecologically justifies the application of biocides (CREMER 1976; PÖTSCH 1976).

Some strains of weeds develop tolerance or resistance to herbicides (see Chap. 11, this Vol.). The gradual development of tolerance is one of the important problems confronting attempts at weed control through biocide application. Weed populations are able to adapt effectively due to maintained genetic heterogeneity, long-term seed dormancy, and perhaps as well due to replenishment by migration from untreated populations in ruderal or other habitats. The most resistant and the most susceptible plants are often morphologically indistinguishable.

The changes in weed density may often have a severe influence on the fauna of the field ecosystems. Young partridges (*Perdix perdix*) rely on an insect diet in their early life. They are affected by the shortage of insects, resulting from herbicide reduction of the insects' food plants in arable fields. Adult birds have been shown to be largely herbivorous, eating the seeds and green parts of a wide range of herbaceous plants. Following herbicidal reduction of weeds such as *Stellaria media,* species of *Polygonum* are a more important item in their diet (POTTS 1970).

Weeds may serve as insect food sources. When the relationship between host plants, pests, predators and parasites is disturbed by man through the use of biocides, pest populations may increase while those of predators and parasites decrease (ZANDSTRA and MOTOOKA 1978). LEIUS (1967) has found that the presence of wild flowers in apple orchards resulted in 18- and 4-fold increases in parasitism of tent caterpillar (*Malacosoma americanum*) pupae and eggs, respectively, over that in non-weedy orchards. Larval parasitism of codling moth (*Carpocapsa pomonella*) also increased five times. Several species of parasitic wasps were responsible. HODEK et al. (1966) reported that natural predators of *Aphis fabae* overwinter in weeds and other plants around fields of sugar beets (*Beta vulgaris*). When white clover (*Trifolium repens*) and red clover (*Trifolium pratense*) were sown under Brussels sprouts (*Brassica oleracea* var. *gemmifera*), the number of cabbage butterflies (*Pieris rapae*) was reduced (DEMPSTER and COAKER 1974). This was due to an increase in predation by ground beetle (*Harpalus rufipes*). By feeding on other larvae in the clover, *Harpalus* increased in numbers to a level where they effectively reduced the butterfly larvae population. The number of cabbage root flies (*Erioischia brassicae*) and the number of eggs they laid were greatly reduced through beetle predation as was the number of cabbage aphids (*Brevicoryne brassicae*) flying to the crop.

The migration of animals between treated and untreated areas is a very important fact in investigation of agro-ecosystems. When biocides are used to control submersed weeds in aquatic ecosystems (KHARE 1977; HOLLINGSWORTH 1978), they sometimes extend their influence on a very large area. They injure not only other macrophytes but also the phyto- and zoo-plankton (BRAGINSKII et al. 1979; JANUSZKO 1979; KRZECZKOWSKA-WOLOSZYN 1979; ACHER and ELGAVISH 1980). SHIARIS et al. (1979) report on the effect of polychlorinated biphenyls on the nitrogen transformation by aquatic microorganisms, MICHAUD et al. (1979) on changes in phosphorus concentrations in an eutrophic lake as a result of macrophyte kill following herbicide application. Because of the severe effects which can arise through the presence of biocides in aquatic ecosystems and in the ground water, biological control of aquatic weeds by using grass carp *Ctenopharyngodon idella* as described by KULSHRESHTHA and SHARMA (1976) may be less dangerous than chemical weed control.

The report of KHAN (1978) shows the trend to incorporate bioactive chemicals into seeds to alleviate environmental stress. Also the use of chemical protectants against insects or other non-parasitic diseases influences the ecosystem. This happens even in cases when we use biocides such as pheremons, antifeedings, hormones or chemosterilants in very small concentrations. We must always anticipate a response of the whole ecosystem.

On summarizing all these facts, limits for chemical crop protection may arise particularly on the cropland itself due to the far-reaching disruptions of the agro-ecosystem. They are manifested by increasing intensiveness of production in the feedback control system: elimination of antagonists → further step up of chemical control → resistance to poison → changing the ecosystem with the disruption in the pathogen balance (DIERCKS 1976). So far it has been possible to defer such ecological limits, which however could at some point develop into real economic limits. A way of warding off this danger is the strategy of Integrated Pest Management (IPM, see preceding Chapter). It utilizes all reasonable economic, ecological, and toxicological methods in order to keep the injurious organisms below the threshold of economic damage by emphasizing exploitation of natural control factors. It needs solid knowledge of the agro-ecosystems and therefore more investigation of this synecological problem than in the past. Sometimes the effects of biocides are already controlled by model ecosystems (TOMIZAWA and KAZANO 1979; ISENSEE et al. 1979). In future it will be necessary to pilot agro-ecosystems from the point of view of economy and ecology (MÜLLER et al. 1980).

12.7 Effects of Biocides and Growth Regulators on Grassland Ecosystems

For several reasons herbicides are seldom used as intensively in grasslands as in cultivated areas. Often a grassland is sprayed in order to eliminate only one species (HARPER 1970). On upland pastures *Molinia coerulea* and *Nardus stricta* can be killed with low doses of dalapon. After herbicide application, re-seeding and fertilizing are usually necessary. Bracken (*Pteridium aquilinum*), one of the most widespread weeds of unimproved grassland, can be controlled very effectively with asulam. In prairies, xerothermic grassland, steppes and pampas, herbicides are used to kill poisonous plants such as *Delphinium, Lupinus,* and *Astragalus*. Arboricides are applied to reduce the number of woody plants, for example *Prosopis* in New Zealand or *Ulex* in Western Europe. In middle European nature reserves with xerothermic vegetation, the natural progression from grassland to scrub or forests has recently accelerated (SCHUBERT 1973). This succession must be controlled through grazing or by mechanical removal, and to a limited extent by biocides, if this type of habitat is to be preserved.

In improved grassland, *Cirsium arvense, Cirsium vulgare,* and large unpalatable species such as *Rumex obtusifolius* are controlled by chemicals. It is necessary to determine the best time of spraying (GRIZHUTINA and MOROZOVA 1970), because seasonal response is found in any mixed vegetation. In recent years the possibility of maintaining grassland through the use of herbicides, either as an alternative to, or in combination with grazing, has been gaining acceptance. Dalapon can reduce the less desirable grasses *Agrostis stolonifera, Poa trivialis* and *Holcus lanatus,* thereby allowing *Lolium perenne* to increase (ALLAN 1968). Glyphosate applied in August to wet upland pastures in the west of Scotland caused a reduction in the dominating *Juncus effusus* population which was replaced by *Deschampsia caespitosa* and *Galium hercynicum* (WILLIAMS 1979).

There is some evidence that herbicides alter the acceptibility and nutritional value of individual plant species to grazing animals. Thus, *Deschampsia caespitosa,* normally unpalatable, is readily eaten after spraying with dalapon. Crude protein content of red and white clover and lucerne was increased as much as 5% by simazine (SOJKOWSKI et al. 1969). On the other hand, phenoxy-acetic herbicides can reduce dry matter yield, digestible crude protein, and starch in grass swards. Poisonous plants, generally avoided by grazing animals due to their unpleasant taste, can be made attractive by spraying, which increases their sugar content, e.g. *Senecio jacobaea*. After treatement with chemicals non-poisonous plants sometimes become less acceptable and toxic due to an increase in the nitrate content of leaves (WAY 1969).

In short-term grassland or in recreational grassland, chemicals are only used to prevent weed invasion during the establishment phase. Herbicides are also recommended to promote the growth and dominance within the community of desirable grass species and to suppress undesirable ones. In extensive areas of grassland adjacent to roads, railways, canals, river banks, airports, and golf courses, chemicals are used more and more often to control the composition of the grass species (HIRANO and KIRITANI 1977). In all these cases we should realize that biocides can detrimentally affect wild animals by killing their preferred food plants or by disturbing predator/prey relationships.

12.8 Effects of Biocides and Growth Regulators on Forest Ecosystems

In forestry management chemicals are mostly used to prevent grassy and herbaceous vegetation from interfering with the early establishment of trees in forest nurseries or in planting schemes (BERGMANN 1968; DEPPENMEIER 1970). There are only few reports about the changes in the flora of the lower layers in forest ecosystems effected by biocides (PREEST 1977; STERRETT and ADAMS 1977). KLÖCK (1969) warns against using Prefix (chlorthiamid) against *Deschampsia flexuosa* on sites poor in nutrition because of the following dominance of this species. In forest nurseries or in planting schemes with a weed-mixture of grasses and herbs a treatment with delapon results in a flora only with herbs, a *Calamagrostis epigeios* – *Rubus idaeus* flora change to a pure *Rubus idaeus* community (PASSARGE 1970). A similar result has been found by FROCHOT 1979 when using dalapon against *Carex brizoides*. This species was replaced by *Rubus* sp. and *Rubus idaeus*. When a combination of dalapon and dichlobenil was given also *Geranium robertianum* becomes dominant.

When application is large scale, e.g. by aerial means, ecological consequences may be extreme, as demonstrated by the military use of biocides (WESTING 1977). Establishment of a monoculture vegetation over wide areas leads to an unstable situation in which a pest species may cause damage of disastrous proportions (WAY and CHANCELLOR 1976). Spraying of 120 ha of a pine forest in Keliya with 2,4-D from the air led to a 50% reduction in the total number of birds and a 50% reduction in the number of bird species, when compared to a similar area in a neighboring unsprayed forest. Terrestrial animals tended to avoid the sprayed area. The remaining birds had to extend their feeding territories as a result of reduction in food supply arising from alteration of the habitat.

Rangeland management practices must be reviewed and made ecologically sound. The present use of 2,4-D or 2,4,5-T or silvex to control shrub growth does not always lead to an improvement in the range grasses nor are the gaps left by the bushes necessarily invaded, as expected, by other desirable or undesirable species (HEDRICK et al. 1966). INGELOG (1978) reports on the control of brushwood by phenoxy acid herbicides and the tolerance and regeneration of grasses after application. In conclusion it can be said, that the more complex the ecosystem, the more severe are the changes that can be expected following the use of growth regulators and biocides. For every biocide it is necessary to have a solid knowledge of its effect on the different plants (MÜNCH 1972). Especially in forests with their numerous plants and animals, the effects of biocides can change the species composition in a serious manner.

12.9 Conclusions

Given the massive present-day use of growth regulators and biocides, it is extremely important to investigate their complex effects on plants, on entire agroecosystems, and on the landscape. As could be shown in this chapter, the ecological impact of these man-made compounds within ecosystems is very serious and not only a good knowledge of the toxicity of biocides is necessary, but also of their ecological influence on organisms and ecosystems. In future it will be important to pilot the agro-ecosystems from the point of view of economy and ecology and to study the effects of biocides by model ecosystems.

One way of warding off the danger of disruption of the agro-ecosystems will be the strategy of Integrated Pest Management, utilizing all reasonable economic, ecological, and toxicological methods to keep the injurious organisms below the threshold of economic damage by emphasizing exploitation of natural control factors.

References

Acher A, Elgavish A (1980) The effect of photochemical treatment of water on algal growth. Water Res 14(5):539–544

Adams JB, Drew ME (1969) Grain aphids in New Brunswick. IV. Effects of malathion and 2,4-D amine on aphid populations and on yields of oats and barley. Can J Zool 47:423–426

Åberg E, Stecko V (1976) Internal factors affecting toxicity. In: Audus LJ (ed) Herbicides Vol II. Academic Press, London New York, pp 175–201

Allan GP (1968) The potential role of selective herbicides in grassland improvement. Proc Br Weed Control Conf 9:1231–1237

Arlt K, Feyerabend G (1972) Herbizide und Kulturpflanzen. Akademie, Berlin

Ashton FM, Crafts AS (ed) (1981) Mode of action of herbicides. Wiley-Interscience, New York

Bachthaler G (1966) Der gegenwärtige Verbreitungsstand von Flughafer (*Avena fatua* L.) in der Bundesrepublik Deutschland. Weed Res 6:193–202

Baker EA (1972) The effect of environmental factors on the development of the leaf wax of *Brassica oleracea* var. *gemmifera*. MS thesis, Univ Bristol, Bristol

Barnes JM (1976) Toxic hazards in the use of herbicides. In: Audus LJ (ed) Herbicides Vol II. Academic Press, London New York, pp 373–391

Baron Le H, Gressel J (eds) (1982) Herbicide resistance in plants. Wiley and Sons, New York

Barralis G, Gasquez J, Jan P, Soffietti S (1979) Comportement ecologique et physiologique des dicotyledones resistantes a l'atrazine en France. Proc EWRS-Symp, Mainz, pp 217–224

Basler E, Todd GW, Meyer RE (1961) Effects of moisture stress on absorption, translocation, and distribution of 2,4-dichlorophenoxacetic acid in bean plants. Plant physic. Lancaster 36:572–576

Beck TH (1970) Der mikrobielle Abbau von Herbiziden und ihr Einfluß auf die Mikroflora des Bodens. Zentralbl Bakteriol Parasitenkd 124:304–313

Ben-Dyke R, Sanderson DM, Noakes DN (1970) Acute toxicity data for pesticides. World Rev Pestic Control 9:119–127

Bengtsson A (1970) 11th Swedish Weed Conference Uppsale, J 76

Bergmann H-J (1968) Die Pflege der europäischen Lärche (*Larix decidua* Miller) mit Herbiziden. Arch Forstwes 17:1033–1058
Bezuglov V, Minenko A, Postdeva R (1976) Herbicides in crop rotations and soil microflora. Khim Sel'sk Khoz 14. 12:45–49
Blake J, Kaufman DD (1975) Characterization of acylanilide-hydrolyzing enzyme(s) from *Fusarium oxysporum* Schlecht. Pestic Biochem Physiol 5:305–313
Blaszyk P (1970) Grundsätzliche Bemerkungen zum Einsatz von Herbiziden in Natur- und Landschaftsschutzgebieten. Z Pflanzenkrankh Sonderh 5:163–166
Braginskii L, Beskaravainaya V, Shcherban E (1979) Responses of fresh water phytoplancton and zoo-plancton to pesticides. Izv Akad Nauk SSSR Ser Biol 4:599–606
Bruinsma J (1967) Uptake and translocation of 4,6-dinitro-o-cresol (DNOC) in young plants of winter rye (*Secale cereale* L.). Acta Bot Neerl 16:73–85
Bukovac MJ (1965) Some factors affecting the absorption of 3-chlorophenoxy-propionic acid by leaves of the peach. Proc. Am Soc Hortic Sci 87:131–138
Bukovac MJ (1976) Herbicide entry into plants. In: Audus LJ (ed) Herbicides Vol I. Academic Press, London New York
Burnside O, Wicks G, Carlson D (1980) Control of weeds in an oat *Avena sativa* and soybean *Glycine max* eco-farming rotation. Weed Sci 28(1):46–50
Crafts AS, Crisp CE (1971) Phloem transport in plants. Freeman, San Francisco
Cramer HH (1977) Herbizide im Spannungsfeld zwischen Ökologie und Ökonomie. Naturwiss Rundsch 30(3):101–102
Cremer J (1976) Siebenjährige Untersuchungen über den Einfluß von Herbiziden auf die Unkrautflora in einer konzentrierten Fruchtfolge. Nachrichtenbl Pflanzenschutzdienst DDR 30:29–32
Curry JP (1970) The effects of different methods of new sward establishment and the effects of the herbicides paraquat and dalapon on the soil fauna. Pedobiologia 10:329–361
Davies PJ, Drennan DSH, Fryer JD, Holly K (1967) The basis of the differential phytotoxicity of 4-hydroxy-3,5-di-iodobenzonitrile. I. The influence of spray retention and plant morphology. Weed Sci 7:220–233
Dempster JP, Coaker TH (1974) Diversification of crops ecosystems as a means of controlling pests. In: Biology in pest and disease control. Blackwell, Oxford, pp 106–114
Deppenmeier E (1970) Bessere Kulturpflege durch chemische Unkrautbekämpfung. Forsttech Inf 9–17
Diercks R (1976) Die ökologischen Grenzen für die Agrarchemie im Sektor Pflanzenbau, bezogen auf den Pflanzenschutz. BASF-Mitt Landbau 2:115–141
Drift van der J (1963) The influence of biocides on the soil fauna. Neth J Plant Physiol 69:188–199
Eble H (1977) Faunistisch-ökologische Untersuchungen der epigäischen Entomofauna einer Agrobiozönose und der Einfluß einer abgestuften Herbizideinwirkung auf die Populationsdynamik. Hercynia 14:107–123
Eijsackers H, Drift van der J (1976) Effects on the soil fauna. In: Audus LJ (ed) Herbicides Vol II. Academic Press, London New York, pp 149–174
Frank H (1970) Das Rebhuhn. Merkbl Niederwildaussch DJV 8:12
Frochot H (1979) Modification de la flore naturelle de Chenates de l'est de la france par de traitements herbicides. Proc EWRS Symp, Mainz, pp 241–247
Fryer JD, Kirkland K (1970) Field experiment to investigate longterm-effects of repeated applications of MCPA, Triallate, Simacine and Linuron: Report after six years. Weed Res 10:133–158
Grizhutina AP, Morozova LK (1970) Poverchnospnoe uluosenie pojnennych lugov v zapoljary pri pomosci. Khim Sel'sk Khoz 8:922–926
Grossbard E (1970) An appraisal of the criteria by which to measure the effect of herbicides on the soil microflora. Meded Fac Landbouwhosch Gent 35:515–530
Grossbard E (1976) Effects on the soil microflora. In: Audus LJ (ed) Herbicides Vol II. Academic Press, London New York, pp 99–148
Gruzdev GS, Mozgovoj AF, Kuz'mina IV (1973) Vlijanie treflana na biologiceskuju aktivnost' pocvy v posevach podsolnecnika. Izv Timirjazevski Skh Akad 96:136–141

Haas H, Streibig JC (1982) Changing patterns of weed distribution as a result of herbicide use and other agronomic factors. In: Baron Le H, Gressel J (eds) Herbicide resistance in plants. Wiley and Sons, New York
Hanf M (1957) Über die Änderung der Blattformen von Dikotyledonen durch Dichlorphenoxyessigsäure (2,4-D). Beitr Biol Pflanz 33:177–218
Hanf M (1979) Einführung zum Einfluß verschiedener Faktoren auf Entwicklung und Bekämpfung von Unkräutern. Proc EWRS Symp, Mainz, pp 5–11
Hanson NS (1972) Weed control practices and research for sugar cane in Hawaii. Weeds 10:192–200
Harper JL (1970) Grazing, fertilizers and pesticides in the management of grasslands. Symp Br Ecol Soc Oxford 11:15–30
Harvey RG, Muzik TJ (1973) Effects of 2,4-D and amino acids on field bindweed in vitro. Weed Sci 21:135–138
Hay JR (1976) Herbicide Transport in Plants. In: Audus LJ (ed) Herbicides Vol I. Academic Press, London New York, pp 365–396
Hedrick DW, Hyder DN, Sneva FA, Poulton CE (1966) Ecological response of sagebrush-grass range in Central Oregon to mechanical and chemical removal of *Artemisia*. *Ecology* 47:432–439
Helmecke K, Hickisch B, Mahn E-G, Prasse J, Sternkopf G (1977) Beiträge zur Wirkung des Herbizideinsatzes auf Struktur und Stoffhaushalt von Agro-Ökosystemen. Hercynia NF 14:375–398
Hirano C, Kiritani K (1977) Paddy ecosystem affected by nitrogenous fertilizer and insecticides. In: Fukushima Yoichi (ed) Proc Int Congr Human Environ (HESC) Kyoto, Japan, 17–26, 1975. JSBNO-08-021948-9, pp 197–206
Hodek J, Holman J, Novak K, Skuhravy V, Stary P, Weisman L, Zeleny J (1966) The present possibilities and prospects of integrated control of *Aphis fabae*. Scop. In: Ecology of aphidophagous insects. Academia, Prague, pp 331–335
Hodgson JM (1964) Variations in ecotypes of Canada thistle. Weeds 12:167–171
Hodgson JM (1973) Lipid deposition on leaves of Canada thistle ecotypes. Weed Sci 21:169–172
Hodgson JM, Moore HD (1972) Stomata variations in Canada thistle and response to herbicides. Weed Sci 20:68–70
Hojden B (1973) Die Reaktion der Ökotypen von *Centaurea cyanus* auf die Herbicide 2,4-D und MCPA. In: Probleme der Agrogeobotanik (P 2). Wiss Beitr MLU, Halle, pp 177–179
Hollingsworth E (1978) A small plot technique for assessing control of submersed weeds. Weed Sci 26(6):554–556
Hull H (1970) Leaf structure as related to absorption of pesticides and other compounds. Residue Rev 31:1–155
Hurle K, Rademacher B (1970) Untersuchungen über den Einfluß langjährig wiederholter Anwendung von DNOC und 2,4-D auf ihren Abbau im Boden. Weed Res 10:159–164
Ingelog T (1978) Effects of the silvicultural use of phenoxy acid herbicides on forest vegetation in Sweden. In: Ramel C (ed) Ecol Bull NRF 27:240–254
Isensee AR, Jones GE, Turner BC (1971) Root absorption and translocation of picloram by oats and soybeans. Weed Sci 19:727–731
Isensee A, Jones G, McCann J, Pitcher F (1979) Toxicity and fate of 9-toxaphene fractions in an aquatic model ecosystem. J Agric Food Chem 27(5):1041–1046
Jacob F, Neumann ST, Strobel U (1973) Studies on mobility of exogen applied substances in plants. In: Proc Res Inst Pomol Skierniewice, Poland, Ser E, Conf Symp 3, pp 315–330
Jacobsohn R, Andersen RN (1968) Differential response of wild oat lines to diallate, triallate and barban. Weed Sci 16:491–494
Januszko M (1979) Development of algae in ponds treated with foschlor and in ponds with controlled time of fill. Pol Arch Hydrobiol 26:247–256
Katan J, Eshel Y (1973) Interaction between herbicides and plant pathogens. Residue Rev 45:145–177
Kaufman DD, Kearney PC (1976) Microbial transformations in the soil. In: Herbicides Vol II. Academic Press, London New York, pp 29–64

Kees H (1979) Beobachtungen über Selektion und Resistenzbildung bei Unkräutern durch Herbizide und Fruchtfolgevereinfachung in Bayern. Proc EWRS-Symp, Mainz, pp 225–232
Khan A (1978) Incorporation of bioactive chemicals into seeds to alleviate environmental stress. Acta Hortic 83:225–234
Khare L (1977) Effect of 4 herbicides on some aquatic weeds. Geobios (Jodhpur) 4(6):271–272
Klöck W (1969) Herbizidanwendung und waldbauliche Vorbeugungsmaßnahmen bei der Unkrautbekämpfung. Allg Forstztg 24:644–646
Koch W, Hurle K (1977) Grundlagen der Unkrautbekämpfung. Uni-Taschenbücher 513. Ulmer, Stuttgart
Koeman JH (1978) Chemicals in the environment and their effects on ecosystems. In: Geissbühler H (ed) Advances in pesticide science, part 1. Pergamon, Oxford, pp 25–38
Krzeczkowska-Woloszyn L (1979) Phyto-plankton in fish ponds treated with an organophosphorus insecticide neguvon. Acta Hydrobiol 21(2):139–148
Kulshreshtha S, Sharma K (1976) Biological control of aquatic weeds in the chambal commanded area of kota using grass carp *Ctenopharyngodon idella*. J Inl Fish Soc India 8:113–114
Kurth H (1968) Chemische Unkrautbekämpfung Vol VII. 3 Aufl. Fischer, Jena
Laudin H (1972) Einfluß der Herbicidanwendung auf die freilebende Tierwelt. Z Pflanzenkrankh Sonderh 6:127–131
Leius K (1967) Influence of wild flowers on parasitism of tent caterpillar and codling moth. Can Entomol 99:444–446
Levis J, Papavizas G, Hora T (1978) Effects of some herbicides on microbial activity in soil. Soil Biol Biochem 10(2):137–142
Linder PL, Mitchell IW, Freeman GD (1964) Persistence and translocation of exogenous regulating compounds that exude from roots. J Agric Food Chem 12:437–438
Luckwill LC, Lloyd-Jones CP (1962) The absorption, translocation and metabolism of 1-naphthaleneacetic acid applied to apple leaves. J Hortic Sci 37:190–206
Mahn E-G, Helmecke K (1979) Effects of herbicide treatment on the structure and functioning of agro-ecosystems. II. Structural changes in the plant community after the application of herbicides over several years. Agro-Ecosystems 5:159–179
Marsh J, Wingfield G, Davies H, Grossbard E (1978) Simultaneous assessment of various responses of the soil microflora to bentazone. Weed Res 18(5):293–300
McWhorter CG (1971) Control of johnsongrass ecotypes. Weed Sci 19:229–233
Michajlova EI, Kruglov JV (1973) Vlijanie nekotorych gerbicidov na al'goflory pocvy. Pocvovedenie 8:81–85
Michaud M, Atchison G, McIntosh A, Mayes R, Nelson D (1979) Changes in phosphorus concentrations in an eutrophic lake as a result of macrophyte kill following herbicide application. Hydrobiologia 66(2):105–112
Mittnacht A, Eberhardt CH, Koch W (1979) Wandel in der Getreideunkrautflora seit 1948, untersucht an einem Beispiel in Südwestdeutschland. Proc EWRS-Symp, Mainz, pp 209–216
Müller G (1972) Faunistisch-ökologische Untersuchungen der Coleopterenfauna der küstennahen Kulturlandschaft bei Greifswald. Pedobiologia 12:169–211
Müller G, Klammt E, Thiele A, Wiese U (1980) Faunistisch-ökologische Untersuchungen an Indikatorgruppen der Makrofauna im Küstenbezirk. In: Forschungsbeiträge zur Ökologie und zum Umweltschutz. EMA Univ Greifswald, pp 127–143
Müller JP (1974) Die modernen Aufgaben der Faunistik für die angewandte Entomologie und die Biosphärenforschung. Folia Entomol Hung 27:Suppl 15–28
Münch WD (1972) Änderung der Artenzusammensetzung von Unkräutern in Forstkulturen nach Anwendung von Herbiziden. Z Pflanzenkrankh 8/9:485–497
Muzik TJ (1976) Influence of environmental factors on toxicity to plants. In: Audus LJ (ed) Herbicides Vol II. Academic Press, London New York, pp 203–247
Muzik TJ, Mauldin WG (1964) Influence of environment on the response of plants to herbicides. Weeds 12:142–145
Overbeck van J (1956) Absorption and translocation of plant regulators. Annu Rev Plant Physiol 7:355–372

Paeschke R, Heitefuss R (1978) The influence of herbicides on metabolic active bacteria and fungi in soil after single or multiannual application. Z Pflanzenkr Pflanzenschutz 85:8–9, 471–481
Passarge H (1970) Zur Kenntnis der Vegetationsfolge nach Kahlschlag, eine Voraussetzung für die rationelle Unkrautbekämpfung. Arch Forstwes 19(3):269–276
Pimentel D (ed) (1981) CRC handbook of pest management in agriculture, vol I–III. CRC Press, Boca Raton
Pötsch J (1976) Ökologische Aspekte der Ackerunkrautverbreitung in ihrer Bedeutung für die Bekämpfungsentscheidung. Wiss Z Paed Hochsch Potsdam 20(1):47–55
Potts GR (1970) The effects of the herbicides in cereals on the feeding ecology of partridges. In: Proc 10th Br Weed Contr Conf, pp 299–302
Prasse J (1975) Investigations concerning the effect of the herbicides 2,4-D and simazin on the coenosis of the collembola and of the acari of a arable soil. In: Proc 5th Int Coll Soil Zool, Prag 1973. Acad Czech Acad Sci, Prague, pp 469–480
Preest D (1977) Long-term growth response of douglas-fir to weed control. N Z J For Sci 7(3):329–332
Rao BHK, Prasad SK (1968) EPTC as a selective nematicide. Nematologia 14:312–313
Reid CPP, Hurtt W (1970) Root exudation of herbicides by woody plants: Allelopathic implications. Nature (London) 225:291
Roche BF, Muzik TJ (1964) Ecological and physiological study of *Echinochloa crus galli* (L.) BEAUV. and the response of its biotypes to sodium 2,2-dichloropropionate. Agron J 56:155–160
Rochecouste E (1962) Studies on the biotypes of *Cynodon dactylon* (L.) PERS. II. Growth response to trichloroacetic and 2,2-dichloropropionic acids. Weed Res 2:136–145
Rola J (1973) Der Einfluß der Intensivierung der Landwirtschaft auf die Segetalgemeinschaften. In: Probleme der Agrogeobotanik (P2). Wiss Beitr MLU, Halle, pp 139–146
Rola J (1979) Das Zusammenwirken von Kulturpflanzen und der angewendeten Herbizide auf die Verunkrautung. Proc EWRS Symp, Mainz, pp 281–285
Sandmann ER (1974) Aromatic metabolism by phenoxy herbicide degrading soil bacteria. M Sc thesis, Univ Natal, South Africa
Sargent JA, Blackman GE (1972) Studies on foliar penetration. IX. Patterns of penetration of 2,4-dichlorophenoxyacetic acid into the leaves of different species. J Exp Bot 23:830–841
Schmidt RR (1979) Änderungen der Unkrautflora auf drei Bodenarten durch unterschiedlichen Fruchtwechsel und durch Herbizide. Proc EWRS Symp, Mainz, pp 281–285
Schooler AB, Bell AR, Nalewaja JD (1972) Inheritance of siduron tolerance in foxtail barley. Weed Sci 20:167–169
Schubert R (1971) Ökologische Fragen bei der Chemisierung der Landwirtschaft. In: Vortragstagung über Agrochemie. Wiss Beitr MLU, Halle (N1)
Schubert R (1973) Probleme der natürlichen Wiederbewaldung von Naturschutzgebieten mit Xerothermstandorten im südlichen Teil der DDR. Acta Bot Acad Sci Hung T 19(1–4):317–327
Schubert R (1977) Ausgewählte pflanzliche Bioindikatoren zur Erfassung ökologischer Veränderungen in terrestrischen Ökosystemen durch anthropogene Beeinflussung unter besonderer Berücksichtigung industrieller Ballungsgebiete. Hercynia NF 14(4):399–412
Schubert R (1980) Neue Ergebnisse aus der herbologischen Forschung unter Berücksichtigung ökologischer und technologischer Aspekte der Herbizidanwendung. Tagungsber 182 Akad Landwirtsch DDR Berlin 182:213–217
Schubert R, Hilbig W (1969) Verbreitungskarten mitteldeutscher Leitpflanzen. 12. Reihe. Wiss. Z Univ Halle, Math-Nat R Jg 18(2):163–210
Sharma MP, Chang FY, Born Vanden WH (1971) Penetration and translocation of picloram in Canada thistle. Weed Sci 19:349–355
Shiaris M, Holt W, Sayler G (1979) The effect of polychloraminated biphenyls on nitrogen transformations by aquatic microorganisms. Abstr Annu Meet Am Soc Microbiol 79:225
Sojkowski Z, Staszewski Z, Sojkowska T (1969) Wplyw simazinu na zawartosc bialka surowego u niektorych roslin uprawnych. Hodwola rosl aklim nasienn 13:433–436

Steche W (1972) Unkrautbekämpfung und Bienenschutz. Z Pflanzenkrankh Sonderh 6:133–135
Sterrett J, Adams R (1977) The effect of forest conversion with herbicides on pine *Pinus*-ssp. establishment soil moisture and understory vegetation. Weed Sci 25(6):521–523
Stryckers JMT (1979) Veränderungen in der Unkrautflora durch Herbizidanwendung. Proc EWRS Symp, Mainz, pp 25–38
Stugren B (1978) Grundlagen der allgemeinen Ökologie. Fischer, Jena
Tomizawa C, Kazano H (1979) Environmental fate of rice paddy pesticides in a model ecosystem. J Environ Sci Health, part B. Pestic Food Contam Agric Wastes 14(2):121–152
Wallnöfer P, Poschenrieder G, Engelhardt G (1979) Kombination von Pflanzenbehandlungsmitteln – ökologische und umwelttoxikologische Aspekte. Gesunde Pflanz 31(8):202–204
Wax LM, Bernard RL, Hayels RM (1974) Response of soybean cultivars to bentazon, bromoxnil, chloroxuron and 2,4-DB. Weed Sci 22:35–41
Way JM (1969) Toxicity and hazards to man, domestic animals and wildlife from some commonly used auxin herbicides. Residue Rev 26:37–62
Way JM, Chancellor RJ (1976) Herbicides and higher plant ecology. In: Audus LJ (ed) Herbicides Vol II. Academic Press, London New York, pp 345–372
Westing A (1977) Ecological effects of the military use of herbicides. In: Perring FH, Mellanby (eds) Linnean Soc Symp Serv No 5 JSBN 0-12-55 1350-X, pp 89–94
Whitecross MJ, Armstrong DJ (1972) Environmental effects on epicuticular waxes of *Brassica napus* L. Aust J Bot 20:87–95
Williams GH (1979) Long-term effects of herbicide applications in hill areas. Proc EWRS Symp, Mainz, pp 233–240
Zandstra BH, Motooka PS (1978) Beneficial effects of weeds in pest management – a review. Pans (3):333–338
Zemánek J (1970) Pokusy s citlivosti vybranych odrud ozime psenice k herbicidum. Ochrana Rostlin Praha 6:261–268
Zemánek J (1979) Einfluß der langjährigen Herbizidanwendung auf die Verunkrautung bei verschiedenen Fruchtfolgen. Proc EWRS Symp, Mainz

13 Eutrophication Processes and Pollution of Freshwater Ecosystems Including Waste Heat

A. KOHLER and B.C. LABUS

CONTENTS

13.1 Introduction 413

13.2 Eutrophication 414
13.2.1 General Eutrophication Processes in Still and Running Waters . . . 415
13.2.2 Eutrophication and Freshwater Vegetation 417
13.2.3 Ecophysiological Studies in Eutrophication 421

13.3 Toxic Pollutants 424
13.3.1 Heavy Metals 425
13.3.1.1 Heavy Metals and Freshwater Vegetation 425
13.3.1.2 Toxicity of Heavy Metals 426
13.3.1.3 Bioaccumulation of Heavy Metals and Adaptation 428
13.3.1.4 Environmental Interactions and Heavy Metal Toxicity . . . 429
13.3.2 Pesticides and Related Organic Compounds (Phenolics, PCBs) . . . 430
13.3.3 Halogens (with Reference to Chlorine) 432
13.3.4 Surfactants 433
13.3.5 Radioactive Wastes 434
13.3.6 Salinity (with Reference to Chlorides) 435
13.3.7 Freshwater Food Chains 437

13.4 Acidification 437

13.5 Waste Heat 439
13.5.1 Waste Heat and Freshwater Vegetation 439
13.5.2 Temperature and Plant Sensitivity to Selected Chemicals 442

13.6 Biological Indicators 442

13.7 Conclusions 443

References 444

13.1 Introduction

Man's increasing utilization of aquatic systems as suppliers and recipients of water used for industrial or municipal purposes, combined with undesirable contributions from cultivated land and streets, contaminated atmospheric fallout, transport as well as excessive recreational use have subjected lentic and lotic waters to perturbations of considerable magnitude within the recent decades. The degree of disturbance is dependent on the ability of a particular water ecosystem to withstand displacement of structure and function induced by adverse conditions (assimilatory capacity) and the ability of the system to snap back after displacement (elasticity), as indicated by CAIRNS (1976).

Freshwater lakes and rivers, estuaries, and oceans display quite distinct characteristics in terms of vulnerability to irreversible effects, severity of environmental stress present, and frequency of man-imposed adverse conditions.

Based on these characteristics a rank-ordering system suggested by CAIRNS and DICKSON (1977) establishes rivers as the presently most severely afflicted systems, followed by estuaries, lakes, and oceans in declining order (see Table 13.1).

Table 13.1. Relative rank-ordering system for lakes, rivers, estuaries, and oceans in respect to irreversibility, severity, and frequency of man-induced perturbations (1 representing highest, 100 lowest irreversiblity, severity or frequency). (After CAIRNS and DICKSON 1977)

	Lakes	Rivers	Estuaries	Oceans
Irreversibility of effects	20	80	100	1
Stress severity	80	50	1	100
Frequency of perturbations	80	1	40	100
Total	180	131	151	201

Changes in the biological, chemical and physical properties of a freshwater ecosystem, resulting from a given pollution situation, basically may be linked to (1) excesses of organic substances, (2) excesses of inorganic substances, (3) specific toxicants, (4) excesses of suspended solids, (5) thermal increases, or possibly (6) radionuclides (CAIRNS et al. 1972). Although a pollution situation may be governed by one of the above influences, the sole presence of just one cause of pollution is rarely the case. This can impose severe problems in establishing specific causality between observed effects on the ecosystem and single pollutants.

13.2 Eutrophication

The term eutrophication originally served to describe the natural phenomenon of increasing production in lakes (THIENEMANN 1931; LINDEMAN 1942). Current consensus among many limnologists allows for a more extended understanding of the term. According to ELSTER (1977) and VOLLENWEIDER (1968) eutrophication is seen as increasing enrichment of surface waters with plant nutrients, with all implied consequences. Increasing nutrient supply results in accelerated productivity that exceeds the rate of production that would have occurred in the absence of the perturbations of the system, affecting all compartments of a limnetic system (ELSTER 1974; WETZEL 1976). WUHRMANN (1976) defines eutrophication as development of an aquatic ecosystem into a state in which aerobic microbial decomposition of organic matter consumes more oxygen than is naturally introduced to the system, resulting in a steadily increasing oxygen deficit (see also SCHMASSMANN 1955).

Numerous factors are involved in eutrophication. In general, geochemical, climatic-hydrological characteristics of the catchment area of a freshwater system, its limnological properties, the properties of the water and anthropogenic alterations determine its productivity (VOLLENWEIDER 1979).

In the current concept of nutrient loading, both lentic and lotic waters are considered open systems; their productivity is seen as resultant of a dynamic equilibrium that describes the systems' metabolism as balance between import and export of matter (VOLLENWEIDER 1979). The concept differentiates between "internal" and "external" loading. External loading, which is of main concern, falls into an inorganic and an organic component. While inorganic loading directly feeds into the available nutrient pool, organic loading affects the system over the bacteria and detritus pool with secondary effects on the oxygen budget of the system and on transfer to the nutrient pool during decomposition.

Phosphorus is generally considered as the decisive limiting factor of primary production in oligotrophic and mesotrophic freshwater ecosystems (HUTCHINSON 1957, 1967; OHLE 1953; SCHINDLER 1971; THOMAS 1973; VOLLENWEIDER 1979). Nitrogen has been reported as a limiting factor, mainly in eutrophic still waters, with high amounts of available phosphorus (CLAESSON and RYDING 1977; FORSBERG 1977; GOLDMAN and WETZEL 1963). Carbon may become limiting, especially in waters of low alkalinity, with excessive amounts of phosphorus and nitrogen present (KING 1970; LANGE 1970; LIKENS 1972; SAKAMOTO 1971; SCHINDLER and NIGHSWANDER 1970). Micronutrient limitation of primary production has also been demonstrated (GOLDMAN 1972; GOLDMAN and CARTER 1965; HAMILTON 1969; LUND 1965; SCHELSKE 1962). Light may be considered an ultimately limiting factor (HYNES 1969; WETZEL and HOUGH 1973).

Domestic sewage is listed as the major contributor to the external loading of freshwaters, accounting for as much as 67% of the total P input followed by industrial sources (10%) as recently stated for West German surface waters (BERNHARDT 1978), with a large proportion (50%–70%) derived from detergents (ELSTER 1977). Runoff from urban and agricultural land, drainage from agricultural land, soil erosion, springs and groundwater from fens and marshland are contributors of nitrogen and phosphorus with regional significance (BERNHARDT 1978). This also holds true for nutrient contributions from decomposing leaf litter accumulation (COWEN and LEE 1973; HOWARTH and FISHER 1976) and from water fowl (MANNY et al. 1975; SANDERSON 1953). Atmospheric fallout and washout provides 0.01 to 0.1 $g\,m^{-2}\,yr^{-1}$ of phosphorus and 0.1 to 1.0 $g\,m^{-2}\,yr^{-1}$ of nitrogen (WETZEL 1975); though rather low, phosphorus input of this magnitude may already induce remarkable primary production in oligotrophic lentic waters (PEUKERT 1976). Nitrogen fixation by blue-green alga heterocysts and bacteria can add substantial portions to the nitrogen input to a lake, with seasonal values from 30% to 40% not being uncommon (BREZONIK 1972; GRANHALL and LUNDGREN 1971; HORNE and FOGG 1970; HORNE and GOLDMAN 1972; HORNE and VINER 1971; STEWART 1968).

Phosphorus occurs in freshwater systems in dissolved and particulate form. Of inorganic forms only orthophosphate (PO_4^{3-}) is significant. More than 90% of the total phosphorus in lake water is organically bound in organisms or adsorbed to inorganic or organic matter (WETZEL 1975). Nitrogen is represented by nitrate (NO_3^-) and ammonia (NH_4^+) as the most important inorganic forms, nitrite (NO_2^-) and molecular nitrogen, subjected to denitrification and nitrification processes, and a vast variety of N-containing organic compounds (amino acids, amines, polypeptides etc.). Permissible and dangerous loading levels of lakes with total nitrogen and phosphorus are indicated according to mean depth by VOLLENWEIDER (1968), with 2.0 $g\,m^{-2}\,yr^{-1}$ of total N and 0.13 $g\,m^{-2}\,yr^{-1}$ of total P considered dangerous for a lake of 5 m depth. The amount of total phosphorus in running waters with only slight loading is less than 100 $mg\,m^{-3}$, while serious loading may lead to values of several hundred $mg\,m^{-3}$, with dangerous loading exceeding 1,000 $mg\,m^{-3}$, as stated by WETZEL (1975).

13.2.1 General Eutrophication Processes in Still and Running Waters

In an eutrophic lake the total potential concentration of nutrients is high; this may imply extremely low concentrations in the water as the major supply is momentarily locked up in organisms or the sediment (HUTCHINSON 1969; WETZEL 1975). The major consequences of heavy nutrient loading in a lake may be listed as follows (see also ELSTER 1977): (1) There is a massive increase in algal biomass (AHL 1975; GRIM 1967; SCHINDLER and FEE 1973; VOLLENWEIDER 1968, 1979) with blooms of Cyanophyceae appearing (LUND 1969; MORGAN 1970; RIEMANN and MATHIESEN 1977) and rapid growth of bacterial populations with raised PO_4^{3-} levels (THOMAS 1973). Light attenuation by self-shading of phytoplankton limits the trophogenic zone to the upper meters of the epilimnion. (2) Photosynthesis leads to daytime oxygen supersaturation (even over 300%) and carbon dioxide shortage in the epilimnion in the summer (THOMAS 1972), while (3) increased amounts of sedimenting organic matter (dead algal cells etc.) further enlarge the oxygen deficit in the hypolimnion and raise the carbon dioxide levels during decomposition. As sediment and bottom water layers turn anaerobic a critical phase is reached. (4) H_2S production by anaerobic sulfate-reducing bacteria in this zone is stimulated (OHLE 1954), while phosphorus, iron, and manganese are released at increasing rates with lowering of the redox potential and reduction of the oxidized microlayer of the sediment which has acted as a barrier to such release (MORTIMER 1971). (5) As insoluble FeS, formed by Fe^{2+} ions and H_2S in water layers close to the sediment, is precipitated (OHLE 1954), released phosphate remains in solution and can be transported to the trophogenic zone with lake turnover in fall or spring (EINSELE 1936). This clearly indicates not only the importance of the chemical dynamics within a lake but also that an element (sulfur) that is not directly limiting production may have profound impact on the chemical dynamics of a limiting nutrient (HUTCHINSON 1969). Other internal phosphorus cycling processes within the water column of a lake, such as excretions by phytoplankton, bacteria and zooplankton, have to be considered, as they may be an order of magnitude more important than normal return from sediment as nutrient supply for phytoplankton (LEAN 1973a, b; PETERS 1972). HUTCHINSON and BOWEN (1950) indicate the littoral region as a phosphorus source that maintains epilimnetic P concentration during summer stratification. CHAMBERLAIN (1968) emphasizes converse net movements of phosphorus, while HAYES et al. (1952) and RIGLER (1956) view the littoral region and epilimnion as an equilibrium exchange system. As pointed out by CONFER (1972) many authors postulate that P circulation is a steady-state system with emphasis on the importance of P influx (see also GRIM 1967; WETZEL 1975). According to WETZEL (1975, p. 236) there is evidence "that the chemical mass of inorganic phosphorus of the water has relatively little relation to growth kinetics. Of overriding importance is the rapidity with which the phosphorus is cycled and exchanged between the particulate phosphorus and soluble and organic phases". More differentiated aspects of nutrient cycles are discussed in Chapter 8, this Volume. (6) Major quantitative and qualitative changes in macrophyte, algal and bacterial populations are involved in the process of eutrophication.

Running waters generally are more fertile than still waters, as the formation of nutrient depleted zones around plants is prevented by turbulent flow (RUTTNER 1926) and plant metabolism is enhanced by water movement (WESTLAKE 1967; WHITFORD 1960). Along the course of a river the total load of suspended matter and usually the concentration of dissolved solids tends to increase. In spite of higher nutrient load and temperature in the lower reaches, plant growth may not be enhanced as turbidity normally limits light penetration (HYNES 1969). Aside from direct import of inorganic nutrients, degradable organic sewage holds a major role in the enrichment of running waters. As organic wastes are decomposed by bacterial self-purification processes in streams and rivers, several zones can be distinguished below a point discharge (BRINLEY 1942; HYNES 1959; WETZEL 1969): (1) In a zone of active bacterial decomposition (polysaprobic zone) immediately below the effluent, most of the sewage is not yet decomposed, bacterial count and biological oxygen demand (BOD) is high, while oxygen concentrations are less than 3 mg l^{-1} (often zero). The plankton volume is low, submerged macrophytes and other higher organisms are missing. (2) Flourishing sewage-tolerant algae and bacteria characterize a zone of intermediate bacterial decomposition (alpha-mesosaprobic zone) in which oxygen-producing algae and mineralizing aerobic bacteria, providing nutrients, stimulate each other. Dissolved oxygen increases to 3 to 5 mg l^{-1} during the day but drops at night. (3) The subsequent fertile zone of recovery (beta-mesosaprobic zone) shows a decline in bacterial count, but displays abundant growth of eutraphent algae and macrophytes, as increased nutrient supply is available from upstream decomposition of organic matter (EICHENBERGER 1976; KOHLER et al. 1974; THOMAS 1975). While ODUM (1956) rated this zone of recovery as extremely productive according to measurements of oxygen concentration in the water, studies of DUFFER and DORRIS (1966) and others indicate that community respiration may exceed photosynthesis. Diurnal variations of dissolved oxygen are large if compared to non-polluted water (SCHMASSMANN 1955) with supersaturation possible during daytime and sharp drops during the night, though usually not below 5 mg l^{-1}. (4) As the stream or river reaches purification in the subsequent oligosaprobic zone, eutraphent vegetation elements are replaced successively by mesotraphent species which only tolerate lower concentrations of available nutrients. Oxygen concentrations should remain near the saturation point. The length of each zone is determined by temperature, rate of flow, the physical nature of the system as well as the amount and quality of the allochthonous organic matter. Continuous or pulsed external loading with nutrients (including degradable organic matter) is the decisive factor in the enrichment of running waters as there is ongoing "elimination" with downstream transport. Excessive growth of certain submerged macrophytes may alter the flow characteristics and raise the water level of running waters and lead to increased tertiary pollution during decomposition. Also sedimentation of suspended matter is enhanced, with the decrease in turbidity favoring plant growth as light penetration increases. Digestive sludge formation may occur as dense mats of sessile algae in eutrophic stretches of a stream and impede water flow along the bottom zone, with anaerobic conditions leading to H_2S formation at the bottom, while aerobic conditions still prevail in the flow zone (THOMAS 1975).

13.2.2 Eutrophication and Freshwater Vegetation

The qualitative and quantitative vegetation changes that occur with increasing nutrient supply are especially troublesome in oligotrophic lakes and wetlands.

In nutrient deficient soft water lakes of Scandinavia recession of *Isoëtes* lawns (*Isoëtes* spp., *Lobelia dortmanna*), *Myriophyllum alterniflorum, Ranunculus peltatus* and other oligotraphent species of the upper and lower littoral, as well as planktonic Desmidiaceae sets in and rapidly proceeds with maintained fertilization. They are largely replaced by allochthonous eutraphent elements such as *Ceratophyllum demersum, Myriophyllum verticillatum* and lemnids; in extreme cases blooms of Cyanophyceae develop. There is noticeable proliferation of *Typha latifolia* and *Glyceria maxima* in the eulittoral zone as available nutrient levels are raised (KURIMO 1970; SUOMINEN 1968; UOTILA 1971). In oligotrophic carbonate lakes *Chara* lawns and *Najas* communities disappear as total phosphorus in the water exceeds 20 $\mu g\, l^{-1}$ (FORSBERG 1964; MELZER 1976) while submerged macrophyte communities become dominated by broadleafed *Potamogeton* species (*P. lucens, P. perfoliatus*), *P. pectinatus, Zannichellia* lawns (LANG 1973; MELZER et al. 1977) or *Myriophyllum spicatum* (AIKEN et al. 1979; COFFEY and MCNABB 1974; FORSBERG 1964; GOLDSBY et al. 1978) as well as *Elodea canadensis* (FISH 1966; KRAUSE 1972). Submersed macrophytes contribute increasingly to the total primary productivity of lakes until the fertility of the system is increased to a point where intense phytoplankton productivity leads to severe light attenuation (WETZEL and HOUGH 1973; see also GOULDER 1969; JUPP and SPENCE 1977). Inverse correlation between phytoplankton and macrophyte development was also demonstrated in experiments of MULLIGAN and BARANOWSKI (1969), with high nitrogen and phosphorus levels producing best growth of phytoplankton, while lower levels resulted in higher yields of vascular plants. With increasing biogenic turbidity the depth distribution boundary of submersed macrophytes is lifted from the lower to the upper littoral (MELZER in KOHLER 1978; LANG 1973). While self-shading effects control phytoplankton growth at higher population densities, the emergent flora and associated epiphytics and eulittoral algae proliferate (WETZEL and HOUGH 1973). These authors view the algae attached to the sub-strata (WETZEL 1964; HARGRAVE 1969) or macrophytes (ALLEN 1971; HICKMAN 1971) as the dominant, if not major site of autochthonous productivity in less fertile sites. In extremely hypereutrophic waters, however, their contribution to total community productivity has been found small (WESTLAKE 1963).

Antagonism between algae and submersed macrophytes, largely attributed to light competition and different nutrient demands and tolerances, may be further complicated if secretions of suppressive substances by established macrophytes that restrict phytoplankton development are also considered (HASLER and JONES 1949; HOGETSU et al. 1960). The effectiveness of epiphyte and filamentous algae to overcome this restriction will depend on the timing of their development. PHILLIPS et al. (1978) suggest that the production of phytoplankton-suppressive secretions of macrophytes is reduced as their net production becomes restricted. FITZGERALD (1969) proposed from laboratory findings that antagonistic activity of aquatic weeds may be due to a "nitrogen sink" effect which prevents the growth of contaminating algae by competition for the limited nitrogen compounds available.

Increased nutrient supply at first leads to higher reed biomass (HÜRLIMANN 1951; BJÖRK 1967; HASLAM 1971), whereas a decline in *Phragmites* stands, as noted for Lake Constance and elsewhere (KLÖTZLI and GRÜNIG 1976; MARKSTEIN and SUKOPP 1980), is encountered with the transgression of a threshold of 10 mg m^{-3} of total P in the water (KLÖTZLI, personal communication). Thick algal blankets developing in and around reed beds bend smaller and peripheral stems of the stands and pull them below the water surface; as the algal mats block the exchange with pelagic water, land-sided total still water zones are formed in which anaerobic conditions develop. Land-sided decline of *Phragmites* stands is explained as consequence of H_2S and NH_3 poisoning of the rhizomes (DEUFEL 1978). Changes initiated by sewage in wetland helophyte stands are described by KLÖTZLI (1965/66). When gross eutrophication is reached in a lake, large visible aggregations of floating algae bloom extensively, particularly with blue-green forms during the late summer. The most common bloom-forming algae according to PALMER et al. (1977) are *Anacystis* and *Anabaena,* but others such as *Aphanizomenon, Gomphosphaeria, Rivularia,* and *Oscillatoria* may also bloom. It is well known that lakes that commonly have water blooms do not always have them, nor do the blooms always arise in the expected period of the year (LUND 1969). Blooms of blue-green algae may cause taste and odor problems; some may release toxins affecting the fauna in smaller bodies of water (MEBS 1978). The development of large populations of blue-greens may partly be explained by the fact that they are hardly grazed upon by zooplankton. Aside from this, heterocyst-forming Cyanophyceae come to dominate in lakes during the summer, when phosphorus and temperature requirements are met (FOEHRENBACH 1974), while decrease of available combined nitrogen (NH_4^+, NO_3^-) limits their competitors. Their capability to function at lower equilibrium CO_2 levels (KING 1970) and also under anaerobic conditions (PEARSON and TOMLINSON 1975) gives them an extra ecological advantage over other algae. MURPHY et al. (1976) found N-fixing *Anabaena* sp. to overcome iron deficiency by producing a hydroxamate chelator which bound the available iron and made it unavailable to other species. This chelator suppressed growth of *Scenedesmus* even when iron was added in excess, indicating interference competition by antibiotic action (KILHAM and KILHAM, personal communication). Numerous examples of similar inhibitory action to other algal species are known, especially for blue-green and unicellular green algae (FOGG 1971; HARRIS 1971; KEATING 1977, 1978; LEFEVRE 1964; RICE 1954; SCHWARZ 1965); their ecological significance, however, is usually restricted to small still waters. The algae that become prominent in addition to bloom-forming species as eutrophication proceeds may be planktonic, floating, periphytic or live attached to rocks or sediment; *Cladophora, Gleocystis, Mougeotia, Oedogonium, Spirogyra* (SCHLICHTING and GEARHEART 1966) as well as some diatoms (PATRICK 1943) may dominate. SCHINDLER (1971) found that the species changed from Cryptophyceae and Chrysophyceae in unfertilized lakes to Chlorophyceae and Cyanophyceae in a fertilized lake. Various authors have arranged algal groups according to minimum and maximum phosphorus concentrations. FORSBERG (1964) indicated 5 $\mu g\,l^{-1}$ of total P as maximum factor for *Dinobryon divergens* and *Uroglena americana,* while *Scenedesmus quadricauda* required at least 500 $\mu g\,l^{-1}$. A similar value as PO_4-P is indicated for *Scenedesmus* by VOLLENWEIDER (1968), with maximum population densities of the diatoms *Fragilaria* and *Tabellaria* at ca. 45 $\mu g\,l^{-1}$ PO_4-P and of the blue-green alga *Oscillatoria rubescens* at 3,000 $\mu g\,l^{-1}$ PO_4-P. FOGG (1973) grouped freshwater algae according to whether their tolerances for PO_4-P were lower, close to, or higher than 20 $\mu g\ l^{-1}$, relying on laboratory data of Rodhe (1948) and Soeder et al. (1971); see also Chu (1943). But it is well understood that not only the concentration of a particular nutrient is important, but that its supply rate relative to other potentially limiting nutrients has an effect on structuring phytoplankton communities. The most important ratios are Si/P for diatoms and N/P for greens and blue-greens (Kilham and Kilham 1978 and personal communication). Increased eutrophication will lead to progressively lower Si/P ratios, and thus induce species changes in the diatom flora of lakes. This has been verified by Tilman (1976, 1977) who observed and predicted, by applying various models, that in mixed populations of *Asterionella formosa* and *Cyclotella meneghiniana, Asterionella* was competitively dominant when both species were phosphate-limited; *Cyclotella* was dominant when both species were silica-limited, while both species coexisted when each species was growth-rate-limited by a different resource. This is much in accord with classical ecological competition theory which predicts that under idealized conditions the one species enabled to make best use of the limiting resource should displace all other competing species (Tilman 1976, 1977). It further supports the ecological concept that as many competing species are able to coexist as there are many limiting resources. With progressing eutrophication, known seasonal succession in phytoplankton populations (as documented in plankton calendars) may be lost and substituted by new successional patterns (Grim 1967). Reduction of phytoplankton species in a lake may also be indicative of increased eutrophication. Palmer et al. (1977) described a decline from 80 algal genera to 30 genera with increasing fertilization of Lake Erie from 1930 to 1950. The quantity of algal mass of the remaining species in a eutrophic lake, however, increases. High positive correlation between phytoplankton biomass and annual load of phosphorus and nitrogen has been demonstrated by Ahl (1975). Chlorophyll a maxima in oligotrophic lakes investigated in the OECD program ranged from 2.6 to 7.6 $mg\ m^{-3}$, while corresponding values in eutrophic lakes ranged from 16.9 to 107 $mg\ m^{-3}$ (Vollenweider 1979). Total productivity in Lake Trummen, Sweden was raised from less than 50 $g\ m^{-2}\ d^{-1}$ of organic substance to more than 600 $g\ m^{-2}\ d^{-1}$ as it was subjected to continuous sewage-loading for decades (Björk, personal communication).

As outlined, conspicuous floristic and vegetational changes may be observed in oligotrophic to mesotrophic running waters, as they are subjected to ongoing organic and subsequent nutrient loading. Specific formations of submersed macrophytes could be assigned to different eutrophication stages and zones of soft water streams in Northern Bavaria: a *Myriophyllum alterniflorum* formation dominated in oligotrophic to mesotrophic stretches, while a *Myriophyllum spicatum* formation was typical for critically polluted zones (Kohler and Zeltner 1974). In carbonate-rich streams from oligotrophic Bavarian fens correlation between species distribution and NH_4^+ (and PO_4^{3-}) concentrations of the water was established that allowed for a floristic-ecological classification and mapping of the stream system (Kohler 1978; Kohler et al. 1971, 1973, 1974; Kutscher and Kohler 1976). A group with *Potamogeton coloratus* as dominant species was characteristic of zones of low NH_4^+ and PO_4^{3-} concentrations

($\leq$0.05 mg l^{-1}) while *Callitriche obtusangula* and *Ranunculus fluitans* proliferated in zones with average values of NH_4^+ >0.1 mg l^{-1}. According to THOMAS (1975), excessive growth of *Ranunculus fluitans* in the Rhine set in at PO_4^{3-} concentrations in the water >0.2 mg l^{-1}. Species diversity is markedly reduced where eutraphent macrophytes prevail. The speed with which the macrophytic vegetation in a stream reacts to progressive eutrophication with adverse changes is not equalled by recolonization with less sensitive species when nutrient loading is reduced (KOHLER et al. 1980). Similar relationships between distribution patterns of submersed macrophytic vegetation and nutrient concentrations in running waters have been described by other authors (CARBIENER 1969; KRAUSE 1972, 1978; LITAV and AGAMI 1976; PIETSCH 1975; SIEFERT 1976). Benthic algae in running waters display also distinct communities depending on the trophic status. BUTCHER (1947) noted in several streams heavily charged with organic matter that there is a succession of different communities of algae as purification takes place, with *Stigeoclonium tenue* dominating in the polysaprobic zone, *Nitzschia palea* and *Gomphonema parvulum* in the mesosaprobic zone and *Cocconeis placentula-Chaemosiphon* spp. and *Ulvella frequens* in the oligosaprobic zone. Eutrophic stretches of some Scandinavian running waters yielded *Vaucheria* spp., *Cladophora glomerata, Hildenbrandia, Batrachospermum boryanum* and *Chaetophora,* while oligotrophic zones usually exhibited *Spirogyra, Schizochlamys* sp. and *Euglena* ssp. (BLUM 1956). Several species of *Oscillatoria* and *Phormidium* seem always to be associated with particularly rich conditions (HYNES 1959; see also PALMER et al. 1977). There is some evidence that N/P ratios are important in determining species composition in stream diatom communities (STOCKNER and SHORTREED 1978). Many forms of the sewage fungus, *Sphaerotilus natans* and some true fungi which do not normally occur in rivers are closely associated with grossly polluted zones, that are usually stripped of vegetation. However *Sphaerotilus* may be also restricted when sewage contains high loads (3–5 mg l^{-1}) of amino acids (DEMOLL and LIEBMANN 1952). Aside from qualitative changes there are also marked quantitative changes with nutrient loading, as has been demonstrated by SCHWOERBEL (1972) for biomass of algae and bacteria.

13.2.3 Ecophysiological Studies in Eutrophication

Inorganic nutrients in aquatic ecosystems are potentially available to submersed macrophytes from two sources: from the water around the shoots and the sediment in which most plants root. Studies on phosphorus absorption by vascular hydrophytes have revealed that P can be absorbed by the roots and translocated through the shoots of various species (BRISTOW and WHITCOMB 1971; DE MARTE and HARTMAN 1974; FUNDERBURK and LAWRENCE 1963; LITTLEFIELD and FORSBERG 1965). Uptake of ammonium through roots and rhizomes of *Potamogeton nodosus* (COLE and TOETZ 1975) and *Elodea densa* (TOETZ 1974) with translocation to stems and leaves also provides evidence that macrophytic hydrophytes may act as nutrient pumps; however, it is also suggested that sediments may only be an auxiliary source of inorganic N. DENNY (1972) points out that there is a range among aquatic plants from emergent plants that solely rely on roots for nutrient supply to rootless plants that entirely depend on their shoots for nutrient absorption; the site of absorption may alter according to the relative nutrient status in the water and the sediment. Various workers have demonstrated preferential uptake of ammonium from the water compared to nitrate. FORSBERG (1965) indicated growth stimulation of *Chara zeylandica* by ammonium (and arginine and cysteine) compared to nitrate. *Lemna minor* can utilize different N sources but uptake of combined nitrogen (NH_4^+, NO_3^-) was superior to NO_2^- and organic N (SIMS et al. 1968). SCHWOERBEL and TILLMANNS (1972, 1974) found in laboratory experiments that *Fontinalis antipyretica, Ranunculus fluitans* and *Callitriche hamulata* utilize ammonium and nitrate, but prefer ammonium if both nitrogen forms are provided together. Ammonium was shown to suppress the activity of nitrate reductase in *Fontinalis,* subsequently leading to inhibition of nitrate uptake. Reactivation of the enzyme under ammonia deficient conditions (<1 µmol l^{-1}) and high nitrate concentration was shown to be light-dependent (SCHWOERBEL and TILLMANNS 1974, 1977). A *Ceratophyllum*-periphyton community studied in situ in a lake assimilated ammonium both during day and night, while nitrate assimilation was strongly light-dependent, with uptake being substantially lowered during the night (TOETZ 1971). SCHWOERBEL and TILLMANNS (1964a) demonstrated concentration dependent increase in ammonium utilization (up to 9 mg l^{-1} NH_4-N) for *Callitriche hamulata,* with even 22 mg l^{-1} in the medium not yet toxic to the plant. The same authors (1964b) suggest uptake mainly as undissociated NH_4OH, as ammonium uptake by *C. hamulata* increased by a factor of 4.6 when the pH was raised from 3.7 to 7.0. With few exceptions all chlorophyllous algae are able to utilize NH_4-N, NO_3-N and NO_2-N. Similar to the above findings with submersed macrophytes, preferential uptake of ammonium over nitrate is known for a large number of algae (BREZONIK 1972; FOGG 1953; MORRIS and SYRETT 1963; PRATT and FONG 1940; SCHULER et al. 1953; SYRETT 1962). The preference of NH_4-N is explained in that nitrate requires energy for its reduction (VOLLENWEIDER 1968). As mentioned in the review by SYRETT (1962) nitrogen-starved algal cells could assimilate NH_4-N four to five times faster than normal cells, probably because nitrogen-starved cells have a carbohydrate reserve which is lacking in normal cells (see also FITZGERALD 1968). Nitrate, the normally predominant N form in freshwaters, also has been shown to be the preferred N source for some algae (CHU 1943; HUTCHINSON 1957; PROVASOLI and PINTNER 1960). A number of algae are known to be able to utilize organic N when inorganic nitrogen supplies are depleted. Of amino acids, glycine was the most readily assimilated by *Chlorella* (SYRETT 1962). FOGG and WESTLAKE (1955) demonstrated that extracellular amino acids were stimulatory to the phytoplankton of hard water lakes of low productivity (FORSBERG and FORSBERG 1969; WETZEL 1968). In this context it is interesting to note that the release of dissolved organic carbon (DOC) and dissolved organic nitrogen (DON) from phytoplankton is known to increase in more eutrophic lakes. Nutrient regeneration cycles would accelerate therefore with increased eutrophication as bacteria in hard water lakes are then less likely to be substrate-limited (WETZEL and

MANNY 1972). Nitrogen fixation by heterocysts of blue-green algae is inhibited when higher ammonium and nitrate levels are present, as they suppress nitrogenase synthesis and heterocyst formation (FOGG 1962; WOLK 1973). Nitrogen fixation also appears to be reduced at higher oxygen concentrations in the water (STEWART et al. 1969; WHITTON 1972). Nitrogen fixation rates or acetylene reduction, as a measure of nitrogenase activity, in blue-green algae also vary markedly and rapidly with the levels of available phosphorus (STEWART et al. 1970). Aside from orthophosphate as the most readily utilized phosphorus, other forms may be available after hydrolysis by enzymes produced by algae (FOGG 1973). This opens a wide range of organic compounds as ultimate phosphorus sources (GALLOWAY and KRAUSS 1963; MIYACHI et al. 1965). Utilization of phosphorus bound as phosphomonoesters and polyphosphates by alkaline phosphatase-producing *Aphanizomenon flos-aquae* is reported by HEATH and COOKE (1975). FORSBERG et al. (1967) have shown that all polyphosphate-containing detergents could act as P sources for *Scenedesmus quadricauda*. Acid phosphatase activity in intact algal cells was shown by TALPASAYI (1962), indicating that the enzyme was situated at the cell surface (see also BONE 1971). These surface phosphatases of algae are most active when the cells are phosphorus-deficient (FITZGERALD and NELSON 1966; REICHARDT 1971); the synthesis of the enzymes involved, however, ceases when phosphorus is restored to the medium (KUENZLER and PERRAS 1965). RHEE (1973) demonstrated that phosphate uptake kinetics and growth of *Scenedesmus* was a function of both internal cellular phosphorus compounds and the external substrate concentration. NYHOLM (1977b) described phosphate uptake by P-deficient algal cells as following Michaelis-Menten kinetics at concentrations less than 500 $\mu g\ l^{-1}$ of PO_4-P. STRANGE et al. (1975) indicated that ^{32}P uptake in some lentic algae was proportional to the surface: volume ratio; higher ratios coincided with increased uptake of phosphorus. SWEAZY et al. (1975) showed that the rate of phosphate uptake in algal cultures was balanced by the rate of loss. SCHWOERBEL and TILLMANNS (1964c) demonstrated that phosphate uptake by *Ranunculus fluitans* and *Callitriche hamulata* increased with concentrations up to 2 $mg\ l^{-1}$ of PO_4-P. NORMANN (1967) stated concentration dependent increased uptake of PO_4-P from $1\text{–}3 \times 10^{-4}$ $mol\ l^{-1}$ in *Ranunculus fluitans*. The high storage potential for phosphorus of *Callitriche hamulata*, which may be as high as 29 mg per 100 g fr.wt. (SCHWOERBEL 1968) is worth mentioning. Luxury consumption, as excess accumulation of nutrients in plants is frequently referred to, is a well established phenomenon in algae. Diatoms (MACKERETH 1953) and green algae (KETCHUM and REDFIELD 1949) store phosphorus. Accumulated polyphosphate in blue-green algae may suffice for several rapid P cycles in the absence of external supply (STEWART and ALEXANDER 1971). From an ecological standpoint this ability to rely on stored polyphosphate may be the reason why poor correlations are found sometimes between the levels of dissolved phosphorus in the water and the extent of algal development in natural environments (STEWART and ALEXANDER 1971). Excess phosphorus from luxury uptake however has little effect on algal growth (GERLOFF and SKOOG 1954). There is also experimental evidence of luxury consumption of nitrogen (FITZGERALD 1968; GERLOFF and SKOOG 1954; NYHOLM 1977a; RODHE 1948).

Increasing eutrophication has been shown to lead to a reduction in starch, hexoses, malic acid and several aminoacids, while asparagine and sucrose contents were raised with increasing phosphorus and nitrogen levels in *Potamogeton pectinatus* (JANAUER 1977, 1979). Nitrogen fertilization increased the carbohydrate and total N content in *Phragmites australis* (RAGHI-ATRI and BORNKAMM 1979). Nitrogen increases the stem diameter of *Phragmites* while phosphorus decreases it; the thickness of the outer sclerenchyma ring is reduced by large additions of phosphate (BORNKAMM et al. 1980). The reduction of the proportion of sclerenchyma in the lower parts of the stalks from 6% to 1% with increasing nutrient supply weakens the stalks and is considered a major factor in the reduced tolerance of *Phragmites* stands to mechanical stress by wind and wave action as well as by floating materials (KLÖTZLI and GRÜNIG 1976; SUKOPP et al. 1975; TOBLER 1943). RAGHI-ATRI and BORNKAMM (1979) noted a marked reduction in stems per m^2 with increased eutrophication from summer to autumn, which adds to the reduced resistance of *Phragmites* stands to mechanical stress.

Various workers have recognized the potential of higher aquatic vegetation for removal of nutrients from lentic systems (BOYD 1970; CULLEY and EPPS 1973; MEREZHKO 1970; NÜMANN 1970). MCNABB (1976) demonstrated that 20%–25% of total P and 50%–70% of total N in the influent to an aerobic sewage pond was removed with a harvest of *Elodea* after 28 days of growth in the summer. Extreme pH elevation due to photosynthesis in such pools, however, may limit vascular plants in such environments (GREER and ZIEBELL 1972). In natural surface waters the problem of biomass removal arises (SCHWOERBEL 1968). WAGNER (1966) has calculated that nitrogen and phosphorus removal from water of Lake Constance by submerged macrophytes was negligible if compared to the nutrient input to the lake from a major tributary. Duckweed populations (*Lemna gibba* and *L. minor*) have been shown to be very potent removers of orthophosphate in static sewage (SUTTON and ORNES 1975). KICKUTH (1978) states that nutrient uptake by *Phragmites australis* does not necessarily lead to lower nutrient levels in the water, as nutrient uptake is exceeded by far by evapotranspiration, which may even lead to increased concentration of nutrients in the aquatic habitat. However stationary microorganism groups, associated with the rhizosphere of *Phragmites* stands have high denitrification potential. This may lead to the elimination of 10–15 t per 10000 m^2 and year of total N on the average in the active horizon of *Phragmites* rhizosphere; phosphate elimination at a rate of 3 to 5 t of P per 10000 m^2 and year is easily achieved and even surpassed by the *Iris pseudacorus* rhizosphere.

Reversal of the eutrophication processes in lakes is feasible but involves costly measures. Outstanding examples of the recovery and restoration of lakes suffering from eutrophication have been given with Lake Trummen, Sweden by BJÖRK (1973, 1977, 1978) and Lake Washington, Washington USA by EDMONDSON (1970, 1977). Sewage diversion has been demonstrated as necessary first step, with removal (BJÖRK 1973) or special chemical treatment of sedimentary ooze (RIPL 1976b) and aeration of the oxygen-depleted hypolimnion required as follow-up to assure success (BJÖRK 1973).

13.3 Toxic Pollutants

The complex reactions of aquatic ecosystems to the introduction of toxic substances are governed by a few basic factors: (1) the nature of the toxicant, (2) toxicant concentration, (3) time of exposure, (4) environmental features of the affected system, (5) biological characteristics of exposed organisms (condition, age etc.), and (6) presence of other toxic substances (CAIRNS et al. 1972).

13.3.1 Heavy Metals

Excess heavy metals in water and sediments of aquatic systems can frequently be attributed to industrial effluents or mine drainage. Aerial inputs of Cd, Cu, Fe, Hg, Pb or Zn from organic fuel burning emission sources (power plants, industries, automotive traffic) also have to be considered. Copper compounds are freely used for algicidal action in surface waters. Heavy metals adsorbed to eroded materials are introduced to freshwater habitats with surface run-off. Within the aquatic environment soluble metal concentrations are governed by precipitation, complexing, adsorption and hydrolysis (SYLVA 1976). In most natural waters metal ions are precipitated as carbonates, hydroxides or sulfides (all of low solubility) when neutral to alkaline pH conditions prevail. Adsorption to suspended solids increases very rapidly at a pH value about one unit less than the pK value of the first hydrolysis reaction (JAMES and HEALY 1972). These immobilization processes keep the heavy metal ion concentrations in the water at levels usually lower than 10 ppb (FÖRSTNER and MÜLLER 1974). Sediments, however, are sinks for immobilized heavy metals. Highly polluted surface waters exhibit heavy metal concentrations in the clay fraction of the sediment at the ppm level.

13.3.1.1 Heavy Metals and Freshwater Vegetation

As essential trace metals (Co, Cu, Fe, Mn, Mo, Zn) appear in concentrations largely exceeding the requirements of certain aquatic plants for normal growth and reproduction, and metals with no evident physiological function in biological systems (Cd, (Cr), Hg, Pb, (V)) become available in critical concentrations, distinct vegetation patterns may be observed in surface waters along pollution gradients.

CARPENTER (1924, 1926) described succession phases in the flora of a river that had been subjected to drainage water from a lead mine (1) with distance from the outfall, and (2) with elapsed time from the shut-down of the source of pollution. Only limited growth of some mosses, liverworts and red algae began to appear as recovery set in and Pb levels in the water dropped to 400 ppb. Some Chlorophyceae followed, and as the pollution level further decreased there was sparse growth of two submerged vascular plants. Twenty years after most of the mines had closed down, elevated heavy metals concentrations in the water still allowed only a restricted flora of bryophytes and some algae (JONES 1940). *Scapania undulata* was a bryophyte found to be especially tolerant of elevated Cu, Pb and Zn levels in these river systems. Diversity of filamentous green algae was reduced to *Hormidium rivulare* at 40 ppm Zn (MCLEAN and JONES 1975). The algal flora in ponds adjacent to Zn waste tips, with *Mougeotia, Tribonema, Coccomyxa* and *Zygnema*, was quite distinct from that in ponds at greater distance from the tips, which consisted of *Cladophora, Spirogyra* and *Oedogonium* (TROLLOPE and EVANS 1976). HYNES (1960) describes serious effects of elevated Cu levels (1 ppm) on lotic algal community structure, as monitored on glass slides exposed at different distances from the copper effluent and a control above the outfall. The community at the control site, consisting of *Chaemaesiphon, Cocconeis, Gomphonema parvulum, Nitzschia palea* and *Stigeoclonium* totaled 1,000 individual mm^{-2}. Right below the outfall the community was restricted to a mere 150–220 individual mm^{-2} consisting of *Chlorococcum* and *Achnanthes*. Five miles downstream cell numbers increased to over 33,000 mm^{-2} with little change in species diversity. The original algal community only reappeared 30 miles downstream. Relative low species diversity was also observed by WILLIAMS and MOUNT (1965) in a periphyton community exposed to Zn levels up to 9 ppm, with slime-forming bacteria and fungi as dominating species.

Excessive algal blooms, dominated by Cyanophyta are a common phenomenon in still waters in the summer. It is established practice to eliminate such blooms at an early stage, applying low $CuSO_4$ concentrations (DOUDOROFF and KATZ 1953). Blue-greens are already very susceptible to concentrations $\leqq 25$ ppb. Trace amounts of Cu^{2+} substantially suppress nitrogen fixation of heterocystous blue-greens, which otherwise may be of considerable advantage when nitrogen depletion limits their competitors in highly productive waters (HORNE and GOLDMAN 1974). While sensitive blue-greens are removed for the rest of the season, numerous green algae (including *Chara* and *Nitella*) remain unaffected and may be joined by diatoms as dominating yet less noxious species as copper concentrations in the water return to background level within a few weeks (AYLES et al. 1976; CRANCE 1963; WHITAKER et al. 1978). Increasing tolerance of certain blue-green algae is feared if copper treatments are maintained over years. A decrease in bacterial count in sediments of copper treated lakes (WATSON and BOLLEN 1952) and in upper layers of stream sediments (LAUBENBERGER and HARTMANN 1970) due to residual copper may be indicative of long-term effects of unforeseen consequences for the habitat. (For general discussion of toxicity and tolerance in response of plants to metals see Chap. 7, Vol. 12C.)

13.3.1.2 Toxicity of Heavy Metals

Ionic Heavy Metals. Ordinarily the heavy metal ion is recognized as potentially toxic. Ionic Cu^{2+} as low as 1–2 ppb produced signs of inhibition of photosynthesis and growth of *Chlorella pyrenoidosa* and *Nitzschia palea* (STEEMANN NIELSEN and WIUM-ANDERSEN 1970), while GREENFIELD (1942) gives somewhat higher values for *Chlorella*. Bioassay data by HUTCHINSON and STOKES (1975) establish Cu^{2+} concentrations from 0.1 to 0.7 ppm as lethal for *Chlorella, Chlamydomonas, Haematococcus,* and *Scenedesmus* in an order of decreasing sensitivity. Inhibition of growth and photosynthesis and reaction of cupric ions with sulfhydryl groups of *Chlorella sorokiniana* has been demonstrated by CEDENO-MALDANO and SWADER (1974). Cu^{2+} strongly inhibited the uncoupled photosynthetic electron transport in chloroplasts (CEDENO-MALDANO et al. 1972). Changes in respiratory rate and pigment patterns in *Chlorella* were related to excess levels of various metal ions in the medium by GROSS et al. (1970). Growth retardation of *Chlorella* populations by Ag^{2+} was more severe than by Cd^{2+}, Hg^{2+} or Ni^{2+} (HUTCHINSON and STOKES 1975). Two diatoms tolerated 0.032–0.32 ppm of $K_2Cr_2O_7$ (HERVEY 1949), while YOUNG et al. (1979) noted toxicity of concentrations >1 ppb to *Euglena gracilis;* however, temperature was involved in the latter findings. The reduction of cell division of *Chlorella* at 0,32 ppm was more pronounced by Cd^{2+} than Cu^{2+} or Hg^{2+}. There was no effect of 0.1 ppm

Cd^{2+} on cell division, while it was retarded by Hg^{2+} (HANNAN and PATOUILLET 1972). Increase of algal cell size in heavy metal solutions is described by ERICKSON et al. (1970), ROSKO and RACHLIN (1977) and STEEMANN NIELSEN and KAMP-NIELSEN (1970). Electron micrographs revealed modified lamellar organization of chloroplasts, disrupted thylakoid membranes and general degeneration of the cytoplasm, caused by lead, with staining techniques suggesting decreased oxidative and phosphorylative abilities of mitochondrial membranes (SILVERBERG et al. 1977). (See also BRINGMANN and KÜHN 1959; FITZGERALD and FAUST 1963; FITZGERALD et al. 1952; KAMP-NIELSEN 1971; PAWLASCZYK-SZPILOWA et al. 1972; RACHLIN and FARRAN 1974.)

Chelated Heavy Metals. There is no agreement concerning the toxicity of chelated heavy metals. While STEEMANN NIELSEN and WIUM-ANDERSON (1970) maintained that copper exhibits no toxicity when it becomes complexed by organic substances in natural waters, GÄCHTER et al. (1973) find this statement perhaps too exclusive, as many interactions may be involved. Chelating agents may improve the transfer of trace elements from the surrounding medium to aquatic plants (FORSBERG 1968). Physiological availability of inorganic reactive metal ions like Mn and Fe can be expected as complexes form (WETZEL 1975). Several types of autochthonous chelating agents, e.g. amino acids, polypeptides and other metabolites, dissolved humic acids (DIETZ 1977; GÖTZ 1978; WETZEL and ALLEN 1970; WETZEL and MANNY 1972), and possibly synthetic chelators (ERICKSON et al. 1970; FÖRSTNER and MÜLLER 1974) are present in aquatic ecosystems.

FOGG and WESTLAKE (1955) showed that extracellular substances released by algae reduced the toxicity of Cu^{2+} ions. Likewise the toxicity of Cu^{2+} to *Chlorella* decreased as chelation with amino acids occurred (STEEMANN NIELSEN and KAMP-NIELSEN 1970). The synthetic complexing agent EDTA reduced the toxicity of Cu^{2+} to *Anabaena cylindrica* (FOGG and WESTLAKE 1955) and that of Cu, Fe, Mn, Ni, Pb, and Zn ions to *Cladophora* (WHITTON 1967). Extensive studies by NUSCH (1977) with representative organisms of different trophic levels (bacteria, algae, protozoa, lower metazoa, fish) indicate the detoxifying effect of strong chelators such as EDTA and NTA in Cu solutions; it was also shown that stable Cu-chelate complexes (20 ppb) inhibited reproduction and oxygen evolution of exposed algae *Anacystis, Chlorella,* and *Scenedesmus.* KHOBOT'YEV et al. (1975) demonstrated variation in the toxicity of copper containing organic complexes with the types of chemical bonds between the metal and the ligands and nature of the ligands. Toxic effects of NTA-chelated Cd, Cu, and Zn to submersed vascular plants were observed by SCHUSTER (1979). While 5×10^{-5} M of chelated Pb caused no effects, first effects of 10^{-7} M of Cd-NTA, 5×10^{-7} M of Cu-NTA and 10^{-5} M of Zn-NTA were indicated by depression of photosynthesis after 21 days. REBECHINI and HANZELY (1974) reported disruption of thylakoid stacking of chloroplasts in *Ceratophyllum demersum* exposed to 1 ppm of EDTA chelated Pb.

Methylated Heavy Metals (with Reference to Mercury). Mercury is commonly introduced to the environment in inorganic or phenylated form (DER RAT 1978).

Presence of methylated mercurials [CH_3Hg^+, $(CH_3)_2Hg$] in aquatic environs may largely be attributed to biotransformation processes of bacteria and fungi, which use mercuric wastes as substrate (JENSEN and JERNELÖV 1969). Such methylating activity has been described for *Clostridium* (YAMADA and TONOMURA 1972), *Neurospora* (LANDNER 1972) and *Methanobacter* (BILLEN 1973; WOOD et al. 1968). The processes of methylation and mineralization of mercury are governed by redox and pH regime with a small imbalance determining whether the environment acts as source or as sink of methylmercury (WOLLAST et al. 1975).

There is evidence that methylated mercurials are more toxic to aquatic biota than inorganic mercury salts. Restriction of growth of *Elodea canadensis* at 0.2 to 2 ppb of CH_3HgCl was more pronounced than by $HgCl_2$ (MORTIMER and KUDO 1975). GEIKE (1977) subjected *Chlorella* and *Chlamydomonas* populations to 2.5×10^{-8}–10^{-5} M of $HgCl_2$ and various organic mercurials, and came to similar conclusions, establishing methylmercury also as the most toxic form of the organic mercury compounds. *Chlorella* appeared to be less sensitive than *Chlamydomonas* as displayed by different decreases in chlorophyll, carbohydrates, total nitrogen and dry matter. Inhibition of photosynthesis and growth of various phytoplankton species by various organic mercurials occurred at 0.1 ppb Hg, while concentrations about two orders of magnitude greater were needed with inorganic mercury (HARRISS 1971). OVERNELL (1975) concluded that the main effect of CH_3Hg^+ on whole cells of *Chlamydomonas reinhardii* was by blocking the Hill and modified Mehler reactions.

13.3.1.3 Bioaccumulation of Heavy Metals and Adaptation

Acute toxicity of heavy metals in freshwater environments may be assumed as less frequent and attributed to considerably elevated levels of plant available metals at particular sites. Concern, however, rests with sublethal concentrations as well, as accumulation of heavy metals in aquatic plants is observed, which implies the hazard of introducing potentially toxic quantities of a given metal into the food-chain of higher organisms.

Concentration factors for heavy metals and effects on the level of primary production can be substantial. Shoots of various macrophytic water plants, collected by DIETZ (1972) from a mildly polluted river system, revealed concentration factors of 100 to 20,800, based on wet weight. Except for extreme Mn accumulation, the order of accumulation of metals in the plants reflected the proportion of the metals present in the water. Best accumulators were the bryophytes *Fontinalis antipyretica* and *Hygroamblystegium*. MORTIMER and KUDO (1975) reported that *Elodea densa* concentrated organic mercury, present in the water at 0.2–10 ppb, by a factor of up to 110,000. HANNERZ (1968) indicated higher concentration factors for methyl mercury than $HgCl_2$ in some aquatic plants kept at 0.2–10 ppb, whereas mosses exceeded the accumulation of vascular plants. Concentration factors for Cu, Fe, Ni, Pb and Zn in *Cladophora, Spirogyra* and *Oedogonium* ranged from 400 to 80,000, depending on the metal (TROLLOPE and EVANS 1976). In aquatic macrophytes differential uptake and accumulation of heavy metals in different organs was observed. In emergent plants, submersed and underground parts showed higher concentrations (ERIKSSON and MORTIMER 1975; GUILIZZONI 1975; HANNERZ 1968) than emergent leaves and stems; likewise submersed vascular plants stored more Cu, Cd, and Zn in their underground parts (RAY and WHITE 1976).

Further records on heavy metal accumulation by macrophytic aquatic plants and algae are by CEARLEY and COLEMAN (1973), COWGILL (1973, 1974), DOLAR et al. (1971), GERHARDS and WELLER (1977), HUTCHINSON and CZYRSKA (1975), MATZKU and BRODA (1970), MAYES and MCINTOSH (1975), MCLEAN and JONES (1975), MILLER and SINGH (1976), MORTIMER (1974), PAYER et al. (1976), RAGHI-ATRI (1978), RANA and KUMAR (1974), SÄRKKÄ et al. (1978), STOKES et al. (1973), WENTSEL and BERRY (1975).

Some detoxification mechanisms of aquatic plants have been described to explain the tolerance of certain species to elevated levels of plant available metals in the water and the accumulation phenomenon. Tolerance may merely be the result of relatively low uptake rates, as suggested by findings of FRANCKE and HILLEBRAND (1980), HARGREAVES and WHITTON (1976), SAY et al. (1977). Cell wall characteristics are assumed to hold an important role among tolerance determining factors. Thickness of the cellulose membrane of *Chlorella* is believed to have greatly contributed to the relative resistance to mercury compounds compared to *Chlamydomonas* (GEIKE 1977; STUART and GAFFRON 1972). The binding of heavy metals to harmless sites, e.g. the cell wall, is suggested by ANTONOVICS et al. (1971) as a feasible detoxification mechanism. Studies on *Asterionella formosa* indicate that a significant percentage of Cd and As was associated with the frustule and its surrounding organic layer (CONWAY 1978). Complexing of heavy metal ions within the cell by protective chelating agents is considered a major avoidance mechanism (ANTONOVICS et al. 1971; GUILIZZONI 1975). Avoidance through excretion of chelating agents by algae to the medium is indicated by FOGG and WESTLAKE (1955) and SOEDER et al. (1972).

13.3.1.4 Environmental Interactions and Heavy Metal Toxicity

The impact of toxic metals on freshwater biota may be substantially reduced in hard, well-buffered freshwater systems. KINKADE and ERDMAN (1975) found that ^{115}Cd uptake of *Nitella flexilis* and *Elodea canadensis* was lower in hard water culture than under soft water conditions. Increased Ca levels in the water reduced the toxicity of Zn to the green algae *Stigeoclonium* (HARDING and WHITTON 1976) and *Hormidium rivulare* (HARGREAVES and WHITTON 1976; SAY and WHITTON 1977). The relative Zn toxicity to *Hormidium* was more reduced by Ca in zinc-tolerant than in zinc-sensitive populations. Water hardness not only appears to be a controlling factor in the toxicity of heavy metal ions but also of heavy metal complexes, as NUSCH (1977) noted higher inhibition of O_2-consumption by Cu-NTA for bacteria kept in hard water (267 ppm $CaCO_3$) than in soft water (17.6 ppm $CaCO_3$), while the converse was true with Cu-EDTA, suggesting differerent affinities of both chelating agents to Ca^{2+} and Mg^{2+} competing with heavy metal ions for binding sites. There are some indications that phosphate may also be involved in reducing the toxicity of heavy metals to algae (KEULDER 1975; RANA and KUMAR 1974; SAY and WHITTON 1977). Variations in the toxicity of a given heavy metal may also be attributed to the pH of the habitat. Toxicity of Zn to *Hormidium* was shown to be least at pH 3.5–4.0, increasing with higher pH (HARGREAVES and WHITTON 1976), whereas an increase of pH from 6.1 to 7.6 resulted in reduced toxicity to *Stigeoclonium* (HARDING and WHITTON 1976). Field studies by BESCH et al. (1972) on diatom communities in acidic and non acidic, heavy metal polluted freshwaters suggest that acid-tolerant species were also tolerant to high levels of Cu and Zn, while the converse was not true. Synergistic or antagonistic effects of different heavy metals present in the medium have been reported by BARTLETT et al. (1974), HUTCHINSON and CZYRSKA (1975), HUTCHINSON and STOKES (1975), and WISSMAR (1972).

13.3.2 Pesticides and Related Organic Compounds (Phenolics, PCBs)

Pesticides encompass a vast variety of chemical formulations with distinct patterns of toxic action, pathways of decomposition, and environmental persistence. Among herbicides phenoxycarboxylic acid derivates (2,4-D, MCPA etc.) outweigh all other compounds (triazines, carbamates, chlorinated carboxylic acids etc.) in terms of actual tonnage. Of insecticides, persistent organochlorine compounds (DDT, dieldrin, endrin, toxaphen etc.) have widely been substituted by organophosphorus formulations of higher degradability. Polychlorinated biphenyls (PCB's) which are similar to organochlorine insecticides are mainly of industrial origin (LIVINGSTON 1977). A vast variety of phenols and phenolic constituents are released by coal conversion plants, coke plants, oil refineries and petrochemical plants into surface waters (BUIKEMA et al. 1979). Introduction of pesticides to freshwaters may be intentional, as is the case for control of noxious flora and fauna, or unintentional by drift from pesticide application, atmospheric fallout, relocation via erosion, spills and application accidents (HURLE and JOHANNES 1978; WARE and ROAN 1970). Within a freshwater ecosystem, sorption to suspended solids and sediments is a major process in reducing the pesticide level in the aqueous phase, largely governed by the solubility of the compound and amounts of sorptive humic substance and clay (HAMAKER et al. 1966; WERSHAW et al. 1969).

Sorption of pesticides containing acidic functional groups is influenced by pH and increases below the pK of a compound (HAMAKER et al. 1966). While many pesticides are not readily degraded in anaerobic sediments, the converse is true for some persistent compounds like DDT and lindane (HILL and MCCARTY 1967). FRANK et al. (1977) noted up to 70.3 ppb of Σ DDT and 252 ppb of PCB residues in sediments of Lake Erie in 1971; restrictions on these materials are monitored by a decline of these residues from 1970 to 1974. Chronic pollution of river water by phenolic constituents from industrial effluents is usually <10 ppb, while concentrations of chlorophenolic compounds commonly remain <1 ppb, as indicated in the comprehensive review of BUIKEMA et al. (1979). (Physiological bases of the effects of biocides and growth regulators and their ecological implications in terrestrial ecosystems are discussed in Chaps. 11 and 12, this Vol.)

Infestations of aquatic vegetation that clog drainage or irrigation channels and water ways, obstructing economic and recreational use of lakes and reservoirs are frequently controlled with short-lived herbicides, e.g. atrazine, 2,4-D, etc., and persistent diquat, paraquat (BROOKER and EDWARDS 1975; NEWBOLD 1975; WAY et al. 1971). Differences in the susceptibility on plant species (not considering effects of time and mode of application of characteristics of the system) may lead to distinct vegetation changes. Common weed killers, like 2,4-D and 2,4,5-T, used on higher aquatic plants at 20 ppm were without appreciable effect on blue-green algae, which proliferated, while dichlone controlled blue-greens at 15 ppb, leaving vascular plants unharmed at as much as 100 ppb

(FITZGERALD et al. 1952). After bipyridylium treatment of lentic waters nuisance-growth species, *Myriophyllum spicatum, Potamogeton pectinatus* etc., declined while macrophytic anchored algae *Chara* spp. and *Nitella* spp. took their place (BROOKER and EDWARDS 1973; FISH 1966; FRY et al. 1973). Aside from direct effects of herbicides on target and non-target plants, vegetation patterns may respond secondarily by alteration of ecological factors. BROOKER (1976) encountered replacement of *Phragmites australis,* the target species after treatment with dalapon by planktonic and epipelic algae as well as floating and submerged macrophytes with improvement of the light climate for the understory flora. *Lemna* mats which developed after similar treatment of drainage ditches on the other hand depressed development of submerged macroflora by light attenuation. Aquatic angiosperms release noticeable amounts of nitrogen and phosphorus on decay (BROOKER and EDWARDS 1973; DODGE 1971). NEWBOLD (1975) suggests that gaps in the plant community from herbicide action may quickly be filled by less susceptible, opportunistic species that utilize the increased nutrient and CO_2 supply. This is in accord with algal blooms after decay of aquatic macrophytes (WALSH et al. 1971). Likewise, bacterial responses can be related to the release of materials from decomposing macrophytes after herbicide treatment (RAMSAY and FRY 1976). Increased biological oxygen demand (BOD) of large quantities of decomposing plants may exert considerable stress on aquatic fauna (MAUCK et al. 1976; STRANGE 1976). While herbicides mainly affect aquatic fauna indirectly by alteration of the habitat, insecticides may be of indirect consequence to freshwater flora. Algal blooms may result from ecological imbalance, caused by a decline of herbivorous organisms after insecticide application (COOK and CONNERS 1963; HURLBERT 1975; TAUB 1976).

A herbicidal triazine inhibited growth of *Chlorella* at concentrations as low as 5×10^{-6} M (KRATKY and WARREN 1971). Similar results were obtained for *Ankistrodesmus,* whereby impairment of photosynthesis seemed to be less pronounced (RAU and GRIMME 1971). HAWXBY et al. (1977) maintained after tests with different algae that the herbicides most toxic were those which acted by specifically inhibiting photosynthesis rather than through other metabolic processes. Uncoupling of oxidative phosphorylation, inhibition of the Hill reaction and cyclic photophosphorylation are considered major actions of many herbicides and phenolics (DEDONDER and VAN SUMERE 1971; HUANG and GLOYNA 1968; MORELAND and HILL 1962; SIKKA and PRAMER 1968; ZWEIG et al. 1963). No adverse effects of phenol on *Chlorella* were noted below 40 ppm (LUKINA 1970) and *Oscillatoria* below 100 ppm (GONCHOROVA 1975). Inhibition of cyclosis in *Nitella* by various phenolics may be due to phenolic binding of the sulphide and thiol groups (STOM and BEYM 1976). DECLEIRE et al. (1976) demonstrated inhibition of nitrate reductase in *Lemna* by a number of herbicides. Increase of alkyl side chain length or number of substituted chlorine atoms in substituted phenols enhanced chlorosis in *Lemna* (BLACKMAN et al. 1955). Concentrations <10 ppm of 2,4-D stimulated growth and nitrogen fixation of *Anabaenopsis* (DAS and SINGH 1977), while notably higher concentrations were damaging, as confirmed by LUNDKVIST (1970) for other blue-green algae. LAZAROFF (1966) noted inhibition of algal development by 1 ppm of DDT, HCH, and thiocarbamate. Increase of cell number of the diatom *Navicula seminulum* was adversely affected by 1.8 to 32 ppm dieldrin (CAIRNS 1968). BIRMINGHAM and COLMAN

(1977) showed that two organophosphate insecticides reduced growth rate of *Chlorella pyrenoidosa* and *Navicula pelliculosa*, while *Anabaena flos-aquae* and *Chlamydomonas* showed growth at 10–100 ppb. LD_{100} of endrin, aldrin and dieldrin for *Microcystis aeruginosa* was as low as 5 ppm (VANCE and DRUMMOND 1969). The lampricide TFM produced EC_{50} of 4.2–15 ppm for a number of algae (MAKI et al. 1975). Bacteria may be relatively resistant to insecticides (LICHTENSTEIN et al. 1966), but sensitivity to low concentrations is also reported (WILLINGHAM and ANDERSEN 1966). Environmental factors may be influential, as water hardness reduced the phytotoxicity of paraquat to *Lemna* (PARKER 1966) and altered the effectiveness of 2,4-D on *Myriophyllum spicatum* (STANLEY 1975b). Synergistic effects of 2,4-D and chloride salts for *M. spicatum* have also been noted by STANLEY (1974b). Similarly synergistic action of DDT and NaCl resulted in growth inhibition of a blue-green alga, *Anacystis nidulans* (BATTERTON et al. 1972). Rapidity of chlorosis in *Spirodela*, when treated with diquat and paraquat, increased with light intensity (BLACKBURN and WELDON 1965).

Pesticides tend to accumulate in living organisms, partly due to their low water solubilities. *Chlorella* readily adsorbed organophosphate insecticides with only slow concomitant release by living and dead cells (AHMED and CASIDA 1958). DDT and lindane were more strongly sorbed by filamentous *Vaucheria sessilis* than by bentonite clay (HILL and MCCARTY 1967). The algae in the experiments of VANCE and DRUMMOND (1969) showed concentration factors of 120–268. Concentration factors of HCH for *Chlorella* ranged from about 200 (CANTON et al. 1975), with preference of accumulation in the cytoplasm rather than the cell wall, to well over 1,000 (HANSEN 1976). EBERHARDT et al. (1971) applied 200 g of labeled DDT per ha to 30 cm deep water of Lake Erie and noted concentration factors in various submersed macrophytes from 7,100 to 84,000, and *Cladophora* 83,000. Fungi, streptomycetes and bacteria have been shown good accumulators of DDT and dieldrin by CHACKO and LOCKWOOD (1967).

Transport, microbial and photochemical degradation of phenols and phenolic compounds is extensively outlined in the review by BUIKEMA et al. (1979). ELLIS (1977) described catabolism of labeled phenol and catechol by various algae at low concentrations. Water hyacinth, *Eichhornia crassipes*, was shown to remove large quantities of phenol from polluted water (WOLVERTON and MCKNOWN 1976). SEIDEL et al. (1967) documented the phenol elimination potential of *Schoenoplectus lacustris*. Various algae and vascular aquatic plants degraded a diphenamid herbicide to further unknown products which were released by the plants (BINGHAM and SHAVER 1977). Dehydrochlorination of lindane by *Chlorella vulgaris* and *Chlamydomonas reinhardii* to non-toxic metabolites has been demonstrated by SWEENEY (1968). Further aspects of metabolism of insecticides are also discussed in the review by WARE and ROAN (1970).

13.3.3 Halogens (with Reference to Chlorine)

In the question of halogens, the use of chlorine for hygenic conditioning of potable water is common practice and widely accepted, whereas chlorination of cooling water and sewage is subject for controversy (INGOLS 1975). Periodic injection of Cl_2 (max. 1–3 ppm) into cooling water systems of thermal power plants is one approach to control undesirable growth of "Aufwuchs" organisms (WUNDERLICH 1978). There are indications that the concentrations used are frequently in excess of demand (BAKER and COLE 1974). Chlorination in sewage treatment reduces bacterial populations and oxidation of organic materials and is used as a measure to relieve recipient streams of increased biological oxygen demand below the sewage outfall, and for aesthetic reasons (WUNDERLICH 1978). As residual chlorine is introduced to freshwater ecosystems, direct and indirect effects are feared; the latter with respect to possible formation of chlorinated hydrocarbons in the presence of organic substance (JOLLEY 1976; ROOK 1974).

Residual Cl_2 from sewage outfalls is extremely damaging to selfpurification processes in streams. BAITY et al. (1933) described a reduction in bacterial count by 90% in a receiving stream below the outfall. TOETZ et al. (1977) showed that impact of Cl_2 and chloramine reduced the ability of phytoplankton (Desmidiaceae, Chrysophyceae, Diatomeae) to take up nitrate and ammonium. Inhibition of enzymes in the cell wall is held responsible for the decrease in NO_3^- uptake. Chlorine, at an initial concentration of 0.028 mg l^{-1} depressed NO_3^- uptake by 50%. This is assumed to greatly influence phytoplankton productivity in times of nitrogen limitation. Uptake of NO_3^- or NH_4^+ appears to be more sensitive to Cl_2 than photosynthesis. BROOKS and BAKER (1972) showed that photosynthesis and respiration of phytoplankton was reduced to 50% by 0.32 mg Cl_2 l^{-1} (see also HAMILTON et al. 1970). TREMBLEY (1960) related the reduction of algal populations in the Delaware River to a point discharge of chlorinated heated water. BETZER and KOTT (1969), KOTT (1969) and KOTT and EDLIS (1969) noted lethal effects of chlorine and bromine for *Cladophora* and *Chlorella sorokiniana*. Chlorine was algistatic in the light and became algicidic in the dark, whereas the reverse was true with bromine. A mixture of both halogens caused greater mortality than single halogens. BRUNGS (1973) gives a summary of the effects of residual chlorine to aquatic life. Residual Cl_2 toxicity may increase with decreasing pH, but otherwise is not governed by environmental factors. It is suggested that aquatic organisms should not be subjected to concentrations of $Cl_2 \geqq 0.002$ mg l^{-1}.

13.3.4 Surfactants

Of surfactants (non-ionics, cationics, anionics), anionic compounds constitute the major group in terms of tonnage, introduced to surface waters, mainly with household sewage. Concentrations of anionic surfactants in European streams have been markedly reduced since the substitution of biologically non-degradable alkyl benzene sulfonates (ABS) in detergents by highly degradable linear chained ABS. Pollution levels in the Rhine and its tributaries rarely exceed 0.1 ppm (FISCHER and WINKLER 1976). Severe pollution levels (10 ppm) were reached elsewhere (LITAV and AGAMI 1976). Quite low concentrations of residual surfactants are known to interfere with gaseous exchange at the water–air interface (GAMESON et al. 1955).

Growth inhibition of algae exposed to different anion-active surfactants occurs within a wide range (1–120 ppm) for various algae, while cases with stimulated growth are also listed in a review by JANICKE (1973). This range was extended from <1 to >10,000 ppm for non-ionic surfactants, while cation-active compounds produced adverse effects from 0.1 to 10 ppm. A cation-active cetyl-pyridinium-bromide produced first adverse effects on *Chlamydomonas, Scenedesmus* and *Chlorella* at 0.1–0.25 ppm, while anion-active and non-ionic compounds only produced retarding effects from 35 to 350 ppm (MATULOVA 1966). HICKS and NEUHOLD (1966) demonstrated that uptake of ^{14}C by *Vaucheria* communities was reduced within 33 h by 4 ppm of an alkyl benzene sulfonate. While a non-ionic surfactant caused no damage to *Ranunculus pseudofluitans,* this macrophyte and *Potamogeton pectinatus* and *Groenlandia densa* were seriously affected by 2.5 ppm of ABS (HYNES and ROBERTS 1962). LABUS (1979) and LABUS et al. (1977a) have shown that 0.5 ppm of a widely used linear ABS caused reductions of photosynthesis of various submerged vascular plants by 27%–77% within 20 days. Damage by surfactants is attributed to interaction with cellular and subcellular membranes. Surfactants penetrate at the protein lipid interphase into grana (KE and CLENDENNING 1956). JIRGENSONS (1961) suggests interaction of the lipophilic part of the surfactant with ligands of proteins, thereby unfolding their tertiary structure. THOMAS et al. (1953) demonstrated inhibition of the Hill reaction in chloroplasts by 3.3×10^{-4} M of anion-active Na-dodecylsulfate.

13.3.5 Radioactive Wastes

Since the nuclear test ban treaty of 1963, atmospheric fallout of radioisotopes from nuclear weapons tests has become insignificant and major attention in radioecological impact studies has shifted to nuclear power plants, which are known for chronic intermittent releases of radiosiotopes in cooling water. The average annual amounts that reach the environment in normal operation are only percentages of legal permissible concentrations (MUNDSCHENK et al. 1977). Of all radioisotopes produced, those with long half-lives, such as the fission products ^{90}Sr, ^{137}Cs, ^{144}Ce and activated nuclides ^{54}Mn, ^{60}Co and ^{65}Zn, are considered to be of ecological significance.

Their distribution in the aquatic environment is controlled by the same mechanisms as apply to their stable counterparts. Radionuclides can remain in solution or in suspension, precipitate and accumulate in sediments or be taken up by aquatic biota. Water chemistry and composition of the sediments, flow characteristics and turbulence govern the decontamination of the water by sedimentation of radionuclides. RUF (1967) lists concentration factors in sediments for ^{90}Sr (46–800) and ^{137}Cs (23–14,000). The solubility of most fission products is affected by pH. At pH 8.5 only 6.3% of ^{144}Ce are found in solution, at pH 7.0 there is an increase to 29.5% and at pH 2 to 57.5% (BAPTIST 1966), enhancing the availability to aquatic organisms in this sequence.

Some surface waters contain naturally radioactive materials. Strong radioactive springs in Japan exhibited 18 algal species, 12 of them being diatoms (MIFUNE et al., cit. PALMER et al. 1977). Aquatic plants are not known to discriminate between stable and radioactive minerals. They accumulate the radioactive isotope of a given element according to the ratio of the radionuclide to the non-active form in the medium (MCKEE and WOLF 1963). Metabolically useless elements like Sr, Y, Cs and their radioactive isotopes are also taken up by algae (HARVEY and PATRICK 1967; WILLIAMS and PICKERING 1961) and vascular plants (BUCKSTEEG 1965). There is evidence for antagonisms in the uptake of ^{90}Sr, resembling Ca and ^{137}Cs, resembling K in aquatic plants. While TIMOFEEVA and KULIKOV (1967) noted discrimination against Sr relative to Ca, results from OPHEL and FRASER (1970) indicate that a majority of 22 aquatic plants (algae, submerged macrophytes) discriminated against Ca, with enrichment of Sr relative to Ca by a factor of two or more. Concentration factors for radionuclides in freshwater plants are quite variable; they may range for ^{90}Sr from 100 to 500,000 and ^{137}Cs from 50 to 11,300, as indicated in the comprehensive literature review by RUF (1967). Algae appeared to be better accumulators than vascular plants in most cases. This is mainly explained by a favorable surface/volume ratio (GLOYNA et al. 1966), as absorption through cell membranes from the surrounding water and adsorption to exposed surfaces are major uptake mechanisms for radionuclides (PALMER et al. 1977). HARVEY and PATRICK (1967) showed that accumulated ^{137}Cs, ^{85}Sr and ^{65}Zn were quickly lost after transfer of green algae to a non-radioactive medium. Release of concentrated radionuclides from algal structures, however, may be retarded by low pH values (MACKENTHUN 1969). The effects of radiation on organisms closely correlate to the dose rate (time-intensity factor), with a dose being generally more effective in causing damage if given within a short period of time than over a longer period. Algae, like other primitive organisms, are usually more resistant to ionizing radiation than higher organisms (CAIRNS et al. 1972). DONALDSON and FOSTER (1957) noted a 50% mortality in algae at 8,000–100,000 roentgens, while fish and mammals were sensitive to values of one to two orders of magnitude less.

13.3.6 Salinity (with Reference to Chlorides)

Chloride concentrations in freshwaters are governed by the geological composition of the catchment area, balance of precipitation and evaporation, and largely by man-originated inputs. In unpolluted freshwaters of temperate zones chloride concentrations usually vary within 2 to 20 ppm Cl^- (WETZEL RG 1975) and hardly exceed 50 ppm. Chloride pollution displays seasonal characteristics, when originating from the use of deicing salts in the winter (BROD and PREUSSE 1975; HUTCHINSON 1968; KLEE 1975). Temporary Cl^- peaks in stretches of a river may exceed 3,000 ppm, when rapid thawing sets in, but return to background level commonly occurs within days. Cl^- pollution levels have been raised considerably in many surface waters; they have tripled to 23 ppm in 1965 in Lake Erie within 50 years (OWNBEY and KEE 1967) and have been raised eightfold within 30 years to 40 ppm in Lake Wingra (SCHRAUFNAGEL 1967). Large chronic chloride loads in streams originate from potassium mines and related industries and coal mines; affected Rhine water averages 170–300 ppm Cl^- (DER RAT 1976). Extreme Cl^- averages of 1,200 and 1,700 ppm are stated for the German rivers Leine and Werra, respectively (MEINCK et al. 1968). Chloride pollution is frequently referred to as salt pollution, as it usually correlates with marked increases of sodium levels (see FÖRSTNER and MÜLLER 1974). High salt accumulation may induce significant alterations in the physical properties of standing waters due to formation of high density water layers on the bottom, obstructing spring circulation and delaying stratification, also contributing to prevailing anaerobic conditions on the bottom (BUBECK et al. 1971; JUDD 1970).

The importance of chloride in determining the composition of algal populations is widely recognized (SUGIHARA and KILHAM 1977). A "halobic system" for diatoms suggested by KOLBE (1927, 1932) arranges algae groups in reference to Cl^- concentrations of the environment. At concentrations $\leqq 100$ ppm Cl^-, indifferent forms constitute 80%–95% of the phytoplankton (PETERSEN 1943). Beyond this threshold halophilous and mesohalobous species increase, corresponding to a reduction of salt-indifferent forms by 56%–70%. Qualitatively similar phytoplankton populations are described by TALLING and TALLING (1965). Findings of CHOLNOKY (1968) suggesting that chloride is an unimportant factor in determining algal distribution, with greater significance of osmotic parameters, are not supported by data of SCHMITZ (1959) and SUGIHARA and KILHAM (1977). According to STEINER and ESCHRICH (1958) osmotic effects exert their main influence on freshwater diatoms at higher salt concentrations, while specific ion effects of Na^+ and Cl^- are of importance at lower concentrations. Extreme alteration of phytoplankton communities in some calcium carbonate streams loaded with alkali chlorides are described by HUSTEDT (1926, 1957), VON TÜMPLING (1960) and ZIEMANN (1967). Under a poikilohaline regime, blooms of halophytic diatoms (e.g. *Amphiprora paludosa, Cyclotella meneghiniana, C. nana, Melosira nummuloides, Thalassiosira fluviatilis*) appeared. A list of algae (largely diatoms) indicative of salt brine (principally NaCl) is provided by PALMER et al. (1977). WETZEL RG (1975) considers the lower flora and fauna to be quite adaptable to a wide range of salinity, although some groups of bacteria and algae are relatively homoiosmotic. Disturbance of self-purification processes appears at 1,200 ppm Cl^- (MEINCK et al. 1968).

Field data from brackish waters indicate that some freshwater macrophytes may tolerate salinities from 1.5%–3.0% (≙800–1,600 ppm Cl^-) (ANDERSON 1964; LUTHER 1951). In greenhouse experiments 50% inhibition of growth occurred at a salinity range from 0.38% (*Ceratophyllum*) to 3.5% (*Potamogeton perfoliatus*) (BOURN 1932). HALLER et al. (1974) who used seawater dilutions showed that *Hydrilla verticillata* and *Vallisneria americana* failed to grow at 0.66% salinity, while *Najas guadalupensis* and *Myriophyllum spicatum* tolerated 1.0% and 1.33% respectively. MCGAHEE and DAVIS (1971) demonstrated that respiration of *Myriophyllum spicatum* shoots was not affected after 10 days at a salinity of 1.6%, whereas photosynthesis was drastically reduced. Floating water plants (*Pistia stratiotes, Eichhornia crassipes*) were already sensitive to 0.17%–0.25% salinity; salt concentrations toxic to emergent *Myriophyllum brasiliense* ranged from 1.0% to 1.33%, while *Lemna* showed increased growth up to 0.33% salinity, suggesting relatively high Cl^- and Na^+ requirements for maximum growth (HALLER et al. 1974). No effects of 1.0% NaCl on *Lemna trisulca* were observed by VIVIER and NISBET (1965). Within *Typha* populations, *T. angustifolia* appeared to be more resistant to gradually increasing NaCl concentrations (0.25%–2.0%) than *T. latifolia* (MCMILLAN 1959). High susceptibility of *Potamogeton alpinus* to Cl^- concentrations as low as 50 ppm (≙0.082% NaCl) was indicated by reduced photosynthesis after 3 weeks of exposure in soft water experiments (NOBEL 1980). Total damage occurred at 1,000 ppm Cl^- whithin two weeks. *Elodea canadensis* and *Myriophyllum alterniflorum* appeared to be less sensitive, as first signs of photosynthetic depression were noticed at 250 and 100 ppm Cl^-, respectively, after 3 weeks. In contrast to terrestrial plants, uptake of Cl^- by the submerged vascular plants in these experiments was outweighed by Na^+ uptake.

According to COWGILL (1974), accumulation of Cl^- in submersed macrophytes is higher than in floating or emergent species. RAVEN (1967) has shown that in the case of *Hydrodictyon africanum* Cl^- is the major intracellular anion.

Calcium is considered a meliorating antagonist in NaCl pollution, as was already observed by GAHLEN (1935) with *Elodea canadensis*. DAVIS et al. (1974) could reduce inhibition of photosynthesis of *Myriophyllum spicatum* exposed to NaCl with the addition of $CaCl_2$. This is in accord with NOBEL (1980) who demonstrated differences in the susceptability of submerged macrophytes to Cl^- with variation of the cation. *Elodea canadensis* was less sensitive to Cl^-, when applied as $CaCl_2$ than was the case with NaCl, KCl or $MgCl_2$, indicated by increased reduction of photosynthesis. Experimental results of CHIMIKLIS and KARLANDER (1973) with *Chlorella* conform to these findings, as NaCl toxicity observed at 0.3 M was reduced in the presence of 10 mM calcium salts. With this Ca^{2+} addition growth could be extended to 0.5 M NaCl. This mainly seems to be due to maintained membrane integrity of cell organelles by Ca^{2+}, whereas disruption was observed in sole presence of elevated NaCl concentrations. Anion ratios may also be involved in tolerance to different salinity levels, as maximum growth rates of the halophilic alga *Thalassiosira fluviatilis* varied with changing Cl^- to SO_4^{2-} ratios of the medium (SCHMITZ 1959), see also Chapter 3 of Vol. 12C.

Extrapolation of laboratory results to natural waters is difficult. Special caution is advised when sea water dilutions are used in laboratory experiments, while focus is intended on the effects of Cl^- pollution of freshwaters. NaCl toxicity is meliorated if seawater dilutions are used, as seawater is an equilibrated solution that is favorable to plants compared to ill-balanced artificial NaCl solutions. STANLEY (1974a) indicated a close to twofold higher toxicity of NaCl in artificial medium for *Myriophyllum spicatum* compared to a tolerance level of this species given by ANDERSON (1964) for brackish water. In seawater dilutions osmotic effects can be assumed to outweigh the effects of single ions such as Na^+ or Cl^-. It is generally agreed that only under extreme conditions, i.e. in hypersaline freshwaters, does salinity become a limiting factor of productivity (LUND 1965; STENGEL and SOEDER 1973).

13.3.7 Freshwater Food Chains

Passing on of toxicants with aquatic food chains and accumulation in organisms along the trophic structure of the habitat is of great concern. The problem of bioaccumulation of toxic chemicals, to which food-chain transmission greatly contributes, lies in the fact that toxic levels for an organism may be reached even though the concentration of a toxicant in the medium is minute and may not be of direct concern. Usually organisms towards the end of the food chain are more sensitive to the toxicant than the organisms they feed on. These lower organisms of the trophic structure may already be potent accumulators. Numerous studies demonstrate bioaccumulation and biomagnification of persistent polychlorinated hydrocarbons, other pesticide ingredients as well as polychlorinated biphenyls (CANTON et al. 1975; EBERHARDT et al. 1971; GREICHUS et al. 1977; HANSEN 1976; ISENSEE and JONES 1975; METCALF 1977) and of heavy metals (BOMBOSCH and PETERS 1975; GREICHUS et al. 1977; TERHAAR et al. 1977; THOMANN et al. 1974).

13.4 Acidification

Most freshwaters display circumneutral to moderate alkaline conditions. High natural H^+ ion activity is encountered in bog lakes as a consequence of active cation exchange in *Sphagnum* cell walls with concomitant release of H^+. Natural low pH values (1.3

to 4.0) are mainly found in surface waters of geothermal and vulcanic regions, as strong mineral acids are present (WETZEL RG 1975). Man originated acid mine drainage renders streams extremely acidic with pH dropping to as low as 1.8–4.0 (HARGREAVES et al. 1975; SAND-JENSEN and RASMUSSEN 1978). Large-scale afforestation of agricultural land with spruce may contribute to acidification of soils and water (BJÖRK 1978). Acid precipitation, originating from sulfur and nitrogen oxide emission from high-rising industrial smoke stacks has become a transregional problem. In northern Europe and America, soft water lakes of low buffering capacity, affected by winds from industrial areas are in a stage of progressive acidification, as the pH of precipitation has dropped below 4.5 due to increased sulfuric and nitric acid contents (ALMER et al. 1974; BJÖRK 1974; LIKENS et al. 1979). Acid precipitation has reduced the pH of some Swedish lakes by as much as 1.8 units over 30 years, leaving many lakes with pH values below 6.0 or 5.0 (ALMER et al. 1974). Also indirect effects of acidification have to be considered as H^+ ion activity is a governing factor in many chemical equilibria.

Hydrogen ion concentration, indicated by pH, may be regarded as a toxicant in the broadest sense (CAIRNS et al. 1975). Acidification affects freshwater ecosystems on all trophic levels, generally indicated by retarded productivity and reduced species diversity. In a stream receiving acid mine drainage bacterial and fungal counts decreased markedly with pH falling from 4.0 to 2.0 (JOSEPH 1953). In acidified lakes the decomposition of organic substance is diminished and fixation of nitrogen is inhibited. Increased transparency, due to precipitation of humic substances and phosphates and restricted algal growth, as well as disappearance of *Phragmites australis* stands leave the impression that the water has become oligotrophic (ALMER et al. 1974). Macrophytic vegetation in waters acidified below pH 4.0 is very limited. Plants which solely rely on dissolved CO_2 as C-source are favored: *Juncus bulbosus, Sparganium simplex, Glyceria fluitans, Eleocharis acicularis* and bryophytes of genus *Polytrichum, Anisothecium, Fontinalis* and *Catharinea* (HARGREAVES et al. 1975; LACKEY 1938; OHLE 1936; PIETSCH 1973; SAND-JENSEN and RASMUSSEN 1978). Species diversity always increased with rising pH and alkalinity. *Juncus bulbosus,* characterized by great phenotypic plasticity (HINNERI 1976) has been described by IVERSEN and OLSEN (1943) as strongly acidotrophic; *Typha latifolia* and *Juncus effusus* were found to be tolerant of pH values as low as 2.8, whereas softwater species *Isoëtes lacustris* and *Lobelia dortmanna* were not found in water of $pH<4$ (HINNERI 1976) and sediments of $pH<5.3$ (MISRA 1938). While in acidified lakes peat-moss swards spread on bottoms that were formerly free from it (ALMER et al. 1974), duckweed covers (*Lemna, Wolffia*) are adversely affected at pH values <4 (MCLAY 1976).

Acidification also affects algal community structure. Blue-green algae are absent under acidic conditions (WHITTON 1972). This is confirmed by SHAPIRO (1973), as lowering the pH in lake water experiments resulted in a shift in dominance from blue-greens to greens. Specialists, tolerant of extremely low pH, are found mainly among flagellates and diatoms (PALMER et al. 1977). Extreme tolerance is noted for *Chlamydomonas acidophila* (pH 1.0), *Euglena mutabilis, Chromulina,* and *Navicula* (all pH 1.8), likewise *Desmidium* and the green algae *Mougeotia* (pH 2.4) and *Ulothrix zonata* (pH 2.6) by LACKEY (1938). See also BENNETT (1969), JOSEPH (1953), HARGREAVES et al. (1975). A pH classification system for diatoms has been suggested by HUSTEDT (1939).

Promising counteraction to acidification of lakes is under way in Sweden, making use of the good ion-exchange potential of humic substances, charged with Na^+, which can be exchanged for H^+ ions of the acidic aquatic environment (Björk 1978; Ripl 1976a).

13.5 Waste Heat

Thermal discharges into aquatic environments have come to be of concern only within the past 25 years or so. The term "thermal pollution", believed to be coined by Moore (1958), has been widely replaced by "waste heat" in the literature. About 80% to 90% of all thermal additions to British and U.S. waterways in 1968 originated from thermal electricity producing power plants (O'Sullivan and Collinson 1976). Nuclear power plants of the 1,000 megawatt category require 45–60 $m^3 s^{-1}$ of cooling water, equivalent to the amount used by a city of the size of Chicago (Belter 1975). Power plant heating effects of 8°–10 °C, expressed as difference between intake and outlet water temperatures (Δt °C), have not been uncommon (Belter 1975), but legislative regulations in many countries limit thermal additions to 3°–5 °C depending on the type of habitat, with certain maximum temperatures not to be exceeded. Rising energy requirements presumably will mainly be met by thermal power plants; although a shift in cooling technology will favor "closed" over "once through" systems by a ratio of 70 by 30 in the year 2000 (Belter 1975), it appears to be unlikely that thermal loading of aquatic systems on the whole will be significantly reduced or stopped.

The assimilatory capacity of a freshwater ecosystem for waste heat is governed by the amount of discharged heat, the mode of release, the properties of the receiving system, also taking the prevailing climatic conditions into account. It has become evident with the multitude of factors involved that the effects of each thermal discharge must be examined on an individual basis, as findings from one freshwater habitat may not necessarily be applicable to another (Cairns and Dickson 1974; O'Sullivan and Collinson 1976).

Temperature correlates highly with many physical and chemical properties of water, e.g. the viscosity, density, ionization, and the solubility of salts and gases. As the temperature increases, the viscosity decreases, leading to enhanced sedimentation of suspended materials. Seasonal temperature characteristics of surface waters may be severely disturbed by thermal discharges. Additions of large amounts of heated water to smaller lentic waters may cause prolonged thermal stratification and delay turnover of the water body in the fall; in extreme cases, when the hypolimnion is also heated in shallow lakes, premature turnover in the summer may be encountered, with nutrient transport from the bottom to the trophogenic zone, stimulating productivity at an undesirable time (Eidg Dep 1968). Thermal discharges will reduce the capacity of the water to keep oxygen in solution. The resulting oxygen deficit in rivers that had been oxygen-saturated before cannot be compensated by artificial aeration, as excess oxygen at the new saturation level rapidly disappears downstream. Aeration, however, may be useful in water that was already oxygen deficient prior to waste heat input. Uptake of oxygen from the air is also reduced as the oxygen saturation deficit decreases with increasing temperature (Eidg Dep 1968). At the same time the situation may be made more difficult as the metabolic activity of microorganisms and biochemical oxygen demand (BOD) is raised (Ross 1977; Stangenberg 1977). This usually decreases the waste assimilatory capacity of the aquatic ecosystem.

13.5.1 Waste Heat and Freshwater Vegetation

Aquatic organisms display certain genetically fixed temperature requirements, commonly outlined by cardinal temperatures: a minimum temperature, below which there is no growth, an optimum temperature with stimulated growth and metabolism, and a maximum temperature above which growth is stopped (see Chaps. 12 and 14, Vol. 12A). Thermal death usually occurs just a few degrees above the maximum growth temperature. While suboptimal temperatures merely slow growth, supraoptimal temperatures generally reduce growth quickly to zero. BROCK (1975) further indicates that the optimum temperature is always closer to the maximum than to the minimum, resulting in a more narrow tolerance margin for temperature additions before death sets in. Organisms in natural environments usually live somewhat below their optimum temperature, so that moderate temperature increases will normally result in accelerated growth, as the temperature will be closer to the optimum. Thus even larger temperature increases (10 °C) in temperate habitats may not necessarily result in changes of population structure, if the organisms have existed far below their optimum temperature (BROCK 1975). Certain variability of the cardinal temperatures of species must be allowed for in different development stages and modes of exposure. The upper temperature limit is usually lower for growing than for resting organisms, with exceptions noticed for some blue-greens and some extreme thermophilic bacteria (LEVITT 1972).

Growth enhancement and shift in population structure of submersed vascular plants in reservoirs receiving waste heat are common phenomena. YOUNG (1974) observed a more than two-fold increase in standing crop of *Myriophyllum spicatum* in ambient water of 25 °C as compared to 15 °C. In similar investigations of GRACE and TILLY (1976) proliferation of this species dwindled again as temperatures of up to 33.4 °C were reached and *Najas guadalupensis* came to dominate. HAAG and GORHAM (1977) also observed changes in population structure in the mixing zone of a thermal effluent into Lake Wabamum, as *Myriophyllum exalbescens* and *Chara globularis* were replaced by *Potamogeton pectinatus* and largely *Elodea canadensis*. ANDERSON (1969) noticed proliferation of *Potamogeton perfoliatus* in 35 °C water below a waste heat discharge in Patuxent River. His data suggest that *P. perfoliatus* is capable of physiological adjustment to higher temperatures as the leaf matures. Fragments of *Myriophyllum spicatum* drawn through cooling systems of an electricity generating station were stimulated or partially inhibited when exposed to temperatures from ambient up to 44 °C for 2.5 to 10 min (STANLEY 1975a).

CAIRNS (1956) has demonstrated pronounced shifts in algal groups collected on glass slides in 18°–20 °C river water, as they were exposed to gradual increases of temperature up to 40 °C. Diatoms remained dominant within 20°–30 °C, giving way to dominance of green algae between 30°–35 °C, while blue-green algae came to prevail at temperatures greater than 35 °C; when the temperature was lowered, green algae and diatoms reappeared at their respective temperatures. Similar shifts have been noticed by BEER and PIPES (1969), BUCK (1970), PATRICK (1969) and TREMBLEY (1960, 1965). Cool stenotherm diatoms are generally more sensitive to temperature change but may withstand a 10 °C temperature change, while diatoms adjusted to warmer temperatures can tolerate changes from 15° to 20 °C (PATRICK 1969). Some caution in the interpretation of slide experiments is suggested by BROCK (1975), as they appear to be selective for smaller algae such as diatoms and blue-greens. Studies by BOYLEN and BROCK (1973) in geothermal habitats indicate that in most cases blue-green algae will only become dominant if the water temperature exceeds 40 °C.

Various authors have focused on the seasonality of waste heat effects. Stimulation of plankton and periphyton productivity by artificial temperature increases in winter and early arrival of spring seasonal succession in periphyton due to heated effluent discharges are noted phenomena (FORESTER 1969; MOTHES et al. 1976). At higher summer temperatures (15°–20 °C) slight increases in temperature may still increase production, whereas higher increases result in depression (WARINNER and BREHMER 1966). Thus phytoplankton photosynthesis was adversely affected at temperatures greater than 30 °C (WUNDERLICH and MÜLLER 1976). While in temperate habitats, waste heat may well be beneficial in winter and harmful in summer, the effects in arctic systems will most likely always be beneficial. Problems are to be expected in warm-water habitats where temperatures are already closer to the upper temperature limit of species (BROCK 1975). Fluctuating temperatures, a common characteristic of power plant effluents inhibit growth of phytoplankton. Even if constant temperatures are higher than the average temperatures of a fluctuating habitat, fluctuation will depress algal growth more than constant high temperatures (BROCK 1975). No significant adverse effects on phytoplankton on passage through condensers of power plants were noted by HOWELLS (1969) and KHALANSKI (1975). Findings by PATRICK (1969) indicate that only little harm, if any, is done if the temperature on passage does not exceed 34°–34.5 °C. MORGAN and STROSS (1970) reported that an increase of 8 °C stimulated algal photosynthesis on passage when the water temperature was cooler than 16 °C, while a similar increase inhibited photosynthesis at temperature $\geqq$20 °C. Although large-scale temperature adaptation by the photosynthetic apparatus of algae appears to be ruled out at physiological temperatures from 1° to 30 °C, there is evidence for adaptive changes in response to seasonal changes in the temperature optimum of photosynthesis (ARUGA 1965). Algae undergo changes in their cellular chlorophyll and protein levels at different temperatures; both appear to increase as the temperature decreases (MORRIS and GLOVER 1974). Protein denaturation and lipid melting are the commonly postulated mechanisms of heat damage to cells (HEBER and SANTARIUS 1973). Alteration of cell membranes (BOWLER et al. 1973; CHAPMAN 1967; DANIELL et al. 1969) may result in their increased leakiness (BENZIONI and ITAL 1973). LANZA and CAIRNS (1972) investigated physiomorphological effects of abrupt thermal stress on diatoms utilizing fluorescent patterns of diatom cells as evaluation criterium. They suggest that a substantial amount of intracellular lipid or its degradation products could be exuded out into the surrounding environment following the disruption of cell membranes by certain temperature stresses. Thereby vertical migration of some species of algae, including diatoms, may be affected as it is in part dependent upon specific gravity changes from cell oil globules.

If no other factors are involved an increase in temperature in general results in "heterotrophication" of the aquatic biocenosis, implying a relative increase of the intensity of respiratory over assimilatory processes (KNÖPP 1973; MOTHES et al. 1976). WUNDERLICH and MÜLLER (1976) indicate inhibition of dehydrogenase activity in psychrophilous cold-adapted saprophytic bacteria after a short temperature increase to 30 °C, while thermotolerant bacteria showed increased activity even beyond 40 °C. Species diversity and biomass of aquatic fungi was reduced after short heating to 30 °C. Variability of bacterial population response to temperature increases was explained by other habitat factors, mainly the amount of available nutrient substrate (RHEINHEIMER 1965). Shifts in microbial groups of river water upon passage through cooling systems were observed by VERSTRAETE et al. (1975).

13.5.2 Temperature and Plant Sensitivity to Selected Chemicals

Literature on the problem of temperature effects on the toxicity of pollutants on aquatic vegetation is limited. Zinc toxicity to *Nitzschia linearis* in soft water increased with a temperature increase from 22° to 30 °C (PATRICK 1971). Similarly, zinc toxicity to *Cyclotella meneghiniana* was shown to increase when the temperature was raised from 15° to 20 °C, while the converse was true for *Scenedesmus quadricauda* and *Chlamydomonas* (CAIRNS et al. 1978a). LABUS (1979) demonstrated increased sensitivity of some submerged macrophytes to a linear alkyl benzene sulfonate as the temperature was raised from 8.5° to 11.5 °C. However, there exists no "universal law" to describe combined chemical and temperature effects as the influence of temperature upon chemical toxicity is both chemical and species specific (CAIRNS et al. 1978a). STANLEY and MADEWELL (1976) demonstrated protection of *Lemna minor* from heat stress by 3 mM of calcium nitrate. Light has been shown as detrimental to *Lemna* in combination with chronic exposures to high temperatures (PORATH and BEN-SHAUL 1971).

13.6 Biological Indicators

Biological indicators, according to ELLENBERG (personal communication) are organisms and/or communities, which by their presence or reactions are indicative of defined ecological factors or pollutants. Similar to terrestrial studies, botanical limnology applies three categories of indicators: (1) aquatic plant species and their communities, (2) monitoring plants which indicate the state of pollution by accumulation of toxicants and/or vitality changes (with plants collected from natural communities in "passive monitoring" and standardized introduction of plants to polluted waters and subsequent analysis in "active monitoring"), and (3) test plants and organisms used for toxicity tests (such as bio-assay) in the laboratory, which display measurable changes when exposed to pollutants (DER RAT 1978).

The indicator value of single species and communities increases with the degree of accuracy attained in knowledge of their ecological potential. Calibration to as many environmental factors as possible, as measured in the natural habitat, is a prerequisite in establishing a biological indicator. Presence of a species yields higher information value than absence, as a species may not only be absent due to unsuitable environmental conditions, but merely because it never had the opportunity to be introduced to the habitat area or because its functional niche has already been occupied by another species (CAIRNS 1974). Community concepts for biological evaluation of water quality based on microphyton are known from COSTE (1974), FJERDINGSTAD (1964), LIEBMANN (1962) dating back to KOLKWITZ and MARSSON (1908), and PATRICK (1949). Standardized species and groups of macrophytes with different tolerance levels for single factor and more complex nutrient pollution have been established for still and running waters (DESCY 1975; EMPAIN 1973; FRAHM 1974; GLÄNZER et al. 1977; KOHLER and ZELTNER 1974; KOHLER et al. 1971, 1973, 1974; KURIMO 1970; KUTSCHER and KOHLER 1976; LANG 1973; MELZER 1976; SUOMINEN 1968; WIEGLEB 1978). Macrophyte communities are used for the classification and mapping of the state of nutrient loading of surface waters (MELZER et al. 1977; HABER and KOHLER 1972; LACHAVANNE and WATTENHOFER 1975). While macrophytic vegetation responds rather quickly to increasing pollution (eutrophication), it does not as rapidly reflect reduction of nutrient load in a once polluted habitat (KOHLER et al. 1980).

Monitoring of pollution through changes in diatom community structure (PATRICK 1977) on exposed glass slides may gain in importance as new techniques (DICKSON et al. 1977) facilitate identification and enumeration. Macrophytes have been used as monitors for accumulation of heavy metals (BENSON-EVANS and WILLIAMS 1976; DIETZ 1972; ERNST and MARQUENIE-VAN DER WERFF 1978; MCLEAN and JONES 1975).

There are numerous approaches towards toxicity testing of chemicals, some of which are based on algal tests (BRINGMANN and KÜHN 1959, 1979; FORSBERG et al. 1978; FRIEDRICH 1975; GELLER and MÄCKLE 1977; HUTCHINSON and STOKES 1975; MALONEY and MILLER 1975; MALONEY and PALMER 1956; MILLER et al. 1974), others on submerged macrophytes (AUERBACH et al. 1973; FRAHM 1975; LABUS and KOHLER 1981; LABUS et al. 1977b; WEISE et al. 1976). Microorganisms have been employed extensively (BRINGMANN and KÜHN 1959, 1979; MÜLLER and WELLNER 1975; PLÖTZ 1974; ROBRA 1979). General and specific aspects of estimating the hazards of chemical substances to aquatic life are presented in CAIRNS and DICKSON (1976), CAIRNS et al. (1977, 1978b) and MAYER and HAMELINK (1977).

13.7 Conclusions

Worldwide freshwater ecosystems are stressed by an increasing number of foreign substances. These substances are leading to far-reaching qualitative and quantitative alterations in habitats and their biocoenosis. Nationally and internationally, efforts to protect and improve water ecosystems have been successful. An example is the restoration of some Swedish lakes. In many areas the water quality of overstressed rivers has been improved.

But new problems of water pollution are continually arising; they may not all be solved by protection management approaches. One serious problem is the preservation of natural oligotrophic water ecosystems in intensively used agricultural areas. The destruction and endangering of natural lake ecosystems with less buffering capacity by acid precipitation, caused by sulfur and nitrogen oxides emissions transported over long distances, is a real and very serious problem. These emissions do not only threaten water ecosystems but also coniferous forests. In order to solve acidification problems we need quick international actions.

From the point of ecological research there are many unsolved questions concerning stress in freshwater ecosystems. For example, we know little about the long-term effects of chronic stress on water ecosystems, or about the combined effects of different pollutants on organisms and ecosystems. In this context we need a better appreciation of biological indicators.

Indeed, the improvement and restoration of our freshwater ecosystems is less a problem of ecological research, than a question of translating ecological research into political decisions and environmental protection. The present level of research in our opinion is sufficient to plan and arrange our water ecosystems in a biologically optimal way. However, in order to apply this research, we need completely new concepts and solutions.

References

Ahl T (1975) Effects of man-induced and natural loading of phosphorus and nitrogen on large Swedish lakes. Verh Int Ver Limnol 19:1125–1132

Ahmed MK, Casida JE (1958) Metabolisms of some organophosphate insecticides by microorganisms. J Econ Entomol 51:59–63

Aiken SG, Newroth PR, Wile I (1979) The biology of Canadian weeds. 34. *Myriophyllum spicatum* L. Can J Plant Sci 59:201–215

Allen HL (1971) Primary productivity, chemo-organotrophy, and nutritional interactions of epiphytic algae and bacteria on macrophytes in the littoral of a lake. Ecol Monogr 41:97–127

Almer B, Dicksen W, Ekström C, Hörnström E, Miller U (1974) Effects of acidification on Swedish lakes. Ambio 3(1):30–36

Anderson RR (1964) Ecology and mineral nutrition of *Myriophyllum spicatum* L. MS thesis, Univ Maryland, College Park

Anderson RR (1969) Temperature and rooted aquatic plants. Chesapeake Sci 10(3, 4):157–164

Antonovics J, Bradshaw AD, Turner RG (1971) Heavy metal tolerance in plants. Adv Ecol Res 7:1–85

Aruga Y (1965) Ecological studies of photosynthesis and matter production of phytoplankton. I. Seasonal changes of photosynthesis in natural phytoplankton. Bot Mag 78:360–365

Auerbach S, Prüfer P, Weise G (1973) Gasstoffwechselphysiologische Schädigungskriterien bei submersen Makrophyten vom Typ *Fontinalis antipyretica* L. unter Einwirkung von Schwermetallen oder Phenol. Int Rev Ges Hydrobiol 58:19–32

Ayles GB, Lark JGI, Barica J, Kling H (1976) Seasonal mortality of rainbow trout (*Salmo gairdneri*) planted in small eutrophic lakes of central Canada. J Fish Res Board Can 33:647–655

Baity HG, Merryfield F, Uzzle AB (1933) Some effects of sewage chlorination upon the receiving stream. Sew Works J 5:429–446

Baker R, Cole S (1974) Residual chlorine: Something new to worry about. Ind Water Eng 11:10–21

Baptist JP (1966) Uptake of mixed fission products by marine fishes. Trans Am Fish Soc 95:145–152

Bartlett L, Rabe FW, Funk WH (1974) Effects of copper, zinc and cadmium on *Selenastrum capricornutum*. Water Res 8:179–185

Batterton JC, Boush GM, Matsumura F (1972) DDT: Inhibition of sodium chloride tolerance by the blue-green alga *Anacystis nidulans*. Science 176:1141–1143

Beer LO, Pipes WO (1969) The effect of discharge of condenser water into the Illinois River. Industrial Bio-Test Laboratories, Northbrook, III

Belter WG (1975) Management of waste heat at nuclear power stations, its possible impact on the environment, and possibilities of its economic use. In: International Atomic Energy Agency (eds) Environmental effects of cooling systems at nuclear power plants. IAEA, Vienna, pp 1–23

Bennett HD (1969) Algae in relation to mine water. Castanea 34:306–328

Benson-Evans K, Williams PF (1976) Transplanting bryophytes to assess river pollution. J Bryol 9:81–91

Ben-Zioni A, Ital C (1973) Short- and long-term effects of high temperatures (47°–49 °C) on tobacco leaves. III. Efflux and ^{32}P incorporation into phospholipids. Physiol Plant 28:493–497

Bernhardt H (ed) (1978) Phosphor. Wege und Verbleib in der Bundesrepublik Deutschland. Chemie, Weinheim

Besch WK, Ricard M, Cantin R (1972) Benthic diatoms as indicators of mining pollution in the Northwest Miramichi river system, New Brunswick, Canada. Int Rev Ges Hydrobiol 57:39–74

Betzer N, Kott Y (1969) Effects of halogens on algae-II. *Cladophora sp*. Water Res 3:257–264

Billen G (1973) Etude de l'ecometabolisme du mercure dans un milieu d'eau douce. Hydrobiol Bull 7:60–68
Bingham SW, Shaver RL (1977) Diphenamid removal from water and metabolism by aquatic plants. Pestic Biochem Physiol 7:8–15
Birmingham BC, Colman B (1977) The effect to two organophosphate insecticides on the growth of freshwater algae. Can J Bot 55:1453–1456
Björk S (1967) Ecological investigations of *Phragmites communis*. Studies in theoretic and applied limnology. Folia Limnol Scand 14:5–248
Björk S (1973) Restoring lakes in Sweden. Tech Rev 76:1–10
Björk S (1974) European Lake rehabilitation activities. Plenary lecture. Conference on Lake Protection and Management, Madison, Wisc, USA, Oct 21–23
Björk S (1977) Recovery and restoration of damaged lakes in Sweden. In: Cairns J, Dickson KL, Herricks EE (eds) Recovery and restoration of damaged ecosystems. Univ Press Virginia, Charlottesville, pp 110–133
Björk S (1978) Restoration of degraded lake ecosystems. Coden Sunbds/(NBLI-3008):1–24 Lund, Sweden
Blackburn RD, Weldon LW (1965) The sensitivity of duckweeds (*Lemnaceae*) and *Azolla* to diquat and paraquat. Weeds 13:147–149
Blackman GE, Parke MH, Garton G (1955) The physiological activity of substituted phenols II. Relationships between physical properties and physiological activity. Arch Biochem Biophys 54:55–71
Blum JL (1956) The ecology of river algae. Bot Rev 22:291–341
Bombosch S, Peters L (1975) Quecksilber in Lebensräumen Süd-Niedersachsens. Naturwissenschaften 62:575–576
Bone DH (1971) Relationship between phosphates and alkaline phosphatase of *Anabaena flos-aquae* in continuous culture. Arch Mikrobiol 80:147–153
Bornkamm R, Raghi-Atri F, Koch M (1980) Effects of eutrophication on *Phragmites australis*. Garten Landsch H1:15–19
Bourn WS (1932) Ecological and physiological studies on certain aquatic angiosperms. Contrib Boyce Thompson Inst 4:425
Bowler K, Duncan CJ, Gladwell RT, Davison TF (1973) Cellular heat injury. Comp Biochem Physiol 45A:441–450
Boyd CE (1970) Vascular aquatic plants for mineral nutrient removal from polluted waters. Econ Bot 24:95–103
Boylen CW, Brock ML (1973) Effects of thermal additions from the Yellowstone geyser basins on the benthic algae of the Firehole River. Ecology 54:1282–1291
Brezonik PL (1972) Nitrogen: Sources and transformations in natural waters. In: Allen HE, Kramer JR (eds) Nutrients in natural waters. Wiley and Sons, New York, pp 1–50
Bringmann G, Kühn R (1959) Vergleichende wassertoxikologische Untersuchungen an Bakterien, Algen und Kleinkrebsen. Gesund Ing 80:115–120
Bringmann G, Kühn R (1979) Vergleich der toxischen Grenzkonzentrationen wassergefährdender Stoffe gegen Bakterien, Algen und Protozoen im Zellvermehrungshemmtest. Gesundheitsring 10(8):249–252
Brinley FJ (1942) Stream pollution. Sewage Works J 14:147–152
Bristow JM, Whitcomb M (1971) The role of roots in the nutrition of aquatic vascular plants. Am J Bot 58:8–13
Brock TD (1975) Predicting the ecological consequences of thermal pollution from observations on geothermal habitats. In: International Atomic Energy Agency (ed) Environmental effects of cooling systems at nuclear power plants. IAEA, Vienna, pp 599–621
Brod HG, Preusse HU (1975) Einfluß von Auftausalzen auf Boden, Wasser und Vegetation. II. Beeinflussung des Wassers. Rasen-Turf-Gazon 6:46–50
Brooker MP (1976) The ecological effects of the use of dalapon and 2,4-D for drainage channel management. Arch Hydrobiol 78:396–412
Brooker MP, Edwards RW (1973) Effects of the herbicide paraquat on the ecology of a reservoir. I. Botanical and chemical aspects. Freshwater Biol 3:157–175
Brooker MP, Edwards RW (1975) Aquatic herbicides and the control of water weeds. Water Res 9:1–15

Brooks AJ, Baker AL (1972) Chlorination at power plants: Impact on phytoplancton productivity. Science 176:1414–1415
Brungs WA (1973) Effects of residual chlorine on aquatic life. J Water Pollut Control Fed 45:2188–2193
Bubeck RC, Diment WH, Deck BL, Baldwin AL, Lipton SD (1971) Runoff of deicing salt: effect on Irondeqouit Bay, Rochester, New York. Science 172:1128–1131
Buck JD (1970) Summary of Connecticut River microbiology study. In: Merriman DE (ed) The Connecticut River Investigation. 10th Semiannu Prog Rep Conn Water Resour Comm
Bucksteeg W (1965) Radiologische Überwachung der Oberflächengewässer im Einzugsgebiet der Ruhr im Hinblick auf die Trinkwasserversorgung – Voraussetzungen und Erfahrungen. GWF Wasser Abwasser 106:1157–1164
Buikema AL Jr, McGinniss MJ, Cairns J Jr (1979) Phenolics in aquatic ecosystems: A selected review of recent literature. Mar Environ Res 2:87–181
Butcher RW (1947) Studies in the ecology of rivers. VII. The algae of organically enriched waters. J Ecol 35:186–191
Cairns J Jr (1956) Effects of increased temperature on aquatic organisms. Ind Wastes 1:150–152
Cairns J Jr (1968) The effects of dieldrin on diatoms. Mosquito News 28:177–179
Cairns J Jr (1974) Indicator species vs. the concept of community structure as an index of pollution. Water Res Bull 10:338–347
Cairns J Jr (1976) Estimating the assimilative capacity of water ecosystems. In: Sharma RK, Buffington JD, McFadden JT (eds) The biological significance of environmental impacts. Natl Tech Inf Serv, Springfield, Virginia, pp 173–189
Cairns J Jr, Dickson KL (1974) Guidelines for developing studies to determine the biological effects of thermal discharges. Aware Mag April 1974:9–12
Cairns J Jr, Dickson KL (eds) (1976) Biological methods for the assessment of water quality. Am Soc Test Mater, Philadelphia
Cairns J Jr, Dickson KL (1977) Recovery of streams from spills of hazardous materials. In: Cairns J Jr, Dickson KL, Herricks EE (eds) Recovery and restoration of damaged ecosystems. Univ Press Virginia, Charlottesville, pp 24–42
Cairns J Jr, Lanza GR, Parker BC (1972) Pollution related structural and functional changes in aquatic communities with emphasis on freshwater algae and protozoa. Proc Natl Acad Sci USA 124:79–127
Cairns J Jr, Heath AG, Parker BC (1975) Temperature influence on chemical toxicity to aquatic organisms. J Water Pollut Control Fed 47:267–280
Cairns J Jr, Dickson KL, Westlake GF (eds) (1977) Biological monitoring of water and effluent quality. ASTM, Philadelphia
Cairns J Jr, Buikema AL Jr, Heath AG, Parker BC (1978a) Effects of temperature on aquatic organisms sensitivity to selected chemicals. Virginia Water Resour Res Cent Bull 106, Blacksburg, Virginia
Cairns J Jr, Dickson KL, Maki AW (eds) (1978b) Estimating the hazard of chemical substances to aquatic life. Am Soc Test Mater, Philadelphia
Canton JH, Greve PA, Slooff W, Esch van GJ (1975) Toxicity, accumulation and elimination studies of α-hexachlorocyclohexane (α-HCH) with freshwater organisms of different trophic levels. Water Res 9:1163–1169
Carbiener R (1969) Aperçu sur quelques éffets de la pollution des eaux douces de la zone temperée sur les biocénoses aquatiques. Bull Sect Geogr Minist Educ Nation 80:45–132
Carpenter KE (1924) A study of the fauna of rivers polluted by lead-mining in the Aberystwyth district of Cardiganshire. Ann Appl Biol 11:1–23
Carpenter KE (1926) The lead mine as an active agent in river pollution. Ann App Biol 13:395–401
Cearley JE, Coleman RL (1973) Cadmium toxicity and accumulation in southern Najad. Bull Environ Contam Toxicol 9:100–101
Cedeno-Maldano A, Swader JA (1974) Studies on the mechanism of copper toxicity in *Chlorella*. Weed Sci 22:443–449
Cedeno-Maldano A, Swader JA, Heath RL (1972) The cupric ion as an inhibitor of

photosynthesis electron transport in isolated chloroplasts. Plant Physiol 50:698–701
Chacko CI, Lockwood JL (1967) Accumulation of DDT and dieldrin by microorganisms. Can J Microbiol 13:1123–1126
Chamberlain WM (1968) A preliminary investigation of the nature and importance of soluble organic phosphorus in the phosphorus cycle of lakes. PhD thesis, Univ Toronto, Toronto, Ont (Microfilm, Natl Lib Canada)
Chapman D (1967) The effect of heat on membranes and membrane constituents. In: Rose AH (ed) Thermobiology. Academic Press, London New York, pp 123–146
Chimiklis PE, Karlander EP (1973) Light and calcium interactions in *Chlorella* inhibited by sodium chloride. Plant Physiol 51:48–56
Cholnoky BJ (1968) Die Ökologie der Diatomeen in Binnengewässern. Cramer, Lehre
Chu SP (1943) The influence of the mineral composition of the medium on the growth of planctonic algae. II. The influence of the concentration of inorganic nitrogen and phosphate phosphorus. J Ecol 31:109–148
Claesson A, Ryding SO (1977) Nitrogen – a growth limiting nutrient in eutrophic lakes. Prog Water Tech 8:291–299
Coffey BT, McNabb CD (1974) Eurasian Water-Milfoil in Michigan. Mich Bot 13:159–165
Cole BS, Toetz DW (1975) Utilization of sedimentary ammonia by *Potamogeton nodosus* and *Scirpus*. Verh Int Ver Limnol 19:2765–2772
Confer JL (1972) Interrelations among plankton, attached algae, and the phosphorus cycle in artificial open systems. Ecol Monogr 42:1–23
Conway HL (1978) Sorption of arsenic and cadmium and their effects on growth, micronutrient utilization, and photosynthetic pigment composition of *Asterionella formosa*. J Fish Res Board Can 35:286–294
Cook SF Jr, Conners JD (1963) The short-term side effects of the insecticidal treatment of Clear Lake, Lake County, California, in 1962. J Econ Entomol 56:819–824
Coste M (1974) Etude sur la mise au point d'une méthode biologique de détermination de la qualité des eaux en milieu fluvial. Division qualité des eaux, pêche et pisciculture. CTGREF, Paris
Cowen WF, Lee GF (1973) Leaves as source of phosphorus. Environ Sci Technol 7:853–854
Cowgill UM (1973) Determination of all detectable elements in the aquatic plants of Linsley Pond and Cedar Lake (North Branford, Conn.) by X-ray emission and optical emission spectroscopy. Appl Spectrosc 27:5–9
Cowgill UM (1974) The hydrogeochemistry of Linsley Pond, North Branford, Connecticut. II. The chemical composition of the aquatic macrophytes. Arch Hydrobiol Suppl 45:1–119
Crance JH (1963) The effects of copper sulphate on *Microcystis* and zooplankton in ponds. Prog Fish Cult 25:198–202
Culley D, Epps E (1973) Use of Duckweed for waste treatment and animal feed. J Water Pollut Control Fed 45:337
Daniell JW, Chapell WE, Couch HB (1969) Effect of sublethal and lethal temperatures on plant cells. Plant Physiol 44:1684–1689
Das B, Singh PK (1977) The effect of 2,4-dichlorophenoxyacetic acid on growth and nitrogen-fixation of blue-green alga *Anabaenopsis raciborskii*. Arch Environ Contam Toxicol 5:437–445
Davis GJ, Fones MN, Lunney CZ, Clark GM (1974) Inhibition of sodium chloride toxicity in seedlings of *Myriophyllum spicatum* L. with calcium. Plant Cell Physiol 15:557–571
Decleire M, Cat de W, Basin R (1976) Rapid detection of various kinds of herbicides in water by the in vivo determination of nitrate reductase activity in *Lemna minor* L. Z Pflanzenphysiol 77:315–322
Dedonder A, Sumere van CF (1971) The effect of phenolics and related compounds on the growth and the respiration of *Chlorella vulgaris*. Z Pflanzenphysiol 65:70–80
DeMarte JA, Hartman RT (1974) Studies on absorption of ^{32}P, ^{59}Fe, and ^{45}Ca by water-milfoil (*Myriophyllum exalbescens* Fernald). Ecology 55:188–194

Demoll R, Liebmann H (1952) Über die Verteilung von *Sphaerotilus natans* im Fluß. Schweiz Z Hydrol 14:289–297
Denny P (1972) Sites of nutrient absorption in aquatic macrophytes. J Ecol 60:819–829
Der Rat von Sachverständigen für Umweltfragen (1976) Umweltprobleme des Rheins. Kohlhammer, Stuttgart
Der Rat von Sachverständigen für Umweltfragen (1978) Umweltgutachten 1978. Kohlhammer, Stuttgart
Descy JP (1975) Value of aquatic plants in the characterization of water quality and principles of the methods used. In: Commission of the European Communities Principles and methods for determining ecological criteria on hydrobiocenoses. Pergamon, New York, pp 157–183
Deufel J (1978) Veränderung der Schilf- und Wasserpflanzenbestände im Bodensee während der Eutrophierung und ihre Auswirkungen auf die Fische. Arb Dtsch Fischereiverb H25:30–34
Dickson KL, Slocomb JP, Cairns J Jr, Almeida SP, Eu JKT (1977) A laser-based optical filtering system to analyze samples of diatom communities. In: Cairns J Jr, Dickson KL, Westlake GF (eds) Biological monitoring of water and effluent quality. ASTM, Philadelphia, pp 121–132
Dietz F (1972) Die Anreicherung von Schwermetallen in submersen Pflanzen. GWF Wasser Abwasser 113:269–273
Dietz F (1977) Zusammenfassende Bewertung der Ergebnisse der methodischen Untersuchungen zur Bestimmung des Remobilisierungsvermögens von Komplexbildnern. Z Wasser Abwasserforsch 1:20–24
Dodge AD (1971) The mode of action of the bipyridylium herbicides, paraquat and diquat. Endeavour 30:130–135
Dolar SG, Keeney DR, Chesters G (1971) Mercury accumulation by *Myriophyllum spicatum* L. Environ Lett 1:191–198
Donaldson LR, Foster RF (1957) Effects of radiation on aquatic organisms. Natl Acad Sci Natl Res Counc Publ 551:96
Doudoroff P, Katz M (1953) Critical review of literature on toxicity of industrial wastes and their components to fish. Sewage Ind Wastes, 25:802–834
Duffer WR, Dorris TC (1966) Primary productivity in a Southern Great Plains stream. J Ecol 46:295–322
Eberhardt LL, Meeks RL, Peterle TJ (1971) Food chain model for DDT kinetics in a freshwater marsh. Nature (London) 230:60–62
Edmondson WT (1970) Phosphorus, nitrogen, and some algae in Lake Washington after diversion of sewage. Science 169:690–691
Edmondson WT (1977) Recovery of Lake Washington from Eutrophication. In: Cairns J Jr, Dickson KL, Herricks EE (eds) Recovery and restoration of damaged ecosystems. Univ Press Virginia, Charlottesville, pp 102–109
Eichenberger E (1976) Biologische und ernährungsphysiologische Gesichtspunkte zur Ausbreitung des Flutenden Hahnenfußes im Oberrhein. Wasser Energie Luft 68:234–239
Eidgenössisches Departement des Innern (1968) Gewässerschutztechnische Gesichtspunkte im Zusammenhang mit der Kühlwasserentnahme und -rückgabe bei konventionell- und nuklearthermischen Kraftwerken. Bern
Einsele W (1936) Über die Beziehungen des Eisenkreislaufs zum Phosphatkreislauf im eutrophen See. Arch Hydrobiol 29:664–686
Ellis BE (1977) Degradation of phenolic compounds by freshwater algae. Plant Sci Lett 8:213–216
Elster HJ (1974) Wechselwirkungen zwischen Chemie und Biologie bei der Eutrophierung des Bodensees. Vom Wasser 43:1–14
Elster HJ (1977) Die Eutrophierung unserer Gewässer. Hohenheimer Arb 88:50–66
Empain A (1973) La végétation bryophytique aquatique et subaquatique de la Sambre belge, son déterminisme écologique et ses relations avec la pollution des eaux. Lejeunia Nouv Ser 69:58
Erickson SJ, Maloney TE, Gentile JH (1970) Effect of nitrilotriacetic acid on the growth and metabolism of estuarine phytoplankton. J Water Pollut Control Fed 42:329–335

Eriksson C, Mortimer DC (1975) Mercury uptake in rooted higher aquatic plants; laboratory studies. Verh Int Ver Limnol 19:2087–2093
Ernst WHO, Marquenie-van der Werff M (1978) Aquatic angiosperms as indicators of copper contamination. Arch Hydrobiol 83:356–366
Fischer WK, Winkler K (1976) Detergentienuntersuchungen im Stromgebiet des Rheins 1958–1975. Vom Wasser 47:81–129
Fish GR (1966) Some effects of the destruction of aquatic weeds in Lake Rotoiti, New Zealand. Weed Res 6:350–358
Fitzgerald GP (1978) Detection of limiting or surplus nitrogen in algae and aquatic weeds. J Phycol 4:121–126
Fitzgerald GP (1969) Some factors in the competition or antagonism among bacteria, algae, and aquatic weeds. J Phycol 5:351–359
Fitzgerald GP, Faust SL (1963) Factors affecting the algicidal and algistatic properties to copper. Appl Microbiol 11:344–351
Fitzgerald GP, Nelson TC (1966) Extractive and enzymatic analyses for limiting or surplus phosphorus in algae. J Phycol 2:32–37
Fitzgerald GP, Gerloff GC, Skoog F (1952) Studies on chemicals with selective toxicity of blue-green algae. Sewage Ind Wastes 24:888–896
Fjerdingstad E (1964) Pollution of streams estimated by benthal phytomicroorganisms. I.A. saprobic system based on communities of organisms and ecological factors. Int Rev Ges Hydrobiol 49:63–131
Foehrenbach J (1974) Eutrophication. J Water Pollut Control Fed 46:1350–1355
Förstner U, Müller G (1974) Schwermetalle in Flüssen und Seen als Ausdruck der Umweltverschmutzung. Springer, Berlin Heidelberg New York
Fogg GE (1953) The metabolism of algae. Wiley and Sons, New York
Fogg GE (1962) Nitrogen Fixation. In: Lewin RA (ed) Physiology and biochemistry of algae. Academic Press, London New York, pp 161–170
Fogg GE (1971) Extracellular products of algae in freshwater. Arch Hydrobiol Beih 5:1–25
Fogg GE (1973) Phosphorus in primary aquatic plants. Water Res 7:77–91
Fogg GE, Westlake DF (1955) The importance of extracellular products of algae in freshwater. Verh Int Ver Limnol 12:219–232
Forester JW (1969) A discussion of heat and freshwater algae. In: Krenkel PA, Parker FL (eds) Biological aspects of thermal pollution. Vanderbilt Univ Press, Nashville, Tenn
Forsberg C (1964) Phosphorus, a maximum factor in the growth of *Characeae*. Nature (London) 201:517–518
Forsberg C (1965) Nutritional studies of *Chara* in axenic cultures. Physiol Plant 18:275–290
Forsberg C (1968) Effects of nitrilotriacetate (NTA) on ^{14}C-assimilation and growth of algae. Vatten 24:339–347
Forsberg C (1977) Nitrogen a growth factor in fresh water. Water Prog Tech 8:275–290
Forsberg C, Forsberg B (1969) Some limnological features of Lake Långsjön, Björklinge. Sven Bot Tidskr 63:49–71
Forsberg C, Jinnerot D, Davidsson L (1967) The influence of synthetic detergents on the growth of algae. Vatten 1:2–16
Forsberg C, Ryding SO, Claesson A, Forsberg A (1978) Water chemical analyses and/or algal assay? – Sewage effluent and polluted lake water studies. Mitt Int Ver Limnol 21:352–363
Frahm JP (1974) Wassermoose als Indikatoren für die Gewaesserverschmutzung am Beispiel des Niederrheins. Gewaesser Abwaesser 53/54:91–106
Frahm JP (1975) Toxitoleranzversuche an Wassermoosen. Gewaesser Abwaesser 57/58:59–66
Francke J, Hillebrand H (1980) Effects of copper on some filamentous freshwater Chlorophyta. Aquat Bot 8:285–289
Frank R, Holdrinet M, Braun HE, Thomas RL, Kemp ALW, Jaquet JM (1977) Organochlorine insecticides and PCBs in sediments of Lake St. Clair (1970 and 1974) and Lake Erie (1971). Sci Total Environ 8:205–227

Friedrich G (1975) Der Algenschnelltest – Ein Beitrag zur Toxizitätsmessung. Gewaesser Abwaesser 57/58:117–120
Fry JC, Brooker MP, Thomas PL (1973) Changes in the microbial populations of a reservoir treated with the herbicide paraquat. Water Res 7:395–407
Funderburk HH Jr, Lawrence JM (1963) Absorption and translocation of radioactive herbicides in submersed and emersed aquatic weeds. Weed Res 3:304–311
Gächter R, Chan KL, Chau YK (1973) Complexing capacity of the nutrient medium and its relation to inhibition of algal photosynthesis to copper. Schweiz Z Hydrol 35:252
Gahlen K (1935) Beiträge zur Physiologie der Blattzellen von *Helodea canadensis*. Protoplasma 22:337–361
Galloway RA, Krauss RW (1963) Utilization of phosphorus sources by *Chlorella*. In: (ed) Microalgae and photosynthetic bacteria. Jpn Soc Plant Physiol, Tokyo, pp 569–575
Gameson ALH, Truesdale GA, Downing AL (1955) Re-aeration studies in a lakeland beck. J Inst Water Eng 9:571
Geike F (1977) Effect of some mercury compounds on freshwater algae. Z Pflanzenkr Pflanzenschutz 84:84–94
Geller W, Mäckle H (1977) Überwachung von Roh- und Trinkwasser. Teil 2: Kontinuierlicher Biotest zur Toxizitätsüberwachung von Trinkwasser. DVGW-Schriftenr Wasser 14:173–181
Gerhards U, Weller H (1977) The uptake of mercury, cadmium and nickel by *Chlorella pyrenoidosa*. Z Pflanzenphysiol 82:292–300
Gerloff GC, Skoog F (1954) Cell contents of nitrogen and phosphorus as a measure of their availability for growth of *Microcystis aeruginosa*. Ecology 35:348–353
Glänzer U, Haber W, Kohler A (1977) Experimentelle Untersuchungen zur Belastbarkeit submerser Fließgewässer-Makrophyten. Arch Hydrobiol 79:193–232
Gloyna EF, Bhagat SK, Yousef Y, Shih C (1966) Transport of radionuclides in a model river. In: IAEA Disposal of radioactive wastes into seas, oceans and surface waters. IAEA, Vienna
Götz R (1978) Die Mobilisierung von Schwermetallen (Pb, Zn, Cu und Ni) im Wasser durch Huminsäuren aus schwerlöslichen Bodenkörpern. Forum Staedte Hyg 29:33–37
Goldman CR (1972) The role of minor nutrients in limiting the productivity of aquatic ecosystems. Nutrients and eutrophication. Spec Symp Am Soc Limnol Oceanogr 1:21–33
Goldman CR, Carter RC (1965) An investigation by rapid carbon-14 bioassay of factors affecting the cultural eutrophication of Lake Tahoe, California-Nevada. J Water Pollut Control Fed 37:1044–1059
Goldman CR, Wetzel RG (1963) A study of the primary productivity of Clear Lake, Lake County, California. Ecology 44:283–294
Goldsby TL, Bates AL, Stanley RA (1978) Effect of water level fluctuation and herbicide on Eurasian Watermilfoil in Melton Hill Reservoir. J Aquat Plant Manag 16:34–38
Goncharova AV (1975) Effect of phenol on the blue-green algae *Oscillatoria splendida* Grew. Biol Vnutr Vod 28:19–21. Chem Abstr 87:7927 6x 1977 (in Russian)
Goulder R (1969) Interactions between rates of production of a freshwater macrophyte and phytoplankton in a pond. Oikos 20:300–309
Grace JB, Tilly LJ (1976) Distribution and abundance of submerged macrophytes, including *Myriophyllum spicatum* L. (Angiospermae), in a reactor cooling reservoir. Arch Hydrobiol 77:475–487
Granhall U, Lundgren A (1971) Nitrogen fixation in Lake Erken. Limnol Oceanogr 16:711–719
Greenfield SS (1942) Inhibitory effects of inorganic compounds of photosynthesis in *Chlorella*. Am J Bot 29:121–131
Greer DE, Ziebell CD (1972) Biological removal of phosphates from water. J Water Pollut Control Fed 44:2342–2348
Greichus YA, Greichus A, Amman BD, Call DJ, Hamman CD, Pott RM (1977) Insecticides, polychlorinated biphenyls and metals in African lake ecosystems. 1. Hartbees-

poort Dam, Transvaal, and Voëlvlei Dam, Cape Province, Republic of South Africa. Arch Environ Contam Toxicol 6:371–383
Grim J (1967) Der Phosphor und die pflanzliche Produktion im Bodensee. GWF Wasser Abwasser 108:1261–1271
Guilizzoni P (1975) Manganese, copper and chromium content in macrophytes of Lake Endine (Northern Italy). Mem Ist Ital Idrobiol 32:313–332
Haag RW, Gorham PR (1977) Effects of thermal effluent on standing crop and net production of *Elodea canadensis* and other submerged macrophytes in Lake Wabamum, Alberta. J Appl Ecol 14:835–851
Haber W, Kohler A (1972) Ökologische Untersuchungen und Bewertung von Fließgewässern mit Hilfe höherer Wasserpflanzen. Landschaft Stadt 4:153–168
Haller WT, Sutton DL, Barlowe WC (1974) Effects of salinity on growth of several aquatic macrophytes. Ecology 55:891–894
Hamaker JW, Goring CAI, Youngson CR (1966) Sorption and leaching of 4-amino-3,5,6-trichloropicolinic acid in soils. In: American Chemical Society (ed.) Organic pesticides in the environment. Adv Chem Ser 60:23–37
Hamilton DH Jr (1979) Nutrient limitation of summer phytoplankton growth in Cayuga Lake. Limnol Oceanogr 14:579–590
Hamilton DH Jr, Flemer DA, Keefe CW, Hursky JAM (1970) Power plants: Effects of chlorination on estuarine primary production. Science 169:197–198
Hannan PJ, Patouillet (1972) Effect of mercury on algal growth rates. Biotech Bioeng 14:93–101
Hannerz L (1968) Experimental investigations on the accumulation of mercury in water organisms. Rep Inst Freshwater Res Drottningholm 48:120–176
Hansen PD (1976) Versuche zur experimentellen Anreicherung von Lindan (−HCH) in einer Nahrungskette aus *Chlorella spec.-Daphnia magna* (Straus) und *Gasterosteus aculeatus* (Linné). Dissertation, Univ Hamburg
Harding JPC, Whitton BA (1976) Resistance to zinc of *Stigeoclonium tenue* in the field and the laboratory. Br Phycol J 11:417–426
Hargrave BT (1969) Epibenthic algal production and community respiration in the sediments of Marion Lake. J Fish Res Board Can 26:2003–2026
Hargreaves JW, Whitton BA (1976) Effect of pH on tolerance of *Hormidium rivulare* to zinc and copper. Oecologia 26:235–243
Hargreaves JW, Lloyd EJH, Whitton BA (1975) Chemistry and vegetation of highly acidic streams. Br Phycol J 11:215–223
Harris DO (1971) Growth inhibitors produced by green-algae (*Volvocaceae*). Arch Mikrobiol 76:47–50
Harriss RC (1971) Ecological implications of mercury pollution in aquatic systems. Biol Conserv 3:279–283
Harvey RS, Patrick R (1967) Concentration of ^{137}Cs, ^{65}Zn, and ^{85}Sr by fresh-water algae. Biotech Bioeng 9:449–456
Haslam SM (1971) Shoot height and density in *Phragmites* stands. Hidrobiologia 12:113–119
Hasler AD, Jones E (1949) Demonstration of the antagonistic action of large aquatic plants on algae and rotifers. Ecology 30:359–364
Hawxby K, Tubea B, Ownby J, Basler E (1977) Effects of various classes of herbicides on four species of algae. Pest Biochem Physiol 7:203–209
Hayes FR, McCarter JA, Cameron ML, Livingstone DA (1952) On the kinetics of phosphorus exchange in lakes. J Ecol 40:202–216
Heath RT, Cooke GD (1975) The significance of alkaline phosphatase in a eutrophic lake. Ver Int Ver Limnol 19:959–965
Heber U, Santarius KA (1973) Cell death by cold and heat and resistence to extreme temperatures. Mechanism of hardening and dehardening. In: Precht HJ, Christopherson J, Hensel H, Larcher W (eds) Temperature and life. Springer, Berlin Heidelberg New York, pp 232–263
Hervey RJ (1949) Effect of chromium on the growth of unicellular *Chlorophyceae* and Diatoms. Bot Gaz 111:1–11

Hickman M (1971) The standing crop and primary productivity of the epiphyton attached to *Equisetum fluviatile* L. in Priddy Pool, North Somerset. Br Phycol J 6:51–59
Hicks CE, Neuhold JM (1966) Alkyl benzene sulfonate effects on stream algae communities. Bull Environ Contam Toxicol 1:225–237
Hill DW, McCarty PC (1967) Anaerobic degradation of selected chlorinated hydrocarbon pesticides. J Water Pollut Control Fed 39:1259–1277
Hinneri S (1976) On the ecology and phenotypic plasticity of vascular hydrophytes in a sulphate-rich-acidotrophic freshwater reservoir, SW coast of Finland. Ann Bot Fenn 13:97–105
Hogetsu K, Okanishi Y, Sugawara H (1960) Studies on the antagonistic relationship between phytoplankton and aquatic plants. Jpn J Limnol 21:124–129
Horne AJ, Fogg GE (1970) Nitrogen fixation in some English lakes. Proc R Soc London Ser B 175:351–366
Horne AJ, Goldman CR (1972) Nitrogen fixation in Clear Lake, California. I. Seasonal variation and the role of heterocysts. Limnol Oceanogr 17:678–692
Horne AJ, Goldman CR (1974) Suppression of nitrogen fixation by blue-green algae in a eutrophic lake with trace additions of copper. Science 183:409–411
Horne AJ, Viner AB (1971) Nitrogen fixation and its significance in tropical Lake George, Uganda. Nature (London) 232:417–418
Howarth RW, Fisher SG (1976) Carbon, nitrogen, and phosphorus dynamics during leaf decay in nutrient enriched stream microecosystems. Freshwater Biol 6:221–228
Howells GP (1969) Hudson River at Indian Point. Annu Rep 4/16/68 to 5/15/69. Inst Environ Med, NY Univ Med Center, New York
Huang J, Gloyna EF (1968) Effect of organic compounds on photosynthetic oxygenation. I. Chlorophyll destruction and suppression of photosynthetic oxygen production. Water Res 2:347–366
Hürlimann H (1951) Zur Lebensgeschichte des Schilfs an den Ufern der Schweizer Seen. Dissertation, Eidg Tech Hochsch, Zürich
Hurlbert SH (1975) Secondary effects of pesticides on aquatic ecosystems. Residue Rev 59:81–148
Hurle K, Johannes H (1979) Eintrag von Pflanzenschutzmitteln in Gewässer. Schr Dtsch Verb Wasserwirtsch Kulturbau H 40:118–144
Hustedt F (1926) *Thalassiosira fluviatilis* nov. spec., eine Wasserblüte im Wesergebiet. Ber Dtsch Bot Ges 43:565–567
Hustedt F (1939) Systematische und ökologische Untersuchungen über die Diatomeen-Flora von Java, Bali und Sumatra. Allgemeiner Teil. Arch Hydrobiol Suppl 16:274–394
Hustedt F (1957) Die Diatomeenflora des Flußsystems der Weser im Gebiet der Hansestadt Bremen. Abh Naturwiss Ver Bremen 34:181–440
Hutchinson FE (1968) Effect of highway salting on the concentration of sodium and chloride in rivers. Res Life Sci 68:12–15
Hutchinson GE (1957) A treatise in limnology. Vol 1. Geography, physics and chemistry. Wiley and Sons, New York
Hutchinson GE (1969) Eutrophication, past and present. In: Eutrophication: Causes, consequences, correctives. Natl Acad Sci Washington DC, pp 17–25
Hutchinson GE, Bowen VT (1950) A direct demonstration of the phosphorus cycle in a small lake. Proc Natl Acad Sci 33:148–153
Hutchinson TC, Czyrska H (1975) Heavy metal toxicity and synergism to floating aquatic weeds. Verh Int Ver Limnol 19:2102–2111
Hutchinson TC, Stockes PM (1975) Heavy metal toxicity and algal bioassay. In: ASTM (ed) Water quality parameters. Am Soc Test Mater, Philadelphia, pp 320–343
Hynes HBN (1959) The biological effects of water pollution. In: Yapp WB (ed) Symp Inst Biol, London, pp 11–24
Hynes HBN (1960) The biology of polluted waters, 5th edn. Liverpool Univ Press, Liverpool
Hynes HBN (1969) The enrichment of streams. In: Eutrophication: causes, consequences, correctives. Natl Acad Sci, Washington DC, pp 188–196

Hynes HBN, Roberts FW (1962) The biological effects of synthetic detergents in the River Lee, Hertfordshire. Ann Appl Biol 50:779–790
Ingols RS (1975) Chlorination of water – potable, possible: wastewater, no! Water Sewage Works: 82–83
Isensee AR, Jones GE (1975) Distribution of 2,3,7,8-tetrachlorodibenzo-p-dioxin (TCDD) in aquatic model ecosystem. Environ Sci Tech 9:668–672
Iversen J, Olsen S (1943) Die Verbreitung der Wasserpflanzen in Relation zur Chemie des Wassers. Bot Tidskr 46:136–145
James RO, Healy TW (1972) Adsorption of hydrolyzable metal ions at the oxide-water interface. J Colloid Interface Sci 40:42
Janauer GA (1977) Amino acids in aquatic macrophytes. Z Pflanzenphysiol 82:45–50
Janauer GA (1979) Metabolic changes in *Potamogeton pectinatus* L. by enhanced eutrophication. Flora 168:344–351
Janicke W (1973) Schadwirkung von Tensiden unter wasserwirtschaftlichem Gesichtspunkt (Eine Übersicht). Bundesgesundheitsblatt 16:258–263
Jensen S, Jernelöv A (1969) Biological methylation of mercury in aquatic organisms. Nature (London) 223:753–754
Jirgensons B (1961) Effects of detergents on the conformation of proteins. I. An abnormal increase of the optical rotatory dispersion constant. Arch Biochem Biophys 94:59–67
Jolley RL (1976) The environmental impact of water chlorination. Proc Conf Environ Impact Water Chlorination, Oak Ridge Nat Lab CONF-751096 UC-11, 41, 48, pp 115–151
Jones JRE (1940) A study of the zinc polluted river Ystwyth in North Cardiganshire, Wales. Ann Appl Biol 27:368–378
Joseph JM (1953) Microbiological study of acid mine waters: preliminary report. Ohio J Sci 53:123–127
Judd JH (1970) Lake stratification caused by runoff from street deicing. Water Res 4:521–532
Jupp BP, Spence DHN (1977) Limitations on macrophytes in a eutrophic lake, Loch Leven. I. Effects of phytoplankton. J Ecol 65:175–186
Kamp-Nielsen L (1971) The effect of deleterious concentrations of mercury on the photosynthesis and growth of *Chlorella pyrenoidosa*. Physiol Plant 24:556–561
Ke B, Clendenning KA (1956) Properties of chloroplast dispersion in the presence of detergents. Biochim Biophys Acta 19:74–83
Keating KI (1977) Allelopathic influence on blue-green bloom sequence in a eutrophic lake. Science 196:885–887
Keating KI (1978) Blue-green algal inhibition of diatom growth: Transition from mesotrophic to eutrophic community structure. Science 199:971–973
Ketchum BH, Redfield AC (1949) Some physical and chemical characteristics of algae grown in mass culture. J Cell Comp Physiol 33:281–300
Keulder PC (1975) Influence of the clay types illite and montmorillonite on the uptake of ^{65}Zn by *Scenedesmus obliquus*. J Limnol Soc S Afr 1:33–35
Khalanski M (1975) Etudes réalisées en France sur les conséquences écologiques de la réfrigération des centrales thermiques en circuit ouvert. In: IAEA-SM-187/36 (ed) Environmental effects of cooling systems at nuclear power plants. IAEA, Vienna, pp 461–477
Khobot'yev VG, Kapkov VI, Rukhadze YG, Turunina NV, Shidlovskaya NA (1975) The toxic effect of copper complexes on algae. Hydrobiol J 11:33–38
Kickuth R (1978) Elimination gelöster Laststoffe durch Röhrichtbestände. Arb Dtsch Fischereiverb H 25:57–70
Kilham SS, Kilham P (1978) Natural community bioassay: Predictions of results based on nutrient physiology and competition. Verh Int Ver Limnol 20:68–74
King DL (1970) The role of carbon in eutrophication. J WPCF 42:2035–2051
Kinkade ML, Erdman HE (1975) The influence of hardness components (Ca^{2+} and Mg^{2+}) in water on the uptake and concentration of cadmium in a simulated freshwater ecosystem. Environ Res 10:308–313
Klee O (1975) Hydrobiologie. Dtsch Verlags-Anstalt, Stuttgart

Klötzli F (1965/66) Umwandlung von Moor- und Sumpfpflanzengesellschaften durch Abwässer des Neeracher Rieds. Veroeff Geobot Inst Zuerich 37:104–112
Klötzli F, Grünig A (1976) Seeufervegetation als Bioindikator. Daten Dok Umweltschutz Univ Hohenheim 19:109–131
Knöpp H (1973) Biologisch-ökologische Konsequenzen der Kühlwassereinleitung. In: Braun H, Heuck F, Ladner HA, Messerschmidt O, Niklas K, Rausch L (eds) Umweltschutz bei nuklearer und konventioneller Energiegewinnung. Strahlenschutz in Forschung und Praxis Vol XII. Thieme, Stuttgart, pp 130–141
Kohler A (1978) Methoden der Kartierung von Flora und Vegetation von Süßwasserbiotopen. Landschaft Stadt 10:73–85
Kohler A, Zeltner GH (1974) Verbreitung und Ökologie von Makrophyten in Weichwasserflüssen des Oberpfälzer Waldes (Naab, Pfreimd, Schwarzach). Hoppea Denkschr Regensburg Bot Ges 33:171–232
Kohler A, Vollrath H, Beisl E (1971) Zur Verbreitung, Vergesellschaftung und Ökologie der Gefäßmakrophyten im Fließwassersystem Moosach (Münchener Ebene). Arch Hydrobiol 69(3):333–365
Kohler A, Wonneberger R, Zeltner GH (1973) Die Bedeutung chemischer und pflanzlicher „Verschmutzungsindikatoren" im Fließgewässersystem Moosach (Münchener Ebene). Arch Hydrobiol 72:533–549
Kohler A, Brinkmeier R, Vollrath H (1974) Verbreitung und Indikatorwert der submersen Makrophyten in den Fließgewässern der Friedberger Au. Ber Bayer Bot Ges 45:5–36
Kohler A, Pensel T, Zeltner GH (1980) Veränderungen von Flora und Vegetation in den Fließgewässern der Friedberger Au (bei Augsburg) zwischen 1972 und 1978. Verh Ges Oekol 8 Freising-Weihenstephan 1979:343–350
Kolbe RW (1927) Zur Ökologie, Morphologie und Systematik der Brackwässer-Diatomeen. Die Kieselalgen des Sperenberger Salzgebiets. Pflanzenforsch 7:1–146
Kolbe RW (1932) Grundlinien einer allgemeinen Ökologie der Diatomeen. Ergeb Biol 8:221–348
Kolkwitz R, Marsson M (1908) Ökologie der pflanzlichen Saprobien. Ber Dtsch Bot Ges 26a:505–519
Kott Y (1969) Effect of halogens on algae. – III. Field experiment. Water Res 3:265–271
Kott Y, Edlis J (1969) Effects of halogens on algae. – I. *Chlorella sorokiniana*. Water Res 3:251–256
Kratky BA, Warren GF (1971) The use of three simple, rapid bioassays on forty-two herbicides. Weed Res 11:257–662
Krause A (1972) Einfluß der Eutrophierung und anderer menschlicher Einwirkungen auf die Makrophytenvegetation der Oberflächengewässer. Ber Landwirtschaft 50:140–146
Krause A (1978) Wasservegetation. In: Olschowy G (ed) Natur- und Umweltschutz in der Bundesrepublik Deutschland. Parey, Hamburg Berlin, pp 277–281
Krause W, Carbiener R (1975) Die Chloridkonzentration in den Gewässern der Oberrheinebene und ihrer Randgebirge. Erdkunde 29:267–277
Kuenzler EJ, Perras JP (1965) Phosphatases of marine algae. Biol Bull 128:271–284
Kurimo U (1970) Effect of pollution on the aquatic macroflora of the Varkaus area, Finnish Lake District. Ann Bot Fenn 7:213–254
Kutscher G, Kohler A (1976) Verbreitung und Ökologie submerser Makrophyten in Fließgewässern des Erdinger Mooses (Münchener Ebene). Ber Bayer Bot Ges 47:175–228
Labus BC (1979) Der Einfluß des Waschrohstoffs Marlon A (anionenaktives Tensid) auf das Wachstum und die Nettophotosynthese verschiedener submerser makrophytischer Wasserpflanzen unter besonderer Berücksichtigung primärer Standortsfaktoren. Dissertation, Univ Hohenheim, Stuttgart
Labus BC, Kohler A (1981) Die Rolle einiger primärer ökologischer Faktoren bei Nettophotosynthesemessungen zur Prüfung der Wirkung von Umweltchemikalien auf submerse Makrophyten. Limnologica 13:373–398
Labus BC, Schuster H, Nobel W, Kohler A (1977a) Wirkung von toxischen Abwasserkomponenten auf submerse Makrophyten. Angew Bot 51:17–36

Labus BC, Nobel W, Smetana R, Kohler A (1977b) Der Einfluß der Abwassersubstanz Marlon A (anionenaktives Tensid) und Bor auf die Photosyntheserate einiger submerser Makrophyten. Verh Ges Oekol, Göttingen 1976:325–333
Lachavanne JB, Wattenhofer R (1975) Contribution d'étude des macrophytes du Léman. Comm Int Pur Protect Faux Léman Rhône Contre Pollut, Genf
Lackey JB (1938) The flora and fauna of surface waters polluted by acid mine drainage. Publ Health Rep 53:1499–1507
Landner L (1972) Biochemical model for the biological methylation of mercury suggested from methylation studies in vivo with *Neurospora crassa*. Nature (London) 230:452–454
Lang G (1973) Die Makrophytenvegetation in der Uferzone des Bodensees. Int Gewaesserschutzkomm Bodensee Ber 12
Lange W (1970) Cyanophyta-bacteria systems: effects of added carbon compounds or phosphate on algal growth at low nutrient concentrations. J Phycol 6:230–234
Lanza GR, Cairns J Jr (1972) Physio-morphological effects of abrupt thermal stress on diatoms. Trans Am Microsc Soc 91:276–298
Laubenberger G, Hartmann L (1970) Speicherung von Schwermetallsalzen in Flußsedimenten und ihr Einfluß auf die Schlammbiocoenose. Wasserwirtschaft 11:372–375
Lazaroff N (1967) Algal response to pesticide pollutants. Bacteriol Proc G 149:48
Lean DRS (1973a) Movements of phosphorus between its biologically important forms in lake water. J Fish Res Board Can 30:1525–1536
Lean DRS (1973b) Phosphorus dynamics in lake water. Science 179:678–680
Lefevre M (1964) Extracellular products of algae. In: Jackson DF (ed) Algae and man. Plenum, New York, pp 337–367
Levitt J (1972) Responses of plants to environmental stresses. Academic Press, London New York
Lichtenstein RF, Skrentny RF, Tsukano Y (1966) Toxicity and fate of insecticide residues in water. Arch Environ Health 12:199
Liebmann H (1962) Handbuch der Frisch- und Abwasserbiologie Vol I, 2nd edn. Oldenbourg, München
Likens GE (ed) (1972) Nutrients and eutrophication: The limiting – nutrient controversy. Spec Symp Am Soc Limnol Oceanogr 1:328
Likens GE, Wright RF, Galloway JN, Butler TJ (1979) Acid rain. Sci American 241(4):39–47
Lindeman RL (1942) The trophic-dynamic aspect of ecology. Ecology 23:399–418
Litav M, Agami M (1976) Relationship between water pollution and the flora of two coastal rivers of Israel. Aquat Bot 2:23–41
Littlefield L, Forsberg C (1965) Absorption and translocation of phosphorus-32 by *Chara globularis* Thuill. Physiol Plant 18:291–296
Livingston RJ (1977) Review of current literature concerning the acute and chronic effects of pesticides on aquatic organisms. CRC 7:325–351
Lukina GA (1970) Effect of phenol on the photosynthesis and respiration of *Chlorella*. Chem Abstr 72:107938a (in Russian)
Lund JWG (1965) The ecology of freshwater phytoplankton. Biol Rev 40:231–293
Lund JWG (1969) Phytoplankton. In: National Academy of Sciences (ed) Eutrophication: causes, consequences, correctives. Natl Acad Sci, Washington DC, 306–330
Lundkvist I (1970) Effect of herbicides on nitrogen fixation by blue-green algae. Sven Bot Tidsk 64:460
Luther H (1951) Verbreitung und Ökologie der höheren Wasserpflanzen im Brackwasser der Ekenäs-Gegend in Südfinnland. Acta Bot Fenn 49/50
MacKenthun KM (1969) Radioactive wastes. In: US Dep Inter Fed Water Pollut Control Admin Div Tech Support (ed) The practice of water pollution biology. US Printing Office, Washington DC
Mackereth FJ (1953) Phosphorus utilization by *Asterionella formosa* Hass. J Exp Bot 4:296–313
Maki AW, Geissel LD, Johnson HE (1975) Toxicity of the lampricide TFM to ten species of algae. Invest Fish Control US Dep Fish Wildl Serv 56:1–17

Maloney TE, Miller WE (1975) Algal assays: Development and application. In: ASTM (ed) Water quality parameters. ASTM, Philadelphia, pp 344–355
Maloney TE, Palmer CM (1956) Toxicity of six chemical compounds to thirty cultures of algae. Water Sewage Works 103:509–513
Manny BA, Wetzel RG, Johnson WC (1975) Annual contribution of carbon, nitrogen and phosphorus by migrant Canada geese to a hardwater lake. Verh Int Ver Limnol 19:949–951
Markstein B, Sukopp H (1980) The waterside vegetation of the Berlin Havel 1962–1977. Garten Landschaft H 1:30–36
Matulová D (1966) The effect of surfactants on water algae. Verh Int Ver Limnol 16:958–962
Matzku S, Broda E (1970) Die Zinkaufnahme in das Innere von *Chlorella*. Planta 92:29–40
Mauck WL, Mayer FL, Holz DD (1976) Simazine residue dynamics in small ponds. Bull Environ Contam Toxicol 16:1–8
Mayer FL, Hamelink JL (eds) (1977) Aquatic toxicology and hazard evaluation. ASTM, Philadelphia
Mayes R, McIntosh A (1975) The use of aquatic macrophytes as indicators of trace metal contamination in fresh water lakes. Proc Univ Miss Annu Conf Trace Subst Environ 15:157–167
McGahee CF, Davis GJ (1971) Photosynthesis and respiration in *Myriophyllum spicatum* as related to salinity. Limnol Oceanogr 16:826–829
McKee JE, Wolf HW (1963) Radioactivity. In: McKee, Wolf HW (eds) Water quality criteria. The Resources Agency of California. State Water Qual Control Board Publ 3-A
McLay CL (1976) The effect of pH on the population growth of three species of duckweed: *Spirodela oligorrhiza, Lemna minor* and *Wolffia arrhiza*. Freshwater Biol 6:125–136
McLean RO, Jones AK (1975) Studies of tolerance to heavy metals in the flora of the rivers Ystwyth and Clarach, Wales. Freshwater Biol 5:431–444
McMillan C (1959) Salt tolerance within a *Typha* population. Am J Bot 46:521–526
McNabb CD Jr (1976) The potential of submersed vascular plants for reclamation of wastewater in temperate zone ponds. In: Tourbier J, Pierson RW Jr (eds) Biological control of water pollution. Univ Pennsylvania Press, Pennsylvania, pp 123–132
Mebs D (1978) Algen-Toxine. Naturwiss Rundsch 31:508–510
Meinck F, Stooff H, Kohlschütter H (1968) Industrie-Abwässer. Fischer, Stuttgart
Melzer A (1976) Makrophytische Wasserpflanzen als Indikatoren des Gewässerzustandes oberbayerischer Seen. Diss Bot 34:1–195
Melzer A, Haber W, Kohler A (1977) Floristisch-ökologische Charakterisierung und Gliederung der Osterseen (Oberbayern) mit Hilfe submerser Makrophyten. Mitt Flor Soz Arbeitsgem NF 19/20:130–151
Merezhko AI (1973) Role of higher aquatic plants in the self-purification of lakes and streams. Hydrobiol J 9:103–109
Metcalf RL (1977) Model ecosystem studies of bioconcentration and biodegradation of pesticides. In: Khan MAQ (ed) Pesticides in aquatic environments. Plenum, New York, pp 127–144
Miller JM, Singh J (1976) AGC/MS Investigation of organic chemical residues found in agricultural soils. Bull Environ Contam Toxicol 16:383–391
Miller WE, Maloney TE, Greene JC (1974) Algal productivity in 49 lake waters as determined by algal assays. Water Res 8:667–679
Misra RD (1938) Edaphic factors in the distribution of aquatic plants in English lakes. J Ecol 26:411–511
Miyachi S, Miyachi S, Benson AA (1965) Metabolism of phospholipids and galactolipids by *Chlorella* cells. Plant Cell Physiol 6:789–792
Moore EW (1958) "Thermal pollution" of streams. Ind Eng Chem 50(1–4):87A
Moreland DE, Hill KL (1962) Interference of herbicides with the Hill reaction of isolated chloroplasts. Weeds 10:229–236

Morgan NC (1970) Changes in the fauna and flora of a nutrient enriched lake. Hydrobiologia 35:545–553
Morgan RP, Stross RG (1970) Destruction of phytoplankton in condensers of an electric power generating station. Chesapeake Sci 10:165
Morris I, Glover HE (1974) Questions on the mechanisms of temperature adaptation in marine phytoplankton. Mar Biol 24:147–154
Morris I, Syrett PJ (1963) The development of nitrate reductase in *Chlorella* and its repression by ammonium. Arch Mikrobiol 47:32–41
Mortimer CH (1971) Chemical exchanges between sediments and water in the Great Lakes – speculations on probable regulatory mechanisms. Limnol Oceanogr 16:387–404
Mortimer DC (1974) The higher aquatic plants as accumulators of mercury and DDT. In: Miller D (ed) Distribution and transport of pollutants in flowing water ecosystems. Ottawa River project. Univ Ottawa, Natl Res Counc Can Rep No 2, Feb 1974
Mortimer DC, Kudo A (1975) Interaction between aquatic plants and bed sediments in mercury uptake from flowing waters. J Environ Qual 4(4):491–495
Mothes G, Koschel R, Scheffler W (1976) Einfluß der Aufwärmung eines Sees durch ein Kraftwerk auf Stoffproduktion und Phosphorhaushalt. Limnologica 10:489–501
Müller D, Wellner B (1975) Zur kontinuierlichen Messung der Schädigung von Mikroorganismen. Dtsch Gewaesserkundl Mitt 19:120–123
Mulligan HF, Baranowski A (1969) Growth of phytoplankton and vascular aquatic plants at different nutrient levels. Verh Int Ver Limnol 17:802–810
Mundschenk H, Engel H, Hanisch HH, Tippner M (1977) Einleit- und Austauschvorgänge im Vorfluter. Modellstudie „Radioökologie Biblis" Teilstudie 2. Bundesanst Gewaesserkd T1/275 KKW Biblis, Koblenz
Murphy TP, Lean DRS, Nalewajko C (1976) Blue-green algae: Their excretion of iron-selective chelators enables them to dominate the algae. Science 192:190–192
Newbold C (1975) Herbicides in aquatic systems. Biol Conserv 7:97–118
Nobel W (1980) Der Einfluss der Belastungsstoffe Chlorid, Borat und Phosphat auf die Photosyntheseleistung submerser Weichwasser-Makrophyten. Dissertation, Univ Hohenheim, Stuttgart
Normann HD (1967) Versuche zur Aufnahme von Phosphat durch *Ranunculus fluitans* Lam. Arch Hydrobiol Suppl 33:243–254
Nümann W (1970) Die Möglichkeit der Gewässerreinigung mit höheren Pflanzen nach den bisherigen Untersuchungsergebnissen und theoretischen Überlegungen. Int Rev Ges Hydrobiol 55:149–158
Nusch EA (1977) Beurteilung der biologischen Schadwirkung organischer Komplexbildner und Komplexe aufgrund toxikologischer Testverfahren mit Bakterien, Algen, Protozoen, niederen Metazoen und Fischen. Z Wasser Abwasserforsch 10:49–61
Nyholm N (1977a) Kinetics of nitrogen limited algal growth. Prog Water Tech 8:347–358
Nyholm N (1977b) Kinetics of phosphate limited algal growth. Biotech Bioeng 19:467–492
Odum HT (1956) Primary production in flowing waters. Limnol Oceanogr 1:102–117
Ohle W (1936) Der schwefelsaure Tonteich bei Reinbek. Arch Hydrobiol 30:604–662
Ohle W (1953) Phosphor als Initialfaktor der Gewässereutrophierung. Vom Wasser 20:11–23
Ohle W (1954) Sulfat als „Katalysator" des limnischen Stoffkreislaufes. Vom Wasser 21:13–32
Ophel IL, Fraser CD (1970) Calcium and strontium discrimination by aquatic plants. Ecology 51:324–327
O'Sullivan AJ, Collinson RI (1976) Wasted heat and aquatic ecosystems. Effluent Water Treat J 16:15–29
Overnell J (1975) The effect of some heavy metal ions on photosynthesis in a freshwater alga. Pestic Biochem Physiol 5:19–26
Ownbey CR, Kee DA (1967) Chlorides in Lake Erie. Proc Conf Great Lakes Res Int Assoc Great Lakes Res 10:382–389

Palmer CM, Walter HJ, Adams S (1977) Algae and water pollution. In: Lewis RL Algae and water pollution. US Government Printing Office, Washington DC
Parker C (1966) Influence of water hardiness on the phytotoxicity of paraquat. Nature (London) 212:1465–1466
Patrick R (1943) The diatoms of Linsley Pond. Proc Acad Natl Sci USA 95:53–110
Patrick R (1949) A proposed biological measure of stream conditions, based on a survey of the Conestoga Basin, Lancaster County, Pennsylvania. Proc Acad Natl Sci USA 101:277–341
Patrick R (1969) Some effects of temperature on freshwater algae. In: Krenkel PA, Parker FL (eds) Biological aspects of thermal pollution. Vanderbilt Univ Press, USA, pp 161–198
Patrick R (1971) Report in water quality criteria 1972. In: Cairns J Jr, Buikema AL Jr, Heath AG, Parker BC (eds) Effects of temperature on aquatic organisms sensitivity to selected chemicals. Virginia Wat Res Res Center Bull 106:1–88
Patrick R (1977) The importance of monitoring change. In: Cairns J Jr, Dickson KL, Westlake GF (eds) Biological monitoring of water and effluent quality. ASTM, Philadelphia, pp 157–190
Pawlaczyk-Szpilowa M, Moskal M, Weretelnik J (1972) The usefullness of biological tests for determining the toxicity of some chemical compounds in waters. Acta Hydrobiol 14:115–127
Payer HD, Runkel KH, Schramel P, Stengel E, Bhumiratana A, Soeder CJ (1976) Environmental influences on the accumulation of lead, cadmium, mercury, antimony, arsenic, selenium, bromine and tin in unicellular algae cultivated in Thailand and Germany. Chemosphere 6:413–418
Pearson HW, Tomlinson DL (1975) Some effects of sulphide on the growth and physiology of blue-green algae. Br Phycol Soc Bull 10:313
Peters RH (1972) Phosphorus regeneration by zooplankton. PhD thesis, Univ Toronto
Petersen JB (1943) Some halobian spectra (diatoms). K Dan Vidensk Selsk Biol Medd 17(9):3–95
Peukert V (1976) Untersuchungen über den Einfluß atmosphärischer Verunreinigungen auf die Eutrophierung der Gewässer. Limnologica 10:483–488
Phillips GL, Eminson D, Moss B (1978) A mechanism to account for macrophyte decline in progressively eutrophicated freshwaters. Aquat Bot 4:102–126
Pietsch W (1973) Vegetationsentwicklung und Gewässergenese in den Tagebauseen des Lausitzer Braunkohlen-Revieres. Arch Naturschutz 13:187–217
Pietsch W (1974) Ökologische Untersuchung und Bewertung von Fließgewässern mit Hilfe höherer Wasserpflanzen – ein Beitrag zur Belastung aquatischer Ökosysteme. Mitt Sekt Geobot Phytotaxon Biol Ges DDR 1975
Plötz J (1974) Ein einfaches Verfahren zur Bestimmung der Gesamttoxizität von Wasser- und Abwasserproben für Bakterien. Dtsche Gewaesserkundl Mitt 3:77–79
Porath D, Ben-Shaul Y (1971) Structural and physiological changes during "heat bleaching" in *Spirodela oligorrhiza*. Isr J Bot 20:157–168
Pratt R, Fong J (1940) Studies on *Chlorella vulgaris*, III. Am J Bot 27:735–743
Provasoli L, Pintner IJ (1960) Artificial media for freshwater algae: Problems and suggestions. In: Tryon CA Jr, Hartman RT (eds) Ecology of algae. Univ Press, Pittsburgh
Rachlin JW, Farran M (1974) Growth response of the green algae *Chlorella vulgaris* to selective concentrations of zinc. Water Res 8:575–577
Raghi-Atri F (1978) Einfluß von Schwermetallen (Pb, Hg) im Substrat auf *Glyceria maxima* (Hartm.) Holmbg. 1. Blei. Angew Bot 52:185–192
Raghi-Atri F, Bornkamm R (1979) Wachstum und chemische Zusammensetzung von Schilf (*Phragmites australis*) in Abhängigkeit von der Gewässereutrophierung. Arch Hydrobiol 85:192–228
Ramsay AJ, Fry JC (1976) Response of epiphytic bacteria to the treatment of two aquatic macrophytes with the herbicide paraquat. Water Res 10:453–459
Rana BC, Kumar HD (1974) Eco-physiological studies on the uptake of the pollutants, copper, zinc and phosphate, by certain algae. Indian J Ecol 1:1–11
Rau I, Grimme LH (1971) Zum Einfluß verschieden substituierter s-Triazine auf Stoffwechselreaktionen der Grünalge *Ankistrodesmus braunii*. Z. Naturforsch 26b:919–921

Raven JA (1967) Ion transport in *Hydrodictyon africanum*. J Exp Bot 19:193–206
Ray S, White W (1976) Selected aquatic plants as indicator species for heavy metal pollution. J Environ Sci Health A 11:717–725
Rebechini HM, Hanzely L (1974) Lead-induced ultrastructural changes in chloroplasts of the hydrophyte *Ceratophyllum demersum*. Z Pflanzenphysiol 73:377–386
Reichardt W (1971) Catalytic mobilization of phosphate in lake water and by Cyanophyta. Hydrobiologia 38:377–394
Rhee GY (1973) A continuous culture study of phosphate uptake, growth rate and polyphosphate in *Scenedesmus sp.* J Phycol 9:495–506
Rheinheimer G (1965) Der Einfluß der Temperatur auf das Bakterienleben in den Gewässern. Naturwiss Rundsch 18:476–481
Rice TR (1954) Biotic influence affecting population growth of plankton algae. Fish Bull US 87:227–245
Riemann B, Mathiesen H (1977) Danish research into phytoplankton productivity. Folia Limnol Scand 17:49–54
Rigler FH (1956) A tracer study of the phosphorus cycle in lake water. Ecology 37:550–561
Ripl W (1976a) Prozesssteuerung in geschädigten See-Ökosystemen. Vierteljahrsschr Naturforsch Ges Zuerich 121:301–308
Ripl W (1976b) Biochemical oxidation of polluted lake sediments with nitrate – A new lake restoration method. Ambio 5:132–135
Robra KH (1979) Akute Bakterientoxizität: Auswertung von Ringversuchen mit einer Reinkultur im Vergleich zu Untersuchungen an Mischpopulationen. Vom Wasser 53:267–282
Rodhe W (1948) Environmental requirements of freshwater plankton algae. Symp Bot Ups 10:1–149
Rook JJ (1974) Formation of haloforms during chlorination of natural waters. Water Treat Exam 23:234–243
Rosko JJ, Rachlin JW (1977) The effect of cadmium, copper, mercury, zinc and lead on cell division, growth, and chlorophyll a content of the chlorophyte *Chlorella vulgaris*. Bull Torrey Bot Club 104:226–233
Ross FF (1977) Summary review of the effects of thermal discharge into fresh water. In: Marois M (ed) Biological balance and thermal modifications. Oxford New York Toronto, pp 169–177
Ruf M (1967) Radiobiologische Untersuchungen über die Konzentration und Verteilung des langlebigen Kernwaffen-fallout in pflanzlichen und tierischen Organismen sowie in den Grundsedimenten von Oberflächengewässern. Habilitationsschrift, Univ München
Ruttner F (1926) Bemerkungen über den Sauerstoffhaushalt der Gewässer und dessen respiratorischen Wert. Naturwissenschaften 14:1237–1239
Särkkä J, Hattula ML, Janatuinen J, Paasivirta J (1978) Chlorinated hydrocarbons and mercury in aquatic vascular plants of Lake Päijänne, Finland. Bull Environ Contam Toxicol 20:361–368
Sakamoto M (1971) Chemical factors involved in the control of phytoplankton production in the Experimental Lakes Area, northwestern Ontario. J Fish Res Board Can 28:203–213
Sanderson WW (1953) Studies of the character and treatment of wastes from duck farms. Proc 8th Ind Wastes Conf Purdue Univ Ext Ser 83:170
Sand-Jensen K, Rasmussen L (1978) Macrophytes and chemistry of acidic streams from lignite mining areas. Bot Tidsskr 72:105–112
Say PJ, Whitton BA (1977) Influence of zinc on lotic plants. II. Environmental effects on toxicity of zinc to *Hormidium rivulare*. Freshwater Biol 7:377–384
Say PJ, Diaz BM, Whitton BA (1977) Influence of zinc on lotic plants I. Tolerance of *Hormidium* species to zinc. Freshwater Biol 7:357–376
Schelske CL (1962) Iron, organic matter, and other factors limiting primary productivity in a marl lake. Science 136:45–46
Schindler DW (1971) Carbon, nitrogen and phosphorus and the eutrophication of fresh water lakes. J Phycol 7:321–329

Schindler DW, Fee E (1973) The roles of nutrient cycling and radiant energy in aquatic communities. In: Cooper JP (ed) Photosynthesis and productivity in different environments. Cambridge Univ Press, Cambridge, pp 323–343
Schindler DW, Nighswander JE (1970) Nutrient supply and primary production in Clear Lake, eastern Ontario. J Fish Res Board Can 27:2009–2036
Schlichting HE Jr, Gearheart RA (1966) Some effects of sewage effluent upon phycoperiphyton in Lake Murray, Oklahoma. Proc Okla Acad Sci 46:19–24
Schmassmann H (1955) Die Haushalts-Typen der Fließgewässer. Arch Hydrobiol Suppl 22:504–509
Schmitz W (1959) Zur Frage der Klassifikation der binnenländischen Brackwässer. Arch Oceanogr Limnol Suppl 11:179–226
Schraufnagel FH (1967) Pollution aspects associated with chemical deicing. Highw Res Board Res Rec 193:22–33
Schuler JF, Diller VM, Kersten HJ (1953) Preferential assimilation of ammonium ion by *Chlorella vulgaris*. Plant Physiol 28:299–303
Schuster HK (1979) Experimentelle Untersuchungen zur Schwermetallresistenz von submersen Makrophyten. Diss Bot 50:1–156
Schwarz D (1965) Der Einfluß von Wirkstoffen auf das Wachstum und die Vermehrung von Algen (Literaturübersicht). Ver Hydrobiol Forschungsabt Dortmunder Stadtwerke 8:206
Schwoerbel J (1968) Untersuchungen über die Rolle der submersen Wasserpflanzen bei der Eliminierung von Phosphaten. Muench Beitr Abwasser Fisch Flussbiol 5:361–374
Schwoerbel J (1972) Produktionsbiologische Aspekte in Fließgewässern. Verh Dtsch Zool Ges 65:57–64
Schwoerbel J, Tillmanns GC (1964a) Untersuchungen über die Stoffwechseldynamik in Fließgewässern. I. Die Rolle höherer Wasserpflanzen: *Callitriche hamulata* Kütz. Arch Hydrobiol Suppl 28:245–258
Schwoerbel J, Tillmanns GC (1964b) Untersuchungen über die Stoffwechseldynamik in Fließgewässern. II. Experimentelle Untersuchungen über die Ammoniumaufnahme und pH-Änderung im Wasser durch *Callitriche hamulata* Kütz und *Fontinalis antipyretica* L. Arch Hydrobiol Suppl 28:259–267
Schwoerbel J, Tillmanns GC (1964c) Konzentrationsabhängige Aufnahme von wasserlöslichem PO_4-P bei submersen Wasserpflanzen. Naturwissenschaften 51:319–320
Schwoerbel J, Tillmanns GC (1972) Ammonium-Adaptation bei submersen Phanerogamen in situ. Arch Hydrobiol Suppl 42:139–141
Schwoerbel J, Tillmanns GC (1974) Stickstoffaufnahme aus dem Wasser und Nitratreduktase-Aktivität bei submersen Wasserpflanzen: *Fontinalis antipyretica* L. Arch Hydrobiol Suppl 47:282–294
Schwoerbel J, Tillmanns GC (1977) Nitrataufnahme aus dem Wasser und Nitratreduktase-Aktivität bei *Fontinalis antipyretica* L. im Hell-Dunkel-Wechsel. Arch Hydrobiol Suppl 48:412–423
Seidel K, Scheffer F, Kickuth R, Schlimme E (1967) Aufnahme und Umwandlung organischer Stoffe durch die Flechtbinse. Gas Wasserfach 6:138–139
Shapiro J (1973) Blue-green algae: Why they become dominant. Science 179:382–384
Siefert A (1976) Über die Verschmutzung von Fließgewässern im südniedersächsischen Raum und ihr Einfluß auf Vorkommen und Verbreitung einiger Makrophyten, Diatomeen und Bakterien. Dissertation, Univ Göttingen
Sikka HC, Pramer D (1968) Physiological effects of fluometuron on some unicellular algae. Weed Sci 16:296–299
Silverberg BA, Wong PTS, Chau YK (1977) Effect of tetramethyl lead on freshwater green algae. Arch Environ Contam Toxicol 5:305–313
Sims AP, Folkes BF, Bussey AH (1968) Mechanisms involved in the regulation of nitrogen assimilation in microorganisms and plants. In: Hewitt EJ, Cutting CV (eds) Recent aspects of nitrogen metabolism in plants. Academic Press, London New York, pp 91–114
Soeder CJ, Müller H, Payer HD, Schulle H (1971) Mineral nutrition of planktonic algae: some considerations, some experiments. Mitt Int Ver Limnol 19:39–58
Soeder CJ, Sternik KH, Trültsch (1972) Einfluß der Stoffwechselaktivität von Plankton-

algen auf physikalische Eigenschaften des waessrigen Milieus. Gewaesserschutz Wasser Abwasser 8:89–110
Stangenberg M (1977) Some new views in the problem of hot-water discharge to inland waters. In: Marois M (ed) Biological balance and thermal modification. Oxford New York Toronto, pp 181–186
Stanley RA (1974a) Toxicity of heavy metals and salts to Eurasian Watermilfoil (*Myriophyllum spicatum* L.). Arch Environ Contam Toxicol 2:331–341
Stanley RA (1974b) Effect of 2,4-D and various salts on Eurasian Watermilfoil. Weed Sci 22:591–594
Stanley RA (1975a) Response of Eurasian Watermilfoil to heat. Hyacinth Control J 13:62–63
Stanley RA (1975b) Interaction of calcium and 2,4-D on Eurasian Watermilfoil. Weed Sci 23:182–184
Stanley RA, Madewell CE (1967) Thermal tolerance of *Lemna minor* L. Tenn Val Auth Circ 2–13, November: 1–16
Steemann Nielsen E, Kamp-Nielsen L (1970) Influence of deleterious concentrations of copper on the growth of *Chlorella pyrenoidosa*. Physiol Plant 23:828–840
Steemann Nielsen E, Wium-Andersen S (1970) Copper ions as poison in the sea and in freshwater. Mar Biol 6:93–97
Steiner M, Eschrich W (1958) Die osmotische Bedeutung der Mineralstoffe. In: Ruhland W (ed) Encyclopedia of plant physiology Vol IV. Springer, Berlin Göttingen Heidelberg, pp 334–354
Stengel E, Soeder CJ (1973) Control of photosynthetic production in aquatic ecosystems. In: Cooper JP (ed) Photosynthesis and productivity in different environments. Cambridge Univ Press, Cambridge London New York Melbourne, pp 645–660
Stewart WDP (1968) Nitrogen input into aquatic ecosystems. In: Jackson DF (ed) Algae, man, and the environment. Syracuse Univ Press, Syracuse, pp 53–73
Stewart WDP, Alexander G (1971) Phosphorus availability and nitrogenase activity in aquatic blue-green algae. Freshwater Biol 1:389–404
Stewart WDP, Haystead A, Pearson HW (1969) Nitrogenase activity in heterocysts of blue-green algae. Nature (London) 224:228–266
Stewart WDP, Fitzgerald GP, Burris RH (1970) Acetylene reduction assay for determination of phosphorus availability in Wisconsin lakes. Proc Natl Acad Sci USA 66:1104–1111
Stockner JG, Shortreed KRS (1978) Enhancement of autotrophic production by nutrient addition in a coastal rain forest stream on Vancouver Island. J Fish Res Board Can 35:28–34
Stokes PM, Hutchinson TC, Krauter K (1973) Heavy metal tolerance in algae isolated from contaminated lakes near Sudbury, Ontario. Can J Bot 11:2155–2168
Stom DI, Beym AM (1976) Effects of phenols on various algal species. Hydrobiol J 12:43–46
Strange JR, Williamson GD, Fletcher DJ (1975) P-32 uptake in lentic algae. Environ Lett 9:1–5
Strange RJ (1976) Nutrient release and community metabolism following application of herbicide to macrophytes in microcosms. J Appl Ecol 13:889–897
Stuart TS, Gaffron H (1972) The mechanism of hydrogen photoproduction by several algae I. The effect of inhibitors of photophosphorylation. Planta 106:91–100
Sugihara G, Kilham P (1977) The ecological importance of major ions for the diatoms *Thalassiosira pseudonana* Hasle and Heimdahl and *Navicula elkab* O. Müller. Mich Acad 9:345–356
Sukopp H, Markstein B, Trepl L (1975) Röhrichte unter intensivem Großstadteinfluß. Beitr Naturkd Forsch Suedwestdschl 34:371–385
Suominen J (1968) Changes in the aquatic macroflora of the polluted Lake Rautavesi, SW-Finland. Ann Bot Fenn 5:65–81
Sutton DL, Ornes WH (1975) Phosphorus removal from static sewage effluent using duckweed. J Environ Qual 4:367–370
Sweazy RM, Lockett MC, Reid GW, Robertson JM (1975) The exchange and growth potential of phosphorus in algae cultures. Int J Environ Stud 7:193–200

Sylva RN (1976) The environmental chemistry of copper(II) in aquatic systems. Water Res 10:789–792
Syrett PJ (1962) Nitrogen assimilation. In: Lewin RA (ed) Physiology and biochemistry of algae. Academic Press, London New York, pp 171–188
Talling JF, Talling IB (1965) The chemical composition of African lake waters. Int Rev Ges Hydrobiol 50:421–463
Talpasayi ERS (1962) Acid phosphatase activity of some algae and its inhibition by molybdenum. Biochim Biophys Acta 59:710–712
Taub FB (1976) Demonstration of pollution effects in aquatic microcosms. Int J Environ Stud 10:23–33
Terhaar CJ, Ewell WS, Dziuba SP, White WW, Murphy PJ (1977) A laboratory model for evaluating the behaviour of heavy metals in an aquatic environment. Water Res 11:101–110
Thienemann A (1931) Der Produktionsbegriff in der Biologie. Arch Hydrobiol 22:616–622
Thomas EA (1972) Fischsterben in Seeabflüssen durch Hyperphotosynthese. Verh Int Ver Limnol 18:454–460
Thomas EA (1973) Phosphorus and eutrophication. In: Griffith EJ et al. (eds) Environmental phosphorus handbook. Wiley and Sons, New York, pp 585–611
Thomas EA (1975) Gewässerfeindliche Wirkungen von Phosphaten in Flüssen und Bächen. Schweiz Z Hydrol 37:273–288
Thomas JB, Blaauw OH, Duysens NM (1953) On the relation between size and photochemical activity of fragments of spinach grana. Biochim Biophys Acta 10:230–240
Thomann RV, Szumski DS, DiTiro DM, O'Connor DJ (1974) A food chain model of cadmium in western Lake Erie. Water Res 8:841–849
Tilman D (1976) Ecological competition between algae: Experimental confirmation of resource-based competition theory. Science 192:463–465
Tilman D (1977) Resource competition between planktonic algae: An experimental and theoretical approach. Ecology 58:338–348
Timofeeva NA, Kulikov NV (1967) The role of freshwater plants in accumulation of Sr-90 and its distribution over the components of a reservoir. In: Åberg B, Hungate FP (eds) Radioecological concentration processes. Pergamon, Oxford, pp 835–841
Tobler F (1943) Stengelbau, Festigkeits- und Verwertungsunterschiede beim Schilfrohr (*Phragmites communis* Trin.). Angew Bot 25:165–177
Toetz DW (1971) Diurnal uptake of NO_3 and NH_4 by a *Ceratophyllum-periphyton* community. Limnol Oceanogr 16:819–822
Toetz DW (1974) Uptake and translocation of ammonia by freshwater hydrophytes. Ecology 55:199–201
Toetz D, Varga L, Pierce M (1977) Effects of chlorine and chloramine on uptake of inorganic nitrogen by phytoplankton. Water Res 11:253–258
Trembley FJ (1960) Research project on effects of condenser discharge water on aquatic life. Progress report 1956 to 1959. Water Pollut Abstr 34:2157
Trembley FJ (1965) Effects of cooling water from steam-electric power plants on stream biota. Publ Health Serv Publ 999-WP-25:334–345
Trollope DR, Evans B (1976) Concentrations of copper, iron, lead, nickel and zinc in freshwater algal blooms. Environ Pollut 11:109–116
Tümpling von W (1960) Probleme, Methoden und Ergebnisse biologischer Güteuntersuchungen an Vorflutern, dargestellt am Beispiel der Werra. Int Rev Ges Hydrobiol 45:513
Uotila P (1971) Distribution and ecological features of hydrophytes in the polluted Lake Vanajavesi, S-Finland. Ann Bot Fenn 8:257–295
Vance BD, Drummond W (1969) J Am Water Works Assoc 61:360–362
Verstraete W, Voets JP, Vanstaen H (1975) Shifts in microbial groups of river water upon passage through cooling systems. Environ Pollut 8:275–281
Vivier P, Nisbet M (1965) Toxicity of some herbicides, insecticides and industrial wastes. In: Biological problems in water pollution, 3rd Seminar, 1962. US Public Health Serv Publ No 999-WP-25:167–169

Vollenweider RA (1968) Scientific fundamentals of the eutrophication of lakes and flowing waters, with particular reference to nitrogen and phosphorus as factors in eutrophication. Rep Org Econ Coop Dev DAS/CSI/68.27, Paris
Vollenweider RA (1979) Das Nährstoffbelastungskonzept als Grundlage für den externen Eingriff in den Eutrophierungsprozeß stehender Gewässer und Talsperren. Z Wasser Abwasserforsch 12 (2):10–20
Wagner G (1966) Phosphatspeicherung und N/P-Quotient bei Laichkräutern in abwasserbelasteten und unbelasteten Ufergebieten des Bodensees. Int Rev Ges Hydrobiol 51:229–235
Walsh GE, Miller CW, Heitmüller PT (1971) Uptake and effects of dichlobenil in a small pond. Bull Environ Contam Toxicol 6:279–288
Ware GW, Roan CC (1970) Interaction of pesticides with aquatic microorganisms and plankton. Residue Rev 33:15–45
Warinner JE, Brehmer ML (1966) The effect of thermal effluents on marine organisms. Int J Air Water Pollut 10:277
Watson GH, Bollen WB (1952) Effect of coppersulfate weed treatment on bacteria in lake bottoms. Ecology 33:522–529
Way JM, Newman JF, Moore NW, Knaggs FW (1971) Some ecological effects of the use of paraquat for the control of weeds in small lakes. J Appl Ecol 8:509–532
Weise G, Horbach W, Hornig L, Gnauck AH (1976) Der Kohlendioxidumsatz submerser Makrophyten als Indikationskriterium des Wassergütezustandes. Acta Hydrochim hydrobiol 4:95–101
Wentsel RS, Berry JW (1975) Cadmium and lead levels in Palestine Lake, Palestine, Indiana. Indiana Acad Sci 84:481–490
Wershaw RL, Burcar PJ, Goldberg MC (1969) Interaction of pesticides with natural organic matter. Environ Sci Technol 3:271–279
Westlake DF (1963) Comparison of plant productivity. Biol Rev 38:385–415
Westlake DF (1967) Some effects of low-velocity currents on the metabolism of aquatic macrophytes. J Exp Bot 18:187–205
Wetzel A (1969) Technische Hydrobiologie. Geest and Portig, Leipzig
Wetzel RG (1964) A comparative study of the primary productivity of higher aquatic plants periphyton and phytoplankton in a large shallow lake. Int Rev Ges Hydrobiol 49:1–64
Wetzel RG (1968) Dissolved organic matter and phytoplankton productivity in marl lakes. Mitt Int Ver Limnol 14:261–270
Wetzel RG (1975) Limnology. Saunders, Philadelphia London Toronto
Wetzel RG, Allen HL (1970) Functions and interactions of dissolved organic matter and the littoral zone in lake metabolism and eutrophication. In: Kajak Z, Hillbricht-Ilkowska A (eds) Productivity problems of freshwaters. Polish Scientific Publ, Warsaw, pp 333–347
Wetzel RG, Hough RA (1973) Productivity and role of aquatic macrophytes in lakes: An Assessment. Pol Arch Hydrobiol 20:9–19
Wetzel RG, Manny BA (1972) Secretion of dissolved organic carbon and nitrogen by aquatic macrophytes. Verh Int Ver Limnol 18:162–170
Whitaker J, Barica J, Kling H, Buckley M (1978) Efficacy of copper sulphate in the suppression of *Aphanizomenon flos-aquae* blooms in prairie lakes. Environ Pollut 15:185–194
Whitford LA (1960) The current effect and growth of freshwater algae. Am Microscop Soc Trans 79:302–309
Whitton BA (1967) Studies in the growth of riverian *Cladophora* in culture. Arch Mikrobiol 58:21–29
Whitton BA (1972) Environmental limits of plants in flowing waters. Symp Zool Soc Lond 29:3–19
Wiegleb G (1978) Untersuchungen über den Zusammenhang zwischen hydrochemischen Umweltfaktoren und Makrophytenvegetation in stehenden Gewässern. Arch Hydrobiol 83:443–484
Williams LG, Mount DI (1965) Influence of zinc on periphyton communities. Am J Bot 52:26–34

Williams LG, Pickering Q (1961) Direct and food-chain uptake of cesium-137 and strontium-85 in bluegill fingerlings. Ecology 42:205–206
Willingham CA, Andersen KJ (1966) Use of microorganisms for detecting toxic materials in water. Water Sew Works 113:464–467
Wissmar RC (1972) Some effects of mine drainage on primary production in Cœur d'Alene River and Lake, Idaho. PhD Thesis, Univ Idaho, Moscow Idaho
Wolk CP (1973) Physiology and cytological chemistry of blue-green algae. Bacteriol Rev 37:32–101
Wollast R, Billen G, Mackenzie FT (1975) Behaviour of mercury in natural systems and its global cycle. In: McIntyre AD, Mills CF (eds) Ecological toxicology research. Effects of heavy metal and organohalogen compounds. Plenum, New York London, pp 145–166
Wolverton BC, McKnown MM (1976) Water hyacinths for removal of phenols from polluted waters. Aquat Bot 2:191–201
Wong PTS, Chau YK, Luxon PL (1975) Methylation of lead in the environment. Nature (London) 253:263–264
Wood JM, Kennedy FS, Rosen CG (1968) Synthesis of methylmercury compounds by extracts of a methanogenic bacterium. Nature (London) 220:173–174
Wuhrmann K (1976) Chemical impact on inland aquatic ecosystem. Pure Appl Chem 45:193–198
Wunderlich M (1978) Study of the influence of cooling water additives on surface waters. UFOPLAN-Nr 102 04 033 Az.: 25 105/9. Bundesanst Gewaesserkd, Koblenz
Wunderlich M, Müller D (1976) Untersuchungen über den Einfluß von Warmwassereinleitungen auf die Gewaesser – chemisch-biologische Auswirkungen. Abschlußbericht T2/011-Wasser 8/69-/6917, Teil C, Bundesanst Gewaesserkd, Koblenz
Yamada M, Tonomura K (1972) Formation of methylmercury compounds from inorganic mercury by *Clostridium cochlearium*. J Ferment Technol 50:159–166
Young CA (1974) The effects of temperature and other environmental factors on standing crop and phenological development of *Myriophyllum spicatum* L. Master's Thesis, Univ Tennessee
Ziemann H (1967) Die Wirkung der Kaliabwässer auf die Flora und Fauna der Gewässer unter besonderer Berücksichtigung der Werra und Wipper. Fortschr Wasserchem 7:50–80
Zweig G, Tamas I, Greenberg E (1963) The effect of photosynthesis on oxygen evolution and fluorescence of illuminated *Chlorella*. Biochim Biophys Acta 66:196–205

14 Ecophysiological Effects of Atmospheric Pollutants

K.J. LENDZIAN and M.H. UNSWORTH

CONTENTS

14.1 Introduction 466

14.2 Pollutants in the Atmosphere 466
14.2.1 Specifications of Concentrations 467
14.2.2 Sulphur 467
14.2.2.1 Sulphur Dioxides 467
14.2.2.2 Other Gaseous Sulphur Compounds 468
14.2.2.3 Removal of Sulphur from the Atmosphere 468
14.2.3 Oxidants 468
14.2.3.1 Atmospheric Photo-Chemistry 469
14.2.3.2 Oxides of Nitrogen 469
14.2.3.3 Ozone 469
14.2.3.4 Peroxyacetyl Nitrate (PAN) 469
14.2.4 Localised Pollutants 470
14.2.4.1 Fluorides 470
14.2.4.2 Particulate Matter 470

14.3 Exposure of Plants to Pollutants 470

14.4 Action of Pollutants on Intact Plants 471
14.4.1 Sulphur 471
14.4.1.1 Sulphur Dioxide 471
14.4.1.2 Other Sulphur Compounds 474
14.4.2 Oxidants 474
14.4.2.1 Symptoms of Acute and Chronic Injury 474
14.4.2.2 Action on Stomata 474
14.4.2.3 Action on Photosynthesis 475
14.4.2.4 Action on Growth and Development 475
14.4.3 Fluorides 475
14.4.3.1 Symptoms of Acute and Chronic Injury 475
14.4.3.2 Action on Stomata 476
14.4.3.3 Action on Gas Exchange 476
14.4.3.4 Action on Growth and Development 476
14.4.4 Particulates 476

14.5 Action of Pollutant Combinations 476

14.6 Acid Precipitation 477

14.7 The Action of Air-Polluting Compounds at the Cellular and the Enzymatic Level 477
14.7.1 Sulphur Dioxide and Hydrogen Sulphide 478
14.7.1.1 Chemical Reactions 478
14.7.1.2 Action on Enzymes and on Cellular Metabolism 479
14.7.2 Ozone and Peroxyacetyl Nitrate 484
14.7.2.1 Chemical Reactions 484
14.7.2.2 Action on Cell Structures 484
14.7.2.3 Action on Enzymes and on Cellular Metabolism 486

14.7.3 Oxides of Nitrogen 486
14.7.3.1 Chemical Reactions 486
14.7.3.2 Action on Enzymes and on Cellular Metabolism 486
14.7.4 Carbon Monoxide 487
14.7.5 Fluoride 487
14.7.5.1 Chemical Reactions 487
14.7.5.2 Action on Enzymes and on Cellular Metabolism 487
14.7.5.3 Subcellular and Cellular Damage 489
14.7.5.4 Metabolism of Fluoride 489
14.7.6 Combined Action of Pollutants 490
14.7.7 Heavy Metals 490

14.8 Conclusion 491

References 491

14.1 Introduction

In a brief review article it is not possible to cover all aspects of a topic such as this, nor can a full sense of historical perspective be given. We have attempted to survey the major sources of pollutants, pollutant uptake or deposition on plants and the physiological and biochemical responses to pollutant exposure. Interactions with diseases and pests or with plant nutrition are scarcely treated, and the interested reader is referred to articles by HEAGLE (1982) or COWLING and KOZIOL (1982) for those subjects.

Responses to air pollution have tended to be studied by pathologists and physiologists on whole plants and by biochemists in vitro. Pathologists have been concerned mainly with acute injury where tissue is severely damaged; perhaps the most informative physiological responses occur when there is less visible injury, and there is a general trend towards experiments at lower gas concentrations. Another recent development is the interpretation of responses in relation to pollutant uptake rather than pollutant concentration. This development seems a necessary step in our understanding of plant response, but should not obscure the importance of developmental stage, edaphic effects, and genetic variation within species, factors which also have a major impact on plant response to pollution. Experimental work on this subject draws together the disciplines of the environmental scientist who defines the pollutant concentration(s) to which plants are exposed, the plant physiologist who measures and analyses the responses and the biochemist who investigates the processes affected. We hope that this review emphasises these links.

14.2 Pollutants in the Atmosphere

Man's activities, industrial, agricultural and domestic, result in the release of many phyto-toxic compounds into the atmosphere. Only those of major significance are discussed here: more detailed reviews of air pollutants, their transformation and transport in the atmosphere may be found in STERN (1977). Examples of concentrations typically found in Europe and North America were given by FOWLER and CAPE (1982). A distinction may be made between *primary pollutants*, i.e. those which affect plants in the form in which they are emitted, and *secondary pollutants*, i.e. those resulting from reactions in the atmosphere. However, for discussions of physiological consequences it seems more convenient to classify pollutants under headings corresponding to the major sources.

14.2.1 Specifications of Concentrations

Two systems are commonly used; gravimetric concentration ($\mu g\, m^{-3}$, $mg\, m^{-3}$) defined as the mass of pollutant per unit volume of air, and volumetric concentration (ppm, i.e. parts in 10^6 by volume, ppb – parts in 10^9). The gravimetric system is equally appropriate for gases and solids, but care must be taken to specify the temperature and pressure because of their influence on the density of gases. Conversion from one system to another may be made using the gas equation, e.g. convert from X ppm to Y $\mu g\, m^{-3}$

$$Y = X \times M \times 10^3 / V$$

where M is the molecular weight and V is the molar volume (24.0 l/mol at 20 °C, 101.3 kPa). Table 14.1 gives conversion factors for some common gaseous pollutants.

Table 14.1. Conversion factors between gravimetric and volumetric concentrations of pollutant gases at 20 °C, 101.3 kPa pressure. (After UNSWORTH and ORMROD 1982)

Gas	Molecular wt $g\, mol^{-1}$	To convert from ppb to $\mu g\, m^{-3}$, multiply ppb by	To convert from $\mu g\, m^{-3}$ to ppb multiply $\mu g\, m^{-3}$ by
Fluorine, F_2	38.0	1.58	0.633
Hydrogen fluoride, HF	20.0	0.83	1.202
Hydrogen sulphide, H_2S	34.1	1.42	0.705
Nitric oxide, NO	30.0	1.25	0.801
Nitrogen dioxide, NO_2	46.0	1.91	0.523
Ozone, O_3	48.0	2.00	0.501
Peroxyacetyl nitrate (PAN), $CH_3COO_2NO_2$	105.0	4.37	0.229
Sulphur dioxide, SO_2	64.1	2.67	0.375

14.2.2 Sulphur

14.2.2.1 Sulphur Dioxides

Combustion of fossil fuels has led to large emissions of sulphur dioxide in urban areas for many hundreds of years. The industrial revolution gave rise to major point sources such as smelters, chemical works, and refineries. Some of the first reports of air pollution damage came from close to point sources (HASELHOFF and LINDAU 1903; NATIONAL RESEARCH COUNCIL OF CANADA 1939) where SO_2 concentrations in the range 1–10 ppm could persist for several hours. With the development of tall stacks to disperse pollutants more effectively, acute damage is now very rare. In many industrialised countries the general background of SO_2 from urban and industrial sources gives annual average SO_2 concentrations in the range 10–30 ppb. A maximum daily average could typically be about four or five times the annual average, and a maximum 3-hour average could be four or five times greater again. MARTIN and BARBER (1981) recently published data for a typical rural site in central England. Natural background levels of SO_2 are below 2 ppb (GRANAT et al. 1976).

14.2.2.2 Other Gaseous Sulphur Compounds

Hydrogen sulphide arises naturally from decomposing vegetation, and from industrial processes such as paper manufacture, coking, and the petroleum industry. In the atmosphere the gas usually oxidises within a few hours to SO_2. Other reduced sulphur compounds such as dimethyl sulphide and methyl mercaptan are believed to play an important part in the global sulphur cycle (GRANAT et al. 1976; GARLAND 1978) but are not known to be generated in phytotoxic concentrations by man's activities.

14.2.2.3 Removal of Sulphur from the Atmosphere

The sulphur cycle has been studied more widely than cycles of other pollutant elements. GARLAND (1978) reviewed the main pathways. Uptake of SO_2 by plants and soils is important near sources (FOWLER and UNSWORTH 1973, 1979; GARLAND 1978); oxidation in the atmosphere mainly by an indirect photochemical reaction (EGGLETON and COX 1978) or in cloud droplets leads to sulphuric acid mists and rains or to sulphate aerosols where the neutralising ion is often ammonium (GRANAT et al. 1976).

14.2.3 Oxidants

The original "smogs" arose in London when SO_2 and smoke combined with fog. Smogs were common in the nineteenth century. From about 1950 onwards, a different type of smog became a problem, initially in North American cities where its occurrence was associated with sunlight and high temperatures, hence the term "photochemical smog". The smog consists of oxidant materials generated by complex chemical reactions in the atmosphere between oxides of nitrogen and hydrocarbons from vehicle exhausts (HAAGEN-SMIT and WAYNE 1976). There are three main phytotoxic constituents, oxides of nitrogen, ozone, and peroxyacyl nitrates (PANs). Although each is discussed separately it should be recognised that they are intimately associated.

14.2.3.1 Atmospheric Photo-Chemistry

In an atmosphere containing oxides of nitrogen but no hydrocarbons, a series of cyclic reactions are set up, driven by energy from sunlight, leading to steady state concentrations of nitric oxide, NO, nitrogen dioxide, NO_2, and ozone, O_3. When hydrocarbons are present, they disrupt the cyclic reaction by combining first with O_2 to form a peroxyacyl radical and then with NO to form PAN. This removal of NO from the system allows a build-up of ozone. In urban areas, emissions are often associated with motor vehicles during "rush-hours" and this dictates the time course of pollutants, provided that meteorological conditions are suitable (stagnant air, bright sunshine). The precursors of photochemical smogs may be transported long distances, at least 1,000 km (Cox et al. 1975), or entrained in quasi-stationary air masses so that oxidant injury to plants may occur on wide regional scales.

14.2.3.2 Oxides of Nitrogen

There are only two oxides of nitrogen known to elicit direct responses from plants, nitric oxide, NO, and nitrogen dioxide, NO_2. Both are produced at the temperatures reached when fuels are burnt in air. Generally the emission ratio is about 9:1, NO/NO_2 by volume (DERWENT and STEWART 1973). However, in the atmosphere, measured ratios of NO to NO_2 are much less than 9:1 and very variable, largely because NO is rapidly oxidised by ozone (generated artificially or present naturally near the surface of the earth at concentrations from 20–40 ppb). Nitrogen dioxide is removed from the atmosphere mainly by oxidation to nitric acid or nitrate and removal in rain or by gaseous uptake by plants and soil. Concentrations of NO and NO_2 vary considerably annually and diurnally: the proportion of NO_2 to total oxides of nitrogen (NO_x) is greater in summer month when the rate of conversion $NO \rightarrow NO_2$ is greater (MARTIN and BARBER 1981; FOWLER and CAPE 1982). Typical annual mean NO_x concentrations in large cities are 50–100 ppb. In severe pollution episodes hourly concentrations may exceed 1,000 ppb (HAAGEN-SMIT and WAYNE 1976).

14.2.3.3 Ozone

Normal background ozone concentrations at the ground are 20–40 ppb. During pollution episodes, urban ozone concentrations generated photochemically may reach 150–300 ppb. High concentrations may also be found in remote regions downwind of local or regional sources of photochemical pollution (Cox et al. 1975). Ozone is removed from the atmosphere mainly by gaseous uptake by plants and soil (LEUNING et al. 1979a, b) and by chemical reactions with other air pollutants (COX and PENKETT 1972).

14.2.3.4 Peroxyacetyl Nitrate (PAN)

Sources of PAN are the same as those for ozone, but much less is known about the general occurrence of PAN. In urban areas of California, daily average PAN concentrations are typically 10–20 ppb with hourly maxima up to about 50 ppb.

14.2.4 Localised Pollutants

The range of pollutants from specific sources which elicit physiological or biochemical responses in plants is limitless. Only those commonly experienced are discussed here.

14.2.4.1 Fluorides

Fluorine is a relatively common and abundant element in the earth's crust. It is emitted in gaseous (hydrogen fluoride, HF) or particulate form during manufacture of bricks, aluminium, fertilisers, steel, and other smelting and combustion processes. Concentrations vary considerably with site. In generally industrialised regions a background concentration of about 0.2–2 ppb (total water-soluble fluoride) is typical (McCUNE et al. 1976).

14.2.4.2 Particulate Matter

This heading covers a vast range of chemical compositions and of particle sizes. Particles larger than about 30 μm diameter generally fall from the atmosphere fairly close to the site of emission. In contrast, particles of sub-micrometer diameter may be dispersed on regional or global scales. The efficiency of capture of particles by vegetation varies considerably with particle size density and windspeed (CHAMBERLAIN 1975; CHAMBERLAIN and LITTLE 1981). Particulate concentrations in urban areas are typically 50–500 $\mu g\ m^{-3}$, and in rural areas are roughly an order of magnitude smaller. Close to motorways and freeways where traffic stirs up much dust and releases particulate lead, concentrations may be large enough to have physiological effects.

14.3 Exposure of Plants to Pollutants

Before physiological or biochemical responses in plants can occur, pollutants must be transferred from the atmosphere to the surface. There are recent reviews describing mechanisms of gas exchange (BENNETT and HILL 1975; UNSWORTH 1981a) and particle transfer (CHAMBERLAIN and LITTLE 1981) to vegetation. Models taking into account various factors influencing uptake of gaseous pollutants were described by BENNETT et al. (1973), FOWLER and UNSWORTH (1979), LEUNING et al. (1979a, b), BLACK and UNSWORTH (1980a) and UNSWORTH (1981a). For single leaves, pollutant molecules must traverse a relatively still layer of air close to the leaf, the "boundary layer", whose thickness depends on leaf shape and structure and on air movement. Further resistances to uptake are imposed by stomata which limit diffusion to the mesophyll, and by the cuticle. Additionally, rates of uptake may be limited by chemical reactions taking place on the cuticle or on cell walls.

In early experiments in controlled environments, the resistance to uptake of pollutants imposed by the boundary layer was sometimes considerable

(MANSFIELD 1976). Consequently much larger pollutant concentrations or longer exposures were required to produce the same damage in the laboratory as that observed in the field where turbulent air flow gave low boundary layer resistances. ASHENDEN and MANSFIELD (1977) showed that yields of ryegrass were significantly reduced by SO_2 in a wind tunnel at 0.4 m s^{-1}, but at a lower windspeed there were no significant reductions. Responses of stomata to pollutants may also influence further uptake of a gas, loss of water, and CO_2 exchange. UNSWORTH (1982) reviewed methods of exposing plants to gaseous pollutants.

Deposition of particles on plants depends strongly on particle size and density, wind speed and surface shape and structure (CHAMBERLAIN 1975; LITTLE and WHIFFEN 1977). There is some evidence that liquid droplets are more effectively captured than dry particles of the same size (CHAMBERLAIN 1975).

Transfer of pollutants to crop canopies can be described by similar concepts as those used for single plants (FOWLER and UNSWORTH 1973, 1979; WESELY et al. 1978; LEUNING et al. 1979a, b; BELOT and GAUTHIER 1975).

14.4 Action of Pollutants on Intact Plants

In this section the action of a range of pollutants on intact plants is reviewed. First, for individual pollutants there are reviews of sites and symptoms of acute and chronic injury, action on stomata, on carbon dioxide exchange and on growth and development.

14.4.1 Sulphur

14.4.1.1 Sulphur Dioxide

Historically, plant responses to sulphur dioxide have been studied for longer than for any other air pollutant, and there have been many reviews of the literature (e.g. recently MUDD 1975a; HÄLLGREN 1978). The gas enters leaves primarily by diffusion through stomata. Oxidation takes place to highly toxic sulphite and then to less toxic sulphate (NIEBOER et al. 1976). When the rate of uptake of SO_2 is high, the relatively slow rate of oxidation from sulphite to sulphate allows sulphite concentrations to rise, producing acute injury. Plants exposed to lower rates of uptake of SO_2 may develop chronic injury, symptoms associated with the accumulation of large concentrations of sulphate in cells (ENVIRONMENTAL PROTECTION AGENCY 1976). At even lower exposures, no visible symptoms develop, but physiological responses can be observed (UNSWORTH 1981b; BLACK 1982).

Symptoms of Acute and Chronic Injury. Details and colour plates of symptoms are given by VAN HAUT and STRATMANN (1970), JACOBSON and HILL (1970), ENVIRONMENTAL PROTECTION AGENCY (1976) and MALHOTRA and BLAUEL (1980). In general, acute injury is characterised by bleaching of tissue, especially marginal and intercostal areas of broadleaf species. The bleached areas are typically ivory or white-coloured, and veins often remain uninjured. In conifers and a few other species (e.g. bracken) the colour of injured tissue is orange-red. Sensitive species such as alfalfa, cucumber, bracken and pine may show visible injury after a few hours exposure to 100–500 ppb SO_2 (depending on conditions of exposure, species, stage of growth, etc.), whereas resistant species require typically 2,000 ppb. Lists of species ranked for sensitivity to acute injury have been published (ENVIRONMENTAL PROTECTION AGENCY 1976), but care should be taken in applying them, as some experiments were performed in unnaturally still air. It is becoming increasingly clear that sensitivity assessed by acute injury does not necessarily correlate with sensitivity in other responses (HORSMAN et al. 1979; AYAZLOO and BELL 1981).

Chronic injury may range from general interveinal chlorosis to more severe areas of white or reddish-brown colouration. The chlorosis arises from lysis of chloroplasts in mesophyll cells leading to destruction of chlorophyll. Injured leaves or needles often abscise prematurely.

Action on Stomata. The subject was reviewed by HÄLLGREN (1978), UNSWORTH and BLACK (1981) and BLACK (1982). Early experiments where acute and chronic injury was produced did not show direct stomatal responses. More recent work indicates that high concentrations of SO_2 induce stomatal closure (VAN HAUT and STRATMANN 1967; SIJ and SWANSON 1974; BONTE et al. 1975; KONDO and SUGAHARA 1978) and low concentrations generally induce stomatal opening (MAJERNIK and MANSFIELD 1971; UNSWORTH et al. 1972; BECKERSON and HOFSTRA 1979a, b; BLACK and UNSWORTH 1980b; TAYLOR et al. 1981). The magnitude of opening appears to increase stomatal conductance typically by about 20% compared with clean air values, independent of SO_2 concentration (BLACK and UNSWORTH 1980b). The response is sometimes reported as reversible (MAJERNIK and MANSFIELD 1971) and sometimes as irreversible (BLACK and UNSWORTH 1980b). This probably depends on SO_2 uptake and plant susceptibility. In species with stomata known to respond to atmospheric humidity, the response to low concentrations of SO_2 (<20 ppb) is complex. At low relative humidity, stomata close in response to SO_2 (MAJERNIK and MANSFIELD 1971; BLACK and UNSWORTH 1980b). Above a humidity threshold, depending on species, typically about 40% r.h. for *Vicia faba*, the "normal" opening response is observed. Such responses persist for at least 3 days, but there have been few reports of direct long-term measurements of stomatal responses. Indirect observations such as increase in transpiration in response to SO_2 have been reported for *Pinus sylvestris* and other tree species (KELLER and MÜLLER 1958).

At higher SO_2 concentrations, stomatal closure is usually reported (SIJ and SWANSON 1974; KONDO and SUGAHARA 1978), although brief stomatal opening may occur on initial exposure. It is not known whether the closure response is reversible; if it is, it would confer some protection in plants.

Several species may be exceptions to these general forms of response, e.g. *Pelargonium hortorum* (BONTE et al. 1975), *Diplacus aurantiacus* and *Heteromeles* (WINNER and MOONEY 1980) and indeed it is not yet clear whether the general distinction between directions of stomatal responses at low and high concentrations will hold when more species have been studied. The responses are probably primarily related to differential sensitivity of cells and to preferential pathways of SO_2 and its derivatives in leaf tissue.

The action of SO_2 on stomata appears to arise in two ways. First, at low SO_2 concentrations, epidermal cells adjacent to stomata lose turgor (BLACK and BLACK 1979a, b) and stomata open. At higher concentrations there is swelling of guard cell chloroplasts (WELLBURN et al. 1972) and ultimately extensive disorganisation of guard cell structure causing stomatal closure. Second, SO_2 interferes with CO_2 exchange in the mesophyll (HÄLLGREN 1978; BLACK and UNSWORTH 1979) and so stomata may also respond to changes in internal CO_2 concentration.

Action on CO_2 Exchange. Inhibition of net photosynthesis by SO_2 was first reported for alfalfa by THOMAS and HILL (1937) and is often regarded as the first sign of SO_2 action on plants, although actions on enzyme systems may be precursors (ZIEGLER 1975; HÄLLGREN 1978). UNSWORTH and BLACK (1981) reported the action of SO_2 on light response curves for net photosynthesis in *Vicia faba*. At all SO_2 concentrations exceeding 35 ppb, net photosynthesis was inhibited compared with clean air. At light saturation the magnitude of inhibition depended on SO_2 concentration, but at low irradiances the inhibition was independent of concentration. Dark respiration rates increased substantially independent of concentration. When exposures continued for up to 3 days, net photosynthesis returned to clean air values about 1 h after fumigation ceased: dark respiration recovered after one photoperiod. Reductions in photosynthesis at low SO_2 concentrations have also been reported by BULL and MANSFIELD (1974) for peas, KELLER (1978, 1981) for conifers, and TANIYAMA (1972) for rice. Magnitudes of inhibition, and recovery rates vary with species and age. Mechanisms of action of SO_2 on carbon dioxide exchange were discussed in detail by ZIEGLER (1975) and HÄLLGREN (1978).

Action on Growth and Development. Much early work, summarised by THOMAS (1951) and HECK and BRANDT (1977), suggested that growth was not affected until visible injury to foliage was observed. Many experiments were, and still are, based on unrealistically high SO_2 concentrations and used exposure systems with poor air exchange. More recently, at lower SO_2 concentrations, below about 150 ppb, numerous experiments have shown reductions in yield of *Lolium perenne* (BELL and CLOUGH 1973; LOCKYER et al. 1976; ASHENDEN and MANSFIELD 1977; HORSMAN et al. 1979), in other grasses (ASHENDEN 1979; HEITSCHMIDT et al. 1978; DAVIES 1980; AYAZLOO and BELL 1981), conifers (FARRAR et al. 1977; GARSED and RUTTER 1982) and broad leaved tree species (GARSED et al. 1979). It seems that much early work implying a threshold of SO_2 concentration below which no responses occurred must now be questioned. It is generally accepted that SO_2 increases rates of leaf senescence. There is some evidence that it also influences numbers of tillers in grasses and cereals and dry-matter partitioning between roots and shoots.

When plants are growing in a medium deficient of sulphur, SO_2 may act as a nutrient (FALLER 1971, 1972; COWLING and LOCKYER 1976, 1978). Implications for plants in their natural environment depend on rates of supply and demand.

14.4.1.2 Other Sulphur Compounds

Hydrogen sulphide induces visible injury at concentrations exceeding about 2 ppm (Thompson and Kats 1978; Bennett et al. 1979). At lower concentrations, yields of pods and dry weight of *Phaseolus vulgaris* in the field were reduced by H_2S (Bennett et al. 1979). Stomatal responses and reductions in photosynthesis rates were also observed (Coyne and Bingham 1978). There are several reports that plants emit H_2S in response to exposure to SO_2 (De Cormis 1968; Wilson et al. 1978; Hällgren et al. 1982) or to excess sulphur available at the roots (Spaleny 1977).

Rain acidified with sulphuric acid has been reported to injure plants, although the pH in laboratory experiments has often been lower than observed in the field. Jacobson (1980) reviewed the subject. Responses include formation of lesions, often near stomata, trichomes and vascular tissue (Evans et al. 1977), premature leaf abscision (Ferenbaugh 1976) and reduction in growth, development and yield (Ferenbaugh 1976; Wood and Bormann 1974). Enhanced leaching of cations from leaves is also reported (Fairfax and Lepp 1975).

14.4.2 Oxidants

Physiological responses to oxidants were reviewed by Heck (1968), Dugger and Ting (1970), Heath (1975), National Academy of Sciences (1977), and Tingey and Taylor (1982).

14.4.2.1 Symptoms of Acute and Chronic Injury

Details and colour plates are found in Jacobson and Hill (1970) and Environmental Protection Agency (1976). Uptake through stomata leads to damage in mesophyll cells. For ozone, palisade cells, where present, are particularly susceptible; PAN preferentially attacks spongy mesophyll. Injury often begins near leaf tips, and successive exposures to PAN may thus result in bands of injury tissue. Ozone injury often consists of flecking or stipples in broadleaf plants; in severe cases the whole leaf may appear chlorotic or bronzed.

14.4.2.2 Action on Stomata

Stomata of many species respond to ozone by partially closing (Hill and Littlefield 1969; Rich and Turner 1972). Because closure decreases the rate of ozone uptake, some species or varieties in which closure takes place appear to be resistant (Rich and Turner 1972; Adedipe et al. 1973; Knudson et al. 1979). It is not clear whether the response is reversible (but see 14.3.2.3). Mechanisms of action in the intact plant are not fully understood although there have been many studies of biochemical action (Mudd 1982; Tingey and Taylor

1982). In particular it is not known whether closure arises independently or as a result of an increase in intercellular CO_2 concentration (MANSFIELD 1976). Stomatal responses to PAN do not seem to have been reported. Closure in response to NO_2 occurred only at concentration exceeding 7,000 ppb, and was probably due to action on photosynthesis (SRIVASTAVA et al. 1975).

14.4.2.3 Action on Photosynthesis

Net photosynthesis is decreased by ozone (HILL and LITTLEFIELD 1969), PAN (MUDD 1975b), NO_2 (TINGEY et al. 1971; BULL and MANSFIELD 1974; SRIVASTAVA et al. 1975; TAYLOR et al. 1975) and NO (HILL and BENNETT 1970; CAPRON and MANSFIELD 1976). It is seldom clear whether stomatal closure accounts partially or totally for inhibition in net photosynthesis, but in vitro it is well established that oxidants interfere with photosynthetic pathways (MUDD 1982). At low dosages, the inhibition is usually reversible.

14.4.2.4 Action on Growth and Development

HECK and BRANDT (1977) and NATIONAL ACADEMY OF SCIENCES (1977) provide tabulations showing reductions in yields of many species exposed to oxidants. There is not a good correlation between foliar injury and reductions in yield (JACOBSON 1982). It is becoming clear that substantial yield reductions by ozone in the field may occur, sometimes without visible injury (HEGGESTAD et al. 1980; JACOBSON 1982). Alterations in assimilate partitioning indicate greater inhibition of root growth than shoot growth (TINGEY 1977; BLUM and HECK 1980). There have been no longterm studies of effects of PAN on growth and yield. Oxides of nitrogen reduce growth in a number of species at NO_x concentrations exceeding about 100 ppb (TAYLOR et al. 1975). Reductions in growth in glasshouses, where the generation of elevated CO_2 concentrations may also lead to high NO_x concentrations, may be substantial (LAW and MANSFIELD 1982). FEDER (1968) reported that ozone decreases germination of pollen in tobacco and petunia, and flower production in carnation (FEDER and CAMPBELL 1968).

14.4.3 Fluorides

Most experimental work has used hydrogen fluoride gas. NATIONAL ACADEMY OF SCIENCES (1971), CHANG (1975), MCCUNE et al. (1976), WEINSTEIN and ALSCHER-HERMAN (1982) and DAVISON (1982) reviewed plant responses.

14.4.3.1 Symptoms of Acute and Chronic Injury

Symptoms vary considerably with species and are described and portrayed in JACOBSON and HILL (1970) and ENVIRONMENTAL PROTECTION AGENCY (1976). On broadleaf species, marginal and intercostal necrosis is observed. *Gladiolus* shows very characteristic necrosis of leaf tips, sharply delineated from live tissue. Conifers also show tipburn.

14.4.3.2 Action on Stomata

Stomatal responses to fluoride have seldom been reported; WOLTZ and LEONARD (1964) measured marked stomatal opening on orange trees. JACOBSON et al. (1966) showed that fluoride can be adsorbed onto leaf surfaces as well as accumulated internally. Fluoride on leaf surfaces remained in soluble form and could be exchanged with the interior of leaves. Consequently stomatal uptake may be less important than for other gases.

14.4.3.3 Action on Gas Exchange

Both stimulation and inhibition of respiration by fluoride have been reported, and there is action on many enzyme systems (CHANG 1975). Fluoride inhibits net photosynthesis in many species at concentrations from 10–50 ppb (THOMAS and HENDRICKS 1956), and inhibits the Hill reaction in chloroplasts of *Phaseolus vulgaris* (BALLANTYNE 1972) and *Beta vulgaris* (SPIKES et al. 1955). WEINSTEIN (1961) found inhibition of formation of chlorophylls a and b in beans and tomato fumigated with 1–12 ppb fluoride, and there was recovery after fumigations ceased.

14.4.3.4 Action on Growth and Development

Responses included substantial reductions in leaf size and decreases in yield, flower or fruit number (WEINSTEIN and ALSCHER-HERMAN 1982). Accelerated ageing in corn roots and inhibited root growth were observed by CHANG and THOMPSON (1966). Effects on pollination and fruit quality are reported for peach, apricot, pear, cherry and citrus (HECK and BRANDT 1977) and for pear and strawberry (BONTE 1982).

14.4.4 Particulates

The wide range of particle sizes and compositions precludes any general conclusions. LERMAN and DARLEY (1975) reviewed the topic. Stomata are often observed blocked with dust (CZAJA 1966; RICKS and WILLIAMS 1974) which may restrict transpiration by day but enhance it by night if stomata cannot shut. FLÜCKIGER et al. (1979) showed that particle size had a significant influence on the magnitude of stomatal response. Photosynthesis may also be influenced by blocked stomata or by reduced light when leaves are heavily dusted. Growth of trees is reported reduced (BOHNE 1963; DARLEY 1966) and pollen germination may be inhibited by cement dust (ANDERSON 1914).

14.5 Action of Pollutant Combinations

Pollutants rarely exist alone. Responses to combinations may be additive (equal to the sum of responses to the pollutants individually), more than additive or less than additive. REINERT et al. (1975) and ORMROD (1982) reviewed the literature. Stomatal responses in a range of species were studied by BECKERSON and HOFSTRA (1979a, b) for mixtures containing ozone and SO_2. Stomata opened in SO_2, closed in O_3 and closed further in SO_2+O_3. Similar responses were reported by COYNE and BINGHAM (1978) for *Phaseolus vulgaris* exposed to H_2S and O_3. Numerous reports of responses of growth and development exist, with and without visible injury (see REINERT et al. 1975; ORMROD 1982), but there are relatively few studies of physiological or biochemical responses. The inhibition of apparent photosynthesis by SO_2 and NO_2 was reported by WHITE et al. (1974), by NO_2 and NO by CAPRON and MANSFIELD (1976), and by O_3 and SO_2 by ORMROD et al. (1981).

14.6 Acid Precipitation

Combustion of fossil fuel has led to large emissions of nitrogen and sulphur oxides which interact with moisture in the atmosphere to give high concentrations of H^+, NO_3^-, and SO_4^{2-}. Precipitation in Europe has an average annual pH of approximately 3.0, with individual pH values as low as pH 2.4 (LIKENS et al. 1979; cf. also WINKLER, 1982; FOWLER et al. 1982). This increased acidity in precipitation ("acid rain") has become a dominant feature of anthropogenic changes in ecological belances. Both aquatic and terrestrial ecosystems are severely affected. Oligotrophic rivers and lakes modify the ionic forms of metals if pH values are lowered. At pH values below 5.0 many of fish species will die (SCHOFIELD 1976; HAINES 1981) mainly due to acidity and aluminum toxicity. Aluminum toxicity (cf. Volume 12C, chapt. 6, this series) has also been a feature when discussing the impact of soil acidification on forest ecosystems (FÖLSTER et al. 1979; ULRICH 1981). Individual plants exposed to acid rainfalls show necrotic lesions on foliage and loss of nutrients (Mg^{2+}, Ca^{2+}) due to leaching from leaves and other foliar organs (EVANS 1980; COWLING 1981; COWLING and LINTHURST 1981).

14.7 The Action of Air-Polluting Compounds at the Cellular and the Enzymatic Level

At their site of action on cellular metabolism, air-polluting gases have already been dissolved in the plasmalemma and within the cell. To elucidate the damaging effect of pollutants, therefore, organelles and enzymes are commonly incubated with the respective ionic forms. Such in vitro experiments may mimic toxic effects during exposure to the pollutants. In vivo fumigation, however, may induce further metabolic changes, including enhanced synthesis of enzymes. These enzymes persist even when fumigation has ceased, and further effects are often completely different from the actual toxic effect during fumigation.

A great deal of research has been carried out concerning the in vitro effect of pollutants on enzymes. For that purpose, almost any testable enzyme has been checked. However, many experiments have been designed lacking a cohesive working hypothesis. Moreover, plants of different and sometimes unknown sensitivity and of different physiological conditions have been used as the source of enzymes. In addition, they have been submitted to various fumigation conditions. Thus, a decrease or increase in enzyme activity (sometimes both responses are reported in different experiments on the same enzyme) or changes in pool sizes of metabolites, in many cases, seem to follow a more statistical rather than logical order.

We have endeavoured to select those cases which may provide some characteristic features of the pollutant's action at the enzymatic level, an action which then results in characteristic changes in cell metabolism.

14.7.1 Sulphur Dioxide and Hydrogen Sulphide

14.7.1.1 Chemical Reactions

When dissolved in water, SO_2 is hydrated and charged molecules are rapidly formed (SCHROETER 1966). Depending on the pH of the water phase, a number of ionic states can be assumed:

$$SO_{2\,gas} + H_2O \rightleftharpoons SO_2 \cdot H_2O \quad (14.1)$$

$$SO_2 \cdot H_2O + H_2O \rightleftharpoons HSO_3^- + H_3O^+ \quad (14.2)$$

$$HSO_3^- + H_2O \rightleftharpoons SO_3^{2-} + H_3O^+ \quad (14.3)$$

$$2\,HSO_3^- \rightleftharpoons S_2O_5^{2-} + H_2O. \quad (14.4)$$

Their relative concentration, expressed as mole fractions, may be calculated using Henry's law constant (1.24 M atm^{-1} at 25 °C; GMELIN 1963) and the K_a values for the equilibria (Eq. 14.2) (1.76×10^{-2}) and (Eq. 14.3) (6.24×10^{-8}) at 25 °C (SILLEN 1964). It can be seen from Fig. 14.1 that at biological pH values there are large changes in the concentrations of HSO_3^- and SO_3^{2-}, with the former predominating at pH values less than 7. There is some doubt that the undissociated acid H_2SO_3 exists in solution (MUDD 1975a) leaving $SO_2 \cdot H_2O$ as the sole unionized sulphur species dissolved in a water phase. It is evident from the graph that SO_2 can act as an oxidizing agent when pH values of the medium are lowered. As the rates of the forward and backward reactions of Eqs. (14.2) and (14.3) are quite rapid (EIGEN et al. 1961; BEILKE and GRAVENHORST 1978), it must be assumed that any disturbance to the concentration of one species results in a fast readjustment of the other species in the equilibria.

Similar variations of the ionic state in aqueous solutions can be described for H_2S:

$$H_2S_{gas} + H_2O \rightleftharpoons (H_2S)_{aq} \quad (14.5)$$

$$(H_2S)_{aq} + H_2O \rightleftharpoons HS^- + H_3O^+ \quad (14.6)$$

$$HS^- + H_2O \rightleftharpoons S^{2-} + H_3O^+ \quad (14.7)$$

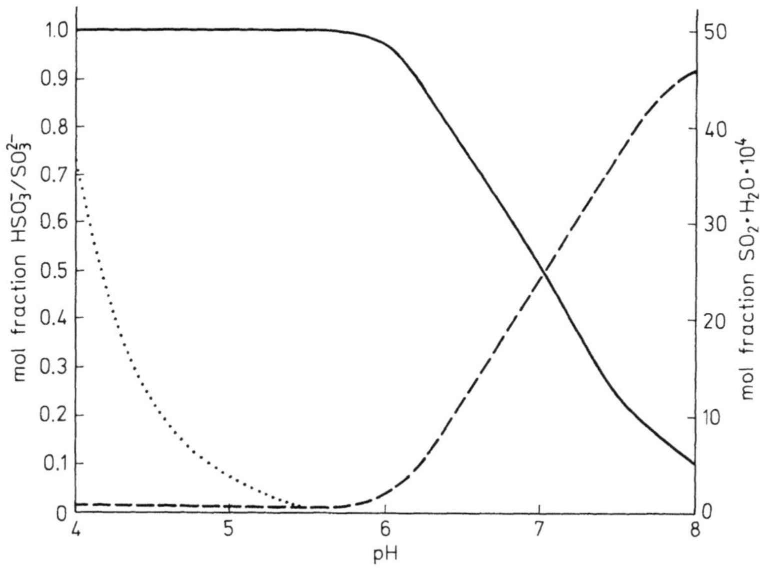

Fig. 14.1. Mole fractions of sulphur species in a solution of S(IV) at different solution pH values. ——— HSO_3^- (scale *left side*); -------- SO_3^{2-} (scale *left side*); ········ $SO_2 \cdot H_2O$ (scale *right side*) (SPEDDING et al. 1980)

Henry's law constant for the solubility of H_2S in water is 35 (WILHELM et al. 1977), indicating a lower solubility of H_2S than is the case for SO_2. With pH values under 6.9, the $(H_2S)_{aq}$ species becomes the predominant one while the HS^- species is found at higher pH values. The mole fractions of HS^- and $(H_2S)_{aq}$ follow opposite curves with respect to pH (SPEDDING et al. 1980). The S^{2-} species will be formed only at pH values higher than 13.

SO_2 is readily oxidized to SO_4^{2-}. Aerobic oxidation in the atmosphere and in aqueous solutions is enhanced by ultraviolet radiation, metal ions and by ozone (FUZZI 1978; BEILKE and GRAVENHORST 1978; HEGG and HOBBS 1978). Oxidation also takes place in plant leaves with illuminated chloroplasts being the predominant location of SO_3^{2-} oxidation (LIBERA et al. 1973; ZIEGLER and LIBERA 1975). There exist several systems that can mediate the oxidation. All systems investigated induce univalent reduction of oxygen and thus initiate univalent oxidation of SO_3^{2-} (ASADA and KISO 1973; LIBERA et al. 1973; ZIEGLER and LIBERA 1975; HÄLLGREN 1978).

14.7.1.2 Action on Enzymes and on Cellular Metabolism

Gaseous pollutants such as SO_2 and H_2S have to overcome several resistances before passing the plasmalemma and entering the cytoplasm. The principal resistances to gas flux are located in the laminar boundary layer, the stomata, the cuticle, the intercellular spaces, and the liquid phase in the cell wall. At low partial pressures SO_2 uptake increases linearly with increasing partial pressure, but this does not seem to be the case at higher partial pressures (GUDERIAN 1970; JENSEN and KOZLOWSKI 1975). The key issue becomes the question of how SO_2 and H_2S gain access to the cytoplasm. It seems that none of the gases or their ionic states are actively transported into plant cells from the intercellular space. It is generally accepted that the gases follow a concentration gradient by means of passive diffusion. The water on the surfaces of the intercellular space and in the cell walls is probably slightly acidic while the pH values in the cytoplasm and the organelles are considered to be around 7 to 8. Considering these conditions and the charged cell wall, only the uncharged species of SO_2 and H_2S ($SO_2 \cdot H_2O$ and H_2S_{aq}) will enter into the cell. Once inside the cell, ionization takes place leading to the formation of protons and the anions (depending on internal pH). As a consequence, the equilibrium within the cell will be disturbed resulting in a further flux of hydrated SO_2 and H_2S into the cell. Thus, it is possible that the ionized species accumulate to significant amounts within the cell compartments.

The interactions between SO_2, H_2S and the cell membrane are not well documented. Transient changes in proton flux across cell membranes are inhibited by glyoxal bisulphite (0.5 mM) or HSO_3^- (BRINKMANN et al. 1971; LÜTTGE et al. 1972). Cl^- transport is affected by the same concentration range and in the same manner. The authors concluded that the effects are nonspecific. The permselectivity of cell membranes is destroyed by SO_2 in lichens, resulting in an efflux of K^+ into the incubation medium (NIEBOER et al. 1976). This effect is most pronounced at low concentrations of SO_2 and with increasing incubation time. It can be assumed that K^+ and Cl^- fluxes are coupled processes.

In general, these data indicate changes of permeability of the limiting plasmalemma towards ions. This phenomenon can be explained by following types of interactions between pollutant and membrane: (a) cleavage of disulphide linkages; (b) free radical reactions; and (c) pH effect.

The cleavage of disulphide linkages by HSO_3^- in proteins, resulting in the production of a thiol and S-sulphonate, is of general interest:

$$RSSR + HSO_3^- \rightleftharpoons RSH + RSSO_3^- .$$

A reaction of this nature leads to destruction of the ternary and quaternary conformation of structural proteins and enzymes. CECIL and WAKE (1962) have demonstrated that disulphide bonds in proteins are easily broken by sulphite in cystine residues while other disulphide-containing residues are quite resistant. GRILL and ESTERBAUER (1973) reported an increased content of thiols in SO_2-injured pine needles. However, this increase has been shown to be due to an increase in the glutathione (reduced) pool; the cysteine level was only slightly elevated (GRILL et al. 1979). Thus, a direct cleaving effect in cell membranes has not yet been clearly demonstrated.

Another observation may be important in the discussion of possible interactions between SO_2 and the cell membrane, that is the reaction of sulphite with methionine and tryptophan (INOUE and HAYATSU 1971) and indole-3-acetic acid (YANG and SALEH 1973). These reactions proceed by a free radical mechanism, initiated by metals such as Mn^{2+} (YANG 1970). Other components, such as olefinic compounds or aldehydes and ketones, are also degraded, resulting in the production of sulphonates (LEHMANN and BENSON 1964; MUDD 1968). A product of the interaction between sulphite and aldehydes, α-hydroxy-sulphonate, has received some interest as this class of compound attacks membranes (MURRAY and BRADBEER 1971). As the free radical mechanism proceeds at sulphite concentrations far below 1 mM, it can be assumed that it plays a specific role during the degradation of membranes after SO_2 fumigation.

The buffer capacity of the cytoplasm is a central question as sulphurous and sulphuric acid have acidifying properties (PAHLICH 1975). THOMPSON and PACE (1962) have shown that the buffer capacity in animal cells increases with the serum content. Presumably also in leaf cells, the proteins and certain amino acids like cysteine are responsible for buffering (THOMAS et al. 1944; RAVEN and SMITH 1976). It is evident that more acid can be buffered at a low concentration fumigation for an extended time than at a high, sudden concentration. This conclusion is most important when comparing the results of two experiments with quite different concentrations used in fumigation. It is generally accepted that the buffer capacity of cells is usually high enough to handle long term fumigation at low SO_2 concentrations. THOMAS et al. (1944) have shown that a fumigation period as long as 5.2 h day^{-1} for 60 days with 0.19 ppm SO_2 scarcely changes the pH of the leaves. Some studies have indicated a correlation between the buffer capacity of lichens and the sensitivity of these plants to SO_2 fumigation (SKYE 1968). However, it has been shown that decreasing the pH value per se does not result in the same damaging effect as SO_2 does (HILL 1971).

A central question in studies dealing with the effect of sulphur compounds on enzymatic reactions is that of the actual concentrations within the cell and the cell compartments. The species distribution of hydrated SO_2 and H_2S and their ionized forms is determined by the pH (SEEL 1967; SPEDDING et al. 1980). Since the average pH of the cytoplasm is around 6.8, approximately 70% is found as HSO_3^- and 55% as H_2S_{aq}, whereas at the optimum pH for Calvin enzymes (pH 7.8 to 8.2), the ratios HSO_3^-/SO_3^{2-} and H_2S_{aq}/HS^{2-} are about 1/9.5 and 1/10, respectively. ZIEGLER (1975) estimated the sulphite concentration after long-term fumigation with 0.2 to 0.3 ppm SO_2 to be up to 1 mM. STEINHART et al. (1976; cited in HÄLLGREN 1978) pointed out that saturation concentration in the mesophyll of alfalfa can be reached within 1 h. The question of the local concentrations of any sulphur species within the plant demands resolution.

Movement of sulphur species into the cell has not been as well characterized as that into chloroplasts (HAMPP and ZIEGLER 1977; ZIEGLER and HAMPP 1977; SPEDDING et al. 1980). The inner envelope of spinach chloroplasts is permeable to sulphate, sulphite and sulphide. Both sulphate and sulphite, but not sulphide, are exchanged for inorganic phosphate, phosphoglyceric acid and dihydroxyacetone phosphate with rates up to 14 nmol mg^{-1} chlorophyll min^{-1}, indicating that the phosphate translocator (HELDT and RAPLEY 1970) is, at least in part, involved in transport across the envelope. Permeability of the S(IV) and the S(II) state is increased by decreasing the pH of the suspension medium from 8 to 5. This observation has been related to diffusion of hydrated SO_2 and H_2S. With all sulphur compounds investigated an increase in uptake with increase in radiation intensity is found. These findings clearly indicate that one has to differentiate between a specific deleterious effect of a specific sulphur compound and a higher permeability of these species.

Once inside plant cells and their organelles, the different species of SO_2 and H_2S show two different effects: a nutritive one and a damaging one. Low dose fumigation over a prolonged period of time may cause an increase in yield (THOMAS et al. 1943; FALLER et al. 1970) and an enhanced increase in yield at sulphur deficiency (COWLING et al. 1973; COWLING and LOCKYER 1978). However, improvement of sulphur supply is not the sole reason for growth promotion. In isolated chloroplasts, reduction of ferricyanide, ferredoxin and CO_2 are enhanced by sulphite concentrations up to 1 mM upon illumination (LIBERA et al. 1973). In vivo, the upper limit of the positive effects may lie around 0.3 ppm SO_2 (BRANDT and HECK 1968; GUDERIAN and VAN HAUT 1970). For further information on the metabolic fate of sulphur the reader is referred to ZIEGLER (1973a) and HÄLLGREN (1978).

The damaging effects of sulphur species result in structural destruction of organelle membranes and in inhibition and decomposition of enzymes. Submicroscopic effects on chloroplast membranes are observed before visible injury of the leaf occurs. In *Vicia faba* chloroplasts (after 3 h of fumigation with 2 ppm SO_2), the stroma becomes granulated and the thylakoids begin to swell (FISCHER et al. 1973). The final phase of damage results in a destruction of the membrane system and a shrinkage of the chloroplasts. At this final stage visible injury of the leaf occurs. Photosynthesis rates are reduced by more than 90%. Swelling of the thylakoids is reversible after short-term and low-concentration fumigation (WELLBURN et al. 1972). Changes in chloroplast organization are also reported for pine needle chloroplasts (MALHOTRA 1976). The microscopic picture of SO_2 damage is expressed in plasmolysis symptoms of mesophyll cells, which lead to destruction of the whole leaf (BRANDT and HECK 1968).

Again, the interactions between sulphur species and organelle membranes may be described by the three types of reactions mentioned above. Sulphonates have been introduced as model substances to study the mechanism of destruction of chloroplast membranes (MURRAY and BRADBEER 1971). Photosynthesis is generally depressed by sulphonates in C_4 plants (ZELITCH 1973), but is unaffected or slightly stimulated in C_3 plants (ZELITCH 1969). ZIEGLER (1975), however, has questioned the nature of sulphonates as model substances for SO_2 action. Furthermore, sulphonate formation seems to be unfavoured at the sulphite concentrations present in cells of fumigated plants.

It is generally accepted that the pH values of the cytoplasm and the interior of the organelles lie around 7 ± 1. SO_2 and H_2S are active mainly in their HSO_3^- or SO_3^{2-} and H_2S_{aq} or HS^- form, respectively. Thus, enzyme studies with dissolved sulphite and sulphide are justified. In this way, pH changes can be eliminated. Studies of air pollution effects on plants have mainly centered on photosynthetic reactions as these reactions are intimately linked to plant growth and productivity. There are at least three central points of attack leading to changes in CO_2 metabolism: chlorophyll, photosynthetic electron transport and photophosphorylation, and reductive pentose phosphate cycle.

The decomposition of chlorophyll to phaeophytin has been observed in numerous plants fumigated with SO_2 (INGLIS and HILL 1974; HÄLLGREN and HUSS 1975; LE BLANC and RAO 1975). However, phaeophytinization is only rapid in experiments using high SO_2 concentrations and at low pH values. Thus, the destruction of chlorophyll proceeds too slowly in vivo to explain the sudden decline of photosynthesis. Besides the concept that acidification causes chlorophyll decomposition there exists substantial evidence that chlorophyllase activity is significantly increased after a low dose fumigation (MALHOTRA 1977). Details of this have been discussed at length by HÄLLGREN (1978).

Information on the interactions between the various sulphur anions and primary reactions of photosynthesis is still contradictory (HÄLLGREN 1978). It has been demonstrated that sulphur species affect oxygen evolution (BALDRY et al. 1968; SILVIUS et al. 1975), photophosphorylation (DATTA et al. 1974; LIBERA et al. 1973; PICK and AVRON 1973) and the reducing site of photosystem I (SHIH and PETERING 1973; LIBERA et al. 1973). In most cases the mechanism of interaction has not been resolved. In one case, however, the interaction is understood quite well. The reaction of SO_3^{2-} with $NADP^+$ resulting in hydroxypyridine-4-sulphonic acid which forms an undissociable complex with dehydrogenases. This complex loses its hydrogen transfer capability (PFLEIDERER and HOHNHOLZ 1959; SHIH and PETERING 1973).

CO_2 fixation by illuminated isolated chloroplasts declines very rapidly at sulphite concentrations above 1 mM. The inhibition is competitive with respect to CO_2 (LIBERA et al. 1975). The same type of inhibition is found in those key enzymes which catalyze CO_2 incorporation. Ribulosebisphosphate (RuBP) carboxylase, phosphoenolpyruvate (PEP) carboxylase and malic enzyme are inhibited competitively by sulphite with respect to HCO_3^- and non-competitively with respect to their substrates and cosubstrates (ZIEGLER 1972, 1973b, 1974, 1977). Therefore the Ki value of sulphite increases with decreasing Km values. PEP carboxylase is far less sensitive to sulphite than RuBP carboxylase. It has been suggested that the competitive type of interaction between SO_2, CO_2 and CO_2-fixing enzyme is a characteristic and specific phenomenon (ZIEGLER 1975). These results also suggest that there should be a remarkable difference in tolerance for SO_2 fumigation between C_3 and C_4 plants. That has been shown recently (CARLSON and BAZZAZ 1982).

Fumigation or incubation with sulphur compounds not only results in inhibition processes but also affects regulatory properties of Calvin cycle enzymes. The biphasic kinetics of chloroplast glyceraldehyde-3-phosphate dehydrogenase versus PGA are altered to normal substrate kinetics in the presence of sulphite (ZIEGLER et al. 1976). Both $NADP^+$-dependent malic dehydrogenase and glucose-6-phosphate dehydrogenase lose their regulatory properties (light-dark regulation) when the light effect mediator has been changed by SO_3^{2-} treatment (ANDERSON and AVRON 1976). In this context it can be speculated that other chloroplast enzymes, regulated by vicinal-thiols, are affected.

A number of investigations have dealt with the action of hydroxysulphonates on chloroplast enzymes. RuBP carboxylase is not effected by 1 mM sulphonate (MURRAY and BRADBEER 1971). There are marked differences in inhibition pattern of sulphonate (OSMOND and AVADHANI 1970) and sulphite (ZIEGLER 1973b) with respect to PEP carboxylase.

The effects of sulphur compounds on light and dark reactions of photosynthesis induce changes in intermediates of plant metabolism (Libera et al. 1975). These changes probably reflect impaired photosynthesis and cannot be explained by specific interactions with sulphur species. There is some evidence for direct interaction with other enzymes. Glutamate dehydrogenase is stimulated by SO_2 towards reductive amination while oxidative deamination is inhibited (Pahlich et al. 1972). SO_2 may be a potent inhibitor of glycolate oxidase (Zelitch 1957) and an activator for oxidase synthesis (Soldatini and Ziegler 1979). Observations of SO_2 effects on dark respiration are still scarce (Keller 1957; Syratt and Wanstall 1969).

Recently it has been demonstrated that leaf tissues of cucumber injured with SO_2 gas or bisulphite ion in solution emit ethylene and ethane (Bressan et al. 1979). These findings indicate that SO_2 injuries (and ozone injuries; Tingey et al. 1976) also affect the hormone metabolism and the membrane structure of living plant cells.

14.7.2 Ozone and Peroxyacetyl Nitrate

14.7.2.1 Chemical Reactions

Ozone and PAN are strong oxidants with standard redox potentials of +2.07 and +0.4 volts, respectively (Dugger and Ting 1970).

The ozone molecule has several intrinsic resonance structures, leading to a slightly ionic form. Its solubility lies around 0.25 ml O_3 dissolved per ml H_2O at 25 °C (Hoather 1948). It is rapidly decomposed in air and in water, releasing molecular oxygen and water products (Kilpatrick et al. 1956). Among the products H_2O_2, hydroxyl radicals and hydroxyl ions are formed resulting in an alkalinization of the solution (Heath 1979). Ozone is relatively more stable under acidic conditions than under alkaline ones (Kilpatrick et al. 1956).

PAN is formed in the atmosphere (Leighton 1961). Its breakdown in aqueous solution is rapid. Decomposition yields acetate, nitrite, singlet oxygen and water (Stephens 1969). The half-life is a function of pH and lies in the order of 6 to 10 min within a pH range from 8.0 to 6.8 (Mudd 1975b).

14.7.2.2 Action on Cell Structures

A central point of attack for metabolic and morphologic changes is the oxidizing effect of both compounds on unsaturated fatty acids (Criegee 1975) and on sulphydryl (Heath et al. 1974) and ring-containing components (Mudd et al. 1969). All three classes of compounds are constituents of the plasmalemma, the outer barrier of protoplasts first in contact with the pollutants. Many of the observed primary damages induced by ozone or PAN treatment have been attributed to these chemical interactions. Ozone causes leakage of trapped glucose from liposomes made from egg yolk phosphatidylcholine (Teige et al. 1974). Results indicate that attack of ozone has produced molecules in which the unsaturated fatty acid in position 2 is shortened at the double bond. Sulfhydryls are oxidized to disulphides or to sulphoxides at very high ozone concentrations (HEATH et al. 1974). Reduced pyridine nucleotides are oxidized by ozone and PAN (MUDD et al. 1974; DUGGER and TING 1968). In addition, PAN oxidizes indole-3-acetic acid resulting in the formation of 3-hydroxymethyloxindole (HALL et al. 1971).

There have been several attempts to demonstrate the primary actions of ozone and PAN on plasmalemma properties. Experiments have been mainly centered on membrane permeability and mechanical integrity. Permeability for K^+ is drastically increased after introduction of ozone into a culture of *Chlorella* cells (CHIMIKLIS and HEATH 1975; HEATH and FREDERICK 1980). Solute permeability of bean leaf cells is also enhanced by ozone (EVANS and TING 1973; PERCHOROWICZ and TING 1974). The effects are rapidly reversible when ozone is removed from the solution. Prolonged incubation leads to irreversible damage. Thus, it can be assumed that initial interactions are not general but take place with specific areas of the membrane. Evidence for the repair of ozone-injured membranes is accumulating (SUTTON and TING 1977). The interaction of ozone with the plasmalemma also weakens its mechanical properties. This results in the rupturing of protoplasts (CAILLOUX et al. 1978).

It is evident that water potential is also affected. A net water loss occurs immediately upon exposure to ozone (EVANS and TING 1973). Water potential is substantially decreased by ozone in certain cultivars of *Petunia hybrida*. SO_2 has no such effect (ELKIEY and ORMROD 1979).

Similar visible results of the effect of oxidants on membrane permeability are reported for mitochondria and chloroplasts. Isolated tobacco mitochondria swell irreversibly after short-term exposure to ozone (LEE 1968). At the same time, protein and nucleotides are released and oxidative phosphorylation ceases. Reflection coefficients for neutral compounds, such as glycerol and erythritol, decrease when isolated pea chloroplasts are exposed to ozone (NOBEL and WANG 1973). Interestingly, the chloroplast envelopes remain impermeable to sucrose. Addition of sulfhydryl reagents does not prevent or reverse the observed effects.

The first damage after ozone or PAN treatment observed by means of electron microscopy is a granulation of the stroma. In the second phase of damage there is a general disintegration of thylakoids and cell membranes (THOMSON et al. 1965, 1966). CHANG (1971) observed a disintegration of chloroplast polysomes to ribosomes. HEATH (1975) suggests a common mechanism for metabolic (and structural) changes in oxidant-induced and desiccation injury.

The inactivating effect of ozone and PAN on phosphoglucomutase and on the UDPG-utilizing, polysaccharide synthesizing system of the cell wall is probably due to its oxidizing effect on sulfhydryl groups (ORDIN and HALL 1967). Fumigating with PAN reduces incorporation of $^{14}CO_2$ into cellulose and glucan, when sections of *Avena* coleoptiles are used. Cell elongation ceases (ORDIN et al. 1967). Growth is further blocked by the interaction of PAN with IAA (HALL et al. 1971). The effect of ozone is much weaker (ORDIN and HALL 1967). Clearly, these findings suggest interactions between the oxidants and the cell wall-synthesizing machinery, however, a direct interaction with either the cell wall or its components has not been demonstrated.

Considering the extreme reactivity of ozone and PAN to the constituents of plant cells, resulting in the production of free radicals, it must be questioned if these compounds pass deeply into the cell. It is very likely, that the primary reactions, and thus toxic effects, are mainly exerted at the plasmalemma or near the cell surface.

14.7.2.3 Action on Enzymes and on Cellular Metabolism

Reactions of ozone and PAN with enzymes are mainly due to their oxidizing effects on sulfhydryl groups. Cellular metabolic changes are induced by alteration of enzyme activities and of membrane permeability. Actions on enzymes have been reviewed at length by ZIEGLER (1973a), HEATH (1975), MUDD (1975b) and HEATH (1980).

14.7.3 Oxides of Nitrogen

14.7.3.1 Chemical Reactions

NO is less water soluble than other air pollutants (7.34 ml per 100 ml H_2O; BOND et al. 1980). NO_2 decomposes rapidly in water and forms a stoichiometric mixture of nitrous and nitric acid.

14.7.3.2 Action on Enzymes and on Cellular Metabolism

As has been pointed out by HEATH (1980), reactions of $(NO)_x$ are poorly characterized but they may involve interactions with free radical reactions. NO_2 may react with C=C bonds of fatty acids and cholesterol in the presence of ozone (KAMEL et al. 1971). OLAJOS (1977) suggests that N-nitroso compounds may be formed when NO_2 is present during ozone attack on uncharged amines.

Nitrous acid (0.25 mM) reduces CO_2 fixation in *Chlorella*. Electron transport is hardly affected. Characteristic changes of Calvin cycle intermediates indicate that the bisphosphatases become inhibited (HILLER and BASSHAM 1965).

Highly toxic concentrations of nitrogen oxides cause degradation of chlorophyll and, in contrast to SO_2 effects, carotene (KÄNDLER and ULLRICH 1964).

Evidence is now accumulating that plants can use NO_2, within the toxic limit, as a source of nitrogen. In *Citrus* seedlings amino acid levels are increased by a factor of 2 after NO_2 fumigation (MATSUSHIMA 1972). When leaves of *Helianthus annuus* and *Spinacia oleracea* have been exposed to $^{15}NO_2$ (4 to 6 ppm for 20 min/day) about 85%–99% of ^{15}N is found in reduced, organic nitrogen compounds (KAJI et al. 1980). The other labelled compounds are nitrate and nitrite. The analysis of ^{15}N content in amino acids reveals that glutamine has the highest content followed by glutamate, aspartate and alanine. It is concluded that NO_2 is converted to nitrite and nitrate and further assimilated into amino acids via the glutamine synthetase/glutamate synthase system. WELLBURN et al. (1976) have assayed several enzymes involved in N-metabolism. On the whole, glutamate dehydrogenase (GDH) and glutamate pyruvate transaminase activity are stimulated upon NO_2 exposure, while glutamate oxaloacetate transaminase activity is variable and glutamine synthetase (GS) activity is unaffected. Upon NO exposure only peroxidase and nitrite reductase activities are significantly increased.

Three clones of *Lolium perenne* show slight elevation in the GDH/GS activity ratio after low level and long-term NO_2 fumigation (WELLBURN 1982). Nitrite reductase activities are significantly elevated. Under the same experimental conditions ATP formation is enhanced and the $NAD(P)H/NAD(P)^+$ ratio is increased.

Further experimentation on enzymatic processes, accumulation of the toxicant and on unusual physiological changes is needed.

14.7.4 Carbon Monoxide

It has been suggested that biological production of CO on a global level may be larger than that by combustion engines (MCCONNALL et al. 1971), even though the latter may cause increased local CO concentrations. The main sources seem to be rice paddy fields and swamps. The methane produced can be oxidized to CO by photolysis in the atmosphere.

CO is taken up by plants both in the light and in the dark (BIDWELL and BEBEE 1974). In darkness, it is oxidized to CO_2 and is released. In the light, CO is incorporated into serine, glycine and subsequently into sucrose. Photosynthetic reducing power is necessary for CO fixation. FISCHER and LÜTTGE (1979) have characterized the light-dependent CO production by several C_3 and C_4 plants.

Cytochrome oxidase is probably the only CO-binding compound in plant mitochondria (PLESNICAR et al. 1967). In young leaves respiration can be inhibited, but not in old ones.

N_2 fixation is extremely sensitive to CO. That holds true for non-symbiotic bacteria as well as for symbiotic root nodule bacteria (BURRIS 1966).

14.7.5 Fluoride

14.7.5.1 Chemical Reactions

HF is readily soluble in water, forming a weak acid (pK = 3.17) and its salts with alkaline earth metal ions are practically insoluble.

14.7.5.2 Action on Enzymes and on Cellular Metabolism

The binding of Mg^{2+} or divalent heavy metals causes the inhibitory action of F (THOMAS and ALTHER 1966), as has been found in vitro for many enzymes, e.g. enolase (WARBURG and CHRISTIAN 1942) or phosphoglucomutase (NAJJAR 1948). The stimulatory action, found for enolase after HF-fumigation of *Phaseolus* or *Sorghum* (MCCUNE et al. 1964) demonstrates that, compared with in vitro incubation, the exposure of a total plant entails a different situation that has not yet been fully analyzed. Moreover, the reaction chain, resulting in an increase in activity of catalase, cytochrome oxidase, or ascorbic acid oxidase (LEE et al. 1966) after fumigation is scarcely understood; the increase seems to be rather unspecific, since it also occurs as a consequence of wounding. This is especially true for peroxidase activity (LEE et al. 1966; KELLER and SCHWAGER 1971), which is known to be stimulated by a variety of toxic compounds or by various infections.

Remarkably, after a low concentration HF-fumigation, the sensitive *Gladiolus* variety Snow Princess shows a decrease in peroxidase, alkaline phosphatase and pyruvate kinase activity, whereas those enzymes of the resistant variety Flowersong are stimulated. With respect to some enzymes, an interconnection with characteristic metabolic changes can be established. The strong inhibition of enolase by fluoride affects only the decomposition of fructose-bisphosphate in seeds, where the glycolytic pathway operates exclusively. On the contrary, the decomposition of fructose-bisphosphate is not reduced in mature leaves, where it proceeds largely to glucose-6-phosphate which is further metabolized by the oxidative pentose phosphate cycle. Moreover, in extracts of the tips of germinating *Pisum sativum* plants, which had been supplied with HF or in plants which had been fumigated with HF, the activity of glucose-6-phosphate dehydrogenase is increased. As a final consequence, fluoride fumigation of *Chenopodium murale* or *Polygonum orientale* does not reduce the glucose metabolism but leads to substantially more $^{14}CO_2$ being released from C_1 of glucose than from C_6 (ROSS et al. 1962).

The inhibitory action of fluoride on phosphoglucomutase may be the cause of some characteristic metabolic changes. In *Glycine max* less saccharose but more reducing sugar was found (YANG and MILLER 1963) after exposure to HF. Moreover, IAA-stimulated *Avena* coleoptiles incorporated very little ^{14}C-glucose into cellulose and into components of the water and acid soluble cell wall fraction. This resulted in strongly inhibited growth (ORDIN and SKOE 1963). Finally, synthesis of starch in pine needles also decreases after HF-fumigation (ADAMS and EMERSON 1961).

The enhanced activity of PEP-carboxylase, found in extracts of fluoride-necrotic leaves of *Glycine max,* seemingly causes the increase in dark fixation of CO_2 (YANG and MILLER 1963). Furthermore, the authors concluded that the accumulation of organic acids, as well as various amino acids, was a result of this. In HF-treated *Sorghum* plants, which showed no visible injury or growth reduction, PEP-carboxylase activity was unaffected (MCCUNE et al. 1964a). This demonstrates that primary enzyme changes by pollutants may differ from subsequent and less specific changes.

With respect to photosynthesis, addition of KF in the mM range was found to decrease the Hill reaction in blue-green algae (VENNESLAND and TURKINGTON 1966) and in *Euglena* (SATOH et al. 1970). During HF fumigation, however, the loss of photosynthetic O_2 evolution only parallels the appearance of visible necrosis (THOMAS and HENDRICKS 1956). Only in cases of high pollutant concentration does a reduction of photosynthesis, of about 40%, occur before visible damage appears.

Whether respiration is enhanced or reduced seems to be mainly dependent on the concentration of fluoride (YU and MILLER 1967). The stimulatory effect, e.g. in healthy *Gladiolus* tissue, adjacent to fluoride-injured tissue (McNulty and Newman 1957), or in beans before visible damage occurs (Applegate and Adams 1960), may be due to higher permeability of glucose; on the other hand the stimulated respiration of *Chlorella pyrenoidosa* by fluoride (McNulty and Lords 1960) was traced back to an interference with regulation by means of the PO_4-ADP-ATP system.

The growth-retarding effect of fluoride on corn seedling roots, brought about by reduction of cell multiplication and cell elongation (Chang and Thompson 1966), seems to be associated with its action on protein synthesis. Total RNA and protein of ribosomes per unit fresh weight was decreased; moreover, due to a reduction of ribosomal RNA, especially in bound ribosomes but also in free ones, the RNA/protein ratio was decreased (Chang 1970). The ribosome distribution was shifted from polysomes to smaller ribosomal particles. These changes were attributed to fluoride-induced accumulation of ATP and activation of ribonuclease.

Fluoride-induced chlorosis seems to be due to an inhibition of chlorophyll synthesis, taking place very early during this process. After Mg^{2+} is incorporated into the ring structure no further inhibition occurs. Thus, the light-dependent transformation of protochlorophyll to chlorophyll was not affected (McNulty and Newman 1961).

14.7.5.3 Subcellular and Cellular Damage

Since carotene formation ceases along with synthesis of chlorophyll, it is suggested that, in leaf tissue treated with NaF, the chloroplast structure was decomposed. Indeed, the disintegration of chloroplast structure has been described in leaves of higher plants (Solberg and Adams 1956) and in various mosses (Comeau and Le Blanc 1972). Membrane disruption, in general, is one among the various electron microscopic changes observed in fluoride-treated leaf cells of soybean (Miller and Wei 1971).

14.7.5.4 Metabolism of Fluoride

Fluoride taken up by the plant finally collects in the cell vacuole, as Ca- or Al-salts, or in the cell wall (see Hill et al. 1959). The content in plants averages 7 to 16 ppm (based on dry weight). However, a number of naturally growing plants are known to contain fluoro-organic compounds up to 1,370 ppm, e.g. *Camellia* (Thomas and Alther 1966). The majority of these plants are toxic to animals. With regard to HF pollution, among fluoro-organic compounds, fluoroacetate and fluorocitrate are the most interesting ones. In plants, other than those known to contain fluoro-organic compounds (e.g. soybean and forage crops), both compounds were traced after HF fumigation (Cheng et al. 1968; Yu and Miller 1970; Lovelace et al. 1968). The metabolic pathways, leading from F^- to both fluoroacids are not yet understood; they seem to be important for detoxification during HF pollution, though the capacity seems to be very limited.

14.7.6 Combined Action of Pollutants

As mentioned above, there has been comparatively little research into combined effects of air pollutants on biochemical reactions in plants, despite the obvious practical significance.

CONSTANTINIDOU and KOZLOWSKI (1979) have shown that carbohydrate, protein, and lipid contents of both actively growing and quiescent seedlings of *Ulmus americana* are reduced most by SO_2+O_3, an intermediate amount by SO_2 and least by O_3 treatment.

Parallel studies on physiological and enzymatic effects of a mixture of NO_2 and SO_2 gases (high concentrations) have been reported (BULL and MANSFIELD 1974; HORSMAN and WELLBURN 1975). At a lower and more physiological range of concentrations (WELLBURN et al. 1976) glutamate dehydrogenase (GDH) activity is markedly increased. This stimulation is an effect of SO_2 in the presence of NO_2. The same is true for peroxidase and glutamate pyruvate transaminase (GPT). RuBP carboxylase activity is depressed by SO_2 but there is no synergistic effect. Glutamine synthetase (GS) activity is not affected by SO_2 and NO_2 singly or in combination. GDH activity is stimulated by an SO_2 effect, by an NH_3 effect (in the presence of SO_2) and by a synergistic interaction between the two gases. GPT activity is unaffected by SO_2 and NH_3 alone, but is significantly inhibited after exposure to SO_2+NH_3.

GDH/GS activity ratios are significantly raised by SO_2 or SO_2+NO_2 in an SO_2-sensitive and by SO_2+NO_2 in an SO_2-resistant cloned cultivar of *Lolium perenne* (WELLBURN 1982). All grasses under investigation respond to NO_2 by increased nitrite reductase activities. This increase is suppressed in clean air or by SO_2 and SO_2+NO_2 treatments.

Explanations for synergistic effects are still poor because mechanisms of interactions between plant constituents and single air pollutants are not well understood.

14.7.7 Heavy Metals

Most investigations have been concerned with the correlation of traffic density and lead accumulation in plant tissues (LEE 1972; BRIGGS 1972; HÖLL and HAMPP 1975). Recently, other heavy metals like cadmium, copper, mercury, and zinc are taken into consideration discussing damaging effects on plant metabolism.

Lead inhibits CO_2 fixation and ATP formation in isolated spinach chloroplasts at concentrations above 2 μM (HAMPP et al. 1973). Photosystem II (but not photosystem I) is inhibited in the presence of Pb^{2+} (MILES et al. 1972). Lead strongly reduces the oxidation of succinate in corn mitochondria. This effect is enhanced in phosphate deficient plants since lead phosphate is almost insoluble in water (KOEPPE and MILLER 1970). Further details have been reviewed (HÖLL and HAMPP 1975).

Information on and conclusive models of the mechanism of action of heavy metals on cellular metabolism are scarce. Special binding sites seem to exist within the cell causing changes in permeability properties of cell membranes. Some effects of heavy metals are described by KAMP-NIELSEN (1971), VALLEE and ULMER (1972) and FATHI and LORENZ (1980).

14.8 Conclusion

The emphasis of this brief review article was to survey the major sources of pollutants, pollutant uptake and desposition on plants and the physiological and biochemical responses to pollutant exposure. We have endeavoured to describe the experimental work which may provide some characteristic features of the pollutant's action on plants. For future experimentation much more emphasis should be drawn on the combined physical, physiological and biochemical interactions between pollutants and plants. For that purpose, it is absolutely necessary to define a testable working hypothesis, to use well defined plant material and adequate pollutant concentrations, and to investigate combined actions of air pollutants.

References

Adams DF, Emerson MT (1961) Variations in starch and polysaccharide content of *Pinus ponderosa* needles with fluoride fumigation. Plant Physiol 36:261–265

Adedipe NO, Khatamian H, Ormrod DP (1973) Stomatal regulation of ozone phytotoxicity in tomato. Z Pflanzenphysiol 68:323–328

Anderson LE, Avron M (1976) Light modulation of enzyme activity in chloroplasts. Generation of membrane-bound vicinal-dithiol groups by photosynthetic electron transport. Plant Physiol 57:209–213

Anderson PJ (1914) The effect of dust from cement mills on the setting of fruit. Plant World 17:57–68

Applegate HG, Adams DF (1960) Effect of atmospheric fluoride in bush beans. Bot Gaz 121:223–227

Asada, K, Kiso K (1973) Initiation of aerobic oxidation of sulfite by illuminated spinach chloroplasts. Eur J Biochem 33:253–257

Ashenden TW (1979) The effects of long-term exposures to SO_2 and NO_2 pollution on the growth of *Dactylis glomerata* and *Poa pratensis*. Environ Pollut 18:249–258

Ashenden TW, Mansfield TA (1977) Influence of wind speed on the sensitivity of ryegrass to SO_2. J Exp Bot 28:729–735

Ayazloo M, Bell JN (1981) Studies on the tolerance to sulphur dioxide of grass populations in polluted areas. I. Identification of tolerant populations. New Phytol 88:203–222

Baldry CW, Cockburn W, Walker DA (1968) Inhibition by sulfate of the oxygen evolution associated with photosynthetic carbon assimilation. Biochim Biophys Acta 153:476–483

Ballantyne D (1972) Fluoride inhibition of the Hill reaction in bean chloroplasts. Atmos Environ 6:267–273

Beckerson DW, Hofstra G (1979a) Stomatal responses of white bean to O_3 and SO_2 singly or in combination. Atmos Environ 13:533–535

Beckerson DW, Hofstra G (1979b) Response of leaf diffusive resistance of radish, cucumber and soybean to O_3 and SO_2 singly or in combinations. Atmos Environ 13:1263–1268

Beilke S, Gravenhorst G (1978) Heterogenous SO_2 oxidation in the droplet phase. Atmos Environ 12:231–239
Bell JN, Clough WS (1973) Depression in yield of ryegrass exposed to SO_2. Nature (London) 241:47–49
Belot Y, Gauthier D (1975) Transport of micronic particles from the atmosphere to foliar surfaces. In: Vries de DA, Afgan NH (eds) Heat and mass transfer in the biosphere. Scripta, Washington
Bennett JH, Hill AC (1975) Interactions of air pollutants with canopies of vegetation. In: Mudd JB, Kozlowski TT (eds) Responses of plants to air pollution. Academic Press, London New York
Bennett JH, Hill AC, Gates DM (1973) A model for gaseous pollutant sorption by leaves. J Air Pollut Control Assoc 23:957–962
Bennett JP, Barnes K, Shinn JH (1979) Interactive effects of H_2S and O_3 on yield of snap beans. Dept Veg Crops, Univ California, Davis
Bidwell RGS, Bebee GP (1974) Carbon monoxide fixation by plants. Can J Bot 52:1841–1847
Black CR, Black VJ (1979a) The effect of low concentrations of sulphur dioxide on stomatal conductance and epidermal cell survival in field bean (*Vicia faba* L.). J Exp Bot 30:291–298
Black CR, Black VJ (1979b) Light and scanning electron microscopy of SO_2-induced injury to leaf surfaces of field bean (*Vicia faba* L.). Plant Cell Environment 2:329–333
Black VJ (1982) Effects of sulphur dioxide on physiological processes in plants. In: Unsworth MH, Ormrod DP (eds) Effects of gaseous air pollution in agriculture and horticulture. Butterworth, London
Black VJ, Unsworth MH (1979) Effects of low concentrations of sulphur dioxide on net photosynthesis and dark respiration of *Vicia faba*. J Exp Bot 30:473–483
Black VJ, Unsworth MH (1980a) Resistance analysis of sulphur dioxide fluxes to *Vicia faba*. Nature (London) 282:68–69
Black VJ, Unsworth MH (1980b) Stomatal responses to sulphur dioxide and vapour pressure deficit. J Exp Bot 31:667–677
Blanc le F, Rao DN (1975) Effects of air pollutants on lichens and bryophytes. In: Mudd JB, Kozlowski TT (eds) Responses of plants to air pollution. Academic Press, London New York
Blum U, Heck WW (1980) Effects of acute ozone exposures on snap bean (*Phaseolus vulgaris* cv. BBL-290) at various stages of its life cycle. Environ Exp Bot 20:73–86
Bohne H (1963) Schädlichkeit von Staub aus Zementwerken für Waldbestände. Allg Forstztg 18:107–111
Bond RG, Straub CP, Prober R (1980) CRC Handbook of environmental control Vol I. Air pollution. CRC Press, Boca Raton
Bonte J (1982) Effects of air pollutants on flowering and fruiting. In: Unsworth MH, Ormrod DP (eds) Effects of gaseous air pollution in agriculture and horticulture. Butterworth, London
Bonte J, Cormis de L, Louguet P (1975) Influence d'une pollution par le dioxyde de soufre sur le degré d'ouverture des stomates du *Pelargonium hortorum*. C R Acad Sci Ser D 280:2377–2380
Brandt CS, Heck WW (1968) Effects of air pollution on vegetation. In: Stern AC (ed) Air pollution Vol I. Academic Press, London New York
Bressan RA, LeCureux L, Wilson LG, Filner P (1979) Emission of ethylene and ethane by leaf tissue exposed to injurious concentrations of SO_2 or bisulfite ion. Plant Physiol 63:924–930
Briggs D (1972) Population differentiation in *Marchantia polymorpha* in various lead pollution levels. Nature (London) 238:166–167
Brinkmann E, Lüttge U, Fischer K (1971) Die Wirkung von SO_2 und Bisulfitverbindungen auf Photosynthese, Ionentransport und Spaltöffnungsregulation bei Blättern von höheren Pflanzen. Ber Dtsch Bot Ges 84:523–524
Bull JN, Mansfield TA (1974) Photosynthesis in leaves exposed to SO_2 and NO_2. Nature (London) 250:443–444

Burris RH (1966) Biological nitrogen fixation. Annu Rev Plant Physiol 17:155–179
Cailloux M, Phan CT, Chung YS (1978) Damage by ozone to the mechanical integrity of the protoplast plasmalemma. Experientia 34:730–731
Capron TM, Mansfield TA (1976) Inhibition of net photosynthesis in tomato in air polluted with NO and NO_2. J Exp Bot 27:1181–1186
Carlson RW, Bazzaz FA (1982) Photosynthetic and growth response to fumigation with SO_2 at elevated CO_2 for C_3 and C_4 plants. Oecologia 54:50–54
Cecil R, Wake RG (1962) The reactions of inter- and intra-chain disulphide bonds in proteins with sulphite. Biochem J 82:401–406
Chamberlain AC (1975) The movement of particles in plant communities. In: Monteith JL (ed) Vegetation and the atmosphere, Vol I. Academic Press, London New York
Chamberlain AC, Little P (1981) Transport and capture of particles by vegetation. In: Grace J, Ford ED, Jarvis PG (eds) Plants and their atmospheric environment. Blackwell, Oxford
Chang CW (1970) Effect of fluoride on ribosomes and ribonuclease from corn roots. Can J Biochem 48:450–454
Chang CW (1971) Effect of ozone on ribosomes in pinto bean leaves. Phytochemistry 10:2863–2868
Chang CW (1975) Fluorides. In: Mudd JB, Kozlowski TT (eds) Responses of plants to air pollution. Academic Press, London New York
Chang CW, Thompson CR (1966) Effect of fluoride on nucleic acids and growth in germinating corn seedling roots. Physiol Plant 19:911–918
Cheng JYO, Yu MH, Miller GW, Welkie GW (1968) Fluoroorganic acids in soybean leaves exposed to fluoride. Environ Sci Technol 2:367–370
Chimiklis PE, Heath RL (1975) Ozone-induced loss of intracellular potassium ion from *Chlorella sorokiniana*. Plant Physiol 56:723–727
Comeau G, Blanc le F (1972) Influence du fluor sur le *Funaria hygrometrica* et l'*Hypogymnia physodes*. Can J Bot 50:847–856
Constantinidou HA, Kozlowski TT (1979) Effects of sulfur dioxide and ozone on *Ulmus americana* seedlings. II. Carbohydrate, proteins and lipids. Can J Bot 57:176–184
Cormis de L (1968) Degagement d'hydrogene sulfate par des plantes soumises a une atmosphere contenant de l'anhydride sulfureux. C R Acad Sci Ser D 273:683–685
Cowling EB (1981) An historical résumé of progress in scientific and public understanding of acid precipitation and its biological consequences. Res Report, SNSF Project, As, Norway
Cowling EB, Linthurst RA (1981) The acid precipitation phenomenon and its ecological consequences. Bioscience 31:649–654
Cowling DW, Koziol MJ (1982) Mineral nutrition and plant response to pollutants. In: Unsworth MH, Ormrod DP (eds) Effects of gaseous air pollution in agriculture and horticulture. Butterworth, London
Cowling DW, Lockyer DR (1976) Growth of perennial ryegrass exposed to a low concentration of sulphur dioxide. J Exp Bot 27:411–417
Cowling DW, Lockyer DR (1978) The effect of SO_2 on *Lolium perenne* L. grown at different levels of sulphur and nitrogen nutrition. J Exp Bot 29:257–265
Cowling DW, Jones LHP, Lockyer DR (1973) Increased yield through correction of sulfur deficiency in ryegrass exposed to sulphur dioxide. Nature (London) 243:479
Cox RA, Penkett SA (1972) Aerosol formation from sulphur dioxide in the presence of ozone and olefinic hydrocarbons. J Chem Soc Faraday Trans I 68:1735–1753
Cox RA, Eggleton AEJ, Derwent RG, Lovelock JE, Pack DH (1975) Long-range transport of photochemical ozone in north-western Europe. Nature (London) 255:118–121
Coyne PI, Bingham GE (1978) Photosynthesis and stomatal light responses in snap beans exposed to hydrogen sulfide and ozone. J Air Pollut Control Assoc 28:1119–1123
Criegee R (1975) Mechanism of ozonolysis. Angew Chem Int Ed 14:745–760
Czaja AT (1966) Über die Einwirkung von Stäuben, speziell von Zementofenstaub auf Pflanzen. Angew Bot 40:106–120
Darley EF (1966) Studies on the effect of cement kiln dust on vegetation. J Air Pollut Control Assoc 16:145–150

Datta DB, Ryrie IJ, Jagendorf AT (1974) Light-dependent modification of spinach chloroplast coupling factor 1 by permanganate ion. J Biol Chem 249:4404–4411

Davies T (1980) Increased sensitivity of grasses to SO_2 pollution when grown under conditions of low irradiance and short day length. Nature (London) 284:483–485

Davison A (1982) The effects of fluorides on plant growth and forage quality. In: Unsworth MH, Ormrod DP (eds) Effects of gaseous air pollution in agriculture and horticulture. Butterworth, London

Derwent RG, Stewart HNM (1973) Air pollution from oxides of nitrogen in the UK. Atmos Environ 7:385–401

Dugger WM, Ting IP (1968) The effect of PAN on plants: Photoreductive reactions and susceptibility of bean plants to PAN. Phytopathology 58:1102–1107

Dugger WM, Ting IP (1970) Air pollution oxidants – their effects on metabolic processes in plants. Annu Rev Plant Physiol 21:215–234

Eggleton AEJ, Cox RA (1978) Homogeneous oxidation of sulphur compounds in the atmosphere. Atmos Environ 12:227–230

Eigen M, Kustin K, Maass G (1961) Die Geschwindigkeit der Hydration von SO_2 in wässriger Lösung. Z Phys Chem 30:130–133

Elkiey T, Ormrod DP (1979) Ozone and sulphur dioxide effects on leaf water potential. Z Pflanzenphysiol 91:177–181

Environmental Protection Agency (1976) Diagnosing vegetation injury caused by air pollution. EPA 68-02-1344, EPA Res Triangle Park, North Carolina

Evans LS, Ting IP (1973) Ozone-induced membrane permeability changes. Am J Bot 60:155–162

Evans LS, Gmur NF, Costa da F (1977) Leaf surface and histological perturbations of leaves of *Phaseolus vulgaris* and *Helianthus annuus* after exposure to simulated acid rain. Am J Bot 64:903–913

Evans LS (1980) Foliar responses that may determine plant injury by simulated acid rain. In: Toribara TY, Miller MW (eds) Polluted rain. Plenum Publ Corp, New York

Fairfax JAW, Lepp NW (1975) Effects of simulated 'acid rain' on cation loss from leaves. Nature (London) 255:324–325

Faller N (1971) Plant nutrient sulphur – SO_2 vs SO_4. Sulphur Inst J 7:5–6

Faller N (1972) In: Isotopes and radiation in soil-plant relationships including forestry. IAEA, Vienna

Faller N, Herwig K, Kühn H (1970) Die Aufnahme von Schwefeldioxyd ($S^{35}O_2$) aus der Luft. I. Einfluß auf den pflanzlichen Ertrag. Plant Soil 33:177–191

Farrar JF, Relton J, Rutter AJ (1977) Sulphur dioxide and the growth of *Pinus sylvestris*. J Appl Ecol 14:861–875

Fathi M, Lorenz H (1980) Bindungsformen von Quecksilber, Cadmium und Blei in Biotopen, Verhalten in der Nahrungskette und Vorkommen in Nahrungsmitteln. Metabolismus in Pflanze, Tier und Mensch. ZEBS-Berichte. Reimer, Berlin

Feder WA (1968) Reduction in tobacco pollen germination and tube elongation induced by low levels of ozone. Science 160:1122

Feder WA, Campbell FJ (1968) Influence of low levels of ozone on flowering of carnations. Phytopathology 58:1038–1039

Ferenbaugh RW (1976) Effects of simulated acid rain on *Phaseolus vulgaris* L. Am J Bot 63:283–288

Fischer K, Lüttge U (1979) Lichtabhängige CO-Bildung grüner Pflanzen und ihre Bedeutung für den CO-Haushalt der Atmosphäre. Flora 168:121–137

Fischer K, Kramer D, Ziegler H (1973) Elektronenmikroskopische Untersuchungen SO_2-begaster Blätter von *Vicia faba*. Protoplasma 76:83–96

Flückiger W, Oertli JJ, Flückinger H (1979) Relationship between stomatal diffusive resistance and various applied particle sizes on leaf surfaces. Z Pflanzenphysiol 91:173–179

Fölster H, Hase H, Ulrich B (1979) Freisetzung von Aluminium in mitteldeutschen sauren Braunerden aus Löss-Sandstein-Fließerden. Z Pflanzenernährung und Bodenkde 142:185–194

Fowler D, Cape JN (1982) Air pollutants in agriculture and horticulture. In: Unsworth

MH, Ormrod DP (eds) Effects of gaseous air pollution in agriculture and horticulture. Butterworth, London
Fowler D, Unsworth MH (1973) Dry deposition of sulphur dioxide on winter wheat. Nature (London) 249:389–390
Fowler D, Unsworth MH (1979) Turbulent transfer of sulphur dioxide to a wheat crop. Q J R Meteorol Soc 105:767–783
Fowler D, Cape JN, Leith ID, Paterson IS, Kinnaird JW, Nicholson IA (1982) Rainfall acidity in northern Britain. Nature 297:383–386
Fuzzi S (1978) Study of iron (III) catalyzed sulphur dioxide oxidation in aqueous solution over a wide range of pH. Atmos Environ 12:1439–1442
Garland JA (1978) Dry and wet removal of sulphur from the atmosphere. Atmos Environ 12:349–362
Garsed SG, Rutter AJ (1982) The relative sensitivities of conifer populations to SO_2 in screening tests with different concentrations of sulphur dioxide. In: Unsworth MH, Ormrod DP (eds) Effects of gaseous air pollution in agriculture and horticulture. Butterworth, London
Garsed SG, Farrar JF, Rutter AJ (1979) The effects of low concentrations of sulphur dioxide on the growth of four broadleaved tree species. J Appl Ecol 16:217–226
Gmelin L (1963) Handbuch Schwefel, 8th edn, part B-3. Dtsch Chem Ges, Frankfurt
Granat L, Rodhe H, Hallberg RO (1976) The global sulphur cycle. Ecol Bull 22:89–123 (SCOPE Rep No 7. R Swed Acad Sci, Orsundsbre, Sweden)
Grill D, Esterbauer H (1973) Quantitative Bestimmung wasserlöslicher Sulfhydrilverbindungen in gesunden und SO_2-geschädigten Nadeln von *Picea abies*. Phyton 15:87–101
Grill D, Esterbauer H, Klösch U (1979) Effect of sulfur dioxide on glutathione in leaves of plants. Environ Pollut 19:187–194
Guderian R (1970) Untersuchungen über quantitative Beziehungen zwischen dem Schwefelgehalt von Pflanzen und dem Schwefeldioxidgehalt der Luft. Z Pflanzenkr Pflanzenschutz 77:289–308
Guderian R, Haut van H (1970) Nachweis von Schwefeldioxid-Wirkungen an Pflanzen. Staub 30:22–35
Haagen-Smit AJ, Wayne LG (1976) Atmospheric reactions and scavenging processes. In: Stern AJ (ed) Air pollution Vol I. 3rd edn. Academic Press, London New York
Haines TA (1981) Acid precipitation and its consequences for aquatic ecosystems: A review. Transact Amer Fisheries Soc 110:669–707
Hällgren J-E (1978) Physiological and biochemical effects of sulphur dioxide on plants. In: Nriagu JO (ed) Sulphur in the environment. part 2. Environmental impacts. Wiley and Sons, New York
Hällgren J-E, Huss K (1975) Effects of SO_2 on photosynthesis and nitrogen fixation. Physiol Plant 34:171–176
Hällgren J-E, Linder S, Richter A, Troeng E, Granat L (1982) Uptake of SO_2 in shoots of Scots Pine: field measurements of net flux of sulphur in relations to stomatal conductance. Plant Cell Environ 5:75–83
Hall MA, Brown RL, Ordin L (1971) Inhibitory products of the action of PAN upon indole-3-acetic acid. Phytochemistry 10:1233–1238
Hampp R, Ziegler I (1977) Sulfate and sulfite translocation via the phosphate translocator of the inner envelope membrane of chloroplasts. Planta 137:309–312
Hampp R, Ziegler H, Ziegler I (1973) The effect of lead ions on the CO_2 fixation and on the ATP synthesis by spinach chloroplasts. Biochem Physiol Pflanz 164:126–134
Haselhoff E, Lindau G (1903) Die Beschädigung der Vegetation durch Rauch. Bornträger, Berlin
Haut van H, Stratmann H (1967) Schriftenreihe der Landesanstalt für Immissions- und Bodennutzungsschutz des Landes Nordrhein-Westfalen, No 7, Essen
Haut van H, Stratmann H (1970) Color-plate atlas of the effects of sulfur dioxide on plants. Girardet, Essen
Heagle AS (1982) Interactions between air pollutants and parasitic plant diseases. In: Unsworth MH, Ormrod DP (eds) Effects of gaseous air pollutants in agriculture and horticulture. Butterworth, London

Heath RL (1975) Ozone. In: Mudd JB, Kozlowski TT (eds) Responses of plants to air pollution. Academic Press, New York
Heath RL (1979) Breakdown of ozone and formation of H_2O_2 in aqueous solutions of amine buffers exposed to ozone. Toxicol Lett 4:449–453
Heath RL (1980) Initial events in injury to plants by air pollutants. Annu Rev Plant Physiol 31:395–431
Heath RL, Frederick PE (1980) Ozone alteration of membrane permeability in *Chlorella*. Plant Physiol 64:455–459
Heath RL, Chimiklis P, Frederick P (1974) Role of potassium and lipids in ozone injury to plant membranes. In: Dugger WM (ed) Air pollution related to plant growth. Am Chem Soc Symp Ser, Washington DC
Heck WW (1968) Factors influencing expression of oxidant damage to plants. Annu Rev Phytopathol 6:165–188
Heck WW, Brandt CS (1977) Effects on vegetation: native, crops, forest. In: Stern AC (ed) Air pollution Vol II. 3rd edn. Academic Press, London New York
Hegg DA, Hobbs PV (1978) Oxidation of sulfur dioxide in aqueous systems with particular reference to the atmosphere. Atmos Environ 12:241–253
Heggestadt HE, Heagle AH, Nennett JH, Koch EJ (1980) The effects of photochemical oxidants on the yield of snap beans. Atmos Environ 14:317–326
Heitschmidt RK, Lauenroth WK, Dodd JL (1978) Effects of controlled levels of sulphur dioxide on western wheatgrass in a south-eastern Montana grassland. J Appl Ecol 15:859–868
Heldt HW, Rapley L (1970) Specific transport of inorganic phosphate, 3-phosphoglycerate, and dihydroxyacetone phosphate and of dicarboxylates across the inner membrane of spinach chloroplasts. FEBS Lett 10:143–148
Hill AC, Bennett JH (1970) Inhibition of apparent photosynthesis by nitrogen oxides. Atmos Environ 4:341–348
Hill AC, Littlefield N (1969) Ozone: effect on apparent photosynthesis, rate of transpiration and stomatal closure in plants. Environ Sci Technol 3:52–56
Hill AC, Pack MR, Transtrum LG, Winters WS (1959) Effects of atmospheric fluorides and various types of injury on the respiration of leaf tissue. Plant Physiol 34:11–16
Hill DJ (1971) Experimental study of the effect of sulphite on lichens with reference to atmospheric pollution. New Phytol 70:831–836
Hiller RG, Bassham JA (1965) Inhibition of CO_2 fixation by nitrous acid. Biochim Biophys Acta 109:607–610
Hoather RC (1948) Some experimental observations on the solubility of ozone and the ozonization of a variety of water. J Inst Water Eng 2:358–368
Höll W, Hampp R (1975) Lead and plants. Residue Rev 54:79–111
Horsman DC, Wellburn AR (1975) Synergistic effect of SO_2 and NO_2 polluted air upon enzyme activity in pea seedlings. Environ Pollut 8:123–133
Horsman DC, Roberts TM, Lambert M, Bradshaw AD (1979) Studies on the effect of sulphur dioxide on perennial ryegrass (*Lolium perenne* L.). J Exp Bot 30:485–493
Inglis F, Hill DJ (1974) The effect of sulphite and fluoride on CO_2 uptake by mosses in the light. New Phytol 73:1207–1213
Inoue M, Hayatsu H (1971) The interactions between bisulfite and amino acids. The formation of methionine sulfoxide from methionine in the presence of oxygen. Chem Pharm Bull 19:1286–1289
Jacobson JS (1980) The influence of rainfall composition on the yield and quality of agricultural crops. In: Drablos D, Tollan A (eds) Ecological impact of acid precipitation. SNSF, Oslo-As, Norway
Jacobson JS (1982) Ozone and the growth and productivity of agricultural crops. In: Unsworth MH, Ormrod DP (eds) Effects of gaseous air pollution in agriculture and horticulture. Butterworth, London
Jacobson JS, Hill AC (1970) Recognition of air pollution injury to vegetation: a pictorial atlas. Air Pollut Control Assoc, Pittsburgh, PA
Jacobson JS, Weinstein LH, McCune DC, Hitchcock AE (1966) The accumulation of fluorine by plants. J Air Pollut Control Assoc 16:412–417

Jensen KF, Kozlowski TT (1975) Absorption and translocation of sulfur dioxide by seedlings of four forest tree species. J Environ Qual 4:379–382
Kändler U, Ullrich H (1964) Detection of NO_2 damage to leaves. Naturwissenschaften 51:518
Kaji M, Yoneyama T, Tosuka T, Iwaki H (1980) Absorption of atmospheric NO_2 by plants and soils. Res Rep Natl Inst Environ Stud 11:51–58
Kamel AM, Weiner ND, Felmeister A (1971) Identification of cholesteryl nitrate as a product of the reaction between NO_2 and cholesterol monomolecular films. Chem Phys Lipid 6:225–234
Kamp-Nielsen L (1971) The effect of deleterious concentrations of mercury on the photosynthesis and growth of *Chlorella pyrenoidosa*. Physiol Plant 24:556–561
Keller H (1957) Beiträge zur Erfassung der durch schweflige Säure hervorgerufenen Rauchschäden an Nadelhölzern. Dissertation, Univ München
Keller H, Müller J (1958) Beiträge zur Erfassung der durch schweflige Säure hervorgerufenen Rauchschäden an Nadelhölzern. Forstwiss Forsch 10:5–63
Keller T (1978) Einfluß niedriger SO_2-Konzentrationen auf die CO_2-Aufnahme von Fichte und Tanne. Photosynthetica 12:316–322
Keller T (1981) Folgen einer winterlichen SO_2-Belastung für die Fichte. Gartenbauwissenschaft 46:170–178
Keller T, Schwager H (1971) Der Nachweis unsichtbarer Fluor-Immissionsschädigungen an Waldbäumen durch eine einfache kolorimetrische Bestimmung der Peroxidase-Aktivität. J For Pathol 1:6–18
Kilpatrick ML, Herrick CC, Kilpatrick M (1956) The decomposition of ozone in aqueous solution. J Am Chem Soc 78:1784–1789
Knudson Butler L, Tibbitts TW (1979) Stomatal mechanisms determining resistance to ozone in *Phaseolus vulgaris* L. J Am Soc Hortic Sci 104:213–216
Koeppe DE, Miller RJ (1970) Lead effects on corn mitochondrial respiration. Science 167:1376–1378
Kondo N, Sugahara K (1978) Changes in transpiration rate of SO_2-resistant and -sensitive plants with SO_2 fumigation and the participation of abscisic acid. Plant Cell Physiol 19:365–373
Law RM, Mansfield TA (1982) Oxides of nitrogen and the greenhouse atmosphere. In: Unsworth MH, Ormrod DP (eds) Effects of gaseous air pollution in agriculture and horticulture. Butterworth, London
Lee CJ, Miller GW, Welkie GW (1966) The effects of hydrogen fluoride and wounding on respiratory enzymes in soybean leaves. Int J Air Water Pollut 10:169–181
Lee JA (1972) Lead pollution from a factory manufacturing antiknock compounds. Nature (London) 238:165–166
Lee TT (1968) Effect of ozone on swelling of tobacco mitochondria. Plant Physiol 43:133–139
Lehmann J, Benson AA (1964) The plant sulfolipid. IX Sulfosugar synthesis from methyl hexoseenides. J Am Chem Soc 86:4469–4472
Leighton PA (1961) Photochemistry of air pollution. Academic Press, London New York
Lerman SL, Darley EF (1975) Particulates. In: Mudd JB, Kozlowski TT (eds) Responses of plants to air pollution. Academic Press, London New York
Leuning R, Neumann HH, Thurtell GW (1979a) Ozone uptake by corn: a general approach. Agric Meteorol 20:115–136
Leuning R, Unsworth MH, Neumann HH, King KM (1979b) Ozone fluxes to tobacco and soil under field conditions. Atmos Environ 13:1155–1163
Libera W, Ziegler H, Ziegler I (1973) Förderung der Hill-Reaktion und der CO_2-Fixierung in isolierten Spinatchloroplasten durch niedere Sulfitkonzentrationen. Planta 109:269–279
Libera W, Ziegler I, Ziegler H (1975) The action of sulfite on the HCO_3-fixation and the fixation pattern of isolated chloroplasts and leaf tissue slices. Z Pflanzenphysiol 74:420–433
Likens GE, Wright RF, Galloway JN, Butler TJ (1979) Acid rain. Sci Am 241:43–51
Little P, Whiffen RD (1977) Emission and deposition of petrol engine exhaust lead: I. Deposition of exhaust Pb to plant and soil surfaces. Atmos Environ 11:437–447

Lockyer DR, Cowling DW, Jones LHP (1976) A system for exposing plants to atmospheres containing low concentrations of sulphur dioxide. J Exp Bot 27:397–409
Lovelace J, Miller GW, Welkie GW (1968) The accumulation of fluoroacetate and fluorocitrate in forage crops collected near a phosphate plant. Atmos Environ 2:187–190
Lüttge U, Osmond CB, Ball E, Brinkmann E, Kinze G (1972) Bisulfite compounds as metabolic inhibitors: Nonspecific effects on membranes. Plant Cell Physiol 13:505–514
Majernik O, Mansfield TA (1971) Effects of SO_2 pollution on stomatal movements in *Vicia faba*. Phytopathol Z 71:123–128
Malhotra SS (1976) Effects of sulphur dioxide on biochemical activity and ultrastructural organization of pine needle chloroplasts. New Phytol 76:239–245
Malhotra SS (1977) Effects of aqueous sulphur dioxide on chlorophyll destruction in *Pinus contorta*. New Phytol 78:101–109
Malhotra SS, Blauel TA (1980) Diagnosis of air pollutant and natural stress symptoms on forest vegetation in western Canada. Rep NOR-X-228. North For Res Cent, Can For Serv, Edmonton, Alberta
Mansfield TA (1976) In: Smith H (ed) Commentaries in plant science. Pergamon, Oxford
Martin A, Barber FR (1981) Sulphur dioxide, oxides of nitrogen and ozone measured continuously for 2 years at a rural site. Atmos Environ 15:567–578
Matsushima J (1972) Influence of SO_2 and NO_2 on assimilation of amino acids, organic acids, and saccharoid in *Citrus natsudaidai* seedlings. Bull Fac Agric Mie Univ 44:131–139
McConnall JC, McElroy MB, Wofsy SC (1971) Natural sources of atmospheric CO. Nature (London) 233:187–188
McCune DC, Weinstein LH, Jacobson JS, Hitchcock AE (1964) Some effects of atmospheric fluoride on plant metabolism. J Air Pollut Control Assoc 14:465–468
McCune DC, Maclean DC, Schneider RE (1976) Experimental approaches to the effects of airborne fluorides on plants. In: Mansfield TA (ed) Effects of air pollutants on plants. Soc Exp Biol Sem Ser 1. Cambridge Univ Press, Cambridge
McNulty JB, Lords JL (1960) Possible explanation of fluoride-induced respiration in *Chlorella pyrenoidosa*. Science 132:1553–1554
McNulty JB, Newman DW (1957) Effects of atmospheric fluoride on the respiration rate of bush bean and gladiolus leaves. Plant Physiol 32:121–124
McNulty JB, Newman DW (1961) Mechanism of fluoride induced chlorosis. Plant Physiol 36:385–388
Miles CD, Brandle JR, Daniel DJ, Chu-Der O, Schnare PD, Uhlik DJ (1972) Inhibition of photosystem II in isolated chloroplasts by lead. Plant Physiol 49:820–825
Miller GW, Wei LL (1971) The fine structure of leaves from *Glycine max* in relation to fluoride injury. 4th Annu Conf Int Soc Fluoride Res, New York
Mudd JB (1968) Determination of sulfite in biological fluids: The usefulness of radioactive N-ethyl maleimide. Anal Biochem 22:242–248
Mudd JB (1975a) Sulfur dioxide. In: Mudd JB, Kozlowski TT (eds) Responses of plants to air pollution. Academic Press, London New York
Mudd JB (1975b) Peroxyacyl nitrates. In: Mudd JB, Kozlowski TT (eds) Responses of plants to air pollution. Academic Press, London New York
Mudd JB (1982) Effects of oxidants on metabolic function. In: Unsworth MH, Ormrod DP (eds) Effects of gaseous air pollution in agriculture and horticulture. Butterworth, London
Mudd JB, Leavitt R, Ongun A, McManus TT (1969) Reaction of ozone with amino acids and proteins. Atmos Environ 3:669–682
Mudd JB, Leh F, McManus TT (1974) Reaction of ozone with nicotinamide and its derivatives. Arch Biochem Biophys 161:408–419
Murray DR, Bradbeer JW (1971) Inhibition of photosynthetic CO_2 fixation in spinach chloroplasts by α-hydroxy-2-pyridine-methanesulphonate. Phytochemistry 10:1999–2003
Najjar VA (1948) The isolation and properties of phosphoglucomutase. J Biol Chem 175:281–290

National Academy of Sciences (1971) Biologic effects of air pollutants: fluorides. NAS, Washington DC
National Academy of Sciences (1977) Biologic effects of air pollutants: Ozone and other photochemical oxidants. NAS, Washington DC
National Research Council of Canada (1939) Effects of sulphur dioxide on vegetation. Publ NRC815, Ottawa
Nieboer E, Richardson DHS, Puckett KJ, Tomassini FD (1976) The phytotoxicity of sulphur dioxide in relation to measurable responses in lichens. In: Mansfield TA (ed) Effects of air pollutants on plants. Cambridge Univ Press, Cambridge
Nobel PS, Wang CT (1973) Ozone increases the permeability of isolated pea chloroplasts. Arch Biochem Biophys 157:388–394
Olajos EJ (1977) Biological interactions of N-nitroso compounds: a review. Ecotoxicol Environ Saf 1:175–196
Ordin L, Hall MA (1967) Studies on cellulose synthesis by a cell-free oat coleoptile enzyme system: Inactivation by airborne oxidants. Plant Physiol 42:205–212
Ordin L, Skoe BP (1963) Inhibition of metabolism in *Avena* coleoptile tissue by fluoride. Plant Physiol 38:416–421
Ordin L, Hall MA, Katz M (1967) PAN-induced inhibition of cell wall metabolism. J Air Pollut Control Assoc 17:811–815
Ormrod DP (1982) Air pollutant interactions in mixtures. In: Unsworth MH, Ormrod DP (eds) Effects of gaseous air pollution in agriculture and horticulture. Butterworth, London
Ormrod DP, Black VJ, Unsworth MH (1981) Depression of net photosynthesis in *Vicia faba* L. exposed to sulphur dioxide and ozone. Nature (London) 291:585–586
Osmond CB, Avadhani PN (1970) Inhibition of the β-carboxylation pathway of CO_2 fixation by bisulfite compounds. Plant Physiol 45:228
Pahlich E (1975) Effect of SO_2 pollution on cellular regulation. A general concept of the mode of action of gaseous air contamination. Atmos Environ 9:261–263
Pahlich E, Jäger HJ, Steubing L (1972) Beeinflussung der Aktivitäten von Glutamatdehydrogenase und Glutaminsynthase aus Erbsenkeimlingen durch SO_2. Angew Bot 46:183–197
Perchorowicz JT, Ting IP (1974) Ozone effects on plant cell permeability. Am J Bot 61:787–793
Pfleiderer G, Hohnholz E (1959) Zum Wirkungsmechanismus von Dehydrogenasen. Biochem Z 331:245–253
Pick U, Avron M (1973) Inorganic sulfate and selenate as energy transfer inhibitors of photophosphorylation. Biochim Biophys Acta 325:297–303
Plesnicar M, Bonner WD, Storey BT (1967) Peroxidase associated with higher plant mitochondria. Plant Physiol 42:366–370
Raven JA, Smith FA (1976) Current advances in plant science. 8:649–656
Reinert RA, Heagle AS, Heck WW (1975) Plant responses to pollutant combinations. In: Mudd JB, Kozlowski TT (eds) Responses of plants to air pollution. Academic Press, London New York
Rich S, Turner NC (1972) Importance of moisture on stomatal behaviour of plants subjected to ozone. J Air Pollut Control Assoc 22:718–721
Ricks GR, Williams RJH (1974) Effects of atmospheric pollution on deciduous woodland. II. Effects of particulate matter upon stomatal diffusion resistance in leaves of *Quercus petraea*. Environ Pollut 6:87–109
Ross CW, Wiebe HH, Miller GW (1962) Effect of fluoride on glucose catabolism in plant leaves. Plant Physiol 37:305–309
Satoh K, Katoh S, Takamiya A (1970) Effects of chlorid ion on Hill reaction in *Euglena* chloroplasts. Plant Cell Physiol 11:453–466
Schofield CS (1976) Effects of acid precipitation on fish. Ambio 5:228–230
Schroeter LC (1966) Sulfur dioxide. Applications in foods, beverages, and pharmaceuticals. Pergamon, Oxford
Seel F (1967) Grundlagen der analytischen Chemie. Chemie, Weinheim
Shih NT, Petering DH (1973) Model reaction for the study of the interaction of sulfur dioxide with mammalian organisms. Biochem Biophys Res Commun 55:1319

Sij JW, Swanson CA (1974) Short term kinetics on the inhibition of photosynthesis by sulfur dioxide. J Environ Qual 3:103–107
Sillen LG (1964) Stability constants of metal-ion complexes. Sect-I-Inorg, 2nd edn, Spec Publ 17. Chem Soc, London
Silvius JE, Ingle M, Baer CH (1975) Sulfur dioxide inhibition of photosynthesis in isolated spinach chloroplasts. Plant Physiol 56:434–437
Skye E (1968) Lichens and air pollution. Acta Phytogeogr Suec 52:1–123
Solberg RA, Adams DF (1956) Histological responses of some plant leaves to hydrogen fluoride and sulfur dioxids. Am J Bot 43:755–760
Soldatini GF, Ziegler I (1979) Induction of glycolate oxidase by SO_2 in *Nicotiana tabacum*. Phytochemistry 18:21–22
Spaleny J (1977) Sulphate transformation to hydrogen sulphide in spruce seedlings. Plant Soil 48:557–563
Spedding DJ, Ziegler I, Hampp R, Ziegler H (1980) Effect of pH on the uptake of ^{35}S-sulfur from sulfate, sulfite, and sulfide by isolated spinach chloroplasts. Z Pflanzenphysiol 96:351–364
Spikes JD, Lumry R, Rieske JS (1955) Inhibition of the photochemical activity of isolated chloroplasts. Arch Biochem Biophys 55:25–37
Srivastava HS, Jolliffe PA, Runeckles VC (1975) Inhibition of gas exchange in bean leaves by NO_2. Can J Bot 53:466–474
Stephens ER (1969) Chemistry of atmospheric oxidants. J Air Pollut Control Assoc 19:181–185
Stern AC (1977) Air pollution, 3rd edn, Vol I. Air pollutants, their transformation and transport. Academic Press, London New York
Sutton R, Ting IP (1977) Evidence for the repair of ozone-induced membrane injury. Am J Bot 64:404–411
Syratt WJ, Wanstall PJ (1969) The effects of SO_2 on epiphytic bryophytes. In: Air pollution. Proc 1st Eur Congr Influence Air Pollut Anim Plants, Wageningen
Taniyama T (1972) Bull Fac Agric Mie Univ 44:111–130
Taylor JS, Reid DM, Pharis RP (1981) Mutual antagonism of sulphur dioxide and abscisic acid in their effect on stomatal aperture in broad bean (*Vicia faba* L) epidermal strips. Plant Physiol 68:1504–1507
Taylor OC, Thompson CR, Tingey DT, Reinert RA (1975) Oxides of Nitrogen. In: Mudd JB, Kozlowski TT (eds) Responses of plants to air pollution. Academic Press, London New York
Teige B, McManus TT, Mudd JB (1974) Reaction of ozone with phosphatidylcholine liposomes and the lytic effect of products on red blood cells. Chem Phys Lipids 12:153–171
Thomas MD (1951) Gas damage to plants. Annu Rev Plant Physiol 2:293–322
Thomas MD, Alther EW (1966) The effects of fluoride on plants. In: Heffter A (ed) Handbuch der experimentellen Parmakologie Vol XX/1. Springer, Berlin Heidelberg New York
Thomas MD, Hendricks RH (1956) Effect of air pollution on plants. In: Magill PL, Holden FR, Ackley C (eds) Air pollution handbook, Sect 9. McGraw-Hill, New York
Thomas MD, Hill GR (1937) Relation of sulphur dioxide in the atmosphere to photosynthesis and respiration of alfalfa. Plant Physiol 12:309–383
Thomas MD, Hendricks RH, Collier TR, Hill GR (1943) The utilization of sulphate and sulphur dioxide for the sulphur nutrition of alfalfa. Plant Physiol 18:345–371
Thomas MD, Hendricks RH, Hill GR (1944) Some chemical reactions of sulphur dioxide after absorption by alfalfa and sugar beets. Plant Physiol 19:212–226
Thompson CR, Kats G (1978) Effects of continuous H_2S fumigation on crop and forest plants. Environ Sci Technol 12:550–553
Thompson CR, Pace DM (1962) The effects of SO_2 upon established cell lines cultivated in vitro. Can J Biochem Physiol 40:207–217
Thomson WW, Dugger WM, Palmer RL (1965) Effects of PAN on ultrastructure of chloroplasts. Bot Gaz 126:66–72

Thomson WW, Dugger WM, Palmer RL (1966) Effect of ozone on the fine structure of the palisade parenchyma cells of bean leaves. Can J Bot 44:1677–1682
Tingey DT (1977) In: Dimitriades B (ed) Photochemical oxidant pollution and its control. US Environ Protect Agency, Res Triangle Park, NC
Tingey DT, Taylor GE (1982) Variation in plant response to ozone: a conceptual model of physiological events. In: Unsworth MH, Ormrod DP (eds) Effects of gaseous air pollution in agriculture and horticulture. Butterworth, London
Tingey DT, Reinert RA, Dunning JA, Heck WW (1971) Vegetation injury from the interaction of nitrogen dioxide and sulphur dioxide. Phytopathology 61:1506–1511
Tingey DT, Standley C, Field RW (1976) Stress ethylene evolution: a measure of ozone effects on plants. Atmos Environ 10:969–974
Ulrich B (1981) Eine ökosystemare Hypothese über die Ursachen des Tannensterbens. Forstwiss Centralbl 100:228–236
Unsworth MH (1981a) Exchange of carbon dioxide and air pollutants between vegetation and the atmosphere. In: Grace J, Ford ED, Jarvis PG (eds) Plants and their atmospheric environment. Blackwell, Oxford
Unsworth MH (1981b) Air pollution and plant productivity. In: Johnson CB (ed) Physiological processes limiting plant productivity. Butterworth, London
Unsworth MH (1982) Exposure to gaseous pollutants and uptake by plants. In: Unsworth MH, Ormrod DP (eds) Effects of gaseous air pollution in agriculture and horticulture. Butterworth, London
Unsworth MH, Black VJ (1981) In: Jarvis PE, Mansfield TA (eds) Stomatal physiology. Cambridge Univ Press, Cambridge
Unsworth MH, Ormrod DP (eds) (1982) Effects of gaseous air pollution in agriculture and horticulture. Butterworth, London
Unsworth MH, Biscoe PV, Pinckney HR (1972) Stomatal responses to sulphur dioxide. Nature (London) 239:458–459
Vallee BL, Ulmer DD (1972) Biochemical effects of mercury, cadmium, and lead. Annu Rev Plant Physiol 41:91–128
Vennesland B, Turkington E (1966) The relationship of the Hill reaction to photosynthesis: Studies with fluoride-poisoned bluegreen algae. Arch Biochem Biophys 116:153–161
Warburg O, Christian W (1942) Isolierung und Kristallisation des Gärungsferments Enolase. Biochem Z 310:385–421
Weinstein LH (1961) Effects of atmospheric fluoride on metabolic constituents of tomato and bean leaves. Contrib Boyce Thompson Inst 21:215–231
Weinstein LH, Alscher-Herman R (1982) Physiological responses of plants to fluorine. In: Unsworth MH, Ormrod DP (eds) Effects of gaseous air pollution in agriculture and horticulture. Butterworth, London
Wellburn AR (1982) Effects of SO_2 and NO_2 on metabolic function. In: Unsworth MH, Ormrod DP (eds) Effect of gaseous air pollution in agriculture and horticulture. Butterworth, London
Wellburn AR, Majernik O, Wellburn FAM (1972) Effects of SO_2 and NO_2 polluted air on the ultrastructure of chloroplasts. Environ Pollut 3:37–49
Wellburn AR, Capron TM, Chan HS, Horsman DC (1976) Biochemical effects of atmospheric pollutants on plants. In: Mansfield TA (ed) Effects of air pollutants on plants. Cambridge Univ Press, Cambridge
Wesely ML, Eastman JA, Cook DR, Hicks BB (1978) Daytime variations in ozone fluxes to maize. Boundary-Layer Meteorol 15:361–373
White KL, Hill AC, Bennett JH (1974) Synergistic inhibition of apparent photosynthesis rate of alfalfa by combinations of sulfur dioxide and nitrogen dioxide. Environ Sci Technol 8:574–576
Wilhelm E, Battino R, Wileock RJ (1977) Low pressure solubility of gases in liquid water. Chem Rev 77:219–262
Wilson LG, Bressan RA, Filner P (1978) Light-dependent emission of hydrogen sulphide from plants. Plant Physiol 61:184–189
Winkler P (1982) Zur Trendentwicklung des pH-Wertes des Niederschlags in Mitteleuropa. Z Pflanzenernaehr Bodenk 145:576–585

Winner WE, Mooney HA (1980) Ecology of SO_2 resistance: II. Photosynthetic changes of shrubs in relation to SO_2 absorption and stomatal behaviour. Oecologia 44:296–302
Woltz SS, Leonard CD (1964) Effect of atmospheric fluorides upon certain metabolic processes in Valencia orange leaves. Proc Fla State Hortic Soc 77:9–15
Wood T, Bormann FH (1974) The effects of an artificial acid mist on the growth of *Betula alleghaniensis* Britt. Environ Pollut 7:259–268
Yang SF (1970) Sulfoxide formation from methionine or its sulfide analogs during aerobic oxidation of sulfite. Biochemistry 9:5008–5014
Yang SF, Miller GW (1963) Biochemical studies on the effect of fluoride on higher plants. Biochem J 88:505–516
Yang SF, Saleh MA (1973) Destruction of indole-3-acetic acid during the aerobic oxidation of sulfite. Phytochemistry 12:1463–1466
Yu MH, Miller GW (1967) Effect of fluoride on the respiration of leaves from higher plants. Plant Cell Physiol 8:483–493
Yu MH, Miller GW (1970) Gas chromatographic identification of fluoro-organic acids. Environ Sci Technol 4:492–495
Zelitch I (1957) alpha-hydroxysulfonates as inhibitor of the enzymic oxidation of glycolic and lactic acids. J Biol Chem 224:251–260
Zelitch I (1969) Increased rate of net photosynthetic carbon dioxide uptake caused by the inhibition of glycolate oxidase. Plant Physiol 41:1623–1631
Zelitch I (1973) Alternate pathways of glycolate synthesis in tobacco and maize leaves in relation to rates of photorespiration. Plant Physiol 51:299–305
Ziegler I (1972) The effect of SO_3^{2-} on the activity of RuDP carboxylase in isolated spinach chloroplasts. Planta 103:155–163
Ziegler I (1973a) Effect of air polluting gases on plant metabolism. In: Coulson F, Korte F (eds) Environmental quality and safety Vol II. Thieme, Stuttgart
Ziegler I (1973b) Effect of sulphite on PEP carboxylase and malate formation in extracts of *Zea mays*. Phytochemistry 12:1027–1030
Ziegler I (1974) Malate dehydrogenase in *Zea mays*: Properties and inhibition by sulfite. Biochim Biophys Acta 364:28–37
Ziegler I (1975) The effect of SO_2 pollution on plant metabolism. Residue Rev 56:79–105
Ziegler I (1977) Sulfite action on RuDP carboxylase in the lichen *Pseudevernia furfuracea*. Oecologia 29:63–66
Ziegler I, Hampp R (1977) Control of $^{35}SO_4^{2-}$ and $^{35}SO_3^{2-}$ incorporation into spinach chloroplasts during photosynthetic CO_2 fixation. Planta 137:303–307
Ziegler I, Libera W (1975) The enhancement of CO_2 fixation in isolated chloroplasts by low sulfite concentrations and by ascorbate. Z Naturforsch 30c:634–637
Ziegler I, Marewa A, Schoepe E (1976) Action of sulphite on the substrate kinetics of chloroplastic NADP-dependent glyceraldehyde-3-phosphate dehydrogenase. Phytochemistry 15:1627–1632

15 Ecophysiological Effects of Changing Atmospheric CO_2 Concentration

D.M. Gates, B.R. Strain, and J.A. Weber

CONTENTS

15.1 Introduction 503

15.2 The Atmospheric Heat Balance 504

15.3 The Global Carbon Cycle 504
15.3.1 The Atmosphere 505
15.3.2 Terrestrial Ecosystems 507
15.3.3 Aquatic Ecosystems 508

15.4 Climate Change 509

15.5 Anticipated Plant Responses to CO_2 Enrichment 510
15.5.1 CO_2 as a Limiting Factor 511
15.5.2 Modeling of Carbon Fixation 512
15.5.3 Aquatic Plant Response 514
15.5.4 Plant Growth and Development 515
15.5.5 Cytological Changes 515
15.5.6 Morphological Changes 517
15.5.7 Plant-Water Relations 517
15.5.8 Phenology and Reproduction 518
15.5.9 Heterotrophic Tissues and Organisms 518
15.5.10 Nitrogen Fixation 520
15.5.11 Genetic Variation 520

15.6 Conclusions: Possible Ecosystem Reponses to CO_2-Induced Plant Changes 521

References 521

15.1 Introduction

The earth's atmosphere and biosphere evolved together over time, the one affecting the other, such that the composition of the atmosphere was strongly influenced by the exchange of gases among them, lithosphere, and hydrosphere. Green plants, through photosynthesis and respiration, have had significant influence on the carbon dioxide, oxygen, and water budgets of the atmosphere.

The primordial earth contained carbon in the igneous rocks estimated at 1.2×10^{18} metric tons (10^3 kg) of carbon. From 1% to 5% of this carbon was released to the atmosphere as CO_2, CO, and COS by venting through fumaroles, hot springs, and various vulcanisms. Much of the volatile carbon released to the atmosphere has been dissolved in the oceans and precipitated in sedimentary limestone. Dynamic tectonic activity of the earth's crust has produced a recycling of some of this carbon in the form of CO_2 to the atmosphere. LI (1972) suggests that 95% of the volatiles, including H_2O, CO_2, CO and H_2S which were contained in the primordial rocks were released during the first billion years of the earth's history. Periods of unusual volcanic activity may have released substantial amounts of CO_2 causing impulsive build-up of this gas in the atmosphere until equilibrium could be reestablished with the oceans and sediments.

The carbon cycle of the earth according to our current understanding suggests that a near equilibrium existed among sources and sinks for carbon involving the atmosphere, soils, vegetation, animals, oceans, and sediments prior to the industrial age in about 1860. Since that time there has been a rapidly increasing amount of CO_2 in the atmosphere, released from storage reservoirs by the propensity of mankind for burning fossil fuels (oil, gas, and coal), cutting of forests, plowing of soils, and manufacturing of cement from limestone.

15.2 The Atmospheric Heat Balance

Carbon dioxide gas in the atmosphere, along with water vapor, plays a critical role in the thermal balance of the earth. These two minor constituents of the atmosphere are transparent to sunlight but have strong infrared absorption bands, beyond the wavelength range of solar energy, which cause these gases to exchange infrared radiation with the ground surface. If the atmosphere were completely transparent, the longwave radiation emitted by the earth's surface would flow unimpeded to space and the surface would equilibrate at -20 °C instead of $+15$ °C. By absorbing some of the outward streaming, infrared radiation water vapor and carbon dioxide warm the atmosphere. In turn these gases reradiate this energy both to space and toward the ground surface, thereby raising the temperature of the ground and lower atmosphere. Mankind, by increasing the atmospheric CO_2 concentration, is enhancing the radiative exchange between the atmosphere and the ground surface, thereby potentially producing a change of the earth's climate. Since photosynthetic productivity responds to both the atmospheric concentration of CO_2 and the climate, one may expect substantial changes in all ecosystems, both the managed and unmanaged, as a result.

15.3 The Global Carbon Cycle

Carbon is continually exchanged between the biota, oceans, and atmosphere, and each represents a large carbon reservoir. Our knowledge concerning the global carbon cycle is not complete and our estimates of reservoir amounts and fluxes between parts of the system are poor. Figure 15.1 shows the amounts of carbon estimated to be in the atmospheric, terrestrial, oceanic, and fossil fuel reservoirs and the fluxes between them. The carbonate or limestone rocks of the world retain an enormous reservoir of carbon, about 30×10^{15} t, and

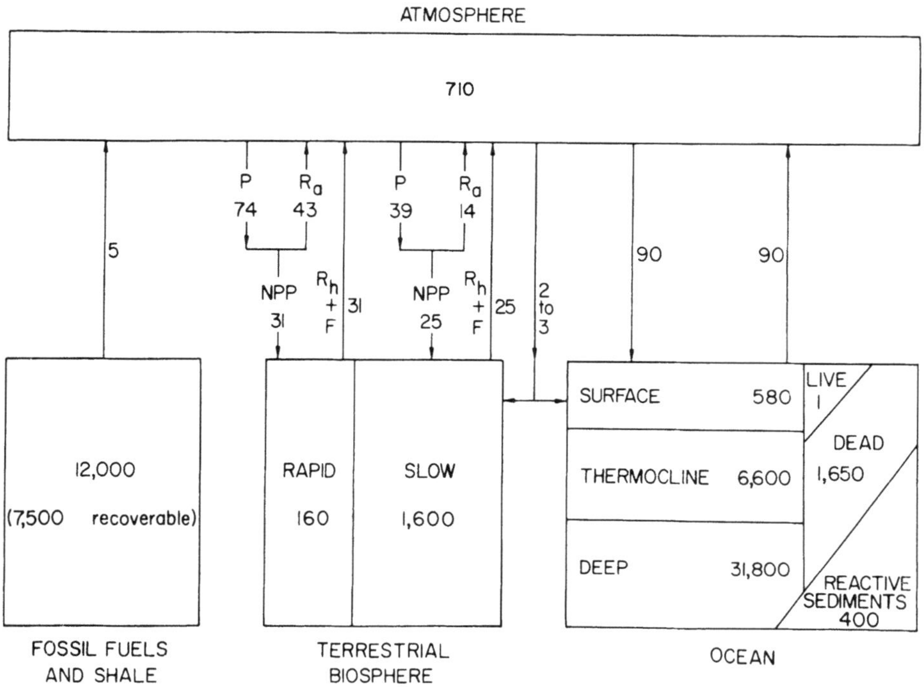

Fig. 15.1. The global carbon cycle (pool sizes in 10^9 t and fluxes in 10^9 t per year). (OLSON et al. 1978)

fossil fuel and shale reservoirs about 12×10^{12} t of which 7.5×10^{12} t may be recoverable. At any given moment the total amount of carbon contained in and exchanged among all reservoirs is a very small fraction (<0.1%) of that in the sedimentary rocks.

15.3.1 The Atmosphere

The atmosphere in 1979 contained a total of 710×10^9 t of carbon at a concentration of 335 ppm of CO_2 by volume. The preindustrial carbon concentration about 1850 is estimated by STUIVER (1978a) to have been about 268 ppm from ^{13}C and ^{14}C isotopic data from tree rings. BROWN and ESCOMBE (1905) reported 91 measurements taken at Kew, England, between 1898 and 1901. According to KEELING (1978), when this data is reanalyzed, discarding some of the data as suspect of industrial contamination, the most probable value in 1900 is 290 ppm. The BROWN and ESCOMBE data indicate a preindustrial value as low as 274 ppm (STUIVER 1978b).

During the International Geophysical Year program accurate measurements of the atmospheric carbon dioxide concentration were begun at Mauna Loa, Hawaii and at the South Pole, Antarctica. These showed a concentration in 1958 of about 314 ppm or an atmospheric carbon load of 665×10^9 t. The

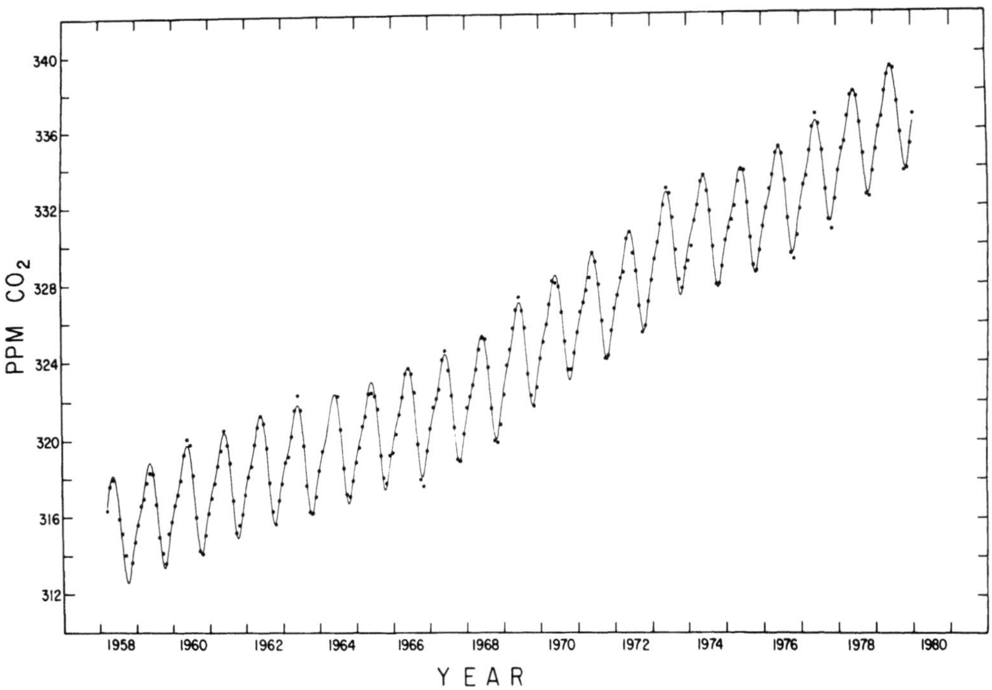

Fig. 15.2. The atmospheric concentration of carbon dioxide measured at Mauna Loa, Hawaii from 1958 through 1979

recorded carbon dioxide concentrations at Mauna Loa from 1958 to 1979 are shown in Fig. 15.2. Depending on which preindustrial levels one selects this gives an increase of atmospheric CO_2 concentration between 1979 and 1900 of $335-290=45$ ppm or 15.5%, or between 1979 and 1850 of $335-268=67$ ppm or 25%.

The amount of carbon released from the burning of fossil fuels and cement production today seems to be well estimated at 5×10^9 t per year (BAES et al. 1977). The rate of release has been increasing at about 4.3% per year and has followed an exponential growth since the beginning of the century. However, estimates of the amount of carbon released from the biosphere by the cutting of forests and the disturbance of terrestrial ecosystems is poorly known. WOODWELL et al. (1978) indicate that the terrestrial biosphere release may currently exceed the fossil fuel release and could be as high as 6×10^9 t of carbon per year. STUIVER (1978a) shows from carbon isotopic data that the rate of biospheric release may have exceeded the rate of fossil fuel release during the 19th century but that since 1930 the rates of release and uptake by the biosphere are nearly balanced. BROECKER et al. (1979) support this contention. Best estimates show that of all the carbon released to the atmosphere about 50% has remained airborne and that the other 50% has gone to a sink, presumed to be the oceans.

It is difficult to predict the future rates of CO_2 release from human activities. If the current rate of increased release continues until the fossil fuel supply is exhausted, about 200 years from now, the atmosphere will have gained 12 times its preindustrial level of CO_2. However, our understanding of all sources and sinks for CO_2 is so poor that this estimate is highly uncertain. No one believes that the maximum fuel use scenario will occur. However, if continued forest destruction, soil disturbance, and cement manufacture add to this, then high rates of CO_2 release will occur. Estimates presently indicate that it is highly probable to have a doubling of the preindustrial concentration sometime in the interval 2,035 to 2,065 (see p. 509 and 511).

15.3.2 Terrestrial Ecosystems

Great uncertainties exist concerning the amount of carbon stored in the terrestrial ecosystems of the world. These ecosystems have been disturbed by mankind and it is difficult to know the amount of carbon contained within them prior to disturbance. There are several ways to consider the carbon distribution in the terrestrial biosphere. The total amount of carbon is estimated at $1{,}760 \times 10^9$ t but with minimum and maximum possible values of 1,527 and $3{,}827 \times 10^9$ t. The greatest error is with the amount of carbon in humus. Of the $1{,}760 \times 10^9$ t, 680×10^9 represent carbon in live material and $1{,}080 \times 10^6$ t in dead materials. SCHLESINGER (1977) reviews the global carbon content of detrital carbon and concludes the total amount is about $1{,}456 \times 10^9$ t. One can partition the biosphere into northern woods above latitude 30 °N, southern woods below 30 °N, and non-woods to get 615, 565, and 580×10^9 t respectively including all above and below ground carbon. Another partitioning is between the pools of rapidly exchanging carbon consisting of leaves, branches, roots, litter, invertebrates, and decomposers totaling 160×10^9 t and the slowly exchanging pools of humus, peat, large branches, trunks and roots, and most animals totaling $1{,}600 \times 10^9$ t. Humus and peat represent about $1{,}000 \times 10^9$ t of the slowly exchanging pool.

All woods together total an area of 47×10^6 km^2 for $1{,}180 \times 10^9$ t of carbon or about 32% of the land surface of 149×10^6 km^2 and consist of forests, open woodlands, and woody swamps. Non-woods cover 83×10^6 km^2 for 580×10^9 t of carbon and are composed of grassland, desert, semidesert, tundra, agriculture, suburban, and urban areas. A detailed analysis deriving these numbers is given by OLSON et al. (1978). A more accurate analysis is not possible until new surveys are made of the vegetated regions of the world using satellite imagery. WHITTAKER and LIKENS (1973) give somewhat larger numbers for the wooded regions, but these may justifiably be lowered when allowing for deforestation.

Shown in Fig. 15.1 are the estimated annual rates of carbon exchange between the terrestrial biosphere and the atmosphere. Gross photosynthetic production between the atmosphere and the terrestrial biosphere is 113×10^9 t per year (74×10^9 t by the rapid pool and 39×10^9 t by the slow pool). Respiration by green vegetation releases 43×10^9 t per year from the rapid pool and 14×10^9 t per year from the slow pool for a total net primary production of 56×10^9 t per year. In addition, respiration by heterotrophic organisms and fires release an equal amount of 56×10^9 t per year to maintain a balance.

15.3.3 Aquatic Ecosystems

The total CO_2 content of the world's oceans, seas, lakes, and rivers is governed by a complex of reactions involving free CO_2, H_2CO_3, HCO_3^-, and CO_3^{2-} ions, solid carbonates in sediments and suspension, other dissolved compounds, pH, ionic strength, and temperature. As one might expect the effect of increased atmospheric CO_2 on this system is quite complex. A detailed discussion of this subject would require many pages and would duplicate much of what has been written (e.g. KEELING 1973; OESCHGER et al. 1975; ELLIOTT and MACHTA 1979). Here we will give a brief overview of the processes involved in CO_2 exchange between the atmosphere and fresh and saltwater bodies of the world. On a global scale freshwater may be neglected because of the small amount present in lakes, ponds, rivers, etc.

Total inorganic carbon dissolved in the oceans is estimated to be $36–39 \times 10^{12}$ t ($3–3.25 \times 10^{18}$ mol C). Of this amount nearly 90% is HCO_3^-, about 10% CO_3^{2-}, and less than 1% is free CO_2. Carbonic acid concentration is over two orders of magnitude smaller than that of free CO_2 (STUMM and MORGAN 1970). While the total inorganic carbon content of freshwaters is negligibly small on a global scale, the distribution of that carbon among the species of dissolved total CO_2 is much more variable than in the oceans. The primary cause of this variability is the wide range of pH found in freshwaters from very acidic (<pH 4) in some volcanic lakes to very alkaline (>pH 10) in soda lakes. The free CO_2 content is controlled largely by the atmospheric partial pressure with variation caused by photosynthetic and respiratory activity within the water. This restraint means that as the pH increases the total inorganic carbon content increases with the free CO_2 level remaining approximately fixed.

There is a cycling of CO_2 between oceans and atmosphere since equatorial regions of the oceans have CO_2 partial pressure higher than atmospheric, and some areas nearer the poles lower than atmospheric (TAKAHASHI 1979). One determinant of the rate at which atmospheric CO_2 can be incorporated into the inorganic carbon of water is the amount of $CaCO_3$ available for the following reaction:

$$CaCO_3 + CO_2 + H_2O \leftrightarrows Ca^{2+} + 2HCO_3^-$$

As more CO_2 dissolves into the waters, the amount of $CaCO_3$ decreases reducing the ability of surface waters to take up free CO_2 (BROECKER 1973). Surface waters are presently several times supersaturated with calcite and aragonite, the principle form of solid carbonate in water. Even a doubling of present atmospheric CO_2 content would not lead to levels below two to four times saturation (ELLIOTT and MACHTA 1979). Deep waters, however, are not saturated with $CaCO_3$, so that the above reaction will be constrained by the rate of dissolution of $CaCO_3$ from the sediments which is a slow process.

As the CO_2 partial pressure increases the pH will become more acidic. In oceanic waters the pH shift will be about 0.3 units (pH 8.1 to 7.8) for a doubling of the CO_2 partial pressure (ELLIOTT and MACHTA 1979, p. 18).

The time required for attaining a new equilibrium among the CO_2 containing components of the biosphere (i.e. oceanic, freshwater, atmospheric, and terrestrial) will be determined by many things besides the rate of CO_2 release by burning of fossil fuels. As noted above equilibration of the increased CO_2 partial pressure with the deep oceanic water and sediment will be constrained by the rate of dissolution of $CaCO_3$ from the sediments (BROECKER 1973). Furthermore, the rate of mixing of surface waters, with a volume of about 0.0224×10^9 km^3, into the deep water, with a volume of about 1.1×10^9 km^3 is quite slow, about 0.5–2×10^6 km^3 yr^{-1} (BROECKER and LI 1970; OESCHGER et al. 1975). Eddy diffusion may also be important in movement of CO_2 to the depths through the thermocline (volume about 0.25×10^9 km^3) at a depth between 70 and 1,000 m. OESCHGER et al. (1975) proposed a model which includes eddy diffusivity based on the rate of movement through the water column of radioactive isotopes produced in atomic bomb tests. This model successfully simulates the increase in atmospheric CO_2 partial pressure. Although movement of CO_2 from the atmosphere to the surface water might be a constraint, TAKAHASHI (1979) could not find a delay in CO_2 partial pressure reaching equilibrium in surface waters. The normal yearly variations obscured any delay.

15.4 Climate Change

The dynamics of climate change are not sufficiently understood to allow accurate predictions of cause and effect. However, certain mechanisms can lead to climate change and one of these is the influence of the atmospheric carbon dioxide concentration on the earth's radiation balance as explained earlier. All modeling of the affect of this mechanism on the atmospheric temperature indicates a doubling of atmospheric CO_2 to give a mean global temperature increase of 2.0 °C (MANABE and WETHERALD 1975; MANABE and STOUFFER 1979). Furthermore, they predict that high latitude regions will have mean temperature rises as high as 7 °C and low latitude regions about 1.5 °C. The anticipated temperature rise at high latitudes is expected to have strong affect on snow cover and sea ice thickness which in turn reduces the albedo of these regions and exacerbates the temperature increase (MANABE and STOUFFER 1979). SCHNEIDER (1975) has reviewed the various climate models and concludes that a doubling of atmospheric CO_2 concentration would produce a mean global warming of 1.5° to 3.0 °C. The MANABE and WETHERALD (1975) model also shows the hydrologic cycle to increase precipitation worldwide by 7% but that certain continental regions (central U.S.) will become drier. It is estimated that a doubling of the preindustrial CO_2 concentration may occur between the years 2,035 and 2,050 and that before the year 2,100 a threefold increase will occur.

Some meteorologists (see BRYSON and WENDLAND 1970) suggest that dust and aerosols in the atmosphere may produce a cooling by reflecting more sunlight to space. ROBOCK (1978) shows that volcanic dust has been an important cause of climate change, mainly cooling, and that anthropogenic factors such as CO_2 and tropospheric dust have not had discernable influence up to the present time. Most low-level dust washes out of the atmosphere relatively quickly, but stratospheric dust from volcanoes persists (MENDONCA et al. 1978). The role of dust in climate is ambiguous and IDSO and BRAZEL (1977) show that low-level dust may produce a warming but above a critical level could produce a cooling. MITCHELL (1971) demonstrated earlier that tropospheric aerosols may either warm or cool the atmosphere near the surface, depending on the properties of the aerosols and the albedo of the surface. Chemical aerosols from air pollution may affect world climate. Aerosols act as condensation nuclei for water droplets and increase the amount of cloud cover, thereby reflecting more sunlight and causing a cooling according to HOBBS et al. (1974). Although aerosols may have a mitigating effect against a warming trend the probability of a warming climate in the 21st century is great.

HANSEN et al. (1981) have modeled the global climate by including radiative and convective energy exchange involving clouds, aerosols, surface albedo, water vapor and carbon dioxide concentrations, and solar activity. Their model has a 100-m mixed layer ocean plus thermal diffusion down to 1,000 m. With this model they can fit very well the observed global mean air temperature with 5-year running means for the period 1880 to 1980. They show that the global mean air temperature has been warming since 1965 by more than 0.3 °C, that all latitude belts between 23.6 °N and 90 °S have been persistently warming throughout this century, and that northern latitudes (23.6 to 90 °N) started a warming trend after 1970. They suggest that the effect of CO_2 on climate is already discernable.

KUKLA and GAVIN (1981) show that the pack ice around Antarctica during summers has decreased during the 1970's compared with its position during the 1930's. Between 1973 and 1980 satellite photographs reveal that the Antarctic summer pack ice decreased by about 35%. KUKLA and GAVIN also found that the mean surface air temperatures along selected northern latitudes between 55° and 80 °N during spring and summer were higher by 0.9 °C in the period 1974 to 1978 than during 1934 to 1938. Although the Arctic ice pack was not yet shrinking, these positive temperature changes appear to be associated with a shifting belt of melting snow. The pattern of change is not yet well understood, however, the suggestion that the climate has been responding to increasing amounts of atmospheric CO_2 is becoming increasingly plausible.

15.5 Anticipated Plant Responses to CO_2 Enrichment

Essentially all of the known physiological responses of plants to carbon dioxide concentration have been presented in prior chapters and volumes of this encyclopedia (see e.g. Vol. 6; Chaps. 7 and 16, Vol. 12B). In addition, the extensive bibliography (STRAIN 1978) prepared for a workshop on the subject should be consulted for references to specific aspects of the problem. This section addresses the integration of known metabolic and growth responses to CO_2 concentration as plants develop through entire life cycles and through several generations. Finally a brief section speculating on ecosystem response to chronically increasing CO_2 is presented.

As stated above, it is assumed that atmospheric CO_2 will double sometime during the first half of the 21st century. It is possible that CO_2 could increase to three times the preindustrial level within a century. Consequently we will consider plant response to atmospheric CO_2 concentration of up to 1,000 ppm. Information on greatly enriched atmospheres (>1,200 ppm) will not be considered in this chapter.

Perhaps the single most important resource of plants to increasing CO_2 concentration is net photosynthesis. Growth, flower, fruit and seed production, photosynthate allocation, nodulation, and symbiotic nitrogen fixation are all changed when plants are grown under high CO_2 concentrations. These changes are, however, largely secondary, and are affected by increasing carbohydrate production caused by increasing net photosynthesis.

15.5.1 CO_2 as a Limiting Factor

Although it has generally been assumed that carbon dioxide is seldom a limiting factor in plant growth and production in unmanaged ecosystems (ATTIWELL 1971), this assumption is definitely false. Research with wheat plants grown at 350 and at 1,000 ppm with and without water stress demonstrated that increased CO_2 completely compensated for water stress (SIONIT et al. 1980a). Thus, even during periods of water stress sufficient to decrease production at normal atmospheric CO_2 concentration, plants grown under high CO_2 produced equally to the unstressed control plants. HART and WRIGHT (1976) demonstrated that pine trees growing in situ responded positively to increasing CO_2 concentration even when soil water potentials were below -15 bars. They determined that drought tolerance is increased as carbon dioxide increases. FORD and THORNE (1967) reported that both C_3 and C_4 plants increased net photosynthesis at 1,000 ppm relative to 300 ppm control plants and that the percentage increase was about the same at high and low irradiance levels. The irradiance interaction is a complicated one, however, perhaps involving leaf temperature or water stress interaction. GIFFORD (1977) reported that wheat grown in low light responded relatively more to CO_2 enrichment than plants grown at higher irradiance.

NEVINS and LOOMIS (1970) fertilized sugar beets with nitrogen to determine the interaction between CO_2 and nitrogen concentrations. Photosynthesis responded linearly as CO_2 was increased from 300–500 ppm. Added nitrogen steepened the response slope but the CO_2 effect was still linear within the concentration ranges tested. WONG (1979) reported the results of growing cotton and maize plants under full sunlight in glass houses containing 330 ppm and 640 ppm CO_2 with four levels of nitrogen nutrient. Cotton plants (*Gossypium hirsutum*), 40 days old, grown in high CO_2, had a 2.0-fold and 1.6-fold increase in dry weight and leaf area respectively to those grown in normal CO_2 concentration. Maize plants (*Zea mays*), 30 days old, showed only a 1.2-fold increase in dry weight and no increase in leaf area for the high to normal CO_2 concentrations. Dry weight and leaf area in both species decreased proportional to nitrogen decreases. According to WONG (1979), "Cotton and maize grown in high CO_2

had a lower assimilation rate in ambient CO_2 compared to plants grown in normal ambient air. This difference was due to the reduction in RuBP carboxylase activity. Water use efficiency was doubled in both cotton and maize plants grown at high CO_2 in all nutrient treatments."

The conclusion from the above studies and a great deal of other literature (see STRAIN 1978) is that carbon dioxide is limiting in net photosynthesis and plant growth. As global atmospheric carbon dioxide concentration increases to two times and even three times the preindustrial level, we anticipate that plants will respond by additional net photosynthesis and consequent changes in growth and development.

It is believed by many investigators (WITTWER 1979) that carbon dioxide has all of the characteristics of a fertilizer and that agricultural production should increase as global carbon dioxide increases. WITTWER (1980) has stated that agricultural technology has already demonstrated the ability to respond to environmental changes of the same order of magnitude as those predicted for the next century. Consequently, it is assumed that future changes in CO_2 will present interesting technical challenges, but challenges which we should expect to be adequately met by agrotechnology. This optimistic view may well be realized in highly managed ecosystems. What of the less managed natural and marginal lands, however?

15.5.2 Modeling of Carbon Fixation

It is not only important to have measured responses of photosynthesis and respiration to increases of atmospheric CO_2 concentration, but to formulate analytical models of these responses so that predictions can be made involving the multitude of variables which affect these responses. These processes are dependent on light intensity, temperature, carbon dioxide and oxygen concentrations, and water and nutrient availability. A large literature exists concerning whole leaf modeling of energy exchange, photosynthesis, and respiration (see Chap. 16, Vol. 12B and GATES 1968; LOMMEN et al. 1971; TENHUNEN et al. 1976a, b, 1977, 1979a, b; TENHUNEN and WESTRIN 1979). Here we summarize only the results.

Leaf temperature and transpiration rate are the direct result of energy exchange between a leaf and its environment through the processes of radiation exchange, convection, and evaporation. The energy budget couples the leaf temperature and transpiration rate to the environmental variables of air temperature, humidity, wind speed, and shortwave and longwave radiation by means of properties of the leaf such as absorptance and emittance, leaf size, shape, and orientation, and stomatal resistance (see Chap. 1, Vol 12A, Chaps. 7 and 17, Vol. 12B).

The exchange of CO_2 by gas diffusion determines the availability of CO_2 to the plant and chemical kinetics determines the photosynthetic and respiratory responses to light, temperature, and concentrations of CO_2 and O_2 at the chloroplasts, mitochondria, and peroxisomes. One can model all of these functional dependencies including competitive inhibition between O_2 and CO_2 for the enzyme ribulose bisphosphate carboxylase. The models referenced above utilize Fick's Law for gas diffusion and for the photosynthetic response, a Michaelis-Menten function to CO_2 concentration at the chloroplasts, and a Smith function to the light intensity. An empirical function involving enzyme kinetics is used for the temperature response of photosynthesis and respiration. The photosynthetic-respiration model includes the maximum photosynthetic and respiration capacities of the leaf, apparent affinity coefficients for CO_2 and O_2, apparent activation energies for temperature, and the mesophyll resistance. The energy budget and the photosynthesis-respiration models are coupled together to give the net photosynthesis response of the leaf to the environmental factors.

Using the whole leaf photosynthetic-respiration model one can get an indication of the net photosynthetic response to increased levels of CO_2. Table 15.1 shows the expected percentage increases of net photosynthesis of leaves to increases of atmospheric CO_2 concentration at various light levels and leaf temperatures using approximately the present CO_2 concentration of 334 ppm as a reference level. It is evident that at low leaf temperatures the percentage increase of net photosynthesis is small even for very large increases of the CO_2 concentration but that at high leaf temperatures rather substantial increases of net photosynthesis are expected. These numbers will be slightly reduced when the

Table 15.1. Calculated rates of whole leaf net photosynthesis ($\mu mol\ m^{-2}\ s^{-1}$) in response to CO_2 concentration as a function of leaf temperature and incident light intensity. Percentage increase of net photosynthesis to CO_2 increase using 334 ppm as a reference level are given in parentheses

Temperature °C	$[CO_2]$ ppm	Light intensity, $mE\ m^{-2}\ s^{-1}$ 0.20	0.45	0.90	1.80
10	334	5.04	5.85	6.05	6.10
	400	5.15 (2.0)	5.99 (2.3)	6.19 (2.4)	6.25 (2.4)
	500	5.26 (4.3)	6.12 (4.6)	6.32 (4.6)	6.38 (4.7)
	600	5.32 (5.6)	6.20 (6.0)	6.42 (6.1)	6.47 (6.1)
20	334	7.09	10.71	12.22	12.70
	400	7.36 (3.7)	11.30 (5.4)	13.01 (6.5)	13.58 (6.9)
	500	7.61 (7.2)	11.83 (10.4)	13.74 (12.4)	14.37 (13.2)
	600	7.78 (9.5)	12.15 (13.4)	14.16 (15.9)	14.84 (16.9)
30	334	6.66	11.58	14.43	15.48
	400	7.10 (7.0)	12.89 (11.3)	16.64 (15.3)	18.13 (17.1)
	500	7.56 (14.0)	14.26 (23.1)	19.21 (33.1)	21.39 (38.2)
	600	7.87 (18.6)	15.15 (30.8)	20.99 (45.5)	23.80 (53.7)
40	334	3.78	6.71	8.49	9.22
	400	4.44 (17.1)	8.04 (19.8)	10.36 (22.0)	11.34 (22.9)
	500	5.17 (36.7)	9.65 (43.7)	12.74 (50.0)	14.11 (52.9)
	600	5.71 (51.1)	10.87 (62.0)	14.64 (72.4)	16.38 (77.5)

The following values were used for the parameters of the photosynthesis model (TENHUNEN and WESTRIN 1979): $P_{MLT} = 40.0\ \mu mol\ m^{-2}\ s^{-1}$; $\alpha = 50$ mol/mE; $R_m = 586\ s\ m^{-1}$; $KC_3 = -1.4789 + 0.015361\ T_K - 0.53188 \times 10^{-4}\ T_K^2 + 0.61405 \times 10^{-7}\ T_K^3$ (units – mol m^{-3}); $K_0 = 0.29068 - 0.67253 \times 10^{-3}\ T_K$ (units – volume ratio); $H_1 = 47700$ cal mol^{-1}; $H = 14{,}200$ cal mol^{-1}; $C_2 = e^{22.47}$; $S = 153.4$ (entropy units).

respiration response of roots, trunks, and other plant parts to increased levels of CO_2 are taken into consideration. These calculations assume a maximum photosynthetic capacity of 40 μmol m^{-2} s^{-1} and an optimum temperature near 40 °C.

15.5.3 Aquatic Plant Response

The question we wish to address now is the response of photosynthesis in the waters of the world to the expected changes in these waters induced by changing atmospheric CO_2 level. The prediction of the effect of increased atmospheric CO_2 on the photosynthetic rate in aquatic environments is quite difficult because of the interaction of several factors: CO_2 concentration, carbonate equilibrium, pH, temperature, and mineral nutrition. As atmospheric CO_2 partial pressure increases, the level of CO_2 in the surface waters will likewise increase. However, if a general warming occurs, the solubility (mol/volume) of CO_2 in water will decrease. Thus the actual concentration of free CO_2 will depend on changes in both CO_2 partial pressure and water temperature.

Terrestrial photosynthesis is limited by the CO_2 concentration in the atmosphere; however, in the aquatic environment CO_2 may be at or near saturating levels, particularly for unicellular algae (VIDAVER 1972). However, many of the studies, particularly in the older literature, do not give light intensities in units which can be compared to environmental levels or to other data. Data on freshweater algae and vascular hydrophytes (STEEMAN-NIELSEN 1947; LUCAS 1975; RAVEN 1968) indicate that increasing CO_2 concentration could increase the rate of photosynthesis with a saturating concentration of $0.5\text{–}1 \times 10^{-3}$ M free CO_2 (at 330 ppm pCO_2 and 25 °C the free CO_2 concentration is about 1×10^{-5} M). From these data an increase in CO_2 concentration would lead to increased photosynthesis.

As the pCO_2 increases in the water the concentration of HCO_3^- will change, depending on both the increase in pCO_2 and pH. With no changes in pH, HCO_3^- concentration will increase; with a larger change in pH, it probably will decrease, the actual change being dependent on the interaction of pH and pCO_2. For those organisms which can utilize HCO_3^- and at HCO_3^- concentration sufficient to provide a significant amount of fixable carbon, changes in HCO_3^- concentration will probably affect net photosynthesis, although much less so than the same changes in free CO_2 (STEEMAN-NIELSEN 1947; LUCAS 1975; RAVEN 1968; LEHMAN 1978).

As noted above, the expected change in the pH of the oceans is about 0.3 units which should be well within tolerances of most if not all marine organisms. The shift in pH in freshwaters will probably be of the same magnitude or less, depending on the buffering capacity of the waters. The shift in the carbonate equilibria following a decrease in pH may pose a hazard to those organisms which depend on the supersaturation of waters with calcite and aragonite for skeletal material. However, only those organisms which require $MgCO_3$ (e.g. coralline algae) would be threatened by any decrease in the supersaturation with precipitated carbonates (ELLIOTT and MACHTA 1979).

There are two further aspects of the aquatic environment which will strongly influence the photosynthetic response to increased CO_2 availability. First, light intensity within the water column is strongly affected by depth and suspended matter. As one goes deeper, light intensity and not CO_2 availability becomes the controlling factor. Increased growth of algae near the surface will further attenuate the flux at lower levels. Second, most waters are deficient in mineral nutrients, particularly fixed nitrogen, including large areas of the pelagic ocean. Where mineral nutrient deficiencies exist, availability of CO_2 will have only a marginal effect.

It would appear then that there are several factors which will restrict the rate of photosynthesis in aquatic plants sufficiently to nullify any increase in the potential rate of photosynthesis following an increase in CO_2 concentration. Where favorable conditions exist, however, photosynthetic rate will likely increase.

15.5.4 Plant Growth and Development

The remainder of this chapter will be devoted to a consideration of whole plant responses of a secondary nature. If carbon accumulation increases, plant growth and developmental changes will be a consequence. What is known about these possible responses?

Undoubtedly the most completely studied species, as far as carbon dioxide related responses are concerned, is the garden tomato. Soybean and maize are perhaps the next best known species in this regard. No non-economic species has been well studied although several have been analyzed from one aspect or another. Unfortunately most studies of plant response to CO_2 enrichment have been of short duration. Plants are usually grown in a normal greenhouse or growth chamber and then tested to determine physiological or growth response to suddenly increased CO_2 levels. The few studies which have been done on plants germinated and grown to maturity under high CO_2 indicate morphogenetic changes which affect subsequent plant response. A brief review of known structural responses is in order.

15.5.5 Cytological Changes

Changes in cellular structure and development have been reported in the algae *Chlorella* spp. (Soeder et al. 1966) and *Gonium* (Vazart and Henry 1972), and in the grass *Lolium* (Wilson and Cooper 1969). The most complete study on a vascular plant is reported by Madsen (1968, 1971, 1973a, b) on tomato. In his extensive studies (summarized in his dissertation, Madsen 1976) he reported changes in cell volume (Madsen 1968) and disruption of chloroplast structure by excessive starch grain accumulation (Madsen 1971).

Hesketh and Hellmers (1973) reported similar cytological changes in cotton (*Gossypium hirsutum*) grown at 1,000 ppm CO_2. Starch grains reaching excessive size and disrupting normal chloroplast structure and other organelle arrangements seems to be the most common cytological problem, see Fig. 15.3. Similar results were obtained by Hofstra and Hesketh (1975) with soybean.

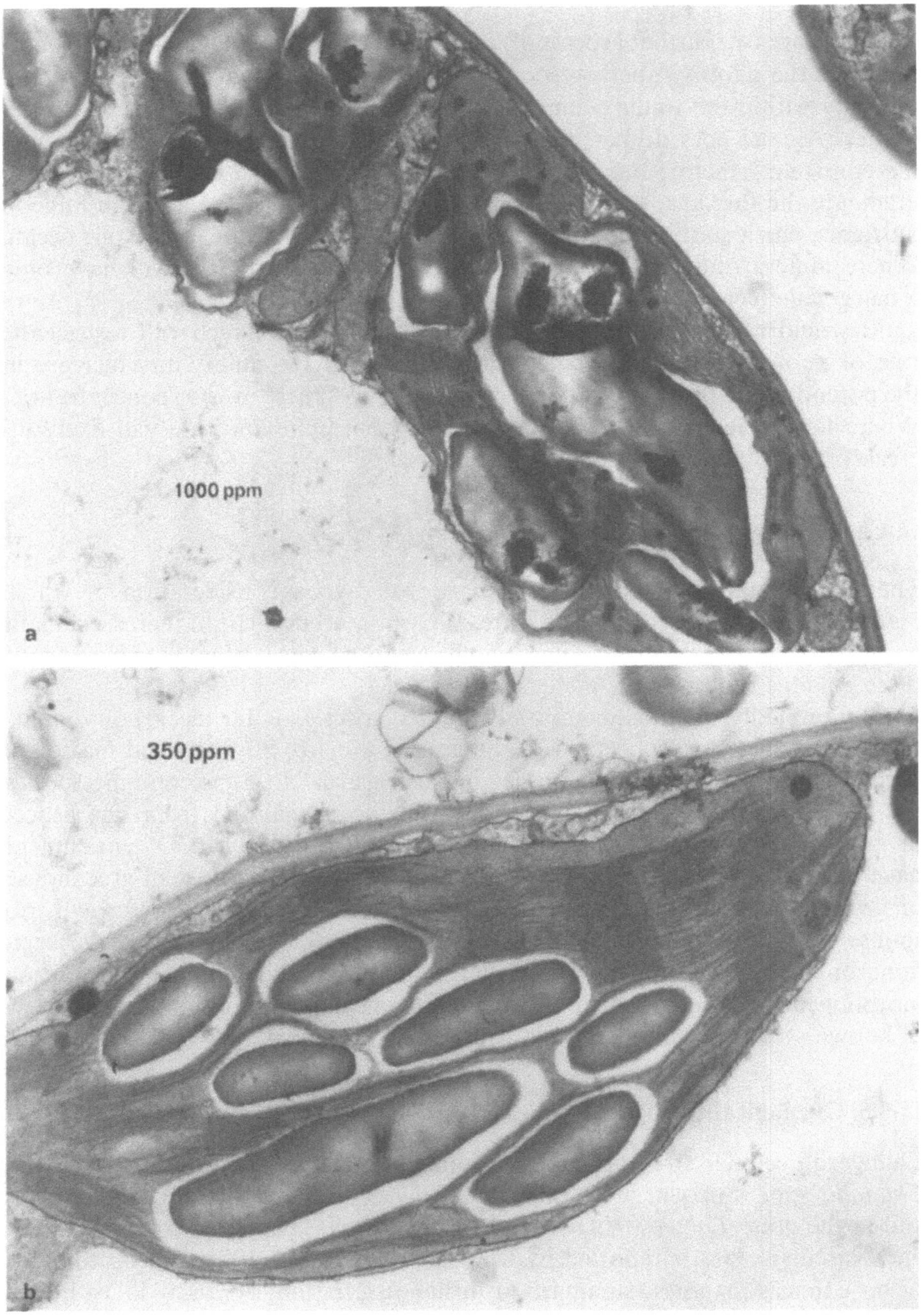

Fig. 15.3 a, b. Electron micrographs of chloroplasts of *Trifolium subterraneum*. Starch grains of plants grown at high CO_2 (**a**) are very large, irregular in shape, and cause disruption of chloroplast grana while chloroplasts grown in a normal atmosphere (**b**) maintained normal granal structure in spite of significant starch grain formation. **a** is from a leaf grown at 1,000 ppm CO_2. **b** is from a leaf grown at 350 ppm. Note that the magnifications differ with **a** being ×21,000 while **b** was photographed at ×42,000. We thank Mr. George Cave, III for allowing us to use this previously unpublished information and these figures

15.5.6 Morphological Changes

Reports of morphological effects of increasing CO_2 are legion. Since BROWN and ESCOMBE (1902) reported changes in the growth of plants as CO_2 concentration increased, more than 100 research papers have appeared which establish without doubt that plant growth and development changes will occur as carbon dioxide increases to two times or three times the preindustrial level.

It is clear that leaf size (FORD and THORNE 1967; HARDY and HAVELKA 1977; HO 1977; KNECHT and O'LEARY 1972; NEWTON 1965) is frequently increased when CO_2 is increased during leaf expansion phase of growth. This increase in leaf area has been shown to increase leaf area index (GIFFORD et al. 1973). Production may increase (HARDY and HAVELKA 1977; HUGHES and COCKSHULL 1969; KRIZEK et al. 1969) because of increasing leaf area. In ecosystems in which the canopy does not close, increasing leaf area could be expected to significantly increase biomass. In closed canopy systems, on the other hand, an increase in leaf area in the canopy and therefore increased self-shading would increase attenuation of radiation penetration and reduce the radiation available for photosynthesis by sub-canopy species or by leaves closer to the ground (GIFFORD et al. 1973). This effect would decrease the impact of CO_2 enhancement on production.

In addition to changes in leaf area and possible changes in the shape of plants and, therefore, the distribution of photosynthetic tissue, other changes in plant form can be expected. There is a great deal of data which indicate that shoot:root ratios will change. Most studies indicate that relatively more carbon enters root tissues when plants are grown under high CO_2 than under current average levels (FORD and THORNE 1967; KRIEDEMANN et al. 1976). One study, however, found that *Rosa hybrida* plants grown at 2,000 ppm had smaller root systems than normal (MATTSON and WIDMER 1971).

15.5.7 Plant-Water Relations

Changes in shoot root ratios should cause changes in plant water and nutrient relations. Increasing root growth relative to top growth could increase water and nutrient availability. On the other hand, a larger root system could deplete the soil reserves more quickly and generate soil water shortages more quickly.

It is much more clear that plant water relations will be affected by a differential response of water and carbon dioxide diffusion to stomatal aperture. Stomates close when atmospheric carbon dioxide concentrations increase (AKITA and MOSS 1972; EGLI et al. 1970; LUDLOW and WILSON 1971; PALLAS 1965; RASCHKE 1972; REGEHR et al. 1975; VAN BAVEL 1974). Wheat, for example, decreased stomatal conductance from 0.82 cm s^{-1} to 0.33 cm s^{-1} when CO_2 was increased from 350 to 1,000 ppm CO_2 (SIONIT et al. 1980b). This decrease in rate of water loss caused by stomatal closure while photosynthesis increases due to an increased concentration gradient leads to increasing water use efficiency (see Chaps. 7 and 17, Vol. 12B and HURT and WRIGHT 1976; MOSS et al. 1961). Field trials have shown that increasing atmospheric carbon dioxide concentration will increase net photosynthesis and decrease water use (TAKAMI and VAN BAVEL 1976).

Partial closing of stomates at high CO_2 while photosynthesis remains high due to increasing CO_2 availability has an interesting and potentially important effect on tolerance to volatile air pollutants. Alfalfa was much less damaged by NO_2 and SO_2 when grown at enriched CO_2 levels (HOU et al. 1977; MAJERNIK and MANSFIELD 1972). Decreased net photosynthesis in tree species, caused by atmospheric ozone and particulate matter, was compensated for by increases in carbon dioxide concentration (GREEN and WRIGHT 1977).

15.5.8 Phenology and Reproduction

The rate of plant growth and development is definitely affected by atmospheric carbon dioxide concentration. Tomato plants grow more rapidly (MADSEN 1976), the flowering date of the first truss may be advanced (CALVERT 1972), the date of first anthesis is accelerated (HAND and POSTLETHWAITE 1971), and fruit mature earlier and may be harvested earlier (HAND and POSTLETHWAITE 1971) at concentrations up to 1,000 ppm.

Similar results have been reported in carnation (GOLDSBERRY 1963), *Callistephus chivensis* (HUGHES and COCKSHULL 1969), wheat (SIONIT et al. 1980b), and Alaska pea (PAEZ et al. 1980). Faster development of flowers and fruits is clearly important in economic crops. Such a response in native plants in unmanaged ecosystems would be equally as important. Many pollinators and herbivorous animals are co-evolved with their host plants and phenologically timed in their own development. If significant changes in phenology do occur a concommitant phenological change will have to occur in the co-adapted animals.

Not all experiments have shown an acceleration in development under high CO_2, however. HESKETH and HELLMERS (1973) reported a significant delay in floral initiation of four cultivars of *Sorghum,* when they were tested at 1,000 ppm in the greenhouses of the Duke Phytotron.

Seed production also is known to be affected by carbon dioxide enrichment (COCK and YOSHIDA 1973; GIFFORD 1977; HARDY and HAVELKA 1977; SIONIT et al. 1980b; YOSHIDA 1973) in grain crops including rice and several varieties of wheat. Number, size and dry weight of grains grown at high CO_2 exceed those grown at normal concentration. As reviewed above, many studies have shown increasing numbers of flowers and fruits and increasing dry weights of fruits in crop plants but few studies could be found of seed number per fruit or plant. It is assumed that seed production increases as fruits increase in size and numbers. Except for the grains reported above, however, no measurements of seed production could be found.

No measurements on germination or establishment and growth of plants from seeds of plants grown at high CO_2 concentration could be found. Such measurements are currently underway with Alaska pea, spring wheat, *Trifolium* species and varieties, and *Desmodium paniculatum,* in the Duke Phytotron, but no results are yet available for publication.

15.5.9 Heterotrophic Tissues and Organisms

Culturing tissues in high CO_2 atmosphere has been reported to enhance growth (LEONARD and PINCKARD 1946). Perhaps the clearest study of the positive effect

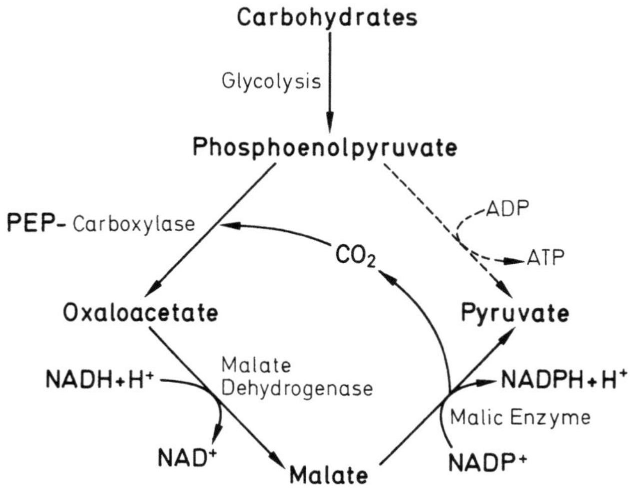

Fig. 15.4. The dark fixation and release of CO_2 with the net transfer of hydrogen from $NADH+H^+$ to $NADP^+$ to form $NADPH+H^+$. *Dashed arrow* indicates the direct pathway for the utilization of PEP. (Modified from TING and DUGGER 1965)

of CO_2 enhancement on tissue growth was reported by SPLITTSTOESSER (1966) on root tissues of maize, carrot, and tomatoes. KUNITAKE et al. (1959) examined dark CO_2 fixation in tobacco leaves and concluded that the reaction is mediated by PEP-carboxylase. Consequently, dark fixation presumably occurs in most plant tissues.

TING and DUGGER (1965, 1966, 1967) believe that the primary effect is a transhydrogenation from $NADH+H^+$ to $NADP^+$ – via formation and degradation of malate (Fig. 15.4). This additional source of reduced $NADPH+H^+$ may be significant for synthetic events in heterotrophic tissues because the $NADPH+H^+$ is available for such reactions as lipogenesis, nitrate reduction, sulfate reduction, and indoleacetic acid metabolism (TING and DUGGER 1965).

Sporulation and the germination of spores of fungi has also been reported to respond positively (IOANNOW et al. 1977) or negatively (BARNETT and LILLY 1955; MUJUMDAR and BHIDE 1970; TAN et al. 1972) to CO_2 enrichment. One study in a freshwater stream (GRAY 1975) found that the bacterium *Escherichia coli* ceased to grow if air without CO_2 was supplied. Lower numbers of *E. coli* cells were associated with plant blooms in the stream. The author suggested that the numbers of *E. coli* cells may be controlled by the levels of dissolved CO_2 which, in turn, depends on the volume of photosynthesizing green cells.

Increasing the growth of bacterial and fungal cultures by CO_2 enrichment of the culture medium is a well known technique. Usually the atmosphere of culture chambers is greatly enriched (2%–5%) but a few studies exist which indicate that the response is also occurring in the range 330 to 1,000 ppm CO_2 (UTTER and WOOD 1951; WOOD and UTTER 1965).

Such reactions have also been discussed for animal tissues (UTTER 1959; UTTER and WOOD 1951). Excellent reviews of the assimilation of CO_2 by dark reactions in heterotrophic organisms were provided by WOOD and STJERNHOLM

(1962) and WOOD and UTTER (1965). Possible significant pathways are the synthesis of acetate from CO_2, the conversion of CO_2 to methane, and the pathways in which the coenzyme biotin is involved (WOOD and STJERNHOLM 1962).

The study of dark CO_2 fixation in autotrophic and heterotrophic tissues should be intensified. The possible increased growth of heterotrophic tissues and organisms is one of the least understood but potentially most significant effects of CO_2 enrichment. Will root systems increase in size? Will heterotrophic decomposers and pathogens respond directly to increased CO_2? What effect would subsequent increased growth have on nutrient cycles?

15.5.10 Nitrogen Fixation

Symbiotic and non-symbiotic nitrogen fixation increases in atmospheres enriched to 1,000 ppm CO_2 (BRUN 1976; BRYAN and WRIGHT 1976; HARDY et al. 1977; HARDY and HAVELKA 1975, 1977; LAWN and BRUN 1974; MARX 1974; MINCHIN and PATE 1973; MULDER and VAN VEEN 1960; QUEBEDEAUX et al. 1975).

PHILLIPS et al. (1976) reported a very significant increase in nitrogen reduction by *Pisum sativum* infected with *Rhizobium leguminosarum* when plants were grown in light for 6 h of CO_2 enrichment (see Chap. 8, Vol. 12C). Plants grown for four weeks at 1,200 ppm CO_2 increased in growth when compared to control plants. They concluded that long-term CO_2 enrichment promotes N_2 reduction by increasing total plant and root nodule development. Such growth and N_2 fixation responses have been reported to increase from ambient to 10,000 ppm CO_2 with no increase above 10,000 ppm (WILSON et al. 1933).

In general, it is believed that N_2 fixation increases with increasing CO_2 concentration because of increasing net photosynthesis (MARX 1974) and carbohydrate available for the growth of nodules and of the bacteria. The additional possibility of direct effects of CO_2 via dark fixation reactions in root tissues and in the bacteria has not been examined and is a very important question which should be addressed.

15.5.11 Genetic Variation

Almost every species or cultivar series which has been examined has shown apparent genetic variation in plant response to increasing CO_2 (BALLARD 1958; GARDNER 1963, 1966; HANAN 1973; HAND and SOFFE 1971; HEDLEY and HARVEY 1975; HEICHEL and HEYNES 1974; HEICHEL and MUSGRAVE 1969; HUGHES and COCKSHULL 1971). The variations include effects on net photosynthesis, growth rates, phenology, and germination of seeds.

Sub-specific variations in response will potentially effect gene frequencies and therefore populations. Differences between interacting species may cause vegetational changes as suggested by BRYAN and WRIGHT (1976).

15.6 Conclusions: Possible Ecosystem Reponses to CO_2 Induced Plant Changes

Ecosystems are integrated units of one or more species functioning in space and through time (ODUM 1971). There is a well-known interaction between the abiotic and biotic components of ecosystems. The concept of holocoenosis described by BILLINGS (1974) is a good way to consider ecosystem structure and function. This concept integrates the first and second laws of thermodynamics with biological principles of selection and evolution. Basically, inorganic mass and energy are transformed via enzymes and organisms into organic compounds. The chemical potential energy which accumulates as the biomass of the primary producers is arrayed vertically and horizontally in patterns which are unique to the various major ecosystems of the world. The study of the details of the holocoenotic behavior of various ecosystems was the object of the International Biological Program. From these system studies and others like them, we now know that according to Newton's third law of motion, "To every action there is always opposed an equal reaction", there is an analogy in ecology. For ecosystems, John MUIR (1911) expressed the concept as "When we try to pick out anything by itself, we find it hitched to everything else in the universe."

Since carbon is the basic building block of life, as we know it, and since carbon dioxide is the primary inorganic source of carbon, it is probable that increasing the atmospheric supply of the gas will have effects on global ecosystems. Although it was the specific objective of this chapter to consider the possible responses of plants to an increase in carbon dioxide, it is assumed that significant changes in the primary producers will cause changes in consumers and decomposers. Long-term chronically increasing carbon dioxide concentrations probably will cause successional changes in global ecosystems. The degree of such change cannot be predicted at this time.

References

Akita S, Moss DN (1972) Differential stomatal response between C_3 and C_4 species to atmospheric CO_2 concentration and light. Crop Sci 12:789–793

Attiwell PM (1971) Atmospheric carbon dioxide and the biosphere. Environ Pollut 1:249–261

Baes CF, Goeller HE, Olson JS, Rotty RM (1977) Carbon dioxide and climate: the uncontrolled experiment. Am Sci 65:310–320

Ballard L (1958) Studies of dormancy in the seeds of subterranean clover (*Trifolium subterraneum* L.) I. Breaking of dormancy by carbon dioxide and by activated carbon. Aust J Biol Sci 11:246–260

Barnett HL, Lilly VG (1955) The effects of humidity, temperature, and carbon dioxide on the sporulation of *Choanephora cucurbitarum*. Mycologia 47:26–29

Bavel van C (1974) Antitranspirant action of carbon dioxide on intact sorghum plants. Crop Sci 14:208–212

Billings WD (1974) Environment: Concept and reality. In: Strain B, Billings WD (eds) Vegetation and environment. Handbook of vegetation science, Vol VI. Junk, The Hague, pp 9–35
Broecker WS (1973) Factors controlling CO_2 content in the oceans and atmosphere. In: Noodwell GM, Pecan EV (eds) Carbon and the biosphere. Proc 24th Brookhaven Symp Biol, pp 32–50
Broecker WS, Li Y-H (1970) Interchange of water between the major oceans. J Geophys Res 75:3545–3552
Broecker WS, Takahashi T, Simpson HJ, Peng TH (1979) Fate of fossil fuel carbon dioxide and the global carbon budget. Science 206:409–418
Brown HT, Escombe F (1902) The influence of varying amounts of carbon dioxide in the air on the photosynthetic process of leaves and on the mode of growth of plants. Proc R Soc London Ser B 70:397–413
Brown HT, Escombe F (1905) Researches on some of the physiological processes of green plants with special reference to the interchange of energy between the leaf and its surroundings. Proc R Soc Lond Ser B 76:29–111
Brun WA (1976) The relation of N_2 fixation to photosynthesis. In: Hill LD (ed) World Soybean Res Conf. Danville, Il, pp 135–150
Bryan W, Wright RD (1976) The effect of enhanced CO_2 levels and variable light intensities on net photosynthesis in competing mountain trees. Am Midl Nat 95:446–450
Bryson RA, Wendland WM (1970) Climatic effects of atmospheric pollution. In: Reidel D (ed) Global effects of environmental pollution. Springer, Berlin Heidelberg New York, pp 130–138
Calvert A (1972) Effects of day and night temperatures and carbon dioxide enrichment on yield of glasshouse tomatoes. J Hortic Sci 47:231–247
Cock JH, Yoshida S (1973) Changing sink and source relations in rice (*Oryza sativa* L.) using carbon dioxide enrichment in the field. Soil Sci Plant Nutr 19:229–234
Egli DB, Pendleton JW, Baker DB (1970) Photosynthetic rate of three soybean communities as related to carbon dioxide levels and solar radiation. Agron J 62:411–414
Elliott WP, Machta L (1979) Carbon dioxide effects research and assessment Program. Workshop on the global effects of carbon dioxide from fossil fuels. Miami Beach, Florida, March 7–11, 1977. CONF 77-0385, US Dep Energy
Ford MA, Thorne GN (1967) Effect of carbon dioxide concentration on growth of sugarbeet, barley, kale, and maize. Ann Bot NS (London) 31:629–644
Gardner R (1963) Response differs with variety. Grower 59:813–814
Gardner R (1966) Effects of carbon dioxide enrichment on crops of tomatoes, lettuce, and chrysanthemums as grown commercially on the N.A.A.S. experimental horticultural stations in England and Wales. Proc 17th Hortic Congr Md (1):347
Gates DM (1968) Transpiration and leaf temperature. Annu Rev Plant Physiol 19:211–238
Gifford RM (1977) Growth pattern, carbon dioxide exchange and dry weight distribution in wheat growing under different photosynthetic environments. Aust J Plant Physiol 4:99–110
Gifford RM (1979) Growth and yield of carbon dioxide enriched wheat under water-limited conditions. Aust J Plant Physiol 6:367–378
Gifford RM, Bremmer PM, Jones DB (1973) Assessing photosynthetic limitation to grain yield in a field crop. Aust J Agric Res 24:297–307
Goldsberry KL (1963) The effects of carbon dioxide on carnation growth. Proc Am Soc Hortic Sci 83:753–760
Gray EA (1975) Survival of *Escherichia coli* in stream water in relation to carbon dioxide and plant photosynthesis. J Appl Bacteriol 39:47–54
Green K, Wright R (1977) Field response of photosynthesis to CO_2 enhancement in ponderosa pine. Ecology 58:687–692
Hanan JJ (1973) Effect of elevated temperatures and carbon dioxide levels on rose production. Hortscience 8:266–267
Hand DW, Postlethwaite JD (1971) The response of CO_2 enrichment of capillary-watered single-truss tomatoes at different plant densities and seasons. J Hortic Sci 46:461–470

Hand DW, Soffe RW (1971) Light-modulated temperature control and the response of greenhouse tomatoes to different CO_2 regimes. J Hortic Sci 46:381–396
Hansen J, Johnson D, Lacis A, Lebedeff S, Lee T, Rind D, Russell G (1981) Climate impact of increasing atmospheric carbon dioxide. Science 213:957–966
Hardy R, Havelka UD (1975) Photosynthate as a major factor limiting nitrogen fixation by field-grown legumes with emphasis on soybeans. In: Nutman PS (ed) Symbiotic nitrogen fixation in plants. Cambridge Univ Press, London, pp 421–439
Hardy R, Havelka UD (1977) Possible routes to increase the conversion of solar energy to food and feed by grain legumes and cereal grains (crop production): CO_2 and N_2 fixation, foliar fertilization, and assimilate partitioning. In: Mitsui A, Miyachi S, San Pietro A, Tamura S (eds) Biological solar energy conversion. Academic Press, London New York, pp 299–322
Hardy R, Criswell JG, Havelka UD (1977) Investigations of possible limitations of nitrogen fixation by legumes: (1) methodology, (2) identification, and (3) assessment of significance. In: Newton EW, Postgate JR, Rodriquez-Barrueco C (eds) Proceedings 2nd International Symposium on Nitrogen Fixation. Proc and Int Symp Nitrogen Fixation. Academic Press, London New York
Hart P, Wright RD (1976) CO_2 compensation point for photosynthesis: effect of variable CO_2 and soil moisture levels. Am Midl Nat 95:450–455
Hedley CL, Harvey DM (1975) The involvement of CO_2 uptake in the flowering behavior of two varieties of *Antirrhinum majus*. In: Marcelle R (ed) Environmental and biological control of photosynthesis. Junk, The Hague, pp. 408
Heichel GJ, Haynes RA (1974) Stimulating emergence and growth of *Kalmia* genotypes with carbon dioxide. Hortscience 9:60–62
Heichel GH, Musgrave RB (1969) Relation of CO_2 compensation concentration to apparent photosynthesis in maize. Plant Physiol 44:1724–1728
Hesketh JD, Hellmers H (1973) Floral initiation in four plant species growing in CO_2 enriched air. Environ Control Biol 11:51–53
Ho LC (1977) Effects of CO_2 enrichment on the rates of photosynthesis and translocation of tomato leaves. Ann Appl Biol 87:191–200
Hobbs PV, Harrison H, Robinson E (1974) Atmospheric effects of pollutants. Science 183:909–915
Hofstra G, Hesketh JD (1975) The effects of temperature and CO_2 enrichment on photosynthesis in soybean. In: Marcelle R (ed) Environmental and biological control of photosynthesis. Junk, The Hague, pp 71–80
Hou L, Hill AC, Soleimani A (1977) Influence of CO_2 on the effects of SO_2 and NO_2 on alfalfa. Environ Pollut 12:7–16
Hughes AP, Cockshull KE (1969) Effects of carbon dioxide concentration on the growth of *Callistephus chinensis* cultivar Johannistag. Ann Bot 33:351–365
Hughes AP, Cockshull KE (1971) The variation in response to light intensity and carbon dioxide concentration shown by two cultivars of *Chrysanthemum morifolium* grown in controlled environments at two times of year. Ann Bot (London) 35:933–945
Hurt P, Wright RD (1976) CO_2 compensation point for photosynthesis: Effect of variable CO_2 and soil moisture levels. Am Midl Nat 95:450–455
Idso SB, Brazel AJ (1977) Planetary radiation balance as a function of atmospheric dust: climatological consequences. Science 198:731–733
Ioannow N, Schneider RW, Grogan RG (1977) Effect of oxygen, carbon dioxide and ethylene on growth, sporulation and production of microsclerotia by *Verticillium dahliae*. Phytopathology 67:645–650
Keeling CD (1973) The carbon dioxide cycle: reservoir models to depict the exchange of atmospheric carbon dioxide with the oceans and land plants. In: Rasool SI (ed) Chemistry of the lower atmosphere. Plenum Press, New York
Keeling CD (1978) Atmospheric carbon dioxide in the 19th century. Science 202:1109
Knecht GN, O'Leary JW (1972) Components of increased leaf growth resulting from carbon dioxide enrichment. J Colo-Wyo Acad Sci 7:2–3
Kriedemann PE, Sward RJ, Downtown WJS (1976) Vine response to carbon dioxide enrichment during heat therapy. Aust J Plant Physiol 3:605–618

Krizek DT, Zimmerman RH, Klueter HH, Bailey WA (1969) Accelerated growth of birch and crabapple seedlings under CO_2 enriched atmospheres. Abstr Plant Physiol Suppl 44:15
Kukla G, Gavin J (1981) Summer ice and carbon dioxide. Science 214:497–503
Kunitake G, Stitt C, Saltman P (1959) Dark fixation of CO_2 by tobacco leaves. Plant Physiol 34:123–127
Lawn RJ, Brun WA (1974) Symbiotic nitrogen fixation in soybeans. I. Effect of photosynthetic source sink manipulation. Crop Sci 14:22–25
Lehman JT (1978) Enhanced transport of inorganic carbon into algal cells and its implications for the biological fixation of carbon. J Phycol 14:33–42
Leonard OA, Pinckard JA (1946) Effects of various oxygen and carbon dioxide concentrations on cotton root development. Plant Physiol 21:18–36
Li Y-H (1972) Geochemical mass balance among lithosphere, hydrosphere, and atmosphere. Am J Sci 272:119–137
Lommen PW, Schwintzer CR, Yocum CW, Gates DM (1971) A model describing photosynthesis in terms of gas diffusion and enzyme kinetics. Planta 98:195–220
Lucas WJ (1975) Photosynthetic fixation of ^{14}C by internodal cells of *Chara corallina.* J Exp Bot 26:331–346
Ludlow MM, Wilson GL (1971) Photosynthesis of tropical pasture plants. I. Illuminance, carbon dioxide concentration, leaf temperature, and leaf-air vapour pressure difference. Aust J Biol Sci 24:449–470
Madsen E (1968) Effect of CO_2-concentration on the accumulation of starch and sugar in tomato leaves. Physiol Plant 21:168–175
Madsen E (1971) Cytological changes due to the effect of carbon dioxide concentration on the accumulation of starch in chloroplasts of tomato leaves. R Vet Agric Univ Yearb, Copenhagen, pp 191–194
Madsen E (1973a) The effect of CO_2-concentration on development and dry matter production in young plants. Acta Agric Scand 23:235–240
Madsen E (1973b) Effect of CO_2-concentration on the morphological, histological and cytological changes in tomato plants. Acta Agric Scand 23:241–246
Madsen E (1976) Effect of CO_2-concentration on the morphological, histological, and cytological and physiological processes in tomato plants. Dissertation, State Seed Test Stan, Lyngby, Denmark
Majernik O, Mansfield TA (1972) Stomatal responses to raised atmospheric CO_2 concentration during exposure of plants to SO_2 pollution. Environ Pollut 3:1–7
Manabe S, Stouffer RJ (1979) CO_2-climate sensitivity study with a mathematical model of the global climate. Nature (London) 282:491–493
Manabe S, Wetherald RT (1975) The effects of doubling the CO_2 concentration on the climate of a general circulation model. J Atmos Sci 32:3–15
Marx J (1974) Nitrogen fixation: research efforts intensify. Science 185:132–136
Mattson RH, Widmer RE (1971) Year-round effects of carbon dioxide supplemented atmospheres on greenhouse rose (*Rosa hybrida*). J Am Soc Hortic Sci 86:487–488
Mendonca BG, Hanson KJ, DeLuisi JD (1978) Volcanically related secular trends in atmospheric transmission at Mauna Loa Observatory, Hawaii. Science 202:513–515
Minchin FR, Pate JS (1973) The carbon balance of a legume and the functional economy of its root nodules. J Exp Bot 24:259–271
Mitchell JM (1971) The effect of atmospheric aerosols on climate with special reference to temperature near the earth's surface. J Appl Meteorol 10:703–714
Moss DN, Musgrave RB, Lemon ER (1961) Photosynthesis under field conditions. III. Some effects of light, carbon dioxide, temperature, and soil moisture on photosynthesis, respiration, and transpiration of corn. Crop Sci 1:83–87
Muir J (1911) My first summer in the Sierra. Houghton Mifflin, Boston
Mujumdar SB, Bhide JP (1970) Effect of carbon dioxide concentration on fungi from rhizosphere and non rhizosphere soil of sugarcane. J Univ Poona 38:137–141
Mulder EG, Veen van WL (1960) The influence of carbon dioxide on symbiotic nitrogen fixation. Plant Soil 13:265–278
Nevins DJ, Loomis RS (1970) Nitrogen nutrition and photosynthesis in sugar beet (*Beta vulgaris* L.) Crop Sci 10:21–25

Newton P (1965) Growth of *Cucumis sativus*, variety Butcher's Disease Resister, with two concentrations of carbon dioxide. Ann Appl Biol 56(1):55–64
Odum E (1974) Fundamentals of ecology, 3rd edn. Saunders, Philadelphia
Oeschger H, Siegenthaler U, Schotterer U, Gugelmann A (1975) A box diffusion model to study the carbon dioxide exchange in nature. Tellus 27:168–192
Olson JS, Pfuderer HA, Chan Y-H (1978) Changes in the global carbon cycle and the biosphere. ORNL/EIS-109. Oak Ridge Natl Lab, Oak Ridge, TN
Paez A, Hellmers H, Strain B (1980) Apical dominance in pea plants unaffected by CO_2 enrichment. Physiol Plant 50:43–46
Pallas JE Jr (1965) Transpiration and stomatal opening with changes in carbon dioxide content of the air. Science 147:171–173
Phillips DA, Newell KD, Hassell SA, Felling CE (1976) The effect of CO_2 enrichment on root nodule development and symbiotic N_2 reduction in *Pisum sativum* L. Am J Bot 63:356–362
Quebedeaux B, Havelka UD, Livak KL, Hardy R (1975) Effect of altered CO_2 in the aerial part of soybean on symbiotic N_2 fixation. Plant Physiol 56:761–764
Raschke K (1972) Saturation kinetics of the velocity of stomatal closing in response to CO_2. Plant Physiol 49:229–234
Raven JA (1968) The mechanism of photosynthetic use of bicarbonate by *Hydrodictyon africanum*. J Exp Bot 19:193–208
Regehr DL, Bazzaz FA, Boggess WR (1975) Photosynthesis, transpiration and leaf conductance of *Populus deltoides* in relation to flooding and drought. Photosynthetica 9:52–61
Robock A (1978) Internally and externally caused climate change. J Atmos Sci 35:1111–1122
Schlesinger WH (1977) Carbon balance in terrestrial detritus. Annu Rev Ecol Syst 8:51–81
Schneider SH (1975) On the carbon dioxide climate confusion. J Atmos Sci 32:2060–2066
Sionit N, Hellmers H, Strain B (1980a) Growth and yield of wheat under CO_2 enrichment and water stress conditions. Crop Sci 20:687–690
Sionit N, Strain B, Hellmers H, Kramer P (1980b) Effects of atmospheric CO_2 concentration and water stress on water relations of wheat. Bot Gaz 142:191–196
Soeder CJ, Strotmann H, Gallough RA (1966) Carbon-dioxide-induced delay of cellular development in two chlorella species. J Phycol 2:117–120
Splittstoesser W (1966) Dark CO_2 fixation and its role in the growth of plant tissue. Plant Physiol 41:755–759
Steeman-Nielsen E (1947) Photosynthesis of aquatic plants with special reference to carbon sources. Dan Bot Ark 12:1–71
Strain BR (ed) (1978) Report of the workshop on anticipated plant responses to global carbon dioxide enrichment. Dep Bot, Duke Univ, Durham, NC
Stuiver M (1978a) Atmospheric carbon dioxide in the 19th century. Science 202:1109
Stuiver M (1978b) Atmospheric carbon dioxide and carbon reservoir changes. Science 199:253–258
Stumm W, Morgan JJ (1970) Aquatic chemistry: An introduction emphasizing chemical equilibria in natural waters. Wiley-Interscience, New York
Takahashi T (1979) Carbon dioxide chemistry in ocean water. In: Elliot WP, Machta L (eds) Carbon Dioxide Effects Research and Assessment Program. Workshop on the global effects of carbon dioxide from fossil fuels. US Dep Energy, CONF 77-0385, pp 63–71
Takami S, Bavel van CHN (1976) Numerical experiments on influence of CO_2 release at ground level on crop assimilation and water use. Agric Meteorol 15:193–203
Tan CK, Badour SS, Waygood ER (1972) Delayed zoospore release by *Chlamydomonas segnis* at high carbon dioxide tension. J Phycol Suppl 8:4
Tenhunen JD, Westrin SS (1979) Development of a photosynthesis model with an emphasis on ecological applications. IV. Wholephot – Whole leaf photosynthesis in response to four independent variables. Oecologia 41:145–162
Tenhunen JD, Yocum CS, Gates DM (1976a) Development of a photosynthesis model with an emphasis on ecological applications. I. Theory. Oecologia 26:89–100

Tenhunen JD, Weber JA, Yocum CS, Gates DM (1976b) Development of a photosynthesis model with an emphasis on ecological applications. II. Analysis of a data set describing the P_M surface. Oecologia 26:101–119
Tenhunen JD, Weber JA, Filipek LH, Gates DM (1977) Development of a photosynthesis model with an emphasis on ecological applications. III. Carbon dioxide and oxygen dependencies. Oecologia 30:189–207
Tenhunen JD, Weber JA, Yocum CS, Gates DM (1979a) Solubility of gases and the temperature dependence of whole leaf affinities for carbon dioxide and oxygen. Plant Physiol 63:916–923
Tenhunen JD, Hesketh JD, Gates DM (1979b) Leaf photosynthesis models. In: Predicting photosynthate production and use for ecosystem models. CRC Press, West Palm Beach
Ting, I.P., Dugger WM Jr (1965) Transhydrogenation in root tissue: mediation by carbon dioxide. Science 150:1727–1728
Ting IP, Dugger WM Jr (1966) CO_2 fixation in *Opuntia* roots. Plant Physiol 41:500–505
Ting IP, Dugger WM Jr (1967) CO_2 metabolism in corn roots. I. Kinetics of carboxylation and decarboxylation. Plant Physiol 42:712–718
Utter MF (1959) The role of CO_2 fixation in carbohydrate utilization and synthesis. Ann N Y Acad Sci 72:451–461
Utter MF, Wood HG (1951) Mechanisms of fixation of carbon dioxide by heterotrophs and autotrophs. Adv Enzymol 12:41–151
Vazart J, Henry Y (1972) The influence of different cultivation conditions on the infrastructure of four species of *Gonium*. H. Microscopy 14:99A
Vidaver, W (1972) Dissolved gases: Plants. In: Kinne O (ed) Marine ecology: a comprehensive, integrated treatise on life in oceans and coastal waters. Wiley-Interscience, New York, pp 1478–1479
Whittaker RH, Likens GE (1973) Carbon in the biota. Brookhaven Symp Biol 24:281–302
Wilson D, Cooper JP (1969) Effects of light intensity and CO_2 on apparent photosynthesis and its relation with leaf anatomy in genotypes of *Lolium perenne* L. New Phytol 68:627–644
Wilson PM, Ford EB, Salmon MR (1933) Relation between carbon dioxide and elemental nitrogen assimilation in leguminous plants. Soil Sci 35:145–165
Wittwer S (1979) Future technological advances in agriculture and their impact on the regulatory environment. BioScience 29:603–610
Wittwer S (1980) Environmental effects on the managed biosphere. In: Ravel R (ed) Environmental and societal consequences of a possible CO_2-induced climate change. AAAS-DOE Workshop Rep, 2–6 April 1979, Annapolis, Maryland
Wong SC (1979) Elevated atmospheric partial pressure of CO_2 and plant growth. I. Interaction of nitrogen nutrition and photosynthetic capacity in C_3 and C_4 plants. Oecologia 44:68–74
Wood HG, Stjernholm RL (1962) Assimilation of carbon dioxide by heterotrophic organisms. In: Gunsalus IC, Stanier RY (eds). The bacteria, vol III. Academic Press, London New York, pp 41–117
Wood HG, Utter MF (1965) The role of CO_2 fixation in metabolism. In: Campbell PN, Greville GD (eds) Essays in biochemistry. Academic Press, London New York, pp 1–27
Woodwell GM, Whittaker RH, Reiners WA, Likens GE, Delwiche CC, Botkin DB (1978) The biota and the world carbon budget. Science 199:141–146
Yoshida S (1973) Effects of carbon dioxide enrichment at different stages of panicle development on yield components and yield of rice (*Oryza sativa*). Soil Sci Plant Nutr 19:311–316

16 Man's Influence on Ecosystem Structure, Operation, and Ecophysiological Processes

W.D. BILLINGS

CONTENTS

16.1 Introduction 528

16.2 Natural Ecosystems 528
16.2.1 Evolution of Natural Ecosystems 528
16.2.2 Structure of Natural Ecosystems 528
16.2.3 Certain Ecosystemic Principles 529
16.2.3.1 The Principle of Limiting Factors 529
16.2.3.2 The Holocoenotic Principle 530
16.2.3.3 The Principle of Trigger Factors 530

16.3 The Origin of Man as a Part of Natural Ecosystems 531

16.4 The Spread of Man into Other Natural Ecosystems 531

16.5 The Origin and Spread of Crop Agricultural Ecosystems 532
16.5.1 Some Ecophysiological Bases of the First Crop Plants 532
16.5.2 Co-Evolution of Plants and Man in Agricultural Ecosystems 535

16.6 The Alteration or Destruction of Natural Ecosystems by Man: The Ecophysiological Bases 536
16.6.1 The Replacement of Natural Ecosystems by Agricultural Systems . . 537
16.6.2 The Replacement of Natural and Agricultural Ecosystems by Urbanization 538

16.7 Some of Man's Effects on Plant Processes, Ecosystems, and the Biosphere 539
16.7.1 Direct Effects 539
16.7.1.1 Effects of Deforestation and Possible Revegetation 539
16.7.1.2 Effects of Feet and Vehicles on Grasslands 540
16.7.1.3 Effects of Herbicides, Pesticides, and Oil on Terrestrial Ecosystems 544
16.7.2 Indirect Effects 546

16.8 Conclusions: Ecophysiology and Ecosystems 546

References 547

16.1 Introduction

Until the mid-Pleistocene the earth's biota evolved and formed communities and ecosystems without man as a constituent or as an influent. Today no part of the biosphere is free of man's effects, great or small. Even on the high, icy plateau of the South Pole, there is the same rate of increase in atmospheric carbon dioxide as on Mauna Loa in Hawaii. This new carbon dioxide is derived mainly from the burning of fossil fuels by industrial peoples.

Tropical forests have been and continue to be destroyed ever more rapidly by clearing, just as have so many other ecosystems. Natural ecosystems on land, untouched by man's effects, probably no longer exist, even though there are still great areas that appear at first glance to be relatively untouched. These are mostly in the boreal taiga and in the arctic; but as with tropical rainforests, what seems at first to be relatively little impact results in the alteration of such systems and their eventual disappearance.

16.2 Natural Ecosystems

16.2.1 Evolution of Natural Ecosystems

Strictly speaking, natural ecosystems do not "evolve" in the sense that plant and animal taxa do or that mankind does. In themselves, ecosystems do not possess genes, and in addition to living organisms, they also include the physical environment. Ecosystems are aggregations of populations of plants, animals, and microorganisms and their encompassing environment. The populations of single taxa, however, do evolve within an ecosystem through time. Populations also migrate into or out of an ecosystem as environments change. Mutations and/or phenotypic adaptations within a population or single organism may allow movement of reproductive units across shorter or longer distances. Ecosystems consist of evolving and migrating populations which come together through time and are continually changing. Change in most ecosystem types is rather rapid. Other ecosystems, such as that dominated by the California coastal redwood (*Sequoia sempervirens*), are relatively stable and change little in biological composition and general environment over long periods of time even though their total area may decrease.

16.2.2 Structure of Natural Ecosystems

Ecosystems consist of (1) a biological community made up of populations of plants, animals, microorganisms, and now, usually, man, and (2) its environment of physical and biological factors. In terms of biomass and physical space occupied, green vegetation is paramount. Animals, whether herbivorous, carnivorous, or omnivorous, are present in much smaller quantities and mass. In natural ecosystems, man is a relatively rare component. But the biomass of microorganisms can be quite great, although not so much as that of green plants unless the vegetation has been destroyed or severely damaged.

Green vegetation is the foundation of any ecosystem: (1) it captures solar energy in photosynthesis, uses it as chemical energy, and passes energy along to other trophic levels with decreasing efficiency, (2) vegetation absorbs mineral nutrients from soil and substratum, fixes these into compounds which are also passed through the other components of the community, and (3) vegetation modifies the physical environment, acts as a pathway for water from soil to air, and determines, directly or indirectly, what other organisms can exist in the ecosystem.

From green vegetation, two main pathways with many interconnections carry energy and nutrients to the ultimate carnivores with loss of energy to the system by respiration all along the way. These two pathways are (1) the "grazing" food chain in which green plants are eaten by herbivorous vertebrates and invertebrates which in turn are eaten by various kinds of carnivores and (2) the "detrital" food chain in which dead plant and animal material is utilized primarily by belowground or surficial "feeding" microorganisms or invertebrates which, of course, are eaten by invertebrate or even vertebrate carnivores which utilize this energy pathway. Many kinds of vertebrates, certain ground-feeding omnivorous birds, for example, operate in both pathways. Microorganisms involved in decomposition, mineral release, or nitrogen-fixation play a large role in the operation of the total ecosystem itself. In many, if not most, ecosystems there is a greater weight of these small microorganisms, including fungi, than exists in the "grazing" food web.

16.2.3 Certain Ecosystemic Principles

Ecosystems operate according to the Laws of Thermodynamics. The trophic structure and trophic dynamics of such systems obey these laws. In addition, it has become obvious in the last century or so that certain other principles govern the structure and operation of populations, communities, and ecosystems. Among the most important of these are the following.

16.2.3.1 The Principle of Limiting Factors

It has been realized for some time that growth and reproduction of organisms, the potential geographic ranges of species, and the productivities of crop ecosystems are limited by the interactions of environmental components and the gene systems of the organisms involved. This principle, first suggested by LIEBIG in the mid-19th century for crops, is now generally recognized to be applicable to all organisms and even to natural ecosystems (BILLINGS 1952, 1974). It is the Principle of Limiting Factors. It works most clearly through individual phenotypes and local populations. Limiting factors are fairly easy to determine by experimentation for single-variety crop ecosystems, but are difficult to determine for multi-specied natural ecosystems because of the very diversity of such systems. Each species in the system has its own tolerance range and needs in regard to environmental components. A factor limiting one species may have little effect on another in the same ecosystem. Also, factors which may be limiting at one time may not be limiting at another. As a factor ceases to be limiting, it is soon replaced by another, or becomes limiting again by a change in its own value or by increased demand by the organism or population. Limiting factors are the checks and balances which keep populations and biological communities under control and within bounds.

16.2.3.2 The Holocoenotic Principle

Any change in a limiting factor will eventually be reflected throughout the whole ecosystem as populations grow or decrease in size. Also, anything *new*, either physical or biological, added to the environment or system will also cause far-reaching effects within an ecosystem. Since the biosphere is an ecosystem, this is true of the earth as a whole. Therefore, any change in temperature or carbon dioxide content of the atmosphere, or any addition of radionuclides or toxic chemicals results in complex chains of events within the system. This also applies to the introduction or extinction of a kind of organism. Such ecosystemic changes are almost always irreversible. They also indicate that there are no barriers or cross-walls to impede changes within the system. The ecosystem is "holocoenotic." This has been known informally for a long time. For a modern treatment of the principle, see BILLINGS (1952, 1974, 1978).

16.2.3.3 The Principle of Trigger Factors

Since an ecosystem is holocoenotic, it can be changed quickly or slowly by changes in limiting factors or by additions or subtractions of environmental or community components. Any such changes can trigger intricate chain reactions within ecosystems or the biosphere. Some of these are quick as are the effects of fire in an old forest; some are slow as, for example, the migrations of marine organisms through sea-level canals such as the series of freshwater canals connecting the Atlantic Ocean with the Great Lakes in North America or the Suez Canal in the Middle East. Through the Suez, Indian Ocean biota are slowly invading the Mediterranean, but more rapidly than those from the Mediterranean are moving southward (see POR 1978). Some changes are so subtle that they may not be noticed until the reaction is far along and it is too late to do much about it. Destruction of the stratospheric ozone layer by fluorocarbons and by high-flying aircraft with resultant increases in solar ultraviolet irradiance at the earth's surface is one that might be solved (Chap. 6, Vol. 12A); but what about the steady increases in atmospheric carbon dioxide (Chap. 15, this Vol.)?

Any initiating factor to such chain reactions within an ecosystem, whether it be a change in a limiting factor or the addition or loss of an ecosystemic component, is termed a "trigger" factor (BILLINGS 1952, 1974).

16.3 The Origin of Man as a Part of Natural Ecosystems

Although ancestral hominids date from Pliocene and even late Miocene times in east central Africa, it was not until the latest Pliocene and earliest Pleistocene (3 million to 1 million years B.P.) that *Homo habilis* and eventually *Homo erectus,* the ancestors of modern man (*Homo sapiens*), evolved from these (LEAKEY et al. 1965, 1967, 1971; BRONOWSKI 1973; BOUGHEY 1975). These early men made tools or choppers of pebbles and other stones. They were hunter-gatherers who developed labor-sharing by living and hunting in groups. This required the development of language. Another, and ecologically important trait developed that persists until the present day: habitat degradation or change by hominids which made life rigorous or impossible for their niche competitors (BOUGHEY 1975).

The evolution of man from primitive vegetarianism to omnivorous and carnivorous habits was accompanied by the development of group and cooperative hunting of big game. This was as necessary as hunting tools.

16.4 The Spread of Man into Other Natural Ecosystems

Hunting, in itself, cannot support a growing population of people in one place; productivity of higher trophic levels is not great enough even in relatively rich grassland ecosystems. If the main source of food is wild meat, human populations must either follow the native herbivore herds, or wait for the migrations of these herds to come past. Otherwise, people starve. Since the latter choice is self-defeating, and the second is risky, the best options are to follow the herds ("wild" transhumance) or to move to far and as yet unexploited regions.

Gradual migrations took thousands of years, but in terms of evolutionary time this is short. And yet, man was evolving biologically and culturally faster than his prey. By one million years ago, man was in northern Africa, by 700,000 years ago in Java, and by 400,000 years ago in China and western Europe. However, it was not until the Bering Land Bridge appeared above the sea during late full-glacial times that the Americas began to be populated. The dates for arrival in Alaska of the ancestral American Indians remain uncertain but 30,000 years B.P. is a good estimate. As long as continental ice remained between Alaska and the south, these peoples could not migrate farther; but once a corridor melted just east of the Canadian Rocky Mountains, they moved rapidly the thousands of miles southward to the southern tip of South America. The Eskimos (Inuit) and Aleuts arrived much later, certainly no more than 10,000 years B.P. Amazingly, islands of the South Pacific including New Zealand and Hawaii also were not invaded until recent millenia or even centuries B.P.

At the same time as the northward migrations of man from Africa into Eurasia, the climate cooled several times with continental glaciations advancing and retreating. Man adapted rapidly to cold climates by the use of shelter and fire. Large mammals of many genera (mammoths, mastodons, bison, etc.)

also adapted rapidly to these colder ecosystems and provided a source of food for the people. Hunting techniques improved. Man depended less on plants and more on wild game. New ways of hunting were developed such as following herds, killing in great numbers often by driving the herds over cliffs. The vegetation could support only a relatively few large mammals per km^2, which resulted in migratory herds. Evidence is ample that the extinction of many of these species and genera of large mammals was aided by people. Man may, indeed, have been the main cause (MARTIN and WRIGHT 1967).

Still, some of the large herbivore species survived, as for example, the American caribou and the Eurasian reindeer (both are subspecies of *Rangifer tarandus*) and one species of American bison (other species of American bison were killed off). The northern Eurasian peoples, having little in the way of plant foods, followed migrations of the wild reindeer herds in a form of transhumance. In the process, the reindeer became semi-domesticated. Only the Lapps of northern Scandinavia still continue this mode with relatively wild animals but there are some variations on the theme. Some of the inland Eskimos in northern Alaska wait until caribou herds come through certain mountain passes or along river bluffs and kill animals without having to travel: a sort of "stay-at-home" transhumance. This goes on today. As recently as two centuries ago, this latter technique was practiced by the Indians of the American West with bison, wapiti, and mountain sheep. This tradition went back for thousands of years and in the process, horses and camels were driven to extinction in western North America. But with the re-introduction of the horse by the Spanish (who rode them rather than ate them), the Indians quickly learned a new form of mobile hunting on horseback. This persisted until only a century ago.

16.5 The Origin and Spread of Crop Agricultural Ecosystems

As recently as 20,000 years ago, man was still a hunter and forager in all parts of the earth which he had reached. Only a little over 10,000 years B.P., came a sudden change in the Old World: the beginnings of crop agriculture in the Middle East and elsewhere. While certain groups of nomads and hunters retained the old wild modes, and some still do, the great majority of peoples went the agricultural way. BRONOWSKI (1973) calls this change from nomad to village agriculture the biggest single step in the cultural ascent of man. SOLHEIM (1972) has some evidence that tropical crop agriculture began as early as 13,000 years B.P. in Southeast Asia with entirely different kinds of plants from those of the Middle East. In the Americas, the change came a bit later (8,000 to 9,000 years B.P.) in Mexico and the Andes (HARLAN 1976).

16.5.1 Some Ecophysiological Bases of the First Crop Plants

It is important to note that the first agricultural plants were annuals. Weeds in and around disturbed areas outside of the wet tropics are also mostly annual plants. Man, in coming together in campsites, unconsciously encouraged pioneer weedy annual plants to invade abandoned and temporary sites, and around the edges of occupied ones. Rapidly growing annuals with at least one generation per year and exposed to catastrophic selection by man's impact were candidates for relatively fast co-evolution with *Homo sapiens*.

In the Middle East, such pioneer grasses included species of *Triticum, Aegilops,* and *Agropyron*. Before 10,000 years B.P., wheat was not what it is today but consisted of several species of *Triticum* among the other wild and weedy grasses. Diploid chromosome numbers of these species were 14. In the wild, at least one accidental cross occurred between *Triticum aegilopoides* and an *Agropyron* or *Aegilops* to produce *T. dicoccoides,* wild emmer, the ancestor of *T. dicoccum,* cultivated emmer. The hybrid was a tetraploid with $2n = 28$. Other hybrids occurred between the 14-chromosome grasses. A few of these hybrids were probably between wild species and the early cultivated forms. Some of the crosses resulted in unstable 14- or 21-chromosome offspring. The only ones of these to survive and to provide productive crop plants were those in which the chromosome numbers doubled to form stable polyploids of 28 or 42 chromosomes. Because of their higher yield, they were selected by the primitive farmers. Without such artificial selection, the polyploid wheats would probably have disappeared in natural selection. It was man who kept them going and, in so doing, set in motion one of the many trains of co-evolution between crop plants and man which continue at an even more rapid pace today.

The invention of the plow in the Middle East, coming on the heels of the evolution of the first cultivated wheats and the domestication of draft animals, made possible a series of quantum jumps in productivity of grain crops. Thus, in the temperate climatic regions, was born the single-specied ecosystem or monoculture which with all its advantages and dangers has replaced natural ecosystems over so much of the earth.

As opposed to the situation in the Old World, the old hunting and gathering way of life persisted among the western North American Indians and Eskimos until the 19th and 20th centuries. In some ways, it still persists as with the whale, walrus, seal, and caribou hunting of the Eskimos of the North Slope in Alaska. But many changes have taken place since the airplane became the principal mode of transportation in the Arctic during the 1930's. Since World War II, the changes have been accelerated until now, in this cold region where agriculture is impossible, the old hunting and gathering culture unfortunately may be dying out. It has been replaced by a dependence on food flown in from warmer climates. This sudden jump in a few decades from a culture based upon the productivities of natural ecosystems, of which man was a part, directly to a society based upon industry (oil) and tourism is extremely unusual in the history of man's relationships with ecosystems. The carbohydrate-rich roots and rhizomes of most of the native tundra plants, eaten a generation ago, remain ungathered.

A similar situation prevailed into the early 20th century and into the depression days of the 1930's among the Paiute Indians of the Great Basin in Nevada. Seeds of many desert plants such as *Oryzopsis hymenoides* (Gramineae) and *Mentzelia* spp. (Loasaceae) were gathered and supplemented a diet of fish and meat (WHEAT 1967). Even today, these Indians annually harvest the big seeds or "pine nuts" of the pinyon (*Pinus monophylla*) which grows abundantly on the desert mountainsides.

The agricultural revolution came independently, and not so much later than in the Middle East, to the American Indians in Mexico and the Andean highlands of South America. In these places, the earliest crop agriculture developed between 8,000 and 9,000 years B.P. The first crop plants in the Andes were lima beans and common beans, followed shortly later by potatoes and peanuts (HARLAN 1976). In Mexico, the first crops were probably squash and gourds followed by maize, beans, and tomatoes. In the lowland tropics of the Americas, other plants were domesticated such as avocados, papayas, cassava and sweet potatoes. In the continental parts of North America, sunflower was the first domesticated crop.

Maize eventually was carried south and north from its Mexican origin and played an important role as a cultivated food plant among Indians all the way from southern Canada to south-temperate South America, and it still does. The section Tuberarium of the genus *Solanum* (wild potatoes) ranges all the way from Colorado, New Mexico, and Arizona across Mexico and the Central American highlands to the Andes and southward along these mountains to Argentina. Surprisingly, in North America (including Mexico) it was not a main staple of the native diet. It was used more as a subsistence or emergency food mainly by the Navajo and Hopi tribes in the United States and by Mexican mountain Indians. This may have been due to the relative abundance of maize, beans, and cucurbits (CORRELL 1962). But in the Andes, it became the mainstay of the diet because it could be grown so easily in the cold climates of the high mountains and on the Altiplano.

All primitive American agriculture was done without the plow and without domesticated draft animals to pull such a tool if it had existed. Even the wheel was unknown. Instead, crude wooden foot-plows were used to turn the earth; each foot-plow required two or three people as the source of energy. Also, small wooden hand-hoes were sometimes used. On the steep Andean slopes, these may have been more useful than would have been the Middle Eastern plow.

Furrows for potatoes in the high Andes run up and down the slopes, not along the contours. This pattern is used to keep the potatoes well-drained so that the tubers will not rot. Naturally, this can lead to severe erosion problems. By the time of the Incas, well-drained terraces were built which made cultivation easier and controlled erosion.

Unfortunately, today, population pressure has pushed potato culture higher in elevation, above most of the terraces; on steep slopes, erosion is again a serious problem. While the old Inca potato varieties are still grown in places, much of the potato crop planted by the Andean Indians is now derived from modern hybrid varieties imported from North America or Europe. In Peru, the highest field of potatoes (modern tetraploid hybrids) which I have seen is at an elevation of 4,150 m where night frosts are frequent. WINTERHALDER and THOMAS (1978) say that two native potato types, *S. tuberosum* subsp. *andigena* ("dulce") and *S.* × *curtilobum* ("amarga") produce well at higher elevations: "dulce" up to 4,250 m and "amarga" up to 4,450 m. The scientific nomenclature is that of CORRELL (1962) who reports subsp. *andigena* to be a tetraploid and that *S.* × *curtilobum* is a pentaploid.

In the same region, maize can be planted in contour furrows. However, it cannot be grown much higher than elevations of about 3,300 m.

While the photosynthetic pathways of all early crop plants have not yet been determined, only three appear to utilize the C_4 pathway. These are maize, sugarcane, and sorghum. All the rest are C_3 plants except for one, pineapple, which has the CAM mode. Every crop species originating in the Fertile Crescent of the Middle East uses the Calvin-Benson (C_3) cycle. Only one C_4 species, maize, is derived from the Americas. Sugarcane was domesticated in Southeast Asia and the Southwest Pacific, and sorghum was put into cultivation first in tropical Africa. Except for maize in Mexico, the nutrition of early peoples resulting in rapid population growth was the result of C_3 crops. And even in Mexico, a great deal of the nutrition was derived from C_3 crops such as beans, sweet potatoes, and tropical fruits. Even in Southeast Asia, the main staple, rice, has the C_3 pathway. In the Andes, before the introduction of maize, all crops appear to have been C_3 plants; this is not surprising in a cool, high-elevation climate.

As EVANS (1975) says, considerable effort has been expended on trying to find a C_3 crop plant without photorespiration in the hopes of improving yields. All of this has been without success so far. EVANS continues by pointing out that the relative advantage of C_4 over C_3 plants is less in total crop photosynthesis than at the level of the single leaf because of the shading of the lower leaves in the field and other considerations (resistances) in measuring CO_2 exchange. Similarly, maximum short term crop growth rates (CGR) in g m^{-2} ground day^{-1} show no apparent differences between C_3 and C_4 crops; in fact, sunflower (*Helianthus annuus*) has the highest measured CGR at 68 g m^{-2} day^{-1}. The great photosynthetic advantage of C_4 plants at the carboxylation level under high light intensities is attenuated at the crop photosynthesis level by stomatal and mesophyll resistances, self-shading, and periods of low light (EVANS 1975). It is possible also, according to EVANS, that the increasing carbon dioxide concentration in the earth's atmosphere will further reduce the relative advantage of the C_4 pathway.

Since most plants operate with the C_3 mode and most primitive agriculture originated in relatively high, dry, cool, and cloudy places, perhaps it is to be expected that C_3 plants would have dominated and continue to dominate crop ecosystems. They seem to have supplied sufficient sustinance to most peoples in the past, we still depend upon their relatively high productivities today, and they grow successfully in more kinds of environments than will C_4 plants.

16.5.2 Co-Evolution of Plants and Man in Agricultural Ecosystems

As mentioned earlier, man aided the rapid evolution of crop plants by selecting, preserving, and replanting new hybrids and mutants which showed high yields. Simultaneously, these crop plants changed and speeded the cultural evolution of people, and probably their biological evolution also. This mutual interdependence has been increased by modern crop genetics, by agronomic research, and by new techniques of planting and harvesting.

Accompanying the evolution of crop plants has been the simultaneous evolution of those opportunists of agriculture, the weeds. Therefore, it is appropriate to at least mention a few examples of how quickly the evolution of weeds has kept up with man's agricultural advances.

Some of the weeds became crop plants themselves. Oats (*Avena*) was carried along with emmer as this latter crop spread over Eurasia. Emmer was a common crop in northern Europe during the climatic optimum in post-glacial times. As the climate began to become colder again, oats was better adapted to low temperatures and replaced emmer as the main food crop of the region (see SAMPSON 1954).

One of the classic examples of co-evolution of a weed with a crop plant is that provided by flaxweed (*Camelina sativa*) with flax (*Linum usitatissimum*). Flaxweed evolved ecotypes that mimicked almost every seed characteristic of the flax varieties with which it grew; these included seed weight, seed shape, capsule shattering, and seed dispersal. Also, depending upon the heights of particular forms of flax, *Camelina* evolved similar plant height forms to match. This rapid evolution of mimicry was aided by man's own threshing techniques; it became difficult to separate the crop from the weed at harvest (KING 1966). In fact, *Camelina sativa* produced forms with edible, oily seeds and became a crop plant itself.

Many kinds of plants evolve much more quickly than was once supposed. They can evolve ecotypes that keep up with even the very rapid changes in environment brought about by industrial man. SNAYDON (in BRADSHAW et al. 1965) found in Britain that plants of two grass species, *Festuca ovina* and *Agrostis canina*, growing just underneath a galvanized fence that was only 30 years old were tolerant to the zinc that had weathered from the fence to the soil. Plants of these same species just a few centimeters away were not tolerant to zinc.

16.6 The Alteration or Destruction of Natural Ecosystems by Man: The Ecophysiological Bases

Man can alter or destroy natural ecosystems either directly or indirectly. Direct ways are exemplified by actually cutting down the vegetation, plowing the sod and soil, by human impacts such as trampling, or by destroying both vegetation and soil in the process of modern urbanization and road-building. These impacts are dramatic and readily seen; it is people against plants and soil in direct confrontation.

Indirect effects are more subtle. These involve effects through one or more environmental factors resulting in chain or "shower" reactions within the holocoenotic system. The beginning of these intricate reactions is in the addition or change of an environmental factor. This triggers a series of events whose long-range effects are usually unpredictable. An ecosystem may not visibly show the results of these changes for a long time; in this lies the subtlety. The first observable changes may be shown by only one or a few species and are usually difficult to trace back to a man-induced change in the environment. An ecosystem is picked apart bit by bit, species by species, slowly and then more rapidly; the system may be gone or irretrievably damaged before the danger is realized. One can see and understand direct impacts by people on ecosystems, and perhaps take corrective measures. But continual observation, censusing, monitoring, and even experimentation is necessary to detect and understand what indirect effects are doing to ecosystems and the biosphere. In the past, direct effects have resulted in most of the changes in ecosystems. But in the last century or two, as the industrial revolution has grown, it is the indirect factors which are quietly changing the composition and operation of the biosphere. It is with the impacts of these effects that plant ecophysiologists must increasingly be concerned in their research.

16.6.1 The Replacement of Natural Ecosystems by Agricultural Systems

The replacement and maintenance of natural ecosystems by agricultural ecosystems requires, in addition to solar radiation, energy inputs from a number of sources: people, draft animals, fertilizers, water, and fossil fuels. This energy is needed to keep the agricultural system simple and productive: a harvestable monoculture of relatively short life span. If this energy is not continually put into the system, the simplified biological community will revert rapidly to a more diverse and stable native vegetation provided seed sources still remain nearby.

As long as environmental conditions are stable and moderate in relation to temperature and moisture, diverse ecological systems will successfully compete with and displace simplified agricultural systems (ODUM 1971). This is because simplified systems have diverted energy away from the stabilization of geochemical cycles, soils, and epidemics, all of which are relatively under control in ecosystems of higher diversity. The annual contribution of agriculture to biospheric productivity is usually much less than that of the natural ecosystems which they have replaced because crops are in full leaf for only part of the year. The removal of highly productive natural ecosystems for agriculture may have more effect on the world carbon cycle and additions to atmospheric carbon dioxide levels through increased decomposition of humus than has the burning of fossil fuels (ODUM 1971; WOODWELL et al. 1978); but this remains to be proven.

Primitive agriculture in the moist lowland tropics tends to be rather casual with diversities higher than in the crop monocultures of temperate regions. But modern agriculture in the tropics has tended toward monoculture, especially in regard to certain crops grown in plantation form. A number of highly destructive diseases have threatened such plantation crops as banana, cacao, and rubber, and have cost a considerable amount of money (ORIANS et al. 1974).

One thing seems certain: agricultural monoculture has its problems where it has replaced highly diverse natural ecosystems in the wet tropics.

A very common form of agriculture throughout the wet tropics is "slash and burn," "swidden," or "shifting agriculture." Forests are cleared by cutting or girdling and then burned. The open tracts are put into gardens for short periods of time and after a few years are abandoned to successional forest. As a forest burns, nutrient cations are accumulated in a thin layer of ash on the surface of the soil. This results in a temporary increase in fertility of the soil since most of the nutrient pools in such forested ecosystems remain in the biomass until fire or decomposition frees them (BENNETT et al. 1974). GROSS, et al. (1979) present some quantitative data on this nutrient release in the Amazonian rainforest and the neighboring cerrado in Brazil. They also conclude that the degree of external market participation by four isolated native Indian villages in this region is related to the difficulty of making a living from slash and burn agriculture as measured by the ratio of labor input to food output. This same principle undoubtedly applies to slash and burn or shifting agriculture in other tropical regions. Again, one sees the energy costs of maintaining even relatively low diversity agricultural ecosystems where surrounded by natural ecosystems of high diversity in which the native trees accumulate most of the nutrient pool in their biomass.

16.6.2 The Replacement of Natural and Agricultural Ecosystems by Urbanization

In the last century or so, agriculture has expanded at the expense of natural ecosystems, part of this by the development of irrigation agriculture on a large scale. At the same time, excellent agricultural land has been encroached upon by the growth of cities, factories, and suburbia made possible by the increased use of fossil fuel for modern transportation. Much of the effect has been direct and mechanically caused. In many places, particularly in North America, "islands" of native ecosystems have been left and are surrounded by agricultural and urban areas. These "islands", however, are relatively depauperate in native plant and animal species compared to the biota of such systems in the great expanses which existed two centuries ago.

Many of the effects of the city on surrounding "islands" or "expanses" of native vegetation, however, are *indirect* and come through changes in the atmospheric environment. Most of these changes are the results of atmospheric pollutants added by industry, vehicles, and airplanes. These pollutants include carbon monoxide, carbon dioxide, ozone, nitrogen oxides, photochemical smog, sulfur dioxide, hydrogen chloride, ions of heavy metals, and a host of organic compounds (see Chapt. 14, this Vol. and GUDERIAN 1977). The effects on plants can be dramatic or subtle and differ from species to species. MAUGH (1979) reporting on the work of NOGGLE, for example, says that agricultural crops in the Tennessee Valley may increase their productivities by utilizing atmospheric SO_2 emitted from coal-burning electrical power plants. This would seem to assume, of course, that the levels of SO_2 do not get very high.

16.7 Some of Man's Effects on Plant Processes, Ecosystems, and the Biosphere

16.7.1 Direct Effects

16.7.1.1 Effects of Deforestation and Possible Revegetation

Man-caused deforestation and loss of other kinds of vegetation has been going on at an ever-increasing rate for thousands of years. It started in Africa, the Middle East, and the Mediterranean region. Now, no part of the earth's native vegetation is exempt from destruction and replacement by fallow ground, agriculture, or urbanization. Such destruction leads to increased runoff, soil erosion by water and wind, and loss of nutrients. It also results in losses of species of plants and animals and, therefore, depletion of the gene pool. If the habitat is destroyed, the closely adapted biota become extinct, at least locally. About one-half of the tropical rainforest of the earth has been destroyed by man (WOODWELL et al. 1978; SAGAN et al. 1979), and the destruction continues at an ever-increasing rate. Many of the species are gone, and, in most places, the rainforest ecosystem cannot recover. The gene pool has been destroyed, and the nutrients have gone with the destruction of the phytomass.

At Hubbard Brook in the low mountains of the northeastern United States, a comprehensive study has been made of the ecological effects of "clear-cutting" in a temperate region forest (LIKENS et al. 1977; BORMANN and LIKENS 1979). This large-scale experiment, which has been carried on for many years, is based on the use of the "small watershed technique." This makes possible the measurement of input–output relationships in regard to soil weathering, nitrogen fixation, biomass accumulation, and species "strategies" in relation to time and space. Each of these is being studied in the context of whole ecosystems that consist of small, tight watersheds. These careful measurements and experiments have allowed the authors and their colleagues to construct a model (really a series of models) of what happens to a forested ecosystem in the middle latitudes after clear-cutting and during the recovery times following such a catastrophic event.

The Hubbard Brook model is an excellent example of how a knowledge of the ecophysiological characteristics of the important species in an ecosystem can aid in understanding the roles played by plants and microorganisms in controlling energy capture and biogeochemical cycling (nutrients and water) in an ecosystem recovering from disturbance. One of the most striking effects is the role played by early successional vegetation which through high transpiration rates reduces streamflow. The result is the holding of nutrients, particularly nitrogen compounds, within the aggrading ecosystem. Without this pioneer vegetation, both nutrients and particulates would be lost to the system by relatively high streamflow. If these losses did occur, there would be little chance of recovery of the forest system as it is now constituted. Obviously, clear-cutting, if followed by natural succession in a temperate region forest ecosystem, stands in strong contrast to continued and permanent destruction in the tropics.

An example of just such destruction of an ecosystem exists in the highlands of the Peruvian Andes (ELLENBERG 1979). During many centuries, the sub-alpine woodlands of *Polylepis incana* and *Buddleia coriacea* between elevations of 3,800 m and 4,600 m have been largely destroyed, except for scattered patches, by man and his animals. This destruction has been accomplished mainly by conversion of the land to crop agriculture and browsing of the native woody vegetation by indigenous domestic animals (llamas, alpacas) as well as introduced ones (sheep, goats, cattle, donkeys, horses). Also, cutting of *Polylepis* for firewood and crude timbers has aided in the destruction. The climate and soils of much of this region below about 4,300 m would support such a scrub woodland but the woodland is gone and has been replaced by crops or a partially man-induced tussock grassland of *Stipa* spp., *Festuca*, and *Calamagrostis*. This grassland burns and thus causes further losses of the scattered remnants or "islands" of *Polylepis* woodland. Frost-sensitive *Eucalyptus* spp. have been planted as replacement wood sources in the lower parts of this sub-alpine zone; but *Polylepis* and *Buddleia* are far better adapted to the environment (ELLENBERG 1979; WINTERHALDER and THOMAS 1978). It is improbable that planting of the native trees would bring back the original sub-alpine Andean woodland ecosystem. Even if this were possible, it is probable that *Polylepis* woodland could not co-exist with man, considering the large populations of native peoples in the region today. Man and the induced grasslands have formed a new and mutually dependent sub-alpine community which is the basis of a new ecosystem.

16.7.1.2 Effects of Feet and Vehicles on Grasslands

One can envision in most of the world's grasslands, as they exist today, a dusty road winding across the countryside and far into the distance. This familiar scene is too common and too obvious to require further explanation. It is simply a modern extension, made by four-wheeled vehicles, of the foot and horse trails of the past.

It is sufficient here to mention only two studies of direct impact of people on plants. The first is concerned with a modern phenomenon of tourism: the proliferation of trails in a rather ad hoc manner by hikers across alpine meadows. Similar trails have been made for thousands of years but their impacts were not studied. HARTLEY (1976) has done just that. He made a quantitative study of trail proliferation in the alpine and sub-alpine meadows of Glacier National Park, Montana, by the use of aerial photographs, counting of people, sampling trailside vegetation, and measuring the effects of repeated trampling along trails and in experimental plots on the survival or death of plants of different species in alpine vegetation. It was to be expected that some species would be sensitive to trampling and would soon disappear from the vegetation while plants of other species would profit from this loss. This occurred both along trails and in the experimental plots. Using a "sensitivity index," he found that 35 species decreased as trails grew and only 7 species increased their numbers.

Table 16.1. Carbohydrate content of trail-side *Erythronium grandiflorum* roots and corms in subalpine meadows of the Lewis Range, Glacier National Park, Montana. (HARTLEY 1976)

Date and location	Distance from trail	Sugar[a]	Starch[a]	Total[a] utilizable carbohydrate
7 August 1969	0–0.5 m	2.2	27.3	29.5
Near Logan Pass	Over 2 m	2.1	33.0	35.1
Visitor Center				
12 August 1969	0–0.5 m	1.6	21.0	22.6
Overlook Path	Over 2 m	2.8	41.6	44.4
Near Road Switchback				
13 August 1969	0–0.5 m	2.9	25.8	28.7
Highline Trail	Over 2 m	3.3	43.8	47.1
Mean	$\bar{y}$ 0–0.5 m	2.3	25.4	27.7
Carbohydrate	$\bar{y}$ Over 2 m	2.6	39.0	41.6
Content	Significant Difference	ns	$^{**}P=0.01$	$^{**}P=0.01$

[a] as % of alcohol-insoluble dry wt.

HARTLEY (1976), in seeking the causes for these decreases in abundance and vigor of sensitive species, measured carbohydrate contents of roots and corms of *Erythronium grandiflorum,* one of the sensitive species, on or close to trails and more than 2 m away from trail edges (see Table 16.1). There was not a significant difference in sugar content between those close to the trail and those away. But there was a very significant difference in starch content of belowground parts with those near the trail having barely half as much stored starch as those not subject to foot traffic.

He also measured belowground carbohydrate contents of plants of *Erythronium grandiflorum* and *Claytonia lanceolata,* two species apparently equal in sensitivity to trampling, on his experimental trampling plots. *Erythronium* again showed a highly significant decrease in stored total utilizable carbohydrate accompanying an increase in trampling intensity. Surprisingly, however, belowground carbohydrates in *Claytonia* plants showed no decrease under increasing trampling intensity even though it is equally sensitive along trailsides. Thus, while disappearance of a species may be caused by decrease in carbohydrate synthesis, it apparently is not the only reason.

HARTLEY (1976) concluded his ecophysiological studies on alpine meadow species by constructing a box, arrow, and controller model of how *man* along with water, nutrients, diseases, and herbivores determines the physiological status of alpine plants under conditions of human foot impact. This model is unique, as far as I know, in depicting the pathways of energy flow in alpine meadow vegetation as stress is applied. It appears, in slightly modified form, as Fig. 16.1. As in other ecosystems, world-wide, man is now a strong controller

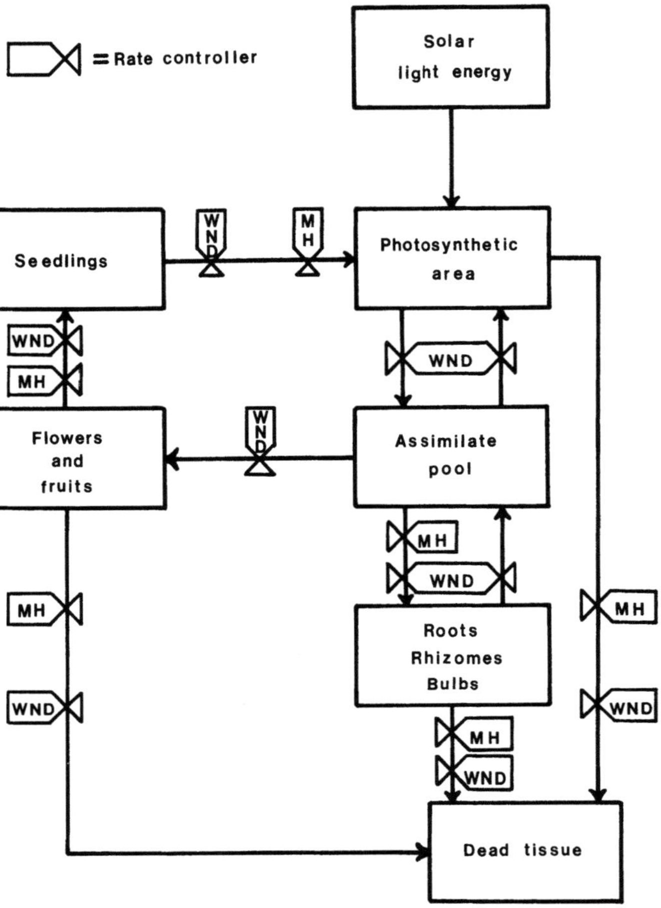

Fig. 16.1. Rate controller model of the impact of man on photosynthesis and carbohydrate storage in subalpine and alpine vegetation in Glacier National Park, Montana, U.S.A. (Slightly modified and redrawn from HARTLEY 1976). *W* water status; *N* nutrient status; *D* disease; *H* herbivores; *M* man

in the physiological processes that determine vegetational diversity, composition, and productivity.

It is but a step from feet to wheeled vehicles. The first roads in a region were, and are, rather casual affairs. Their direct effects on plants are probably similar to those caused by heavy foot traffic. However, if not graveled or paved, soil erosion is likely to result, especially in hilly or mountainous regions, with consequent damage to the ecosystem. The modern proliferation of paved highways, carefully engineered, may not cause erosion but does increase hydrologic runoff into nearby stream systems. Such increase in flooding may affect species composition of downstream ecosystems, but data on this effect are not readily available even though speculative estimates have been made.

However, as a second example of the effects of people on natural ecosystems, I shall use an extreme case: the impact of vehicles on unprotected wet tundra in the Arctic. We and others (HOK 1969; BILLINGS and PETERSON 1980) have studied this situation in the wet coastal tundra near Barrow, Alaska (71°18′N lat.). The undisturbed vegetation is dominated by C_3 graminoid plants. The soil parent material is a gray clay which is overlain with 12–15 cm of peat. The soil is underlain by ice-rich permafrost to a depth of 400–500 m. The annual depth of thaw ranges from 25–35 cm. These depths are reached in late July and early August; after this time the surface soils begin to refreeze and from September to June the whole soil and geological column is frozen and thinly snow-covered. Within the permafrost, with their tops just under the permafrost table lies a polygoned network of pure ice wedges. Such wedges range in top width from 1 to 6 m across. These wedges cause the soil above the permafrost table to be patterned in a polygonal fashion with the polygons ranging from about 5 to 20 m across.

Because of low population levels, human foot traffic causes little damage even at time of maximum thaw. But in recent decades, the use of heavy tracked or wheeled vehicles over the roadless tundra during the short period of thaw has resulted in extreme damage to the vegetation and soil. The metal tracks and large ribbed tires of the various kinds of vehicles dig into the thawed peat and destroy the tundra vegetation. The surface energy budget changes so that the annual depth of thaw becomes deeper; this is especially true over the tops of the ice wedges. The white ice of the wedge melts and forms a "bead pond" at the intersection of the track and ice wedge where the peat insulation has been disturbed. From the air, the track now resembles a string of beads; thus the name "bead pond." Repeated traffic over the same track results in deep thermokarst erosion due to melting of the top of the ice wedge. The initial bead pond chain may become a shallow canal stretching across the tundra as the bead ponds unite. Such canals are usually the result of more than ordinary vehicle traffic; most were caused by bulldozing or "blading" off the peat to expose the ice so that large sleds could be dragged behind tracked vehicles over the ice.

Whether the thermokarst results in bead ponds, canals, or extremely deep erosion gullys, the change in vegetation and soil is disastrous. Between the bead ponds in the Barrow region, the remaining peat on the track becomes so dry that revegetation is very slow. One of the principal pioneers on such dry peat is the moss *Psilopilum cavifolium* which is quite rare in undisturbed tundra. Lichens, which are abundant in the original tundra vegetation, are very slow to invade the tracks; most lichen species are not adapted to the new environment. Pioneer vascular plant species on the dry peat are *Stellaria edwardsii, Luzula confusa,* and *Eriophorum scheuchzeri* which are not the common dominants of the original tundra vegetation even though present there.

Nutrients are released into the bead ponds from the decomposing wet peat. The growth of algae in the ponds is thus encouraged. Three species of native graminoids also invade the ponds: *Carex aquatilis, Dupontia fischeri,* and *Arctophila fulva*. They, too, are aided by the increased nutrient supply, particularly nitrogen and phosphorus, from the decomposing peat and are much greener than in the surrounding undisturbed vegetation. Also, there is often conspicuous flowering and fruiting of *Eriophorum* spp. along even lightly traveled trails.

Almost all of this damage has occurred within the last 30 years, and most of it is irreversible. It is another example of the direct effect of man on a previously untouched ecosystem. In this case, low soil temperatures and nutrients have been limiting to growth of plants. Disturbance eliminates these limiting factors by warming the soil and releasing nutrients to the native plants (there are no exotic species). But this very amelioration causes changes in vegetational composition and the actual elimination of plant species in an already impoverished flora of only 125 vascular plant species.

16.7.1.3 Effects of Herbicides, Pesticides, and Oil on Terrestrial Ecosystems

In the last 30 to 40 years, there has been a dramatic increase in the use of insecticides such as DDT in all parts of the biosphere. By the mid-1950's, DDT was known to accumulate in cropland soils where it had been applied aboveground to control insect pests. WOODWELL (1961) soon found it in undisturbed forest soils in Maine. Later, it was shown that DDT not only destroyed the target insects but also killed pollinating insects in orchards and more complex ecosystems. WOODWELL and his co-workers found that DDT also entered ecosystemic food webs and accumulated in lethal amounts in carnivores, particularly birds. It was taken in even by vascular plants such as *Spartina patens,* in an estuarine salt marsh, where it was concentrated in the roots (WOODWELL et al. 1967).

In spite of the lethal effects of DDT on animals, the actual pool of DDT residues in the biota is very small compared to those in the atmosphere and oceans (WOODWELL et al. 1971); it is fortunate that this is so, otherwise a biospheric catastrophe could occur. There is no assurance that biotic pools of DDT and other pesticides will not increase in the future if environmental pools become larger. Other biocides such as heavy metals (mercury, lead, zinc), polychlorinated biphenols (PCB's), and chlorinated hydrocarbons have escaped or have been dumped into modified or natural ecosystems. The final effects of these on plants, animals, people, and ecosystems are not yet known in detail; however, they are all known to be toxic. BAZZAZ et al. (1974a, b), for example, have found that the heavy metals Pb, Cd, Ni, and Tl strongly inhibit both photosynthesis and transpiration in corn and sunflower. The strong reduction seems to be triggered by stomatal closure.

One of the strongest direct effects of man on the vegetational basis of an ecosystem is by the use of herbicides (see Chaps. 11, 12, and 13, this Vol.). Within the last 40 years, the use of synthetic auxin herbicides such as 2,4-D, 2,4,5-T, and picloram for killing of dicotyledonous weeds in monocotyledonous crops has had great economic significance. These herbicides can be used also to modify or destroy natural ecosystems. For example, 2,4,5-T has been used to eliminate certain hardwood species from pine stands and woody plants such as mesquite (*Prosopis*) from semi-desert grasslands. They were used also in Vietnam by the United States for clearing tropical forests in warfare zones. Mangrove ecosystems were particularly vulnerable to these herbicides and were largely destroyed, at least temporarily.

With the development of petroleum reserves in the Arctic and other cold regions, there has been considerable research into the effects of oil spills on terrestrial vegetation in such regions. DENEKE et al. (1975) did some pioneer work on this problem in Alaska utilizing plants growing around natural oil seeps near Cape Simpson (71 °N lat.) in the coastal tundra and by directly applying oil to tundra vegetation near Barrow at about the same latitude. They also studied the effects of accidental leaks and spillages in taiga vegetation along a military pipeline from Fairbanks in central Alaska to Haines on the southeast coast. They found that rapid recovery of plant communities subjected to spills will occur only if the root systems remain relatively unaffected. Damage is long-term if the roots are saturated with oil. Also, since it takes time for oil to get to the roots in frozen and cold soils, the effects often do not appear in the plants for several years. It then shows up in a high degree of winterkill. Oil seems to affect the plants through breaking down cold hardiness.

Variation exists in species susceptibility. In the tundra, *Carex aquatilis,* an abundant and highly competitive dominant sedge, is very resistant to crude oil damage; in fact, *Carex* plants grew quite well when surrounded by oil both in natural seeps and when artificially applied. On the other hand, in the taiga, *Picea mariana,* black spruce, is very susceptible. Fertilization enhances recovery from oil spills not only through plant nutrition but by stimulating microbial decomposition of oil. It had been shown earlier that, following application of oil, the total number of microbial species decreases while the total number of microbial organisms increases; i.e., the detrital community becomes simplified in regard to diversity.

LINKINS and his co-workers (LINKINS et al. 1978; ANTIBUS and LINKINS 1978; LINKINS and ANTIBUS 1978) have studied the effects of crude oil on certain physiological responses of tundra plants and soil microbial activity at Barrow. They found that tundra soil microbiota can actively modify oil and use it to support metabolism. Soil respiration rates increased above background after oil was applied. This indicates that certain microbiota can degrade and utilize oil faster than they can utilize normal peat and litter. They also found that ectomycorrhizae on roots of *Salix rotundifolia* in the Barrow tundra were quickly damaged by oil, and that this effect was still apparent three growing seasons later. However, the mycorrhizae on *Salix rotundifolia* growing near the oil seep at Cape Simpson were not so susceptible to oil damage as those at Barrow. Also, at Barrow, they found that oil-treated mycorrhizal roots of this willow species continued to have depressed respiration rates through the whole short growing season; but there was recovery in this regard during the second summer. Viable root biomass declined from year to year in oil-treated soils. Again, as in the work by DENEKE et al., cold acclimation was reduced; in this case, respiration rates of roots lost cold acclimation. There was a lesser effect on root respiration and cold acclimation of the mycorrhizal roots of *Salix* growing near the periphery of the natural oil seep at Cape Simpson. In summary, it seems apparent that oil, in the Arctic, affects root respiration, cold acclimation, cold hardiness, and soil microbial activity. Also, there seems to be some ecotypic adjustment to oil of mycorrhizal roots of *Salix rotundifolia* from populations near natural oil seeps.

16.7.2 Indirect Effects

In a holocoenotic biosphere, changes in environmental components have effects within the environment itself. Such interactions result in indirect and often unexpected impacts on populations and biological communities. These effects are difficult to predict in time, space, and among organisms (BILLINGS 1974).

In regard to terrestrial ecosystems, almost all of the environmental changes which result in indirect effects are due to climatic changes or to additions of natural or synthetic gases, molecules, or particles to the atmosphere or soil. These include: (1) heavy metals from the exhausts of vehicles, smelters, and industrial complexes, (2) hydrocarbons, ozone, sulfur dioxide, nitrogen oxides, carbon dioxide, and other gases derived directly or indirectly (by photo-oxidation, etc.) from the burning of fossil fuels, (3) various organic compounds of industrial origin, or (4) radionuclides and ionizing radiation. These will not be discussed further since they have been introduced in considerable detail earlier in these volumes: stratospheric ozone deterioration and its effects on ultraviolet-B irradiation (Chap. 6, Vol. 12A), ionizing radiation (Chap. 7, Vol. 12A), heavy metals (Chap. 7, Vol. 12B), biocides (Chaps. 11 and 12, this Vol.), atmospheric pollutants (Chap. 14, this Vol.), the increases in atmospheric carbon dioxide concentrations (Chap. 15, this Vol.), and aquatic pollutants (Chap. 13, this Vol.). In addition to these, changing concentrations of dust, volcanic ash, and water vapor in the air can affect solar irradiance in the lower atmosphere and thus change biospheric climates as has been suggested for the increase in atmospheric carbon dioxide. All of these added components have intricate pathways of effects in the biosphere and generally act through other factors which are the agents of direct impact. However, almost all of the added factors also can act directly on plants by causing damage or by increasing growth as seems to be the case with moderate increases in carbon dioxide concentrations (Chap. 15, this Vol.).

16.8 Conclusions: Ecophysiology and Ecosystems

Ecophysiologists and their researches on plant processes must by definition be involved with the ways in which these processes operate in nature. This, of course, means that ecophysiologists must understand the structures, energy and nutrient pathways, productivities, and carrying capacities of both natural and agricultural ecosystems. Since ecosystems are holocoenotic, there is the necessity for considering ecophysiological problems in the context of *whole* ecosystems. Usually, such an approach is beyond the abilities of one person alone. Therefore, ecophysiologists must join forces with other scientists in team approaches to the solutions of ecosystemic problems. This has been successfully demonstrated in much of the work of the International Biological Program (see TIESZEN 1978).

A particularly ecophysiological approach to ecosystemic problems which requires less manpower and can be done at lower cost is to bring relatively undisturbed microcosms of certain ecosystems into phytotrons and carry on experiments under controlled conditions. We have found this to be a very useful technique in studying the effects of increased temperature and atmospheric carbon dioxide on tundra ecosystems (BILLINGS et al. 1982).

With increasing numbers of people on earth and the rapid growth of industry and agriculture, the effects of man on ecosystem composition and ecosystemic tolerance to environmental stress is becoming acute. Since many of these problems concern plant processes, and, thus, ecosystemic productivities and feedback rates, ecophysiologists have an important role to play in designing their researches toward solutions to biospheric problems. The challenge is clear; let us accept it.

References

Antibus RK, Linkins AE (1978) Ectomycorrhizal fungi of *Salix rotundifolia* Trautv. I. Impact of surface applied Prudhoe Bay crude oil on mycorrhizal structure and composition. Arctic 31:366–380

Bazzaz FA, Carlson RW, Rolfe GL (1974a) The effect of heavy metals on plants: Part 1. Inhibition of gas exchange in sunflower by Pb, Cd, Ni, Tl. Environ Pollut 7:241–246

Bazzaz FA, Rolfe GL, Carlson, RW (1974b) Effect of Cd on photosynthesis and transpiration of excised leaves of corn and sunflower. Physiol Plant 32:373–376

Bennett C et al. (1974) Interaction of man and tropical environments. In: Farnsworth EG, Golley FB (eds) Fragile ecosystems: Evaluation of research and applications in the Neotropics. Springer, Berlin Heidelberg New York, pp 139–182

Billings WD (1952) The environmental complex in relation to plant growth and distribution. Q Rev Biol 27:251–265

Billings WD (1974) Environment: concept and reality. In: Strain BR, Billings WD (eds) Vegetation and environment. Handbook of vegetation science, Vol VI. Junk, The Hague, pp 7–35

Billings WD (1978) Plants and the ecosystem, 3rd edn. Wadsworth, Belmont

Billings WD, Peterson KM (1980) Vegetational change and ice-wedge polygons through the thaw-lake cycle in arctic Alaska. Arct Alp Res 12:413–432

Billings WD, Luken JO, Mortensen DA, Peterson KM (1982) Arctic tundra: A source or sink for atmospheric carbon dioxide in a changing environment? Oecologia 53:7–11

Bormann FH, Likens GE (1979) Pattern and process in a forested ecosystem. Springer, Berlin Heidelberg New York

Boughey AS (1975) Man and the environment, 2nd edn. MacMillan, New York

Bradshaw AD, McNeilly TS, Gregory RPG (1965) Industrialization, evolution, and the development of heavy metal tolerance in plants. In: Ecology and the industrial society. 5th Symp Br Ecol Soc, Oxford

Bronowski J (1973) The ascent of man. Little Brown, Boston

Correll DS (1962) The potato and its wild relatives. Tex Res Foundation, Renner

Deneke FJ, McCown BH, Coyne PI, Rickard W, Brown J (1975) Biological aspects of terrestrial oil spills: USA CRREL oil research in Alaska, 1970–1974. US Army CRREL Res Rep 346, Hanover

Ellenberg H (1979) Man's influence on tropical mountain ecosystems in South America. J Ecol 67(2):401–416

Evans LT (ed) (1975) Crop physiology: some case histories. Cambridge Univ Press, London

Gross DR et al.(1979) Ecology and acculturation among native peoples of central Brazil. Science 206:1043–1050

Guderian R (1977) Air pollution. Ecol Stud, Vol 22. Springer, Berlin Heidelberg New York
Harlan JR (1976) The plants and animals that nourish man. Sci Am 235(3):88–97
Hartley EA (1976) Man's effects on the stability of alpine and subalpine vegetation in Glacier National Park, Montana. PhD Dissertation, Duke Univ, Durham
Hok JR (1969) A reconnaissance of tractor trails and related phenomena on the North Slope of Alaska. US Dep Interior, Bur Land Manage, Washington DC
King LJ (1966) Weeds of the world: biology and control. Wiley-Interscience, New York
Leakey LSB, Tobias PV, Leakey MD (1965–1971) Olduvai Gorge. 3 Vols. Cambridge Univ Press, Cambridge
Likens GE, Bormann FH, Pierce RS, Eaton JS, Johnson NM (1977) Biogeochemistry of a forested ecosystem. Springer, Berlin Heidelberg New York
Linkins AE, Antibus RK (1978) Ectomycorrhizal fungi of *Salix rotundifolia* Trautv. II. Impact of surface applied Prudhoe Bay crude oil on mycorrhizal root respiration and cold acclimation. Arctic 31:381–391
Linkins AE, Atlas RM, Gustin P (1978) Effect of surface applied crude oil on soil and vascular plant root respiration, soil cellulase, and hydrocarbon hydroxylase at Barrow, Alaska. Arctic 31:355–365
Martin PS, Wright HE (eds) (1967) Pleistocene extinctions. Yale Univ Press, New Haven
Maugh TH (1979) SO_2 pollution may be good for plants. Science 205:383
Odum TH (1971) Environment, power, and society. Wiley-Interscience, New York
Orians G et al. (1974) Tropical population ecology. In: Farnsworth EG, Golley FB (eds) Fragile ecosystems: Evaluation of research and applications in the Neotropics. Springer, Berlin Heidelberg New York, pp 5–65
Por FD (1978) Lessepsian migration. Ecol Stud, Vol 23. Springer, Berlin Heidelberg New York
Sagan C et al. (1979) Anthropogenic albedo changes and the earth's climate. Science 206:1363–1368
Sampson DR (1954) On the origin of oats. Bot Mus Leaflets Harvard Univ 16 (10):265–303
Tieszen LL (ed) (1978) Vegetation and production ecology of an Alaskan arctic tundra. Ecol Stud, Vol 29. Springer, Berlin Heidelberg New York
Wheat MM (1967) Survival arts of the primitive Paiutes. Univ Nevada Press, Reno
Winterhalder BP, Thomas RB (1978) Geoecology of southern highland Peru: A human adaptation perspective. Inst Arct Alp Res, Univ Colorado Occ Pap No 27
Woodwell GM (1961) The persistence of DDT in a forest soil. For Sci 7:194–196
Woodwell GM, Wurster CF Jr, Isaacson PA (1967) DDT residues in an East Coast estuary: A case of concentration of a persistent insecticide. Science 156:821–823
Woodwell GM, Craig PP, Johnson HA (1971) DDT in the biosphere: Where does it go? Science 174:1101–1110
Woodwell GM, Whittaker RH, Reiners WA, Likens GE, Delwiche CC, Botkin DB (1978) The biota and the world carbon budget. Science 199:141–146

Author Index

Page numbers in *italics* refer to the references

Abeles FB 358, 361, *372*
Aber JD 251, *272*
Åberg E, Stecko V 396, *406*
Acevedo E, see Hsiao TC 140, 141, *147*, 189, *200*, 261, *275*
Acher A, Elgavish A 403, *406*
Ackerman TL, Bamberg SA 214, *226*
Adams DF, Emerson MT 488, *491*
Adams DF, see Applegate HG 489, *491*
Adams DF, see Solberg RA 489, *500*
Adams JB, Drew ME 401, *406*
Adams MS, Strain BR 217, *266*
Adams MS, see Prentki RT 64, *81*
Adams MS, see Smith CS 65, *82*
Adams R, see Sterrett J 405, *411*
Adams S, see Palmer CM 419, 420, 421, 434, 435, 436, 438, *458*
Adams SS, see Smith CS 65, *82*
Addicott FT 361, *372*
Adedipe NO, Khatamian H, Ormrod DP 474, *491*
Agami M, see Litav M 421, 433, *455*
Agee BA, see Powell TM 314, *325*
Ågren GI, Axelsson B 261, *272*
Ågren GI, Axelsson B, Flower-Ellis JGK, Linder S, Persson H, Staaf H, Troeng E 264, 270, *272*
Ahl T 416, 420, *444*
Ahmed MK, Casida JE 432, *444*
Ahmed SI, Kenner RA, Pakkard TT 92, *108*
Aiken SG, Newroth PR, Wile I 418, *444*
Akhtar MA, see Rolfe GL 17, *44*
Akita S, Moss DN 517, *521*
Akoyunoglou GA 336, *372*
Alberda T, Sibma L 165, *170*
Albrektson A 237, 250, *272*
Albrektson A, Aronsson A, Tamm CO 251, *272*
Alexander G, see Stewart WDP 423, *461*
Alexander M, Marshall KC, Hirsch P 58, *75*
Aljaro N, see Montenegro G 220, *229*
Allan GP 404, *406*
Allen HE, Kramer JR 86, *108*
Allen HL 418, *444*
Allen HL, see Wetzel RG 427, *463*
Allen LH, see Lemon E 293, *302*
Allison DA, see Palm HL 354, *384*
Alm AA, see Tappeiner JC 26, *44*
Almeida SP, see Dickson KL 443, *448*
Almer B, Dicksen W, Ekström C, Hörnström E, Miller U 438, *444*
Alscher-Herman R, see Weinstein LH 475, 476, *501*
Alther EW, see Thomas MD 487, 489, *500*
Al-Ugaily L, see Stewart WDP 63, *83*
Alway FJ, Marsh AW, Methley WJ 11, *39*
Amesz J, see Velthuys BR 338, *390*
Amman BD, see Greichus YA 437, *450*
Ammon HU, see Gressel J 366, 368, 369, *378*
Amrhein N, Deus B, Gehrke P, Steinrücken HC 341, 354, *372*
Anderegg BN, Lichtenstein EP, Kemp JD 365, *373*
Andersen JM 58, 65, *75*
Andersen KJ, see Willingham CA 432, *464*
Andersen RN, see Jacobsohn R 397, *408*
Anderson AS 360, 363, *373*
Anderson DM, Morel FMM 103, *108*
Anderson HW, Zsuffa L 236, *272*
Anderson J, see Menzel DW 365, *383*
Anderson JL, Thomson WW 355, *373*
Anderson LE, Avron M 483, *491*
Anderson MC 244, 245, 247, 251, *272*
Anderson PJ 476, *491*
Anderson RR 436, 437, 440, *444*
Anderson SM, Krinsky NJ 351, *373*
Angus JF, see Trenbath TR 164, *172*
Antia NJ, McAllister CD, Parsons TR, Stephens K, Strickland JDH 100, *108*, 321, *323*
Antibus RK, Linkins AE 545, *547*
Antibus RK, see Linkins AE 545, *548*
Antonovics J, Bradshaw AD, Turner RG 429, *444*
Antunes-Madeira MC, Madeira VC 366, *373*
Aoki M, see Yabuki M 293, *303*
Aouni El MH, Mousseau M 260, *272*
Apple JL, Smith RF 333, 371, *373*

Appleby AP, see Stanger CE, 350, 351, *388*
Appleby AP, see Wiedman SJ 355, *391*
Applegate HG, Adams DF 489, *491*
Archer S, Tieszen LL 192, *197*
Archer S, see Tieszen LL 191, *202*
Arkin GF, see Maas SJ 129, *148*
Arlt K, Feyerabend G 396, *406*
Armond PA, Mooney HA 216, 224, *226*
Armond PA, see Björkman O 219, *226*
Armstrong DE, see Li WC 65, *80*
Armstrong DE, see Williams JD 65, *84*
Armstrong DJ, see Whitecross MJ 395, *411*
Armstrong JE, see Estep M 102, *110*
Armstrong R, see Richerson P 105, *114*
Arnold LE, see Rolfe GL 17, *44*
Arntzen CJ, Ditto C, Brewer P 370, *373*
Arntzen CJ, Linck AJ, Cunningham WP 355, *373*
Arntzen CJ, Pfister K, Steinback KE 370, *373*
Arntzen CJ, see Bowes J 338, 342, *375*
Arntzen CJ, see Darr S 370, *376*
Arntzen CJ, see Pfister K 338, 342, 343, 344, 345, 368, 370, *385*
Arntzen CJ, see Radosevich SR 370, *386*
Arntzen CJ, see Souza Machado V 370, *388*
Arntzen CJ, see Steinback KE 342, 370, *388*
Arntzen CJ, see York AC 339, 367, *391*
Aronsson A, see Albrektson A 251, *272*
Aruga Y 441, *444*
Asada K, Kiso K 479, *491*
Aschmann H 206, *226*
Ashenden TW 473, *491*
Ashenden TW, Mansfield TA 471, 473, *491*
Ashton DH 237, 267, *272*
Ashton FM, Bayer DE 354, 355, *373*
Ashton FM, Crafts AS 330, 334, 338, *373*, 394, 396, *406*
Ashton FM, Villiers de OT, Glenn RK, Duke WB 336, 353, 354, *373*
Ashton FM, see Penner D 341, 357, *385*
Ashworth EN, see St John JB 353, *388*
Aspiras RB, see Graetz DA 57, *78*
Assche CJ van 338, 342, *373*
Assche CJ van, see Carles PM 345, *375*
Astier C, Vernotte C, Der-Vartanian M, Joset-Espardellier F 371, *373*
Atchison G, see Michaud M 403, *409*
Atkinson D 268, *272*
Atlas RM, see Linkins AE 539, *548*
Attiwell PM 511, *521*
Attiwill PM 20, *39*, 241, 267, *272*
Attiwill PM, Guthrie HB, Leuning R 267, *272*
Audus LJ 330, 331, 338, 358, 362, *373*
Auerbach S, Prüfer P, Weise G 443, *444*
Auger S, see Thorpe MR 247, *279*
Ausmus BS, Witkamp M 33, *39*
Ausmus BS, see Cox TL 26, *40*
Ausmus BS, see Witkamp M 28, 36, *45*
Austin ER, Lee FG 59, *75*
Austin KH, see Strickland JDH 99, *114*
Avadhani PN, see Osmond CB 483, *499*
Avery RJ, see Kerr MW 354, *380*
Avron M, see Anderson LE 483, *491*
Avron M, see Pick U 483, *499*
Axelsson B, see Ågren GI 261, 264, 270, *272*
Axelsson L, see Klockare B 352, *381*
Axler RP, see Gersberg RM 61, *77*
Ayazloo M, Bell JN 472, 473, *491*
Ayers GP, Gras JL 90, *108*
Ayers GS, see Giannopolitis GN 350, 351, *378*
Ayles GB, Lark JGI, Barica J, Kling H 426, *444*
Azam F, see Fuhrmann JA 52, *77*
Azam F, see Harrison WG 98, *111*

Baalen C van, see Estep M 102, *110*
Baars MA, see Gieskes WWC 322, *323*
Baas Becking HG, see Hodgkinson KC 194, 195, *200*
Baccini P, Suter U 53, *75*
Bach TJ, see Buschmann C 350, *375*
Bach TJ, see Grumbach KH 350, *378*
Bachthaler G 401, *406*
Back E, see Pimentel D 331, *385*
Badger MR, see Björkman O 219, *226*
Badour SS, see Tan CK 519, *525*
Baer CH, see Silvius JE 483, *500*
Baes CF, Goeller HE, Olson JS, Rotty RM 506, *521*
Bailey WA, see Krizek DT 517, *524*
Baillie AC, see Wright BJ 347, 350, *391*
Baity HG, Merryfield F, Uzzle AB 433, *444*
Baker AL, see Brooks AJ 433, *446*
Baker CH, Curry RB 124, 125, *146*
Baker CH, see Curry RB 131, 143, *146*
Baker CH, see Meyer GE 131, *148*
Baker DB, see Egli DB 517, *522*
Baker DN, Lambert JR, Phene CJ, McKinion JM 131, 133, 136, 137, 143, *146*

Baker DN, see Bar-Yosef B 141, *146*
Baker DN, see Lambert JR 131, 139, 141, 142, 145, *148*
Baker DN, see Phene CJ 132, *149*
Baker DN, see Stapleton HN 136, *149*
Baker EA 395, *406*
Baker HG, see Opler PA 295, 296, *303*
Baker JH, see Campbell PGC 52, *76*
Baker R, Cole S 433, *444*
Baldry CW, Cockburn W, Walker DA 483, *491*
Baldwin AL, see Bubeck RC 435, *446*
Baldwin BC, see Dodge AD 345, 346, *376*
Ball E, see Lüttge U 480, *498*
Ballantyne D 476, *491*
Ballard L 520, *521*
Ballio A 362, *373*
Ballio A, see Dohrmann U 362, *376*
Bamber RK 20, *39*
Bamberg SA, see Ackerman TL 214, *226*
Bamberg SA, see Wallace A 213, *231*
Bandeen JD, Stephenson GR, Cowett ER 368, 369, *373*
Bandeen JD, see Jensen KIN 370, *380*
Bandeen JD, see Souza Machado V 370, *388*
Bannister TT, see Laws EA 320, *324*
Banse K 93, *108*
Baptist JP 434, *444*
Baranowski A, see Mulligan HF 418, *457*
Barbaree JM, Payne WJ 96, *108*
Barber DA, Loughman BC 29, *39*
Barber FR, see Martin A 468, 469, *498*
Barber J 336, *373*
Barber LE, Ensign JC 54, *75*
Barber RT 54, *75*, 103, *108*
Barber RT, Dugdale RC, MacIsaac JJ, Smith RL 103, *108*
Barbour MG 211, 212, *226*
Barea JL, see Herrera J 60, *78*
Barica J, see Ayles GB 426, *444*
Barica J, see Whitaker J 426, *463*
Barko JW, Smart RM 64, *75*
Barley KP 20, *39*
Bärlocher F, Mackay RJ, Wiggins GB 53, *75*
Barlow EWR, Boersma L, Young JL 195, *197*
Barlowe WC, see Haller WT 436, *451*
Barnes A 137, *146*
Barnes A, Greenwood DJ, Cleaver TJ 129, 143, *146*
Barnes A, Hole CC 139, *146*, 262, 265, *272*
Barnes A, see Hole CC 263, *274*
Barnes H 86, *108*
Barnes JM 400, *406*
Barnes JR, Ovink R, Cummins KW 53, *76*
Barnes K, see Bennett JP 474, *492*
Barnett HL, Lilly VG 519, *521*
Barney GO 332, 333, *373*
Baron H le, Gressel J 398, *406*
Barralis G, Gasquez J 369, *373*
Barralis G, Gasquez J, Jan P, Soffietti S 397, *406*
Barratt DHP, see Flavell RB *146*
Barrett JR, Peart RM 135, *146*
Barrs HD, see Kriedemann PE 260, *275*
Barsdate RJ, see McRoy CP 64, *80*
Bartell SM, see Gallepp GW 65, *77*
Bartels PG, Hilton JL 341, 355, 356, *373*
Bartels PG, Hyde A 350, 352, 355, *373*
Bartels PG, McCullough C 340, 350, *374*
Bartels PG, Watson CW 350, 352, *374*
Bartholomew WV, Clark FE 29, *40*
Bartle GA, see Carbon BA 250, 251, *273*
Bartlett L, Rabe FW, Funk WH 430, *444*
Barvenik FW, see Walsh JJ 89, *115*, 314, *327*
Bar-Yosef B, Lambert JR, Baker DN 141, *146*
Basilevich NI, see Rodin LE 16, 17, 24, 25, *44*
Basin R, see Decleire M 431, *447*
Basler E, Todd GW, Meyer RE 396, *406*
Basler E, see Hawxby K 431, *451*
Bassata G, see Susmel L 210, *231*
Bassham JA, see Hiller RG 486, *496*
Bates AL, see Goldsby TL 418, *450*
Batra SWT 372, *374*
Batt S, Venis MA 362, *374*
Batterton JC 365, *374*
Batterton JC, Boush GM, Matsumura F 432, *444*
Battino R, see Wilhelm E 479, *501*
Baumann G, see Fischer WR 65, *77*
Bavel C van 517, *521*
Bavel CHN van, see Takami S 517, *525*
Baxter R, see Weeks DP 354, 355, *391*
Bayer DE, see Ashton FM 354, 355, *373*
Bayer DE, see Hess FD 355, 356, *379*
Bayer DE, see Rost TL 355, *386*
Bazilevich NI, see Rodin LE 211, 212, *230*
Bazzaz FA, Carlson RW, Rolfe GL 544, *547*
Bazzaz FA, Rolfe GL, Carlson, RW 544, *547*
Bazzaz FA, see Carlson RW 483, *493*
Bazzaz FA, see Regehr DL 517, *525*
Beadle CL, Neilson RE, Jarvis PG, Talbot H 260, 261, *272*
Beadle CL, Talbot H, Jarvis PG 241, 246, 251, 253, *273*
Beadle NCW 13, *40*
Bearder JR 358, *374*

Beatley JC 213, 214, 216, *226*
Beauchamp RSA 66, *76*
Bebee GP, see Bidwell RGS 487, *492*
Bechard MJ, Rayburn WR 69, *76*
Beck TH 398, *406*
Beckerson DW, Hofstra G 472, 477, *491*
Beek J 142, *146*
Beek J, Frissel MJ 142, *146*
Beer LO, Pipes WO 440, *444*
Beets P, see Madgwick HAI 235, 236, 237, *277*
Begg JE, Torsell BWR 224, *226*
Behrendt S, see Knittel H 361, *381*
Beilke S, Gravenhorst G 479, *492*
Beisl E, see Kohler A 420, 442, *454*
Bell AR, see Schooler AB 397, *410*
Bell JN, Clough WS 473, *492*
Bell JN, see Ayazloo M 472, 473, *491*
Bell LC, see Gillman GP 21, *41*
Bell SL, Schwarz OJ, Hughes KW 354, *374*
Belot Y, Gauthier D 471, *492*
Belter WG 439, *444*
Ben-Aziz A, Koren E 350, *374*
Bender M, see McCaffney RJ 97, *112*
Ben-Dyke R, Sanderson DM, Noakes DN 400, *406*
Benedict HM, Brown GB 195, *197*
Bengtsson A 397, *406*
Benjamin RW, see Keulen H van 131, 142, *147*
Bennett C 538, *547*
Bennett D, Downes PA, Janes AVJ 126, 127, *146*
Bennett HD 438, *444*
Bennett JH, Hill AC 470, *492*
Bennett JH, Hill AC, Gates DM 470, *492*
Bennett JH, see Hill AC 475, *496*
Bennett JH, see White KL 477, *501*
Bennett JP, Barnes K, Shinn JH 474, *492*
Bennett KJ, Rook DA 260, *273*
Ben-Shaul Y, see Porath D 442, *458*
Bensink J, see Horie T 144, *147*
Benson AA, see Lehmann J 481, *497*
Benson AA, see Miyachi S 423, *456*
Benson-Evans K, Williams PF 443, *444*
Benz J, see Rüdiger W 352, *387*
Ben-Zioni A, Ital C 441, *444*
Berg MA, see McNeal FH 166, 167, *171*
Berger A, see Thorpe MR 247, *279*
Bergersen FJ, Kennedy GS, Wittman W 12, *40*
Bergfeld R, see Frosch S 350, *378*
Bergmann H-J 405, *407*
Bergmann W 53, *76*
Berman T 51, 63, *76*
Berman T, see Williams PJLeB 95, *115*
Bernard RL, see Wax LM 397, *411*
Berner RA, see Martens CS 97, *112*
Bernhard F *300*
Bernhard-Reversat F 25, *40*
Bernhard-Reversat F, see Huttel Ch 283, 291, 294, 295, 297, *301*
Bernhardt H 415, *444*
Bernice R 61, *76*
Berry JA, Downton WJS 258, 260, *273*
Berry JA, see Björkman O 219, *226*
Berry JA, see Ehleringer JR 216, 217, *227*
Berry JA, see Mooney HA 215, 216, 219, 224, *229*
Berry JA, see Pearcy RW 219, *230*
Berry JW, see Wentsel RS 429, *463*
Besch WK, Ricard M, Cantin R 430, *444*
Beschel RE 174, *197*
Besemer AFH, see Voerman S 364, *390*
Beskaravainaya V, see Braginskii L 403, *407*
Best D, Mantai KE 60, *76*
Best GR, Monk CD 37, 38, *40*
Beste CE, Schreiber MM 355, *374*
Betzer N, Kott Y 433, *444*
Bevege DI, Bowen GD, Skinner MF 23, *40*
Bevege DI, Richards BN, Moore AW 12, *40*
Bevenue A 364, *374*
Beyer EM, Johnson AL, Sweetser PB 360, *374*
Beyers RJ, Odum HT 322, *323*
Beym AM, see Stom DI 431, *461*
Bezuglov V, Minenko A, Postdeva R 398, *407*
Bhagat SK, see Gloyna EF 435, *450*
Bhagsari AS, see Phillips BA 367, 368, *385*
Bhide JP, see Mujumdar SB 519, *524*
Bhumiratana A, see Payer HD 429, *458*
Bidwell RGS, Bebee GP 487, *492*
Biggs DC, Rowland RG, Wurster CF 365, *374*
Bilinski H, see Stumm W 71, *83*
Bille J-C, Poupon H 282, *300*
Billen G 428, *445*
Billen G, see Wollast R 428, *464*
Billings WD 521, *522*, 529, 530, 546, *547*
Billings WD, Luken JO, Mortensen DA, Peterson KM *547*
Billings WD, Peterson KM 543, *547*
Binder BJ, Chisholm SW 70, *76*
Bingham FT, see Jenny H 34, *42*
Bingham GE, see Coyne PI 474, 477, *493*
Bingham SW, Shaver RL 432, *445*
Birch HF 32, *40*
Birmingham BC, Colman B 365, *374*, 431, *445*
Biscoe PV, see Gallagher JN 238, *274*

Biscoe PV, see Marshall B 258, *277*
Biscoe PV, see Unsworth MH 472, *501*
Bjerregaard RS 213, *226*
Björk S 418, 420, 424, 438, 439, *445*
Björkman O, Badger MR, Armond PA 219, *226*
Björkman O, Boynton J, Berry JA 219, *226*
Björkman O, Ludlow MM, Morrow PH 288, 290, 291, *300*
Björkman O, Mooney HA, Ehleringer J 221, 223, *226*
Björkman O, Pearcy RW, Harrison AT, Mooney HA 217, 219, *226*
Björkman O, see Ehleringer J 163, *171*
Björkman O, see Ehleringer JR 217, 219, 222, 223, *227*, 240, 261, *273*
Björkman O, see Mooney H 287, *302*
Björkman O, see Mooney HA 215, 217, 219, 221, 222, 223, *229*
Björkman O, see Pearcy RW 219, 223, *230*
Blaauw OH, see Thomas JB 434, *462*
Black CC 176, 179, 184, *197*
Black CR, Black VJ 473, *492*
Black JN 235, *273*
Black VJ 471, 472, *492*
Black VJ, Unsworth MH 470, 472, 473, *492*
Black VJ, see Black CR 473, *492*
Black VJ, see Ormrod DP 477, *499*
Black VJ, see Unsworth MH 472, 473, *501*
Blackburn RD, Weldon LW 432, *445*
Blackburn WM, Petr T 53, *76*
Blackman GE, Parke MH, Garton G 431, *445*
Blackman GE, see Sargent JA 394, *410*
Blackmon WJ, see Moreland DE 349, *383*
Blad BL, see Lemeur R 164, *171*
Blair AM, Parker C, Kassasian L 367, *374*
Blake J, Kaufman DD 398, *407*
Blanc F le, Rao DN 483, *492*
Blanc F le, see Comeau G 489, *493*
Blasco F, Tassy B 294, *300*
Blaszyk P 400, *407*
Blauel TA, see Malhotra SS 471, *498*
Bledsoe CS, see Johnson DW *42*
Bliss LC 174, *198*
Blum JL 421, *445*
Blum U, Heck WW 475, *492*
Boardman NK 179, *198*, 258, *273*
Bodungen B von, see Smetacek V 97, *114*
Boehle J, see Ellis JF 367, *377*
Boehler-Kohler BA, Schopf M, Böger P 340, 353, *374*
Boersma L, see Barlow EWR 195, *197*
Böger P, Kunert KJ 342, 346, 351, 353, *374*
Böger P, Sandmann G, Miller R 371, *374*
Böger P, Schlue U 352, *374*
Böger P, see Boehler-Kohler BA 340, 353, *374*
Böger P, see Böhme H 336, *374*
Böger P, see Kunert KJ 340, 350, 351, 352, *381*
Böger P, see Lambert R 340, 347, 351, 352, *381*
Böger P, see Sandmann G 340, 349, 350, 351, 352, 353, *387*
Boggess WR, see Regehr DL 517, *525*
Böhme H, Kunert KJ, Böger P 336, *374*
Böhme H, Reimer S, Trebst A 339, 345, *374*
Bohne H 476, *492*
Bokhari UG, see Dyer MI 194, *199*
Bollard EG, see Ferguson AR 60, *77*
Bollen WB, see Watson GH 426, *463*
Bolton JR, see Harbour JR 353, *379*
Bombosch S, Peters L 437, *445*
Bond RG, Straub CP, Prober R 486, *492*
Bone DH 423, *445*
Bonham CD 191, *198*
Bonner WD, see Plesnicar M 487, *499*
Bonotto S, see Borghi H 365, *374*
Bonte J 476, *492*
Bonte J, Cormis L de, Louguet P 472, *492*
Borghi H, Puiseux-Dao S, Bonotto S, Hoursiangou-Neubrunn D 365, *374*
Borghi H, see Puiseux-Dao S 365, *385*
Borman FH, see Marks PL 250, *277*
Bormann FH, Likens GE 6, 7, 8, 20, *40*, 539, *547*
Bormann FH, see Eaton JS 24, *41*
Bormann FH, see Gosz JR 24, *41*
Bormann FH, see Likens GE 545, *548*
Bormann FH, see Wood T 474, *502*
Born WH van den, see Sharma MP 396, *410*
Bornkamm R, Raghi-Atri F, Koch M 424, *445*
Bornkamm R, see Raghi-Atri F 424, *458*
Boroschewski G, see Trebst A 339, *390*
Borovskaya TA, see Kurochkina LY 213, *228*
Bosch R van der, see Flint ML 371, *377*
Botkin DB, see Woodwell GM 262, 264, *280*, 506, *526*, 537, 539, *548*
Boughey AS 531, *547*
Bourn WS 436, *445*
Boush GM, see Batterton JC 432, *444*
Bowen GD, Theodorou C 21, 27, *40*
Bowen GD, see Bevege DI 23, *40*
Bowen VT, see Hutchinson GE 416, *452*
Bowes GW 365, *374*
Bowes J, Crofts AR, Arntzen CJ 338, 342, *375*

Bowes PM, see Carlucci AF 95, 103, *109*
Bowler K, Duncan CJ, Gladwell RT, Davison TF 441, *445*
Bowman MJ, Esaias WE 89, *108*
Bowonder B, see Menon KAP *148*
Boyd CE 424, *445*
Boyer JS 195, *198*
Boyle EA 103, *108*
Boylen CW, Brock ML 440, *445*
Boynton J, see Björkman O 219, *226*
Boynton WR, Keefe C, Kemp WM 307, 309, 319, *323*
Boynton WR, Kemp WM, Osborne CG 306, *323*
Boynton WR, see Mihursky JA 315, *325*
Boysen-Jensen P 164, *170*, 256, *273*
Bradbeer JW, see Murray DR 481, 482, 483, *498*
Bradbury IK, Malcolm DC 261, *273*
Bradkowski TA, see Hilton BD 366, *379*
Bradshaw AD, McNeilly TS, Gregory RPG 536, *547*
Bradshaw AD, see Antonovics J 429, *444*
Bradshaw AD, see Horsman DC 472, 473, *496*
Braginskii L, Beskaravainaya V, Shcherban E 403, *407*
Bramley PM, see Sandmann G 350, *387*
Brand T von, see Rakestraw MW 95, *113*
Brandle JR, see Miles CD 490, *498*
Brandt CS, Heck WW 482, *492*
Brandt CS, see Heck WW 473, 475, 476, *496*
Bransford AV, see Fuhr I 11, *41*
Braun HE, see Frank R 430, *449*
Braun R 66, *76*
Bray JR, Gorham E 25, 26, *40*, 266, 267, *273*, 292, 294, 295, *300*
Bray JR, see Struik GJ 195, *202*
Bray JT, Bricker OP, Troup BM 100, *109*
Brazel AJ, see Idso SB 510, *523*
Brehmer ML, see Warinner JE 441, *463*
Breitner NF, see Davis CO 101, *110*
Bremmer PM, see Gifford RM 517, *522*
Bressan RA, LeCureux L, Wilson LG, Filner P 484, *492*
Bressan RA, see Wilson LG 474, *501*
Brewer P, see Arntzen CJ 370, *373*
Brewer R, see Stace HCT 15, *44*
Brezonik PL 415, 422, *445*
Brezonik PL, Lee GF 58, *76*
Brezonik PL, see Brooks RH 56, *76*, 90, *109*
Brezonik PL, see Keirn MA 56, 57, *79*
Brian RC, Homer RF, Stubbs J, Jones RL 331, 339, 345, *375*
Bricker OP, see Bray JT 100, *109*
Briggs D 490, *492*
Briggs DR, see Siminovitch D 296, *303*
Briggs WR, see Cross JW 362, *376*
Bringmann G, Kühn R 427, 443, *445*
Brinkmann E, Lüttge U, Fischer K 480, *492*
Brinkmann E, see Lüttge U 480, *498*
Brinkmeier R, see Kohler A 417, 420, 442, *454*
Brinley FJ *445*
Brinson MM, see Twilley RR 64, *83*
Bristow JM 57, *76*
Bristow JM, Whitcomb M 421, *445*
Bristow JM, Whitcombe M 64, *76*
Britton G 350, *375*
Brix H 261, *273*
Brock ML, see Boylen CW 440, *445*
Brock TD 440, 441, *445*
Brockel K von, see Parsons TR 96, *113*
Brockington NR 125, 129, *146*
Brod HG, Preusse HU 435, *445*
Broda E, see Matzku S 429, *456*
Brödsel K von, see Smetacek V 97, *114*
Brody R, see Mandelli EF 311, *324*
Broecker WS 508, 509, *522*
Broecker WS, Li Y-H 509, *522*
Broecker WS, Takahashi T, Simpson HJ, Peng TH 506, *522*
Broecker WS, see Schindler DW 49, *82*
Bronowski J 531, 532, *547*
Brooker MP 431, *445*
Brooker MP, Edwards RW 430, 431, *445*
Brooker MP, see Fry JC 431, *450*
Brooks AJ, Baker AL 433, *446*
Brooks GT 364, *375*
Brooks RH, Brezonik PL, Putman HD, Keirn MA 56, *76*, 90, *109*
Brougham RW 163, *170*
Brouwer R 187, *198*
Brouwer R, Wit CT de 137, 141, *146*
Brouwer R, see Wit CT de 137, 144, *150*
Brown AHF, see Carlisle A 25, *40*
Brown AWA 364, *375*
Brown EJ, Harris RF 64, *76*
Brown EJ, Harris RF, Koonce JK 64, *76*
Brown GB, see Benedict HM 195, *197*
Brown GW 205, *226*
Brown HT, Escombe F 505, 517, *522*
Brown J, see Deneke FJ 545, *547*
Brown LF, Trlica MJ 178, 179, *198*
Brown LL, see Zimmermann MH 287, *303*
Brown RH 176, 191, *198*
Brown RH, Simmons RE 176, *198*

Brown RL, see Hall MA 485, *495*
Brown V, Ducker SC, Rowan KS 104, *109*
Bruinsma J 396, *407*
Brun C, see Puiseux-Dao S 365, *385*
Brun R 283, 291, *300*
Brun WA 520, *522*
Brun WA, see Lawn RJ 520, *524*
Brungs WA 433, *446*
Bruno SF, Staker RD 103, *109*
Brunskill GJ, see Schindler DW 49, *82*
Brunsting AB, see Penning de Vries FWT 138, *149*
Brunsting AHM, see Penning de Vries FWT 160, 161, *171*, 240, 265, *278*
Bryan W, Wright RD 520, *522*
Brylinsky M, Mann KH 305, 313, 314, 319, *323*
Bryson RA, Wendland WM 509, *522*
Bual SW, see Wells EG 24, 25, *45*
Bubeck RC, Diment WH, Deck BL, Baldwin AL, Lipton SD 435, *446*
Bucha HC, Todd CW 331, *375*
Buchanan GA, see Burns ER 340, 350, 352, *375*
Büchel KH 330, 331, 342, 348, 364, *375*
Büchel KH, Draber W 342, *375*
Büchel KH, Draber W, Trebst A, Pistorius E 348, *375*
Büchel KH, see Draber W 339, 342, *376*
Buck JD 440, *446*
Bucklay B, see Oviatt CA 316, *325*
Buckley M, see Whitaker J 426, *463*
Bucksteeg W 434, *446*
Bugg MW, Whitmarsh J, Rieck CE, Cohen WS 339, 345, 352, *375*
Buikema AL Jr, McGinniss MJ, Cairns J Jr 430, 432, *446*
Buikema AL Jr, see Cairns J Jr 442, *446*
Bukovac MJ 394, 395, *407*
Bula RJ, see Holt DA 131, *147*
Bula RJ, see Schreiber MM 124, 131, *149*
Bull JN, Mansfield TA 473, 475, 490, *492*
Bullock HC, see Kramer PJ 21, *42*
Bunce JA *198*, 951
Bunt JS 315, *323*
Burcar PJ, see Wershaw RL 430, *463*
Burges A, see Kendrick WB 31, *42*
Buringh P, Heemst HDJ van, Staring GJ 134, *146*
Burkard G, see Lichtenthaler HK 352, *382*
Burkard G, see Meier D 352, *383*
Burkholder LM, see Burkholder PR 103, *109*
Burkholder PR, Burkholder LM 103, *109*
Burkholder PR, see Mandelli EF 311, *324*
Burnett WC, see Veeh MM 100, *115*
Burnison BK 64, *76*
Burns ER, Buchanan GA, Carter MC 340, 350, 352, *375*
Burns NM, Nriagu JO 72, *76*
Burnside O, Wicks G, Carlson D 401, *407*
Burris RH 487, *493*
Burris RH, see Stewart WDP 423, *461*
Burrows WH 212, 213, *226*
Burrows WJ 355, *375*
Burton JD, Liss PS 101, *109*
Buschbom U, see Evenari M 213, 214, *227*
Buschbom U, see Kappen L 214, 222, 223, *228*
Buschbom U, see Lange OL 219, 221, *228*
Buschmann C, Grumbach KH, Bach TJ 350, *375*
Buschmann C, see Pfister K 339, 343, 344, *385*
Büsgen M, Münch E 257, *273*
Bush LP, see Carringer RD 367, *375*
Bussey AH, see Sims AP 422, *460*
Buta JG, see Steffens GL 362, *388*
Butcher RW 421, *446*
Butler DR, Landsberg JJ 262, 264, 265, 266, *273*
Butler EI, Corner EDS, Marshall SM 96, 100, *109*
Butler EI, Knox S, Liddicoat MI 93, *109*
Butler GL 364, *375*
Butler HS, see Duke WB 341, 355, *376*
Butler TJ, see Likens GE 438, *455*, 477, *497*
Buxton DR, see Stapleton HN 136, *149*
Byrne RH, Kester DR 102, *109*

Cabanettes A 207, 210, *226*
Cadenas E, see Serra JL 95, *114*
Caemmerer S von, Farquhar GD 258, *273*
Cailloux M, Phan CT, Chung YS 485, *493*
Cain SA, Oliveira Castro GM de, Murca Pires J, Tomas de Silva N 285, 291, *300*
Cairns J Jr 413, 431, *446*
Cairns J Jr, Buikema AL Jr, Heath AG, Parker BC 442, *446*
Cairns J Jr, Dickson KL 414, 439, 443, *446*
Cairns J Jr, Dickson KL, Maki AW 443, *446*
Cairns J Jr, Dickson KL, Westlake GF 443, *446*
Cairns J Jr, Heath AG, Parker BC 438, *446*
Cairns J Jr, Lanza GR, Parker BC 414, 424, 435, *446*
Cairns J Jr, see Buikema AL Jr 430, 432, *446*
Cairns J Jr, see Dickson KL 443, *448*
Cairns J Jr, see Lanza GR 441, *455*
Calderbank A, Slade P 345, *375*
Caldwell MM, Johnson DA, Fareed M 187, *198*
Caldwell MM, Osmond CB, Nott DL 219, *226*
Caldwell MM, White RS, Moore RT, Camp LB 212, 213, *226*

Caldwell MM, see Deputit EJ 219, *227*
Caldwell MM, see Johnson DA 187, *200*
Call DJ, see Greichus YA 437, *450*
Calvayrac R, Ledoigt G, Laval-Martin D 371, *375*
Calvayrac R, see Laval-Martin D 371, *382*
Calvert A 518, *522*
Calvert SE 101, *109*
Calvert SE, Price NB 95, *109*
Cameron ML, see Hayes FR 416, *451*
Camp LB, see Caldwell MM 212, 213, *226*
Campbell FJ, see Feder WA 475, *494*
Campbell PGC, Baker JH 52, *76*
Campbell RS, see McNaughton SJ 187, *201*
Campbell WF, see Singh B 355, *388*
Canelli E, see Fuhs GW 64, *77*
Cantin R, see Besch WK 430, *444*
Canton JH, Greve PA, Slooff W, Esch GJ van 432, 437, *446*
Cape JN, see Fowler D 467, 469, 477, *494*, *495*
Caperon J 91, *109*
Caperon J, Cattell SA, Krasnick G 320, *323*
Capinera JL, Roltsch WJ 194, *198*
Cappenberg ThE 68, *76*
Capron TM, Mansfield TA 475, 477, *493*
Capron TM, see Wellburn AR 486, 490, *501*
Carbiener R 421, *446*
Carbiener R, see Krause W *454*
Carbon BA, Bartle GA, Murray AM 250, 251, *273*
Card KA, see Troughton JH 298, *303*
Carles PM, Assche CJ van 345, *375*
Carlisle A, Brown AHF, White EJ 25, *40*
Carlson D, see Burnside O 401, *407*
Carlson PS, see Meredith CP 366, *383*
Carlson RW, Bazzaz FA 483, *493*
Carlson RW, see Bazzaz FA 544, *547*
Carlucci AF, Bowes PM 103, *109*
Carlucci AF, Hartwig EO, Bowes PM 95, *109*
Carlucci AF, Strickland JDH 95, *109*
Carlucci AF, see Provasoli L 103, *113*
Carmiggelt CJW, see Horne AJ 56, *78*
Carpenter EJ, Guillard RRL 93, *109*
Carpenter EJ, McCarthy JJ 314, *323*
Carpenter EJ, see Gonye ER 102, *111*
Carpenter JH, Pritchard DW, Whaley RC 90, *109*
Carpenter KE 425, *446*
Carr JF, see Ogawa RE 56, *81*
Carringer RD, Riek CE, Bush LP 367, *375*
Carter MC, see Burns ER 340, 350, 352, *375*
Carter RC, see Goldman CR 415, *450*
Cartwright PM 341, 356, 357, 358, *375*
Casida JE 364, *375*
Casida JE, Kimmel EC, Lay MM, Ohkawa H, Rodebush JE, Gray RA, Teseng CK, Tilles H 334, 367, *375*
Casida JE, see Ahmed MK 432, *444*
Casida JE, see Lay MM 367, *382*
Casida JE, see Pallos FM 367, *384*
Castri F di, Mooney HA 205, *226*
Cat W de, see Decleire M 431, *447*
Čatský J, see Šesták Z 237, *279*
Catsky J, see Sestark Z 216, 217, *230*
Cattell SA, see Caperon J 320, *323*
Cearley JE, Coleman RL 429, *446*
Cecil R, Wake RG 480, *493*
Cedeno-Maldano A, Swader JA 426, *446*
Cedeno-Maldano A, Swader JA, Heath RL 426, *446*
Cernusca A, see Larcher W 181, *200*
Cha JW, see Wallace A 213, *231*
Chacko CI, Lockwood JL 432, *447*
Challa H 138, *146*
Challa H, Vooren J van de 136, *146*
Chamberlain AC 470, 471, *493*
Chamberlain AC, Little P 470, *493*
Chamberlain WM 416, *447*
Champ MA, see Ewald WG 365, *377*
Chan HS, see Wellburn AR 486, 490, *501*
Chan KL, see Gächter R 427, *450*
Chan Y-H, see Olson JS 505, 507, *525*
Chancellor RJ, see Way JM 405, *411*
Chandorkar KR, see Khan MU 353, *380*
Chang CW 475, 476, 485, 489, *493*
Chang CW, Thompson CR 476, 489, *493*
Chang FY, see Sharma MP 396, *410*
Chang FY, see Stephenson GR 367, *388*
Chapell WE, see Daniell JW 441, *447*
Chapin FS 21, *40*
Chapin FS, III 196, *198*
Chapin FS, III, Cleve K van, Tieszen LL 190, *198*
Chapin FS, see Tieszen LL 176, 177, 181, *202*
Chapman ARO, Craigie JS 104, *109*
Chapman D 441, *447*
Chapman RC, see Kinerson RS 252, *275*
Charley JL, Cowling SW 212, 213, *226*
Charley JL, West NE 15, 31, *40*

Charley JL, see Hurditch WJ 33, *42*
Charley JL, see Richards BN 32, *44*
Chase VC, see Strain BR 219, 221, *231*
Chatarpaul L, Robinson JB, Kaushik NK 58, *76*
Chau YK, see Gächter R 427, *450*
Chau YK, see Silverberg BA 427, *460*
Chau YK, see Wong PTS *464*
Chave KE 49, *76*
Chen C, see Lewin J 102, *112*
Chen KL, see McGregor AN 56, *80*
Chen LG, Switzer CM, Fletcher RA 341, 354, 355, 357, *375*
Chen M, see Fuhs GW 64, *77*
Chen RI, see Keeney DR 58, *79*
Chen RL, Keeney DR, Graetz DA 57, 58, *76*
Cheng JYO, Yu MH, Miller GW, Welkie GW 489, *493*
Cherry JH 341, 354, 355, 357, *375*
Chervin MB, see Malone TC 314, *324*
Chester R, see Riley JP 86, 100, *114*
Chesters G, see Dolar SG 429, *448*
Chew AE, see Chew RM 212, *226*
Chew RM, Chew AE 212, *226*
Child GI, see Golley FB 17, 25, *41*, 283, 294, *300*
Chimiklis P, see Heath RL 484, 485, *496*
Chimiklis PE, Heath RL 485, *493*
Chimiklis PE, Karlander EP 437, *447*
Chisholm SW, see Binder BJ 70, *76*
Chmiel H, see Ries SK 355, *386*
Cholnoky BJ 436, *447*
Christensen O 241, 267, *273*
Christian W, see Warburg O 487, *501*
Christiansen MN, see St John JB 353, *388*
Christie EK 176, 194, *198*
Chu C, see Mooney HA 207, 208, 210, *229*
Chu SP 420, 422, *447*
Chu-Der O, see Miles CD 490, *498*
Chun ChH, see Talpaz H 127, *150*
Chung YS, see Cailloux M 485, *493*
Cintrón B, see Lugo A 287, 293, 295, *302*
Cintrón G, see Odum HT 287, 293, *303*
Claesson A, Ryding SO 415, *447*
Claesson A, see Forsberg C 443, *449*
Clark EA, Loomis RS 164, *170*
Clark FE, see Bartholomew WV 29, *40*
Clark GM, see Davis GJ 437, *447*
Clark HE 156, *170*
Clark JG, see Ledig FT 240, 263, 265, *276*
Cleaver TJ, see Barnes A 129, 143, *146*
Cleaver TJ, see Greenwood DJ 127, *147*
Clements RG, see Golley FB 17, 25, *41*, 283, 294, *300*
Clendenning KA, see Ke B 434, *453*
Cleve K van, see Chapin FS, III 190, *198*
Clifford CH, see Rowe GT 98, *114*, 306, 314, *326*
Clough WS, see Bell JN 473, *492*
Coaker TH, see Dempster JP 403, *407*
Coatsworth JL, see Eppley RW 92, 94, 107, *110*
Coble HD, Slife FW 357, *375*
Cock JH, Wholey D, Gutierrez de la Casas O 160, *170*
Cock JH, Yoshida S 518, *522*
Cockburn W, see Baldry CW 483, *491*
Cockshull KE, see Hughes AP, 517, 518, 520, *523*
Cocucci SM, see Dohrmann U 362, *376*
Coffey BT, McNabb CD 418, *447*
Cohen WS, see Bugg MW 339, 345, 352, *375*
Coker RD, see Nriagu JO 68, *81*
Cole B, see Toetz DW 59, *83*
Cole BS, Toetz DW 421, *447*
Cole CV, Innis GS, Steward JWB 129, 142, *146*
Cole DR, Plapp FW 365, *375*
Cole DW, Rapp M 18, *40*
Cole DW, see Johnson DW 37, 39, *42*
Cole S, see Baker R 433, *444*
Coleman DC 33, *40*
Coleman DC, see Dyer MI 191, *199*
Coleman DC, see Weigert RG 28, *45*
Coleman RL, see Cearley JE 429, *446*
Collatz GJ, see Mooney HA 217, 219, 221, 222, *229*
Collier TR, see Thomas MD 482, *500*
Collins DC, see Meisel WS 134, *148*
Collins NJ, see Oechel WC 179, 181, 182, *201*
Collins NM 294, *300*
Collinson RI, see O'Sullivan AJ 439, *457*
Colman B, see Birmingham BC 365, *374*, 431, *445*
Colman RL, Lazenby A 190, *198*
Comeau G, Blanc le F 489, *493*
Confer JL 416, *447*
Connell GH, see Hall AE 165, *171*
Conners JD, see Cook SF Jr 431, *447*
Conrad CE, see DeBano LF 208, *227*
Conrad SG, Radosevich SR 369, *376*
Constantinidou HA, Kozlowski TT 490, *493*
Conway HL 92, *109*, 429, *447*
Conway HL, Harrison PJ 95, *109*
Conway HL, Harrison PJ, Davis CO 101, *109*
Conway HL, see Harrison PJ 101, *111*
Conway HL, see Parker JI 69, *81*

Cook DR, see Wesely ML 471, *501*
Cook SF Jr, Conners JD 431, *447*
Cooke GD, see Heath RT 63, *78*, 423, *451*
Cooper JM, see Miller HG 251, *277*
Cooper JP 160, *170*, 236, *273*
Cooper JP, see Wilson D 179, *203*, 515, *526*
Copeland BJ 315, *323*
Corbett JR 330, 331, 335, 348, 349, 364, *376*
Corbett JR, see Wright K 337, 338, 339, 351, *391*
Corley RHV 157, 161, *171*
Cormis L de 474, *493*
Cormis L de, see Bonte J 472, *492*
Corner EDS, Davies AG 96, *109*
Corner EDS, see Butler EI 96, 100, *109*
Cornforth JS 294, *300*
Correll DS 534, 535, *547*
Correll RL, see Sands R 195, *202*
Corson MJ, see Rook DA 262, 264, 265, *278*
Costa F da, see Evans LS 474, *494*
Coste M 442, *447*
Couch HB, see Daniell JW 441, *447*
Coulston F, Korte F 364, *376*
Coupland RT 191, *198*
Cours G 160, *171*
Coutinho LM 291, *300*
Cowan IR, see Wong SC 140, *150*
Cowen WF, Lee GF 415, *447*
Cowett ER, see Bandeen JD 368, 369, *373*
Cowgill UM 429, 436, *447*
Cowling DW, Jones LHP, Lockyer DR 482, *493*
Cowling DW, Koziol MJ 466, *493*
Cowling DW, Lockyer DR 473, 482, *493*
Cowling DW, see Lockyer DR 473, *498*
Cowling EB 477, *493*
Cowling EB, Linthurst RA 477, *493*
Cowling SW, see Charley JL 212, 213, *226*
Cox MC, see Nicholls KH 10, *43*
Cox RA, Eggleton AEJ, Derwent RG, Lovelock JE, Pack DH 469, *493*
Cox RA, Penkett SA 469, *493*
Cox RA, see Eggleton AEJ 468, *494*
Cox TL, Harris WF, Ausmus BS, Edwards NT 26, *40*
Coyne PI, Bingham GE 474, 477, *493*
Coyne PI, see Deneke FJ 545, *547*
Crafts AS, Crisp CE 396, *407*
Crafts AS, see Ashton FM 330, 334, 338, *373*, 394, 396, *406*
Craig PP, see Woodwell GM 544, *548*
Craigie JS, see Chapman ARO 104, *109*
Cramer HH 331, *376*, 398, *407*
Crance JH 426, *447*
Cremer J 402, *407*
Criegee R 484, *493*
Crisp CE, see Crafts AS 396, *407*
Crist JB, Dawson DH 236, *273*
Criswell JG, see Hardy R 520, *523*
Crocker RL, Major J 11, *40*
Croft BA, see Welch SM 136, *150*
Crofts AR, see Bowes J 338, 342, *375*
Cromack K Jr, see Fogel R 34, *41*
Cromack K, see Johnson DW *42*
Cronin LE, see Neilson BJ 305, *325*
Cross FA, Sunda WG 103, *109*
Cross JW, Briggs WR 362, *376*
Cross R, see Robb F 53, *82*
Crossley DA Jr 28, *40*
Crossley DA Jr, Witkamp M 28, *40*
Crossley DA Jr, see Edwards CA 28, *41*
Croze E, Kelly M, Horton P 342, *376*
Cullen D, see McCaffney RJ 97, *112*
Cullen JJ, see Eppley RW 89, *110*
Culley D, Epps E 424, *447*
Cummins KW, see Barnes JR 53, *76*
Cummins KW, see Suberkropp K 53, *83*
Cunningham GL, Strain BR 223, 224, *226*
Cunningham GL, see Syvertsen JP 215, *231*
Cunningham WP, see Arntzen CJ 355, *373*
Curl HC Jr, see Iverson RL 314, 319, *324*
Curry JP 399, *407*
Curry RB, Baker CH, Streeter JG 131, 143, *146*
Curry RB, see Baker CH 124, 125, *146*
Curry RB, see Meyer GE 131, *148*
Curtis EJC, Durrant K, Harman MMI 57, *76*
Cushing DH 100, *109*, 316, *323*
Czaja AT 476, *493*
Czyrska H, see Hutchinson TC 429, 430, *452*

Dacey JWH, Klug MJ 54, *77*
Dale RF, see Nelson WL 127, *148*
Daley RJ, see Overbeck J 52, *81*
Dalrymple DG 165, *171*
Daniel DJ, see Miles CD 490, *498*
Daniell JW, Chapell WE, Couch HB 441, *447*
Darley EF 476, *493*
Darley EF, see Lerman SL 476, *497*
Darr S, Souza Machado V, Arntzen CJ 370, *376*
Das B, Singh PK *447*
Datta DB, Ryrie IJ, Jagendorf AT 483, *494*
Davey E, see McCaffney RJ 97, *112*
Davidson JL, Milthorpe FL 194, *198*
Davidson JM, see Mapp HP 136, *148*
Davidsson L, see Forsberg C 423, *449*

Davies AG, see Corner EDS 96, *109*
Davies BR, see Robb F 53, *82*
Davies H, see Marsh J 398, *409*
Davies PJ, Drennan DSH, Fryer JD, Holly K 395, *407*
Davies T 473, *494*
Davis AW, see Polya GM 363, *385*
Davis CO, Breitner NF, Harrison PJ 101, *110*
Davis CO, see Conway HL 101, *109*
Davis CO, see Harrison PJ 106, *111*
Davis DE, see Mellis JM 341, 353, *383*
Davis GJ, Fones MN, Lunney CZ, Clark GM 437, *447*
Davis GJ, see McGahee CF 436, *456*
Davis GJ, see Twilley RR 64, *83*
Davison A 475, *494*
Davison TF, see Bowler K 441, *445*
Dawson DH, see Crist JB 236, *273*
Dawson WA, see Welch EB 315, *327*
Day BE, see Mann JD 341, 354, *382*
Day JW, see Hall CAS 125, *147*
Day W, see Legg BJ 238, 239, *276*
Dayan E, Dovrat A 138, *146*
Dean W, see Lundgren DG 72, 73, *80*
Deans JD 21, *40*, 241, 268, 269, *273*
Deans JD, see Ford ED 269, *274*
DeBach P 371, *376*
DeBano LF, Conrad CE 208, *227*
DeBoer JA, see D'Elia CF 104, *110*
Deck BL, see Bubeck RC 435, *446*
Decleire M, Cat W de, Basin R 431, *447*
Dedonder A, Sumere CF van 431, *447*
Delaney ME, Owen WJ, Rogers LJ 365, *376*
Delaney ME, see Owen WJ 365, *384*
Delaney ME, see Rogers LJ 365, *386*
D'Elia CF, DeBoer JA 104, *110*
D'Elia CF, see Webb KL 314, *327*
DeLuisi JD, see Mendonca BG 510, *524*
Delwiche CC, see Woodwell GM 506, *526*, 537, 539, *548*
DeMarte JA, Hartman RT 421, *447*
Demmerle SD, see Fuhs GW 64, *77*
Demoll R, Liebmann H 421, *448*
Dempster JP, Coaker TH 403, *407*
Denaeyer-De Smet S, see Duvigneaud P 17, *41*
Deneke FJ, McCown BH, Coyne PI, Rickard W, Brown J 545, *547*
Dennis JG 196, *198*
Denny P 64, *77*, 422, *448*
Deppenmeier E 405, *407*
Depuit EJ, Caldwell MM 219, *227*
Der Rat von Sachverständigen für Umweltfragen 427, 435, 442, *448*
Derby SB, Ruber E 365, *376*
Der-Vartanian M, see Astier C 371, *373*
Derwent RG, Stewart HNM 469, *494*
Derwent RG, see Cox RA 469, *493*
Descy JP 442, *448*
Despain DG, see Mayo JM 178, 181, 182, *200*, *201*
Detling JK 179, 195, *198*
Detling JK, Dyer MI, Procter-Gregg C, Winn DT 194, *199*
Detling JK, Dyer MI, Winn DT 193, 194, 195, 196, *199*
Detling JK, Klikoff LG 189, *199*
Detling JK, Parton WJ, Hunt HW 178, 179, *199*
Detling JK, see Dyer MI 191, *199*
Detling JK, see Painter EL 193, *201*
Deufel J 419, *448*
Deus B, see Amrhein N 341, 354, *372*
Devilliers OT, see Radosevich SR 370, *386*
Devlin RM, Kisiel MJ, Kostusiak AS 350, *376*
Diaz BM, see Say PJ 429, *459*
Dickinson CE, Dodd JL 176, 185, 186, *199*
Dickoré K, see Draber W 339, 342, *376*
Dicksen W, see Almer B 438, *444*
Dickson KL, Slocomb JP, Cairns J Jr, Almeida SP, Eu JKT 443, *448*
Dickson KL, see Cairns J Jr 414, 439, 443, *446*
Dieckert GB, Graf EG, Thauer RK 70, *77*
Diercks R 393, 403, *407*
Dietz F 427, 429, 443, *448*
Digiano FA, see Ku WC 65, *79*
Diller VM, see Schuler JF 422, *460*
Dilley DR, see Ries SK 355, *386*
Dilley RA, see Pritchard MK 352, *385*
Dillon TM, see Powell TM 314, *325*
Diment WH, see Bubeck RC 435, *446*
Dina SJ, Klikoff LF, Keddington MB 189, *199*
Dinger BE, see Reed KL 258, 259, 260, 261, *278*
Diskin S, see Orshan G 212, *230*
DiTiro DM, see Thomann RV 437, *462*
Ditto C, see Arntzen CJ 370, *373*
Dodd JL, Lauenroth WK 176, *199*
Dodd JL, see Dickinson CE 176, 185, 186, *199*
Dodd JL, see Heitschmidt RK 473, *496*
Dodge AD 431, *448*
Dodge AD, Harris N, Baldwin BC 345, 346, *376*

Dodge AD, Lawes GB 350, *376*
Dodge AD, see Harris N 340, 346, 351, 353, *379*
Dodge AD, see Pallet KE 350, 351, 353, *384*
Dodge AD, see Youngman RJ 346, 351, 353, *391*
Dodson AN, see Thomas WH 90, 94, 101, *115*
Doheny TE, see Mandelli EF 311, *324*
Dohnt G, see Renger G 342, *386*
Dohrmann U, Hertel R, Kowalik H 362, *376*
Dohrmann U, Hertel R, Pesci P, Cocucci SM, Marré E, Randazzo G, Ballio A 362, *376*
Dolar SG, Keeney DR, Chesters G 429, *448*
Dommergues Y 12, *41*
Donaldson LR, Foster RF 435, *448*
Doorenbos J, Kassam AH 140, 141, *146*
Dorris TC, see Duffer WR 417, *448*
Doudoroff P, Katz M 426, *448*
Dougherty PM, Teskey RO, Phelps JE, Hinckley TM 260, 264, 265, *273*
Dougherty PM, see Hinckley TM 268, 269, *274*
Dovrat A, see Dayan E 138, *146*
Dovrat A, see Williams WA 163, *172*
Downes PA, see Bennett D 126, 127, *146*
Downing AL, see Gameson ALH 433, *450*
Downton WJS, see Berry JA 258, 260, *273*
Downtown WJS, see Kriedemann PE 517, *523*
Dowsett JR, see Wright BJ 347, 350, *391*
Dozier BJ, see Powell TM 314, *325*
Draayder HA, see Greichus YA 364, *378*
Draber W 344, 358, 360, 362, *376*
Draber W, Büchel KH, Dikoré K, Trebst A, Pistorius E 339, 342, *376*
Draber W, Fedtke C 341, 345, 350, 356, *376*
Draber W, see Büchel KH 342, 348, *375*
Draber W, see Trebst A 339, 342, 345, 348, *390*
Drennan DSH, see Davies PJ 395, *407*
Drew AP, see Ledig FT 240, 263, 265, *276*
Drew ME, see Adams JB 401, *406*
Drift J van der 399, *407*
Drift J van der, see Eijsackers H 399, *407*
Droop MR 91, *110*
Droop, MR, McLaughlin JJA, Pinter IJ, Provasoli L 103, *110*
Drummond W, see Vance BD 432, *462*
Dubertret G, see Laval-Martin D 371, *382*
Duc SK le, see Steyeart LT 126, *149*
Ducker SC, see Brown V 104, *109*
Duever MJ, see Golley FB 17, 25, *41*, 283, 294, *300*
Duff RB, see Webley DM 13, *45*
Duffer WR, Dorris TC 417, *448*
Dugdale RC 86, 91, 93, 102, *110*, 314, *323*
Dugdale RC, Goering JJ 85, 89, 98, *110*
Dugdale RC, Menzel DW, Ryther JH 90, *110*
Dugdale RC, see Barber RT 103, *108*
Dugdale RC, see Goering JJ 96, *111*
Dugdale RC, see Harrison PJ 101, *111*
Dugdale RC, see MacIsaac JJ 94, *112*
Dugger K, see Lugo A 287, 293, 295, *302*
Dugger WM, Ting IP 474, 484, 485, *494*
Dugger WM Jr, see Ting IP 519, *526*
Dugger WM, see Thomson WW 485, *500, 501*
Duhme F, see Hinckley TM 260, *274*
Duke WB, Slife FW, Hanson JB, Butler HS 341, 355, *376*
Duke WB, see Ashton FM 336, 353, 354, *373*
Dumford SW, see Ebert E 339, *376*
Dunbar MJ 320, *323*
Duncan CJ, see Bowler K 441, *445*
Duncan DA, Woodmansee RG 209, 211, *227*
Duncan WG 164, *171*
Duncan WG, Loomis RS, Williams WA, Hanau R 164, *171*
Duncan WG, see Williams WA 163, *172*
Dunn EL 217, 218, *227*
Dunn EL, see Mooney HA 206, 209, *229*
Dunn EL, see Schulze E-D 261, *279*
Dunning JA, see Tingey DT 475, *501*
Durrant K, see Curtis EJC 57, *76*
Dustin P 356, *376*
Duvigneaud P, Denaeyer-De Smet S 17, *41*
Duvigneaud P, see Kestemont P 236, *275*
Duysens NM, see Thomas JB 434, *462*
Dye AJ 187, 188, *199*
Dyer MI, Bokhari UG 194, *199*
Dyer MI, Detling JK, Coleman DC, Hilbert DW 191, *199*
Dyer MI, see Detling JK 193, 194, 195, 196, *199*
Dyke GV, see Garner HV 153, *171*
Dykstra GF 260, *273*
Dziuba SP, see Terhaar CJ 437, *462*

Early RW, see Penner D 354, 355, *385*
Eastman JA, see Wesely ML 471, *501*
Eaton JS, Likens GE, Bormann FH 24, *41*
Eaton JS, see Gosz JR 24, *41*

Eaton JS, see Likens GE 545, *548*
Eberhardt CH, see Mittnacht A 401, *409*
Eberhardt LL, Meeks RL, Peterle TJ 432, 437, *448*
Ebert E 367, *376*
Ebert E, Dumford SW 339, *376*
Ebert E, see Paradies I 367, *384*
Ebert M, see Farrington JA 346, 351, *377*
Eble H 401, *407*
Eckardt F, Heim G, Methy M, Sauvezon R 208, 209, 218, *227*
Edelmann M, see Mattoo AK 342, *383*
Eder FA 350, 353, *376*
Edlis J, see Kott Y 433, *454*
Edminster TW 118, *146*
Edmonds RL, see Grier CC 241, 269, 271, *274*
Edmonds RL, see Johnson DW *42*
Edmondson WT 424, *448*
Edwards CA 364, *377*
Edwards CA, Heath GW 28, *41*
Edwards CA, Reichle DE, Crossley DA Jr 28, *41*
Edwards NT, see Cox TL 26, *40*
Edwards NT, see Harris WF 26, *42*, 269, *274*
Edwards PJ 294, *300*
Edwards PJ, Grubb PJ 283, *300*
Edwards RW, Rolley HLJ 58, *77*
Edwards RW, see Brooker MP 430, 431, *445*
Eggleton AEJ, Cox RA 468, *494*
Eggleton AEJ, see Cox RA 469, *493*
Egli DB, Pendleton JW, Baker DB 517, *522*
Egmond F van 143, *146*
Ehleringer J, Björkman O 163, *171*
Ehleringer J, see Björkman O 221, 223, *226*
Ehleringer J, see Mooney HA 215, 216, 219, 223, *229*
Ehleringer JR 176, *199*
Ehleringer JR, Björkman O 217, 219, 222, 223, *227*, 240, 261, *273*
Ehleringer JR, Björkman O, Mooney HA 223, *227*
Ehleringer JR, Mooney HA 223, 224, *227*
Ehleringer JR, Mooney HA, Berry JA 216, 217, *227*
Ehleringer JR, see Forseth I 224, *227*
Ehleringer JR, see Miller PC 187, *201*
Ehleringer JR, see Mooney HA 216, 219, 224, *229*
Eichenberger E 417, *448*
Eickmeier WF 176, *199*
Eickmeier WG 215, *227*
Eidgenössisches Departement des Innern 439, *448*
Eidman VR, see Mapp HP 136, *148*
Eigen M, Kustin K, Maass G 478, *494*
Eijsackers H, Drift J van der 399, *407*
Einsele W 65, *77*, 416, *448*
Ekström C, see Almer B 438, *444*
Eleftheriadis D, see Schweisfurth R 73, *82*
Elgavish A, see Acher A 403, *406*
Elkiey T, Ormrod DP 485, *494*
Ellen J, see Spiertz JHJ 164, *172*
Ellenberg H 540, *547*
Elliott WP, Machta L 508, 514, *522*
Ellis BE 432, *448*
Ellis JF, Peek JW, Boehle J, Müller G 367, *377*
Ellis RP, see Vogel JC 176, *202*
Elster HJ 414, 415, 416, *448*
Elstner EF 340, 346, 351, 353, *377*
Elstner EF, Fischer HP, Osswald W, Kwiatkowski G 351, *377*
Elstner EF, Lengfelder E, Kwiatkowski G 346, 351, *377*
Elstner EF, Osswald W 340, 346, 350, 351, 353, *377*
Elstner EF, Pils J 350, 351, 353, *377*
Elstner EF, see Paradies I 367, *384*
Emerson MT, see Adams DF 488, *491*
Emerson S, see Schindler DW 49, *82*
Emery KO, Orr WK, Rittenberg SC 88, *110*
Emery RM, see Welch EB 315, *327*
Eminson D, see Phillips GL 418, *458*
Emmingham WH, Waring RH 239, 261, *273*
Emmingham WH, see Waring RH 250, 251, *280*
Empain A 442, *448*
Engel H, see Mundschenk H 434, *457*
Engelhardt G, see Wallnöfer P 396, *411*
Ensign JC, see Barber LE 54, *75*
Environmental Protection Agency 471, 472, 474, 475, *494*
Eppley RW 315, *323*
Eppley RW, Coatsworth JL 94, *110*
Eppley RW, Coatsworth JL, Solórzano Lucia 92, 107, *110*
Eppley RW, Peterson BJ 96, *110*, 314, 319, *323*
Eppley RW, Renger EH, Harrison WG 89, *110*
Eppley RW, Renger EH, Harrison WG, Cullen JJ 89, *110*
Eppley RW, Rogers JN, McCarthy JJ 85, 93, 106, 107, *110*
Eppley RW, Sloan PR 321, *323*
Eppley RW, Strickland JDH 91, *110*
Eppley RW, Thomas WH 91, *110*
Eppley RW, Weiler CS 104, *110*
Eppley RW, see Harrison WG 98, *111*
Epps E, see Culley D 424, *447*
Erdman HE, see Kinkade ML 429, *453*
Erickson RO 144, *146*
Erickson SJ 103, *110*

Erickson SJ, Maloney TE, Gentile JH 427, *448*
Eriksson C, Mortimer DC 429, *449*
Erion JL, see Fox JE 363, *378*
Ernst WHO, Marquenie van der Werff M 443, *449*
Esaias WE, see Bowman MJ 89, *108*
Esaias WE, see Walsh JJ 314, *327*
Escalante G, see Jordan CF 37, 42, 283, 297, *301*
Escarre J, see Houssard C 252, *275*
Esch GJ van, see Canton JH 432, 437, *446*
Eschrich W, see Steiner M 436, *461*
Escobar DE, see Waller H 361, *391*
Escombe F, see Brown HT 505, 517, *522*
Eshel Y, see Katan J 399, *408*
Estep M, Armstrong JE, Baalen C van 102, *110*
Esterbauer H, see Grill D 480, *495*
Estermann EF, McLaren AD 29, *41*
Etienne AL, see Lavorel J 338, *382*
Eu JKT, see Dickson KL 443, *448*
Evans B, see Trollope DR 425, 428, *462*
Evans C, see O'Reilly JE 311, *325*
Evans GC 284, *300*
Evans LS 477, *494*
Evans LS, Gmur NF, Costa F da 474, *494*
Evans LS, Ting IP 485, *494*
Evans LT 157, *171*, 535, *547*
Evans PS 194, *199*
Evenari M, Schulze ED, Lange OL 212, 213, *227*
Evenari M, Schulze ED, Lange OL, Kappen L, Buschbom U 213, 214, *227*
Evenari M, see Kappen L 214, 222, 223, *228*
Evenari M, see Lange OL 219, 221, *228*
Ewald WG, French JE, Champ MA 365, *377*
Ewell WS, see Terhaar CJ 437, *462*
Ezra G, Gressel J 367, *377*

Fagerström T 143, *146*
Fairfax JAW, Lepp NW 474, *494*
Falkner R, Simonis W 364, *377*
Falkowski PG 92, *110*, 317, *323*
Falkowski PG, Hopkins TS, Walsh JJ 320, *323*
Falkowski PG, Rivkin RB 92, *110*
Falkowski PG, Stone DP 92, *111*, 311, *323*
Faller N 473, *494*
Faller N, Herwig K, Kühn H 482, *494*
Faludi-Daniel A, see Muschinek Gy 350, *383*
Fang SC, see Lee SS 343, *382*, *455*
FAO 157, 158, 159, 162, *171*
FAO Agricultural Series 10 331, *377*
Fareed M, see Caldwell MM 187, *198*
Farmer FS, see Moreland DE 349, *383*
Farquhar GD, see Caemmerer S von 258, *273*
Farquhar GD, see Wong SC 140, *150*
Farran M, see Rachlin JW 427, *458*
Farrar JF, Relton J, Rutter AJ 473, *494*
Farrar JF, see Garsed SG 473, *495*
Farrington JA, Ebert M, Land EJ, Fletcher K 346, 351, *377*
Fathi M, Lorenz H 491, *494*
Faulkner JS 366, *377*
Faust SL, see Fitzgerald GP 427, *449*
Feder WA 475, *494*
Feder WA, Campbell FJ 475, *494*
Fedtke C 330, 338, 352, 355, 357, *377*
Fedtke C, see Draber W 341, 345, 350, 356, *376*
Fedtke C, see Kiermayer O 356, *380*
Fee EJ 305, 314, 319, *323*
Fee EJ, see Schindler DW 48, 49, *82*, 416, *460*
Feierabend J, Schubert B 352, *377*
Feierabend J, Schulz U, Kemmerich P, Lowitz T 352, *377*
Feller MC 21, 38, *41*
Felling CE, see Phillips DA 520, *525*
Felmeister A, see Kamel AM 486, *497*
Feng TH, see Ku WC 65, *79*
Ferenbaugh RW 474, *494*
Fereres E, see Hsiao TC 140, 141, *147*, 261, *275*
Ferguson AR, Bollard EG 60, *77*
Ferrante JG 63, *77*
Ferrar PJ, see Slatyer RO 260, *279*
Feyerabend G, see Arlt K 396, *406*
Fiance S, see Pimentel D 331, *385*
Fick GW 124, 138, *146*
Field RW, see Tingey DT 484, *501*
Filipek LH, see Tenhunen JD, 512, *526*
Filner P, see Bressan RA 484, *492*
Filner P, see Ries SK 355, *386*
Filner P, see Wilson LG 474, *501*
Finenko ZZ 89, *111*
Finke R, Seeley HW Jr 56, *77*
Finlayson DG, MacCarthy HR 364, *377*
Finstein MS, Strom PF, Matulewich VA 58, *77*
Fischer HP, see Elstner EF 351, *377*
Fischer K, Kramer D, Ziegler H 482, *494*
Fischer K, Lüttge U 487, *494*
Fischer K, see Brinkmann E 480, *492*
Fischer NS, Wurster CF 365, *377*
Fischer RA 238, *273*
Fischer RA, Turner NC 169, *171*
Fischer WK, Winkler K 433, *449*

Fischer WR, Baumann G 65, *77*
Fish GR 418, 431, *449*
Fisher NS 104, *111*
Fisher SG, Likens GE 52, *77*
Fisher SG, see Howarth RW 415, *452*
Fitz FK, see Gholz HL 251, *274*
Fitzgerald GP 91, *111*, 418, 422, 423, *449*
Fitzgerald GP, Faust SL 427, *449*
Fitzgerald GP, Gerloff GC, Skoog F 431, *449*
Fitzgerald GP, Nelson TC 423, *449*
Fitzgerald GP, see Stewart WDP 423, *461*
Fjerdingstad E 442, *449*
Flavell RB, Barratt DHP *146*
Flemer DA 309, *323*
Flemer DA, see Hamilton DH Jr 433, *451*
Flemer DA, see Heinle DR 316, *324*
Fletcher DJ, see Strange JR 423, *461*
Fletcher K, see Farrington JA 346, 351, *377*
Fletcher RA, see Chen LG 341, 354, 355, 357, *375*
Flett RJ, see Rudd JWM 54, *82*
Flint ML, Bosch R van der 371, *377*
Flower-Ellis JGK, Persson H 253, *273*
Flower-Ellis JGK, see Ågren GI 264, 270, *272*
Flückiger W, Oertli JJ, Flükkinger H 476, *494*
Flückinger H, see Flückiger W 476, *494*
Focht DD, Verstraete W 58, *77*
Foehrenbach J 419, *449*
Fogel R, Cromack K Jr 34, *41*
Fogelfors H, see Gressel J 366, 368, 369, *378*
Fogg GE 51, *77*, 419, 422, 423, *449*
Fogg GE, Westlake DF 422, 427, 429, *449*
Fogg GE, see Horne AJ 56, *78*, 415, *452*
Folkes BF, see Sims AP 422, *460*
Fölster H, Hase H, Ulrich B 477, *494*
Fones MN, see Davis GJ 437, *447*
Fong J, see Pratt R 422, *458*
Fong KL, see Lai EK 352, *381*
Fonteyn PJ, Mahall BE 211, *227*
Foote CS 353, *378*
Foote KC, Schaedle M 264, *273*
Ford EB, see Wilson PM 520, *526*
Ford ED 235, 236, 250, *273*
Ford ED, Deans JD 269, *274*
Ford ED, Newbould PJ 246, 251, *274*
Ford MA, Thorne GN 511, 517, *522*
Forester JW 441, *449*
Fork DC, see Pearcy RW 219, *230*
Forrester JW 121, *146*
Forsberg A, see Forsberg C 443, *449*
Forsberg B, see Forsberg C 422, *449*
Forsberg C 60, 61, *77*, 415, 418, 419, 422, 427, *449*
Forsberg C, Forsberg B 422, *449*
Forsberg C, Jinnerot D, Davidsson L 423, *449*
Forsberg C, Ryding SO, Claesson A, Forsberg A 443, *449*
Forsberg C, see Littlefield L 64, *80*, 421, *455*
Forseth I, Ehleringer JR 224, *227*
Förstner U, Müller G 425, 427, 435, *449*
Foster RB, see Haines B 294, *300*
Foster RF, see Donaldson LR 435, *448*
Fowler D, Cape JN 467, 469, *494*
Fowler D, Cape JN, Leith ID, Paterson IS, Kinnaird JW, Nicholson IA 477, *495*
Fowler D, Unsworth MH 468, 470, 471, *495*
Fox JE, Erion JL 363, *378*
Fox JE, see Pratt HM 363, *385*
Foy CL 341, *378*
Frahm JP 442, 443, *449*
Francke J, Hillebrand H 429, *449*
Francko DA, Wetzel RG 64, *77*
Frank H 400, *407*
Frank R, Holdrinet M, Braun HE, Thomas RL, Kemp ALW, Jaquet JM 430, *449*
Frankie GW, see Opler PA 295, 296, *303*
Franklin JF, see Waring RH 235, *280*
Franklin RT 268, *274*
Franz JM, Krieg A 372, *378*
Fraser AB, see Norman JM 247, *278*
Fraser CD, see Ophel IL 434, *457*
Frear DS 341, 357, *378*
Frederick P, see Heath RL 484, 485, *496*
Frederick PE, see Heath RL 485, *496*
Freed VH, see Haque R 364, *379*
Freed VH, see Lee SS *455*
Freeman GD, see Linder PL 396, *409*
French JE, see Ewald WG 365, *377*
French NR 176, 177, 191, *199*
Fresson R, see Malaisse F 282, 294, *302*
Frey BE, Small LF 103, *111*
Freyer RA, see McNaughton SJ 187, *201*
Fridovich I 353, *378*
Friedrich G 443, *450*
Frissel MJ, see Beek J 142, *146*
Frochot H 405, *407*
Froelich P, see McCaffney RJ 97, *112*
Frosch S, Jabben M, Bergfeld R, Kleinig H, Mohr H 350, *378*
Fry B, see Thayer GW 306, *326*
Fry JC, Brooker MP, Thomas PL 431, *450*
Fry JC, see Ramsay AJ 431, *458*

Fryer JD, Kirkland K 398, *407*
Fryer JD, Matsunaka S 371, *378*
Fryer JD, see Davies PJ 395, *407*
Fuchs M, Schulze ED, Fuchs MI 261, *274*
Fuchs M, see Schulze E-D 251, 254, 260, 270, *279*
Fuchs MI, see Fuchs M 261, *274*
Fuchs MI, see Schulze E-D 251, 254, 260, 270, *279*
Fuhr I, Bransford AV, Silver SD 11, *41*
Fuhrmann JA, Azam F 52, *77*
Fuhs GW, Demmerle SD, Canelli E, Chen M 64, *77*
Fujimori J, Kawanabe S, Saito H, Grier CC, Shidei T 235, *274*
Fujimori T 235, 236, *274*
Füller F, see Kopecek K 365, *381*
Fuls A, see Vogel JC 176, *202*
Funderburk HH Jr, Lawrence JM 421, *450*
Funk WH, see Bartlett L 430, *444*
Furutani A, see Rudd JWM 54, *82*
Fuzzi S 479, *495*

Gaarder T, Gran HH 321, *323*
Gabbott PA 342, *378*
Gächter R, Chan KL, Chau YK 427, *450*
Gadgil MD, see Schaffer WM 206, *230*
Gaffron H, see Stuart TS 429, *461*
Gahlen K 437, *450*
Gallagher JN, Biscoe PV 238, *274*
Gallahan D, see Pimentel D 331, *385*
Gallapher S, see Madgwick HAI 235, 236, 237, *277*
Gallepp GW, Kitchell JF, Bartell SM 65, *77*
Gallough RA, see Soeder CJ, 515, *525*
Galloway JN, see Likens GE 438, *455*, 477, *497*
Galloway RA, Krauss RW 423, *450*
Galloway RA, see Gemmill ER 104, *111*
Gamble DS, Schnitzer M 53, *77*
Gameson ALH, Truesdale GA, Downing AL 433, *450*
Gandar PW, Tanner CB 195, *199*
Gannon JJ, see Werzenak CT 57, *84*
Garab GyI, see Muschinek Gy 350, *383*
Garcia-Moya E, McKell CM 213, *227*
Gardner G, see Kende H 362, *380*
Gardner G, see Pfister K 342, *385*
Gardner R 520, *522*
Garland JA 468, *495*
Garland JHN 57, *77*
Garner HV, Dyke GV 153, *171*
Garrett SD 30, *41*
Garsed SG, Farrar JF, Rutter AJ 473, *495*
Garsed SG, Rutter AJ 473, *495*
Garside C, Malone TC, Roels OA, Shartstein BA 90, *111*
Garton G, see Blackman GE 431, *445*
Gasquez J, see Barralis G 369, *373*, 397, *406*
Gasquez J, see Gressel J 366, 368, 369, *378*
Gast A, Knuesli E, Gysin H 331, *378*
Gates DM 512, *522*
Gates DM, see Bennett JH 470, *492*
Gates DM, see Lommen PW, 512, *524*
Gates DM, see Tenhunen JD 512, *525*, *526*
Gausman HW, see Waller H 361, *391*
Gauthier D, see Belot Y 471, *492*
Gavin J, see Kukla G 510, *524*
Gavis J, see Pasciak WJ 94, *113*
Gearheart RA, see Schlichting HE Jr 419, *460*
Gehrke P, see Amrhein N 341, 354, *372*
Geike F 428, 429, *450*
Geissbühler H 330, *378*
Geissel LD, see Maki AW 432, *455*
Geissler AE, see Katekar GF 360, *380*
Geller W, Mäckle H 443, *450*
Gemmill ER, Galloway RA 104, *111*
Gentile JH, see Erickson SJ 427, *448*
Gentner WA, see Hilton JL 339, *379*
Gerakis PA, see Loomis RS 157, 158, 160, *171*
Gerasimenko TV 184, *199*
Gerasimenko TV, Zalenskiy LV *199*
Gerber HR, see Nyffeler A 367, *384*
Gerhards U, Weller H 429, *450*
Gerloff GC 142, *147*
Gerloff GC, Skoog F 423, *450*
Gerloff GC, see Fitzgerald GP 431, *449*
Gersberg RM, Axler RP, Krohn K, Peek N, Goldman CR 61, *77*
Gersper PL, Holowaychuk N 15, 24, *41*
Gersper PL, see Ulrich A 189, *202*
Gessel SP, see Jenny H 34, *42*
Gessel SP, see Johnson DW 37, 39, *42*
Gholz HL, Fitz FK, Waring RH 251, *274*
Gholz HL, see Waring RH 250, 251, *280*
Giannopolitis GN, Ayers GS 350, 351, *378*
Gibb MJ, see Welbank PJ 167, *172*
Gieskes WWC, Kraay GW, Baars MA 322, *323*
Giesy JP Jr 54, *77*
Gifford RM 511, 518, *522*
Gifford RM, Bremmer PM, Jones DB 517, *522*
Gifford RM, Marshall C 193, 194, 195, *199*

Gigon A 215, 220, *227*
Giliberto J, see Mooney HA 207, 208, 210, *229*
Gill DS, see Schlesinger WH 208, 210, *230*
Gillman GP, Bell LC 21, *41*
Giovando LF, see Parsons TR 313, *325*
Givnish TJ 286, *300*
Givnish TJ, Vermeij GJ 285, 291, *300*
Gjessing ET 53, *77*
Gladwell RT, see Bowler K 441, *445*
Glänzer U, Haber W, Kohler A 442, *450*
Glenn RK, see Ashton FM 336, 353, 354, *373*
Glooschenko WA, see MacFarlane RB 365, *382*
Glover HE 103, *111*
Glover HE, see Morris I 441, *457*
Gloyna EF, Bhagat SK, Yousef Y, Shih C 435, *450*
Gloyna EF, see Huang J 431, *452*
Gmelin L 478, *495*
Gmur NF, see Evans LS 474, *494*
Gnauck AH, see Weise G 443, *463*
Gocke K 61, *77*
Godden DA, see Powell TM 314, *325*
Godshalk GL, Wetzel RG 53, *77*
Goeller HE, see Baes CF 506, *521*
Goering JJ 61, *78*, 94, *111*
Goering JJ, Dugdale RC 96, *111*
Goering JJ, Pamatmat MM 97, *111*
Goering JJ, see Dugdale RC 85, 89, 98, *110*
Goering JJ, see Nelson DM 101, 102, *113*
Goffinet G, see Malaisse F 282, 294, *302*
Goldberg MC, see Wershaw RL 430, *463*
Goldman CR 415, *450*
Goldman CR, Carter RC 415, *450*
Goldman CR, Wetzel RG 415, *450*
Goldman CR, see Gersberg RM 61, *77*
Goldman CR, see Horne AJ 56, *78*, 415, 426, *452*
Goldman CR, see Kaplan LA 65, *79*
Goldman CR, see Kellar PE 56, *79*
Goldman CR, see Richerson P 105, *114*
Goldman JC 99, *111*, 320, *323*
Goldman JC, McCarthy JJ, Peavey DG 93, *111*, 322, *323*
Goldman JC, see McCarthy JJ 95, *112*, 317, 322, *325*
Goldsberry KL 518, *522*
Goldsby TL, Bates AL, Stanley RA 418, *450*
Golley FB 282, *300*
Golley FB, McGinnis JT, Clements RG, Child GI, Duever MJ 17, 25, *41*, 283, 294, *300*
Golterman HL 63, *78*
Goncharova AV 431, *450*
Gonye ER, Carpenter EJ 102, *111*
Gonzalez-Liboy JA, see Lugo A 287, 293, 295, *302*
Good NE 347, *378*
Good NE, see Izawa S 337, 338, *380*
Goodwin PB, see Letham DS 330, 358, *382*
Gordon JC 235, 236, *274*
Gorham E, Vitousek PM, Reiners WA 13, *41*
Gorham E, see Bray JR 25, 26, *40*, 266, 267, *273*, 292, 294, 295, *300*
Gorham PR, see Haag RW 440, *451*
Goring CAI, see Hamaker JW 430, *451*
Gosz JR, Likens GE, Eaton JS, Bormann FH 24, *41*
Götz R 427, *450*
Goudriaan J, Laar HH van 140, *147*
Goudriaan J, see Horie T 144, *147*
Goudriaan J, see Keulen H van 135, 140, 142, *147*
Goudriaan J, see Wit CT de 121, 125, *150*
Goulder R 418, *450*
Govindjee 336, *378*
Govindjee, Papageorgiou G 338, *378*
Govindjee, see Rensen JJS van 348, 349, *386*
Grabherr G, see Larcher W 181, *200*
Grace J 258, *274*
Grace J, Marks TC 182, *199*
Grace J, Woolhouse HW 180, 181, 184, *199*, *200*
Grace JB, Tilly LJ 440, *450*
Gradwell GR, see Varley GC 268, *280*
Graebe JE, Ropers HJ 361, *378*
Graetz DA, Keeney DR, Aspiras RB 57, *78*
Graetz DA, see Chen RL 57, 58, *76*
Graetz DA, see Keeney DR 58, *79*
Graf EG, see Dieckert GB 70, *77*
Gran HH, see Gaarder T 321, *323*
Granat L, Rodhe H, Hallberg RO 468, *495*
Granat L, see Hällgren J-E 474, *495*
Granhall U, Lundgren A 56, *78*, 415, *450*
Grant MC, see Lewis WM 37, *43*
Gras JL, see Ayers GP 90, *108*
Gräser H 330, 357, *378*
Gravenhorst G, see Beilke S 479, *492*
Gray EA 519, *522*
Gray RA, see Casida JE 334, 367, *375*
Gray TRG, see Williams ST 34, *45*
Greaves MP, Webley DM 29, *41*
Green GH, see Hargrave BT 63, *78*
Green K, Wright R 518, *522*
Greenberg E, see Zweig G 431, *464*
Greene JC, see Miller WE 443, *456*
Greenfield SS 426, *450*
Greenland DJ 32, *41*
Greenland DJ, Kowal JL 297, *300*

Greenwood DJ, Cleaver TJ, Loquens SMH, Niendorf KB 127, *147*
Greenwood DJ, see Barnes A 129, 143, *146*
Greer DE, Ziebell CD 424, *450*
Gregory E, Perry RS, Staley JT 73, 74, *78*
Gregory LE, see Steffens GL 362, *388*
Gregory RPG, see Bradshaw AD 536, *547*
Gregory SV 65, 66, *78*
Greichus A, see Greichus YA 364, *378*, 437, *450*
Greichus YA, Greichus A, Amman BD, Call DJ, Hamman CD, Pott RM 437, *450*
Greichus YA, Greichus A, Draayder HA, Marshall B 364, *378*
Gressel J, Ammon HU, Fogelfors H, Gasquez J, Kay QON, Kees H 366, 368, 369, *378*
Gressel J, Segel LA 369, *378*
Gressel J, see Baron H le 398, *406*
Gressel J, see Ezra G 367, *377*
Gressel J, see LeBaron H 368, *382*
Greve PA, see Canton JH 432, 437, *446*
Grewe F 330, *378*
Grier CC 26, *41*
Grier CC, Logan RS 235, 264, 271, *274*
Grier CC, Running SW 251, *274*
Grier CC, Vogt KH, Keys MR, Edmonds RL 241, 269, 271, *274*
Grier CC, see Fujimori J 235, *274*
Grier CC, see Henderson GS 37, 38, *42*
Grier CC, see Johnson DW *42*
Grier CC, see Keyes MR 237, 241, *275*
Grier CC, see Waring RH 250, 251, *280*
Griffin DM 34, *41*
Griffith FB, see Kirkman H 64, *79*
Grill D, Esterbauer H 480, *495*
Grill D, Esterbauer H, Klösch U 480, *495*
Grill EV 101, *111*
Grill EV, Richards FA 102, *111*
Grim J 416, 420, *451*
Grimme LH, see Kümmel HW 350, *381*
Grimme LH, see Rau I 431, *458*
Grizhutina AP, Morozova LK 404, *407*
Grizzard T, see Henderson GS 24, *42*
Grogan RG, see Ioannow N 519, *523*
Gromet-Elhanan Z 339, 340, 348, *378*
Gross DR 538, *547*
Gross et al. 426, *451*
Grossbard E 398, *407*
Grossbard E, see Marsh J 398, *409*
Groth P 71, *78*
Grubb PJ 283, 284, 286, 291, 294, *300*
Grubb PJ, see Edwards PJ 283, *300*
Gruenhagen RD, Moreland DE 354, 355, *378*
Gruenhagen RD, see Moreland DE 341, 354, 355, *383*
Grumbach KH, Bach TJ 350, *378*
Grumbach KH, see Buschmann C 350, *375*
Grumbach KH, see Lichtenthaler HK 352, *382*
Grünig A, see Klötzli F 418, 424, *454*
Gruzdev GS, Mozgovoj AF, Kuz'mina IV 398, *407*
Guderian R 479, *495*, 538, *548*
Guderian R, Haut H van 482, *495*
Gugelmann A, see Oeschger H 508, 509, *525*
Guilizzoni P 429, *451*
Guillard RRL, Kilham P, Jackson TA 94, *111*
Guillard RRL, see Carpenter EJ 93, *109*
Guillard RRL, see Nelson DM 102, *113*
Guillard RRL, see Sunda W 53, *83*, 103, *114*
Gulmon SL, see Mooney HA 206, 220, *229*
Gundlach H, see Schweisfurth R 73, *82*
Gustin P, see Linkins AE 539, *548*
Guthrie HB, see Attiwill PM 267, *272*
Gutierrez AP, Wang Y 130, *147*
Gutierrez de la Casas O, see Cock JH 160, *170*
Gysin H, see Gast A 331, *378*

Haag RW, Gorham PR 440, *451*
Haagen-Smit AJ, Wayne LG 468, 469, *495*
Haan H de 54, *78*
Haas H, Streibig JC 401, *408*
Haas LW, Webb KL 104, *111*
Haber W, Kohler A 442, *451*
Haber W, see Glänzer U 442, *450*
Haber W, see Melzer A 418, 442, *456*
Hagem O 261, *274*
Hagemann R, see Renger G 342, *386*
Haines B, Foster RB 294, *300*
Haines EB 306, *323*
Haines KC, Wheeler PA 104, *111*
Haines TA 477, *495*
Hale SS, see Nixon WS 97, *113*
Halim Y, Morcos SA 90, *111*
Hall A, see Mooney H 287, *302*
Hall AD 152, 155, *171*
Hall AE, Connell GH, Lawton HW 165, *171*
Hall CAS 309, *324*
Hall CAS, Day JW 125, *147*
Hall CAS, Moll R 306, 322, *324*
Hall DO, see Reeves SG 337, *386*
Hall M, see Zeroni M 358, 363, *391*
Hall MA, Brown RL, Ordin L 485, *495*
Hall MA, see Ordin L 485, *499*

Hallberg RO, see Granat L 468, *495*
Haller WT, Sutton DL, Barlowe WC 436, *451*
Hällgren J-E 471, 472, 473, 479, 481, 482, 483, *495*
Hällgren J-E, Huss K 483, *495*
Hällgren J-E, Linder S, Richter A, Troeng E, Granat L 474, *495*
Hallsworth EG, see Stace HCT 15, *44*
Halstead RL, McKercher R 29, 30, *41*
Halvorson WL, Patten DT 189, *200*
Hamaker JW, Goring CAI, Youngson CR 430, *451*
Hamann O 286, *300*
Hamelink JL, see Mayer FL 443, *456*
Hamerly ER, see Reed KL 258, 259, 260, 261, *278*
Hamilton DH Jr 415, *451*
Hamilton DH Jr, Flemer DA, Keefe CW, Hursky JAM 433, *451*
Hamilton PL, see Rowe GT 306, 314, *326*
Hamilton RD, see Rudd JWM 54, *82*
Hamman CD, see Greichus YA 437, *450*
Hamner CL, Tukey HB 331, 358, *379*
Hampp R, Ziegler H, Ziegler I 490, *495*
Hampp R, Ziegler I 481, *495*
Hampp R, see Höll W 490, *496*
Hampp R, see Spedding DJ 479, 481, *500*
Hampp R, see Ziegler I 481, *502*
Hanan JJ 520, *522*
Hanau R, see Duncan WG 164, *171*
Hand DW, Postlethwaite JD 518, *522*
Hand DW, Soffe RW 522, *523*
Handley WRC 34, *41*
Hanf M 396, *408*
Hanisch HH, see Mundschenk H 434, *457*
Hanks RJ 127, *147*
Hanley DP 236, *274*
Hannan PJ, Patouillet 427, *451*
Hannerz L 428, 429, *451*
Hansch C 342, *379*
Hansen GK, Jensen CR 140, *147*
Hansen J, Johnson D, Lacis A, Lebedeff S, Lee T, Rind D, Russell G 510, *523*
Hansen PD 432, 437, *451*
Hanson JB, see Duke WB 341, 355, *376*
Hanson KJ, see Mendonca BG 510, *524*
Hanson NS 402, *408*
Hanson RS, see Harrits SM 54, *78*
Hanzely L, see Rebechini HM 427, *459*
Haque R, Freed VH 364, *379*
Harbour JR, Bolton JR 353, *379*
Harding JPC, Whitton BA 429, 430, *451*
Harding LW, Phillips JH 364, 365, *379*
Hardy R, Criswell JG, Havelka UD 520, *523*
Hardy R, Havelka UD 517, 518, 520, *523*
Hardy R, see Quebedeaux B 520, *525*
Hardy RWF 360, 363, *379*
Hargrave BT 418, *451*
Hargrave BT, Green GH 63, *78*
Hargreaves JW, Lloyd EJH, Whitton BA 438, *451*
Hargreaves JW, Whitton BA 429, 430, *451*
Hari P, Leikola M *127*, 147
Harlan JR 532, 534, *548*
Harley JL 22, 23, *41*
Harley JL, Lewis DH 23, *41*
Harman MMI, see Curtis EJC 57, *76*
Harmsen R, see Rose MR 125, 134, *149*
Harper JL 191, *200*, 404, *408*
Harris DO 419, *451*
Harris GP 51, *78*
Harris N, Dodge AD 340, 346, 351, 353, *379*
Harris N, see Dodge AD 345, 346, *376*
Harris RF, see Brown EJ 64, *76*
Harris RF, see Li WC 65, *80*
Harris RF, see Williams JD 65, *84*
Harris WF, Kinerson RS, Edwards NT 26, *42*, 269, *274*
Harris WF, see Cox TL 26, *40*
Harris WF, see Henderson GS 24, *42*
Harrison A, Small E, Mooney HA 215, *227*
Harrison AT 211, *227*
Harrison AT, see Björkman O 217, 219, *226*
Harrison AT, see Mooney HA 206, 218, 220, *229*
Harrison AT, see Pearcy RW 217, 219, 223, *230*
Harrison H, see Hobbs PV 510, *523*
Harrison PJ, Conway HL, Dugdale RC 101, *111*
Harrison PJ, Davis CO 106, *111*
Harrison PJ, see Conway HL 95, 101, *109*
Harrison PJ, see Davis CO 101, *110*
Harrison PJ, see Turpin DH 92, 106, 107, 108, *115*
Harrison W 59, *78*
Harrison WG 94, 95, *111*
Harrison WG, Azam F, Renger EH, Eppley RW 98, *111*
Harrison WG, see Eppley RW 89, *110*
Harriss RC 419, 428, *451*
Harriss RC, see MacFarlane RB 365, *382*
Harrits SM, Hanson RS 54, *78*
Hart P, Wright RD 511, *523*
Harth E, see Trebst A 342, *390*
Hartley EA 540, 541, 542, *548*
Hartley W 176, *200*
Hartman RT, see DeMarte JA 421, *447*
Hartmann L, see Laubenberger G 426, *455*
Hartmeyer H, see Urbach W 371, *390*

Hartney VJ, see Neales TF 223, *229*
Hartsock TL, Nobel PS 217, *227*
Hartsock TL, see Nobel PS 219, *229*
Hartwig EO 97, *111*
Hartwig EO, see Carlucci AF 95, *109*
Harvey DM, see Hedley CL 520, *523*
Harvey RG, Muzik TJ 397, *408*
Harvey RS, Patrick R 434, 435, *451*
Hase H, see Fölster H 477, *494*
Haselhoff E, Lindau G 468, *495*
Haslam SM 418, *451*
Hasler AD, Jones E 418, *451*
Hassell MP, see Varley GC 268, *280*
Hassell SA, see Phillips DA 520, *525*
Hastings JR, Humphrey RR 206, *227*
Hathaway RL 236, *274*
Hatilos KK, Penner D 367, *379*
Hattori A, Wada E 95, *111*
Hattula ML, see Särkkä J 429, *459*
Haun JR 127, *147*
Hauska G, Trebst A, Kötter C, Schulz H 340, 347, *379*
Haut H van, Stratmann H 471, 472, *495*
Haut H van, see Guderian R 482, *495*
Havelka UD, see Hardy R 517, 518, 520, *523*
Havelka UD, see Quebedeaux B 520, *525*
Havranek WM 264, *274*
Hawxby K, Tubea B, Ownby J, Basler E 431, *451*
Hay JR 396, *408*
Hayashi F, see Robinson SJ 339, 345, *386*
Hayatsu H, see Inoue M 480, *496*
Haydock DP, see Wilson JR 190, 195, *203*
Hayels RM, see Wax LM 397, *411*
Hayes FR, McCarter JA, Cameron ML, Livingstone DA 416, *451*
Hayes JD, see Owen WJ 365, *384*
Hayes RM, Yeargan KV, Witt WW, Raney HG 364, *379*
Haynes RA, see Heichel GJ 520, *523*
Hays RI, see Mooney HA 207, 208, 210, *229*
Haystead A, see Stewart WDP 423, *461*
Head GC 268, *274*
Heagle AH, see Heggestadt HE 475, *496*
Heagle AS 466, *495*
Heagle AS, see Reinert RA 476, 477, *499*
Heal OW, see Rosswall T 173, *201*
Healey FP 91, 92, *112*
Healy TW, see James RO 425, *453*
Heath AG, see Cairns J Jr 438, 442, *446*
Heath GR, see Hinga KR 322, *324*
Heath GW, see Edwards CA 28, *41*
Heath RL 474, 484, 485, 486, *496*
Heath RL, Chimiklis P, Frederick P 484, 485, *496*
Heath RL, Frederick PE 485, *496*
Heath RL, see Cedeno-Maldano A 426, *446*
Heath RL, see Chimiklis PE 485, *493*
Heath RT, Cooke GD 63, *78*, 423, *451*
Heber U, Santarius KA 441, *451*
Heck WW 474, *496*
Heck WW, Brandt CS 473, 475, 476, *496*
Heck WW, see Blum U 475, *492*
Heck WW, see Brandt CS 482, *492*
Heck WW, see Reinert RA 476, 477, *499*
Heck WW, see Tingey DT 475, *501*
Hedley CL, Harvey DM 520, *523*
Hedrick DW, Hyder DN, Sneva FA, Poulton CE 405, *408*
Heemst HDJ van, see Buringh P 134, *146*
Heemst HDJ van, see Wit CT de 162, *172*
Hegg DA, Hobbs PV 479, *496*
Heggestadt HE, Heagle AH, Nennett JH, Koch EJ 475, *496*
Heichel GH, Musgrave RB 520, *523*
Heichel GJ, Haynes RA 520, *523*
Heim G, see Eckardt F 208, 209, 218, *227*
Heinle DR 316, *324*
Heinle DR, Flemer DA 316, *324*
Heinonen S, see Kallio P 181, 184, *200*
Heinrichs H, Mayer R 37, 39, *42*
Heitefuss R, see Paeschke R 398, *410*
Heitkamp D, see Ovington JD 26, *43*
Heitmüller PT, see Walsh GE 431, *463*
Heitschmidt RK, Lauenroth WK, Dodd JL 473, *496*
Heldt HW, Rapley L 481, *496*
Hellebust JA 51, *78*
Hellmers H, see Hesketh JD 515, 518, *523*
Hellmers H, see Paez A 518, *525*
Hellmers H, see Sionit N 511, 517, 518, *525*
Hellmuth EO 219, 223, *227*
Helmecke K, Hickisch B, Mahn E-G, Prasse J, Sternkopf G 398, 399, 402, *408*
Helmecke K, see Mahn E-G 402, *409*
Helms JA 260, *274*
Helvey JD, Patric JH 23, *42*
Henderson DW, see Hsiao TC 140, 141, *147*, 261, *275*
Henderson GS, Harris WF, Todd DE, Grizzard T 24, *42*
Henderson GS, Swank WT, Waide JB, Grier CC 37, 38, *42*

Hendricks RH, see Thomas MD 476, 481, 482, 488, *500*
Hendy CH, see Troughton JH 298, *303*
Henry Y, see Vazart J 515, *526*
Hensley JR, see Nyffeler A 367, *384*
Hepler PK, Jackson WT 356, *379*
Herbert TY, see O'Brien RD 366, *384*
Herbicide Handbook of the Weed Science Society of America 330, 343, *379*
Herczeg T, Lehoczki E, Szalay L 339, *379*
Hermann RK 22, *42*
Hermann RK, see Sanantonio D 21, *44*, 269, *279*
Herrera J, Paneque A, Maldonado JM, Barea JL, Losada M 60, *78*
Herrera R, Jordan CF, Klinge H, Medina E 284, 290, *300*
Herrera R, see Klinge H 283, 285, 287, 290, *301*
Herrick CC, see Kilpatrick ML 484, *497*
Herrmann A, see Steinberg C 54, 72, *83*
Herrmann R, see Neuland H 74, *80*
Herrmann V, Jüttner F 61, *78*
Hertel R, Thomson KS, Russo, VEA 360, 362, *379*
Hertel R, see Dohrmann U 362, *376*
Hertel R, see Lembi CA 360, 362, *382*
Hertel R, see Thomson KS 360, *362*
Hervey RJ 426, *451*
Herwig K, see Faller N 482, *494*
Hesketh JD, Hellmers H 515, 518, *523*
Hesketh JD, Jones JW 125, *471*
Hesketh JD, see Hofstra G 515, *523*
Hesketh JD, see Tenhunen JD 512, *526*
Hess FD 336, *379*
Hess FD, Bayer DE 355, 356, *379*
Heuveldop J 285, *300*
Hickisch B, see Helmecke K 398, 399, 402, *408*
Hicklenton P, see Oechel WC 178, *201*
Hickman JC 189, *200*
Hickman M 418, *452*
Hicks BB, see Wesely ML 471, *501*
Hicks CE, Neuhold JM 434, *452*
Higginbotham KO, see Kinerson RS 252, *275*
Higgins TJV, Jacobsen JV 363, *379*
Higgins TJV, see Letham DS 330, 358, *382*
Hilbert DW, see Dyer MI 191, *199*
Hilbig W, see Schubert R 401, *410*
Hill AC, Bennett JH 475, *496*
Hill AC, Littlefield N 474, 475, *496*
Hill AC, Pack MR, Transtrum LG, Winters WS 489, *496*
Hill AC, see Bennett JH 470, *492*
Hill AC, see Hou L 518, *523*
Hill AC, see Jacobson JS 471, 474, 475, *496*
Hill AC, see White KL 477, *501*
Hill AJ, see Woodell SRJ 211, *231*
Hill DJ 481, *496*
Hill DJ, see Inglis F 483, *496*
Hill DW, McCarty PC 430, 432, *452*
Hill GR, see Thomas MD 473, 481, 482, *500*
Hill GW, see Thomas WD 11, *44*
Hill KL, see Moreland DE 431, *456*
Hillebrand H, see Francke J 429, *449*
Hiller RG, Bassham JA 486, *496*
Hilton BD, Bradkowski TA, Yamada M, Narahashi T, O'Brien RD 366, *379*
Hilton BD, O'Brien RD 366, *379*
Hilton BD, see O'Brien RD 366, *384*
Hilton JL, Monaco TJ, Moreland DE, Gentner WA 339, *379*
Hilton JL, Scharen AL, St. John JB, Moreland DE, Norris KH 339, 340, 350, *379*
Hilton JL, see Bartels PG 341, 355, 356, *373*
Hilton JL, see Moreland DE 339, 340, 348, *383*
Hilton JL, see St John JB 340, 350, 352, 353, 354, *388*
Hinckley TM, Dougherty PM, Lassoie JP, Roberts JE, Teskey RO 268, 269, *274*
Hinckley TM, Teskey RO, Duhme F, Richter H 260, *274*
Hinckley TM, see Dougherty PM 260, 264, 265, *273*
Hinga KR, Sieburth JMcN, Heath GR 322, *324*
Hinneri S 438, *452*
Hirano C, Kiritani K 405, *408*
Hirsch P, see Alexander M 58, *75*
Hitchcock AE, see Jacobson JS 476, *496*
Hitchcock AE, see McCune DC 487, 488, *498*
Hitchcock GL 320, *324*
Hitchcock GL, Smayda TJ 313, *324*
Ho LC 517, *523*
Ho LC, Thornley JHM 138, *147*
Hoather RC 484, *496*
Hobbie JE, see Wright RT 52, *84*
Hobbs PV, Harrison H, Robinson E 510, *523*
Hobbs PV, see Hegg DA 479, *496*
Hodek J, Holman J, Novak K, Skuhravy V, Stary P, Weisman L, Zeleny J 402, *408*
Hodgkinson KC, Baas Becking HG 194, 195, *200*
Hodgkinson KC, Smith NG, Miles GE 193, *200*
Hodgson JM 397, *408*
Hodgson JM, Moore HD 397, *408*
Hof J van't, see Webster PL 355, *391*

Hoffmann A, see Mooney HA 207, 208, 210, *229*
Hoffmann AE, see Montenegro G 220, *229*
Hoffmann AJ, see Montenegro G 220, *229*
Hoffmann G 21, *42*
Hoffmann G, see Lyr H 21, *43*, 268, *276*
Hoffmann OL 367, 368, *379*
Hoffmann-Falk H, see Mattoo AK 342, *383*
Hofstra G, Hesketh JD 515, *523*
Hofstra G, see Beckerson DW 472, 477, *491*
Hogen D, see Lakowicz JR 366, *381*
Hogetsu K, Okanishi Y, Sugawara H 418, *452*
Hogeweg P 144, *147*
Hohnholz E, see Pfleiderer G 483, *499*
Hojden B 397, *408*
Hok JR 543, *548*
Holdrinet M, see Frank R 430, *449*
Hole CC, Barnes A 263, *274*
Hole CC, see Barnes A 139, *146*, 262, 265, *272*
Höll W, Hampp R 490, *496*
Hollibaugh JT 95, *112*
Hollingsworth E 403, *408*
Holly K 334, *380*
Holly K, see Davies PJ 395, *407*
Holman J, see Hodek J 402, *408*
Holmes EB, Wilson LA 160, *171*
Holm-Hansen O, see Mague TM 90, *112*
Holm-Hansen O, see Parsons TR 96, *113*
Holm-Hansen O, see Williams PJLeB 95, *115*
Holowaychuk N, see Gersper PL 15, 24, *41*
Holt DA, Bula RJ, Miles GE, Schreiber GE, Peart RM 131, *147*
Holt DA, see Schreiber MM 124, 131, *149*
Holt JS, see Radosevich SR 369, 370, *385*
Holt W, see Shiaris M 403, *410*
Holz DD, see Mauck WL 431, *456*
Homer RF, see Brian RC 331, 339, 345, *375*
Hopkins B 294, *300*
Hopkins TS, see Falkowski PG 320, *323*
Hopton JW, see Melchiorri-Santolini U 52, *80*
Hora T, see Levis J 398, *409*
Horbach W, see Weise G 443, *463*
Horie T 262, 263, 265, *275*
Horie T, Wit CT de, Goudriaan J, Bensink J 144, *147*
Horne AJ 56, *78*
Horne AJ, Carmiggelt CJW 56, *78*
Horne AJ, Fogg GE 56, *78*, 415, *452*
Horne AJ, Goldman CR 56, *78*, 415, 426, *452*
Horne AJ, Sandusky JC, Carmiggelt CJW 56, *78*
Horne AJ, Viner AB 415, *452*
Horne AJ, see Roth JC 59, *82*
Hornig L, see Weise G 443, *463*
Hörnström E, see Almer B 438, *444*
Horowitz M 336, *380*
Horsman DC, Roberts TM, Lambert M, Bradshaw AD 472, 473, *496*
Horsman DC, Wellburn AR 490, *496*
Horsman DC, see Wellburn AR 486, 490, *501*
Horton P, see Croze E 342, *376*
Hou L, Hill AC, Soleimani A 518, *523*
Hough J, see Pimentel D 331, *385*
Hough RA, Wetzel RG 51, *78*
Hough RA, see Wetzel RG 415, 418, *463*
Hoursiangou-Neubrunn D, see Borghi H 365, *374*
Houssard C, Escarre J 252, *275*
Howard-Williams C , see Robb F 53, *82*
Howarth RW, Fisher SG 415, *452*
Howe SO, see Walsh JJ 89, *115*, 314, *327*
Howells GP 441, *452*
Hozumi K, Yoda K, Kira T 283, 285, 287, 292, 293, *300*
Hozumi K, see Kira T 249, 260, *275*, 284, 285, *301*
Hozumi K, see Yoda K 264, *280*
Hsiao TC, Acevedo E 189, *200*
Hsiao TC, Acevedo E, Fereres E, Henderson DW 261, *275*
Hsiao TC, Fereres E, Acevedo E, Henderson DW 140, 141, *147*
Huang J, Gloyna EF 431, *452*
Hubble GD, see Stace HCT 15, *44*
Huber O 290, 291, *301*
Hudson JP 360, 363, *380*
Huffaker CB, Messenger PS 371, 372, *380*
Hughes AP, Cockshull KE 517, 518, 520, *523*
Hughes DJ, see Sharp JH 104, *114*
Hughes KW, see Bell SL 354, *374*
Hull H 395, *408*
Humphrey RR, see Hastings JR 206, *227*
Hunt HW, see Detling JK 178, 179, *199*
Hunt R 195, *200*
Hunt WF, Loomis RS 124, 139, *147*
Huntsman SA, see Walsh JJ 89, *115*, 314, *327*
Hurd DC 101, *112*
Hurditch WJ, Charley JL, Richards BN 33, *42*
Hurlbert SH 364, *380*, 431, *452*
Hurle K, Johannes H 430, *452*
Hurle K, Rademacher B 398, *408*
Hurle K, see Koch W 398, *409*
Hürlimann H 418, *452*
Hursky JAM, see Hamilton DH Jr 433, *451*
Hurt P, Wright RD 517, *523*
Hurtt W, see Reid CPP 396, *410*

Huss K, see Hällgren J-E 483, *495*
Hustedt F 436, 438, *452*
Hutchinson FE 415, 416, 422, 435, *452*
Hutchinson GE 11, *42*, 48, *78*, 105, *112*
Hutchinson GE, Bowen VT 416, *452*
Hutchinson GL, Millington RJ, Peters DB 11, *42*
Hutchinson TC, Czyrska H 429, 430, *452*
Hutchinson TC, Stokes PM 426, 430, 443, *452*
Hutchinson TC, see Stokes PM 429, *461*
Hutchison BA, Matt DR 244, 245, 249, 250, 254, 256, *275*
Huttel Ch, Bernhard-Reversat F 283, 291, 294, 295, 297, *301*
Hutzinger O, Safe ASW, Zitko V 364, *380*
Hyde A, see Bartels PG 350, 352, 355, *373*
Hyder DN, see Hedrick DW 405, *408*
Hynes HBN 415, 417, 421, 425, *452*
Hynes HBN, Roberts FW 434, *453*
Hynes HBN, see Kaushik NK 59, *79*

IBM 125, *147*
Idso SB, Brazel AJ 510, *523*
Ikeda T, see Takahashi M 96, *114*
Ikuma H, see Robinson SJ 339, 345, *386*
Imbamba SK, see Tieszen LL 176, 177, *202*
Ingelog T 405, *408*
Ingham G 11, *42*
Ingle M, see Silvius JE 483, *500*
Inglis F, Hill DJ 483, *496*
Ingols RS 433, *453*
Innis GS 124, 125, 141, *147*
Innis GS, see Cole CV 129, 142, *146*
Innis GS, see Reuss JO 129, 142, *149*
Inoue M, Hayatsu H 480, *496*
Ioannow N, Schneider RW, Grogan RG 519, *523*
Isaacson PA, see Woodwell GM 544, *548*
Isensee A, Jones G, McCann J, Pitcher F 403, *408*
Isensee AR, Jones GE 437, *453*
Isensee AR, Jones GE, Turner BC 395, *408*
Ital C, see Ben-Zioni A 441, *444*
Iversen J, Olsen S 438, *453*
Iverson RL, Curl HC Jr, O'Connors HB Jr, Kirk D, Zakar K 314, 319, *324*
Iwaki H, see Kaji M 486, *497*
Izawa S, Good NE 337, 338, *380*

Jabben M, see Frosch S 350, *378*
Jackman M, see Specht R 208, *231*
Jackson GA 93, *112*
Jackson TA, Schindler DW 63, *78*
Jackson TA, see Guillard RRL 94, *111*
Jackson WT, Stetler DA 356, *380*
Jackson WT, see Hepler PK 356, *379*
Jacob F, Neumann ST, Strobel U 396, *408*
Jacobs M, see Schweisfurth R 73, *82*
Jacobsen JV 363, *380*
Jacobsen JV, see Higgins TJV 363, *379*
Jacobsohn R, Andersen RN 397, *408*
Jacobson JS 474, 475, *496*
Jacobson JS, Hill AC 471, 474, 475, *496*
Jacobson JS, Weinstein LH, McCune DC, Hitchcock AE 476, *496*
Jacobson JS, see McCune DC 487, 488, *498*
Jagendorf AT, see Datta DB 483, *494*
Jäger HJ, see Pahlich E 484, *499*
James GB, see Jarvis PG 256, 257, 264, *275*
James RO, Healy TW 425, *453*
Jameson D 192, *200*
Jan P, see Barralis G 397, *406*
Janatkova H, Wildner GF 371, *380*
Janatuinen J, see Särkkä J 429, *459*
Janauer GA 424, *453*
Janes AVJ, see Bennett D 126, 127, *146*
Janicke W 433, *453*
Jansson M *78*, 100, *112*
Jansson SL 32, *42*
Januszko M 403, *408*
Janzen DH 296, *301*
Janzen DH, Wilson DE 296, *301*
Jaquet JM, see Frank R 430, *449*
Jarvis PG 235, 236, 258, 259, 260, *275*
Jarvis PG, James GB, Landsberg JJ 256, 257, 264, *275*
Jarvis PG, see Beadle CL 241, 246, 251, 253, 260, 261, *272*, *273*
Jarvis PG, see Leverenz JW 256, 257, 258, 260, *276*
Jarvis PG, see Ludlow MM 260, *276*
Jarvis PG, see Neilson RE 260, *277*
Jarvis PG, see Norman JM 245, 246, 247, 249, 252, 258, 264, *278*
Jarvis PG, see Reed KL 258, 259, 260, 261, *278*
Jarvis PG, see Šesták Z 237, *279*
Jarvis PG, see Sestark Z 216, 217, *230*
Jarvis PG, see Watts WR 260, *280*
Jarvis PG, see Whitehead D 251, 260, *280*
Jaworski EG 341, 354, *380*
Jeanne N 365, *380*
Jeanne-Levain N, see Puiseux-Dao S 365, *385*
Jenkins D, Medsker LL, Thomas JF 69, *78*
Jenkinson DS 153, *171*
Jenny H 6, *42*, 152, *171*
Jenny H, Gessel SP, Bingham FT 34, *42*
Jensen CR, see Hansen GK 140, *147*
Jensen HL 12, *42*

Jensen KF, Kozlowski TT 479, *497*
Jensen KIN, Bandeen JD, Souza-Machado V 370, *380*
Jensen KIN, see Souza Machado V 370, *388*
Jensen S, Jernelöv A 428, *453*
Jernelöv A, see Jensen S 428, *453*
Jinnerot D, see Forsberg C 423, *449*
Jirgensons B 434, *453*
Jitts HR, see Turner RE 315, *327*
Johannes H, see Hurle K 430, *452*
Johannes RE 63, *79*
Johannes RE, see Pomeroy LR 322, *325*
Johannes RL, see Webb KL 96, *115*
Johanningmeier U, see Oettmeier W 342, 343, 345, *384*
Johansson N 263, *275*
John DM 294, 295, *301*
Johnson AL, see Beyer EM 360, *374*
Johnson AW, see Mooney HA 207, 208, 210, *229*
Johnson D, see Hansen J 510, *523*
Johnson DA, Caldwell MM 187, *200*
Johnson DA, Tieszen LL 178, 194, *200*
Johnson DA, see Caldwell MM 187, *198*
Johnson DW, Cole DW, Bledsoe CS, Cromack K, Edmonds RL, Gessel Sp, Grier CC, Richards BN, Vogt KA *42*
Johnson DW, Cole DW, Gessel SP 37, *42*
Johnson DW, Cole DW, Gessel SP, Singer MJ, Mind RV 39, *42*
Johnson FL, Risser PG 17, *42*
Johnson HA, see Woodwell GM 544, *548*
Johnson HB, see Szarek SR 218, 223, 224, *231*
Johnson HE, see Maki AW 432, *455*
Johnson NM, see Likens GE 545, *548*
Johnson WC, see Manny BA 415, *456*
Jolley RL 433, *453*
Jolliffe PA, see Srivastava HS 475, *500*
Jon W, see Kummerow J 207, 211, *228*
Jones AK, see McLean RO 425, 429, 443, *456*
Jones AM, see Melhado LL 363, *383*
Jones DB, see Gifford RM 517, *522*
Jones E, see Hasler AD 418, *451*
Jones G, see Isensee A 403, *408*
Jones GE, see Isensee AR 395, *408*, 437, *453*
Jones HG, see Landsberg JJ 260, *276*
Jones JRE 425, *453*
Jones JW, see Hesketh JD 125, *471*
Jones JW, see Potter JR 195, *201*
Jones LHP, see Cowling DW 482, *493*
Jones LHP, see Lockyer DR 473, *498*
Jones RL 358, 361, *380*
Jones RL, see Brian RC 331, 339, 345, *375*
Jordan CF 23, *42*, 282, 294, 295, *301*
Jordan CF, Escalante G 283, 297, *301*
Jordan CF, Kline J 287, *301*
Jordan CF, Murphy PG 294, 295, *301*
Jordan CF, Todd RL, Escalante G 37, *42*
Jordan CF, Uhl Ch 283, 287, 292, *301*
Jordan CF, see Herrera R 284, 290, *300*
Jordan CF, see Odum HT 287, 293, *303*
Jordan CF, see Stark N 37, *44*
Jordan DC 12, *42*
Jordan LS, see Mann JD 341, 354, *382*
Jordan MJ, Likens GE 52, *79*
Jorgensen JR, see Penning de Vries FWT 143, *149*
Joseph JM 438, *453*
Joset-Espardellier F, see Astier C 371, *373*
Judd JH 435, *453*
Jung W, see Schweisfurth R 73, *82*
Junge CE 69, *79*
Jupp BP, Spence DHN 418, *453*
Jüttner F 51, *79*
Jüttner F, Matuschek T 51, *79*
Jüttner F, see Herrmann V 61, *78*

Kabanova JG, see Koblentz-Mishke OJ 88, *112*, 305, 308, *324*
Kaji M, Yoneyama T, Tosuka T, Iwaki H 486, *497*
Kallio P, Heinonen S 181, 184, *200*
Kallio P, Karenlampi K 181, *200*
Kalmaz EV, Kalmaz GD 364, *380*
Kalmaz GD, see Kalmaz EV 364, *380*
Kaltwasser H, see Zaiss U 54, *84*
Kamel AM, Weiner ND, Felmeister A 486, *497*
Kamp-Nielsen L 59, 65, 69, *79*, 427, *453*, 491, *497*
Kamp-Nielsen L, see Steemann Nielsen E 427, *461*
Kändler U, Ullrich H 486, *497*
Kapkov VI, see Khobotyev VG 427, *453*
Kaplan LA, Goldman CR, Knight AW 65, *79*
Kappen L, Lange OL, Schulze ED, Evenari M, Buschbom U 214, 222, 223, *228*
Kappen L, Oertli JJ, Lange OL, Schulze E-D, Evenari M, Buschbom U 214, *228*
Kappen L, see Evenari M 213, 214, *227*
Kappen L, see Lange OL 219, 221, 223, *228*
Karenlampi K, see Kallio P 181, *200*
Karlander EP, see Chimiklis PE 437, *447*
Karolin A, see Moldau H 139, *148*

Kassam AH, see Doorenbos J 140, 141, *146*
Kassasian L, see Blair AM 367, *374*
Katan J, Eshel Y 399, *408*
Katekar GF 362, *380*
Katekar GF, Geissler AE 360, *380*
Katoh S, see Satoh K 488, *499*
Kats G, see Thompson CR 474, *500*
Katz M, see Doudoroff P 426, *448*
Katz M, see Ordin L 485, *499*
Kaufman DD, Kearney PC 398, *408*
Kaufman DD, see Blake J 398, *407*
Kaufman DD, see Kearney PC 330, 334, 338, *380*
Kaushik NK 52, 53, *79*
Kaushik NK, Hynes HBN 59, *79*
Kaushik NK, see Chatarpaul L 58, *76*
Kawanabe S, see Fujimori J 235, *274*
Kay QON, see Gressel J 366, 368, 369, *378*
Kazano H, see Tomizawa C 403, *411*
Ke B, Clendenning KA 434, *453*
Kearney PC, Kaufman DD 330, 334, 338, *380*
Kearney PC, see Kaufman DD 398, *408*
Keating KI 419, *453*
Keddington MB, see Dina SJ 189, *199*
Kee DA, see Ownbey CR 435, *457*
Keefe C, see Boynton WR 307, 309, 319, *323*
Keefe CW, see Hamilton DH Jr 433, *451*
Keeley S, see Mooney HA 207, 208, 210, *229*
Keeling CD 505, 508, *523*
Keeney DR 58, *79*
Keeney DR, Chen RI, Graetz DA 58, *79*
Keeney DR, see Chen RL 57, 58, *76*
Keeney DR, see Dolar SG 429, *448*
Keeney DR, see Graetz DA 57, *78*
Keeney DR, see McGregor AN 56, *80*
Keeney DR, see Nichols DS 60, *80*
Kees H 397, *409*
Kees H, see Gressel J 366, 368, 369, *378*
Kefeli VJ 362, *380*
Keirn MA, Brezonik PL 56, 57, *79*
Keirn MA, see Brooks RH 56, *76*, 90, *109*
Kellar PE, Goldman CR 56, *79*
Keller H 484, *497*
Keller H, Müller J 472, *497*
Keller T 473, *497*
Keller T, Schwager H 488, *497*
Kelley JC, see Walsh JJ 89, *115*, 314, *327*
Kelly M, see Croze E 342, *376*
Kelly Robertson C 54, *79*
Kemmerich P, see Feierabend J 352, *377*
Kemp ALW, see Frank R 430, *449*
Kemp ALW, see Nriagu JO 68, *81*
Kemp JD, see Anderegg BN 365, *373*
Kemp PR, Williams GJ, III 185, 195, *200*
Kemp RP, see Williams GJ III 179, *203*
Kemp WM 315, *324*
Kemp WM, see Boynton WR 306, 307, 309, 319, *323*
Kemp WM, see Ulanowicz RE 306, *327*
Kende H, Gardner G 362, *380*
Kendrick WB, Burges A 31, *42*
Kennedy FS, see Wood JM 428, *464*
Kennedy GS, see Bergersen FJ 12, *40*
Kenner RA, see Ahmed SI 92, *108*
Kenyon C, see Robb F 53, *82*
Kerr MW, Avery RJ 354, *380*
Kerr MW, Wain RL 339, 348, *380*
Kersten HJ, see Schuler JF 422, *460*
Kessel JF van 58, *79*
Kestemont P 236, *275*
Kestemont P, Duvigneaud P, Paulet E 236, *275*
Kester DR, see Byrne RH 102, *109*
Ketchum BH 90, 91, *112*, 314, 318, *324*
Ketchum BH, Redfield AC 423, *453*
Ketchum BH, see Kuenzler EJ 91, *112*
Ketchum BH, see Redfield AC 86, 95, 97, *114*
Keulder PC 429, *453*
Keulen H van 134, 140, 141, *147*
Keulen H van, Laar HH van, Louwerse W, Goudriaan J 135, 140, *147*
Keulen H van, Seligman NG, Benjamin RW 131, 142, *147*
Keulen H van, Seligman NG, Goudriaan J 142, *147*
Keulen H van, Wit CT de, Lof H 127, *147*
Keulen H van, see Penning de Vries FWT 143, *149*
Keulen H van, see Seligman N 142, 143, *149*
Keulen H van, see Wit CT de 130, 131, 132, 135, 136, 137, 138, 139, 140, *150*
Keyes MR, Grier CC 237, 241, *275*
Keys MR, see Grier CC 241, 269, 271, *274*
Khalanski M 441, *453*
Khan A 403, *409*
Khan MA, Tsunoda S 120, *148*
Khan MAQ, see Neudorf S 364, *384*
Khan MU, Lem NW, Chandorkar KR, Williams JP 353, *380*
Khare L 403, *409*
Khatamian H, see Adedipe NO 474, *491*
Khobotyev VG, Kapkov VI, Rukhadze YG, Turunina NV, Shidlovskaya NA 427, *453*
Kickuth R 424, *453*

Kickuth R, see Seidel K 432, *460*
Kiermayer O, Fedtke C 356, *380*
Kiermayer O, Meindl U 365, *380*
Kilham P, see Guillard RRL 94, *111*
Kilham P, see Kilham SS 419, 420, *453*
Kilham P, see Sugihara G 435, 436, *461*
Kilham SS, Kilham P 419, 420, *453*
Kilham SS, see Nelson DM 102, *113*
Kilpatrick M, see Kilpatrick ML 484, *497*
Kilpatrick ML, Herrick CC, Kilpatrick M 484, *497*
Kimmel EC, see Casida JE 334, 367, *375*
Kinerson RS 252, 265, *275*
Kinerson RS, Higginbotham KO, Chapman RC 252, *275*
Kinerson RS, Ralston CW, Wells CG 264, 271, *275*
Kinerson RS, see Harris WF 26, *42*, 269, *274*
King DL 415, 419, *453*
King KM, see Leuning R 469, 470, 471, *497*
King LJ 536, *548*
Kinkade ML, Erdman HE 429, *453*
Kinnaird JW, see Fowler D 477, *495*
Kinze G, see Lüttge U 480, *498*
Kira T 235, 236, 237, *275*, 283, 291, 296, 297, 298, *301*
Kira T, Ogawa H 292, *301*
Kira T, Ogawa H, Yoda K, Ogino K 292, 298, *301*
Kira T, Shidei T 235, 237, 250, 263, 267, *275*, 284, 298, *301*
Kira T, Shinozaki K, Hozumi K 249, 260, *275*, 284, 285, *301*
Kira T, Yabuki K 243, 271, *275*, 296, *301*
Kira T, see Hozumi K 283, 285, 287, 292, 293, *300*
Kira T, see Ogawa H 283, *303*
Kira T, see Shinozaki K 252, *279*
Kira T, see Yoda K 264, *280*
Kirita H, see Nishioka M 267, *277*
Kiritani K, see Hirano C 405, *408*
Kirk D, see Iverson RL 314, 319, *324*
Kirkland K, see Fryer JD 398, *407*
Kirkman H, Griffith FB, Parker RR 64, *79*
Kirkwood RC 348, 349, *380*
Kisiel MJ, see Devlin RM 350, *376*
Kiso K, see Asada, K 479, *491*
Kitchell JF, see Gallepp GW 65, *77*
Kitchen LM, Witt WW, Rieck CE 352, *380*
Kittredge J 207, 210, *228*
Klammt E, see Müller G 399, 403, *409*
Klaveness D 74, *79*
Klee O 435, *453*
Kleinig H, see Frosch S 350, *378*
Klemmedson JO, see Tiedemann AR 15, *44*
Kleudgen HK 352, *381*
Kleudgen HK, see Lichtenthaler HK 352, *382*
Klikoff LF, see Dina SJ 189, *199*
Klikoff LG, see Detling JK 189, *199*
Kline J, see Jordan CF 287, *301*
Kling H, see Ayles GB 426, *444*
Kling H, see Whitaker J 426, *463*
Klinge H 283, 293, 294, *301*
Klinge H, Herrera R 283, 285, 287, 290, *301*
Klinge H, Rodrigues WA 294, *301*
Klinge H, see Herrera R 284, 290, *300*
Klinkhammer G, see McCaffney RJ 97, *112*
Klöck W 405, *409*
Klockare B, Axelsson L, Ryberg H, Sandelius AS, Widell KO 352, *381*
Klösch U, see Grill D 480, *495*
Klötzli F 419, *454*
Klötzli F, Grünig A 418, 424, *454*
Klueter HH, see Krizek DT 517, *524*
Klug MJ, see Dacey JWH 54, *77*
Klug MJ, see Suberkropp K 53, *83*
Kluge M, Ting IP 223, 224, 225, *228*
Knaggs FW, see Way JM 430, *463*
Knapp WW, see Russo JM 129, *149*
Knauer GA, see Reshkin SJ 98, *114*
Knecht GN, O'Leary JW 517, *523*
Knight AW, see Kaplan LA 65, *79*
Knittel H, Behrendt S, Schott PE 361, *381*
Knöpp H 441, *454*
Knops HJ, see Trebst A 345, *390*
Knox S, see Butler EI 93, *109*
Knudson Butler L, Tibbitts TW 474, *497*
Knuesli E, see Gast A 331, *378*
Koblentz-Mishke OJ, Volkovinsky VV, Kabanova JG 88, *112*, 305, 308, *324*
Koch EJ, see Heggestadt HE 475, *496*
Koch M, see Bornkamm R 424, *445*
Koch W, Hurle K 398, *409*
Koch W, see Lange OL 223, *228*
Koch W, see Mittnacht A 401, *409*
Koch W, see Schulze ED 219, *230*
Koeller P, see Parsons TR 96, *113*
Koeman JH 400, *409*
Koeppe DE, Miller RJ 490, *497*
Kohler A 418, 420, *454*
Kohler A, Brinkmeier R, Vollrath H 417, 420, 442, *454*
Kohler A, Pensel T, Zeltner GH 421, *454*

Kohler A, Vollrath H, Beisl E 420, 442, *454*
Kohler A, Wonneberger R, Zeltner GH 420, 442, *454*
Kohler A, Zeltner GH 420, 442, *454*
Kohler A, see Glänzer U 442, *450*
Kohler A, see Haber W 442, *451*
Kohler A, see Kutscher G 420, 442, *454*
Kohler A, see Labus BC 434, 443, *454*
Kohler A, see Melzer A 418, 442, *456*
Kohlschütter H, see Meinck F 435, 436, *456*
Kolbe RW 435, *454*
Kolkwitz R, Marsson M 442, *454*
Kondo N, Sugahara K 472, *497*
Koonce JK, see Brown EJ 64, *76*
Kopecek K, Füller F, Ratzmann W, Simonis W 365, *381*
Koren E, see Ben-Aziz A 350, *374*
Korte F, see Coulston F 364, *376*
Korycka A, see Niewolak S 59, *81*
Koschel R, see Mothes G 441, *457*
Kostusiak AS, see Devlin RM 350, *376*
Kott Y 433, *454*
Kott Y, Edlis J 433, *454*
Kott Y, see Betzer N 433, *444*
Kötter C, see Hauska G 340, 347, *379*
Kowal JL, see Greenland DJ 297, *300*
Kowalik H, see Dohrmann U 362, *376*
Koyama H 287, 288, 290, 293, *301*
Koziol F, see Pimentel D 331, *385*
Koziol MJ, see Cowling DW 466, *493*
Kozlowski TT, see Constantinidou HA 490, *493*
Kozlowski TT, see Jensen KF 479, *497*
Kozlowski TT, see Kramer PJ 261, *275*
Kraay GW, see Gieskes WWC 322, *323*
Krambeck C 52, *79*
Kramer D, see Fischer K 482, *494*
Kramer HJM, see Rensen JJS van 342, *386*
Kramer JR, see Allen HE 86, *108*
Kramer P, see Sionit N 517, 518, *525*
Kramer PJ, Bullock HC 21, *42*
Kramer PJ, Kozlowski TT 261, *275*
Krämer W 330, *381*
Krasnick G, see Caperon J 320, *323*
Kratky BA, Warren GF 336, *381*, 431, *454*
Krause A 418, 421, *454*
Krause D, see Kummerow J 207, 211, *228*
Krause W, Carbiener R *454*
Krauss RW, see Galloway RA 423, *450*
Krauter K, see Stokes PM 429, *461*
Kremer DJ, see Penning de Vries FWT 137, 138, *149*, 240, *278*
Kremer JN, see Nixon SW 313, *325*
Kriedemann PE, Barrs HD 260, *275*
Kriedemann PE, Sward RJ, Downtown WJS 517, *523*
Krieg A, see Franz JM 372, *378*
Krinsky NJ 350, 351, 352, *381*
Krinsky NJ, see Anderson SM 351, *373*
Krishnakumari MK 365, *381*
Krizek DT, Zimmerman RH, Klueter HH, Bailey WA 517, *524*
Krohn K, see Gersberg RM 61, *77*
Kruger FJ 208, *228*
Kruglov JV, see Michajlova EI 398, *409*
Krul J, see Penning de Vries FWT 143, *149*
Krummel J, see Pimentel D 331, *385*
Krzeczkowska-Woloszyn L 403, *409*
Ku WC, Digiano FA, Feng TH 65, *79*
Kudo A, see Mortimer DC 428, *457*
Kuenzler EJ 91, 100, 104, *112*
Kuenzler EJ, Ketchum BH 91, *112*
Kuenzler EJ, Perras JP 63, *79*, 99, *112*, 423, *454*
Kühn H, see Faller N 482, *494*
Kühn R, see Bringmann G 427, 443, *445*
Kukla G, Gavin J 510, *524*
Kulikov NV, see Timofeeva NA 434, *462*
Kulshreshtha S, Sharma K 403, *409*
Kumar HD, see Rana BC 429, *458*
Kümmel HW, Grimme LH 350, *381*
Kummerow J, Krause D, Jon W 207, 211, *228*
Kummerow J, see Mooney HA 207, 208, 210, *229*
Kunert KJ, Böger P 340, 350, 351, 352, *381*
Kunert KJ, see Böger P 342, 346, 351, 353, *374*
Kunert KJ, see Böhme H 336, *374*
Kunert KJ, see Lambert R 340, 347, 351, 352, *381*
Kunert KJ, see Sandmann G 350, *387*
Kunitake G, Stitt C, Saltman P 519, *524*
Künstle E, Mitscherlich G 260, *275*
Kurimo U 418, 442, *454*
Kurochkina LY, Osmanova LT, Borovskaya TA 213, *228*
Kurth H 330, 331, *381*, 394, *409*
Kuruoiwa S 247, 257, *276*
Kusnetsov SI 57, 59, *79*
Kusnezov SI 50, 55, 67, 68, 73, *79*
Kustin K, see Eigen M 478, *494*
Kusumoto T 260, *276*

Kutscher G, Kohler A 420, 442, *454*
Kuz'mina IV, see Gruzdev GS 398, *407*
Kwiatkowski G, see Elstner EF 346, 351, *377*

Laar HH van, see Goudriaan J 140, *147*
Laar HH van, see Keulen H van 135, 140, *147*
Laar HH van, see Penning de Vries FWT 125, 138, 139, *149*, 160, 161, *171*, 240, 265, *278*
Laar HH van, see Wit CT de 130, 131, 132, 135, 136, 137, 138, 139, 140, *150*
Laasch H, Pfister K, Urbach W 339, 343, 344, 348, *381*
Labus BC 434, 442, *454*
Labus BC, Kohler A 443, *454*
Labus BC, Nobel W, Smetana R, Kohler A 443, *454*
Labus BC, Schuster H, Nobel W, Kohler A 434, *454*
Lachavanne JB, Wattenhofer R 442, *455*
Lacis A, see Hansen J 510, *523*
Lackey JB 438, *455*
LaCroix MW, see Thayer GW 306, *326*
Lai EK, Fong KL, McCay PB 352, *381*
Lakowicz JR, Hogen D 366, *381*
Lakowicz JR, Hogen D, Omann G 366, *381*
Lam CWY, Vincent WF, Silvester WB 56, *79*
Lamarra VA Jr 63, *79*
Lambers HT 140, *148*
Lambert JR, Baker DN, Phene CJ 131, 139, 141, 142, 145, *148*
Lambert JR, see Baker DN 131, 133, 136, 137, 143, *146*
Lambert JR, see Bar-Yosef B 141, *146*
Lambert JR, see Phene CJ 132, *149*
Lambert M, see Horsman DC 472, 473, *496*
Lambert R, Kunert KJ, Böger P 340, 347, 351, 352, *381*
Lambert R, see Sandmann G 351, *387*
Lampert W 51, *79*
Land EJ, see Farrington JA 346, 351, *377*
Landivar JA 135, *148*
Landner L 428, *455*
Landsberg JJ, Jones HG 260, *276*
Landsberg JJ, see Butler DR 262, 264, 265, 266, *273*
Landsberg JJ, see Jarvis PG 256, 257, 264, *275*
Landsberg JJ, see Seaton KA 141, *149*
Landsberg JJ, see Thorpe MR 259, *279*
Landsberg JJ, see Watson RL 258, *280*
Lang G 418, 442, *455*
Lange OL, Kappen L, Schulze E-D 219, 223, *228*
Lange OL, Koch W, Schulze ED 223, *228*
Lange OL, Schulze ED, Evenari M, Kappen L, Buschbom U 219, 221, *228*
Lange OL, Schulze ED, Kappen L, Evenari M, Buschbom U 221, *228*
Lange OL, see Evenari M 212, 213, 214, *227*
Lange OL, see Kappen L 214, 222, 223, *228*
Lange OL, see Schulze ED 219, *230*
Lange W 415, *455*
Lännergren C, Lundgren A 56, *79*
Lanza GR, Cairns J Jr 441, *455*
Lanza GR, see Cairns J Jr 414, 424, 435, *446*
Larcher W 261, *276*, 287, 288, *302*
Larcher W, Cernusca A, Schmidt L, Grabherr G, Nötzel, Smeets N 181, *200*
Larcher W, see Pisek A 181, *201*
Lark JGI, see Ayles GB 426, *444*
Larow EJ, McNaught DC 63, *79*
Larsson CM, Tillberg JE 365, *381*
Lassoie JP, see Hinckley TM 268, 269, *274*
Laubenberger G, Hartmann L 426, *455*
Laudelout H, Meyer J 294, *302*
Laudin H 400, *409*
Lauenroth WK 174, 186, *200*
Lauenroth WK, Sims PL 174, 187, *200*
Lauenroth WK, see Dodd JL 176, *199*
Lauenroth WK, see Heitschmidt RK 473, *496*
Lauenroth WK, see Sala OE 189, *202*
Lauenroth WK, see Sims PL 191, *202*
Laval-Martin D, Dubertret G, Calvayrac R 371, *382*
Laval-Martin D, see Calvayrac R 371, *375*
Lavigine P, see Souza-Machado V 370, *388*
Lavorel J, Etienne AL 338, *382*
Law RM, Mansfield TA 475, *497*
Lawes GB, see Dodge AD 350, *376*
Lawlor DW, see Legg BJ 238, 239, *276*
Lawn RJ, Brun WA 520, *524*
Lawrence DB, see Ovington JD 26, *43*
Lawrence JM, see Funderburk HH Jr 421, *450*
Laws EA, Bannister TT 320, *324*
Laws EA, Wong DCL 95, *112*
Lawton HW, see Hall AE 165, *171*
Lay MM, Casida JE 367, *382*
Lay MM, see Casida JE 334, 367, *375*
Lazaroff N 431, *455*
Lazenby A, see Colman RL 190, *198*
Leakey LSB, Tobias PV, Leakey MD 531, *548*
Leakey MD, see Leakey LSB 531, *548*
Lean DRS 62, *79*, 100, *112*, 416, *455*
Lean DRS, see Murphy TP 51, 61, *80*, 102, *113*, 419, *457*

Lean DRS, see Nalewajko C 64, *80*
Lean DRS, see Paerl HW 64, *81*
Lean DRS, see Peters RH 63, *81*
Lean DRS, see Schindler DW 48, 49, *82*
Leavitt JRC, Penner D 367, *382*
Leavitt R, see Mudd JB 484, *498*
LeBaron H, Gressel J 368, *382*
Lebedeff S, see Hansen J 510, *523*
Lebedjantzev AN 32, *43*
LeBrasseur RJ, see Parsons TR 313, *325*
LeCureux L, see Bressan RA 484, *492*
Ledig FT, Drew AP, Clark JG 240, 263, 265, *276*
Ledoigt G, see Calvayrac R 371, *375*
Lee CJ, Miller GW, Welkie GW 488, *497*
Lee FG, see Austin ER 59, *75*
Lee GF, see Brezonik PL 58, *76*
Lee GF, see Cowen WF 415, *447*
Lee JA 490, *497*
Lee K, Nalewajko C 51, *79*
Lee KE, Wood TG 31, *43*
Lee SS, Fang SC 343, *382*
Lee SS, Fang SC, Freed VH *455*
Lee T, see Hansen J 510, *523*
Lee TT 485, *497*
Lee-Kaden J, see Simonis W 365, *388*
Lefevre M 419, *455*
Leftley JW, see Syrett PJ 61, *83*
Legg BJ, Day W, Lawlor DW, Parkinson KJ 238, 239, *276*
Leh F, see Mudd JB 485, *498*
Lehman JT 514, *524*
Lehmann J, Benson AA 481, *497*
Lehoczki E, see Herczeg T 339, *379*
Leighton PA 484, *497*
Leighty DA, see Rice HV 103, *114*
Leikola M, see Hari P *127*, 147
Leinweber CL, see Reardon PO 194, *201*
Leith ID, see Fowler D 477, *495*
Leius K 402, *409*
Lem NW, see Khan MU 353, *380*
Lembi CA, Morre DJ, Thomson KS, Hertel R 360, 362, *382*
Lemeur R, Blad BL 164, *171*
Lemon E, Allen LH, Muller L 293, *302*
Lemon ER, see Moss DN 517, *524*
Lempert U, see Rüdiger W 352, *387*
Lengfelder E, see Elstner EF 346, 351, *377*
Leonard CD, see Woltz SS 476, *502*
Leonard NJ, see Melhado LL 363, *383*
Leonard OA, Pinckard JA 518, *524*
Leopold AC, see Thomson KS 362, *389*
Lepley CR, see Williams WA 163, *172*
Lepp NW, see Fairfax JAW 474, *494*
Lerman JC, see Ode D 176, 186, *201*
Lerman SL, Darley EF 476, *497*
Letham DS 358, 361, 362, *382*
Letham DS, Goodwin PB, Higgins TJV 330, 358, *382*
Leuning R, Neumann HH, Thurtell GW 469, 470, 471, *497*
Leuning R, Unsworth MH, Neumann HH, King KM 469, 470, 471, *497*
Leuning R, see Attiwill PM 267, *272*
Leverenz JW 260, *276*
Leverenz JW, Jarvis PG 256, 257, 258, 260, *276*
Levine S 62, *80*
Levis J, Papavizas G, Hora T 398, *409*
Levitt J 440, *455*
Lewin J, Chen C 102, *112*
Lewin JC 102, *112*
Lewis AG, see Whitfield PH 103, *115*
Lewis DH, see Harley JL 23, *41*
Lewis MC, see Tieszen LL 176, 177, 181, *202*
Lewis WM, Grant MC 37, *43*
Li WC, Armstrong DE, Williams JDH, Harris RF, Syers JK 65, *80*
Li WKW 317, *324*
Li Y-H *524*
Li Y-H, see Broecker WS 509, *522*
Libera W, Ziegler H, Ziegler I 479, 482, 483, *497*
Libera W, Ziegler I, Ziegler H 483, 484, *497*
Libera W, see Ziegler I 479, *502*
Lichtenstein EP, see Anderegg BN 365, *373*
Lichtenstein RF, Skrentny RF, Tsukano Y 432, *455*
Lichtenthaler HK 352, *382*
Lichtenthaler HK, Burkard G, Grumbach KH, Meier D 352, *382*
Lichtenthaler HK, Kleudgen HK 352, *382*
Lichtenthaler HK, see Meier D 352, *383*
Lichtenthaler HK, see Pfister K 339, 343, 344, *385*
Liddicoat MI, see Butler EI 93, *109*
Lieberman M 358, 361, *382*
Liebmann H 442, *455*
Liebmann H, see Demoll R 421, *448*
Lien S, McBridge JC, Togasaki RK, San Pietro A 371, *382*
Lieth H 252, *276*, 287, 290, *302*, 309, *324*
Lignowski EM, Scott EG 355, *382*
Likens GE 48, 49, *80*, 415, *455*
Likens GE, Bormann FH, Pierce RS, Eaton JS, Johnson NM 545, *548*
Likens GE, Wright RF, Galloway JN, Butler TJ 438, *455*, 477, *497*
Likens GE, see Bormann FH 6, 7, 8, 20, *40*, 539, *547*

Likens GE, see Eaton JS 24, *41*
Likens GE, see Fisher SG 52, *77*
Likens GE, see Gosz JR 24, *41*
Likens GE, see Jordan MJ 52, *79*
Likens GE, see Whittaker RH 158, *172*, 174, *202*, 507, *526*
Likens GE, see Woodwell GM 506, *526*, 537, 539, *548*
Lilly VG, see Barnett HL 519, *521*
Linck AJ 354, 356, 357, *382*
Linck AJ, see Arntzen CJ 355, *373*
Lindau G, see Haselhoff E 468, *495*
Lindeman RL 414, *455*
Linder PL, Mitchell IW, Freeman GD 396, *409*
Linder S 261, *276*
Linder S, Lohammar T 271, *276*
Linder S, Rook DA 269, *276*
Linder S, Troeng E 255, 262, 264, 265, *276*
Linder S, see Ågren GI 264, 270, *272*
Linder S, see Hällgren J-E 474, *495*
Linder S, see Troeng E 255, 260, 261, *280*
Lindström K 70, *80*
Link K, see Oettmeier W 339, 342, 345, *384*
Linkins AE, Antibus RK 545, *548*
Linkins AE, Atlas RM, Gustin P 539, *548*
Linkins AE, see Antibus RK 545, *547*
Linscott DL, see Muslih RK 353, *383*
Linthurst RA, see Cowling EB 477, *493*
Lipton SD , see Bubeck RC 435, *446*
Liss PS, see Burton JD 101, *109*
Litav M, Agami M 421, 433, *455*
Little CMA, Loach K 260, *276*
Little P, Whiffen RD 471, *497*
Little P, see Chamberlain AC 470, *493*
Littlefield L, Forsberg C 64, *80*, 421, *455*
Littlefield N, see Hill AC 474, 475, *496*
Littler M, see Wheeler P 52, *84*
Litvinova NP 212, 213, *228*
Livak KL, see Quebedeaux B, 520, *525*
Livingston RJ 430, *455*
Livingstone DA, see Hayes FR 416, *451*
Llama MJ, see Serra JL 95, *114*
Lloyd EJH, see Hargreaves JW 438, *451*
Lloyd-Jones CP, see Luckwill LC 394, *409*
Loach K, see Little CMA 260, *276*
Lockett MC, see Sweazy RM 423, *461*
Lockwood JL, see Chacko CI 432, *447*
Lockyer DR, Cowling DW, Jones LHP 473, *498*
Lockyer DR, see Cowling DW 473, 482, *493*
Lof H, see Keulen H van 127, *147*
Logan RS, see Grier CC 235, 264, 271, *274*
Lohammar T, see Linder S 271, *276*
Löhr E 287, 288, *302*
Lommen PW, Schwintzer CR, Yocum CW, Gates DM 512, *524*
Long JN, Turner J 250, 251, *276*
Longstreth DL, see Nobel PS 219, *229*
Loomis RS, Gerakis PA 157, 158, 160, *171*
Loomis RS, Ng E 124, 131, *148*
Loomis RS, Rabbinge R, Ng E 122, 125, 131, 135, 136, 137, *148*, 169, *171*, 238, *276*
Loomis RS, Williams WA 163, 164, *171*
Loomis RS, see Clark EA 164, *170*
Loomis RS, see Duncan WG 164, *171*
Loomis RS, see Hunt WF 124, 139, *147*
Loomis RS, see McDermitt DK 161, *171*
Loomis RS, see Nevins DJ 511, *524*
Loomis RS, see Williams WA 163, *172*
Loos MA 341, 357, *382*
Loquens SMH, see Greenwood DJ 127, *147*
Lords JL, see McNulty JB 489, *498*
Lorenz H, see Fathi M 491, *494*
Lorenzen CH 313, *324*
Losada M, see Herrera J 60, *78*
Lossaint P 207, 208, 210, *228*
Lossaint P, Rapp M 210, *228*
Loughman BC, see Barber DA 29, *39*
Louguet P, see Bonte J 472, *492*
Louwerse W, see Keulen H van 135, 140, *147*
Lovelace J, Miller GW, Welkie GW 489, *498*
Lovelock JE, Maggs RJ, Rasmussen RA 69, *80*
Lovelock JE, see Cox RA 469, *493*
Lowitz T, see Feierabend J 352, *377*
Lucas WJ 514, *524*
Luckwill LC, Lloyd-Jones CP 394, *409*
Ludlow MM 176, 179, 184, 185, 189, 195, *200*
Ludlow MM, Jarvis PG 260, *276*
Ludlow MM, Ng TT 194, *200*
Ludlow MM, Wilson GL 517, *524*
Ludlow MM, see Björkman O 288, 290, 291, *300*
Ludlow MM, see Neilson RE 260, *277*
Luedtke N, see McCaffney RJ 97, *112*
Lugo A 287, 290, *302*
Lugo A, Gonzalez-Liboy JA, Cintrón B, Dugger K 287, 293, 295, *302*

Lugo A, see Odum HT 287, 293, *303*
Lui NST, Roels OA 92, *112*
Luken JO, see Billings WD *547*
Lukina GA 431, *455*
Lumry R, see Spikes JD 476, *500*
Lund JWG 415, 416, 419, 437, *455*
Lundgren A, see Granhall U 56, *78*, 415, *450*
Lundgren A, see Lännergren C 56, *79*
Lundgren DG, Dean W 72, 73, *80*
Lundkvist I 431, *455*
Lunney CZ, see Davis GJ 437, *447*
Lurz G, see Urbach W 371, *390*
Luther H 436, *455*
Lüttge U, Osmond CB, Ball E, Brinkmann E, Kinze G 480, *498*
Lüttge U, see Brinkmann E 480, *492*
Lüttge U, see Fischer K 487, *494*
Luxon PL, see Wong PTS *464*
Lyr H, Hoffmann G 21, *43*, 268, *276*

Maas SJ, Arkin GF 129, *148*
Maass G, see Eigen M 478, *494*
MacCarthy HR, see Finlayson DG 364, *377*
MacFarlane RB, Glooschenko WA, Harriss RC 365, *382*
Mach W, see Norman JM 247, *278*
Machta L, see Elliott WP 508, 514, *522*
MacIsaac JJ, Dugdale RC 94, *112*
MacIsaac JJ, see Barber RT 103, *108*
Mackay RJ, see Bärlocher F 53, *75*
MacKenthun KM 435, *455*
Mackenzie FT, see Wollast R 428, *464*
Mackereth FJ 423, *455*
Mäckle H, see Geller W 443, *450*
Macko D, see Mesarovic MD 134, *148*
Maclean DC, see McCune DC 470, 475, *498*
Madeira VC, see Antunes-Madeira MC 366, *373*
Madewell CE, see Stanley RA 442, *461*
Madge DS 294, *302*
Madgwick HAI 236, *276*
Madgwick HAI, Beets P, Gallapher S 235, 236, 237, *277*
Madsen E 515, 518, *524*
Maggs RJ, see Lovelock JE 69, *80*
Mague TM, Weane NM, Holm-Hansen O 90, *112*
Mahall BE, see Fonteyn PJ 211, *227*
Mahn E-G, Helmecke K 402, *409*
Mahn E-G, see Helmecke K 398, 399, 402, *408*
Majernik O, Mansfield TA 472, *498*, 518, *524*
Majernik O, see Wellburn AR 482, *501*
Majerus ME 195, *200*
Major J, see Crocker RL 11, *40*
Maki AW, Geissel LD, Johnson HE 432, *455*
Maki AW, see Cairns J Jr 443, *446*
Malaisse F, Fresson R, Goffinet G, Malaisse-Mousset M 282, 294, *302*
Malaisse-Mousset M, see Malaisse F 282, 294, *302*
Malcolm DC, see Bradbury IK 261, *273*
Maldonado JM, see Herrera J 60, *78*
Malhotra SS 482, 483, *498*
Malhotra SS, Blauel TA 471, *498*
Malhotra SS, see Moreland DE 341, 354, 355, *383*
Malo BA, Purvis ER 11, *43*
Malone PLG, Towe KM 100, *112*
Malone TC 310, 311, 314, *324*
Malone TC, Chervin MB 314, *324*
Malone TC, see Garside C 90, *111*
Maloney TE, Miller WE 443, *456*
Maloney TE, Palmer CM 443, *456*
Maloney TE, see Erickson SJ 427, *448*
Maloney TE, see Miller WE 443, *456*
Manabe S, Stouffer RJ 509, *524*
Manabe S, Wetherald RT 509, *524*
Mandava NB, see Steffens GL 362, *388*
Mandelli EF, Burkholder PR, Doheny TE, Brody R 311, *324*
Mann JD, Jordan LS, Day BE 341, 354, *382*
Mann JD, Pu M 353, *382*
Mann KH 104, *112*
Mann KH, see Brylinsky M 305, 313, 314, 319, *323*
Mannion EC, see Voigt GK 14, *45*
Manny BA, Wetzel RG, Johnson WC 415, *456*
Manny BA, see Wetzel RG 61, *84*, 422, 423, 427, *463*
Mansfield TA 471, 475, *498*
Mansfield TA, see Ashenden TW 471, 473, *491*
Mansfield TA, see Bull JN 473, 475, 490, *492*
Mansfield TA, see Capron TM 475, 477, *493*
Mansfield TA, see Law RM 475, *497*
Mansfield TA, see Majernik O 472, *498*, 518, *524*
Mantai KE, see Best D 60, *76*
Mapp HP, Eidman VR, Stone JF, Davidson JM 136, *148*
Marewa A, see Ziegler I 483, *502*
Margaris NS 207, 210, 220, *228*
Mark AF 191, *200*
Markley JL, see Williams GJ III 176, 185, *203*
Marks PL 250, *277*
Marks PL, Borman FH 250, *277*
Marks TC, see Grace J 182, *199*
Markstein B, Sukopp H 418, *456*

Markstein B, see Sukopp H 424, *461*
Marquenie van der Werff M, see Ernst WHO 443, *449*
Marra J 317, *324*
Marré E, see Dohrmann U 362, *376*
Marsh AW, see Alway FJ 11, *39*
Marsh J, Wingfield G, Davies H, Grossbard E 398, *409*
Marsh JE, see Nash TH 214, *229*
Marshall B, Biscoe PV 258, *277*
Marshall B, see Greichus YA 364, *378*
Marshall C, see Gifford RM 193, 194, 195, *199*
Marshall KC, see Alexander M 58, *75*
Marshall SM, see Butler EI 96, 100, *109*
Marsson M, see Kolkwitz R 442, *454*
Martens CS, Berner RA, Rosenfield JK 97, *112*
Martin A, Barber FR 468, 469, *498*
Martin CW 37, *43*
Martin H, Worthing CR 330, *382*
Martin JH 316, *325*
Martin PS, Wright HE 532, *548*
Martinez P 215, 216, 218, *228*
Marx J 520, *524*
Marxsen J 52, *80*
Masson K, see Oettmeier W 342, 343, 348, *384*
Mathews HM, see Pomeroy LH 100, *113*
Mathiesen H, see Riemann B 416, *459*
Matlick HA, see Prézelin BB 320, *325*
Matsuda RI, see Welch EB 315, *327*
Matsumura F, see Batterton JC 432, *444*
Matsunaka S 351, *382*
Matsunaka S, see Fryer JD 371, *378*
Matsunaka S, see Mine A 339, *383*
Matsushima J 486, *498*
Matt DR, see Hutchison BA 244, 245, 249, 250, 254, 256, *275*
Mattoo AK, Pick U, Hoffmann-Falk H, Edelmann M 342, *383*
Mattson RH, Widmer RE 517, *524*
Matulewich VA, see Finstein MS 58, *77*
Matulová D 433, *456*
Matuschek T, see Jüttner F 51, *79*
Matzku S, Broda E 429, *456*
Mauck WL, Mayer FL, Holz DD 431, *456*
Maugh TH 538, *548*
Mauldin WG, see Muzik TJ 395, *409*
Maxwell JO, Redmann RE 189, *200*
Mayer FL, Hamelink JL 443, *456*
Mayer FL, see Mauck WL 431, *456*
Mayer R, see Heinrichs H 37, 39, *42*
Mayer T, see Williams JDH 64, *84*
Mayes R, McIntosh A 429, *456*
Mayes R, see Michaud M 403, *409*
Mayo JC, see Tieszen LL 176, 177, 181, *202*
Mayo JM, Despain DG, Zinderen Bakker EM van Jr 178, 181, 182, *200*, *201*
McAllister CD, Parsons TR, Stephens K, Strickland JDH 321, *325*
McAllister CD, see Antia NJ 100, *108*, 321, *323*
McBee RH 28, *43*
McBridge JC, see Lien S 371, *382*
McCaffney RJ, Myers AC, Davey E, Morrison G, Bender M, Luedtke N, Cullen D, Froelich P, Klinkhammer G 97, *112*
McCann J, see Isensee A 403, *408*
McCarter JA, see Hayes FR 416, *451*
McCarthy JJ 93, 94, *112*
McCarthy JJ, Goldman JC 95, *112*, 317, 322, *325*
McCarthy JJ, see Carpenter EJ 314, *323*
McCarthy JJ, see Eppley RW 85, 93, 106, 107, *110*
McCarthy JJ, see Goldman JC 93, *111*, 322, *323*
McCarty PC, see Hill DW 430, 432, *452*
McCarty RE 347, *383*
McCay PB, see Lai EK 352, *381*
McColl JG 21, 38, 39, *43*
McConnall JC, McElroy MB, Wofsy SC 487, *498*
McCown BH, see Deneke FJ 545, *547*
McCracken MD, Middaugh RE, Middaugh RS 51, *80*
McCree KJ 134, 139, *148*, 240, 262, 263, 265, *277*
McCree KJ, Silsbury JM 265, *277*
McCullough C, see Bartels PG 340, 350, *374*
McCune DC, Maclean DC, Schneider RE 470, 475, *498*
McCune DC, Weinstein LH, Jacobson JS, Hitchcock AE 487, 488, *498*
McCune DC, see Jacobson JS 476, *496*
McDermitt DK, Loomis RS 161, *171*
McElroy MB, see McConnall JC 487, *498*
McEwen FL 331, *383*
McEwen FL, Stephenson GR 364, *383*
McGahee CF, Davis GJ 436, *456*
McGinnies WG 205, *228*
McGinnis JT, see Golley FB 17, 25, *41*, 283, 294, *300*
McGinniss MJ, see Buikema AL Jr 430, 432, *446*
McGregor AN, Keeney DR, Chen KL 56, *80*
McIntosh A, see Mayes R 429, *456*
McIntosh A, see Michaud M 403, *409*
McIntosh L, see Steinback KE 370, *388*
McKee JE, Wolf HW 434, *456*

McKell CM, see Garcia-Moya E 213, *227*
McKercher R, see Halstead RL 29, 30, *41*
McKinion JE, Weaver REC 137, 138, *148*
McKinion JM, see Baker DN 131, 133, 136, 137, 143, *146*
McKinion JM, see Phene CJ 132, *149*
McKnight DM, Morel FMM 53, 61, *80*
McKnight DM, see Swallow KC 53, *83*
McKnown MM, see Wolverton BC 432, *464*
McLachlan J 104, *113*
McLaren AD, Skujins J 29, *43*
McLaren AD, see Estermann EF 29, *41*
McLaughlin JJA, see Droop, MR 103, *110*
McLaughlin SB, Shriner DS 242, 268, *277*
McLay CL 438, *456*
McLean RO, Jones AK 425, 429, 443, *456*
McLeod GC, see Rice HV 103, *114*
McManus TT, see Mudd JB 484, 485, *498*
McManus TT, see Teige B 484, *500*
McMillan C 436, *456*
McNabb CD Jr 424, *456*
McNabb CD, see Coffey BT 418, *447*
McNaught DC, see Larow EJ 63, *79*
McNaughton SJ 192, 194, 196, *201*
McNaughton SJ, Campbell RS, Freyer RA, Mylroie JE, Rodland KD 187, *201*
McNeal FH, Berg MA, Watson CA 166, 167, *171*
McNeilly TS, see Bradshaw AD 536, *547*
McNulty JB, Lords JL 489, *498*
McNulty JB, Newman DW 489, *498*
McQuigg JD, see Steyeart LT 126, *149*
McRoy CP, Barsdate RJ 64, *80*
McWhorter CG 397, *409*
Mead JF 353, *383*
Mebs D 419, *456*
Medawar PB 320, *325*
Medina E 284, 286, 289, 290, 291, *302*
Medina E, Minchin P 299, *302*
Medina E, Zelwer M 294, 295, *302*
Medina E, see Herrera R 284, 290, *300*
Medina E, see Mooney H 287, *302*
Medsker LL, see Jenkins D 69, *78*
Medway Lord 294, 295, *302*
Meeks RL, see Eberhardt LL 432, 437, *448*
Meentemeyer V 35, *43*
Meier D, Lichtenthaler HK, Burkard G 352, *383*
Meier D, see Lichtenthaler HK 352, *382*
Meinck F, Stooff H, Kohlschütter H 435, 436, *456*
Meindl U, see Kiermayer O 365, *380*
Meisel WS, Collins DC 134, *148*
Melchiorri-Santolini U, Hopton JW 52, *80*
Melhado LL, Jones AM, Leonard NJ, Vanderhoef LN 363, *383*
Mellanby K, see Perring FH 364, *385*
Mellis JM, Pillai P, Davis DE, Truelove B 341, 353, *383*
Melzer A 418, 442, *456*
Melzer A, Haber W, Kohler A 418, 442, *456*
Menaut J-C 283, *302*
Mendonca BG, Hanson KJ, DeLuisi JD 510, *524*
Mendt WJ, see Steffens GL 362, *388*
Menon KAP, Bowonder B *148*
Menzel DW, Anderson J, Randtke A 365, *383*
Menzel DW, Ryther JH 103, *113*
Menzel DW, Spaeth JP 90, 103, *113*
Menzel DW, see Dugdale RC 90, *110*
Menzel DW, see Ryther JH 321, *326*
Meredith CP, Carlson PS 366, *383*
Merek EL 74, *80*
Merezhko AI 424, *456*
Merrill A, see Pimentel D 331, *385*
Merrill LB, see Reardon PO 194, *201*
Merryfield F, see Baity HG 433, *444*
Mesarovic MD, Macko D, Takahara Y 134, *148*
Messenger PS, see Huffaker CB 371, 372, *380*
Metcalf RL 437, *456*
Methley WJ, see Alway FJ 11, *39*
Methy M, see Eckardt F 208, 209, 218, *227*
Methy M, see Thorpe MR 247, *279*
Meyer GE, Curry RB, Streeter JG, Baker CH 131, *148*
Meyer J, see Laudelout H 294, *302*
Meyer JL 53, 65, *80*
Meyer RE, see Basler E 396, *406*
Michajlova EI, Kruglov JV 398, *409*
Michaud M, Atchison G, McIntosh A, Mayes R, Nelson D 403, *409*
Middaugh RE, see McCracken MD 51, *80*
Middaugh RS, see McCracken MD 51, *80*
Mihursky JA, Boynton WR, Setzler-Hamilton EM, Wood KV, Polgar TT 315, *325*
Milborrow BV 358, 361, *383*
Miles CD, Brandle JR, Daniel DJ, Chu-Der O, Schnare PD, Uhlik DJ 490, *498*
Miles GE, see Hodgkinson KC 193, *200*
Miles GE, see Holt DA 131, *147*
Miles GE, see Schreiber MM 124, 131, *149*
Miller CW, see Walsh GE 431, *463*
Miller EE, see Norman JM 246, *278*
Miller GW, Wei LL 489, *498*

Miller GW, see Cheng JYO 489, *493*
Miller GW, see Lee CJ 488, *497*
Miller GW, see Lovelace J 489, *498*
Miller GW, see Ross CW 488, *499*
Miller GW, see Yang SF 488, *502*
Miller GW, see Yu MH 488, 489, *502*
Miller HG, Cooper JM, Miller JD 251, *277*
Miller HG, Miller JD 251, *277*
Miller JD, see Miller HG 251, *277*
Miller JM, Singh J 429, *456*
Miller PC 257, *277*
Miller PC, Stoner WA 209, 211, *228*
Miller PC, Stoner WA, Ehleringer JR 187, *201*
Miller PC, Stoner WA, Richards SP 129, 139, 143, *148*
Miller PC, Stoner WA, Tieszen LL 180, 182, 183, 196, *201*
Miller PC, see Oechel WC 178, *201*
Miller PC, see Poole DK 220, *230*
Miller PC, see Tieszen LL 176, 177, 181, *202*
Miller R, see Böger P 371, *374*
Miller RB 25, *43*
Miller RJ, see Koeppe DE 490, *497*
Miller U, see Almer B 438, *444*
Miller WE, Maloney TE, Greene JC 443, *456*
Miller WE, see Maloney TE 443, *456*
Millington RJ, see Hutchinson GL 11, *42*
Milthorpe FL, see Davidson JL 194, *198*
Min HS, see Pomeroy LH 100, *113*
Minchin FR, Pate JS 520, *524*
Minchin P, see Medina E 299, *302*
Mind RV, see Johnson DW 39, *42*
Minderman G 35, *43*, 236, *277*
Mine A, Matsunaka S 339, *383*
Minenko A, see Bezuglov V 398, *407*
Miroshnichenko YM, see Rodin LE 212, *230*
Misra R 283, 294, *302*
Misra RD 438, *456*
Mitchell IW, see Linder PL 396, *409*
Mitchell JM 510, *524*
Mitscherlich G, see Künstle E 260, *275*
Mittnacht A, Eberhardt CH, Koch W 401, *409*
Miyachi S, Miyachi S, Benson AA 423, *456*
Miyachi S, see Miyachi S 423, *456*
Miyake Y, see Sugimura Y 102, *144*
Moaledj K, see Stabel HH 54, *83*
Mohr H, see Frosch S 350, *378*
Moldau H, Karolin A 139, *148*
Moll J, see Schönheit P 70, *82*
Moll R, see Hall CAS 306, 322, *324*
Monaco TJ, see Hilton JL 339, *379*
Monheimer RH 52, *80*
Monk CD, see Best GR 37, 38, *40*
Monsi M, Saeki T 284, *302*
Monteith JL 158, 160, 163, *171*, 234, 238, 243, 255, *277*
Montenegro G, Hoffmann AJ, Aljaro N, Hoffmann AE 220, *229*
Mooney H 289, *302*
Mooney H, Björkman O, Hall A, Medina E, Tomlinson PB 287, *302*
Mooney HA 194, 195, *201*, 205, 207, 208, 209, 215, 219, 222, *229*
Mooney HA, Björkman O, Collatz GJ 217, 219, 221, 222, *229*
Mooney HA, Dunn EL 209, *229*
Mooney HA, Dunn EL, Shropshire F, Song L 206, *229*
Mooney HA, Ehleringer J, Berry JA 216, 219, *229*
Mooney HA, Ehleringer J, Björkman O 215, 223, *229*
Mooney HA, Ehleringer JR 216, 219, 224, *229*
Mooney HA, Gulmon SL, Parsons DJ, Harrison AT 206, 220, *229*
Mooney HA, Harrison AT, Morrow PA 218, 220, *229*
Mooney HA, Kummerow J, Johnson AW, Parsons DJ, Keeley S, Hoffmann A, Hays RI, Giliberto J, Chu C 207, 208, 210, *229*
Mooney HA, Troughton JH, Berry JA 215, 224, *229*
Mooney HA, West M 221, *229*
Mooney HA, see Armond PA 216, 224, *226*
Mooney HA, see Björkman O 217, 219, 221, 223, *226*
Mooney HA, see Castri F di 205, *226*
Mooney HA, see Ehleringer JR 216, 217, 223, 224, *227*
Mooney HA, see Harrison A 215, *227*
Mooney HA, see Morrow PA 220, *229*
Mooney HA, see Pearcy RW 219, 223, *230*
Mooney HA, see Winner WE 472, *502*
Mooney HA, see Woodell SRJ 211, *231*
Mooney MA, see Schulze E-D 261, *279*
Moore AW, see Bevege DI 12, *40*
Moore EW 439, *456*
Moore HD, see Hodgson JM 397, *408*
Moore NW, see Way JM 430, *463*
Moore RT, see Caldwell MM 212, 213, *226*
Morcos SA, see Halim Y 90, *111*
Morel FMM, see Anderson DM 103, *108*

Morel FMM, see McKnight DM 53, 61, *80*
Morel FMM, see Swallow KC 53, *83*
Morel NML, see Swallow KC 53, *83*
Moreland DE 330, 335, 339, 340, 342, 347, 348, 349, 355, 356, *383*
Moreland DE, Blackmon WJ, Todd HG, Farmer FS 349, *383*
Moreland DE, Hill KL 431, *456*
Moreland DE, Hilton JL 339, 340, 348, *383*
Moreland DE, Malhotra SS, Gruenhagen RD, Shokraii EH 341, 354, 355, *383*
Moreland DE, see Gruenhagen RD 354, 355, *378*
Moreland DE, see Hilton JL 339, 340, 350, *379*
Morgan JJ, Stumm W 65, *80*
Morgan JJ, see Stumm W 72, *83*, 508, *525*
Morgan NC 416, *457*
Morgan PW 358, 360, 362, 363, *383*
Morgan RP, Stross RG 441, *457*
Morozova LK, see Grizhutina AP 404, *407*
Morre DJ, see Lembi CA 360, 362, *382*
Morris I 75, *80*
Morris I, Glover HE 441, *457*
Morris I, Syrett PJ 422, *457*
Morris I, see Yentsch CS 315, *327*
Morrison G, see McCaffney RJ 97, *112*
Morrod RS 340, 354, *383*
Morrow PA, Mooney HA 220, *229*
Morrow PA, see Mooney HA 218, 220, *229*
Morrow PH, see Björkman O 288, 290, 291, *300*
Mortensen DA, see Billings WD *547*
Mortimer CH 416, 429, *457*
Mortimer DC, Kudo A 428, *457*
Mortimer DC, see Eriksson C 429, *449*
Moser W 181, *201*
Moshier LJ, see Simkins GS 367, *388*
Moskal M, see Pawlaczyk-Szpilowa M 427, *458*
Moss B, see Phillips GL 418, *458*
Moss DN, Musgrave RB, Lemon ER 517, *524*
Moss DN, see Akita S 517, *521*
Mothes G, Koschel R, Scheffler W 441, *457*
Motooka PS, see Zandstra BH 402, *411*
Mount DI, see Williams LG 426, *463*
Mousseau M, see Aouni El MH 260, *272*
Mozgovoj AF, see Gruzdev GS 398, *407*
Muckenhirn RJ, see Stauffer RS 153, *172*
Mudd JB 471, 474, 475, 478, 481, 484, 486, *498*
Mudd JB, Leavitt R, Ongun A, McManus TT 484, *498*
Mudd JB, Leh F, McManus TT 485, *498*
Mudd JB, see Teige B 484, *500*
Muir J 521, *524*
Mujumdar SB, Bhide JP 519, *524*
Mukammal EI 249, *277*
Mulcahy MJ, see Stace HCT 15, *44*
Mulder EG, Veen WL van 520, *524*
Müller D, Nielsen J 263, *277*, 292, 296, 298, *302*
Müller D, Wellner B 443, *457*
Müller D, see Wunderlich M 441, *464*
Müller G 399, *409*
Müller G, Klammt E, Thiele A, Wiese U 399, 403, *409*
Müller G, see Ellis JF 367, *377*
Müller G, see Förstner U 425, 427, 435, *449*
Müller H, see Soeder CJ 420, *460*
Müller J, see Keller H 472, *497*
Müller JP 400, 401, *409*
Muller L, see Lemon E 293, *302*
Muller S, see Thomson KS 360, *362*
Müller U 70, *80*
Mulligan HF, Baranowski A 418, *457*
Mulroy TW, Rundel PW 215, 217, 223, *229*
Münch E, see Büsgen M 257, *273*
Münch WD 399, 405, *409*
Mundschenk H, Engel H, Hanisch HH, Tippner M 434, *457*
Mur LR, see Zevenboom W 60, *84*
Murata Y 127, *148*, 158, *171*
Murawski DA, see Teeri JA 215, *231*
Murca Pires J, see Cain SA 285, 291, *300*
Murphy CE, see Penning de Vries FWT 143, *149*
Murphy PG 282, *302*
Murphy PG, see Jordan CF 294, 295, *301*
Murphy PJ, see Terhaar CJ 437, *462*
Murphy TP, Lean DRS, Nalewajko C 51, 61, *80*, 102, *113*, 419, *457*
Murphy TP, see Williams JDH 64, *84*
Murray AM, see Carbon BA 250, 251, *273*
Murray DF 173, 174, *201*
Murray DR, Bradbeer JW 481, 482, 483, *498*
Muschinek Gy, Garab GyI, Mustardy LA, Faludi-Daniel A 350, *383*
Musgrave RB, see Heichel GH 520, *523*
Musgrave RB, see Moss DN, 517, *524*
Muslih RK, Linscott DL 353, *383*
Mustardy LA, see Muschinek Gy 350, *383*
Muzik TJ 395, *409*
Muzik TJ, Mauldin WG 395, *409*
Muzik TJ, see Harvey RG 397, *408*
Muzik TJ, see Roche BF 397, *410*
Muzik TJ, see West LD 370, *391*

Myers AC, see McCaffney RJ 97, *112*
Mylroie JE, see McNaughton SJ 187, *201*
Myrup LO, see Powell TM 314, *325*

Naiman RJ, Sibert RJ 89, *113*
Najjar VA 487, *498*
Nalejwajko C, Schindler DW 51, *80*
Nalewaja JD, see Schooler AB 397, *410*
Nalewajko C, Lean DRS 64, *80*
Nalewajko C, see Lee K 51, *79*
Nalewajko C, see Murphy TP 51, 61, *80*, 102, *113*, 419, *457*
Namken LN, see Waller H 361, *391*
Narahashi T, see Hilton BD 366, *379*
NAS 161, *171*
Nash TH, White SL, Marsh JE 214, *229*
National Academy of Sciences 305, *325*, 474, 475, *499*
National Research Council of Canada 468, *499*
Naumann K 364, *383*
Naveh Z 206, *229*
Naylor AW 334, *384*
Neales TF, Patterson AA, Hartney VJ 223, *229*
Nedwell DB 58, *80*
Negisi K 262, 263, *277*
Neilands JB, see Simpson FB 102, *114*
Neilson BJ, Cronin LE 305, *325*
Neilson RE, Ludlow MM, Jarvis PG 260, *277*
Neilson RE, see Beadle CL 260, 261, *272*
Neilson RE, see Watts WR 260, *280*
Nelson D, see Michaud M 403, *409*
Nelson DM, Goering JJ 101, *113*
Nelson DM, Goering JJ, Kilham SS, Guillard RRL 102, *113*
Nelson LE, see Switzer GL 18, 19, *44*, 250, *279*
Nelson TC, see Fitzgerald GP 423, *449*
Nelson WL, Dale RF 127, *148*
Nennett JH, see Heggestadt HE 475, *496*
Neudorf S, Khan MAQ 364, *384*
Neuhold JM, see Hicks CE 434, *452*
Neuland H, Schrimpff E, Herrmann R 74, *80*
Neumann HH, see Leuning R 469, 470, 471, *497*
Neumann ST, see Jacob F 396, *408*
Nevins DJ, Loomis RS 511, *524*
Newbold C 430, 431, *457*
Newbould PJ 252, *277*, 292, *302*
Newbould PJ, see Ford ED 246, 251, *274*
Newell BS, see Tranter DJ 103, *115*
Newell KD, see Phillips DA 520, *525*
Newman DW, see McNulty JB 489, *498*
Newman JF, see Way JM 430, *463*
Newroth PR, see Aiken SG 418, *444*
Newton P 517, *525*
Ng E, see Loomis RS 122, 124, 125, 131, 135, 136, 137, *148*, 169, *171*, 238, *276*
Ng TT, see Ludlow MM 194, *200*
Nicholas AK, see Wells EG 24, 25, *45*
Nicholls KH, Cox MC 10, *43*
Nichols DS, Keeney DR 60, *80*
Nicholson IA, see Fowler D 477, *495*
Nicholson TH 22, *43*
Nickel LG 358, *384*
Nickell GL, see Syvertsen JP 215, *231*
Nieboer E, Richardson DHS, Puckett KJ, Tomassini FD 471, 480, *499*
Nielsen J, see Müller D 263, *277*, 292, 296, 298, *302*
Niendorf KB, see Greenwood DJ 127, *147*
Niering WA, see Whittaker RH 212, *231*
Niewolak S, Korycka A, Potocka E 59, *81*
Nighswander JE, see Schindler DW 415, *460*
Nihlgard B 10, 25, *43*
Nisbet M, see Vivier P 436, *462*
Nishimura M, see Takahama U 352, *389*
Nishioka M, Kirita H 267, *277*
Nixon SW 305, 306, 314, *325*
Nixon SW, Oviatt CA, Kremer JN, Perez K 313, *325*
Nixon SW, see Oviatt CA 316, *325*
Nixon WS, Oviatt CA, Hale SS 97, *113*
Noakes DN, see Ben-Dyke R 400, *406*
Nobel PS 217, 218, 219, 223, 224, *229*
Nobel PS, Longstreth DL, Hartsock TL 219, *229*
Nobel PS, Wang CT 485, *499*
Nobel PS, see Hartsock TL 217, *227*
Nobel W 436, 437, *457*
Nobel W, see Labus BC 434, 443, *454*
Nolting DJ, see Stapleton HN 136, *149*
Nooden LD 362, *384*
Nooden LD, see Upadhyaya MK 341, 356, *390*
Norman JM 245, 247, 248, 256, 259, *277*, *278*
Norman JM, Jarvis PG 245, 246, 247, 249, 252, 258, 264, *278*
Norman JM, Miller EE, Tanner CB 246, *278*
Norman JM, Perry SG, Fraser AB, Mach W 247, *278*
Norman JM, Welles J 247, *278*
Normann HD 423, *457*
Norris DO 11, *43*
Norris KH, see Hilton JL 339, 340, 350, *379*
North B, see Wheeler P 52, *84*
North BB, see Wheeler PA 93, *115*

Northcote KH, see Stace HCT 15, *44*
Nott DL, see Caldwell MM 219, *226*
Nötzel, see Larcher W 181, *200*
Novak K, see Hodek J 402, *408*
Nowak KE 61, *81*
Nriagu JO 68, 72, *81*
Nriagu JO, Coker RD, Kemp ALW 68, *81*
Nriagu JO, see Burns NM 72, *76*
Nümann W 424, *457*
Nunez AF, see Williams WA 163, *172*
Nusch EA 427, 429, *457*
Nye PH 25, *43*
Nye PH, Tinker PB 20, *43*
Nyffeler A, Gerber HR, Hensley JR 367, *384*
Nyholm N 423, *457*

Oberweiss AL, see Wild A 365, *391*
O'Brien RD, Herbert TY, Hilton BD 366, *384*
O'Brien RD, see Hilton BD 366, *379*
O'Connor DJ, see Thomann RV 437, *462*
O'Connors HB Jr, see Iverson RL 314, 319, *324*
Ode D, Tieszen LL, Lerman JC 176, 186, *201*
Odell RT, see Stauffer RS 153, *172*
Odening W, Strain BR, Oechel WC 218, 223, 224, *229*
Odening WR, see Oechel WC 217, *230*
Odum E 521, *525*
Odum EP 6, *43*
Odum EP, see Weigert RG 28, *45*
Odum HT 8, 9, 22, *43*, 283, 285, 286, 287, 293, 294, 297, 298, *303*, 417, *457*
Odum HT, Lugo A, Cintrón G, Jordan CF 287, 293, *303*
Odum HT, see Beyers RJ 322, *323*
Odum TH 537, *548*
Oechel W 211, 216, 218, 219, 221, *230*
Oechel WC, Collins NJ 179, 181, 182, *201*
Oechel WC, Hicklenton P, Sveinbjörnsson, Miller PC, Stoner WL 178, *201*
Oechel WC, Strain BR, Odening WR 217, *230*
Oechel WC, see Odening W 218, 223, 224, *229*
Oechel WC, see Tieszen LL 176, 177, 181, *202*
Oertli JJ, see Flückiger W 476, *494*
Oertli JJ, see Kappen L 214, *228*
Oeschger H, Siegenthaler U, Schotterer U, Gugelmann A 508, 509, *525*
Oettmeier W 342, 348, *384*
Oettmeier W, Johanningmeier U, Trebst A 343, 345, *384*
Oettmeier W, Masson K 342, 343, 348, *384*
Oettmeier W, Masson K, Johanningmeier U 342, *384*
Oettmeier W, Reimer S, Link K 339, 342, 345, *384*
Ogawa H 292, 294, 295, *303*
Ogawa H, Yoda K, Ogino K, Kira T 283, *303*
Ogawa H, see Kira T 292, 298, *301*
Ogawa H, see Yoda K 264, *280*
Ogawa RE, Carr JF 56, *81*
Ogino K, see Kira T 292, 298, *301*
Ogino K, see Ogawa H 283, *303*
Ohad I, see Regitz G 342, *386*
Ohkawa H, see Casida JE 334, 367, *375*
Ohle W 48, 54, 65, 67, *81*, 415, 416, 438, *457*
Okanishi Y, see Hogetsu K 418, *452*
Olajos EJ 486, *499*
O'Leary JW, see Knecht GN, 517, *523*
Oliveira Castro GM de, see Cain SA 285, 291, *300*
Olofimboba MU 296, *303*
Olsen RA 11, *43*
Olsen S 65, *81*
Olsen S, see Iversen J 438, *453*
Olson JS 34, *43*
Olson JS, Pfuderer HA, Chan Y-H 505, 507, *525*
Olson JS, see Baes CF 506, *521*
Omann G, see Lakowicz JR 366, *381*
Ongun A, see Mudd JB 484, *498*
Oorshot JLP van 334, 346, 349, *384*
Ophel IL, Fraser CD 434, *457*
Opler PA, Frankie GW, Baker HG 295, 296, *303*
Ordin L, Hall MA 485, *499*
Ordin L, Hall MA, Katz M 485, *499*
Ordin L, Skoe BP 488, *499*
Ordin L, see Hall MA 485, *495*
Orebamjo TO, Stewart GR 60, *81*
O'Reilly JE, Thomas JP, Evans C 311, *325*
Ören TI 125, *148*
Orians G 537, *548*
Orman HR, Will GM 20, *43*
Ormrod DP 476, 477, *499*
Ormrod DP, Black VJ, Unsworth MH 477, *499*
Ormrod DP, see Adedipe NO 474, *491*
Ormrod DP, see Elkiey T 485, *494*
Ormrod DP, see Unsworth MH 467, *501*
Ornes WH, see Sutton DL 424, *461*
Orr WK, see Emery KO 88, *110*
Orshan G 220, *230*
Orshan G, Diskin S 212, *230*
Orshan G, Zand G 223, *230*
Osborne CG, see Boynton WR 306, *323*
Osborne DJ 361, *384*
Osmanova LT, see Kurochkina LY 213, *228*
Osmond CB, Avadhani PN 483, *499*
Osmond CB, see Caldwell MM 219, *226*
Osmond CB, see Lüttge U 480, *498*
Osswald W, see Elstner EF 340, 346, 350, 351, 353, *377*

O'Sullivan AJ, Collinson RI 439, *457*
Otsuki A, see Wetzel RG 49, *84*
Overbeck J 52, *81*
Overbeck J, Daley RJ 52, *81*
Overbeck J, see Reichardt W 63, *81*
Overbeck J, see Stabel HH 54, *83*
Overbeck J, see Witzel KP 58, *84*
Overbeck J van 395, *409*
Overnell J 428, *457*
Overton WS, see Sanantonio D 21, *44*, 269, *279*
Oviatt CA, Nixon SW, Perez KT, Bucklay B 316, *325*
Oviatt CA, see Nixon SW 313, *325*
Oviatt CA, see Nixon WS 97, *113*
Ovington JD 235, 236, 250, *278*
Ovington JD, Heitkamp D, Lawrence DB 26, *43*
Ovink R, see Barnes JR 53, *76*
Owen WJ, Delaney ME, Rogers LJ 365, *384*
Owen WJ, Rogers LJ, Hayes JD 365, *384*
Owen WJ, see Delaney ME 365, *376*
Owen WJ, see Rogers LJ 365, *386*
Ownbey CR, Kee DA 435, *457*
Ownby J, see Hawxby K 431, *451*

Paasche E 94, 101, 103, *113*
Paasivirta J, see Särkkä J 429, *459*
Paatero G, see Skult H 365, *388*
Pace DM, see Thompson CR 481, *500*
Pack DH, see Cox RA 469, *493*
Pack MR, see Hill AC 489, *496*
Packard TT 87, 92, *113*
Packard TT, see Ahmed SI 92, *108*
Packard TT, see Whitledge TE 102, *115*
Paerl HW 56, *81*
Paerl HW, Lean DRS 64, *81*
Paeschke R, Heitefuss R 398, *410*
Paez A, Hellmers H, Strain B 518, *525*
Pahlich E 481, *499*
Pahlich E, Jäger HJ, Steubing L 484, *499*
Painter EL, Detling JK 193, *201*
Painter HA 57, 58, *81*
Pallas JE Jr 517, *525*
Pallet KE, Dodge AD 350, 351, 353, *384*
Pallos FM, Casida JE 367, *384*
Palm HL, Riggleman JD, Allison DA 354, *384*
Palmer CM, Walter HJ, Adams S 419, 420, 421, 434, 435, 436, 438, *458*
Palmer CM, see Maloney TE 443, *456*
Palmer RL, see Thomson WW 485, *500*, *501*
Pamatmat MM, see Goering JJ 97, *111*
Paneque A, see Herrera J 60, *78*
Papageorgiou G, see Govindjee 338, *378*
Papavizas G, see Levis J 398, *409*
Paradies I, Ebert E, Elstner EF 367, *384*
Parka SJ, Soper OF 341, 356, *384*
Parke MH, see Blackman GE 431, *445*
Parker BC, see Cairns J Jr 414, 424, 435, 438, 442, *446*
Parker C 361, *385*, 432, *458*
Parker C, see Blair AM 367, *374*
Parker J 264, *278*
Parker JI, Conway HL, Yaguchi EM 69, *81*
Parker PL, see Thayer GW 306, *326*
Parker RR, see Kirkman H 64, *79*
Parkinson KJ, see Legg BJ 238, 239, *276*
Parry WH 241, *278*
Parsons DJ, see Mooney HA 206, 207, 208, 210, 220, *229*
Parsons TR 101, 108, *113*
Parsons TR, Brockel K von, Koeller P, Takahashi M, Reeve MR, Holm-Hansen O 96, *113*
Parsons TR, Giovando LF, LeBrasseur RJ 313, *325*
Parsons TR, Strickland JDH 52, *81*
Parsons TR, Takahashi M 108, *113*
Parsons TR, see Antia NJ 100, *108*, 321, *323*
Parsons TR, see McAllister CD 321, *325*
Parton WJ, see Detling JK 178, 179, *199*
Parton WJ, see Sala OE 189, *202*
Pasciak WJ, Gavis J 94, *113*
Passarge H 405, *410*
Passioura JB 135, 141, *148*
Pastac I, see Truffaut G 331, *390*
Pate JS, see Minchin FR 520, *524*
Paterson IS, see Fowler D 477, *495*
Patouillet, see Hannan PJ 427, *451*
Patric JH, see Helvey JD 23, *42*
Patrick R 74, *81*, 419, 440, 441, 442, 443, *458*
Patrick R, see Harvey RS 434, 435, *451*
Patten DT 213, *230*
Patten DT, see Halvorson WL 189, *200*
Patterson AA, see Neales TF 223, *229*
Patterson DT 24, *43*
Paulet E, see Kestemont P 236, *275*
Pawlaczyk-Szpilowa M, Moskal M, Weretelnik J 427, *458*
Payer HD, Runkel KH, Schramel P, Stengel E, Bhumiratana A, Soeder CJ 429, *458*
Payer HD, see Soeder CJ 420, *460*
Payne G, see White E 62, *84*
Payne WJ 58, *81*
Payne WJ, see Barbaree JM 96, *108*
Pearcy RW 221, *230*

Pearcy RW, Berry JA, Fork DC 219, *230*
Pearcy RW, Harrison AT 217, 219, *230*
Pearcy RW, Harrison AT, Mooney HA, Björkman O 219, 223, *230*
Pearcy RW, Troughton J 291, *303*
Pearcy RW, see Björkman O 217, 219, *226*
Pearson HW, Tomlinson DL 419, *458*
Pearson HW, see Stewart WDP 423, *461*
Pearsons JE, see Phene CJ 132, *149*
Peart RM, see Barrett JR 135, *146*
Peart RM, see Holt DA 131, *147*
Peavey DG, see Goldman JC 93, *111*, 322, *323*
Peek JW, see Ellis JF 367, *377*
Peek N, see Gersberg RM 61, *77*
Pemble M, see Stewart WDP 63, *83*
Pendleton JW, see Egli DB 517, *522*
Peng TH, see Broecker WS 506, *522*
Peng TH, see Schindler DW 49, *82*
Penhale PA, Smith WO Jr 51, *81*
Penhale PA, see Wetzel RG 49, *84*
Penkett SA, see Cox RA 469, *493*
Pennefather M, see Schönau APG 236, *279*
Penner D, Ashton FM 341, 357, *385*
Penner D, Early RW 354, 355, *385*
Penner D, see Hatilos KK 367, *379*
Penner D, see Leavitt JRC 367, *382*
Penner D, see Putman AR 364, *385*
Penner D, see Rivera CM 340, 353, 354, *386*
Penning de Vries FWT 128, 136, 139, *148*, *149*, 240, 263, 265, *278*
Penning de Vries FWT, Brunsting AB, Laar HH van 138, *149*
Penning de Vries FWT, Brunsting AHM, Laar HH van 160, 161, *171*, 240, 265, *278*
Penning de Vries FWT, Krul J, Keulen H van 143, *149*
Penning de Vries FWT, Laar HH van 125, 139, *149*
Penning de Vries FWT, Murphy CE, Wells CG, Jorgensen JR 143, *149*
Penning de Vries FWT, Witlage JM, Kremer DJ 137, 138, *149*, 240, *278*
Penning de Vries FWT, see Wit CT de 119, 133, 137, 144, *150*
Penny D, see Penny P 363, *385*
Penny P, Penny D 363, *385*
Pensel T, see Kohler A 421, *454*
Perchorowicz JT, Ting IP 485, *499*
Perez K, see Nixon SW 313, *325*
Perez KT, see Oviatt CA 316, *325*
Perras JP, see Kuenzler EJ 63, *79*, 99, *112*, 423, *454*
Perring FH, Mellanby K 364, *385*
Perry MJ 93, 94, 98, 99, *113*
Perry RS, see Gregory E 73, 74, *78*
Perry SG, see Norman JM 247, *278*
Persson H 241, 269, *278*
Persson H, see Ågren GI 264, 270, *272*
Persson H, see Flower-Ellis JGK 253, *273*
Pesci P, see Dohrmann U 362, *376*
Petering DH, see Shih NT 483, *499*
Peterle TJ, see Eberhardt LL 432, 437, *448*
Peters RH 62, 63, *81*
Peters DB, see Hutchinson GL 11, *42*
Peters L, see Bombosch S 437, *445*
Peters RH 98, *113*, 416, *458*
Peters RH, Lean DRS 63, *81*
Petersen JB 435, *458*
Peterson BJ 306, 321, 322, *325*
Peterson BJ, see Eppley RW 96, *110*, 314, 319, *323*
Peterson KM, see Billings WD 543, *547*
Peterson R 105, *113*
Petr T, see Blackburn WM 53, *76*
Peukert V 415, *458*
Pfister K, Arntzen CJ 338, 344, 345, 368, 370, *385*
Pfister K, Buschmann C, Lichtenthaler HK 339, 343, 344, *385*
Pfister K, Radosevich SR, Arntzen CJ 343, 345, 370, *385*
Pfister K, Schreiber U 338, 342, *385*
Pfister K, Steinback KE, Arntzen CJ 342, 343, *385*
Pfister K, Steinback KE, Gardner G, Arntzen CJ 342, *385*
Pfister K, Watson J, Arntzen CJ 343, 344, 368, 370, *385*
Pfister K, see Arntzen CJ 370, *373*
Pfister K, see Laasch H 339, 343, 344, 348, *381*
Pfister K, see Steinback KE 342, *388*
Pfleiderer G, Hohnholz E 483, *499*
Pfuderer HA, see Olson JS 505, 507, *525*
Phan CT, see Cailloux M 485, *493*
Pharis RP, see Taylor JS 472, *500*
Phelps JE, see Dougherty PM 260, 264, 265, *273*
Phene CJ, Baker DN, Lambert JR, Pearsons JE, McKinion JM 132, *149*
Phene CJ, see Baker DN 131, 133, 136, 137, 143, *146*
Phene CJ, see Lambert JR 131, 139, 141, 142, 145, *148*
Phillips BA, Bhagsari AS 367, 368, *385*
Phillips DA, Newell KD, Hassell SA, Felling CE 520, *525*
Phillips GL, Eminson D, Moss B 418, *458*

Phillips JH, see Harding LW 364, 365, *379*
Pick U, Avron M 483, *499*
Pick U, see Mattoo AK 342, *383*
Pickering Q, see Williams LG 434, *464*
Pierce M, see Toetz D 433, *462*
Pierce RS, see Likens GE 545, *548*
Pierrou U 98, *113*
Pietsch W 421, 438, *458*
Pillai P, see Mellis JM 341, 353, *383*
Pillsbury RD, see Walsh JJ 89, *115*, 314, *327*
Pils J, see Elstner EF 350, 351, 353, *377*
Pimentel D 331, 332, 371, *385*, 393, *410*
Pimentel D, Krummel J, Gallahan D, Hough J, Merrill A, Schreiner I, Vittum P, Koziol F, Back E, Yen D, Fiance S 331, *385*
Pinckard JA, see Leonard OA, 518, *524*
Pinckney HR, see Unsworth MH 472, *501*
Pinter IJ, see Droop, MR 103, *110*
Pintner IJ, see Provasoli L 422, *458*
Pipes WO, see Beer LO 440, *444*
Pisek A, Larcher W, Unterholzner R 181, *201*
Pistorius E, see Büchel KH 348, *375*
Pistorius E, see Draber W 339, 342, *376*
Pistorius E, see Trebst A 339, *390*
Pitcher F, see Isensee A 403, *408*
Pittendrigh CS 292, *303*
Plapp FW, see Cole DR 365, *375*
Platt T 314, *325*
Plesnicar M, Bonner WD, Storey BT 487, *499*
Plötz J 443, *458*
Polgar TT, see Mihursky JA 315, *325*
Polya GM, Davis AW 363, *385*
Pomeroy LH, Mathews HM, Min HS 100, *113*
Pomeroy LR, Johannes RE 322, *325*
Poole DK, Miller PC 220, *230*
Por FD 530, *548*
Porath D, Ben-Shaul Y 442, *458*
Porcella DB, see Tunzi MG 321, *327*
Poschenrieder G, see Wallnöfer P 396, *411*
Postdeva R, see Bezuglov V 398, *407*
Postlethwaite JD, see Hand DW 518, *522*
Postma H 100, *113*
Potocka E, see Niewolak S 59, *81*
Pötsch J 402, *410*
Pott RM, see Greichus YA 437, *450*
Potter JR, Jones JW 195, *201*
Potts GR 400, 402, *410*
Poulton CE, see Hedrick DW 405, *408*
Poupon H, see Bille J-C 282, *300*
Powell CE, see Ryle GJA 194, 195, *202*
Powell TM, Richerson PJ, Dillon TM, Agee BA, Dozier BJ, Godden DA, Myrup LO 314, *325*
Powers CD, Rowland RG, Wurster CF 365, *385*
Pramer D, see Sikka HC 431, *460*
Prasad SK, see Rao BHK 399, *410*
Prasse J 399, *410*
Prasse J, see Helmecke K 398, 399, 402, *408*
Pratt HM, Fox JE 363, *385*
Pratt R, Fong J 422, *458*
Preest D 405, *410*
Prentki RT, Adams MS 64, *81*
Preusse HU, see Brod HG 435, *445*
Prézelin BB, Matlick HA 320, *325*
Price NB, see Calvert SE 95, *109*
Pritchard DW, see Carpenter JH 90, *109*
Pritchard MK, Warren GF, Dilley RA 352, *385*
Prober R, see Bond RG 486, *492*
Procter-Gregg C, see Detling JK 194, *199*
Provasoli L, Carlucci AF 103, *113*
Provasoli L, Pintner IJ 422, *458*
Provasoli L, see Droop, MR 103, *110*
Prüfer P, see Auerbach S 443, *444*
Pu M, see Mann JD 353, *382*
Puckett KJ, see Nieboer E 471, 480, *499*
Puiseux-Dao S, Jeanne-Levain N, Roux F, Ribier J, Borghi H, Brun C 365, *385*
Puiseux-Dao S, see Borghi H 365, *374*
Purvis ER, see Malo BA 11, *43*
Putman AR, Penner D 364, *385*
Putman HD, see Brooks RH 56, *76*, 90, *109*
Putwain PD, see Scott KR 370, *387*

Quebedeaux B, Havelka UD, Livak KL, Hardy R 520, *525*

Rabbinge R, see Loomis RS 122, 125, 131, 135, 136, 137, *148*, 169, *171*, 238, *276*
Rabe FW, see Bartlett L 430, *444*
Rabinowich EI 321, *325*
Rachlin JW, Farran M 427, *458*
Rachlin JW, see Rosko JJ 427, *459*
Rademacher B, see Hurle K 398, *408*
Radosevich SR 370, *385*
Radosevich SR, Devilliers OT 370, *386*
Radosevich SR, Holt JS 369, 370, *385*
Radosevich SR, Steinback KE, Arntzen CJ 370, *386*
Radosevich SR, see Conrad SG 369, *376*
Radosevich SR, see Pfister K 343, 345, 370, *385*

Radosevich SR, see Sims Holt JD 369, *388*
Rafes PM 268, *278*
Raghi-Atri F 429, *458*
Raghi-Atri F, Bornkamm R 424, *458*
Raghi-Atri F, see Bornkamm R 424, *445*
Rakestraw MW, Brand T von 95, *113*
Ralston CW, see Kinerson RS 264, 271, *275*
Ramsay AJ, Fry JC 431, *458*
Rana BC, Kumar HD 429, *458*
Ranawat MPS, Vyas LN 294, *303*
Randazzo G, see Dohrmann U 362, *376*
Randtke A, see Menzel DW 365, *383*
Raney HG, see Hayes RM 364, *379*
Rao BHK, Prasad SK 399, *410*
Rao DN, see Blanc F le 483, *492*
Rapley L, see Heldt HW 481, *496*
Rapp M, see Cole DW 18, *40*
Rapp M, see Lossaint P 210, *228*
Rappaport L 361, *386*
Raschke K 140, *149*, 517, *525*
Rasmussen L, see Sand-Jensen K 438, *459*
Rasmussen RA, see Lovelock JE 69, *80*
Rastorfer JR 182, *201*
Ratzmann W, see Kopecek K 365, *381*
Rau I, Grimme LH 431, *458*
Rauner JL 237, 249, 250, 252, 262, 266, 271, *278*
Raven JA 436, *459*, 514, *525*
Raven JA, Smith FA 481, *499*
Ray S, White W 429, *459*
Ray TB 341, 354, *386*
Rayburn WR, see Bechard MJ 69, *76*
Rayson P, see Specht R 208, *231*
Reardon PO, Leinweber CL, Merrill LB 194, *201*
Reardon PO, Merrill LB 194, *201*
Rebechini HM, Hanzely L 427, *459*
Reddy MM 49, *81*
Redfield AC 93, *113*
Redfield AC, Ketchum BH, Richards FA 86, 95, 97, *114*
Redfield AC, see Ketchum BH 423, *453*
Redmann RE 186, 187, 188, 189, *201*
Redmann RE, see Maxwell JO 189, *200*
Reed KL, Hamerly ER, Dinger BE, Jarvis PG 258, 259, 260, 261, *278*
Rees AR, Thornley JHM 128, *149*
Reeve MR, see Parsons TR 96, *113*
Reeves SG, Hall DO 337, *386*
Regehr DL, Bazzaz FA, Boggess WR 517, *525*
Regitz G, Ohad I 342, *386*
Reichardt W 423, *459*
Reichardt W, Overbeck J, Steubing L 63, *81*
Reichle DE, see Edwards CA 28, *41*
Reid CPP, Hurtt W 396, *410*
Reid DM, see Taylor JS 472, *500*
Reid GW, see Sweazy RM 423, *461*
Reimer S, see Böhme H 339, 345, *374*
Reimer S, see Oettmeier W 339, 342, 345, *384*
Reimer S, see Trebst A 345, *390*
Reiners WA 25, *43*
Reiners WA, see Gorham E 13, *41*
Reiners WA, see Vitousek PM 6, *44*
Reiners WA, see Woodwell GM 506, *526*, 537, 539, *548*
Reinert JC, Tinsley IJ 366, *386*
Reinert RA, Heagle AS, Heck WW 476, 477, *499*
Reinert RA, see Taylor OC 475, *500*
Reinert RA, see Tingey DT 475, *501*
Reinhardt H 127, *149*
Reisberg E, see Urbach W 350, 351, 353, *390*
Relton J, see Farrar JF 473, *494*
Renger EH, see Eppley RW 89, *110*
Renger EH, see Harrison WG 98, *111*
Renger G 342, *386*
Renger G, Hagemann R, Dohnt G 342, *386*
Rensen JJS van 339, 340, 351, 353, *386*
Rensen JJS van, Kramer HJM 342, *386*
Rensen JJS van, Vet E van der, Vliet WPA van 348, 349, *386*
Rensen JJS van, Wong D, Govindjee 348, 349, *386*
Reshkin SJ, Knauer GA 98, *114*
Reuss JO, Innis GS 129, 142, *149*
Rhee GY 64, *82*, 423, *459*
Rheinheimer G 442, *459*
Ribier J, see Puiseux-Dao S 365, *385*
Ricard M, see Besch WK 430, *444*
Rice EL, see Wolfe JM 51, *84*
Rice HV, Leighty DA, McLeod GC 103, *114*
Rice TR 419, *459*
Rich PH, Wetzel RG 52, *82*
Rich PH, see Wetzel RG 49, *84*
Rich S, Turner NC 474, *499*
Richards BN 9, 22, 34, 35, *43*
Richards BN, Charley JL 32, *44*
Richards BN, see Bevege DI 12, *40*
Richards BN, see Hurditch WJ 33, *42*
Richards BN, see Johnson DW *42*
Richards BN, see Voigt GK 14, *45*
Richards FA, see Grill EV 102, *111*
Richards FA, see Redfield AC 86, 95, 97, *114*
Richards PW 294, *303*
Richards SP, see Miller PC 129, 139, 143, *148*
Richardson DHS, see Nieboer E 471, 480, *499*

Richerson P, Armstrong R, Goldman CR 105, *114*
Richerson PJ, see Powell TM 314, *325*
Richter A, see Hällgren J-E 474, *495*
Richter H, see Hinckley TM 260, *274*
Rick JR 135, *149*
Rickard W, see Deneke FJ 545, *547*
Ricks GR, Williams RJH 476, *499*
Ridley J, see Ridley SM 350, *386*
Ridley SM 349, 350, *386*
Ridley SM, Ridley J 350, *386*
Rieck CE, see Bugg MW 339, 345, 352, *375*
Rieck CE, see Kitchen LM 352, *380*
Riek CE, see Carringer RD 367, *375*
Riemann B, Mathiesen H 416, *459*
Ries SK 355, 357, *386*
Ries SK, Chmiel H, Dilley DR, Filner P 355, *386*
Rieske JS, see Spikes JD 476, *500*
Riggleman JD, see Palm HL 354, *384*
Rigler FH 63, *82*, 416, *459*
Riley GA 305, 306, 309, 311, 312, 313, 318, 319, 321, 322, *325*, *326*
Riley JP, Chester R 86, 100, *114*
Riley JP, Skirrow G 86, *114*
Rind D, see Hansen J 510, *523*
Ripl W 424, 439, *459*
Risser PG, see Johnson FL 17, *42*
Rittenberg SC, see Emery KO 88, *110*
Rittig FR, see St John JB 353, *388*
Rittig FR, see Waller H 361, *391*
Rivera CM, Penner D 340, 353, 354, *386*
Rivkin RB, see Falkowski PG 92, *110*
Roan CC, see Ware GW 364, *391*, 430, 432, *463*
Robb F, Davies BR, Cross R, Kenyon C, Howard-Williams C 53, *82*
Roberts FW, see Hynes HBN 434, *453*
Roberts JE, see Hinckley TM 268, 269, *274*
Roberts TM, see Horsman DC 472, 473, *496*
Robertson JM, see Sweazy RM 423, *461*
Robinson E, see Hobbs PV 510, *523*
Robinson JB, see Chatarpaul L 58, *76*
Robinson RA 158, *171*
Robinson SJ, Yocum CF, Ikuma H, Hayashi F 339, 345, *386*
Robock A 509, *525*
Robra KH 443, *459*
Robson MJ 179, *201*
Roche BF, Muzik TJ 397, *410*
Rochecouste E 397, *410*
Rodebush JE, see Casida JE 334, 367, *375*
Rodhe H, see Granat L 468, *495*
Rodhe W 420, 423, *459*
Rodin LE, Basilevich NI 16, 17, 24, 25, *44*
Rodin LE, Bazilevich NI 211, 212, *230*
Rodin LE, Bazilevich NI, Miroshnichenko YM 212, *230*
Rodland KD, see McNaughton SJ 187, *201*
Rodrigues WA, see Klinge H 294, *301*
Rodriguez RR, see Waller H 361, *391*
Roels OA, see Garside C 90, *111*
Roels OA, see Lui NST 92, *112*
Rogers JN, see Eppley RW 85, 93, 106, 107, *110*
Rogers LJ, Owen WJ, Delaney ME 365, *386*
Rogers LJ, see Delaney ME 365, *376*
Rogers LJ, see Owen WJ 365, *384*
Rogers RW, see Westman WE 17, *45*
Rola J 396, 401, *410*
Rolfe GL, Akhtar MA, Arnold LE 17, *44*
Rolfe GL, see Bazzaz FA 544, *547*
Rolley HLJ, see Edwards RW 58, *77*
Roltsch WJ, see Capinera JL 194, *198*
Romanenko VI 52, *82*
Rook DA, Corson MJ 262, 264, 265, *278*
Rook DA, see Bennett KJ 260, *273*
Rook DA, see Linder S 269, *276*
Rook JJ 433, *459*
Ropers HJ, see Graebe JE 361, *378*
Rose MR, Harmsen R 125, 134, *149*
Rosen CG, see Wood JM 428, *464*
Rosenfield JK, see Martens CS 97, *112*
Rosko JJ, Rachlin JW 427, *459*
Ross CW, Wiebe HH, Miller GW 488, *499*
Ross FF 439, *459*
Ross J 164, *172*
Rosswall T, Heal OW 173, *201*
Rost TL 355, *386*
Rost TL, Bayer DE 355, *386*
Roth JC, Horne AJ 59, *82*
Rotty RM, see Baes CF 506, *521*
Roux F, see Puiseux-Dao S 365, *385*
Rovira AD 27, *44*
Rowan KS, see Brown V 104, *109*
Rowe GT, Clifford CH, Smith KL Jr 98, *114*
Rowe GT, Clifford CH, Smith KL Jr, Hamilton PL 306, 314, *326*
Rowland RG, see Biggs DC 365, *374*
Rowland RG, see Powers CD 365, *385*
Ruber E, see Derby SB 365, *376*
Rubery PH 362, *387*
Rudd JWM, Furutani A, Flett RJ, Hamilton RD 54, *82*
Rudd JWM, Hamilton RD 54, *82*

Rüdiger W, Benz J 352, *387*
Rüdiger W, Benz J, Lempert U, Schoch S, Steffens D 352, *387*
Ruehle W, see Wild A 365, *391*
Ruf M 434, 435, *459*
Rukhadze YG, see Khobot'yev VG 427, *453*
Rundel PW, see Mulroy TW 215, 217, 223, *229*
Runeckles VC, see Srivastava HS 475, *500*
Runkel KH, see Payer HD 429, *458*
Running SW, see Grier CC 251, *274*
Russ OG, see Simkins GS 367, *388*
Russell G, see Hansen J 510, *523*
Russo JM, Knapp WW 129, *149*
Russo RC, Smith CF, Thurston RU 57, *82*
Russo, VEA, see Hertel R 360, 362, *379*
Rustanov IG 213, *230*
Rutter AJ, see Farrar JF 473, *494*
Rutter AJ, see Garsed SG 473, *495*
Ruttner F 74, *82*, 416, *459*
Ryan GF 368, 369, *387*
Ryberg H, see Klockare B 352, *381*
Ryding SO, see Claesson A 415, *447*
Ryding SO, see Forsberg C 443, *449*
Ryle GJA, Powell CE 194, 195, *202*
Ryrie IJ, see Datta DB 483, *494*
Ryther JH 309, 318, *326*
Ryther JH, Menzel DW 321, *326*
Ryther JH, Vaccaro RF 321, *326*
Ryther JH, Yentsch CS 311, *326*
Ryther JH, see Dugdale RC 90, *110*
Ryther JH, see Menzel DW 103, *113*
Ryther JH, see Vaccaro RF 95, *115*
Saeki T 285, 293, *303*
Saeki T, see Monsi M 284, *302*
Safe ASW, see Hutzinger O 364, *380*
Sagan C 539, *548*
Saito H 267, *278*
Saito H, see Fujimori J 235, *274*
Sakamoto M 415, *459*
Sala OE, Lauenroth WK, Parton WJ, Trlica MJ 189, *202*
Saleh MA, see Yang SF 480, *502*
Salisbury FB, Spommer GG 181, *202*
Salmon MR, see Wilson PM 520, *526*
Saltman P, see Kunitake G 519, *524*
Salunkhe DD, see Singh B 355, *388*
Sampson AW 208, *230*
Sampson DR 536, *548*
San José JJ 287, 289, *303*
San Pietro A, see Lien S 371, *382*
Sanantonio D 268, 269, *278*
Sanantonio D, Hermann RK, Overton WS 21, *44*, 269, *279*
Sandelius AS, see Klockare B 352, *381*
Sanders HJ 330, *387*
Sanderson DM, see Ben-Dyke R 400, *406*
Sanderson WW 415, *459*
Sand-Jensen K, Rasmussen L 438, *459*
Sand-Jensen K, see Søndergaard M 49, *82*
Sandmann ER 398, *410*
Sandmann G, Böger P 340, 349, 352, 353, *387*
Sandmann G, Bramley PM, Böger P 350, *387*
Sandmann G, Kunert KJ, Böger P 350, *387*
Sandmann G, Lambert R, Böger P 351, *387*
Sandmann G, see Böger P 371, *374*
Sands R, Correll RL 195, *202*
Sandusky JC, see Horne AJ 56, *78*
Santarius KA, see Heber U 441, *451*
Sargent JA 334, *387*
Sargent JA, Blackman GE 394, *410*
Särkkä J, Hattula ML, Janatuinen J, Paasivirta J 429, *459*
Satchell JE 28, 31, *44*
Satoh K, Katoh S, Takamiya A 488, *499*
Satoo T 235, 237, 251, 252, *279*
Sauer RH 129, *149*
Saugier B, see Thorpe MR 247, *279*
Saunders GW 52, *82*
Sauvezon R, see Eckardt F 208, 209, 218, *227*
Say PJ, Diaz BM, Whitton BA 429, *459*
Say PJ, Whitton BA 429, *459*
Sayler G, see Shiaris M 403, *410*
Schaedle M, see Foote KC 264, *273*
Schaffer WM, Gadgil MD 206, *230*
Scharen AL, see Hilton JL 339, 340, 350, *379*
Schauberger CW, Wildman RB 364, 365, *387*
Schefczik K, Simonis W 366, *387*
Schefczik KW 366, *387*
Scheffer F, see Seidel K 432, *460*
Scheffler W, see Mothes G 441, *457*
Scheinpflug H, Schlör H, Widdig A 330, *387*
Schelske CL 415, *459*
Schiff JA, see Vaisberg AJ 350, *390*
Schindler DW 305, 313, 319, *326*, 415, 419, *459*
Schindler DW, Brunskill GJ, Emerson S, Broecker WS, Peng TH 49, *82*
Schindler DW, Lean DRS, Fee EJ 48, 49, *82*, 416, *460*
Schindler DW, Nighswander JE 415, *460*
Schindler DW, see Jackson TA 63, *78*
Schindler DW, see Nalejwajko C 51, *80*
Schindler JE, Williams DJ, Zimmerman AP 54, *82*

Schlesinger WH 507, *525*
Schlesinger WH, Gill DS 208, 210, *230*
Schlichting HE Jr, Gearheart RA 419, *460*
Schlimme E, see Seidel K 432, *460*
Schlör H, see Scheinpflug H 330, *387*
Schlue U, see Böger P 352, *374*
Schmassmann H 414, 417, *460*
Schmidt L, see Larcher W 181, *200*
Schmidt RL 103, *114*
Schmidt RR 396, *410*
Schmidt WD 73, 74, *82*
Schmitt MR, see Smith CS 65, *82*
Schmitz W 436, 437, *460*
Schnare PD, see Miles CD 490, *498*
Schneider EA, Wightman F 341, 360, *387*
Schneider G 360, *387*
Schneider RE, see McCune DC 470, 475, *498*
Schneider RW, see Ioannow N 519, *523*
Schneider SH 509, *525*
Schnitzer M, see Gamble DS 53, *77*
Schoch S, see Rüdiger W 352, *387*
Schoepe E, see Ziegler I 483, *502*
Schofield CS 477, *499*
Schönau APG, Pennefather M 236, *279*
Schönheit P, Moll J, Thauer RK 70, *82*
Schooler AB, Bell AR, Nalewaja JD 397, *410*
Schopf M, see Boehler-Kohler BA 340, 353, *374*
Schott PE, see Knittel H 361, *381*
Schotterer U, see Oeschger H 508, 509, *525*
Schramel P, see Payer HD 429, *458*
Schraufnagel FH 435, *460*
Schreiber GE, see Holt DA 131, *147*
Schreiber K 362, *387*
Schreiber MM, Miles GE, Holt DA, Bula RJ 124, 131, *149*
Schreiber MM, see Beste CE 355, *374*
Schreiber U, see Pfister K 338, 342, *385*
Schreiner I, see Pimentel D 331, *385*
Schrimpf A, see Steinberg C 63, *83*
Schrimpff E, see Neuland H 74, *80*
Schroeter LC 478, *499*
Schubert B, see Feierabend J 352, *377*
Schubert R 393, 397, 402, 404, *410*
Schubert R, Hilbig W 401, *410*
Schuler JF, Diller VM, Kersten HJ 422, *460*
Schulle H, see Soeder CJ 420, *460*
Schulz H 339, *387*
Schulz H, see Hauska G 340, 347, *379*
Schulz H, see Trebst A 339, *390*
Schulz U, see Feierabend J 352, *377*
Schulze E-D 258, 260, 261, 262, 265, 270, *279*
Schulze E-D, Fuchs MI, Fuchs M 251, 254, 260, 270, *279*
Schulze ED, Lange OL, Koch W 219, *230*
Schulze E-D, Mooney MA, Dunn EL 261, *279*
Schulze ED, Schulze I 215, 224, *230*
Schulze ED, see Evenari M 212, 213, 214, *227*
Schulze ED, see Fuchs M 261, *274*
Schulze ED, see Kappen L 214, 222, 223, *228*
Schulze ED, see Lange OL 219, 221, 223, *228*
Schulze I, see Schulze ED 215, 224, *230*
Schuster H, see Labus BC 434, *454*
Schuster HK 427, *460*
Schütz W, see Tranquillini W 262, 264, 265, *280*
Schwager H, see Keller T 488, *497*
Schwarz D 419, *460*
Schwarz OJ, see Bell SL 354, *374*
Schweisfurth R 74, *82*
Schweisfurth R, Eleftheriadis D, Gundlach H, Jacobs M, Jung W 73, *82*
Schwintzer CR, see Lommen PW 512, *524*
Schwoerbel J 421, 423, 424, *460*
Schwoerbel J, Tillmanns GC 60, *81*, 422, 423, *460*
Scott EG, see Lignowski EM 355, *382*
Scott JT, see Walsh JJ 314, *327*
Scott KR, Putwain PD 370, *387*
Scott TK 358, *387*
Seaton KA, Landsberg JJ 141, *149*
Seel F 481, *499*
Seeley HW Jr, see Finke R 56, *77*
Segel LA, see Gressel J 369, *378*
Seginer I 136, *149*
Seibert DLRS, see Thomas WH 90, *115*
Seidel K, Scheffer F, Kickuth R, Schlimme E 432, *460*
Seiwell HR 321, *326*
Seligman N, Keulen H van 142, 143, *149*
Seligman NG, see Keulen H van 131, 142, *147*
Senft WH 99, *114*
Senyimba MM, see Tieszen LL 176, 177, *202*
Serra JL, Llama MJ, Cadenas E 95, *114*
Šesták Z, Jarvis PG, Čatský J 237, *279*
Sestark Z, Catsky J, Jarvis PG 216, 217, *230*
Setlik I *149*
Setzler-Hamilton EM, see Mihursky JA 315, *325*
Shantz HL 174, *202*
Shapiro J 438, *460*
Sharma K, see Kulshreshtha S 403, *409*
Sharma MP, Chang FY, Born WH van den 396, *410*
Sharp JH 50, *82*, 322, *326*
Sharp JH, Underhill PA, Hughes DJ 104, *114*

Sharpe TM, see Wright BJ 347, 350, *391*
Shartstein BA, see Garside C 90, *111*
Shaver GR 220, *230*
Shaver RL, see Bingham SW 432, *445*
Shcherban E, see Braginskii L 403, *407*
Shepard RK 264, *279*
Sheridan RP 67, *82*
Shiaris M, Holt W, Sayler G 403, *410*
Shibles RM, Weber CR 163, *172*
Shidei T, see Fujimori J 235, *274*
Shidei T, see Kira T 235, 237, 250, 263, 267, *275*, 284, 298, *301*
Shidlovskaya NA, see Khobot'yev VG 427, *453*
Shih C, see Gloyna EF 435, *450*
Shih NT, Petering DH 483, *499*
Shinn JH, see Bennett JP 474, *492*
Shinozaki K, Kira T 252, *279*
Shinozaki K, see Kira T 249, 260, *275*, 284, 285, *301*
Shinozaki K, see Yoda K 264, *280*
Shipman LL 345, *387*
Shokraii EH, see Moreland DE 341, 354, 355, *383*
Shortreed KRS, see Stockner JG 421, *461*
Shreve F 205, *231*
Shreve F, Wiggins IL 205, 223, *231*
Shriner DS, see McLaughlin SB 242, 268, *277*
Shropshire F, see Mooney HA 206, *229*
Shvetsova VM, see Zalenskiy OV 181, *203*
Shvetsova WM, Voznessenskiy VL 179, 180, 181, *202*
Sibert RJ, see Naiman RJ 89, *113*
Sibma L 165, *172*
Sibma L, see Alberda T 165, *170*
Sieburth JMcN, see Hinga KR 322, *324*
Siefermann-Harms D 351, *387*
Siefert A 421, *460*
Siegel MR, Sissler HD 330, *387*
Siegenthaler U, see Oeschger H 508, 509, *525*
Sigurdson DC, see Tieszen LL 176, *202*
Sij JW, Swanson CA 472, *500*
Sikka HC, Pramer D 431, *460*
Sillen LG 478, *500*
Silsbury JH 263, *279*
Silsbury JM, see McCree KJ 265, *277*
Silver SD, see Fuhr I 11, *41*
Silverberg BA, Wong PTS, Chau YK 427, *460*
Silvester WB, see Lam CWY 56, *79*
Silvius JE, Ingle M, Baer CH 483, *500*
Siminovitch D, Wilson CM, Briggs DR 296, *303*
Simkins GS, Moshier LJ, Russ OG 367, *388*
Simmons RE, see Brown RH 176, *198*
Simonis W, Lee-Kaden J 365, *388*
Simonis W, see Falkner R 364, *377*
Simonis W, see Kopecek K 365, *381*
Simonis W, see Schefczik K 366, *387*
Simonson RW 12, *44*
Simpson FB, Neilands JB 102, *114*
Simpson HJ, see Broecker WS 506, *522*
Sims AP, Folkes BF, Bussey AH 422, *460*
Sims Holt JD, Stemler AJ, Radosevich SR 369, *388*
Sims PL, Singh JS 192, *202*
Sims PL, Singh JS, Lauenroth WK 191, *202*
Sims PL, see Lauenroth WK 174, 187, *200*
Sinclair TR, Wit CT de 144, *149*, 167, *172*
Singer MJ, see Johnson DW 39, *42*
Singer MJ, see Turner J 26, *44*
Singh B, Campbell WF, Salunkhe DD 355, *388*
Singh J, see Miller JM 429, *456*
Singh JS, see Sims PL 191, 192, *202*
Singh PK, see Das B *447*
Sionit N, Hellmers H, Strain B 511, *525*
Sionit N, Strain B, Hellmers H, Kramer P 517, 518, *525*
Sissler HD, see Siegel MR 330, *387*
Skinner MF, see Bevege DI 23, *40*
Skirrow G, see Riley JP 86, *114*
Skoe BP, see Ordin L 488, *499*
Skoog F 358, *388*
Skoog F, see Fitzgerald GP 431, *449*
Skoog F, see Gerloff GC 423, *450*
Skrentny RF, see Lichtenstein RF 432, *455*
Skuhravy V, see Hodek J 402, *408*
Skujins J, see McLaren AD 29, *43*
Skult H, Paatero G 365, *388*
Skye E 481, *500*
Slade P, see Calderbank A 345, *375*
Slatyer RO, Ferrar PJ 260, *279*
Sleeman JR, see Stace HCT 15, *44*
Slicher van Bath BH 153, *172*
Slife FW, see Coble HD 357, *375*
Slife FW, see Duke WB 341, 355, *376*
Slife FW, see York AC 339, 367, *391*
Sloan PR, see Eppley RW 321, *323*
Slocomb JP, see Dickson KL 443, *448*
Slooff W, see Canton JH 432, 437, *446*
Small E, see Harrison A 215, *227*
Small LF, see Frey BE 103, *111*
Smart RM, see Barko JW 64, *75*
Smayda TJ 97, *114*

Smayda TJ, see Hitchcock GL 313, *324*
Smeets N, see Larcher W 181, *200*
Smetacek V, Bodungen B von, Brödsel K von, Zeitzschel B 97, *114*
Smetana R, see Labus BC 443, *454*
Smith CF, see Russo RC 57, *82*
Smith CS, Adams MS, Schmitt MR, Adams SS 65, *82*
Smith FA, see Raven JA 481, *499*
Smith KL 322, *326*
Smith KL Jr, see Rowe GT 98, *114*, 306, 314, *326*
Smith MS, see Wain RL 334, *391*
Smith NG, see Hodgkinson KC 193, *200*
Smith RF, see Apple JL 333, 371, *373*
Smith RL, see Barber RT 103, *108*
Smith VH 314, 319, *326*
Smith WH 27, *44*
Smith WH, see Switzer GL 250, *279*
Smith WO Jr, see Penhale PA 51, *81*
Sneva FA, see Hedrick DW 405, *408*
Södergren A 365, *388*
Söderlund R, Svensson BH 86, *114*
Soeder CJ, Müller H, Payer HD, Schulle H 420, *460*
Soeder CJ, Sternik KH, Trültsch 429, *460*
Soeder CJ, Strotmann H, Gallough RA 515, *525*
Soeder CJ, see Payer HD 429, *458*
Soeder CJ, see Stengel E 437, *461*
Soffe RW, see Hand DW 522, *523*
Soffietti S, see Barralis G 397, *406*
Sojkowska T, see Sojkowski Z 404, *410*
Sojkowski Z, Staszewski Z, Sojkowska T 404, *410*
Solberg RA, Adams DF 489, *500*
Soldatini GF, Ziegler I 484, *500*
Soleimani A, see Hou L 518, *523*
Solheim 532
Sollns P 267, *279*
Solórzano Lucia, see Eppley RW 92, 107, *110*
Søndergaard M, Sand-Jensen K 49, *82*
Song L, see Mooney HA 206, *229*
Soper OF, see Parka SJ 341, 356, *384*
Sorokin YuI 68, *83*
Sournia A 90, *114*
Soutar A, see Veeh MM 100, *115*
Souza Machado V 366, 370, *388*
Souza Machado V, Arntzen CJ, Bandeen JD, Stephenson GR 370, *388*
Souza Machado V, Bandeen JD, Stephenson GR, Jensen KIN 370, *388*
Souza-Machado V, Bandeen JD, Stephenson GR, Lavigine P 370, *388*
Souza Machado V, see Darr S 370, *376*
Souza-Machado V, see Jensen KIN 370, *380*
Spaeth JP, see Menzel DW 90, 103, *113*
Spaleny J 474, *500*
Specht R, Rayson P, Jackman M 208, *231*
Specht RL 206, 207, 208, 210, *231*
Spedding DJ, Ziegler I, Hampp R, Ziegler H 479, 481, *500*
Spellenberg RW, see Syvertsen JP 215, *231*
Spence DHN, see Jupp BP 418, *453*
Spencer CP 86, 91, *114*
Spiertz JHJ, Ellen J 164, *172*
Spikes JD, Lumry R, Rieske JS 476, *500*
Splittstoesser W 519, *525*
Spommer GG, see Salisbury FB 181, *202*
Spratt M, see Stark N 290, 298, *303*
Springett BP 23, *44*
Srivastava HS, Jolliffe PA, Runeckles VC 475, *500*
Staaf H, see Ågren GI 264, 270, *272*
Stabel HH 51, 52, *83*
Stabel HH, Moaledj K, Overbeck J 54, *83*
Stace HCT, Hubble GD, Brewer R, Northcote KH, Sleeman JR, Mulcahy MJ, Hallsworth EG 15, *44*
Stadelmann E, see Walter H 218, 223, *231*
Staker RD, see Bruno SF 103, *109*
Staley JT, see Gregory E 73, 74, *78*
Standley C, see Tingey DT 484, *501*
Stangenberg M 439, *461*
Stanger CE, Appleby AP 350, 351, *388*
Stanhill G 153, 155, *172*
Stanier RY 30, *44*
Stanley RA 432, 437, 440, *461*
Stanley RA, Madewell CE 442, *461*
Stanley RA, see Goldsby TL 418, *450*
Stapleton HN, Buxton DR, Watson FL, Nolting DJ, Baker DN 136, *149*
Staring GJ, see Buringh P 134, *146*
Stark N, Jordan CF 37, *44*
Stark N, Spratt M 290, 298, *303*
Stary P, see Hodek J 402, *408*
Staszewski Z, see Sojkowski Z 404, *410*
Stauffer RS, Muckenhirn RJ, Odell RT 153, *172*
Steche W 401, *411*
Stecko V, see Åberg E 396, *406*
Steele JH, Yentsch CS 94, *114*
Steemann-Nielsen E 306, 316, 321, *326*, 514, *525*
Steemann Nielsen E, Kamp-Nielsen L 427, *461*
Steemann Nielsen E, Wium-Andersen S 426, 427, *461*,
Steffens D, see Rüdiger W 352, *387*

Steffens GL, Buta JG, Gregory LE, Mandava NB, Mendt WJ, Worley JF 362, *388*
Steinback KE, McIntosh L, Arntzen CJ 370, *388*
Steinback KE, Pfister K, Arntzen CJ 342, *388*
Steinback KE, see Arntzen CJ 370, *373*
Steinback KE, see Pfister K 342, 343, *385*
Steinback KE, see Radosevich SR 370, *386*
Steinberg C 51, 53, 54, 61, *83*
Steinberg C, Herrmann A 54, 72, *83*
Steinberg C, Schrimpf A 63, *83*
Steiner M, Eschrich W 436, *461*
Steinhardt U 10, *44*
Steinrücken HC, see Amrhein N 341, 354, *372*
Steinwandter H 364, *388*
Stemler AJ, see Sims Holt JD 369, *388*
Stengel E, Soeder CJ 437, *461*
Stengel E, see Payer HD 429, *458*
Stephens ER 484, *500*
Stephens G, see Wheeler P 52, *84*
Stephens GC, see Wheeler PA 93, *115*
Stephens GR 251, *279*
Stephens GR, Waggoner PE 287, 288, 289, *303*
Stephens K, see Antia NJ 100, *108*, 321, *323*
Stephens K, see McAllister CD 321, *325*
Stephenson GR , see Souza Machado V 370, *388*
Stephenson GR, Chang FY 367, *388*
Stephenson GR, see Bandeen JD 368, 369, *373*
Stephenson GR, see McEwen FL 364, *383*
Stephenson GR, see Souza Machado V 370, *388*
Stephenson GR, see Souza-Machado V 370, *388*
Stern AC 467, *500*
Sternik KH 63, *83*
Sternik KH, see Soeder CJ 429, *460*
Sternkopf G, see Helmecke K 398, 399, 402, *408*
Sterrett J, Adams R 405, *411*
Stetler DA, see Jackson WT 356, *380*
Steubing L, see Pahlich E 484, *499*
Steubing L, see Reichardt W 63, *81*
Steward JWB, see Cole CV 129, 142, *146*
Stewart GR, see Orebamjo TO 60, *81*
Stewart HNM, see Derwent RG 469, *494*
Stewart WDP 56, *83*, 90, *114*, 415, *461*
Stewart WDP, Alexander G 423, *461*
Stewart WDP, Fitzgerald GP, Burris RH 423, *461*
Stewart WDP, Haystead A, Pearson HW 423, *461*
Stewart WDP, Pemble M, Al-Ugaily L 63, *83*
Steyeart LT, Duc SK le, McQuigg JD 126, *149*
Stitt C, see Kunitake G 519, *524*
Stjernholm RL, see Wood HG, 519, 520, *526*
St John JB 353, *388*
St John JB, Christiansen MN 353, *388*
St John JB, Hilton JL 340, 350, 352, 353, 354, *388*
St John JB, Rittig FR, Ashworth EN, Christiansen MN 353, *388*
St John JB, see Hilton JL 339, 340, 350, *379*
Stocker O 287, 291, *303*
Stockner JG, Shortreed KRS 421, *461*
Stoddart JL 363, *389*
Stoddart JL, Venis MA 358, 363, *389*
Stokes PM, Hutchinson TC, Krauter K 429, *461*
Stokes PM, see Hutchinson TC 426, 430, 443, *452*
Stom DI, Beym AM 431, *461*
Stone DP, see Falkowski PG 92, *111*, 311, *323*
Stone JF, see Mapp HP 136, *148*
Stoner WA, see Miller PC 129, 139, 143, *148*, 180, 182, 183, 187, 196, *201*, 209, 211, *228*
Stoner WL, see Oechel WC 178, *201*
Stooff H, see Meinck F 435, 436, *456*
Storey BT, see Plesnicar M 487, *499*
Stottlemeyer JR, see Stross RG 319, *326*
Stouffer RJ, see Manabe S 509, *524*
Stowe LG, see Teeri JA 176, *202*, 215, *231*
Strain B, see Paez A 518, *525*
Strain B, see Sionit N 511, 517, 518, *525*
Strain BR 217, 218, 219, 221, 223, *231*, 510, 512, *525*
Strain BR, Chase VC 219, 221, *231*
Strain BR, see Adams MS 217, *266*
Strain BR, see Cunningham GL 223, 224, *226*
Strain BR, see Odening W 218, 223, 224, *229*
Strain BR, see Oechel WC 217, *230*
Strange JR, Williamson GD, Fletcher DJ 423, *461*
Strange RJ 431, *461*
Stratmann H, see Haut H van 471, 472, *495*
Straub CP, see Bond RG 486, *492*
Streeter JG, see Curry RB 131, 143, *146*
Streeter JG, see Meyer GE 131, *148*
Streibig JC, see Haas H 401, *408*
Strickland JDH 306, *326*
Strickland JDH, Austin KH 99, *114*
Strickland JDH, see Antia NJ 100, *108*, 321, *323*
Strickland JDH, see Carlucci AF 95, *109*
Strickland JDH, see Eppley RW 91, *110*
Strickland JDH, see McAllister CD 321, *325*

Strickland JDH, see Parsons TR 52, *81*
Strobel U, see Jacob F 396, *408*
Strom PF, see Finstein MS 58, *77*
Stross RG, Stottlemeyer JR 319, *326*
Stross RG, see Morgan RP 441, *457*
Strotmann H, see Soeder CJ, 515, *525*
Strotmann H, see Tischer W 342, 343, 344, *389*
Strube LR, see Yentsch CS 315, *327*
Struik GJ, Bray JR 195, *202*
Stryckers JMT 397, *411*
Stuart TS, Gaffron H 429, *461*
Stubbs J, see Brian RC 331, 339, 345, *375*
Stugren B 393, *411*
Stuiver M 505, 506, *525*
Stumm W, Bilinski H 71, *83*
Stumm W, Morgan JJ 72, *83*, 508, *525*
Stumm W, see Morgan JJ 65, *80*
Suberkropp K, Klug MJ, Cummins KW 53, *83*
Suchanka M, see Urbach D 340, 350, *390*
Sugahara K, see Kondo N 472, *497*
Sugawara H, see Hogetsu K 418, *452*
Sugihara G, Kilham P 435, 436, *461*
Sugimura Y, Suzuki Y, Miyake Y 102, *144*
Sukopp H, Markstein B, Trepl L 424, *461*
Sukopp H, see Markstein B 418, *456*
Sullivan CW 70, *83*
Sumere CF van, see Dedonder A 431, *447*
Sumida S, Ueda M 336, 353, *389*
Sumida S, Yoshida R, Ueda M 353, 355, *389*
Summers LA 339, 345, 346, 351, *389*
Sunda W, Guillard RRL 53, *83*, 103, *114*
Sunda WG, see Cross FA 103, *109*
Suominen J 418, 442, *461*
Susmel L, Viola F, Bassata G 210, *231*
Sutcliffe WH 315, *326*
Suter U, see Baccini P 53, *75*
Sutton DL, Ornes WH 424, *461*
Sutton DL, see Haller WT 436, *451*
Sutton R, Ting IP 485, *500*
Sutton RF 21, *44*
Suzuki Y, see Sugimura Y 102, *144*
Sveinbjörnsson, see Oechel WC 178, *201*
Svensson BH, see Söderlund R 86, *114*
Sverdrup HU 313, 319, *326*
Swader JA, see Cedeno-Maldano A 426, *446*
Swallow KC, Westall JC, McKnight DM, Morel NML, Morel FMM 53, *83*
Swank WT, see Henderson GS 37, 38, *42*
Swanson CA, see Sij JW 472, *500*
Sward RJ, see Kriedemann PE 517, *523*
Sweazy RM, Lockett MC, Reid GW, Robertson JM 423, *461*
Sweeney 432, *461*
Sweetser PB, see Beyer EM 360, *374*
Switzer CM, see Chen LG 341, 354, 355, 357, *375*
Switzer GL, Nelson LE 18, 19, *44*
Switzer GL, Nelson LE, Smith WH 250, *279*
Syers JK, see Li WC 65, *80*
Syers JK, see Williams JD 65, *84*
Sylva RN 425, *462*
Syratt WJ, Wanstall PJ 484, *500*
Syrett PJ 422, *462*
Syrett PJ, Leftley JW 61, *83*
Syrett PJ, see Morris I 422, *457*
Syvertsen JP, Nickell GL, Spellenberg RW, Cunningham GL 215, *231*
Szalay L, see Herczeg T 339, *379*
Szarek SR 212, *231*
Szarek SR, Johnson HB, Ting IP 218, 223, 224, *231*
Szarek SR, Ting IP 217, 218, 224, 225, *231*
Szeicz G 238, 248, *279*
Szumski DS, see Thomann RV 437, *462*

Tadaki Y 250, 251, *279*
Tailor HR, see Talpaz H 127, *150*
Takahama U 351, 352, *389*
Takahama U, Nishimura M 352, *389*
Takahara Y, see Mesarovic MD 134, *148*
Takahashi M, Ikeda T 96, *114*
Takahashi M, see Parsons TR 96, 108, *113*
Takahashi T 508, 509, *525*
Takahashi T, see Broecker WS 506, *522*
Takami S, Bavel CHN van 517, *525*
Takamiya A, see Satoh K 488, *499*
Takegami T, Yoshida K 363, *389*
Talbot H, see Beadle CL 241, 246, 251, 253, 260, 261, *272*, *273*
Talling IB, see Talling JF 436, *462*
Talling JF, Talling IB 436, *462*
Talpasayi ERS 423, *462*
Talpaz H, Chun ChH, Tailor HR 127, *150*
Tamas I, see Zweig G 431, *464*
Tamm CO 251, *279*
Tamm CO, see Albrektson A 251, *272*
Tamura S 362, *389*
Tan CK, Badour SS, Waygood ER 519, *525*
Taniyama T 473, *500*
Tanner CB, see Gandar PW 195, *199*
Tanner CB, see Norman JM 246, *278*
Tappeiner JC, Alm AA 26, *44*
Tassy B, see Blasco F 294, *300*
Taub FB 431, *462*

Tavares JE, see Thomson KS 360, *362*
Taylor GE, see Tingey DT 474, *501*
Taylor JS, Reid DM, Pharis RP 472, *500*
Taylor OC, Thompson CR, Tingey DT, Reinert RA 475, *500*
Taylor PJ, see Welbank PJ 167, *172*
Teeri JA, Stowe LG 176, *202*, 215, *231*
Teeri JA, Stowe LG, Murawski DA 215, *231*
Teige B, McManus TT, Mudd JB 484, *500*
Tenhunen JD, Hesketh JD, Gates DM 512, *526*
Tenhunen JD, Weber JA, Filipek LH, Gates DM 512, *526*
Tenhunen JD, Weber JA, Yocum CS, Gates DM 512, *526*
Tenhunen JD, Westrin SS 512, 513, *525*
Tenhunen JD, Yocum CS, Gates DM 512, *525*
Terhaar CJ, Ewell WS, Dziuba SP, White WW, Murphy PJ 437, *462*
Teseng CK, see Casida JE 334, 367, *375*
Teskey RO, see Dougherty PM 260, 264, 265, *273*
Teskey RO, see Hinckley TM 260, 268, 269, *274*
Tessenow U 65, 69, 74, *83*
Tevis L Jr 214, *231*
Thauer RK, see Dieckert GB 70, *77*
Thauer RK, see Schönheit P 70, *82*
Thayer GW 309, *326*
Thayer GW, Parker PL, LaCroix MW, Fry B 306, *326*
Theodorou C, see Bowen GD 21, 27, *40*
Therriault JC 314, *326*
Thiele A, see Müller G 399, 403, *409*
Thienemann A 414, *462*
Thomann RV, Szumski DS, DiTiro DM, O'Connor DJ 437, *462*
Thomas EA 415, 416, 417, 421, *462*
Thomas JB, Blaauw OH, Duysens NM 434, *462*
Thomas JF, see Jenkins D 69, *78*
Thomas JP, see O'Reilly JE 311, *325*
Thomas MD 473, *500*
Thomas MD, Alther EW 487, 489, *500*
Thomas MD, Hendricks RH 476, 488, *500*
Thomas MD, Hendricks RH, Collier TR, Hill GR 482, *500*
Thomas MD, Hendricks RH, Hill GR 481, *500*
Thomas MD, Hill GR 473, *500*
Thomas PL, see Fry JC 431, *450*
Thomas RB, see Winterhalder BP 534, 540, *548*
Thomas RL, see Frank R 430, *449*
Thomas TH 358, *389*
Thomas WD, Hill GW 11, *44*
Thomas WH 90, *114*, 321, *326*
Thomas WH, Dodson AN 94, 101, *115*
Thomas WH, Seibert DLRS, Dodson AN 90, *115*
Thomas WH, see Eppley RW 91, *110*
Thompson CR, Kats G 474, *500*
Thompson CR, Pace DM 481, *500*
Thompson CR, see Chang CW 476, 489, *493*
Thompson CR, see Taylor OC 475, *500*
Thomson KS, Hertel R, Muller S, Tavares JE 360, *362*
Thomson KS, Leopold AC 362, *389*
Thomson KS, see Hertel R 360, 362, *379*
Thomson KS, see Lembi CA 360, 362, *382*
Thomson WW, Dugger WM, Palmer RL 485, *500, 501*
Thomson WW, see Anderson JL 355, *373*
Thorne GN, see Ford MA 511, 517, *522*
Thornley JHM 125, 137, 138, 139, *150*, 241, 258, 262, 265, *279*
Thornley JHM, see Ho LC 138, *147*
Thornley JHM, see Rees AR 128, *149*
Thorpe MR, Saugier B, Auger S, Berger A, Methy M 247, *279*
Thorpe MR, Warrit B, Landsberg JJ 259, *279*
Thorpe MR, see Watson RL 258, *280*
Thünen JH von 154, *172*
Thurston RU, see Russo RC 57, *82*
Thurtell GW, see Leuning R 469, 470, 471, *497*
Tibbitts TW, see Knudson Butler L 474, *497*
Tiedemann AR, Klemmedson JO 15, *44*
Tieszen LL 176, 177, 178, 179, 180, 181, 182, 183, 184, 185, 187, 191, 192, 196, *202*, 546, *548*
Tieszen LL, Archer S 191, *202*
Tieszen LL, Miller PC, Lewis MC, Mayo JC, Oechel WC, Chapin FS 176, 177, 181, *202*
Tieszen LL, Senyimba MM, Imbamba SK, Troughton JH 176, 177, *202*
Tieszen LL, Sigurdson DC 176, *202*
Tieszen LL, see Archer S 192, *197*
Tieszen LL, see Chapin FS, III 190, *198*
Tieszen LL, see Johnson DA 178, 194, *200*
Tieszen LL, see Miller PC 180, 182, 183, 196, *201*
Tieszen LL, see Ode D 176, 186, *201*
Tijssen SB 322, *326*
Tillberg JE, see Larsson CM 365, *381*
Tilles H, see Casida JE 334, 367, *375*
Tillmanns GC, see Schwoerbel J 60, *81*, 422, 423, *460*
Tilly LJ, see Grace JB 440, *450*

Tilman D 105, 106, *115*, 420, *462*
Timofeeva NA, Kulikov NV 434, *462*
Ting IP, Dugger WM Jr 519, *526*
Ting IP, see Dugger WM 474, 484, 485, *494*
Ting IP, see Evans LS 485, *494*
Ting IP, see Kluge M 223, 224, 225, *228*
Ting IP, see Perchorowicz JT 485, *499*
Ting IP, see Sutton R 485, *500*
Ting IP, see Szarek SR 217, 218, 223, 224, 225, *231*
Tingey DT 475, *501*
Tingey DT, Reinert RA, Dunning JA, Heck WW 475, *501*
Tingey DT, Standley C, Field RW 484, *501*
Tingey DT, Taylor GE 474, *501*
Tingey DT, see Taylor OC 475, *500*
Tinker PB, see Nye PH 20, *43*
Tinsley IJ, see Reinert JC 366, *386*
Tippner M, see Mundschenk H 434, *457*
Tischer W, Strotmann H 342, 343, 344, *389*
Titman D 70, *83*, 105, *115*
Tobias PV, see Leakey LSB 531, *548*
Tobler F 424, *462*
Todd CW, see Bucha HC 331, *375*
Todd DE, see Henderson GS 24, *42*
Todd GW, see Basler E 396, *406*
Todd HG, see Moreland DE 349, *383*
Todd RL, see Jordan CF 37, *42*
Toetz D, Varga L, Pierce M 433, *462*
Toetz DW 60, *83*, 421, 422, *462*
Toetz DW, Cole B 59, *83*
Toetz DW, see Cole BS 421, *447*
Togasaki RK, see Lien S 371, *382*
Tollenaar M 136, *150*
Tomas de Silva N, see Cain SA 285, 291, *300*
Tomassini FD, see Nieboer E 471, 480, *499*
Tomizawa C, Kazano H 403, *411*
Tomlinson DL, see Pearson HW 419, *458*
Tomlinson PB, see Mooney H 287, *302*
Tonomura K, see Yamada M 428, *464*
Topinka JA 104, *115*
Torsell BWR, see Begg JE 224, *226*
Tosuka T, see Kaji M 486, *497*
Towe KM, see Malone PLG 100, *112*
Tranquillini W 261, *279*
Tranquillini W, Schütz W 262, 264, 265, *280*
Transtrum LG, see Hill AC 489, *496*
Tranter DJ, Newell BS 103, *115*
Trebst A 336, 337, 338, 339, 342, 343, 344, 345, *389*
Trebst A, Draber W 339, 342, 345, 348, *390*
Trebst A, Harth E 342, *390*
Trebst A, Pistorius E, Boroschewski G, Schulz H 339, *390*
Trebst A, Reimer S, Draber W, Knops HJ 345, *390*
Trebst A, Wietoska H 339, *390*
Trebst A, see Böhme H 339, 345, *374*
Trebst A, see Büchel KH 348, *375*
Trebst A, see Draber W 339, 342, *376*
Trebst A, see Hauska G 340, 347, *379*
Trebst A, see Oettmeier W 343, 345, *384*
Trembley FJ 433, 440, *462*
Trenbath TR, Angus JF 164, *172*
Trepl L, see Sukopp H 424, *461*
Trlica MJ, see Brown LF 178, 179, *198*
Trlica MJ, see Sala OE 189, *202*
Troeng E, Linder S 255, 260, 261, *280*
Troeng E, see Ågren GI 264, 270, *272*
Troeng E, see Hällgren J-E 474, *495*
Troeng E, see Linder S 255, 262, 264, 265, *276*
Trollope DR, Evans B 425, 428, *462*
Troughton J, see Pearcy RW 291, *303*
Troughton JH, Card KA, Hendy CH 298, *303*
Troughton JH, see Mooney HA 215, 224, *229*
Troughton JH, see Tieszen LL 176, 177, *202*
Troughton JH, see Winter K 215, 224, *231*
Troup BM, see Bray JT 100, *109*
Truelove B, see Mellis JM 341, 353, *383*
Truesdale GA, see Gameson ALH 433, *450*
Truffaut G, Pastac I 331, *390*
Trültsch, see Soeder CJ 429, *460*
Tselniker YL 260, *280*
Tsukano Y, see Lichtenstein RF 432, *455*
Tsunoda S, see Khan MA 120, *148*
Tsutsumi T 17, *44*
Tubea B, see Hawxby K 431, *451*
Tukey HB 24, *44*
Tukey HB, see Hamner CL 331, 358, *379*
Tümpling W von 436, *462*
Tunzi MG, Porcella DB 321, *327*
Turkington E, see Vennesland B 488, *501*
Turner BC, see Isensee AR 395, *408*
Turner J, Singer MJ 26, *44*
Turner J, see Long JN 250, 251, *276*
Turner NC, see Fischer RA 169, *171*
Turner NC, see Rich S 474, *499*

Turner NC, see Waggoner PE 249, *280*
Turner RE, Woo SW, Jitts HR 315, *327*
Turner RG, see Antonovics J 429, *444*
Turpin DH 87, *115*
Turpin DH, Harrison PJ 92, 106, 107, 108, *115*
Turunina NV, see Khobot'yev VG 427, *453*
Turvey ND 25, *44*
Twilley RR, Brinson MM, Davis GJ 64, *83*

Ueda M, see Sumida S 336, 353, 355, *389*
Uhl Ch, see Jordan CF 283, 287, 292, *301*
Uhlik DJ, see Miles CD 490, *498*
Ulanowicz RE, Kemp WM 306, *327*
Ullrich H, see Kändler U 486, *497*
Ulmer DD, see Vallee BL 491, *501*
Ulrich A, Gersper PL 189, *202*
Ulrich B 477, *501*
Ulrich B, see Fölster H 477, *494*
Underhill PA, see Sharp JH 104, *114*
UNESCO 281, 282, 297, *303*
Ungs MJ 249, *280*
Unsworth MH 470, 471, *501*
Unsworth MH, Biscoe PV, Pinckney HR 472, *501*
Unsworth MH, Black VJ 472, 473, *501*
Unsworth MH, Ormrod DP 467, *501*
Unsworth MH, see Black VJ 470, 472, 473, *492*
Unsworth MH, see Fowler D 468, 470, 471, *495*
Unsworth MH, see Leuning R 469, 470, 471, *497*
Unsworth MH, see Ormrod DP 477, *499*
Unterholzner R, see Pisek A 181, *201*
Uotila P 418, *462*
Upadhyaya MK, Nooden LD 341, 356, *390*
Urbach D, Suchanka M, Urbach W 340, 350, *390*
Urbach D, see Urbach W 350, 351, 353, 371, *390*
Urbach W 337, *390*
Urbach W, Lurz G, Hartmeyer H, Urbach D 371, *390*
Urbach W, Urbach D, Reisberg E 350, 351, 353, *390*
Urbach W, see Laasch H 339, 343, 344, 348, *381*
Urbach W, see Urbach D 340, 350, *390*
Utter MF 519, *526*
Utter MF, Wood HG 519, *526*
Utter MF, see Wood HG 519, 520, *526*
Uzzle AB, see Baity HG 433, *444*

Vaccaro RF, Ryther JH 95, *115*
Vaccaro RF, see Ryther JH 321, *326*
Vaisberg AJ, Schiff JA 350, *390*
Vallee BL, Ulmer DD 491, *501*
Vance BD, Drummond W 432, *462*
Vanderhoef LN, see Melhado LL 363, *383*
Vanstaen H, see Verstraete W 442, *462*
Varga L, see Toetz D 433, *462*
Varley GC 241, *280*
Varley GC, Gradwell GR, Hassell MP 268, *280*
Vazart J, Henry Y 515, *526*
Veeh MM, Burnett WC, Soutar A 100, *115*
Veen JA van 140, 142, *150*
Veen R van der, see Wessels JSC 331, 339, *391*
Veen WL van, see Mulder EG, 520, *524*
Velthuys BR 338, 342, *390*
Velthuys BR, Amesz J 338, *390*
Venis MA 354, 355, 363, *390*
Venis MA, see Batt S 362, *374*
Venis MA, see Stoddart JL 358, 363, *389*
Vennesland B, Turkington E 488, *501*
Verduin J 322, *327*
Vermeij GJ, see Givnish TJ 285, 291, *300*
Vernotte C, see Astier C 371, *373*
Verstraete W, Voets JP, Vanstaen H 442, *462*
Verstraete W, see Focht DD 58, *77*
Vet E van der, see Rensen JJS van 348, 349, *386*
Vidaver, W 514, *526*
Villiers de OT, see Ashton FM 336, 353, 354, *373*
Vincent WF, see Lam CWY 56, *79*
Viner AB, see Horne AJ 415, *452*
Viola F, see Susmel L 210, *231*
Vitousek PM 6, 7, *44*
Vitousek PM, Reiners WA 6, *44*
Vitousek PM, see Gorham E 13, *41*
Vittum P, see Pimentel D 331, *385*
Vivier P, Nisbet M 436, *462*
Vliet WPA van, see Rensen JJS van 348, 349, *386*
Voerman S, Besemer AFH 364, *390*
Voets JP, see Verstraete W 442, *462*
Vogel JC 299, *303*
Vogel JC, Fuls A, Ellis RP 176, *202*
Vogt KA, see Johnson DW *42*
Vogt KH, see Grier CC 241, 269, 271, *274*
Voigt GK, Richards BN, Mannion EC 14, *45*
Volkovinsky VV, see Koblentz-Mishke OJ 88, *112*, 305, 308, *324*
Vollenweider RA 58, 61, *84*, 414, 415, 416, 419, 420, 422, *463*
Vollrath H, see Kohler A 417, 420, 442, *454*
Vooren J van de, see Challa H 136, *146*
Vooys CGN de 309, *327*
Voznessenskiy VL, see Shvetsova WM 179, 180, 181, *202*

Voznessenskiy VL, see Zalenskiy OV 181, *203*
Vyas LN, see Ranawat MPS 294, *303*
Vyskot M 235, *280*

Wada E, see Hattori A 95, *111*
Waggoner PE 125, 136, *150*
Waggoner PE, Turner NC 249, *280*
Waggoner PE, see Stephens GR 287, 288, 289, *303*
Wagner G 424, *463*
Waid JS 33, *45*
Waide JB, see Henderson GS 37, 38, *42*
Wain RL 358, 363, *390*
Wain RL, Smith MS 334, *391*
Wain RL, see Kerr MW 339, 348, *380*
Wainwright CM 224, *231*
Wake RG, see Cecil R 480, *493*
Walker DA 337, *391*
Walker DA, see Baldry CW 483, *491*
Wallace A, Bamberg SA, Cha JW 213, *231*
Waller H, Gausman HW, Rittig FR, Namken LN, Escobar DE, Rodriguez RR 361, *391*
Wallnöfer P, Poschenrieder G, Engelhardt G 396, *411*
Wallwork JA 31, *45*
Walsh GE, Miller CW, Heitmüller PT 431, *463*
Walsh JJ 313, 315, 316, 318, *327*
Walsh JJ, Kelley JC, Whitledge TE, Huntsman SA, Pillsbury RD 314, *327*
Walsh JJ, Whitledge TE, Barvenik FW, Wirick CD, Howe SO 89, *115*
Walsh JJ, Whitledge TE, Barvenik FW, Wirick CD, Howe SO, Esaias WE, Scott JT 314, *327*
Walsh JJ, Whitledge TE, Kelley JC, Huntsman SA, Pillsbury RD 89, *115*
Walsh JJ, see Falkowski PG 320, *323*
Walter H 234, *280*, 290, *303*
Walter H, Stadelmann E 218, 223, *231*
Walter HJ, see Palmer CM 419, 420, 421, 434, 435, 436, 438, *458*
Walton DC 358, 361, *391*
Wang CT, see Nobel PS 485, *499*
Wang Y, see Gutierrez AP 130, *147*
Wanstall PJ, see Syratt WJ 484, *500*
Warburg O, Christian W 487, *501*
Wardlaw IF 195, *202*
Ware GW, Roan CC 364, *391*, 430, 432, *463*
Wareing PF 358, *391*
Waring RH, Emmingham WH, Gholz HL, Grier CC 250, 251, *280*
Waring RH, Franklin JF 235, *280*
Waring RH, see Emmingham WH 239, 261, *273*
Waring RH, see Gholz HL 251, *274*
Warinner JE, Brehmer ML 441, *463*
Warren GF, see Kratky BA 336, *381*, 431, *454*
Warren GF, see Pritchard MK 352, *385*
Warrit B, see Thorpe MR 259, *279*
Watanabe I 12, *45*
Water quality data, British Columbia, 1961–1971 90, *115*
Watson CA, see McNeal FH 166, 167, *171*
Watson CW, see Bartels PG 350, 352, *374*
Watson FL, see Stapleton HN 136, *149*
Watson GH, Bollen WB 426, *463*
Watson J, see Pfister K 343, 344, 368, 370, *385*
Watson RL, Landsberg JJ, Thorpe MR 258, *280*
Wattenhofer R, see Lachavanne JB 442, *455*
Watts WR, Neilson RE, Jarvis PG 260, *280*
Wax LM, Bernard RL, Hayels RM 397, *411*
Way JM 401, 404, *411*
Way JM, Chancellor RJ 405, *411*
Way JM, Newman JF, Moore NW, Knaggs FW 430, *463*
Waygood ER, see Tan CK 519, *525*
Wayne LG, see Haagen-Smit AJ 468, 469, *495*
Weane NM, see Mague TM 90, *112*
Weaver GT 17, *45*
Weaver REC, see McKinion JE 137, 138, *148*
Weaver RJ 330, 341, 357, 358, *391*
Webb KL, D'Elia CF 314, *327*
Webb KL, Johannes RL 96, *115*
Webb KL, see Haas LW 104, *111*
Webber PJ 174, 186, *202*
Weber CR, see Shibles RM 163, *172*
Weber JA, see Tenhunen JD, 512, *526*
Webley DM, Duff RB 13, *45*
Webley DM, see Greaves MP 29, *41*
Webster PL, Hof J van't 355, *391*
Weeks DP, Baxter R 354, 355, *391*
Wegler R 330, 331, 364, *391*
Wei LL, see Miller GW 489, *498*
Weigert RG, Coleman DC, Odum EP 28, *45*
Weiler CS, see Eppley RW 104, *110*
Weiner ND, see Kamel AM 486, *497*
Weinstein LH 476, *501*
Weinstein LH, Alscher-Herman R 475, 476, *501*
Weinstein LH, see Jacobson JS 476, *496*
Weinstein LH, see McCune DC 487, 488, *498*
Weise G, Horbach W, Hornig L, Gnauck AH 443, *463*
Weise G, see Auerbach S 443, *444*
Weisman L, see Hodek J 402, *408*
Welbank PJ, Gibb MJ, Taylor PJ, Williams ED 167, *172*
Welch E 309, *327*

Welch EB, Emery RM, Matsuda RI, Dawson WA 315, *327*
Welch LF 153, 154, *172*
Welch SM, Croft BA 136, *150*
Weldon LW, see Blackburn RD 432, *445*
Welkie GW, see Cheng JYO 489, *493*
Welkie GW, see Lee CJ 488, *497*
Welkie GW, see Lovelace J 489, *498*
Wellburn AR 473, 487, 490, *501*
Wellburn AR, Capron TM, Chan HS, Horsman DC 486, 490, *501*
Wellburn AR, Majernik O, Wellburn FAM 482, *501*
Wellburn AR, see Horsman DC 490, *496*
Wellburn FAM, see Wellburn AR 482, *501*
Weller H, see Gerhards U 429, *450*
Welles J, see Norman JM 247, *278*
Wellner B, see Müller D 443, *457*
Wells CG, see Kinerson RS 264, 271, *275*
Wells CG, see Penning de Vries FWT 143, *149*
Wells EG, Nicholas AK, Bual SW 24, 25, *45*
Wendland WM, see Bryson RA 509, *522*
Went FW 214, *231*
Wentsel RS, Berry JW 429, *463*
Weretelnik J, see Pawlaczyk-Szpilowa M 427, *458*
Werner D 70, *84*
Wershaw RL, Burcar PJ, Goldberg MC 430, *463*
Werzenak CT, Gannon JJ 57, *84*
Wesely ML, Eastman JA, Cook DR, Hicks BB 471, *501*
Wessels JSC, Veen R van der 331, 339, *391*
West LD, Muzik TJ, Witters RE 370, *391*
West M, see Mooney HA 221, *229*
West NE, see Charley JL 15, 31, *40*
Westall JC, see Swallow KC 53, *83*
Westing A 405, *411*
Westlake DF 416, 418, *463*
Westlake DF, see Fogg GE 422, 427, 429, *449*
Westlake GF, see Cairns J Jr 443, *446*
Westman WE 25, *45*
Westman WE, Rogers RW 17, *45*
Westman WE, Whittaker RH 235, 236, 251, *280*
Westrin SS, see Tenhunen JD, 512, 513, *525*
Wetherald RT, see Manabe S 509, *524*
Wetzel A 417, 418, *463*
Wetzel RG 48, 49, 50, 51, 55, 61, 66, 67, 68, 71, 72, 73, *84*, 307, 308, 313, *327*, 414, 415, 416, 422, 427, 435, 436, 438, *463*
Wetzel RG, Allen HL 427, *463*
Wetzel RG, Hough RA 415, 418, *463*
Wetzel RG, Manny BA 61, *84*, 422, 423, 427, *463*
Wetzel RG, Otsuki A 49, *84*
Wetzel RG, Penhale PA 49, *84*
Wetzel RG, Rich PH 49, *84*
Wetzel RG, see Francko DA 64, *77*
Wetzel RG, see Godshalk GL 53, *77*
Wetzel RG, see Goldman CR 415, *450*
Wetzel RG, see Hough RA 51, *78*
Wetzel RG, see Manny BA 415, *456*
Wetzel RG, see Rich PH 52, *82*
Whaley RC, see Carpenter JH 90, *109*
Wheat MM 534, *548*
Wheeler P, North B, Littler M, Stephens G 52, *84*
Wheeler PA, North BB, Stephens GC 93, *115*
Wheeler PA, see Haines KC 104, *111*
Whiffen RD, see Little P 471, *497*
Whitaker J, Barica J, Kling H, Buckley M 426, *463*
Whitcomb M, see Bristow JM 421, *445*
Whitcombe M, see Bristow JM 64, *76*
White E, Payne G 62, *84*
White EJ, see Carlisle A 25, *40*
White KL, Hill AC, Bennett JH 477, *501*
White PS 8, *45*
White RS, see Caldwell MM 212, 213, *226*
White SL, see Nash TH 214, *229*
White W, see Ray S 429, *459*
White WW, see Terhaar CJ 437, *462*
Whitecross MJ, Armstrong DJ 395, *411*
Whitehead D, Jarvis PG 251, 260, *280*
Whitfield PH, Lewis AG 103, *115*
Whitford LA 416, *463*
Whitledge TE, Packard TT 102, *115*
Whitledge TE, see Walsh JJ 89, *115*, 314, *327*
Whitledge TE, see Walsh JJ, 314, *327*
Whitmarsh J, see Bugg MW 339, 345, 352, *375*
Whittaker RH 211, *231*, 235, *280*
Whittaker RH, Likens GE 158, *172*, 174, *202*, 507, *526*
Whittaker RH, Niering WA 212, *231*
Whittaker RH, see Westman WE 235, 236, 251, *280*
Whittaker RH, see Woodwell GM 506, *526*, 537, 539, *548*
Whittaker RM, Woodwell GM 263, 264, *280*
Whitton BA 423, 427, 438, *463*
Whitton BA, see Harding JPC 429, 430, *451*
Whitton BA, see Hargreaves JW 429, 430, 438, *451*
Whitton BA, see Say PJ 429, *459*
Wholey D, see Cock JH 160, *170*

Wicks G, see Burnside O 401, *407*
Widdig A, see Scheinpflug H 330, *387*
Widell KO, see Klockare B 352, *381*
Widmer RE, see Mattson RH, 517, *524*
Wiebe H 264, *280*
Wiebe HH, see Ross CW 488, *499*
Wiedman SJ, Appleby AP 355, *391*
Wiegleb G 442, *463*
Wielgolaski FE 174, 175, 191, *202*
Wiese U, see Müller G 399, 403, *409*
Wietoska H, see Trebst A 339, *390*
Wiggins GB, see Bärlocher F 53, *75*
Wiggins IL, see Shreve F 205, 223, *231*
Wightman F, see Schneider EA 341, 360, *387*
Wild A 13, *45*
Wild A, Oberweiss AL, Ruehle W 365, *391*
Wildman RB, see Schauberger CW 364, 365, *387*
Wildner GF, see Janatkova H 371, *380*
Wile I, see Aiken SG 418, *444*
Wileock RJ, see Wilhelm E 479, *501*
Wilhelm E, Battino R, Wileock RJ 479, *501*
Wilkinson RE 341, 353, *391*
Will GM 25, *45*
Will GM, see Orman HR 20, *43*
Willey RW 165, *172*
Williams DJ, see Schindler JE 54, *82*
Williams ED, see Welbank PJ 167, *172*
Williams GH 404, *411*
Williams GJ III 176, 185, 195, *203*
Williams GJ III, Kemp RP 179, *203*
Williams GJ III, Markley JL 176, 185, *203*
Williams GJ, III, see Kemp PR 185, 195, *200*
Williams JD, Syers JK, Harris RF, Armstrong DE 65, *84*
Williams JDH, Murphy TP, Mayer T 64, *84*
Williams JDH, see Li WC 65, *80*
Williams JP, see Khan MU 353, *380*
Williams LG, Mount DI 426, *463*
Williams LG, Pickering Q 434, *464*
Williams PF, see Benson-Evans K 443, *444*
Williams PJLeB, Berman T, Holm-Hansen O 95, *115*
Williams PM 90, *115*
Williams RB 309, *327*
Williams RJH, see Ricks GR 476, *499*
Williams ST, Gray TRG 34, *45*
Williams WA, Loomis RS, Duncan WG, Dovrat A, Nunez AF 163, *172*
Williams WA, Loomis RS, Lepley CR 163, *172*
Williams WA, see Duncan WG 164, *171*
Williams WA, see Loomis RS 163, 164, *171*
Williamson GD, see Strange JR 423, *461*
Willingham CA, Andersen KJ 432, *464*
Wilson AL 71, *84*
Wilson CM, see Siminovitch D 296, *303*
Wilson D, Cooper JP 179, *203*, 515, *526*
Wilson DE, see Janzen DH 296, *301*
Wilson GL, see Ludlow MM, 517, *524*
Wilson JR 176, 190, 191, 195, *203*
Wilson JR, Haydock DP 190, 195, *203*
Wilson LA, see Holmes EB 160, *171*
Wilson LG, Bressan RA, Filner P 474, *501*
Wilson LG, see Bressan RA 484, *492*
Wilson PM, Ford EB, Salmon MR 520, *526*
Wingfield G, see Marsh J 398, *409*
Winkler K, see Fischer WK 433, *449*
Winkler P 477, *501*
Winn DT, see Detling JK 193, 194, 195, 196, *199*
Winner WE, Mooney HA 472, *502*
Winter K, Troughton JH 215, 224, *231*
Winterhalder BP, Thomas RB 534, 540, *548*
Winters WS, see Hill AC 489, *496*
Wirick CD, see Walsh JJ 89, *115*, 314, *327*
Wissmar RC 430, *464*
Wit CT de 125, *150*, 168, *172*, 243, *280*
Wit CT de, Brouwer R, Penning de Vries FWT 137, 144, *150*
Wit CT de, et al. 161, 169, *172*
Wit CT de, Goudriaan J 121, 125, *150*
Wit CT de, Heemst HDJ van 162, *172*
Wit CT de, Laar HH van, Keulen H van 130, 131, 132, 135, 136, 137, 138, 139, 140, *150*
Wit CT de, Penning de Vries FWT 119, 133, *150*
Wit CT de, see Brouwer R 137, 141, *146*
Wit CT de, see Horie T 144, *147*
Wit CT de, see Keulen H van 127, *147*
Wit CT de, see Sinclair TR 144, *149*, 167, *172*
Witkamp M 32, *45*
Witkamp M, Ausmus BS 28, 36, *45*
Witkamp M, see Ausmus BS 33, *39*
Witkamp M, see Crossley DA Jr 28, *40*
Witlage JM, see Penning de Vries FWT 137, 138, *149*, 240, *278*
Witt HT 336, 350, *391*
Witt WW, see Hayes RM 364, *379*
Witt WW, see Kitchen LM 352, *380*

Witters RE, see West LD 370, *391*
Wittman W, see Bergersen FJ 12, *40*
Wittwer S 512, *526*
Wittwer SH 357, 358, 361, 362, *391*
Witzel KP 58, *84*
Witzel KP, Overbeck J 58, *84*
Wium-Andersen S 49, *84*
Wium-Andersen S, see Steemann Nielsen E 426, 427, *461*
Wofsy SC, see McConnall JC 487, *498*
Woledge J 179, *203*
Wolf HW, see McKee JE 434, *456*
Wolfe JM, Rice EL 51, *84*
Wolk CP 423, *464*
Wollast R, Billen G, Mackenzie FT 428, *464*
Woltz SS, Leonard CD 476, *502*
Wolverton BC, McKnown MM 432, *464*
Wong D, see Rensen van JJS 348, 349, *386*
Wong DCL, see Laws EA 95, *112*
Wong PTS, Chau YK, Luxon PL *464*
Wong PTS, see Silverberg BA 427, *460*
Wong SC 511, *526*
Wong SC, Cowan IR, Farquhar GD 140, *150*
Wonneberger R, see Kohler A 420, 442, *454*
Woo SW, see Turner RE 315, *327*
Wood HG, Stjernholm RL 519, 520, *526*
Wood HG, Utter MF 519, 520, *526*
Wood HG, see Utter MF 519, *526*
Wood JM, Kennedy FS, Rosen CG 428, *464*
Wood KV, see Mihursky JA 315, *325*
Wood T, Bormann FH 474, *502*
Wood TG, see Lee KE 31, *43*
Woodell SRJ, Mooney HA, Hill AJ 211, *231*
Woodman JN 260, *280*
Woodmansee RG, see Duncan DA 209, 211, *227*
Woodwell GM 544, *548*
Woodwell GM, Botkin DB 262, 264, *280*
Woodwell GM, Craig PP, Johnson HA 544, *548*
Woodwell GM, Whittaker RH, Reiners WA, Likens GE, Delwiche CC, Botkin DB 506, *526*, 537, 539, *548*
Woodwell GM, Wurster CF Jr, Isaacson PA 544, *548*
Woodwell GM, see Whittaker RM 263, 264, *280*
Woolhouse HW, see Grace J 180, 181, 184, *199*, *200*
Woolley JT 246, *280*
Worley JF, see Steffens GL 362, *388*
Worthing CR 330, *391*
Worthing CR, see Martin H 330, *382*
Wright BJ, Baillie AC, Wright K, Dowsett JR, Sharpe TM 347, 350, *391*
Wright HE, see Martin PS 532, *548*
Wright K, Corbett JR 337, 338, 339, 351, *391*
Wright K, see Wright BJ 347, 350, *391*
Wright MJ 128, *150*
Wright R, see Green K 518, *522*
Wright RD, see Bryan W 520, *522*
Wright RD, see Hart P 511, *523*
Wright RD, see Hurt P 517, *523*
Wright RF, see Likens GE 438, *455*, 477, *497*
Wright RT, Hobbie JE 52, *84*
Wright STC 361, 362, *391*
Wuhrmann K 414, *464*
Wunderlich M 433, *464*
Wunderlich M, Müller D 441, *464*
Wurster CF Jr, see Woodwell GM 544, *548*
Wurster CF, see Biggs DC 365, *374*
Wurster CF, see Fischer NS 365, *377*
Wurster CF, see Powers CD 365, *385*
Yabuki K, see Kira T 243, 271, *275*, 296, *301*
Yabuki M, Aoki M 293, *303*
Yaguchi EM, see Parker JI 69, *81*
Yamada M, Tonomura K 428, *464*
Yamada M, see Hilton BD 366, *379*
Yamaguchi J 139, *150*
Yang SF 480, *502*
Yang SF, Miller GW 488, *502*
Yang SF, Saleh MA 480, *502*
Yeargan KV, see Hayes RM 364, *379*
Yen D, see Pimentel D 331, *385*
Yentsch CM, see Yentsch CS 315, *327*
Yentsch CS 315, *327*
Yentsch CS, Yentsch CM, Strube LR, Morris I 315, *327*
Yentsch CS, see Ryther JH 311, *326*
Yentsch CS, see Steele JH 94, *114*
Yocum CF, see Robinson SJ 339, 345, *386*
Yocum CS, see Tenhunen JD 512, *525*, *526*
Yocum CW, see Lommen PW, 512, *524*
Yoda K 263, 264, *280*, 285, 286, 287, 291, 292, 293, *303*
Yoda K, Shinozaki K, Ogawa H, Hozumi K, Kira T 264, *280*
Yoda K, see Hozumi K 283, 285, 287, 292, 293, *300*
Yoda K, see Kira T 292, 298, *301*
Yoda K, see Ogawa H 283, *303*
Yoneyama T, see Kaji M 486, *497*
York AC, Arntzen CJ, Slife FW 339, 367, *391*
Yoshida K, see Takegami T 363, *389*
Yoshida R, see Sumida S 353, 355, *389*
Yoshida S 518, *526*
Yoshida S, see Cock JH 518, *522*
Young CA 426, 440, *464*

Young JL, see Barlow EWR 195, *197*
Youngman RJ, Dodge AD 346, 351, 353, *391*
Youngson CR, see Hamaker JW 430, *451*
Yousef Y, see Gloyna EF 435, *450*
Yu MH, Miller GW 488, 489, *502*
Yu MH, see Cheng JYO 489, *493*

Zaiss U 63, *84*
Zaiss U, Kaltwasser H 54, *84*
Zakar K, see Iverson RL 314, 319, *324*
Zalenskiy LV, see Gerasimenko TV *199*
Zalenskiy OV, Shvetsova VM, Voznessenskiy VL 181, *203*
Zand G, see Orshan G 223, *230*
Zandstra BH, Motooka PS 402, *411*
Zeitzschel B, see Smetacek V 97, *114*
Zeleny J, see Hodek J 402, *408*
Zelitch I 482, 484, *502*
Zeltner GH, see Kohler A 420, 421, 442, *454*
Zelwer M, see Medina E 294, 295, *302*
Zemánek J 396, 397, *411*
Zeroni M, Hall M 358, 363, *391*
Zevenboom W, Mur LR 60, *84*
Ziebell CD, see Greer DE 424, *450*
Ziegler H 49, *84*, 262, 264, 265, *280*
Ziegler H, see Fischer K 482, *494*
Ziegler H, see Hampp R 490, *495*
Ziegler H, see Libera W 479, 482, 483, 484, *497*
Ziegler H, see Spedding DJ 479, 481, *500*
Ziegler I 473, 481, 482, 483, 486, *502*
Ziegler I, Hampp R 481, *502*
Ziegler I, Libera W 479, *502*
Ziegler I, Marewa A, Schoepe E 483, *502*
Ziegler I, see Hampp R 481, 490, *495*
Ziegler I, see Libera W 479, 482, 483, 484, *497*
Ziegler I, see Soldatini GF 484, *500*
Ziegler I, see Spedding DJ 479, 481, *500*
Ziemann H 436, *464*
Zimmerman AP, see Schindler JE 54, *82*
Zimmerman RH, see Krizek DT 517, *524*
Zimmermann MH, Brown LL 287, *303*
Zinderen Bakker EM van Jr, see Mayo JM 178, 181, 182, *200*, *201*
Zinke PJ 15, *45*
Zitko V, see Hutzinger O 364, *380*
Zohary M 223, *231*
Zsuffa L, see Anderson HW 236, *272*
Zweig G, Tamas I, Greenberg E 431, *464*

Taxonomic Index

Page numbers in **bold face** refer to tables or figures

Abies **17**
Abies amabilis 26, 35, 269
Abies balsamea **260**
Abies grandis **236**
Abies sachalinensis **236**, 264
Abies veitchii 250
Acantholimon diapensioides **213**
Acari 399
Acer 53
Acer saccharum 27
Achnanthes 426
Achromobacter 58
Acinetobacter 57
Acyrthosiphon pisum 401
Adenostoma fasciculatum **210**
Aegilops 533
Agave deserti **217**
Agropyron **188**, 533
Agropyron dasystachyum 187, 189
Agropyron repens 397
Agropyron smithii 185, 187, **190**, 192, **193**
Agrostis canina 536
Agrostis stolonifera 404
Alangium ridleri **288**
Alfalfa, see Medicago sativa
Algae, see Phycophyta
Allium cepa 362
Allopophora caliginosa 399
Alnus crispa 11
Alocasia macrorhiza **288**
Amaranthus hybridus 369
Amaranthus palmeri **217**
Amaranthus retroflexus 397
Ambrosia 211
Amphiprora paludosa 436
Anabaena 56, 419
Anabaena azollae 56
Anabaena cylindrica 427
Anabaena flos-aquae 432
Anabaenopsis 431
Anacardiaceae 246
Anacystis 419, 427
Anacystis nidulans 432
Ananas sativus 535
Andropogon 176
Andropogon gerardii **190**
Andropogon hallii **178**, **184**
Andropogon scoparius **190**, 195
Anisothecium 438
Ankistrodesmus 371, 431
Anthocephalus cadamba 290
Antiaris africana 297
Antocephalus cadamba **288**
Apera spica-venti 397, 401
Aphanizomenon 56, 419
Aphanizomenon flos-aquae 423
Aphis fabae 403
Apis 401
Apple, see Malus
Apricot, see Prunus armeniaca
Arachis hypogaea 534
Arbutus menziesii 220
Arctophila fulva 543
Arctostaphylos 219
Arrhenatherum elatius **208**
Artemisia badhysi **213**
Artemisia frigida 189
Artemisia herba-alba 211, **212**
Artemisia rhodantha **213**
Artemisia sieberi **212**
Artemisia terra-alba **213**
Arthrobacter 54, 58
Aspidosperma cruentum 290
Asterionella 70, 420
Asterionella formosa 69, 420, 429
Astragalus 404
Atriplex confertifolia **212**, **213**
Atriplex hymenelytra 223
Atriplex lentiformis **217**
Atriplex vesicaria **212**, **213**
Avena **154**, 366, 367, 397, 401, 485, 488, 536
Avena fatua 397, 401
Avocado, see Persea
Azolla 56
Azotobacter 56

Bacillariophyceae 69, 70, 74, 92, 95, 100, 101, 102, 103, 104, 106, **107**, 313, 419, 420, 421, 423, 426, 430, 431, 433, 434, 436, 438, 440, 441, 443
Bacillus 58, **73**
Bacteria, see Schizomycetes

Banana, see Musa
Barley, see Hordeum
Batrachospermum boryanum 421
Bean, see Phaseolus vulgaris
Beech, see Fagus
Beet, see Beta
Beggiatoa 68
Bermudagrass, see Cynodon dactylon
Beta saccharifera **131**, 137, 143, **155**, 157, **160**, 164, 166, 362, 397, 511
Beta vulgaris 403, 476
Betula allegheniensis **17**, 27
Betula pendula **260**
Betulaceae 22
Birch, see Betula
Bison bison 186, 194, 532
Black spruce, see Picea mariana
Blue-green algae, see Cyanophyceae
Bouteloua 176
Bouteloua gracilis **178**, 179, 185, 187, **188**, 189, **190**, 194, 195
Bracken, see Pteridium aquilinum
Brassica campestris 369
Brassica napus 364
Brassica oleracea var. acephala **160**
Brassica oleracea var. gemmifera 403
Brevicoryne brassicae 401, 403
Brickellia incana **213**
Bromeliaceae 291, 292
Brown algae, see Phaeophyceae
Brussels sprouts, see Brassica oleracea var. gemmifera
Bryophyta **175**, **177**, 178, 179, 181, 182, 425, 428, 438, 489, 543
Buchloë 176
Buddleia coriacea 540
Bumilleriopsis 371
Byronima crassifolia 289

Cacao, see Theobroma cacao
Cactaceae 291
Calamagrostis 540
Calamagrostis epigeios 405
Calanus 100
California coastal redwood, see Sequoia sempervirens
Callistephus chivensis 518
Callitriche hamulata 422, 423
Callitriche obtusangula 421
Calluna 180, 181, 182
Camelina sativa 536
Camellia 489
Camissonia claviformis **217**, 219
Carex aquatilis 543, 545
Carex brizoides 405
Carica papaya 534
Carpocapsa pomonella 402
Carrot, see Daucus carota
Carthamus tinctorius 166
Cassava, see Manihot esculenta
Castanea sativa **246**
Castanopsis cuspidata **249**, 264
Casuarina equisetifolia 12
Catharinea 438
Ceanothus megacarpus **210**
Cecropia peltata **288**, **289**, 290
Centaurea cyanus 397
Ceratoides lanata **212**, **213**
Ceratoides papposa **213**
Ceratophyllum 422, 436
Ceratophyllum demersum 60, 418, 427
Cercidium **212**, 213
Cercidium floridum **217**
Cercosporella herpotrichoides 399
Chaemosiphon 421, 425
Chaetoceros **107**
Chaetoceros gracilis **94**
Chaetophora 421
Chara 64, 418, 426, 431
Chara globularis 440
Chara zeylandica 422
Chenopodium album 369, 397
Chenopodium murale 488
Chenopodium serotinum 397
Cherry, see Prunus avium
Chilopsis linearis **224**
Chlamydomonas 61, 356, 371, 426, 429, 434, 442
Chlamydomonas acidophila 438
Chlamydomonas reinhardii 428, 432
Chlorella 61, 422, 427, 428, 429, 431, 434, 437, 485, 486, 515
Chlorella pyrenoidosa 426, 432, 489
Chlorella sorokiniana 426, 433
Chlorella vulgaris 432
Chlorobacteriaceae 68
Chlorobium **67**
Chlorococcum 426
Chlorophyceae 69, 419, 420, 422, 423, 425, 426, 429, 435, 438, 440
Chlorophylla maxima 420
Chromatiaceae 68
Chromatium **67**, 68
Chromulina 438
Chroomonas **107**
Chrysophyceae 419, 433
Cinnamomum parthenoxylon **288**
Cirsium arvense 397, 404
Cirsium vulgare 404
Citrus 361, 476, 486
Citrus sinensis 476
Cladocerae 51
Cladophora 419, 425, 427, 428, 432, 433

Cladophora glomerata 421
Claytonia lanceolata 541
Clostridium 56, **73**, 428
Clover, see Trifolium
Coccolithophoraceae 103
Coccomyxa 425
Cocconeis 425
Cocconeis placentubula 421
Cochlospermum vitifolium 289
Coffea **159**
Collembola 399
Coniferae 25, 34, 53, 234, 235, **236**, 241, **249**, 250, 251, 252, 253, 258, 264, 472, 473, 475
Convolvulus arvensis 397
Coralline algae, see Rhodophyceae
Cordia alliodora **288**, **289**
Cordyline rubra **288**
Cotton, see Gossypium
Croton glabellus **288**, **289**
Cryptocarya alba **210**
Cryptomeria japonica **236**, 240
Cryptophyceae 419
Ctenopharyngodon idella 403
Cucumber, see Cucumis sativus
Cucumis sativus 472, 484
Cucurbitaceae 166, 534
Curatella americana 289
Curly-top virus 155
Cyanophyceae 12, 51, 53, **55**, 56, 63, 67, 74, 85, 88, 90, 102, 415, 416, 418, 419, 420, 423, 426, 430, 431, 432, 438, 440, 488
Cyclobalanopsis gilva 264
Cyclotella meneghiniana 70, 420, 436, 442
Cyclotella nana 436
Cylindrotheca **107**
Cynodon dactylon **160**, 397
Cyperaceae 401

Dacryodes excelsa 290
Dactylis glomerata 194
Daphnia 59
Daphnia pulex 51
Daucus carota 364, 519
Delphinium 404
Deschampsia caespitosa 404
Deschampsia flexuosa 405
Desmidiaceae 418, 433
Desmidium 438
Desmodium paniculatum 518
Desulfotomaculum 67
Desulfovibrio **67**, 68, **73**
Diatoms, see Bacillariophyceae
Digitaria sanguinalis 397
Dinobryon divergens 419
Dinoflagellatae 104
Diplacus aurantiacus 472
Dipterocarpaceae 22
Dipterocarpus crinitus **288**
Dipterocarpus sublamellatus **288**
Ditylum brightwellii **94**
Douglas fir, see Pseudotsuga menziesii
Dryas 178, 179, 182
Dryas integrifolia 181
Duckweed, see Lemna
Dunaliella tertiolecta 91
Dupontia **180**, 181
Dupontia fischeri **182**, **184**, 543

Echinochloa crus-galli 397
Eichhornia crassipes 432, 436
Elaeis guineensis 157, 161
Eleocharis acicularis 438
Elm, see Ulmus
Elodea 49, 424
Elodea canadensis 418, 428, 429, 436, 437, 440
Elodea densa 60, 366, 421, 428
Elymus canadensis **178**, **184**
Emmer, see Triticum dicoccum
Encelia californica 219
Encelia farinosa **217**, **222**, 223, **224**
Eperua leucantha 290
Ephedra nevadensis **213**
Equisetum 70
Eremophila gilesii **212**, **213**
Erioischia brassicae 403
Eriophorum scheuchzeri 543
Erythronium grandiflorum **541**
Erythrophleum fordii **288**
Escherichia coli 519
Eucalyptus **17**, **25**, 26, 53, 234, 241, 247, 251, 254, 540
Eucalyptus globulus **236**, 268
Eucalyptus grandis **236**
Eucalyptus maculata **249**
Eucalyptus nitens **236**
Eucalyptus obliqua 20, 267
Eucalyptus pauciflora **260**
Euglena 61, 371, 421, 488
Euglena gracilis 426
Euglena mutabilis 438
Euphausia pacifica 96
Euphorbia 291
Euphorbia exigua 402
Eurotia ceratoides **213**
Eurotia lanata 189, **213**

Fabaceae 11
Fagaceae 22
Fagus **17**, **25**, 53, **296**, 299
Fagus crenata **249**
Fagus grandifolia 27
Fagus sylvatica 254, 261, **270**, 271
Fern, see Pteridophyta

Festuca 540
Festuca ovina 536
Fir, see Abies
Flagellatae 104
Flax, see Linum usitatissimum
Flaxweed, see Camelina sativa
Fontinalis 49, 438
Fontinalis antipyretica 422, 428
Fragaria vesca 476
Fragilaria 419
Franseria **212**
Franseria dumosa **213**
Fucales 104
Fucus 104
Fungi, see Mycophyta

Galium aparine 397, 401
Galium hercynicum 404
Gallionella 72, **73**
Geranium robertianum 405
Geum reptans 181
Gladiolus 475, 488, 489
Gleocystis 419
Gloeotrichia 56
Glyceria borealis 57
Glyceria fluitans 438
Glyceria maxima 418
Glycine max **131**, 138, 144, 154, 155, **159**, **160**, **161**, 367, 368, 397, 488, 489, 515
Goethalsia meiantha **288**, **289**
Gomphonema parvulum 421, 425
Gomphosphaeria 419
Gonium 515
Gonyaulax tamarensis 103
Gossypium **131**, 133, 135, 136, 137, 156, **159**, 168, 333, 361
Gossypium hirsutum 511, 512, 515
Gourd, see Cucurbitaceae
Gramineae 533
Grape, see Vitis
Grayia spinosa **213**
Green algae, see Chlorophyceae
Groenlandia densa 434

Haematococcus 426
Hammada **214**
Hammada scoparia **212**, **221**, 222
Harpalus rufipes 403
Helianthus annuus **208**, 486, 534, 535
Heliconia latispatha **288**
Hemlock, see Tsuga
Hesperochloa kingii **184**
Heteromeles 472
Heteromeles arbutifolia 220
Hevea brasiliensis 362
Hickory, see Hicoria
Hicoria **17**
Hildenbrandia 421
Holcus lanatus 404
Homo erectus 531
Homo habilis 531
Homo sapiens 531, 533
Hordeum 166, 168, 238, 361, 365, 401, 541
Hordeum jubatum 397
Hormidium 430
Hormidium rivulare 425, 429
Hormogonales 56
Hydrilla verticillata 436
Hydrodictyon africanum 436
Hygroamblystegium 428
Hymenoclea salsola **213**

Ipomoea batatas **159**, 534, 535
Iris pseudacorus 424
Isoëtes 418
Isoëtes lacustris 438

Jacquinia pungens 296
Juncus bulbosus 438
Juncus effusus 404, 438

Kale, see Brassica oleracea
Kelp, see Phaeophyceae
Koeleria cristata 189
Krameria grayi **213**

Laminariales 104
Larrea 211, **214**
Larrea divaricata **212**, **213**, **217**, 219, 221, 222, **224**
Leersia oryzoides **178**, **184**
Leguminosae 11
Lemna 431, 432, 438
Lemna gibba 424
Lemna minor 70, 422, 424, 442
Lemna trisulca 436
Leptocylindrus **94**
Leptothrix 72, **73**
Leucopoa kingii **178**
Licania heteromorpha 290
Lichenes 12, **175**, **177**, 178, 181, 480, 481, 543
Lima bean, see Phaseolus lunatus
Linum usitatissimum 536
Liriodendron 27, 269
Liriodendron tulipifera **244**, **249**, 250, **254**, 256, 259, **260**, 264
Loasaceae 533
Lobelia dortmanna 418, 438
Loiseleuria 181
Lolium 471, 515
Lolium multiflorum 193
Lolium perenne **160**, **165**, 179, 404, 473, 487, 490

Lucerne, see Medicago sativa
Lupinus 404
Lupinus arizonicus **224**
Luzula confusa 543
Lycium andersonii **213**
Lycopersicon esculentum 135, 166, 476, 515, 518, 519, 534

Macaranga hypoleuca **288**
Macrocystis 104
Macrosiphum avenae 401
Maize, see Zea mays
Malacosoma americanum 402
Mallomonas 70
Mallotus paniculatus **288**
Malus 262, 264, 397, 402
Mangletia glauca **288**
Manihot esculenta **159**, **160**, 534
Maple, see Acer
Medicago 157, **160**
Medicago sativa **131**, **159**, 166, 168, 169, 404, 472, 473, 481, 518
Melosira nummuloides 436
Mentzelia 533
Mesquite, see Prosopis
Metallogenium 74
Methanobacter 428
Methanobacterium **50**, 68
Methanococcus **50**
Micranda spruceana 287
Micrasterias 356
Micrococcus 58
Microcystis aeruginosa 432
Molinia coerulea 404
Mosses, see Bryophyta
Mougeotia 419, 425, 438
Muhlenbergia 176
Muhlenbergia filiculmis **178**
Musa 537
Mustard, see Sinapis
Mycophyta **22**, 29, **50**, 58, 66, 331, 333, 398, 399, 421, 426, 428, 432
Myriophyllum alterniflorum 418, 420, 436
Myriophyllum brasiliense 436
Myriophyllum exalbescens 440
Myriophyllum spicatum 60, 418, 420, 431, 432, 436, 437, 440
Myriophyllum verticillatum 418
Myrtaceae 22

Najas 418
Najas guadalupensis 436, 440
Napiergrass, see Pennisetum purpureum
Nardus stricta 404
Navicula 438
Navicula pelliculosa 432
Navicula seminulum 431
Neurospora 428
Nicotiana tabacum **159**, 362, 475, 485
Nitella 426, 431
Nitella flexilis 429
Nitrobacter 57
Nitrosomonas 57
Nitzschia **107**
Nitzschia linearis 442
Nitzschia palea 421, 425, 426
Noctiluca 104
Nostoc 56
Nothofagus 53, 234
Notholaena parryi **217**, 219
Nuphar luteum 54

Oak, see Quercus
Oat, see Avena
Ochrobium **73**
Ochromonas 70
Oedogonium 419, 425, 428
Oil palm, see Elaeis guineensis
Onion, see Allium cepa
Opuntia basilaris **217**, 224, **225**
Orange, see Citrus sinensis
Orchidaceae 291
Oryza sativa 132, **133**, 134, 156, **159**, **160**, **162**, **163**, 169, 334, 367, 473, 487, 518, 535
Oryzopsis hymenoides 533
Oscillatoria 90, 419, 421
Oscillatoria agardhii 60
Oscillatoria redekei 51, 63
Oscillatoria rubescens 419
Oxyria digyna 181

Panicum 166
Panicum virgatum **178**, **184**
Papaver rhoeas 401
Papaya, see Carica papaya
Parashorea densiflora **288**
Pea, see Pisum sativum
Peach, see Prunus persica
Peanut, see Arachis hypogaea
Pear, see Pyrus communis
Pedunculate oak, see Quercus robur
Pelargonium hortorum 472
Pennisetum 166
Pennisetum purpureum **160**, 162
Pennisetum typhoides **163**
Perdix perdix 400, 402
Peridinium cinctum 70
Persea 534
Petunia 475
Petunia hybrida 485
Phaeophyceae 104
Phaseanus colchicus 400

Phaseolus 487
Phaseolus lunatus 534
Phaseolus vulgaris 474, 476, 477, 485, 489, 534
Phlomis fruticosa **210**
Phormidium 421
Phragmites 418, 419
Phragmites australis 70, 424, 431, 438
Phycophyta 51, 59, 60, 61, 63, 64, 66, 69, 72, 74, 316, 356, 365, 366, 371, 398, 417, 418, 419, 421, 423, 425, 427, 431, 432, 435, 441, 514, 515, 543
Picea **17**, 24, **25**, 26, 290
Picea abies **236**, **249**, 254, 261, 262, **270**, 271
Picea mariana 545
Picea sitchensis **236**, 246, **249**, 250, 252, 256, **257**, **259**, **260**, 264, 268, 269
Pieris brassicae 400
Pieris rapae 400, 403
Pinaceae 22
Pine, see Pinus
Pineapple, see Ananas sativus
Pinus **17**, **25**, 143, **207**, 211, 472, 480, 482, 488, 544
Pinus aristata 261
Pinus cembra 261
Pinus contorta **260**
Pinus densiflora 240, 250
Pinus halepensis **210**
Pinus monophylla 534
Pinus monticola **38**
Pinus nigra **260**
Pinus nigra var. nigra **236**
Pinus pinea **210**
Pinus radiata 20, **236**, **249**, **260**, 268
Pinus resinosa 26, **249**
Pinus rigida 240, 263, 265
Pinus strobus **249**
Pinus sylvestris 31, **249**, 250, 251, **253**, **255**, 261, 262, 264, 269, **270**, 472
Pinus taeda 19, 250, 264
Pistia stratiotes 436
Pisum 518
Pisum sativum 365, 473, 485, 488, 520
Poa trivialis 404
Polygonum convolvulus 402
Polygonum lapathifolium 397
Polygonum orientale 488
Polygonum persicaria 397
Polylepis incana 540
Polytrichum 181, 438
Polytrichum alpinum 181
Populus balsamifera **236**
Populus deltoides **236**
Populus tremula **249**, **260**
Populus tremuloides 240
Potamogeton alpinus 436
Potamogeton coloratus 420
Potamogeton lucens 418
Potamogeton nodosus 421
Potamogeton pectinatus 60, 418, 424, 431, 434, 440
Potamogeton perfoliatus 418, 436, 440
Potato, see Solanum tuberosum
Prosopis 404, 544
Protium glabrum **288**
Protozoa 72, 100, 427
Prunus armeniaca 222, 476
Prunus avium 476
Prunus persica 476
Pseudomonas 57, 58
Pseudomonas manganoxidans 73
Pseudomonas perfectomarinus 96
Pseudotsuga menziesii 34, **236**, **249**, **261**, 264, 269
Psilopilum cavifolium 543
Pteridium aquilinum 404
Pteridophyta 56, 218, 219, 291, 472
Pyramimonas **107**
Pyrus communis 476

Quercus **17**, **25**, 53, 237, 241
Quercus agrifolia **210**
Quercus alba **260**, 269
Quercus coccifera **210**
Quercus coccinea 264
Quercus ilex **207**, **208**, 209, **210**, 264, 296
Quercus robur **249**, 266, **271**

Radiolaria 100
Rangifer tarandus 532
Ranunculus arvensis 401
Ranunculus fluitans 421, 422, 423
Ranunculus glacialis 181
Ranunculus peltatus 418
Ranunculus pseudofluitans 434
Rape, see Brassica napus
Red algae, see Rhodophyceae
Red clover, see Trifolium pratense
Rhaphanus raphanistrum 400
Rhizobium leguminosarum 520
Rhizoctonia solani 399
Rhizosolenia 90
Rhodophyceae 104, 425
Rhopalosiphum padi 401
Rice, see Oryza sativa
Rivularia 419
Robinia pseudoacacia 296
Rosa hybrida 517
Rubus 182
Rubus idaeus 405
Rumex acetosella 401
Rumex obtusifolius 404
Ryegrass, see Lolium

Saccharum **160**, 362, 401, 402, 535
Safflower, see Carthamus tinctorius
Salicornia fruticosa **208**
Salix 177
Salix matsudana **236**
Salix pulchra 178
Salix purpurea **236**
Salix rotundifolia 545
Salmo gairdneri 57
Sargassum 105
Scapania undulata 425
Scenedesmus 49, 54, 61, 426, 427, 434
Scenedesmus quadricauda 419, 423, 442
Schizochlamys 421
Schizomycetes 12, 29, 36, 49, **50**, 52, **55**, 56, 57, 58, 59, 63, 64, 66, 67, 68, 69, 70, 72, 73, 74, 85, **87**, 88, 90, 95, 96, 97, 98, 103, 104, 398, 415, 416, 417, 422, 426, 427, 428, 431, 432, 441, 487, 519, 520
Schoenoplectus lacustris 432
Scirpus 60
Scleranthus annuus 401
Scots pine, see Pinus sylvestris
Sedges, see Cyperaceae
Senecio jacobaea 404
Senecio vulgaris 368, 369, 370
Sequoia sempervirens **236**, 528
Shorea leprosula **288**
Shorea macroptera **288**
Shorea pauciflora **288**
Siderocapsa 74
Siderocapsa geminata **73**
Siderococcus **73**
Silicoflagellatae 100
Sinapis 364
Sinapis arvensis 400, 401
Sitka spruce, see Picea sitchensis
Skeletonema **107**
Skeletonema costatum 92, **94**, 107
Sloanea berteriana 290
Solanum nigrum 397
Solanum tuberosum **159**, **160**, **163**, 166, 362, 396
Solanum tuberosum ssp. andigena 534, 535
Solanum tuberosum ssp. curtilobum 534, 535
Sorghum **129**, **162**, 166, 334, 367, 487, 488, 518, 535
Sorghum bicolor **160**
Sorghum halepense 397
Sorghum vulgare **163**
Soybean, see Glycine max
Sparganium simplex 438
Spartina patens 544
Spartina pectinata **178**, **184**
Sphaerotilus natans 421
Sphagnum 437, 438
Spinach, see Spinacia oleracea
Spinacia oleracea 481, 486, 490
Spirodela 432
Spirogyra 419, 421, 425, 428
Sporobolus heterolepis 189
Spruce, see Picea
Spumella 70
Squash, see Cucurbitaceae
Stellaria edwardsii 543
Stellaria media 397, 401, 402
Stigeoclonium 425, 429, 430
Stigeoclonium tenue 421
Stipa 540
Stipa viridula 189
Strawberry, see Fragaria vesca
Sudangrass, see Sorghum vulgare
Sugar beet, see Beta saccharifera
Sugar beet yellows virus 155
Sumac, see Anacardiaceae
Sweet chestnut, see Castanea sativa
Sweet potato, see Ipomoea batatas
Synechococcus lividus 67
Synura 70

Tabellaria 419
Tea, see Thea sinensis
Thalassiosira **107**
Thalassiosira fluviatilis **94**, 436, 437
Thalassiosira pseudonana **94**, 103
Thea sinensis **159**
Theobroma cacao **159**, 537
Thiobacillus **67**, 68, 72
Thiorhodaceae 68
Thiothrix 68
Thuja plicata **236**
Tidestromia oblongifolia **217**
Tobacco, see Nicotiana tabacum
Tomato, see Lycopersicon esculentum
Tribonema 425
Trichodesmium 90
Trifolium 153, **154**, 518
Trifolium pratense 403, 404
Trifolium repens 403, 404
Trifolium subterraneum **516**
Triticum 126, **155**, 157, **159**, **160**, **162**, 164, 166, **167**, 168, 169, 238, 361, 397, 511, 517, 518
Triticum aegilopoides 533
Triticum dicoccoides 533
Triticum dicoccum 533, 536
Triticum vulgare 153
Tsuga 26
Tsuga heterophylla **236**
Tulip poplar, see Liriodendron tulipifera
Tulipa 128

Typha angustifolia 436
Typha latifolia 194, 418, 436, 438

Ulex 404
Ulmus 53
Ulmus americana 490
Ulothrix zonata 438
Ulvella frequens 421
Uroglena americana 419

Vallisneria americana 436
Vaucheria 421, 434
Vaucheria sessilis 432
Veronica polita 402
Vicia faba 401, 472, 473, 482
Vigna 166
Virola koschnyi **288**
Vitis 361

Water hyacinth, see Eichhornia crassipes
Wheat, see Triticum
White clover, see Trifolium repens
White pine, see Pinus monticola
Wild emmer, see Triticum dicoccoides
Wolffia 438

Xanthophyceae 69

Yellow birch, see Betula allegheniensis

Zannichellia 418
Zea mays 152, 153, **154**, 155, 156, **159**, **160**, **161**, **162**, **163**, 165, 169, 334, 362, 367, 368, 369, 397, 511, 512, 515, 519, 534, 535
Zygnema 425
Zygophyllum dumosum 211, **212**

Subject Index

General index for all four parts of Volume 12, *Physiological Plant Ecology*, containing a detailed index for the present volume (part IV), with most important keywords from parts I, II, and III. Letters placed before page numbers refer to Volumes 12A–D (cf. Contents). Page numbers in **bold face** refer to tables or figures.

Abscisic acid, absorption spectrum A **171**
affected by water potential B281ff.
effect of environmental conditions B286
effect of water stress B278, B279, B281ff., B**302**ff.
effect on growth B286
effect on hydraulic conductance B285
effect on stomatal closure B18, B281, B288ff., B460
growth inhibition B273
metabolism B284
phloem transport B304
photoperiodism A142
proton transport inhibition B273
quantum yield A172
seed development B305
seed germination B406, B407
use as growth regulator D358, D359, D361
Abscission of leaves, effect of water stress B279, B**302**ff.
Absorption, radiation on plant A18
Acclimation, drought B330, B333
light A94, C451, D258
quantum requirement D258
temperature A283, A299, C449, D222, D223, D545
desert plants D222, D223
effect of oil D545
UV radiation A190
Acid phosphatase D63
Acid rain C202
effect on ecological balance D477
effect on pH of lakes D438
effect on plants D474
Acidic soil, effect on plant life C207ff.
Acidification, effect of ammonium C217
freshwater ecosystems D437, D438
Acidophilic plants, definition C203, C204
resistance to aluminum C213
Actinomycetes, heat resistance A515
Actinorrhiza C302
cross-infection studies C335
culture of the endosymbiont C334
development C336
distribution C337
geochemical significance D11
physiology C335ff.
Actinorrhizal nodules, nitrogen fixation C308, C333ff.
Activation energy A292
and growth A342
for mitochondrial respiration A308
for RuBP carboxylase-oxygenase A304
of glycolysis A310
of membrane-bound enzymes A384
Adaptation, and changes of microbial populations A365
of microorganisms to temperature A347
to fire A515ff.
to secondary thermophily A366
to temperature A299, A366
Aerenchyma, root ventilation B463ff.
Aerial roots, water uptake B119
Agricultural ecosystems D151ff.
co-evolution of plants and man D535, D536
origin and spread D532ff.
Agricultural production, and world population D**332**
global D158ff.
seasonal pattern D165
Agricultural systems, replacement of natural ecosystems D537, D538
Agriculture, technological influences D**155**
Air blockage, vessels B139
Air capacity, soil A546
Air content, soil A550
Air pollution D465ff.
cold hardening A415
effect on lichens C451ff.
Algae, cold resistance A**408**
CO_2 fixation D49
desiccation B357
epilithic D66
heat tolerance A451, A461, A463
photoadaptive responses to white light A266
Alkaline phosphatase D63
Alkaloids, in fruits C643
in seeds C648
Allelopathy C694ff.
after fire A510, A521
Allocation of nutrients, in plant communities D5ff.

Allocation of photosynthates A72, B626ff.
root B516
tundra and grassland D194ff.
Allochthonous carbon, aquatic ecosystems D49, D52
Allophycocyanin, adaptochrome-like function A264
in algae, light-harvesting function A261
Alpha rays, biological effect A201
Alpine plants, frost resistance **A411**
leaf conductance B213
Aluminum, effect on DNA synthesis C213
effect on membranes C213
effect on root branching C**212**
polynuclear pH effect C256, C**257**
Aluminum tolerance C252, C**253**
Aluminum toxicity C260
Amino acid synthesis, effect of herbicides D354
Amino acids, cytoplasmic osmotica B57
floral secretions C585
nectar constituents C602ff.
Ammonification C163, C168, C173, D59
after fire A513
effect of soil moisture C175
temperature dependence C175
Ammonium, absorption by soil and plant D11
uptake by aquatic macrophytes D421ff.
Ammonium nutrition, effect on root growth C181
pH dependence C187, C188
Anaerobis point, soil A551
Anemophilic pollination C560
Aneuploidy, ionizing radiation effect A203
Animal-chloroplast associations C480ff.
Annuals B**619**
blooming C561
carbon allocation B627, B628
competition B639
growth characteristics B634ff.
photoperiodism A142
seed production B636ff.
structural forms B627
vegetative growth B639
Anthesis C565ff.
Anthocyanin pigmentation, photoperiodism A142
Anthocyanins, in flowers C575, C576
Antidotes, for herbicides D366ff.
Aphotic zone, H_2S occurrence D68
Apoplasmic water, drought adaptation B67ff.
tissue water potential B54ff.
water potential isotherm B**46**
Apoplastic water movement B90, B92ff., B**95**, C17
endodermis C**22**
Apparent free space, root B92
Aquatic angiosperms, xylem B141
Aquatic ecosystems, acidification D437, D438
bioindicators for pollutants D442, D443
carbon concentration D508, D509
diversity A250
effect of waste heat D439ff.
eutrophication D414ff.
phytoplankton productivity D305ff., D306ff., D311ff., D316, D321
measurement D321
physiological adaptations D316
regulation D311ff.
seasonal patterns D306ff.
pollution D424ff.
Aquatic environment A229ff., B489ff., D47ff., D85ff., D413ff.
irradiance A238, A242
light spectrum A249
nitrogen fixation A236
photic zones A254, A255
UV radiation A181
Aquatic macrophytes, CO_2 fixation D49
effect of eutrophication D418ff.
excretion of photosynthetic products D51
role in nutrient cycles D104, D105
uptake and translocation of nutrients D421ff.
Aquatic plants, CO_2 compensation point B493
CO_2 concentrating mechanism B491ff.
CO_2 response D514, D515
CO_2 users B492ff.
HCO_3^- users B492ff.
ion pumps C17ff.
light regulation A263ff.
light responses A249ff.
nitrate reductase D60
phosphorus uptake D64
photosynthesis B489ff.
productivity A**252**
uptake and translocation of nutrients D421ff.
Archaebacteria C149
Arctic plants, frost resistance **A411**
leaf conductance B213
Arid conditions, photosynthetic adaptation D220ff.
Aridity gradients D220
Ascospores, germination B438, B440
Ash, deposition A508
Assimilates, allocation B514ff., B627ff., B640ff.
annuals B627, B628
herbaceous perennials B628, B629, B640ff.
woody perennials B629, B630

partitioning B 22, B 514ff.
Assimilation, see Photosynthesis
Atmospheric CO_2 concentration, ecophysiological effects D 503ff.
increase by burning fossil fuels B 499
Atmospheric dust A 14
Atmospheric heat balance D 504
Atmospheric photo-chemistry D 469
Atmospheric pollutants, ecophysiological effects D 465ff.
ATP content, during desiccation B **366**
ATP production B 564ff.
ATP synthesis, effect of herbicides D **340**, D 347, D 355
ATP synthetase content, light regime A 84
Attractants, flowers C 574ff., C 578ff., C 587ff., C 591, C 592, C 593ff.
Autochthonous carbon, aquatic ecosystems D 49
Autochthonous primary producers, aquatic ecosystems D 49
Autotroph-heterotroph symbiosis C 469ff.
Auxin, see Indolacetic acid
Auxin herbicides D **341**
Azimuthal angles, leaf D 245
effect on radiation interception D 245

Bacteria, heat resistance A 461, A 514
nitrifying D 57
Bacterial production, aquatic ecosystems D 49, D 51
measurement D 51
Bacteriocytes, definition C 470
Bacterioplankton D 51
Bacteroids, characterization C 312
Basiphilic plants, definition C 203, C 204
Bead pond formation, in tundra D 543
Begleiter, hydroids of bryophytes B 122
Belowground productivity, desert plants D 211
Benthic macrophytes, zonation A 255
Beta rays, biological effect A 201
Bicarbonate fixation C 232
Biennials B **619**, B 629, B 640
Biliprotein, adaptochrome A 263
chemical structure A 261
in algae A 257, A 261
in aquatic plants A **251**
Biocide effectiveness, influence of environmental conditions D 394ff.
Biocide effects, model ecosystems D 403
Biocide production, and world population D **332**
Biocide resistance D 333, D 397, D **398**
Biocides, ecological implications D 333, D 393ff.
effects in aquatic ecosystems D 403
effects on agro-ecosystems D 401ff.
effects on faunal populations D 399ff.
effects on forest ecosystems D 405
effects on grassland ecosystems D 404
effects on plant populations D 394ff.
effects on soil microbial populations D 398, D **399**
history and development D 330ff.
inhibition of respiratory electron transport D 348
mode of action D 335ff.
physiological effects D 329ff.
plant susceptibility D 396ff.
translocation within plants D 396
uncoupling of mitochondrial phosphorylation D 348
Biocides, see also Herbicides
Bioindicators, for pollutants D 442, D 443
Biological efficiency, radiation A 202
Biomass, daily production B 20
different organs B **647**
different plant life forms B 660ff.
effect of stress treatment B **638**
Biomass distribution, desert plants D 211ff.
mediterranean-climate plants D **207**ff.
tropical forests D 282ff.
Bipyridylium herbicides, mode of action D 345, D **346**
Blooming periods C 562
Blue light, action spectrum A **117**
influence on aquatic plants A 265
influence on fungi A 266
Blue light photoreceptor A 117, A 127
chloroplast movement control A 268
phototaxis A 268
Blue-green algae, associations with other organisms C 338ff.
morphological groups C 352
nitrogen fixation C 302, C **352**ff.
Borderline metals, toxicity and tolerance C 261ff.
Boreal forest, fire effect A 505
Boreal forest zone, soil temperature B 381
Boron content, radiosensitivity A 214
Bound water B 42
Boundary layer, air A 477
benthic A 233
eddy transfer A 478
effect on pollutant uptake D 470, D 471
thickness A 480
water motion A 232
wind effect A 478
Boundary layer resistance, leaf A 23
leaf characteristics B 235
Bryophytes, desiccation B 362ff.
heat tolerance A 452, A 461, A 463
Bulb formation, photoperiodism A 141
Bulk density, soil A 540, A 544

Bundle-sheath cells A 98
Buttresses, wind effect A 493

$\delta^{13}C$ value B 494 ff.
carnivorous plants C 185, C **494**
seasonal progression D **186**
tropical plants D 298, D **299**
Cadmium toxicity C 274, C 275
Calcareous soil, effect on plant life C 222
Calcicole plants, definition C 203, C 204, C **223**
Calcification, corals C 478
Calcifuge plants, definition C 203, C 204, C **223**
Calciophobic plants, definition C 222, C **223**
Calciotrophic plants, definition C 222, C **223**
Calcium, after fire A 508, A 512
effect on photosynthesis C 209, C 225
Calcium deficiency C 223
effect on metabolism C 209
radiosensitivity A 214
Calcium uptake, root C **226**
CAM, effect of salinity C 112, C 116
malate oscillations C 44
CAM induction B **505**, B 529, B **531**
effect of water deficit C 112, C 116
CAM plants, $\delta^{13}C$ value B 529 ff.
characterization B 498, B 504
CO_2 concentrating mechanisms B 488, B 489
CO_2 exchange B 205
CO_2 recycling B 532
cytoplasmic pH regulation C 23
drought avoidance D 223
effect of herbivory D 191
electrolyte concentration B 56
heat tolerance A 455, A 464
photorespiration B 489
photosynthesis B 487 ff.
photosynthesis and leaf conductance D 217
productivity B 519, B 532
stomatal response B 489
water use efficiency D 224
Canopy architecture, calculation models D 164
measurement D 245
Canopy environment A 30 ff.
radiation interception D 243 ff., D **249**
water loss B 253, B 254
Canopy extinction coefficient A 33
Canopy model, photosynthesis A 37, B 581
transpiration A 37
Canopy transmission coefficient A 33, A 34
Carbohydrate content, effect of trampling D **541**
Carbohydrate storage, impact of man D **542**
Carbohydrate transport, cost D 138
external and internal factors D 138
Carbohydrates, synthesis respiration D 138
Carbon, in aquatic ecosystems D 49 ff.
in terrestrial ecosystems D 507
Carbon allocation, after fire A 520
Carbon assimilation, see Photosynthesis
Carbon balance D **121**
Carbon cycle, global D 504 ff.
Carbon cycling, aquatic ecosystems D **50**
Carbon exchange, between atmosphere and biosphere D 507
Carbon isotope discrimination A 94
relation to photosynthetic pathway and environment B 494 ff.
Carbon monoxide, effect on plants D 487
production by plants D 487
Carbon uptake, aquatic ecosystems D 48 ff.
Carbonic anhydrase, effect of water stress B 293
Carnivorous plants C 489 ff.
$\delta^{13}C$ value C **494**
δD value C **494**
digestive enzymes C 504 ff., C 507
genera C **491**
geographical distribution C **491**
nutritional requirements C 492 ff.
β-Carotene, in algae A 260
Carotenoid biosynthesis, effect of herbicides D 349 ff.
Carotenoids, in algae A 257, A 258, A 260
in aquatic plants A **251**
Casparian strips B 90, B 92, C 104, C 107
Cation uptake, mycorrhiza C 398
Cauliflory C 629
Cavitation, vessels B 54, B 144 ff.
Cell elongation, radicle protrusion B 402, B 409, B 410
Cell potential, effect on ion pumps C 14 ff.
Cell size, drought adaptation B 69
mesophyll surface A **90**
Cell wall, elasticity B 49 ff., C 41
Cell wall formation, effect of herbicides D 355
Cell wall relaxation, indolacetic acid B 273
Cephalodium, metabolism C 432, C 433
Chamaephytes B **619**
Chaparral, fire effect A 526
nitrogen volatilization A 509
productivity D **207**, D **208**
soil conditions A 506
Chelator D 53, D 61
Chilling, adaptation A 406
cell ultrastructure A 379
definition A 404
photoreductive capacity A 378
photosynthetic depression A 420
stem extension A 140
Chilling injury A 377 ff., A 405 ff.
Arrhenius relationship 'breaks' A 380 ff.
C_4 plants B 509
mechanism A 379
seed germination B 408

Chilling sensitive plants A 97, A 286, A 382ff., A 405ff., A 419
photosynthesis A 419
potential distribution A 426
protein synthesis A 317
Chironomid larvae D 69
Chloride, in freshwater ecosystems D 435ff.
Chlorophyll, absorption spectra A 259, A **260**
effect of desiccation B 333, B 365, B 369
fluorescence spectra A 259
in algae A 257, A 258, A 259
in aquatic plants A **251**
photoreceptor A 116, A 127
Chlorophyll a/b ratio A 67, A 84
in algae A 259
Chlorophyll bleaching A 185
and light A 62, A 95
effect of biocides D 335, D **340**, D **349**ff.
effect of bipyridylium herbicides D 346
Chlorophyll content, and action spectrum A 47
correlation to light saturated photosynthesis A 78
light regime A **76**, A **77**, A **83**
of algae A **265**
quantum yield A **69**
shade and sun plants A 67
Chlorophyll fluorescence, heat injury effect A 373
Chlorophyll synthesis, effect of herbicides D 352
Chlorophyllase activity, in marine plants A 270
Chloroplast, desiccation damage B 357
effect of CO_2 concentration D **516**
effect of sulphur dioxide D 481ff.
in animal cells C 480ff.
ion concentration C 73
Chloroplast biogenesis, effect of herbicides D 352
Chloroplast formation, heat injury A 457
Chloroplast ribosomes, synthesis inhibition by high temperature A 317
Chlorosis C 186
cadmium-induced C 274
copper-induced C 276
effect of fluoride D 489
lead-induced C 284
lime-induced C 217, C 219, C 229ff., C 230
nickel-induced C 269
zinc-induced C 264
Chromatic adaptation, aquatic plants A 263
gene regulation A 263
to blue green light A 264
Chromium toxicity C 268
Chromophore, UV A **171**
Circadian rhythm, light control A 267
phytochrome A 160
sleep movements of leaves A 161
Clouds, UV spectral irradiance A 179
Clumped vegetation, light penetration A 34
Cobalt toxicity, mechanism C 273
Coevolution, plant-herbivores C 672, C 673
Cohesion theory B 136
Cold acclimation A 390ff., A 410
effect of oil D 545
Cold hardening A **414**, B 69
of juvenile plant stages A 423
Cold resistance A 403ff.
adaptive A 419
specific plant types A 408
Cold susceptibility, definition A 404
Colligative properties, solutions C 36
Compatible solutes, cytoplasm B 43
Compensatory growth, effect of grazing D 192
Competition C 679ff.
C_3/C_4 plants A 289, D 191
definition C 203, C 680
effect on photosynthesis B **609**
effect on plant distribution C 204ff.
effect on transpiration B **609**
for dispersers C 637
for light C 681, C 686ff.
for mineral nutrients C 689
for nodulation C 329
for pollinators C 697
for water C 692ff.
root systems B 608
within plant communities B 624
Competition, see also Interference
Competitive index C 681
Comprehensive model, definition D 126
examples D **131**
productivity D 130ff.
Compressive modulus, definition B 55
Conductance, cell wall B 92
cuticular B 386ff.
hydraulic B 13, B 81, B 85, B 87, B 142, B 275, B 276, B 285, B 380ff., B 382, B 383, B 413ff., C 42, C 113, C 118, D 251
and water flow B 87
coefficient B 81, B 85, B 103
effect of salinity C 118
influence of air bubbles B 382, B 383
influence of soil frost B 380ff.
root B 13, B 285
seed/soil interface B 416, B 417
soil B 413ff.
wood D 251
xylem B 142
leaf A 86, A **87**, B 182ff., B 190ff., B **192**, B 196, B 199, B **202**, B 203ff., B 208ff., B 213ff., B 221, B 222, B 239, B 592, B 605, D 216

Conductance, leaf
age effect B186, B**187**
changes during drought B**200**
cold environments B213, B214
desert plants D216
genotype B187ff.
humidity effect B221
light effect A**87**, B**185**
operational range B182ff., B214ff.
photosynthesis A86, A**87**, B182ff., B**184**, B192, B203ff.
photosynthetic capacity B182ff., B**185**, B186, B**187**, B**202**
mesophyll D258
effect on photosynthesis D258
root/soil interface B100ff.
stomatal B10, B17, B18, B20, B190ff., B**193**, B**194**, B198ff., B202, B244ff., B**302**, B303, B597, D258
and abscisic acid B18
and turgor B10
effect on photosynthesis D258
leaf/air vapor pressure difference B190ff., B**193**, B**194**
relation to mesophyll capacity for carbon fixation B597
relation to photosynthesis B20
threshold responses B244ff.
Conductance, see also Resistance
Coniferous forest, fire effect A506, A510ff., A524
Convergence, morphological structure B616
Conversion efficiency, calculation D138
Convexity coefficient D258
Copper, biochemical effects C**277**
effect on membranes C278, C279
effect on phosphate uptake C**279**, C**280**, C**284**
effect on root growth C**251**, C277, C**278**
role in marine ecosystem D102, D103
Copper tolerance C250, C**251**, C**254**
examples C**281**
Copper toxicity, symptoms and mechanisms C276ff.
Coralloid roots, nitrogen fixation C342
Corals, carbon fixation by the symbiont C478
CO_2 assimilation, see Photosynthesis
CO_2 compensation point, definition B556
effect on photosynthesis D258
CO_2 concentrating mechanisms, aquatic plants B491ff.
CAM plants B485ff., B488, B489
C_4 plants B485ff., B500, B506
CO_2 concentration, atmospheric D503ff., D504, D509ff., D517, D518ff.
ecophysiological effects D503ff.
effect on climate D509, D510
effect on heat balance D504
effect on heterotrophic tissues and organisms D518ff.
increase D505, D506
limiting factor in plant growth D511, D512
plant responses D510ff.
effect on cellular structure D515
effect on chloroplasts D**516**
effect on nitrogen fixation C322
intercellular B182
morphological effects D517
stomatal responses B243, B244, D517
CO_2 dark fixation D**519**
CO_2 diffusion rate, in ice A419
CO_2 exchange, relation to water content B**358**ff.
lichens B**358**ff.
CO_2 fixation, different pathways B479ff.
CO_2 solubility in water D514
CO_2 uptake, see also Photosynthesis
Crassulacean acid metabolism, see CAM
Crop plants, co-evolution with man D533
ecophysiological bases D532ff.
growth rates D163
photosynthesis D535
photosynthetic pathways D535
productivity D157, D533
statistics D157
Crop rotation D153
Cross-protection, viruses C541
Cryptophytes B**619**
Cushion plants, frost resistance A**411**
Cuticle proper B166ff.
Cuticular membrane, chemistry and structure B165ff.
effect on pollutant uptake D470
model B163
permeability and composition B175
permeability and structure B155ff.
resistance B159, B**160**
transport properties B172ff.
water permeability B157, B158, B**160**, B161, B**173**
wettability B113
Cuticular transpiration B153ff., B159
Cutin, cross-linking B166
Cyanelles C470, C471
Cyanobacteria, see Blue-green algae
Cyphellae B126
Cytochrome f A**85**
Cytokinin, effect of herbivory C669
effect of water stress B280, B281, B**302**ff.
effect on mycorrhizal symbiosis C400
production by Rhizobium C313
synthesis in the root apex B459
use as growth regulator D358, D359, D361

Cytokinin content, reduction at high temperatures A 374, A 376
Cytoplasm, buffer capacity D 481
effect of sulphur dioxide D 481
Cytoplasmic osmotica B 57
Cytoplasmic pH regulation, root C 23
Cytoplasmic ribosome, synthesis inhibition by high temperature A 317
Cytoplasmic volume, regulation C 87
C_3 plants, CO_2 concentration and quantum yield A 289
CO_2 exchange B **204** ff.
dominance at high elevation A **288**
drought responses B 203 ff.
leaf conductance B **204** ff.
light acclimation A 60
light efficiency A 70
nitrogen requirement A 289
photosynthesis A **63**, A 301, A **302**, B **216**, B **217**, B **219**, B 569
photosynthetic capacity B 182
quantum yield A 68 ff., A 289, A 419, B 483
temperature response A 287
C_3/C_4 intermediate plants, $\delta^{13}C$ value B 511
C_3/C_4 plants, allocation strategies D 194
competition A 289, D 191
distribution B 526
effect of atmospheric CO_2 concentration D 511
effect of herbivory D 191
effect of sulphur dioxide D 482, D 483
grain yields D **162**
light response D 179
nitrogen use efficiency B 510, D 191
photosynthesis and leaf conductance D 218
productivity B 518 ff., D 158
effect of growing season B 519 ff.
effect of plant competition B 522
quantum efficiency D 163
temperature response B 515, B **517**, D 184 ff., D 191, D 219, D **221**, D **222**
US grassland D **177**
water potential D 189
water use efficiency B 514 ff., D 189
C_3/C_4/CAM plants B 479 ff.
among different life forms D 215 ff.
carbon isotope discrimination B 494 ff.
crops D 535
distribution B 523 ff.
in tropical forests D 291, D **292**
intercellular CO_2 concentration B 489 ff.
light response curves B 505 ff.
photosynthetic CO_2 response curves B **501**, B **503**
quantum yield B 506, B 507
stomatal frequency B 503
temperature response B 508
water use efficiency B 501 ff.
C_4 plants, chilling injury A 406, A 419, A 420, B 509
CO_2 concentrating mechanism A 81, B 485 ff., B 500, B 506
bundle-sheath cells B 486, B 500
productivity B 506
CO_2 exchange B 204 ff.
distribution in North America B **525**
dominance at low elevation A **288**
drought responses B 203 ff.
heat tolerance A 454, A 464
leaf conductance B **204**, B 205
light acclimation A 60
light efficiency A 70
nitrogen requirement A 289
osmotic adjustment B 502
photoinhibition B 488, B 509
photosynthesis A 69, A 301, A **302**, B **218**, B **219**, B 485 ff., B 580
mathematical model B 580
photosynthetic capacity B 182
productivity B 518 ff.
quantum yield A 68 ff., A 289, A 419
summer annuals A 289
temperature response A 287

δD value, carnivorous plants C **494**
C_3/C_4/CAM plants B 498, B **499**
Darcy's Law A 550, B 100, B 103
water flux density B 12
Dark respiration, shade and sun plants A 60, A 71
Daylength A 136 ff.
flowering response A **137**
latitude A 136, A **137**
seed germination A 138
Daylength, see also Photoperiod
Daylight, spectral distribution A 151
Deciduous forests, northern limits A 424
productivity D 233 ff.
Deciduous leaves, characterization B 633, B 634
Deciduous perennials B **619**
geographic distribution B 634
growth characteristics B 647 ff.
Decomposition, affecting factors D 33 ff.
measurement D 32, D 33
regulation D 35, D 36
Defoliation, by insects D 268
effect on below-ground organs D 194
effect on productivity D 268
Defoliators C 660
Deforestation, effects D 539, D 540
Demonstrative model, types D 122
Denitrification A 237, C 166, C 168, C 173, D 58
fire effect A 508

Desert annuals, productivity D213, D**214**
Desert plants, avoidance of heat stress A443
biomass and productivity D211ff.
dew uptake B114
effect of midday stomatal closure D223
frost resistance A**411**
heat injury A442
life forms D216ff.
light response D219
photosynthetic adaptation D221ff.
seedlings A457
temperature response D219
Desiccation, changes in membrane integrity B338, B344ff.
effect on ATP content B**366**
effect on chlorophyll B333, B365, B369
effect on growth B**357**
effect on photosynthesis and respiration B**356**ff.
enzyme activities B349, B**350**
leakage of substances B344ff.
RNA and protein synthesis B351ff.
role in seed development B407
Desiccation avoidance, habitat selection B336
Desiccation injuries, winter B**390**ff.
Desiccation tolerance B70, B325ff., B386ff.
anatomical and morphological adaptations B338, B339
and CO_2 exchange B336, B337
avoidance of mechanical stress B338, B339
characteristics B**331**
desert plants D223
development B333
different species B326ff.
poikilochlorophyllous plants B327, B333, B369
poikilohydrous plants B326, B333
poikiloxerophytes B326
resurrection plants B326, B341
Desiccation tolerance, see also Drought tolerance
Detritus, definition D52
dissolved D53
effect on phosphorus uptake D66
food chain D52, D54
particulate D52
role in aquatic ecosystems D52ff.
Detritus decomposition, affecting factors D33ff.
Dew, effect on transpiration B113
growth effect B112
lichen growth B126
stemflow B113
Dew absorption, through leaves B112, B114ff.
Diastase activity, heat tolerance in frost-hardened plants A457
Dichogamy, explication C571ff.
Diffusion coefficient, cuticular membrane B154
Diffusion resistance, see Resistance
Diffusive transfer of CO_2, leaf A88
Dispersal, of fruits C636
of seeds C646
Distribution of plants, effect of biotic factors C204ff.
effect of chemical factors C206
effect of competition C204ff.
effect of physical factors C206
DNA, composition and growth temperature A**345**
UV radiation absorption A183
DNA damage, by UV radiation A177, A**178**, A**179**
DNA synthesis, effect of aluminum C213
effect of herbicides D354, D355
Dormancy A150, B404, B405, B409
photoperiodism A141
spores B440
Dose-response curves, ionizing radiation A204, A205
Drag, water motion A233
Drag coefficient A486
Drainage, soil A551
Drought, acclimation B330, B333
adaptation B305ff., B311
and turgor maintenance B62
effect on leaf area index D187
progressive B**200**, B**202**
Drought adaptation, evergreen plants D220
Drought avoidance B325, B327
desert plants D223
mechanisms B341
mediterranean-climate plants D220
Drought experience, leaf conductance B203ff.
Drought resistance B67, B325
mycorrhiza C401
Drought tolerance, dry races B332
mechanisms B341
Drought tolerance, see also Desiccation tolerance
Dry matter balance, annual D**270**
Dry matter losses D266ff.
calculation D241, D242
Dry matter partitioning, patterns D166ff.
Dry matter production, maximal rates D**236**
relation to water use B664
temperate forest D**236**
Dynamic model D121ff.
physiological processes D136ff.

Ecophysiological processes, man's influence D527ff.
Ecosystem structure, man's influence D527ff.

Ecosystemic principles D 529ff.
Ecosystems, food chain D 529
Ecotypes, light regime A 68
  photoperiodism A 148
  physiological differences A 286
  shade and sun plants A 61, A 68
Ectomycorrhiza C 395
Ectotrophic mycorrhiza C 392
Eddy diffusion D 509
Eddy diffusion coefficient A 484
Einstein, definition A 43
Elaiophors C 589, C **590**
Elasticity modulus B **45**, B 51, B 60ff.
  marine macrophytes A 235
Electrochemical gradient, sodium transport C 16
Electron transport, inhibition by water stress B 293ff.
  inhibition by herbicides D 347
  photosynthetic D 338, D 343
    inhibition by herbicides D 338, D 343
  respiratory D 348
    inhibition by biocides D 348
Embryogenesis, temperature sensitivity A 422
Emissivity, calculation A 18
Endocyanoses C 339
Endodermis, drop in water potential B 14, B **96**
  ion transport C 21, C 104ff., C 107
  suberization B 93ff., B 98, B 99
  water movement B 93
Endotrophic mycorrhiza C 392
Energy balance, forest B 654
    seasonal variation B 654
  leaf B 193, B 592
    transpiration B 193
  stomatal effects B 182ff.
Energy budget, leaf A 20ff., D 512
Environmental conditions, influence on biocide effectiveness D 394ff.
Enzyme activity, effect of salinity C **74**
Enzyme conformations, temperature-dependence A 296
Enzymes, protection against water stress B 342
  regulation during water stress B 277, B 278
  responses to ions C 73ff.
  stabilizators A 344
  temperature inhibition A 325
  thermostability A 375
Ephemerals, C_3/C_4 pathway D 216ff.
  photosynthesis and leaf conductance D 216ff.
  solar tracking D 224
Epicormic growth, after fire A 517
Ericaceous mycorrhiza C 394
Erosion, by fires A 504, A 506
Essential oils, photoperiod control A 142
Estuarine systems, primary production D 309
Ethylene, affecting growth B 273
  biosynthetic pathway B 280
  effect of water stress B 279, B 280, B **302**ff.
  effect on defloration C 568
  flooding response B 460
  use as growth regulator D 361
Euphotic zone, ammonification D 59
  depth reduction A 254
  H_2S occurrence D 68
  irradiance A 256
Eurythermal microorganisms A 342
Eutrophication D 49
  consequences D 415, D 416
  definition D 414
  ecophysiological studies D 421ff.
  effect of phosphorus D 62, D 64
  effect on freshwater vegetation D 414ff.
Evaporation, soil surface B 23
Evaporation, see also Transpiration
Evergreen forest, productivity D **207**ff., D 233ff.
Evergreen leaves, characterization B 633, B 634
Evergreen perennials B **619**
  drought adaptation D 220
  dry matter production A 421, A 422
  geographic distribution B 634
  growth characteristics B 647ff.
    leaf longevity B 655
    nutrient balance B 649ff.
    water balance B 651, B **652**
Evergreen trees, fruits C 628
Evolution of man, in natural ecosystems D 531
Explanatory model D 122
Extensibility coefficient, cell growth B 270, B 272
Extinction coefficient, in canopy A 33
Extrafloral nectaries C 593, C 594, C 671

Fallow, benefits D 153
Fatty acid composition, chilling sensitivity A 382, A 384
  heat injury accumulation A 374
  temperature responses A 327
Fatty acids, synthesis after rehydration B 347
Ferns, photosynthesis and leaf conductance D 217
Ferredoxin content, light regime A 84
Fertilizer production, and world population D **332**
Fick's law A 235, A 479, A 552, B 237
  cuticular membrane permeability B 154
Fire, ecological factor A 501ff.
  ground surface temperatures A **503**
  nutrient flow A 508ff.
  soil temperature profiles A **503**

Fire intensity, community effects A 525
resprouting ability A 520
Fire temperatures A 502
Flooding B 453 ff.
effect of anoxia B 455, B 465 ff.
effect on leaf water potential B 454
effect on soil aeration B 456
epinasty B 454
hormonal imbalance B 459, B 460
mineral deficiency B 454
morphological adaptations B 458, B **459**
soil toxins B 460 ff.
survival time B 455, B **456**
Flood-tolerant plants, ethanol production B 458
Floral nectaries C **593**, C 594
Floral secretions C 585 ff.
Flower, induction and seasonality C 561
morphogenesis C 562 ff.
postfloration C 569
respiration C 568, C 569
UV absorption and reflection C **576**, C 577
Flower development, chilling susceptibility A 407
photoperiodism A 141
temperature effect A 323
Flower induction, gibberellic acid A 148
Flower sterility, chilling injury A 422
Flowering, fire stimulated A 515
inhibition by low temperatures A 323
photoperiodism A 143, A 146
reproductive development A 138
synchronization A 150
Flowering hormone A 151
Flowering inhibition, red/far-red light ratio A 160
Fluorides, action on enzymes and cellular metabolism D 487 ff.
chemical reactions D 487 ff.
injury symptoms D 475
metabolism D 489
plant responses D 475
role as atmospheric pollutants D 470, D 475, D 476, D 487 ff.
Fog deserts B 126
Fog drip B 113, B 115
Fog moisture, uptake B 115
Foliage distribution, effect of radiation transmission D 256
effect on radiation interception D 245, D 251, D 252
Food chain, detritus D 52, D 54
effect of biocides D 544
effect of organic nitrogen D 61
in ecosystems D 529
in freshwater D 437
Forced convection A 477
leaf A 23
Forest limit, alpine B 381, B 386, B 390
Forests, effect of biocides D 405
Free convection A 479
Freezing avoidance A 404
Freezing injury A 385 ff.
mechanism A 388, A 389
Freezing point, nonhardening species A 410
Freezing point depression, osmotic potential C 37
Freezing process, cellular dehydration A 386
ice formation A 386
Freezing resistance, desiccation-tolerant cormophytes A 410
short term adaptation A 415
specific differences A 409 ff.
Freezing, see also Frost
Freezing-sensitive plants, reduction of photosynthetic activity A 420
Freshwater ecosystems, acidification D 437, D 438
bioindicators for pollutants D 442, D 443
carbon concentration D 508
effect of waste heat D 439 ff.
eutrophication D 414 ff.
food chain D 437
nutrient cycling D 47 ff.
pollution D 424 ff.
Freshwater lake, light penetration A 254
Freshwater vegetation, effect of eutrophication D 417 ff.
Friction velocity, wind in canopy A 31
Frost, influence on water uptake B 380 ff.
Frost drought, ecological significance B 379 ff.
Frost hardening, and photoperiodism A 141, A 150
influence on photosynthetic activity A 421
Frost resistance A 409 ff., B 395
lichens C 436, C 438 ff.
potential A 410
responsive adjustment A 414
specific plant types A 410, A **411**
Frost, see also Freezing
Frost stress, effect on CO_2 exchange A **421**
effect on stomatal resistance A **421**
Frosttrocknis, see Frost drought
Frugivorous animals, seed dispersal C 641
Fruit, attractants C 642
function C 626
nutrient storage C 630, C 635, C 640 ff.
physiological ecology C 625 ff.
protective secondary compounds C 643
Fruit dispersal, animals C 636
Fruit ecology C 625 ff.
Fruit morphology, protective C 633 ff.
Fruits, chilling susceptibility A 409

heat tolerance A 465
Fucoxanthin, in algae A 257, A 258, A 260
in aquatic plants A **251**
Fulvic acid D 53, D 61
Fungi, heat resistance A 452, A 461, A 514
Fungicides, effect on Rhizobium C 331
Fynbos, fire effect A 519, A 526
productivity D **208**

Gall formers C 661
Gamma rays, mutation response A **218**
Garrigue, productivity D **207**, D **208**
Genetic engineering, modeling of results D 135
Genetic variation, effect of atmospheric CO_2 concentration D 520
Geophytes, fire adaptation A 519
frost resistance A **411**
Geotropism, flower C 564, C **565**
Germination, chilling susceptibility A 407
effect of herbicides D 356
fire-stimulated A 520
in canopies A 119
initiation A 321
light sensitivity A 119
phytochrome control A 119, A 322
temperature effects A 319, A 422
Germination, see also Seed or Spore germination
Gibberellic acid, affecting growth B 272 ff.
breaking seed dormancy A 322
effect of water stress B 279, B 280
flower induction A 148
role in seed development B 407
synthesis in the root apex B 459
use as growth regulator D 358 ff.
Global agricultural productivity D 158 ff.
Global radiation, spectrum A 110
Grassland, effect of biocides D 404
effects of trampling D 540 ff.
productivity D 173 ff.
Grassland ecosystems, responses to environmental conditions B **645**
Grassland plants, temperature response D 184
water relations D 187
Grasslands, characterization D 174
Gravitational potential, soil water A 545, A 548, A 555
Grazing, effect in tundra and grassland D 191 ff.
effect on below-ground organs D 194
Green waters, spectral distribution A **253**
Gross primary productivity, definition D 292
tropical forest D 293, D 298
measurement D 293, D 298
Growth, aquatic environment A 242
balanced A 341
cell properties B 50
cell wall synthesis B 276
crops D 163
doubling time A 340
effect of abscisic acid B 286
effect of atmospheric CO_2 concentration D 511, D 512, D 515, D 518
effect of calcium C **209**
effect of fluoride D 476
effect of herbicides D 356, D 357
effect of hydraulic conductance B 275, B 276
effect of ions C 64, C 101 ff., C 214
effect of leaf area index B **632**, D 163, D 164
effect of nutrient availability D 142, D 143
effect of oxidants D 475
effect of PAR A 61, B **239**, D 163, D 164
effect of phosphate C **203**
effect of salinity C 61 ff., C 82, C 85, C 91, C **92**, C **95**, C **98**, C **100**, C **102**, C 112 ff., C 120, C 143
effect of soil pH C **205**, C **216**, C **220**
effect of sulphur dioxide D 473
effect of turgor B 300
effect of water relations B 270 ff., C **113**, D 186 ff.
herbivory response C 667 ff.
influence of phytochrome B 402
interactions between light and salinity C 64
interactions between relative humidity and salinity C 62
kinetics A 340
long-term B 66
microorganisms A 365
temperature limits A 343 ff., A 365
osmotic adjustment B 64
osmotic maintenance B 274
physiological time D 137
potential D **239**
protein synthesis B 276 ff.
relation to mortality B 606 ff.
response to temperature A 316 ff., A 340, A **341**, A **345**
responses to contrasting thermal regimes A 281
short-term B 64
stand D **239**
temperate forest D 235 ff.
Growth characteristics, plant life forms B 634 ff.
Growth coefficient, definition D 265
Growth equations B 63 ff.
Growth factors D 242, D 243, D **269** ff.
for prediction D **270**
Growth forms, definition B 616
effect of abiotic factors D **175**
physiological significance D 176, D 177
Growth forms, see also Life forms
Growth inhibition, ionizing radiation effect A 203, A 205

Growth mechanism B272ff.
Growth model D117ff.
forests D237ff.
Growth of cells, physics B63
Growth optimum C204
Growth rate, crop D535
equation A72, B602, B603, B606, B620, D237
external and internal factors D137
organs D137
calculation D137
Growth reduction, effect of radiation losses D254
Growth regulation, influence of water and nutrient status B623, B624
Growth regulators, classification D**359**
ecological implications D393ff.
effects on agro-ecosystems D401ff.
effects on faunal population D399ff.
effects on forest ecosystems D405
effects on grassland ecosystems D404
effects on plant populations D394ff.
effects on soil microbial populations D398, D**399**
history and development D330ff.
mode of action D335ff., D357ff.
physiological effects D329ff.
Growth respiration B620, D240, D265
equation D265
Growth temperature, acclimation A283
effect on fatty acid composition A**343**
photosynthetic response A**284**

Halobacteria, growth media C158
Halogens, in freshwater ecosystems D432, D433
Halophilic prokaryotes C137ff.
Halophilism, mechanism C155
Halophytes C68ff.
effect of water relations on growth C**113**
frost resistance A**411**
growth C61
heat tolerance A463
ion concentrations C**71**
leaf vapor concentration A22
metabolism C79
osmoregulation C59ff.
osmotic adjustment B511
salt secretion C70
salt tolerance C59ff.
seed germination B409
transpiration B511
water use efficiency C118
Halophytic scrub, productivity D**208**
Halotolerant eucaryotes C59ff.
Hapaxanthic flowers C561
Haploidy, ionizing radiation effect A203

Harvest index B627
forests D235
Heat activation, spores A348ff.
Heat capacity, soil A**555**
Heat exchange, forced convection A477
Heat flow, soil A554
Heat injury, by sunscald A441
natural A441
Heat requirement, degree-days A279
Heat shocks, effects on plants A442, A457
hardening effect A458
Heat storage, leaf A23
Heat stress, effect on seeds and germination B421
lichens B334
Heat tolerance A445ff.
acclimatization A**458**
ecotypes A454
effect of water content A459, A**462**
lichens B334, C440ff.
methods for measuring A447
specific plant types A**460**, A514
terminology A450
Heat tolerance limit, criteria for evaluation A450
main plant groups A**448**, A**449**
of photosynthesis A**445**
Heat transfer, from cylinders and spheres A25
in soil A502
Heath, fire effect A409, A527
productivity D**208**
Heavy metals, bioaccumulation D428, D429
detoxification mechanisms D429
effect on competition C699
effects on plants D490, D544
in freshwater ecosystems D425ff.
role in aquatic plants D70ff.
toxicity D426ff.
Hemicryptophytes B**619**
Hemiparasites C522
Henry's law constant D478
Herbaceous perennials B**619**, B630ff.
carbon allocation B628, B629, B640ff.
competition B646, B654
geographic distribution B616, B629
growth characteristics B639ff.
root/shoot ratio B643
Herbicide absorption, effect on transpiration D395, D**396**
Herbicide effectiveness, influence of environmental conditions D394ff.
Herbicide resistance D344, D368ff.
Herbicide selectivity D333, D334, D366, D368
Herbicide susceptibility D334, D368
Herbicide tolerance D334, D366
Herbicides, competition D344

effect on Rhizobium C 331, C 332
effects on ecosystems D 544, D 545
history and development D 330 ff.
in freshwater ecosystems D 430 ff.
mode of action D 336 ff.
physiological effects D 329 ff.
pre-emergence application D 334
uptake and translocation D 334
Herbicides, see also Biocides
Herbicide-safeners D 366 ff.
Herbivores, classification C 658 ff.
coevolution with plants C 672, C 673
symbiosis with plants C 671
Herbivory C 657 ff.
effect on below-ground organs D 194
effect on phytoplankton productivity D 315, D **316**
in forests D 268
in tundra and grassland D 191 ff.
Heterocysts C 304
nitrogen fixation C 433, D 56
nitrogenase C 352, C 355
Heterotrophic activity, bacterioplankton D 51
High temperature injury A 372 ff.
High temperature resistance, ecological significance A 439 ff.
Hill reaction, effect of fluoride D 488
effect of herbicides D 338
Höfler diagram B **45**
Holocoenotic principle, in ecosystems D 530, D 536
effect of environmental changes D 536
Holoparasites C 522
Homoiohydrous plants, water stress B 326
Host distribution, effect on virus spread C 541, C 542
Host ranges, of viruses C 539
Host recognition, definition C 526 ff.
Host resistance, definition C 526 ff.
Host specificity C 524 ff.
Host-parasite interactions, higher plants C 519 ff.
Humic acid D 53
complexes with aluminum C 258
composition C **258**
effect on soil acidification C 208
Humidity, effect on biocide effectiveness D 395, D **396**
effect on nectar secretion C 597
stomatal response B 191, B 205, B 233, B 242, B 243, B 248
Humus forms, nitrogen production C 175
Hydathodes B 113
Hydraulic architecture, trees B 142
Hydraulic conductance, see Conductance, hydraulic
Hydraulic potential, soil water A 548
Hydrogen fluoride, effect on lichens C 452
Hydrogen isotope discrimination, relation to photosynthetic pathway and environment B 498
Hydrogen sulphide, in lakes D 68
Hydrophilic pollination C 560
Hydrostatic pressure, see Turgor or Pressure potential
Hypersaline lakes, ion concentration C 140
Hypolimnion, denitrification D 58
phosphorus supply D 63

Ice formation, extracellular A 386
intracellular A 386, 390
photosynthesis A 419
water conducting system B 382, B 383
Ideal solutions, osmotic potential B 40
Inclination angles of leaves D 245, D **246**, D 256, D 257
distribution D **246**
effect on light saturation D 256, D 257
effect on radiation interception D 245
Indicator species, for cold resistance A 425
Indolacetic acid, affecting growth B 272 ff.
effect of herbivory C 669
effect of water stress B 279, B **302** ff.
effect on mycorrhizal symbiosis C 400
effect on nectary-specific enzymes C 595
flowers C 564
use as growth regulator D 357 ff.
Insecticides, effect on photosynthesis D 365, D 366
effect on Rhizobium C 331, C 332
effects on plants D 363 ff.
in freshwater ecosystems D 432
Integrated Pest Management D 333, D 363, D 371, D 403
Intercellular CO_2, calculation B 182
C_3/C_4/CAM plants B 489 ff.
photosynthetic response B **575**, B **576**
quantum yield A **70**
stomatal response B 233, B 235
Interference, aboveground C 684
allelopathy C 694 ff.
belowground C 684
definition C 680
effect on relative growth rate C 683
in the field C 680, C 681
mathematical model C 683
Interference, see also Competition
Interference ability C 698 ff.
effect of species combination C 700
effect on plant character C 701 ff.
Ion concentration, cell wall C 102
cells C 24
chloroplast C 73
cytoplasm C 65 ff., C 72

Ion concentration, effect of transpiration C105, C109
effect on growth C101ff., C104, C112
effect on water relations C101ff.
leaves C68
phloem C110
regulation C109
roots C70, C**71**
salt bladders C68
shoot C**69**, C103ff.
vacuole C72
xylem C**106**, C108, C109
Ion influx, active C101
Ion pumps C14ff., C17ff.
Ion selectivity C25
Ion transport, mechanism C6
passive pathways C107
via the phloem C111
Ion uptake, control C24, C25
roots C20
xylem C107
Ionic environment C5ff.
Ionizing radiation, energy range A201, A**202**
genetic effects A203
linear energy transfer A220
long-term effects A220
plant response A199ff.
relative biological efficiency A**219**, A220
units of dose and exposure A202
Iron, flooding injury B461, B**462**
role in aquatic plants D70
role in marine ecosystem D102
Iron availability C229, C230
Iron cycle, aquatic ecosystems D**73**
Iron deficiency, biochemical reactions C230
effect on root growth C**214**
Iron immobilization C233
Iron toxicity C272
Iron uptake, and transport C**26**
root C20
Irradiance, aquatic environment A238, A242
canopy shading A152
clouds A155
definition A42, A238
downwelling A238
in different water types A253
plant surface A15ff.
spectral distribution A151, A**152**, A**153**
underwater A254
upwelling A238
Irradiance, see also Light or Radiation
Irrigation B306
Isothermal radiation A27

Kettletraps C569
Kinetin, effect of water stress B278
K-strategists, definition C547

Lakes, eutrophic D54
primary production D308, D**309**
Laminar flow, mass transfer A24
Latitude, effect on phytoplankton productivity D313
Lead toxicity, mechanisms C252, C284
Leaf, energy budget B235, B592
Leaf abscission, economic consequence B611
photoperiodic response A150
Leaf area, mesophyll volume A90, A91
models D144
phenology D252
Leaf area density D284
Leaf area distribution, tropical forests D284ff.
Leaf area duration, definition D284
Leaf area index, canopy transmission A34
changes with age D**271**
conservative plant B26
definition A32, D284
different vegetation formations B**664**
effect of atmospheric CO_2 concentration D517
effect of drought D187
effect of radiation C687
effect on growth B**632**, D163, D164
effect on production D187
effect on radiation interception B630ff., D245
effect on transpiration B18, B25, B253, B254, B309, D287
effect on water use efficiency B**26**
forests D250ff.
mediterranean-climate ecosystems D**207**
seasonal variation D252ff.
tropical forests D**283**
Leaf biomass density D284
Leaf conductance, see Conductance
Leaf elongation B**65**
Leaf fall, low temperature stress A416
Leaf growth, photoperiodism A139
Leaf inclination, relation to total leaf area B**632**
Leaf life span, tropical forest D294
Leaf miners C660
Leaf morphology, photoperiodism A140
wind effect A492
Leaf senescence D167
Leaf spectral properties, desert plants D223
Leaf surface property, effect on herbicide selectivity D334
Leaf temperature, difference to air temperature A**28**
fundamentals A20
optimal range B210
overtemperature type A441
undertemperature type A441
wind effect A481

Leaf variables, effect on photosynthesis D **259**
Leaf weight ratio A 72
light regime A 75
Lectins, effect on root infection C 311
Leghaemoglobin C 304
Legume nodules, nitrogen fixation C 304, C 308 ff.
Lenticels, permeability B 161
Lichen acids C 434 ff.
Lichen coefficient C 437
Lichen symbiosis C 343, C 423 ff.
Lichenization C 428, C 429
Lichenometry C 456
Lichens, artificial synthesis C 423, C 424, C 435
cold resistance A **408**, C 438 ff.
desert habitat B 358 ff.
desiccation B **358** ff.
growth C 454 ff.
heat tolerance A 452, A 461, C 440 ff.
nitrogen fixation C 344 ff.
photosynthesis C 444 ff., C 449 ff.
water relations B **126**, C 447
sensitivity to air pollution C 451 ff.
water relations C 441
Life form adaptation, winter stress A 417
Life forms, classification B 616 ff.
desert plants D 216 ff.
distribution of different photosynthetic pathways D 215 ff.
mediterranean-climate plants D 215, D **216**
tropical plants D 291
Life forms, see also Growth forms
Light, acclimation A 59 ff., D 258
chloroplast structure A **266**
effect on biocide effectiveness D 395, D **396**
effect on nitrogen fixation C 319
effect on nodulation C 318 ff.
effect on phytoplankton productivity D 313
photoadaptive responses in algae A 266
Light, see also Irradiance or Radiation
Light compensation point, shade plants A 60
Light extinction coefficient D 285
Light interception, annual D 243
canopy C 687
changes with age D **271**
effect on photosynthesis D 238, D **239**
effect on productivity D 234
Light quality, chloroplast orientation A 129
non-photosynthetic responses A 109 ff.
photomorphogenesis A 120
photoperiodism A 128
seed germination A 118 ff.
stomatal aperture A 129
Light requirements, of different photosynthetic pathways B 505 ff.
Light response curves, C_3/C_4/CAM plants B 505 ff.
Light scattering, by water A 239
canopy A 35
Light utilization efficiency A 70
Lignotubers, fire adaptation A 519
Lime-induced chlorosis C 217, C 219, C 229 ff.
effect of iron C 230
Limestone, influence on vegetation C 201 ff.
Limiting factors, principle in ecosystems D 529, D 530
Lipids, effect of herbicides D **340**, D 351 ff.
transport properties of cuticular membranes B 153 ff.
Liquid-phase resistance for CO_2, photosynthesis B 577, B 578
Litter, average rates D 267
decomposition after fire A 513
measurement D 252
mineralization D 27 ff.
nutrient reserves D 16 ff.
role in mineral cycling D 8, D 24 ff.
Litter fall, in evergreen forest D 210
Litter layer, influence on soil heating A 502
Litter production, tropical forest D 292 ff.
Littoral zone, phosphorus supply D 64
Long-day plants A **144**
Longevity, life forms B 620 ff.
Long-wave radiation, computation A 17 ff.
in canopy A 35
Low temperature resistance, ecological significance A 403 ff.
Lysosomes, enzyme secretion C 506

Magnesium, after fire A 508, A 512
Maintenance coefficient, definition D 263
Maintenance costs, tropical forest D 296
Maintenance processes D 139
Maintenance respiration A 312, D 240, D 241, D 263, D 264
calculation D 139, D 263, D 264
Malate oscillations, CAM C 45
Malic dehydrogenase, thermostability A 375
Malthusian fitness B 610
Manganese, role in aquatic plants D 70
Manganese cycle, aquatic ecosystems D **73**
Manganese toxicity, mechanism C 261
Manganous ions, flooding injury B 461
Mangroves A 252
chilling sensitivity A 406
frost resistance A **412**
Marine ecosystem, nitrogen fixation C 347
nutrient cycle D 85 ff.
primary production D 309 ff.
Marine habitat, productivity A 252
Marsh grasses A 252
Mathematical models, photosynthesis B 549 ff.
transpiration B 231 ff.
Matorral, fire effect A 526

Matric potential B36ff., C39, C47
cell wall C42
definition B8
soil A545, A548, A551, A**552**, A553, A555, C42
anaerobis point A551
influence of temperature A553
Maximal photosynthesis, see Photosynthetic capacity
Mediterranean plants, dew uptake B114
heat tolerance A**454**
Mediterranean sclerophylls, frost resistance A410, A**412**
photosynthetic capacity A422
Mediterranean shrubs, fire effect A503, A526
Mediterranean-climate life forms D215, D**216**
Mediterranean-climate plants, biomass and productivity D206ff.
dimorphic leaves D220
drought adaptation D220
light response D219
photosynthetic adaptation D220, D221
stomatal control D220
temperature response D218, D219
Mehler reaction B488
Membrane bilayer formation, temperature dependence A320
Membrane fluidity, fatty acids unsaturation A383
Membrane lesions, freezing injury A387
Membrane lipids, fatty acids composition and temperature A343, A345
Membrane models C**7**ff.
Membrane permeability, effect of lichen substances C436
oxidative phosphorylation B418
rehydration B418
temperature B418
Membrane potential, amino acid uptake C**18**, C19
charge separation C11
proton cotransport C17, C**18**
sugar uptake C**18**, C19
Membrane properties, effect of ions C79
Membrane stability, temperature responses A373
Membrane structure, effect of herbicides D351ff.
Membrane water permeability, flux equations B154, B155
Membranes, composition B345ff.
conformational changes B345ff.
during desiccation and rehydration B345ff.
destruction of permselectivity by sulphur dioxide D480
effect of aluminum C213
effect of calcium C220ff.
effect of herbicides D**340**
thermal behavior A383
Mercury toxicity, mechanism C287
Meromictic lakes, H_2S occurrence D68
Mesophyll, area and volume A90, A91
internal surface A88
photosynthesis A90, A91
Mesophyll resistance to CO_2, crop plants D535
effect of salinity C119
effect on water use efficiency C119
Metal ions, categories C246ff., C256ff.
Metal toxicity and tolerance C245ff.
Metalliferous soils, effect on plant distribution C248, C249
Methane, in aquatic ecosystems D54
Microbial activity, after fire A513
soil D30ff.
Microorganisms, temperature response A339ff.
Microtubule system, effect of herbicides D355, D356
Microtubules, reversible disruption A315
Midday depression of gas exchange B**204**, B**207**, B208, B594
desert plants D223
Middle lamellae, water permeability B174
Mineral cycling, and soil metabolism D8
biochemical D18ff.
geochemical D20ff., D36ff.
in ecosystems D5ff.
Mineral nutrition, carnivorous plants C493ff.
Mineralization, of litter A510
fire effect A510
of litter and soil organic matter D27ff.
soil D32, D33
Minerals, turnover in different vegetation types D**17**
Mitchell's theory C12, D**337**
Mitochondria, oxidase activity A**309**
Arrhenius relationship A**309**
respiratory capacity A308
Mitochondrial phosphorylation, uncoupling by biocides D348
Mitosis, effect of herbicides D355, D356
Modeling, concepts D119
objectives and procedures B232, B254ff.
Models, relative values D**123**
types D121ff.
Moisture compensation point, lichens C445
Moonlight, plant response A153
red/far-red light ratio A156
spectral distribution A112, A151, A**152**, A**153**
Morphogenesis, effect of organ partitioning D168
Mortality, of plants B606ff.

Mutation, ionizing radiation effect A 203
Mutation response, gamma rays A **218**
Mycobiont, lichens C 423
Mycorrhiza C 391 ff.
  classification C 395
  effect of growth substances C 399 ff.
  effect of oil D 545
  effect on lime sensitivity C 219
  effect on nodulation C 324, C 332, C 406
  effect on nutrient uptake C 211, D 22
  effect on phosphorus competition C 690
  effect on phosphorus nutrition C 332
  morphology C 392 ff.
  orchids C 645
  relation to animals C 407
  role in biogeochemical cycling D **22**
  water relations C 401, C 410
Mycorrhizal fungi, anthropogenic influence C 409
Mycorrhizal symbiosis, ecophysiology C 391 ff.
Myxospermy B 418

NADPH production B 563, B 564
Natural ecosystems D 528 ff.
  alteration or destruction by man D 536 ff.
    ecophysiological bases D 536 ff.
  evolution D 528
  structure D 528, D 529
Nectar, caloric value C 607
  energy source C 605 ff.
  floral secretion C 585
  flower attractant C 593 ff.
Nectar constituents C 599 ff.
Nectar quantity C 598
Nectar secretion C 594
Nectaries, physiological explanation C 594
Negative turgor B 48
Net primary productivity, definition D 292
  tropical forest D **297**, D 298
Neutrons, biological effect A 201
Nickel toxicity, symptoms C 269, C 270
Nitrate dissimilation D 58
Nitrate nutrition, chlorosis C 186
  pH dependence C 187, C 188
Nitrate reductase C 181 ff.
  effect of light C 182
  effect of temperature C 182
  effect on nitrogen uptake C 179, C 180
  indicator of nitrogen nutrition C 217
  induction C 184
  temperature sensitivity C 184
  water plants D 60
Nitrate reductase activity, during water stress B 278
  UV radiation A 172
Nitrate uptake C 179 ff.
  dependence on root system size C 181
  effect of nitrate reductase C 180
  phytoplankton D **87**
Nitrate:ammonium ratio C 174 ff., C 183
Nitrification C 163 ff., C 215
  after fire A 508, A 513, A 514
  in aquatic ecosystems D 57, D 58
  marine ecosystem D 95
Nitrifying bacteria, carbon-autotrophic D 57, D 58
  carbon-heterotrophic D 57, D 58
  heat resistance A 515
Nitrogen, after fire A 508
  agricultural soil D 152
  aquatic environment A 236
  competition for C 689 ff.
  effect on growth and nitrogen fixation C 326
  effect on photosynthesis D 191
  mineralization A 236, C 163, C 166 ff., C 169, C 173, C 175 ff., C 215
  role in freshwater ecosystems D 415
  sedimentation A 237
  transport forms C 314
  utilization D 61
  volatilization from soil A **509**
Nitrogen assimilation C 181 ff., C 186
Nitrogen capital, ecosystems C 164 ff.
Nitrogen content, seasonal course D **190**
Nitrogen cycle, aquatic ecosystems D **55**
  ecological aspects D 56 ff.
  marine ecosystems D 86 ff.
  terrestrial ecosystems D 36
Nitrogen fixation C 166, C 219, C 301 ff.
  actinorrhiza C 333 ff., C 336, D 11
  agricultural and ecological systems C 307 ff.
  aquatic environment A 236, A 237, D 55 ff.
  blue-green algae C 352 ff., D 12, D 415, D 423
  cephalodia C 433
  coralloid roots C 342
  effect of atmospheric CO_2 concentration D 520
  effect of carbon monoxide D 487
  effect of mycorrhiza C 406
  energy cost C 315
  epiphytic blue-green algae C 346 ff.
  fire effect A 508, A 510
  free-living bacteria C 356 ff., C 360, D 12
  heterocysts C 433
  heterotrophic C 306
  inhibition in acidified lakes D 438
  legume nodulation C 308 ff., D 11
  legumes D 168, D 169
  lichens C 344 ff.
  measurement C 305 ff., C 348
  nitrate inhibition C 325
Nitrogen forms, assimilation C 185, C 186
  effect of microorganisms C 219
  effect on cation uptake C 218

Nitrogen forms, pH values C185, C187, C215
utilization C184ff.
Nitrogen immobilization C215
Nitrogen mobilization D167
Nitrogen nutrition C163ff.
agricultural systems D167
different ecosystems C174
effect in marine ecosystem D85
effect on growth D164
leaf conductance B185, B**186**
Nitrogen oxides, role as atmospheric pollutants D469, D486, D487
action on enzymes and cellular metabolism D486, D487
chemical reactions D486, D497
Nitrogen reduction C181ff.
Nitrogen regeneration, in marine ecosystem D95ff.
Nitrogen uptake C167
effect of water availability C181
macrophytes D422ff.
mycorrhiza C397, C398
phytoplankton D92ff.
enzyme system D92ff.
Nitrogen utilization, phytoplankton D92ff.
enzyme system D92ff.
Nitrogenase D56
in heterocysts C352
oxygen sensitivity C304, C360
Nitrogenase activity, actinorrhiza C335, C336
blue-green algae C339
effect of light C319, C341, C343, C355
effect of mycorrhizal symbiosis C406
effect of oxygen diffusion C320
effect of soil CO_2 level C323
effect of water potential C320ff., C350
nitrite inhibition C326
prokaryotes C302
sponges C471
Nitrogen-use efficiency, C_3/C_4 plants D191
Nodulation, actinorrhiza C333ff.
competition C329
effect of mycorrhiza C406
effect of nutrient status C323ff.
effect of waterlogging C320, C321
interactions with mycorrhiza C332
legume C308ff.
Non-osmotic volume C82
plant cells C**41**
Nucleic acids, denaturation temperature A344
Nutrient allocation, in plant communities D5ff.
Nutrient budget, oceans D**88**
Nutrient cycling, in ecosystems D5ff., D47ff., D85ff., D416, D539
freshwater D47ff., D416
marine D85ff.
terrestrial D5ff.
Nutrient regeneration, from sediments D97
in water D95ff.
effect on production D**96**
marine ecosystems D86ff., D104, D105
nitrogen D86ff.
phosphorus D98ff., D99
physical advection D88, D89
role of aquatic macrophytes D104, D105
silicon D100ff.
terrestrial and atmospheric sources D89, D90
Nutrient reserves, litter D16ff.
soil D14ff.
Nutrient return, effect of litter fall D24ff.
effect of throughfall and stemflow D23ff.
Nutrient sources, atmospheric D9ff.
soil D12
weathering D13
Nutrient status, effect on root/shoot ratio A286
in tundra and grassland D189
Nutrient storage, fruit C630, C640ff.
Nutrient uptake, effect of mycorrhiza D22
effect of water stress B297ff.
effect on transpiration B297
models D142
phytoplankton D91ff., D316
Nutrients, after fire A513
effect on assimilate partitioning B517
effect on phytoplankton productivity D313ff.
soil colloids A508
volatilization A508
Nutrition, parasitic plants C522, C523
Nuts C641, C647

Odours, flowers C578
Oil, effect on cold hardiness D545
effect on detritus D545
effect on ecosystems D544, D545
effect on mycorrhiza D545
Oligotrophic lakes, effect of eutrophication D417ff.
Optimization of growth, effect of water and nutrient status B623, B624
Optimization of photosynthesis and transpiration B190ff., B589ff.
Optimization of production D135, D136
effect of plant water status B622
Optimization of stomatal behavior B20, B590, B591, B596, B597
Optimization of water use efficiency B193ff., B222, B247ff.
Orchidaceous mycorrhiza C393, C394
Organic matter, soil D**9**, D28ff., D31, D32, D33

decomposition D **9**, D 28 ff.
turnover in different vegetation types D **17**
Ortho-phosphate, see Phosphorus
Osmolality C 52
definition C 36, C 37
Osmophores C 574, C 580, C **582**
definition C 583
heat production C 583
Osmoregulation C 35 ff., C 59 ff., C 85, C 88, C 91
halophytes C 104
microalgae C **83**, C 84
units and terminology C 36 ff.
water relations C 40
Osmotic adjustment B 56 ff., B 202
desiccation-tolerant plants B 341
halophytes B 511
turgor maintenance B 62
Osmotic coefficient, definition C 37
solutes C 44, C 47
Osmotic potential B 36, B 40 ff., B **45**
control C 35
definition C 38
diurnal changes B 59
effect of ion accumulation C 101
effect on anthesis C 568
effect on cell volume C 40
effect on ion uptake C 24, C 27
indicator for water stress B 268
leaf B 202, D 189
effect of water stress D 189
organic solutes C 90
root exudation B 82
seasonal changes B 58
winter B 390 ff.
soil water A 545, A 555
Osmotic potential gradient, volume flow of water B 156
Osmotic pressure, see Osmotic potential
Osmotic solutes, accumulation C 93
balance C 90
effect on photosynthesis C 94
synthesis and degradation C 88, C **89**
Osmotic volume C 82
Oxidants, role as atmospheric pollutants D 468, D 469, D 474, D 475
plant responses D 474, D 475
Oxygen level, soil C 321, C 322
Ozone, atmospheric concentration A 180
resistance of spores B 436
role as atmospheric pollutant D 469, D 474, D 484 ff.
action on cell structures D 484, D 485
action on enzymes and cellular metabolism D 486
chemical reactions D 484, D 485
injury symptoms D 474
Ozone layer, UV radiation attenuation A 174, A 175

PAN, see Peroxyacetyl nitrate
PAR A 13, A 41 ff.
amount with photoperiod A 139
conversion factor A 51
effect on growth D 163, D 164, D **239**
effect on nitrogen fixation C 345
effect on photosynthate accumulation D **255**
effective efficiency D 242
for temperate forest D 234
in canopy B **307**
measurement A 43 ff., A 48, A 50, A 52
ratio to total solar radiation A 13
PAR utilization, changes with age D **271**
Parasitic plants, characterization C 519
disease compensation C 523
interactions with the host C 519 ff.
nutrition C 522, C 523
Parasitism C 520 ff.
Particle radiation A **200** ff.
Particulates, role as atmospheric pollutants D 470, D 476
plant responses D 476
Passive transport, ions C 107
Pasteur effect A 310
Penman evapotranspiration equation A 27, B 18
Penman-Monteith transpiration formula A 27
PEP carboxylase B 56, B 486 ff.
^{13}C discrimination B 497
CO_2 fixation in Phaeophyceae B 492
C_4 pathway B 486
effect of water stress B 293
Perennial herbaceous plants, see Herbaceous perennials
Perennial woody plants, see Woody perennials
Perfume flowers C 591, C **592**
Perianth movements C 569
Periderm membrane B 156
resistance B 161
transport properties B 174 ff.
Periphyton, effect of waste heat D 440, D 441
Permafrost, fire effect A 505
Permeability, hydraulic B 63
Permeability coefficient, cell B 91
cuticular membrane B 154 ff.
polymer matrix membrane B **171**
Peroxisomes B 481, B **483**
Peroxyacetyl nitrate, role as atmospheric pollutant D 469, D 474, D 484, D 485, D 486
action on cell structures D 484, D 485
action on enzymes and cellular metabolism D 486
chemical reactions D 484, D 485
injury symptoms D 474

Pest control, chemicals D 331 ff.
Pesticides, effects on ecosystems D 544, D 545
in freshwater ecosystems D 430 ff.
Pesticides, see also Biocides
pH, effect on growth C **205**, C **216**, C **220**
effect on nitrogen fixation C 323 ff.
effect on potassium content C **221**
influence on vegetation C 201 ff.
soil C 201 ff., C **205**
after fire A 512, A 514, A 515
determination C 202
effect on growth C **205**
influence on vegetation C 201 ff.
pH regulation, cytoplasm C 16, C 23
Phagocytosis C 476
Phanerophytes B **619**
Phenoseason, effect on radiation D **244**, D 245
Pheromones C 583
Phloem transport, abscisic acid B 284
inhibition by low temperature A 315
Phosphatase, acid D 63
Phosphate, soluble A 237
aquatic environment A 237
Phosphate uptake, dependency C **26**
marine microorganisms D 98
mycorrhiza C 396
root C 20, C **24**, C 228, C **229**
Phosphorus, after fire A 508
aquatic ecosystems D 65
competition for C 689 ff.
precipitation D 65
retention D 65
flowing waters D 65
role in aquatic plants D 62
role in freshwater ecosystems D 415
uptake by aquatic macrophytes D 421 ff.
Phosphorus cycling, fire effect A 511
freshwater A 237, D **62**
lake D 416
marine ecosystem D 98 ff.
Phosphorylation, mitochondrial D 348
uncoupling by biocides D 348
Photic zones, distribution of algae A 255
Photoinhibition A 59
by UV radiation A 177, A **178**
carboxylating enzyme A 186
chlorophyll bleaching A 95
C_4 pathway B 488, B 509
disruption of thylakoid membranes A 185
effect of water stress B 295
effect on photosystem II B **484**
interactions with temperature A 97
irradiation during cold A 420
mechanism A 96
plastoquinone destruction A 185
shade and sun plants A 101
Photon flux density, see Quantum flux density
Photoperiod A 135 ff.
effect on heat tolerance A 462
effect on seed ripening B 421
flowering response A 138, A 143, A 146, A 150
interactions with CO_2 A 139
interactions with nutrient levels A 148
interactions with temperature A 146
Photoperiod, see also Daylength
Photoperiodic response A **137**, A **143** ff.
after vernalization A 147
dormancy A 150
frost-hardening A 150
leaf abscission A 150
Photoperiodic sensitivity, in natural ecosystems A 149
Photoperiodism A 135 ff.
critical day and night A 139, A 146, A 149, A 158, A 160
ecological aspects A 148 ff.
effect of altitude A 149
interactions with temperature A 143, A 157
light interruption A **154**, A 157, A 159
mechanism A 151 ff., A 161
phytochrome A 128, A 160
role in ecosystems A 136
Photophosphorylation, effect of herbicides D 347
freezing injury A 387
thermal inhibition A 373
Photoprotection A 184
Photoreceptors A 116 ff.
Photorespiration A 98, B 57, B 481 ff.
crop plants D 535
C_4 pathway B 487
effect on plant productivity B 485, D 136
release of NH_3 B 483
Photosynthate accumulation, effect of PAR D **255**
Photosynthesis, action spectrum A 41, A 46, A **47**, A 257
adaptation and acclimation A 306
affected by osmotic solutes C 94
after frost stress A **421**
animal-chloroplast associations C **483**
annual gain B **655**
arid conditions B 203 ff.
ATP production B 564 ff.
carbon isotope discrimination B 494 ff.
carboxylation efficiency B 573 ff., B 580
carnivorous plants C 493
cold environments B 213, B 214
correlation to mesophyll size A 93
CO_2 response B **575**, B 576, D **513**
C_3/C_4/CAM plants B **501**, B **503**
crop plants D 535

C_3/C_4 plants A69, A301, A**302**, B**216**, B**217**, B**219**, B479ff.
desert and mediterranean-climate plants D214ff.
diurnal course B203ff., B**204**, B211, B222, B223
effect of calcium C209, C225
effect of competition B**609**
effect of desiccation B356ff.
effect of fluoride D476, D488
effect of heavy metals D544
effect of herbicides D336ff.
effect of humidity B222, B223
effect of insecticides D365, D366
effect of leaf age B579
effect of leaf water potential D**224**
desert plants D**224**
effect of oxidants D475
effect of rehydration B364ff.
effect of salinity C99, C117, C119
effect of soil water content D**188**
effect of sulphur dioxide D473, D482ff.
effect of water status B19, B20, B**201**
effect of water stress B215, B291ff., B504, D186ff.
effect of wind speed A483
effect on atmospheric CO_2 concentration D507
equation B593, B620
fruits C627, C628
grassland D178, D**188**
heat tolerance limit A458
high temperatures A300
impact of man D**542**
inhibition by UV radiation A173
intercellular CO_2 resistance B576, B577
lichens B**126**, C444ff.
light reactions D**337**
light regime A75
light response A48, A59, B222, B223, B598, C686, D218, D219, D**257**, D**289**
tropical plants D**289**
low temperatures A306, A419
operational range B190ff., B214ff., B221
optimization B**219**, B247ff., B595, B**604**, B**605**, B622
photoinhibition A59
phytoplankton A254
potential D243, D**261**
seasonal course D**261**
quantum yield A44, A**45**, A**46**, A68, A**301**, B294, B570ff., D219
desert plants D219
spectral A44, A**45**, A**46**
temperature effect A**301**
relation to leaf conductance B**192**, B195, B203ff., B**216ff.**
relation to transpiration B20, B**21**, B235, B**598**, B**601**
seasonal pattern B203ff., B**655**, D178, D**188**, D**261**
seeds C628, C644
shade and sun plants A59, A72, B211ff., B**216**ff.
temperature response A**282**, A**283**, A299, A300ff., B208, B222, B223, B579, D181ff., D218, D219, D**221**, D**222**, D260, D290, D**291**
desert plants D**221**, D**222**
effect on competition D182
species-dependent differences A299
tropical plants D290, D**291**
tropical plants D287ff.
tundra D178
weak light A62ff.
Photosynthesis inhibitors, herbicides D331
Photosynthesis model B549ff., B581, B582, D258, D512, D513
canopy A37, B581
shade and sun leaves A84
Photosynthetic capacity, and leaf conductance B182ff., B**202**, B597
C_3/C_4 plants B182
different species B625, B626
effect of blue-green light A264
effect of growth form D177, D178
effect of photosynthetic pathway D177, D178
inorganic nutrition B185, B**186**
relation to nitrogen content B625, B626
seasonal maximum A**285**
shade and sun leaves A72
Photosynthetic efficiency, dependence on environment D260, D**261**
dependence on leaf photosynthetic properties D258ff.
dependence on structural properties D256ff.
forests D239, D240
Photosynthetic pathways, biochemistry B481ff.
desert and mediterranean-climate plants D215ff.
different life forms D215ff.
effect of nutrition and salinity B510ff.
effect on photosynthetic capacity D178
effect on plant distribution B523ff.
effect on plant productivity B518ff., D158
functional significance B479ff.
light requirements B505ff.
temperature response B508ff.
tropical plants D291, D**292**
water economy B500ff.
Photosynthetic unit size A68
alteration by white light A267
light regime A85

Photosynthetically active radiation, see PAR
Photosystem I, reduction of bipyridylium herbicides D346
Photosystem I, II, in shade and sun leaves A84, A**85**
Photosystem II, effect of lead D490
Photosystem II herbicides, antidotes D368
binding site D**344**, D370
inhibitory efficiency D**343**
mode of action D338ff.
Phototaxis A267
action spectrum A268
ATP supply A269
light control A268
Phototrophic bacteria, thermophilic A356
Phototropism A128
flower C564, C574
Phrygana, fire effect A514
productivity D**207**
Phycobiont, lichens C423, C**424**
Phycocyanin, in algae A259, A261
Phycoerythrin, in algae A259, A261
Phyllotaxis, photoperiodism A142
Phytochrome A113ff.
absorption spectrum A**171**, A172
anthesis induction C565
effect on nodulation C318ff.
influence of hydration B408
influence on growth B402
photoperiodism A128, A142, A157
photoreceptor A116, A127
phototransformation A322
red/far-red effect in algae A269
sensitivity to red/far-red light A125
spectral response A151
Phytohormones, as growth regulators D357ff.
effect of growth regulators D331
effect of herbivory C668ff.
effect of sulphur dioxide D484
influence during water stress B279ff., B301ff.
protection against herbivory C658
regulation of stomatal response B605
transport B304
Phytoplankton A250
absorption spectra A**258**
depth distribution A256
effect of chloride D435
effect of eutrophication D418
effect of halogens D433
effect of phosphorus D63
effect of waste heat D441
excretion of photosynthetic products D50
nitrogen uptake D60
nutrient uptake D91ff.
photosynthesis, quantum efficiency A254
radiation absorption A254
sensitivity to UV radiation A181
Phytoplankton productivity D305ff., D311ff., D316, D321
measurement D321
physiological adaptation D316
regulation D311ff.
seasonal patterns D306ff.
Phytotoxins, effect on mycorrhiza C410
Pigments, effect of herbicides D**340**, D349ff., D351
protection mechanism D351
in flowers C574, C575
in lichens C436
nectar constituents C604
oxydation by bipyridylium herbicides D346
Plant life forms B615ff.
Plant nutrition, agricultural system D152
Plant productivity, importance of aquatic plants A**252**
Plant-water relations, effect of root/shoot ratio D517
Plasmalemma, effect of ozone and peroxyacetyl nitrate D484ff.
Plasmalemma permeability, effect of sulphur dioxide D480
Plasmodesmata, conductance B93
structure B93, B**94**
water movement B93
Plasmolysis, turgor measurement C83
Plasticity, cell wall B49ff.
Plastoquinone content, light regime A84
Podzolization C202, C208
Poikilochlorophyllous plants, desiccation tolerance B327, B333, B369
Poikilohydric, see Poikilohydrous
Poikilohydrous angiosperms, cytoplasmic adjustments B333
foliar water uptake B119
Poikilohydrous ferns, foliar water uptake B120
Poikilohydrous plants, CO_2 exchange B**204**
desiccation B343
desiccation tolerance B326
heat tolerance A464
maximum temperature A441
osmoregulation B341
Poikiloxerophytes, desiccation tolerance B326
Pollen, constituents C588ff.
floral attractant C587ff.
wind dispersal A487
Pollen release, regulation C571
Pollen vectors C562
Pollination A150
ecophysiology C559ff.
effect of competition C697
Pollutant combinations, effects on plants D476, D490

Pollutants, atmospheric D465ff., D470, D538
action on plants D471ff.
ecophysiological effects D465ff.
effect of urbanization D538
particulates D470
specifications of concentrations D**467**
transfer to plant surface D470, D471
effect on enzymes D477
in freshwater ecosystems D442ff.
bioindicators D442, D443
temperature effect D442
Pollution, effect of biocides D333
Polymer matrix, cuticular membranes B156, B170, B**171**
Polyphosphate D63
Polysome formation, effect of temperature A346
Polysomes, changes during desiccation B351ff.
Porous cuticular membrane, model B163, B**164**
Postfloration, regulation C569
Potassium, after fire A508, A511
competition for C692
effect of soil pH on retention C**221**
Potassium accumulation, root C228
Potassium uptake C**25**
root C**226**
Potometer B88
Pre-dawn leaf water potential B202
Predictive model, growth rate D135
plant water use D135
Prefloration C565
Preliminary model, definition D126
examples D**129**
productivity D128ff.
Pressure chamber B48, B**101**, C39
cavitation of xylem B54
Pressure potential B36, B**45**, B47ff.
Pressure probe B36, B48
Pressure-volume curves B43, B**45**
Primary producers, nitrogen nutrition D59
Primary production, aquatic ecosystems D49ff.
equation B620
grassland D**192**
Production, optimization B622
Production levels, plant growth D119ff.
plant summary model D133ff.
Production models D117ff.
Productive structures, mediterranean-climate plants D**207**ff.
Productivity, above and belowground A285
actual D242
agricultural systems D151ff.
aquatic ecosystems D305ff.
aquatic plants A252
deciduous forest D233ff.
definition B20
desert and mediterranean-climate plants D205ff.
different plant life forms B660ff.
effect of atmospheric CO_2 concentration D511, D512
effect of grazing D191ff.
effect of leaf area index D187
effect of photosynthetic pathways B518ff.
effect of radiation D234
effect of soil water B23
effect of water status B9
evergreen forest D233ff.
forests D242, D243
global D158
grassland and tundra D173ff.
models D126ff.
poikilohydrous plants B337
potential D242, D243
prediction B523
seasonal pattern B28
simulation D**133**
temperate forest D233ff.
time scales B6
tropical forest and woodland A252, D281ff., D292ff.
Prokaryotes, nitrogenase activity C302
Proline, cytoplasmic osmoticum B57
cytosolute C49
Protein concentration, temperature response A307
Protein content, blue-green light A264
correlation to light saturated photosynthesis A81
grain D167
shade and sun plants A71, A**76**
Protein synthesis, during water stress B276ff.
effect of fluoride D489
effect of herbicides D354, D355
effect of salinity C79
effect of UV radiation A172
halobacteria C150
in thermophiles A344
inhibition by high temperature A317
sensitivity to drought stress B351ff.
temperature acclimation A318
Proterandry C571, C**572**
Proterogyny C571, C**573**
Proton pumps C21ff.
algae C16
Pseudocyphellae B126
Psychrometer B8, B48, B**101**
use in transpiration measurement B87
Psychrophilic eukaryotes A362, A**363**, A**364**
Psychrophilic microorganisms A342, A360ff.
habitat A361

Psychrophilic prokaryotes A 362, A **363**
Pteridophytes, heat tolerance limit A 452, A 455
Pubescence, leaf A 482, B 209, B 210
heat exchange A 482
light absorption A 66
Purple membrane C **144** ff.

Quantum efficiency, C_3/C_4 plants D 163
Quantum flux density A 43, A 44
definition A 13
effect on water stress D 179
photosynthetic response D 179, D **180**, D **257**
plant responses A 57 ff.
Quantum yield B 483, D 258
C_3/C_4/CAM plants A 68 ff., B 506, B 507
definition A 43, B 570 ff.
desert plants D 219
effect of water stress B 294
photoinhibition A 96
phytochrome A **117**
response to temperature A 289, A 300 ff.
Q_{10}, definition A 293
greening A 316
ion uptake A 313
photoperiodism A 146
protein synthesis A 317
relation to activation energy A 293
seed respiration A 321
seedling growth A 316
translocation A 315
water flow A 314
water use efficiency A 291

Radiant flux, definition A 42, A 238
Radiation, absorption A 18
absorption by leaves A 21
absorption by water A 115, A 239
amplification effect A 177 ff., A **179**
annual D 234
temperate forest D 234
attenuation in water A 253
canopy environment A 32, A 112, C 687, D **254**
classification A **200**
definitions and units A 238
effect on heat tolerance A 462
effect on productivity D 234
forest B **653**
fundamentals A 11 ff.
ionizing A 199 ff.
long-term effects B 183 ff., B 195
long-wave A 12 ff.
monthly D **235**
temperate regions D **235**
short-wave A 12 ff.
spectral distribution A 12
stomatal response B 190, B 241, B 242
Radiation, see also Light or Irradiance
Radiation efficiency, for CO_2 fixation D 238
Radiation frost A 405
Radiation interception, canopy D 243 ff., D **249**
effect of structural properties D 245, D 246
effect on yield D 254
equation D 248 ff., D 253
measurement D 244, D 245
models D 246 ff.
Radiation penetration, measurement D 246
Radiation transmittance, equation D 247
Radioactive carbon method D 306, D 321, D **322**
Radioactive wastes, in freshwater ecosystems D 434, D 435
Radiometry A 42 ff.
Radioresistant cells, quiescent center A 216
Radiosensitivity A 206 ff.
Raunkiaer life forms B 616
Raw humus C 202
ammonium C 215
nitrogen production C 175
Reaction centers, shade and sun plants A 71
Reaction wood, wind effect A 493, A 494
Red/far-red light ratio, apical dominance A 122
beneath canopy A **114**
definition A 113
effect in algae A 269
reduction during twilight A 122
Redistribution of nutrients, models D 143
Reflectance, leaf D 245, D 246
effect on radiation interception D 245, D 246
Reflection, radiation on plant A 18
Regression model, definition D 126
examples D **127**
productivity D 126 ff.
yield prediction D 126
Rehydration, changes in membrane integrity B 344 ff.
effect on ATP content B **366**
effect on enzyme activities B 349, B **350**
effect on leakage of substances B 344 ff.
effect on photosynthesis and respiration B **364** ff.
effect on RNA and protein synthesis B 351 ff.
Relative water content B 44, B **45**
tissue B 267
Reproduction, fire adaptation A 523
Resin, flower attractant C 591, C 592
in fruits C 635
Resistance, cuticular A 22, B 159, B **160**, B 235, B 386 ff.
desiccation B 386 ff.

diffusion B153ff.
endodermis B97
hydraulic B13, B15, B16, B18, B19, B81, B416
root/soil interface B13, B19, B81, B100ff., B103ff., B251
seed/soil interface B416
soil B102, B103
peridermal B386ff.
stomatal A22, A**421**
after frost stress A**421**
crop plants D535
temperature and pressure dependence A24
to vapor transfer B237ff.
to viruses C540
Resistance, see also Conductance
Respiration, biosynthetic A312
calculation D266
different species B626
effect of desiccation B356ff.
effect of fluoride D476, D488
effect of herbicides D348, D349
effect of rehydration B**364**ff.
effect of sulphur dioxide D473, D484
effect of water stress B296
effect on atmospheric CO_2 concentration D507
flowers C568, C569
growth A311, D240, D265
equation D265
lichens C444, C**445**, C**448**
low temperature stress A309
maintenance D240, D241, D263, D264
equation D263, D264
measurement, tropical forests D292, D**293**
osmophores C**584**
roots B22, B456, D139, D545
effect of oil D545
seed germination B403ff.
temperature response A307, A308, D262, D291
equation D262
tropical plants D291
wound A311
Respiration models D512, D513
Resprouting, fire-stimulated A517
Resupination C564
Resurrection plants B70
desiccation tolerance B326, B341
Resurrection plants, see also Poikilohydrous plants
Ribosome content, temperature dependence A319
Ribosomes, denaturation temperature A344
Ribulose-1,5-bisphosphate carboxylase-oxygenase, see RuBP carboxylase-oxygenase
Rivers, primary production D308, D**309**
RNA, UV radiation absorption A183
RNA content, blue-green light A264
RNA synthesis, effect of herbicides D354, D355
inhibition B351ff.
in desiccated mosses B353
Root, air space B**464**, B**465**
cytoplasmic pH regulation C23
distribution B13
effect on interference ability C703
hydraulic resistance B15, B80, B98
interfacial resistance B14, B100ff., B251
ion pumps C20ff.
resistance to water flow A314, B14, B15, B96
resistance to water uptake B384
water sink A551, A556
water supply B23
water uptake and flow B79ff.
Root detritus, nutrient return D26
Root exudates, nutrient return D27
Root exudation, rate B82ff.
Root feeders C660
Root growth D139
effect of aluminum C**212**
effect of copper C**251**
effect of ions C214
effect of iron deficiency C**214**
effect of malate C**232**
effect of relative water content B621, B**622**
effect of soil water potential B24
effect of temperature A554
effect of water stress B24, B308
effect of water uptake D141
response to rewatering B24
temperature optimum A312
Root hairs B104, B105
Root length, effective B13
Root membrane, permeability B384
Root pressure B80, B81
Root production, mediterranean-climate plants D210, D211
tropical forest D297
Root respiration B22, C230ff., D139, D545
Root surface area D144, D145
Root turnover D241, D268
Rootcrowns, fire adaptation A519
Rooting density, crop plants B28
definition B12
Root/shoot ratio, crop plants B28
desert plants D211ff.
effect of leaf and soil water potential B308
effect of salinity C110, C114
effect of temperature A285, B515, B**517**, D196
effect of water availability B514ff., B612
effect of water stress B24, B308, D195

Root/shoot ratio, effect on plant-water relations D517
interactions with nutrient status A286
mediterranean-climate ecosystems D**207**
photoperiodism A140
respiratory cost B24
tropical forests D282, D**284**
UV radiation A187
Root/soil, interfacial resistance B81
Roughness parameter, wind in canopy A31
r-strategists, definition C547
RuBP carboxylase-oxygenase A61, A69, A**76**, A307, B481 ff.
activation energy A304
aquatic plants B493
^{13}C discrimination B497
carboxylation B**578**
changes by blue-green light A264
correlation to light saturated photosynthesis A79
effect of water stress B277, B293
effect on photosynthesis B**554**
kinetics B551 ff.
rate limiting A78
regeneration B558 ff., B568, B574, B575, B**578**
shade and sun plants A**80**
temperature response A**305**
Running water, denitrification D58
nitrification D58

Saline conditions, osmotic adjustment B59
Saline lakes, ion concentration C140
microbiology C139
Salinity, changes in enzyme properties C77
effect on assimilate partitioning B517
effect on ATPase activity C78
effect on CAM C112, C116
effect on cell wall characteristics C95
effect on electrical potential C91
effect on freshwater productivity D436
effect on growth C61 ff., C78, C82, C85, C91, C**92**, C**95**, C**98**, C**100**, C**102**, C112 ff., C120, C**139**
effect on osmotic potential C113
effect on photosynthesis C99, C117, C119
effect on protein synthesis C78
effect on root conductance C118
effect on root/shoot ratio C110
effect on succulence C109, C111, C114, C**115**, C119
effect on turgor C99, C113, C115
effect on water relations C99, C112, C116 ff.
in freshwater ecosystems D435 ff.
Salt bladders C104, C111
ion concentrations C68
Salt excretion C104, C111
Salt glands C104, C111
ion pumps C19
Salt pollution, of freshwater ecosystems D435 ff.
Salt secretion, halophytes C70
Sap feeders C661
Sapwood area, relation to leaf area D251
Savanna, fire effect A509
history of fire A527
Scent emission, flowers C581 ff.
Scent production, flowers C578, C579
Sclerophylly, relation to water content B649, B**650**
Seawater, osmolality C39
Seaweeds, zonation A255
Secondary plant substances, classification C662 ff.
effect on herbivory C662 ff., C672
effect on selectivity and evolution C664 ff.
production under stress conditions B485
Sediment, ammonification D59, D60
denitrification D58
nitrification D57
phosphorus release D65
phosphorus supply D64
Seed, defensive compounds C636, C648
Seed burial B423
Seed coats, diffusivity to water B415
mucilaginous B417, B418
selective permeability B412
Seed dispersal C646
rainfall sensors B425
wind C629
Seed ecology C625 ff.
Seed germination B401 ff.
at suboptimal temperature B409
chilling injury B408
control B635
critical hydration level B412
effect of soil pH C205
geocarpy B422
growth phase B409, B410
halophytes B409
inhibition B405
light sensitivity B403, B408, B422
optimal soil depth B421 ff.
osmotic inhibition B410
oxygen B403, B404
regulatory mechanisms B404, B405, B421 ff.
temperature B404, B422, B424
water relations B401 ff.
water supply B421 ff., B423, B424
Seed predators C631, C636, C641, C646 ff.
Seed production, effect of atmospheric CO_2 concentration D518

Seed reserves, amount C 644
Seed storage, on-plant A 522
fire adaptation A 522
Seed yield, photoperiodism A 141
water stress B 637
Seedling competition, for PAR B 405
Seedling establishment A 487
low temperature effect A 423
Self-pollination C 571
Self-purification, effect of chloride D 436
freshwater ecosystems D 436
Senescence, photoperiodism A 142
retardation by cytokinin A 374
Sewage, effect on eutrophication D 419
Sex expression, dioecious plants C 563
photoperiodism A 141
Shade, red/far-red light ratio A 160
Shade and sun plants A 59 ff., B 211 ff.
chlorophyll content A 67
chloroplast morphology A 67
dark respiration A 60
ecotypes A 61
effects of radiation B 183
leaf growth in relation to light intensity B 514
light acclimation A 59 ff.
light response D 179
photoinhibition A 101
photosynthesis A **58**, B 211 ff., B **219**
photosynthesis model A 84
photosynthetic reaction centers A 71
protein content A 71
RuBP carboxylase-oxygenase activity A **80**
Shifting agriculture D 538
Short-day plants A **143**
Short-long-day plants A **145**
Short-wave radiation, computation A 13, A 15 ff.
definition A 12
in canopy A 34
Silica, role in aquatic plants D 69 ff.
Silicates, influence on vegetation C 201 ff.
Silicon cycle, marine ecosystem D 100 ff.
Silicon metabolism, aquatic plants D 70
Silver toxicity, mechanism C 287
Simulation, models D 122
Sinking rate, marine plants and spores A 233, A **234**
Siphonaxanthin, in algae A 258, A 260
Skylight, spectral distribution A 151, A **152**
Smog, photochemical D 468, D 469
Snow cover, influence on water uptake B 381
UV spectral irradiance A 180
Soil, air capacity A 546
air content A 550
conductivity A **549**, B 413 ff.
consistence A 541
element contents D **15**
heat transfer A 502
ionic composition D **21**
microstructure A **540**
nutrient reserves D 14 ff.
profile A 544, A 546, A 555
structure A **542**, A **543**
temperature, root growth A 554
water A **546**, B 23, B 603
characteristic curves A **546**
critical level B 603
productivity B 23
Soil acidification, effect of acid rain D 477
Soil adsorption, effect on herbicide selectivity D 334
Soil environment A 539 ff.
fire effect A 505, A 506
Soil enzymes D 29
Soil faunal populations, effects of biocides D 399 ff.
Soil frost, duration in different habitats B **382**
influence on water uptake B 380 ff.
Soil microbial populations, effects of biocides D 398
Soil-plant-atmosphere continuum B 5 ff.
water flow B 5 ff., B 196
Solar radiation, see Radiation
Solar tracking, desert ephemerals D 224
Solar zenith angle D 247, D 257
Specific leaf area, definition D 284
tropical forests D 285, D **286**
Spectral absorptance, leaf A **47**
Spectral changes, canopy environment A 157
Spectral distribution, daylight A **111**
Spindle formation, effect of herbicides D 355
Spine, dew uptake B 114
fruit protection C 634
Spore germination A 348 ff., B 433 ff.
effect of temperature A 351
inhibition by smoke A 523
temperature induction A 348, A 350
Spores, activation B 438 ff.
desiccation resistance B 436
dispersal A 486, B 435 ff., B 436
dormancy B 440, B 442
Sporulation, temperature limits A 352
Spread of man, into natural ecosystems D 531, D 532
Starlight, spectral distribution A 112
Stem feeders C 660
Stemflow B 25
effect on nutrient return D 23, D 24
Sterility, ionizing radiation effect A 203
Stomata, control B 181 ff.
effect of fluoride D 476
effect of oxidants D 474, D 475
effect of sulphur dioxide D 472, D 473

Stomata, effect on pollutant uptake D 470
effects on CO_2-exchange B 181 ff., B 286 ff., B 590
effects on leaf energy balance B 182 ff.
effects on transpiration B **21**, B 181 ff., B 590
mechanism B 288 ff.
optimal behavior B 20, B 22, B 590, B 591, B 596, B 597, B 605
water uptake B 113
Stomatal conductance, see Conductance
Stomatal frequency, C_3/C_4/CAM plants B 503
photoperiodism A 142
Stomatal function, dynamic model B 249, B 250
Stomatal mechanics, mathematical models B 232
Stomatal resistance, see Resistance
Stomatal response, and cold hardening B 385
and photosynthesis B 181 ff., B 286 ff., B 590
effect of ground frost B 385, B 386
effect of soil water potential B 605
induced by abscisic acid B 281, B 288 ff., B 385, B 460
leaf water status B 202, B 288 ff., B 308
models B 236 ff.
to CO_2 concentration B 233, B 243, B 244, D 517
to humidity B 117, B 222, B 233, B 242, B 243, B 248, B 287 ff.
to light B 222, B 223, B 241, B 242
to plant water relations B 196 ff., B 208, B 244 ff., B 605
to temperature B 191, B 222, B 223, B 243, B 248, B 385, B 386
to water stress B 287 ff., B 485, B 489
Stratification, decrease of abscisic acid A 321
Stress resistance D 168
Stress-strain relationship, physical model B 264 ff.
Strip miners C 660
Structural types of plants, definition B 616
Suberin, transport properties B 153 ff.
Succession, after clear-cutting D 539, D 540
Succulence, effect of salinity C 109, C 111, C 114, C **115**, C 119
Succulents, effect of herbivory D 191
photosynthesis and leaf conductance D 217
photosynthetic pathway B 488
Sulphur, inorganic D 66 ff.
microbial transformation D 66 ff.
role as atmospheric pollutant D 467, D 468, D 471 ff.
removal D 468
role in aquatic plants D 66
uptake by mycorrhiza C 398
Sulphur cycle, aquatic ecosystems D **67**
biosphere D 468
terrestrial ecosystems D 36, D 37
Sulphur dioxide, absorption by soil and plant D 11
effect on lichens C 452
role as atmospheric pollutant D 467, D 471 ff., D 478 ff.
action on enzymes and cellular metabolism D 479 ff.
chemical reactions D 478, D 479
plant responses D 471 ff.
Sulphur dioxide injury, symptoms D 471, D 472
Sulphur-oxidizing processes, aquatic ecosystems D 68
Sulphur-reducing processes D 67, D 68
Summary model, for production D 133 ff.
Sun plants, see Shade and sun plants
Sunflecks, in plant canopy A 115
spectral distribution A 115
Sunlight, spectral distribution A 151, A **152**, A **153**
Supercooling, xylem A 424
Super-infection immunity, viruses C 541
Surfactants, plant damage D 434
in freshwater ecosystems D 433, D 434
Symbiosis, actinorrhizal C 302
algae-invertebrate C 475 ff.
between plants and animals C 469 ff., C 474, C 477 ff.
cyanelles C 470, C 471
lichens C 343, C 423 ff.
mycorrhizal C 391 ff.
nitrogen fixation C 325 ff., C 330, C 339
plant-herbivore C 671
zooxanthellae C 471 ff.
Symplastic ion transport C 17
root C 22
Symplastic water movement B 90, B 93 ff., B **95**

Taiga, fire effect A 504, A 525
Tearing, leaf A 489
Technological influences on agriculture D **155**
Temperate forest, productivity D 233 ff.
Temperature, acclimation A 390, A 457, A 458, C 449
desert plants D 222, D 223
adaptation and acclimation strategies A 299
effect on assimilate partitioning B 515, B **517**
effect on biocide effectiveness D 395, D **396**
effect on growth C 341, D 181 ff., D 184
grassland plants D 184
tundra plants D 181 ff.
effect on hydrophobic interactions A 297
effect on membrane structure A 298
effect on membrane transport C 7
effect on nectar secretion C 597

effect on nitrogen content of C_3 and C_4 plants A **290**
effect on nodulation C 316 ff.
effect on nutrient requirement A 291
effect on phytoplankton productivity D 315
effect on protein synthesis A 307, A 317
effect on quantum yield A **70**
effect on root/shoot ratio A 285
effect on transpiration A 290
extreme A 371 ff.
in canopy environment A 31
leaf A 27, A 300, A 481
lichens C 449
limits of growth A 343 ff.
photosynthetic response A 282, A 299, A 300 ff., C 449, D 181, D **182**, D 260, D 290, D **291**
tropical plants D 290, D **291**
physiological optima and limits A 280, A 296
responses of C_3/C_4 plants A 287
responses of microorganisms A 339 ff., A 346
Temperature extremes, plant organs A 440
soil A 440
Temperature injury, see Heat or Frost injury
Temperature optimum, desert plants D 222
photosynthesis C 449
Temperature resistance A 403 ff., A 439 ff.
lichens C 437 ff.
Tensile strength, of water B 138
Terrestrial ecosystems, carbon concentration D 507
Thermal diffusivity, soil A 554, A **555**
Thermal shocks, heat resistance A 376
Thermal springs, pH variation A **354**
Thermal tolerance, see Heat or Frost tolerance
Thermoduric microorganisms A 342
Thermodynamic laws, role in ecosystems D 529
Thermophilic microorganisms A 342, A 353 ff.
habitats A 353
Thermoregulation A 278
Therophytes B **619**
Throughfall, effect on nutrient return D 23 ff.
Thylakoids, effect of sulphur dioxide D 482
Timber-line, alpine A **414**, A 420, A 422, B 381, B 656
annual production B 656
influence of snow cover B 381
arid B 658
CO_2 balance during winter A 422
environmental conditions B **657**
frost drought B 393, B 395
frost resistance A **414**, A 420
Tipburn, fluoride injury D 475
Tissue water potential, apoplasmic water B 54 ff.
Tissue-pressure B 49
Toxic chemicals, bioaccumulation in freshwater food chain D 437
Trace elements, effect on primary production D 71
transfer in marine ecosystem D 102 ff.
Trampling sensitivity D 540
Transfer cells, effect on iron uptake C 230
ion transport C **8**, C **9**, C 17
xylem C 22
Translocation, temperature effect A 315
Translocation of herbicides D 334
Transmission, radiation on plant A 18
Transmission coefficient, in canopy A 33, A 35
Transmittance, diffuse D 247, D 248 ff.
equation D 247
extinction coefficient D 248 ff.
leaf D 245, D 246
effect on radiation interception D 245, D 246
Transpiration B 85 ff.
and photosynthesis B 286 ff., B **598**, B **601**
controlling parameters A 551
cuticular B 153 ff., B 166, B 386 ff.
C_3 and C_4 halophytes B 511
diurnal course B **204** ff.
driving force B 154
effect of competition B **609**
effect of heavy metals D 544
effect of leaf abrasion A 489
effect of leaf area index B 18, B 25, B 253, B 254, B 309, D 287
effect of leaf orientation B 235
effect of leaf pubescence B 235
effect of soil water status B 19
effect of temperature A 290
effect of wind A 29, A **30**, A 482, A 489, A 491
effect on ion concentration C 105, C 109
function of soil matric potential A **552**
general equation A 22, A 27, B 85
influence on leaf water potential B 98, B **101**, B 196 ff.
leaf energy balance B 235
lichens C 442
optimization B 595, B 601 ff., B 605, B **606**, B **609**, B 622
physical aspects B 236 ff.
sensitivity to changes in leaf conductance B 195
Transpiration equation B 592
Transpiration model B 231 ff., D 140
canopy A 37, B 253, B 254
Transpiration ratio, C_3/C_4/CAM plants B 501 ff.
Transpiration ratio, see also Water use efficiency
Transpiratory cooling A 445

Transport, apoplastic C 17
symplastic C 17
Traps, carnivorous plants C 490, C **491**, C 497 ff., C 500
Tree mortality D 267, D 268
Trees, growth characteristics B 652 ff.
Trigger factors, principle in ecosystems D 530
Trigger hair, effect of stimulation C 497
Tropical forest, productivity A 252, D 281 ff., D 292 ff.
measurement D 292 ff.
Tropical life forms D 291
Tropical plants, photosynthesis D 287 ff.
photosynthetic pathways D 291, D **292**
respiration D 287 ff.
Tundra, characterization D 173, D 174
fire effect A 505
productivity D 173 ff.
vegetation types **D 175**
Tundra plants, temperature response D 181 ff.
water relations D 187
water stress B 214
Turbulence, water A 230
Turbulence intensity, mass transfer A 25
wind speed variations A 476
Turgor, and cell wall properties B 52
and growth B 269 ff.
and transpiration B 89
changes in guard cells and subsidiary cells B 199
effect of salinity C 99, C 113, C 115
effect on ion uptake C 24, C 27
indicator for water stress B 267, B 268
influence on stomatal conductance B 10
maintenance B 62 ff.
measurement B 268, C 83
negative B 48
regulation C 36
tissue C 97
Turgor, see also Hydrostatic pressure and Pressure potential
Turgor-volume maintenance C 85
Twilight, photoperiodic control A 128
plant response A 153
red/far-red light ratio A 160
spectral changes A 157
spectral distribution A 111, A 136, A 151, A **152**, A **153**, A 155
Two-peaked course of gas exchange, see Midday depression

Upwelling irradiance, definition A 238
Upwelling radiation, in water A 115
Urbanization, atmospheric pollutants D 538
effect on natural ecosystems D 536
replacement of natural and agricultural ecosystems D 538
UV chromophore A **171**
UV radiation, abscission of leaves A 187
absorption and reflection by flowers C **576**, C 577
action spectra A 172
adaptation to A 188 ff.
attenuation in the atmosphere A 175
competitive balance of plant species A 192
damage A **174**
effect on growth A 187
effect on plant production A 192
effect on root/shoot ratio A 187
in canopies A 189
photoprotection A 184
plant response A 169 ff.
reflection A 189
spectrum A 176 ff.

Vapor pressure deficit of air, stomatal response B 192
Vegetation, role in an ecosystem D 529
Vegetation formations, biomass B 660 ff.
leaf area index B **664**
productivity B 660 ff.
water use efficiency B 660 ff.
Velamen B 119
Vernalization, flower initiation A 148
Vesicular-arbuscular mycorrhiza C 393
Vessel length B **137**
Virus ecology C 537 ff.
Virus epidemic forecasting C 546
Virus groups, characterization C 550 ff.
Viruses C 537 ff.
Virus vectors, characteristics C 542 ff., C 552 ff.
Vitamins, nectar constituents C 604
role in marine ecosystem D 102 ff.

Wall softening B 64
Waste heat, effect in freshwater ecosystems D 439 ff.
Waste water technology, nitrification process D 57
Water activity, definition C 37
effect on metabolism C 93
Water balance, effect on growth rate D 140 ff.
Water conduction, bryophytes B 121
efficiency B 138
root B 79 ff.
soil-plant-atmosphere continuum B 11 ff.
structural basis B 136 ff.
Water content, effect on heat tolerance A 459
leaf B 88
measurement by a β gauge B 88
lichen thallus B **125**, B **126**, B 358 ff.
plant B 390 ff.
frost-drought B 390 ff.
seed B 415
soil A 537 ff., B 413, B 415, B 603

Water economy, and photosynthetic pathways B 500 ff.
competition B 599
Water flow, apoplastic B 90
from air to root B 112
regulation B 148
roots B 79 ff., B 90 ff., B **91**
root-soil interface B 13
seed-soil interface B 413, B 415, B 416
soil-plant-atmosphere continuum B 11 ff.
symplastic B 90
vacuolar B 90
Water in tissues and cells B 35 ff.
Water loss, see Transpiration
Water motion, boundary layer A 232
drag A 233
mass transport A 235
plant response A 238
Water plants, see Aquatic plants
Water potential B 36 ff.
and cell expansion B 270 ff.
and radicle elongation B 410
and seed germination B 403, B 406, B 409, B 412, B 416
components B 8, B 9, B 36, B 37
C_3/C_4 plants D 189
definition B 8, B 36 ff., C 38
different species B 143
ecological implication B 55 ff.
effect of nitrogenase activity C 350
effect of ozone and peroxyacetyl nitrate D 485
effect on ion uptake C 27
importance for water transport B 112
indicator for water stress B 267
leaf A 22, B 17, B 87, B **88**, B 98, B **101**, B 196 ff., B 198, B 199, B **200**, B 202, B **206**, B **207**, B 211, B 222, B 235, B 281 ff., B 286 ff., B 308, D 195, D **224**, D **225**
and competition for water C **693**
and transpiration B 17, B **88**, B **89**, B 98, B 101, B 196 ff., B 222, B 235, B 286 ff.
changes during drought B **200**
diurnal course B **206**, B **207**
effect of abscisic acid B 281 ff.
effect of air humidity B **101**
effect of flooding B 454
effect of soil water potential B 605
effect on conductance B 196, B 199, B **200**, B 211, B 308, B 605
effect on leaf elongation B 270 ff., D 195
effect on nocturnal acidification of CAM plants D **225**
effect on photosynthesis B **201**, B 286 ff., D **224**
effect on root/shoot ratio B 308
measurement B 87
pre-dawn B 202
seasonal pattern B **652**
short-term and long-term changes B **197** ff.
threshold for stomatal closure B 199
measurements B 10, B 16
perirhizal drop B 102
radicle elongation B 410
root B **101**
root cortex B 86
soil A 544 ff., B 24, B 81, B 98, B **101**, B 202, B 308, B 411 ff., B 605, B **606**, B **609**, B **659**, C 39, C 169, C 319
soil/seed interface B 403, B 411
total B 9
xylem B 86
Water potential isotherm B 38, B **45**, B **46**
Water relations, and frost-drought B 390 ff.
components B 9
effect of environmental factors B **7**
effect of ion concentration C 99, C 101 ff., C 112, C 116 ff.
effect of physiological processes B **6**
effect on growth B 67, C **113**
effect on leaf conductance B 214 ff.
effect on photosynthesis B 19, B 196 ff., B 208, B 222
halophytes C 102
lichens C 441 ff., C 443
mathematical models B 231 ff.
mycorrhiza C 401
seed B 421
seed germination B 401 ff., B 418
Water status, definitions B 7 ff.
Water status, see also Water relations
Water storage B 143 ff., B 233
Water stress B 233, B 263 ff.
adaptation B 175, B 305 ff., B 325 ff., B 341
and cell wall synthesis B 276
diurnal fluctuation B 58
effect of midday stomatal closure D 223
desert plants D 223
effect of reduced nitrate uptake B 278
effect on biocide effectiveness D 395, D **396**
effect on biomass accumulation D 282, D **283**
effect on carbon balance B 336
effect on cell size B 70
effect on dry matter distribution D 141
effect on electron transport B 293 ff.
effect on growth B 269 ff., D 140, D 186 ff.
effect on guard cell K^+ B 289 ff.
effect on ion uptake and transport B 285
effect on leaf abscission B **302** ff.
effect on leaf elongation D 195
effect on mRNA translation B **353**
effect on mycorrhiza C 401

Water stress, effect on nectar secretion C597
effect on nutrient uptake B297ff.
effect on photophosphorylation B294, B295
effect on photosynthesis and respiration B291ff., B356ff., B504, D179, D186ff.
effect on phytohormones B278, B281, B**302**ff.
effect on productivity B305
effect on proline and betaine accumulation and transport B296, B297
effect on proteins and amino acids B296, B297
effect on quantum yield of photosynthesis B294
effect on root/shoot ratio B308
effect on seed yield B637
effect on stomatal conductance B289ff., B**302**, B303, B485, B489
effect on turgor B296
enzyme activities B277, B278
frost drought B379ff.
growth B286
induced by chilling A420
inhibition of chlorophyll formation B293
leaf expansion B305
osmotic adjustment B272, B292, B308
photoinhibition B295
protection of enzymes B342
protein synthesis B276ff., B351
timberline ecotone B393, B395
Water transport, see Water flow
Water types, irradiance A253
Water uptake, bryophytes B121
by organs other than roots B111ff.
effect on root growth D141
foliar B114ff.
hydraulic conductance B271
lichens B124, C441, C442
reduced by cooled roots A420
root B79ff., B93, B**94**, B380ff., B381
seed B407, B411, B417ff.
trichomes B116
Water use B25ff., B603, D140
and carbon gain B589ff.
long-term regulation B600ff.
prediction B233
short-term regulation B589ff.
Water use efficiency B25ff., B183ff., C118, D517
and leaf area index B**26**
and photosynthetic pathways B500ff.
CAM plants C116, D224
C_3/C_4 plants D189
C_4 plants D166
definition A291, B20
different plant life forms B660ff.
effect of atmospheric CO_2 concentration D512
effect of salinity C118
effect on dry matter production B664
heterotrophic tissues B28
optimization B193ff., B222, B247ff., B589ff., B664
Q_{10} A291
Water use efficiency, see also Transpiration ratio
Waterlogging, effect on nitrogen fixation C320, C321
Weed, biocide resistance D397, D**398**, D402
development D402
co-evolution with crop plants D536
Weed control, chemicals D330ff.
Wind, ecological factor A475ff.
effect on biocide effectiveness D395, D**396**
seed dispersal C629
Wind erosion, after fire A508
Wind pollination A487
Wind speed, free convection A479
growth chambers A476
optimum for plant growth A485
Windbreak A491
Winter dormancy A412, B384
influence on photosynthetic activity A421
Winter drought, reduction by leaf-fall A416
Wood production, tropical forest D**295**
Woody perennials B**619**, B630ff.
carbon allocation B629, B630
geographic distribution B630
growth characteristics B646ff.
Wound respiration A311

Xanthophylls, in algae A260
Xylem, axial resistance B15
capacitance B16
deep-supercooling A424, A426
frost resistance A418
functional adaptation B139
spiral grain B142
Xylem pressure potential, see Water potential

Yield, effect of radiation losses D254
models D126
of organic materials D160, D161
Yield tables, forests D235

Zinc toxicity C252, C264
Zoochlorella C474
Zoophilic pollination C559ff.
Zooplankton, ammonium excretion D59
effect on dissolved organic carbon D51
phosphorus consumption D63
Zooxanthellae C471ff.

Encyclopedia of Plant Physiology

New Series

Editors: A. Pirson, M. H. Zimmermann

Volume 14, Part A
Nucleic Acids and Proteins in Plants I
Structure, Biochemistry and Physiology of Proteins
Editors: **D. Boulter, B. Parthier**
With contributions by numerous experts
1982. 135 figures. XX, 768 pages
ISBN 3-540-11008-9

Volume 14, Part B
Nucleic Acids and Proteins in Plants II
Structure, Biochemistry and Physiology of Nucleic Acids
Editors: **B. Parthier, D. Boulter**
With contributions by numerous experts
1982. 173 figures. XVIII, 774 pages
ISBN 3-540-11140-9

Volume 13, Part A
Plant Carbohydrates I
Intracellular Carbohydrates
Editors: **F. A. Loewus, W. Tanner**
With contributions by numerous experts
1982. 103 figures. XXII, 918 pages
ISBN 3-540-11060-7

Volume 13, Part B
Plant Carbohydrates II
Extracellular Carbohydrates
Editors: **W. Tanner, F. A. Loewus**
With contributions by numerous experts
1981. 124 figures. XXII, 769 pages
ISBN 3-540-11007-0

Volume 11
Hormonal Regulation of Development III
Role of Environmental Factors
Editors: **R. P. Pharis, D. M. Reid**
With contributions by numerous experts
ISBN 3-540-10197-7
In preparation

Volume 10
Hormonal Regulation of Development II
From the Cell to the Whole Plant
Editor: **T. K. Scott**
With contributions by numerous experts
ISBN 3-540-10196-9
In preparation

Volume 9
Hormonal Regulation of Development I
Molecular Aspects of Plant Hormones
Editor: **J. MacMillan**
With contributions by numerous experts
1980. 126 figures. XVII, 681 pages
ISBN 3-540-10161-6

Volume 8
Secondary Plant Products
Editors: **E. A. Bell, B. V. Charlwood**
1980. 176 figures, 44 tables and numerous schemes and formulas. XVI, 674 pages
ISBN 3-540-09461-X

Volume 7
Physiology of Movements
Editors: **W. Haupt, M. E. Feinleib**
1979. 185 figures, 19 tables. XVII, 731 pages
ISBN 3-540-08776-1

Volume 6
Photosynthesis II
Photosynthetic Carbon Metabolism and Related Processes
Editors: **M. Gibbs, E. Latzko**
1979. 75 figures, 27 tables. XX, 578 pages
ISBN 3-540-09288-9

Springer-Verlag
Berlin
Heidelberg
New York
Tokyo

MIX
Papier aus verantwortungsvollen Quellen
Paper from responsible sources
FSC® C105338

If you have any concerns about our products,
you can contact us on
ProductSafety@springernature.com

In case Publisher is established outside the EU,
the EU authorized representative is:
Springer Nature Customer Service Center GmbH
Europaplatz 3, 69115 Heidelberg, Germany

Printed by Libri Plureos GmbH
in Hamburg, Germany